ALAN J. KALAMEJA

The
AutoCAD® 2000
Tutor for
Engineering Graphics

ALAN J. KALAMEJA

The AutoCAD® 2000 Tutor for Engineering Graphics

Autodesk.
Press

Thomson Learning™

Africa • Australia • Canada • Denmark • Japan • Mexico • New Zealand
Philippines • Puerto Rico • Singapore • Spain • United Kingdom • United States

NOTICE TO THE READER

Publisher does not warrant or guarantee any of the products described herein or perform any independent analysis in connection with any of the product information contained herein. Publisher does not assume, and expressly disclaims, any obligation to obtain and include information other than that provided to it by the manufacturer.

The reader is expressly warned to consider and adopt all safety precautions that might be indicated by the activities herein and to avoid all potential hazards. By following the instructions contained herein, the reader willingly assumes all risks in connection with such instructions.

The publisher makes no representation or warranties of any kind, including but not limited to, the warranties of fitness for particular purpose or merchantability, nor are any such representations implied with respect to the material set forth herein, and the publisher takes no responsibility with respect to such material. The publisher shall not be liable for any special, consequential, or exemplary damages resulting, in whole or part, from the readers' use of, or reliance upon, this material. Autodesk does not guarantee the performance of the software and Autodesk assumes no responsibility or liability for the performance of the software or for errors in this manual.

Trademarks

AutoCAD ® and the AutoCAD logo are registered trademarks of Autodesk, Inc. Windows is a trademark of the Microsoft Corporation. All other product names are acknowledged as trademarks of their respective owners.

Autodesk Press Staff
Executive Director: Alar Elken
Executive Editor: Sandy Clark
Developmental Editor: John Fisher
Editorial Assistant: Allyson Powell
Executive Marketing Manager: Maura Theriault
Executive Production Manager: Mary Ellen Black
Production Coordinator: Jennifer Gaines
Art and Design Coordinator: Mary Beth Vought
Marketing Coordinator: Paula Collins
Technology Project Manager: Tom Smith

Cover Design by Scott Keidong's Image Enterprises. Lower right image provided courtesy of Autodesk, Inc.

For more information, contact
Autodesk Press
3 Columbia Circle, Box 15-015
Albany, New York USA 12212-15015;
or find us on the World Wide Web at http://www.autodeskpress.com

Library of Congress Cataloging-in-Publication Data

Kalameja, Alan J.
   The AutoCAD 2000 tutor for engineering graphics / Alan J. Kalameja.
     p. cm.
   ISBN 0-7668-1238-3
   1. Engineering graphics. 2. AutoCAD. I. Title.
T357.K35 1999
604.2'0285'5369—dc21
                                                        99-16749
                                                          CIP

CONTENTS

×

CHAPTER 9 An Introduction to Drawing Layouts

CHAPTER 10 Plotting Your Drawings 577

CHAPTER 11 Dimensioning Basics 637

INTRODUCTION

Engineering graphics is the process of defining an object graphically before it is constructed and used by consumers. Previously, this process for producing a drawing involved the use of drawing aids such as pencils, ink pens, triangles, T-squares, etc. to place an idea on paper before making changes and producing blue-line prints for distribution. The basic principles and concepts of producing engineering drawings have not changed even when using the computer as a tool.

This text uses the basics of engineering graphics to produce 2D drawings and 3D computer models using AutoCAD 2000 and a series of tutorial exercises that follow each chapter. Following the tutorials in most chapters, problems are provided to enhance your skills in producing engineering drawings. A brief description of each chapter follows:

CHAPTER 1 – GETTING STARTED WITH AUTOCAD

This first chapter introduces you to the following fundamental AutoCAD concepts: Screen elements; Use of function keys; Opening up an existing drawing file; Basic drawing techniques using the LINE, CIRCLE, and PLINE commands; Understanding absolute, relative, and polar coordinates; Using the Direct Distance mode for drawing lines; Using Object snaps and the AutoSnap feature, polar and object tracking techniques; Using the Erasing command; Saving a drawing. Drawing tutorials follow along with additional problems at the end of this chapter.

CHAPTER 2 – DRAWING ORGANIZATION

This chapter introduces the concept of drawing in real world units through the setting of drawing units and limits. The importance of organizing a drawing through layers is also discussed through the use of the Layer Properties Manager dialog box. Color, linetype, and lineweight are assigned to layers and applied to drawing objects. The use of various layer express tools are also discussed in this chapter. A tutorial exercise on performing these layer tasks is found at the end of this chapter.

CHAPTER 3 – AUTOCAD DISPLAY AND SELECTION OPERATIONS

This chapter discusses the ability to magnifying a drawing using numerous options of the ZOOM command. The PAN command is also discussed as a means of staying in a zoomed view and moving the display to a new location. Productive uses of realtime zooms and pans along with the effects the Intellimouse has on zoom and pan are included. All object selection set modes are discussed such as Window, Crossing, Fence, and All to name a few. Finally this chapter discusses the ability to save the image of your display through the View Control dialog box. This image can then be retrieved for later use.

CHAPTER 4 – MODIFY COMMANDS

The full range of Modify commands are discussed in this chapter and include ARRAY, BREAK, CHAMFER, COPY, EXPLODE, EXTEND, FILLET, LENGTHEN, MIRROR, MOVE, OFFSET, PEDIT, ROTATE, SCALE, STRETCH, and TRIM. Express tools that concentrate on modify commands are also discussed in this chapter. Tutorial exercises follow along with additional problems at the end of this chapter.

CHAPTER 5 – PERFORMING GEOMETRIC CONSTRUCTIONS

This chapter discusses how AutoCAD 2000 command when used for constructing geometric shapes. The following drawing related commands are included in this chapter: ARC, DONUT, ELLIPSE, MLINE, POINT, POLYGON, RAY, RECTANG, SPLINE, and XLINE. Express tools that deal with drawing commands are also discussed in this chapter. Tutorial exercises follow along with additional problems at the end of this chapter.

CHAPTER 6 – ADDING TEXT TO YOUR DRAWING

Use this chapter for placing text in your drawing. Various techniques for accomplishing this task include the use of the DTEXT and MTEXT commands. The creation of text styles and the ability to edit text once it is placed in a drawing is also included. Express tools that deal with text are also discussed in this chapter. A tutorial exercise is included at the end of this chapter.

CHAPTER 7 – OBJECT GRIPS AND CHANGING THE PROPERTIES OF OBJECTS

The topic of grips and how they are used to enhance the modification of a drawing is presented. The ability to modify objects through the Properties dialog box is discussed in great detail. Tutorial exercises are included at the end of this chapter to reinforce the importance of grips and changing the properties of objects.

CHAPTER 8 – SHAPE DESCRIPTION/MULTIVIEW PROJECTION

Describing shapes and producing multiview drawings using AutoCAD 2000 are the focus of this chapter. The basics of shape description are discussed along with proper use of linetypes, fillets, rounds, chamfers, and runouts. A tutorial exercise is available at the end of this chapter and outlines the steps used for creating a multiview drawing using AutoCAD 2000. A series of freehand sketching exercises along with additional drawing problems are provided at the end of the chapter.

CHAPTER 9 – BEGINNING DRAWING LAYOUTS

This chapter deals with the creation of layouts before a drawing is plotted out. A layout takes the form of a sheet of paper and is referred to as Paper Space. A wizard to assist in the creation of layouts is also described. Once a layout of an object is created, scaling through the Viewports toolbar is discussed. Numerous layouts for the same drawing is also introduced including a means of freezing layers only in certain layouts. Various tutorials are provided at the end of this chapter to reinforce the importance of layouts.

CHAPTER 10 – PLOTTING YOUR DRAWING

Beginning and advanced of outputting your drawings to a print or plot device are discussed in this chapter through a series of tutorial exercises. One tutorial demonstrates the use of the Add-A-Plotter Wizard to configure a new plotter. Plotting from a layout is discussed through a tutorial. This includes the assignment of a sheet size. Tutorial exercises are also provided to create color-dependent and named plot styles. These allow you to control the appearance of your plot.

CHAPTER 11 – DIMENSIONING BASICS

Basic dimensioning rules are introduced in the beginning of this chapter before all AutoCAD 2000 dimensioning commands are discussed. Placing linear, diameter, and radius dimensions are then demonstrated. The powerful QDIM command allows you to place baseline, continuous and other dimension groups in a single operation. A tutorial exercise follows along with additional dimensioning problems at the end of this chapter.

CHAPTER 12 – THE DIMENSION STYLE MANAGER

A thorough discussion of the use of the Dimension Styles Manager dialog box is included in this chapter. The ability to create, modify, and override dimension styles are discussed. A detailed tutorial exercise follows along with additional dimensioning problems at the end of this chapter.

CHAPTER 13 – ANALYZING 2D DRAWINGS

This chapter provides information on analyzing a drawing for accuracy purposes. The AREA, ID, LIST, and DIST commands are discussed in detail in addition to how they are used to determine the accuracy of various objects in a drawing. A series of tutorial exercises follow that allows the user to test their drawing accuracy. Numerous problems along with questions on each problem are provided at the end of this chapter that further test a user's drawing accuracy.

CHAPTER 14 – SECTION VIEWS

The basics of section views are described in this chapter including full, half, assembly, aligned, offset, broken, revolved, removed, and isometric sections. Hatching techniques through the use of the Boundary Hatch dialog box are also discussed. Material on the use of the SUPERHATCH command which is part of the Express tools is also provided. Tutorial exercises follow along with additional problems at the end of the chapter that deal with the topic of section views.

CHAPTER 15 – AUXILIARY VIEWS

Producing auxiliary views using AutoCAD 2000 is discussed in this chapter. Items discussed include rotating the snap at an angle to project lines of sight perpendicular to a surface to be used in the preparation of the auxiliary view. One tutorial exercise and additional problems follow in this chapter.

CHAPTER 16 – ISOMETRIC DRAWINGS

This chapter discusses constructing isometric drawings with particular emphasis on using the SNAP command along with the Style option in AutoCAD 2000 used to create an isometric grid. Methods of toggling between right, top and left isometric modes will be explained. In addition to isometric basics, creating circles and angles in isometric are also discussed. Tutorial exercises on producing isometric drawings follow along with additional problems at the end of this chapter.

CHAPTER 17 – BLOCK CREATION, AUTOCAD DESIGNCENTER, AND MDE

This chapter covers the topic of creating blocks in AutoCAD 2000. Creating local and global blocks such as doors, windows, and pipe symbols will be demonstrated. The Insert dialog box is discussed as a means of inserting blocks into drawings. The chapter continues by explaining the many uses of the AutoCAD DesignCenter. This feature allows the user to display a browser containing blocks, layers, and other named objects that can be dragged into the current drawing file. This chapter also discusses the ability to open numerous drawings through the Multiple Document Environment and transfer objects and properties between drawings. Tutorial exercises and drawing problems can be found at the end of this chapter.

CHAPTER 18 – EXTERNAL REFERENCES

The chapter discusses the use of External References in drawings. An external reference is a drawing file that can be attached to another drawing file. Once the referenced drawing file is edited or changed, these changes are automatically seen once the drawing containing the external reference is opened again. Performing in-place editing of external references will also be demonstrated. A tutorial exercise follows at the end of this chapter to practice this important design concept.

CHAPTER 19 – MULTIPLE VIEWPORT DRAWING LAYOUTS

This very important chapter is designed to utilize advanced techniques used in laying out a drawing before it is plotted. The ability to layout a drawing consisting of various images at different scales will also be discussed. The ability to create user-defined rectangular viewports through the MVIEW command and is discussed. Non-rectangular viewports will also be demonstrated. A series of tutorial exercises follow to practice this advanced layout mode.

CHAPTER 20 – SOLID MODELING FUNDAMENTALS

This chapter begins with a comparison between isometric, extruded, wireframe, surfaced, and solid model drawings. The chapter continues with a detailed discussion on the use of the User Coordinate System and how it is positioned to construct objects in 3D. The display of 3D images through the 3DORBIT, VPOINT and 3DCORBIT commands is discussed. Creating various solid primitives such as boxes, cones and cylinders is discussed in addition to the ability to construct complex solid objects through the use of the Boolean operations of union, subtraction, and intersection. The chapter continues on by discussing extruding and rotating operations for creating solid models in addition to filleting and chamfering solid models. Analyzing solid models is also covered in this chapter. Because of the importance of this design paradigm taking you from 2D to 3D, four tutorial exercises follow along with additional problems at the end of this chapter.

CHAPTER 21 – EDITING SOLID MODELS

The ALIGN and ROTATE3D commands are discussed as a means of introducing the editing capabilities of AutoCAD 2000 on solid models. Modifications can also be made to a solid model through the use of the SOLIDEDIT command, which is discussed in this chapter. The ability to extrude existing faces, imprint objects, and create thin wall shells will be demonstrated throughout this chapter. Tutorial exercises can be found at the end of this chapter.

CHAPTER 22 – CREATING ORTHOGRAPHIC VIEWS FROM A SOLID MODEL

Once the solid model is created, the SOLVIEW command is used to layout 2D views of the model and the SOLDRAW command is used to draw the 2D views. Layers are automatically created to assist in the annotation of the drawing through the use of dimensions. A lengthy tutorial exercise is available at the end of this chapter along with instructions on how to apply the techniques learned in this chapter to other solid models.

ONLINE COMPANION

This new edition contains a special Internet companion piece. The online Companion is your link to AutoCAD on the Internet. We've compiled supporting resources with links to a variety of sites. Not only can you find out about training and education, industry sites, and the online community, we also point to valuable archives compiled for AutoCAD users from various Web sites. In addition, there is information of special interest to users of *The AutoCAD Tutor for Engineering Graphics*. These include updates, information about the author, and a page where you can send us your comments. You can find the Online Companion at:

http://www.autodeskpress.com/onlinecompanion.html

When you reach the Online Companion page, click on the title *The AutoCAD Tutor for Engineering Graphics*.

ACKNOWLEDGEMENTS

I wish to thank the staff at Autodesk Press for their assistance with this document, especially John Fisher, Sandy Clark, Jennifer Gaines, and Mary Beth Vought. Thanks also go out to John Shanley of Phoenix Creative Graphics for his part in the desktop publishing aspects of this document. I would also like to thank Don Baer of Trident Technical College for performing the technical edit on the entire manuscript. Special thanks go to Kevin Lang for his expertise in completing Chapters 17, 18, 19, and 20.

The publisher and author would line to thank and acknowledge the many professionals who reviewed the manuscript to help us publish this AutoCAD 2000 text. A special acknowledgement is due to the following instructors who reviewed the chapters in detail:

Don Baer, Trident Technical College, Charleston, SC

George W. Baggs, Augusta Technical Institute, Augusta, GA

Edith Hansen, Big Bend Community College, Moses Lake, WA

Bill Julian, Hudson Valley Community College, Wynantskill, NY

Kevin Lang, Trident Technical College, Charleston, SC

Gerald Vinson, Texas A & M University, College Station, TX

DEDICATION

This book is dedicated in memory to Frank Dagostino for his commitment to education and dedication to quality of instruction in the field of Engineering Design Graphics.

ABOUT THE AUTHOR

Alan J. Kalameja is the Department Head of Computer-Integrated Manufacturing at Trident Technical College located in Charleston, South Carolina. He has been at the College for over 18 years and has been using AutoCAD since 1984. He directs the Premier AutoCAD training Center at Trident, which is charged with providing industry training to local and regional companies. Currently, he is a Technical Specialist with Autodesk, is a member of the AutoCAD Certification Exam Board, and has authored AutoCAD Certification Exam Preparation Manuals in Release 12, 13, and 14. He is also the author of the AutoCAD Tutor for Engineering Graphics R14. All of his books are published by Delmar/Autodesk Press.

CONVENTIONS

All tutorials in this publication use the following conventions in the instructions:

Whenever you are told to enter text, the text appears in boldface type. This may take the form of entering an AutoCAD command or entering such information as absolute, relative or polar coordinates. You must follow these and all text inputs by pressing the ENTER key to execute the input. For example, to draw a line using the LINE command from point 3,1 to 8,2, the sequence would look like the following:

Command: **L** *(For LINE)*
Specify first point: **3,1**
Specify next point or [Undo]: **8,2**
Specify next point or [Undo]: *(Press ENTER to exit this command)*

Instructions for selecting objects are in italic type. When instructed to select an object, move the pickbox on the object to be selected and press the pick button on the mouse or digitizing puck.

If you enter the wrong command for a particular step, you may cancel the command by pressing the ESC key. This key is located in the upper left hand corner of any standard keyboard.

Instructions in this tutorial are designed to enter all commands, options, coordinates, etc., from the keyboard. You may use the same commands by selecting them from the screen menu, digitizing tablet, pulldown menu area, or from one of the any floating toolbars.

NOTES TO THE STUDENT AND INSTRUCTOR CONCERNING THE USE OF TUTORIAL EXERCISES

Various tutorial exercises have been designed throughout this book and can be found at the end of each chapter. The main purpose of each tutorial is to follow a series of steps towards the completion of a particular problem or object. Performing the tutorial will also prepare you to undertake the numerous drawing problems also found at the end of each chapter.

As you work on the tutorials, you should follow the steps very closely, taking care not to make a mistake. However, most individuals rush through the tutorials to get the correct solution in the quickest amount of time only to forget the steps used to complete the tutorial. A typical comment made by many is "I completed the tutorial…but I don't understand what I did to get the correct solution."

It is highly recommended to both student and instructor that all tutorial exercises be performed two or even three times. Completing the tutorial the first time will give you the confidence that it can be done; however, you may not understand all of the steps involved. Completing the tutorial a second time will allow you to focus on where certain operations are performed and why things behave the way they do. This still may not be enough. More complicated tutorial exercises may need to be performed a third time. This will allow you to anticipate each step and have a better idea what operation to perform in each step. Only then will you be comfortable and confident to attempt the many drawing problems that follow the tutorial exercises.

The CD-ROM in the back of the book contains AutoCAD drawing files for the tutorial exercises. To use drawing files, copy files to your hard drive, then remove their read-only attribute. Files cannot be used without AutoCAD. Files are located in the /Drawing Files/ directory.

SUPPLEMENTS

E.RESOURCE

e.resource™ — This is an educational resource that creates a truly electronic classroom. It is a CD-ROM containing tools and instructional resources that enrich your classroom and make your preparation time shorter. The elements of *e.resource* link directly to the text and tie together to provide a unified instructional system. Spend your time teaching, not preparing to teach.

ISBN 0-7668-1239.1

Features contained in *e.resource* include:

- **Syllabus:** Lesson plans created by chapter. You have the option of using these lesson plans with your own course information..

- **Chapter Hints:** Objectives and teaching hints that provide the basis for a lecture outline that helps you to present concepts and material.

- **Answers to Review Questions:** These solutions enable you to grade and evaluate end of chapter tests.

- **PowerPoint® Presentation:** These slides provide the basis for a lecture outline that helps you to present concepts and material. Key points and concepts can be graphically highlighted for student retention. There are over 300 slides, covering every chapter in the text.

· **Computerized Test Bank:** Over 600 questions of varying levels of difficulty are provided in true/false and multiple-choice formats. Exams can be generated to assess student comprehension or questions can be made available to the student for self-evaluation.

· **CADD Drawing Files:** Drawing files that link to figures and exercises in the text are provided. These drawings can be launched and manipulated to illustrate the use of commands or the completion of an exercise.

· **World Class Manager:** This allows instructors to automatically extract and track data from on-line tests, practice tests and tutorials created by World Class Test. Instructors can also create reports such as individual student reports, missing assignment reports, a summary of grade distribution, class lists and more. All statistics are updated automatically as data is entered or changed.

· **Video and Animation Resources:** These AVI files graphically depict the execution of key concepts and commands in drafting, design, and AutoCAD and let you bring multi-media presentations into the classroom.

Spend your time teaching, not preparing to teach!

CHAPTER 1

Getting Started with AutoCAD 2000

THE AUTOCAD 2000 DRAWING SCREEN

When you first launch AutoCAD 2000, you will see the drawing screen illustrated in Figure 1–1. At the top of this screen is the menu bar, with the pull-down menus containing various commands under such categories as File, Edit, View, and Insert (to name a few). Directly beneath the menu bar is the Standard toolbar containing buttons for such commands as NEW and OPEN in addition to displaying two flyouts supporting the Zoom and Object Snap features. The Object Properties toolbar (below the Standard toolbar) is designed for manipulating layers, color, linetypes, and line weights. Two additional toolbars appear on the left side of Figure 1–1. The Draw toolbar contains buttons for such commands as LINE, CIRCLE, ARC, and BHATCH. The Modify toolbar contains FILLET, CHAMFER, ERASE, and ARRAY, to name just a few. The User Coordinate System icon is displayed in the lower corner of the drawing editor to alert you to the coordinate system currently in use. The "W" in the icon indicates that you are in the World Coordinate System.

At the lower part of the display screen is the command prompt area, where AutoCAD 2000 prompts you for input depending on which command is currently in progress. At the very bottom of the screen is the status bar. Use this area to toggle on or off the following modes: Coordinate Display, Snap, Grid, Ortho, Polar Snap, Object Snap, Object Tracking, Line Weight, and Model/Paper. Click once on the button to turn the mode on or off. Scroll bars are present just below and to the right of the drawing editor screen. You can use these to pan to a different screen location especially when the screen has been magnified using one of the ZOOM command options. You can use the graphics cursor to pick points on the screen or select objects to edit.

When you begin a new drawing, you perform the drawing inside Model mode. To plot the drawing out, you switch to one of the Layout modes presented by tabs at the bottom of the display screen.

2

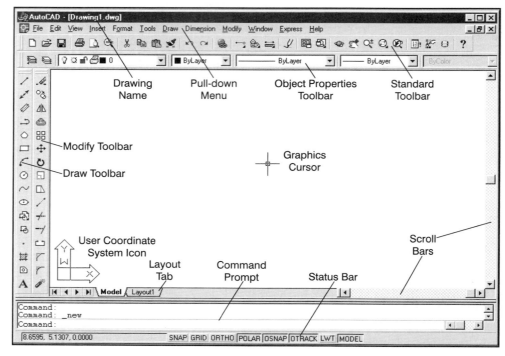

Figure 1–1

USING KEYBOARD FUNCTION KEYS

Figure 1–2

Once you open a drawing file, you have additional aids to control such settings as grid and snap. Illustrated in Figure 1–2 is a group of function keys, which are found above the alphanumeric keys of a typical keyboard. These function keys are labeled F1 through F12; only keys F1 through F11 will be discussed here. Software companies commonly program certain functions into these keys to assist you in their application, and AutoCAD 2000 is no exception. Most keys act as switches, which turn functions on or off (see Table 1–1).

Table 1–1

Function Key	Definitions
F1	AutoCAD Help Topics
F2	Toggle Text/Graphics Screen
F3	Object Snap settings On/Off
F4	Toggle Tablet Mode On/Off
F5	Toggle Isoplane Modes
F6	Toggle Coordinates On/Off
F7	Toggle Grid Mode On/Off
F8	Toggle Ortho Mode On/Off
F9	Toggle Snap Mode On/Off
F10	Toggle Polar Mode On/Off
F11	Toggle Object Snap Tracking On/Off

When you press F1, the AutoCAD Help Topics dialog box displays.

Pressing F2 takes you to the text screen consisting of a series of previous prompt sequences. This may be helpful for viewing the previous command sequence in text form.

Pressing F3 toggles the current Object Snap settings on or off. This will be discussed later in this chapter.

Use F4 to toggle Tablet mode on or off. This mode is only activated when the digitizing tablet has been calibrated for the purpose of tracing a drawing into the computer.

Pressing F5 scrolls you through the three-supported Isoplane modes used to construct isometric drawings (Right, Left, and Top).

F6 toggles the coordinate display, located in the lower left corner of the status bar, on or off. When the coordinate display is off, the coordinates update when you pick an area of the screen with a mouse or digitizer puck. When the coordinate display is on, the coordinates dynamically change with the current position of the cursor.

F7 turns the display of grid on or off. The actual grid spacing is set by the GRID command and not by this function key.

F8 toggles ORTHO mode on or off. Use this key to force objects such as lines to be drawn horizontally or vertically.

Use F9 to toggle Snap mode on or off. The SNAP command sets the current snap value.

Use F10 to toggle the Polar Tracking on or off. Use F11 to toggle Object Snap Tracking on or off. Both of these function keys will be discussed in greater detail in this chapter.

The status bar, illustrated in Figure 1–3, holds most commands controlled by the function keys. For example, click once on the GRID button to turn the grid on; click again on the GRID button to turn the grid off. You can control Object Snap (F3), the coordinate display (F6), Grid (F7), Ortho (F8), Snap (F9), Polar Tracking (F10), and Object Snap Tracking (F11) in the same way. Additional buttons such as LWT turn Lineweight mode on or off. Clicking on the MODEL button enters the Page Setup mode. Both the LWT and the MODEL buttons will be discussed in greater detail in the chapters that follow.

Figure 1–3

Right-clicking on one of the button displays the shortcut menu in Figure 1–4. Click on Settings... to access various dialog boxes that control certain features associated with the button.

Figure 1–4

METHODS OF CHOOSING COMMANDS

AutoCAD 2000 provides numerous ways to enter or choose commands for constructing or editing drawings. The Draw pull-down menu, shown in Figure 1–5, is very popular with many users. It allows you to choose most AutoCAD 2000 commands. You can also enter commands directly from the keyboard. This practice is popular for users who are already familiar with the commands.

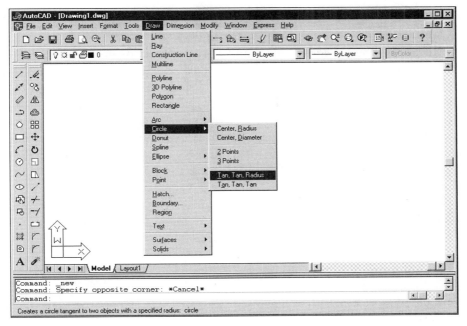

Figure 1–5

TOOLBARS

Illustrated in Figure 1–6B is an additional method for initiating commands, namely, through a toolbar. The toolbar illustrated in Figure 1–6A gives you access to Zoom commands. You can find other toolbars by choosing Toolbars... from the View pull-down menu. By default, the toolbar is displayed flat. When the cursor rolls over a tool, a 3D border displays along with a tooltip that explains the purpose of the command (see Figure 1–6B).

Figure 1–6A

Figure 1–6B

ACTIVATING TOOLBARS

Many toolbars are available to assist you in using other types of commands. Choosing Toolbars...from the View pull-down menu, as shown in Figure 1–7A, displays the Toolbars dialog box, illustrated in Figure 1–7B. By default, four toolbars are already active or displayed in all drawings; these are the Object Properties, Standard, Draw, and Modify toolbars. To make another toolbar active, click in the empty box next to the name of the toolbar. This will display the toolbar and allow you to preview it before making it a part of the display screen. If this toolbar is correct, click on the Close button. This will close the Toolbars dialog box but leave the selected toolbar. Illustrated in Figure 1–7C is the Dimension toolbar, which is activated when a check is placed in the box next to Dimension in the Toolbars dialog box.

Figure 1–7A

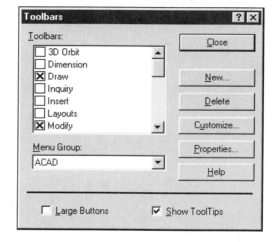

Figure 1–7B

Figure 1–7C

RIGHT-CLICK SHORTCUT MENUS

Many shortcut or cursor menus have been developed to assist with the rapid access of commands. Clicking on the right mouse button activates these commands. The Default shortcut menu is illustrated in Figure 1–8. It displays whenever you right-click in the drawing area and no command or selection set is in progress.

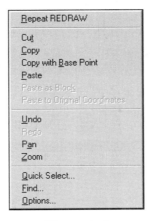

Figure 1–8

Illustrated in Figure 1–9 is an example of the Edit shortcut menu. This shortcut consists of numerous editing and selection commands. This menu activates whenever you right-click in the display screen with an object or group of objects selected but no command is in progress.

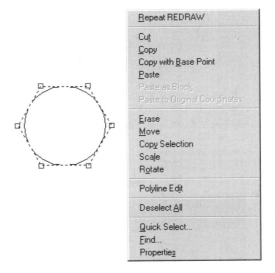

Figure 1–9

Right-clicking in the Command prompt area of the display screen activates the shortcut menu in Figure 1–10. This menu provides for a quick way to access the Options dialog box. Also, a record of the six most recent commands is kept, which allows you to select from this group of previously used commands.

Illustrated in Figure 1–11 is an example of a Command-Mode shortcut menu. When you enter a command and right-click, this menu will display options of the command. This menu supports a number of commands. In the figure, the 3P, 2P, and Ttr (tan tan radius) listings are all options of the CIRCLE command.

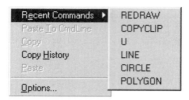

Figure 1–10   **Figure 1–11**

ICON MENUS

Icon menus display graphical representations of commands or command options. The icon menu in Figure 1–12 shows numerous hatch patterns. The graphic images of the patterns allow you preview the pattern before you place it in a drawing. Icon menus allow you to select items quickly because of their graphical nature.

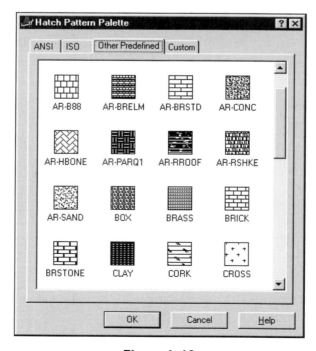

Figure 1–12

COMMAND ALIASES

You can enter commands directly from the keyboard. However, you must know the exact command name, including its exact spelling. To assist with the entry of AutoCAD 2000 commands from the keyboard, numerous commands are available in shortened form, and these shorter forms are called *aliases*. The list in Table 1–2 shows the command aliases that are located in the support file ACAD.PGP. Once you are comfortable with the keyboard, command aliases provide a fast and efficient method of activating AutoCAD 2000 commands.

Table 1–2 Command Aliases

Alias	Command	Alias	Command
3A,	*3DARRAY	-CH,	*CHANGE
3DO,	*3DORBIT	CHA,	*CHAMFER
3F,	*3DFACE	COL,	*COLOR
3P,	*3DPOLY	CP,	*COPY
A,	*ARC	D,	*DIMSTYLE
ADC,	*ADCENTER	DAL,	*DIMALIGNED
AA,	*AREA	DAN,	*DIMANGULAR
AL,	*ALIGN	DBA,	*DIMBASELINE
AP,	*APPLOAD	DBC,	*DBCONNECT
AR,	*ARRAY	DCE,	*DIMCENTER
ATT,	*ATTDEF	DCO,	*DIMCONTINUE
-ATT,	*-ATTDEF	DDI,	*DIMDIAMETER
ATE,	*ATTEDIT	DED,	*DIMEDIT
-ATE,	*-ATTEDIT	DI,	*DIST
ATTE,	*-ATTEDIT	DIV,	*DIVIDE
B,	*BLOCK	DLI,	*DIMLINEAR
-B,	*-BLOCK	DO,	*DONUT
BH,	*BHATCH	DOR,	*DIMORDINATE
BO,	*BOUNDARY	DOV,	*DIMOVERRIDE
-BO,	*-BOUNDARY	DR,	*DRAWORDER
BR,	*BREAK	DRA,	*DIMRADIUS
C,	*CIRCLE	DS,	*DSETTINGS
CH,	*PROPERTIES	DST,	*DIMSTYLE

Alias	Command	Alias	Command
DT,	*DTEXT	-LA,	*-LAYER
DV,	*DVIEW	LE,	*QLEADER
E,	*ERASE	LEN,	*LENGTHEN
ED,	*DDEDIT	LI,	*LIST
EL,	*ELLIPSE	LO,	*-LAYOUT
EX,	*EXTEND	LS,	*LIST
EXIT,	*QUIT	LT,	*LINETYPE
EXP,	*EXPORT	-LT,	*-LINETYPE
EXT,	*EXTRUDE	LTS,	*LTSCALE
F,	*FILLET	LW,	*LWEIGHT
FI,	*FILTER	M,	*MOVE
G,	*GROUP	MA,	*MATCHPROP
-G,	*-GROUP	ME,	*MEASURE
GR,	*OPTIONS	MI,	*MIRROR
H,	*BHATCH	ML,	*MLINE
-H,	*HATCH	MO,	*PROPERTIES
HE,	*HATCHEDIT	MS,	*MSPACE
HI,	*HIDE	MT,	*MTEXT
I,	*INSERT	MV,	*MVIEW
-I,	*-INSERT	O,	*OFFSET
IAD,	*IMAGEADJUST	OP,	*OPTIONS
IAT,	*IMAGEATTACH	OS,	*OSNAP
ICL,	*IMAGECLIP	-OS,	*-OSNAP
IM,	*IMAGE	P,	*PAN
-IM,	*-IMAGE	-P,	*-PAN
IMP,	*IMPORT	PA,	*PASTESPEC
IN,	*INTERSECT	PE,	*PEDIT
INF,	*INTERFERE	PL,	*PLINE
IO,	*INSERTOBJ	PO,	*POINT
L,	*LINE	POL,	*POLYGON
LA,	*LAYER	PR,	*OPTIONS

Alias	Command	Alias	Command
PRE,	*PREVIEW	SU,	*SUBTRACT
PRINT,	*PLOT	T,	*MTEXT
PS,	*PSPACE	-T,	*-MTEXT
PU,	*PURGE	TA,	*TABLET
R,	*REDRAW	TH,	*THICKNESS
RA,	*REDRAWALL	TI,	*TILEMODE
RE,	*REGEN	TO,	*TOOLBAR
REA,	*REGENALL	TOL,	*TOLERANCE
REC,	*RECTANGLE	TOR,	*TORUS
REG,	*REGION	TR,	*TRIM
REN,	*RENAME	UC,	*DDUCS
-REN,	*-RENAME	UCP,	*DDUCSP
REV,	*REVOLVE	UN,	*UNITS
RM,	*DDRMODES	-UN,	*-UNITS
RO,	*ROTATE	UNI,	*UNION
RPR,	*RPREF	V,	*VIEW
RR,	*RENDER	-V,	*-VIEW
S,	*STRETCH	VP,	*DDVPOINT
SC,	*SCALE	-VP,	*VPOINT
SCR,	*SCRIPT	W,	*WBLOCK
SE,	*DSETTINGS	-W,	*-WBLOCK
SEC,	*SECTION	WE,	*WEDGE
SET,	*SETVAR	X,	*EXPLODE
SHA,	*SHADE	XA,	*XATTACH
SL,	*SLICE	XB,	*XBIND
SN,	*SNAP	-XB,	*-XBIND
SO,	*SOLID	XC,	*XCLIP
SP,	*SPELL	XL,	*XLINE
SPL,	*SPLINE	XR,	*XREF
SPE,	*SPLINEDIT	-XR,	*-XREF
ST,	*STYLE	Z,	*ZOOM

STARTING A NEW DRAWING

To begin a new drawing file, select the NEW command with the button shown in the following command sequence or click on the New command on the File pull-down menu, as shown in Figure 1–13A. You can also enter the NEW command at the command prompt, similar to the following command sequence:

Command: **NEW**

Entering the NEW command displays the Create New Drawing dialog box illustrated in Figure 1–13B. Available in this dialog box are the following buttons: Start from Scratch, Use a Template, Use a Wizard, and Tips on each of the options.

Clicking on the Start from Scratch button begins the AutoCAD 2000 drawing file with such default settings as a sheet of paper 12 units by 9 units and four-place decimal precision.

Clicking on the Use a Template button displays an edit box containing various files that have an extension of .DWT (see Figure 1–13C). These represent template files and are designed to conform to various standard drawing sheet sizes. Associated with each template file is a corresponding title block that is displayed in the Preview area. You may scroll through the various template files and get a glimpse of the title block matched to the template file.

The Use a Wizard button has two options associated with it: Quick Setup and Advanced Setup. Both these options will be discussed in greater detail in Chapter 2, "Drawing Setup and Organization."

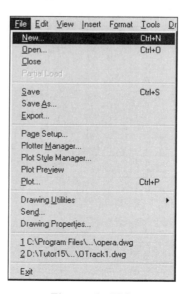

Figure 1–13A

Figure 1–13B

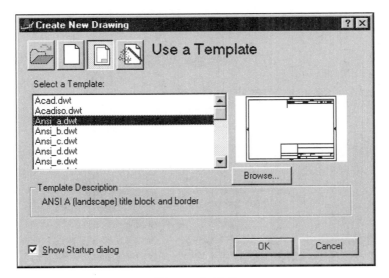

Figure 1–13C

OPENING AN EXISTING DRAWING

The OPEN command is used to edit a drawing that has already been created. Choose this command from the File menu as shown in Figure 1–14. When you select this command, a dialog box appears similar to Figure 1–15. Listed in the edit box area are all files that match the type shown at the bottom of the dialog box. Since the file type is .DWG, all drawing files supported by AutoCAD 2000 are listed. To choose a different subdirectory, use standard Windows file management techniques by clicking in the Look in edit box at "A." This will display all subdirectories associated with the drive. Clicking on the subdirectory will display any drawing files contained in it.

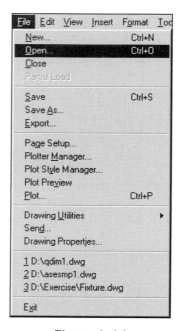

Figure 1–14

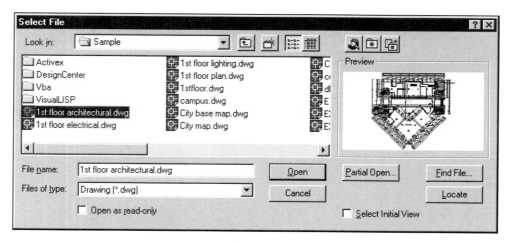

Figure 1–15

When you first enter AutoCAD 2000, the Startup dialog box in Figure 1–16 displays. In addition to the buttons for Use a Wizard, Use a Template, Start from Scratch, there is an Open a Drawing button, which allows you to open an existing drawing. The Select a File edit box displays the existing drawing files in the current subdirectory. To perform a search for drawing files in a different subdirectory, choose the Browse button. As you choose drawing files, valid AutoCAD Release 13, AutoCAD LT 95, Release 14, and AutoCAD 2000 files will display in the Preview area of the dialog box. Drawing files created in versions prior to Release 13 must be opened in AutoCAD 2000 and saved before they preview.

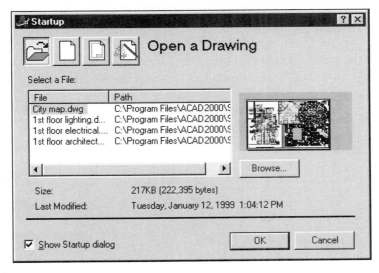

Figure 1–16

A view is a previously saved portion of the drawing display screen, which will be explained later in this chapter. Placing a check in the Select Initial View check box of the enlarged area of the Select File dialog box in Figure 1–17A causes the Select Initial View dialog box to display, illustrated in Figure 1–17B, before the drawing file loads. Listed are a series of view names that were already created. Instead of opening a drawing file where the entire drawing is displayed, you have the option of opening the drawing file and automatically going to one of the views. This is considered good practice; rather than bring up the entire drawing and then zoom into a specific area of the screen, you bypass this zoom operation by going directly to the named view. Before the drawing file loads, the dialog box in Figure 1–17B displays on the screen, alerting you to select the desired view.

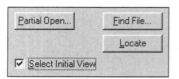

Figure 1–17A   Figure 1–17B

If the Open as read-only check box of the Select File dialog box is checked as in Figure 1–18A, a drawing file displays on the screen and objects can be drawn or edited as with all drawing files. However, none of the changes may be saved because the drawing may only be viewed; this is the purpose of read-only mode. If any changes are made and a save is attempted, the alert box illustrated in Figure 1–18B informs you that the current file is write protected and no permanent changes may be made.

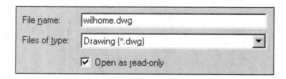

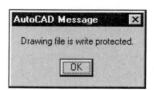

Figure 1–18A   Figure 1–18B

THE BROWSE TAB

Click on the Find File button of the Select File dialog box as shown in Figure 1–19A to display the Browse/Search dialog box illustrated in Figure 1–19B. Use the Browse tab to preview groups of drawings; select the desired file to open as the current drawing file. Click on the Search tab to activate a dialog box that enables you to enter drive information, file type information, and dates and times of the file in order to perform a search.

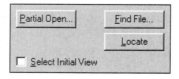

Figure 1–19A

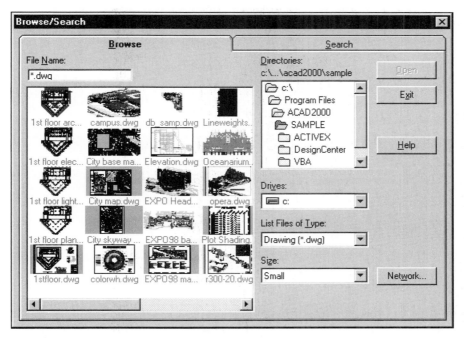

Figure 1–19B

BASIC DRAWING TECHNIQUES THE LINE COMMAND

Use the LINE command to construct a line from one endpoint to the other. As the first point of the line is marked, the rubber-band cursor is displayed along with the normal crosshairs to help you see where the next line segment will be drawn. The LINE command stays active until you either use the Close option or issue a null response by pressing ENTER at the prompt "To point." Study Figure 1–20 and the following command sequence for using the LINE command.

Command: **L** *(For LINE)*
Specify first point: *(Mark a point at "A")*
Specify next point or [Undo]: *(Pick a point at "B")*
Specify next point or [Undo]: (Pick a point at "C")
Specify next point or [Close/Undo]: *(Pick a point at "D")*
Specify next point or [Close/Undo]: *(Pick a point at "E")*
Specify next point or [Close/Undo]: *(Pick a point at "F")*
Specify next point or [Close/Undo]: C *(To close the shape and exit the command)*

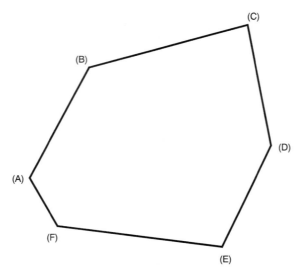

Figure 1–20

From time to time, you might make mistakes in the LINE command by drawing an incorrect segment. As illustrated in Figure 1–21, segment DE is drawn incorrectly. Instead of exiting the LINE command and erasing the line, you can use a built-in Undo within the LINE command. This removes the previously drawn line while still remaining in the LINE command. Refer to Figure 1–21 and the following prompts to use the Undo option of the LINE command.

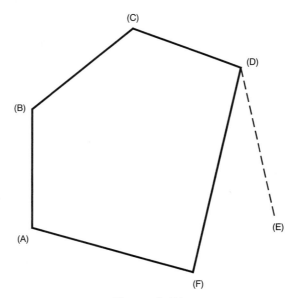

Figure 1–21

Command: **L** *(For LINE)*
Specify first point: *(Mark a point at "A")*
Specify next point or [Undo]: *(Pick a point at "B")*
Specify next point or [Undo]: *(Pick a point at "C")*
Specify next point or [Close/Undo]: *(Pick a point at "D")*
Specify next point or [Close/Undo]: *(Pick a point at "E")*
Specify next point or [Close/Undo]: U *(To undo or remove the segment from "D" to "E"*
   and still remain in the LINE command)
Specify next point or [Close/Undo]: *(Pick a point at "F")*
Specify next point or [Close/Undo]: End *(For Endpoint mode)*
of *(Select the endpoint of the line segment at "A")*
Specify next point or [Close/Undo]: *(Press ENTER to exit this command)*

Another option of the LINE command is the Continue option. The dashed line segment in Figure 1–22 was the last segment drawn before the LINE command was exited. To pick up at the last point of a previously drawn line segment, type the LINE command and press ENTER. This will activate the Continue option of the LINE command.

Command: **L** *(For LINE)*
Specify first point: *(Press ENTER to activate Continue Mode)*
Specify next point or [Undo]: *(Pick a point at "B")*
Specify next point or [Undo]: *(Pick a point at "C")*
Specify next point or [Close/Undo]: End *(For Endpoint mode)*
of *(Select the endpoint of the vertical line segment at "A")*
Specify next point or [Close/Undo]: *(Press ENTER to exit this command)*

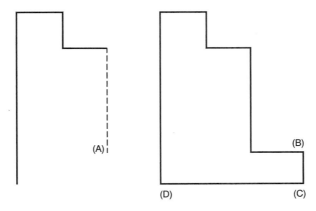

Figure 1–22

THE DIRECT DISTANCE MODE FOR DRAWING LINES

Another method is available for constructing lines, and it is called drawing by Direct Distance mode. In this method, the direction a line will be drawn is guided by the location of the cursor. You enter a value, and the line is drawn at the specified distance at the angle specified by the cursor. This mode works especially well when drawing horizontal and vertical lines. Illustrated in Figure 1–23 is an example of how the Direct Distance mode is used. To begin, Ortho mode is turned on. Use the following command sequence to construct the line segments.

Command: **L** *(For LINE)*
Specify first point: **2.00,2.00**
Specify next point or [Undo]: *(Move the cursor to the right and enter a value of 7.00 units)*
Specify next point or [Undo]: *(Move the cursor up and enter a value of 5.00 units)*
Specify next point or [Close/Undo]: *(Move the cursor to the left and enter a value of 4.00 units)*
Specify next point or [Close/Undo]: *(Move the cursor down and enter a value of 3.00 units)*
Specify next point or [Close/Undo]: *(Move the cursor to the left and enter a value of 2.00 units)*
Specify next point or [Close/Undo]: C *(To close the shape and exit the command)*

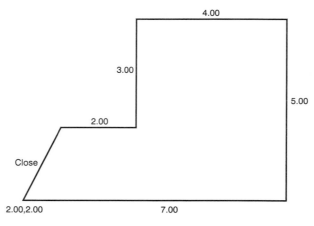

Figure 1–23

Illustrated in Figure 1–24 is another example of an object drawn with Direct Distance mode. Each angle was constructed from the location of the cursor. In this example, Ortho mode is turned off.

Command: **L** *(For LINE)*
Specify first point: *(Pick a point at "A")*
Specify next point or [Undo]: *(Move the cursor and enter 3.00)*
Specify next point or [Undo]: *(Move the cursor and enter 2.00)*

Specify next point or [Close/Undo]: *(Move the cursor and enter 1.00)*
Specify next point or [Close/Undo]: *(Move the cursor and enter 4.00)*
Specify next point or [Close/Undo]: *(Move the cursor and enter 2.00)*
Specify next point or [Close/Undo]: *(Move the cursor and enter 1.00)*
Specify next point or [Close/Undo]: *(Move the cursor and enter 1.00)*
Specify next point or [Close/Undo]: C *(To close the shape and exit the command)*

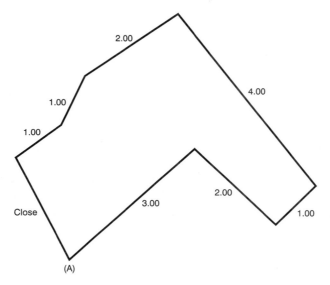

Figure 1–24

USING OBJECT SNAP FOR GREATER PRECISION

The most important mode used for locking onto key locations of objects is Object Snap. Figure 1–25 is an example of constructing a vertical line connecting up the endpoint of the fillet with the endpoint of the line at "A." The LINE command is entered and the Endpoint mode activated. When the cursor moves over a valid endpoint, an Object Snap symbol appears along with a tooltip telling you which OSNAP mode is currently being used. Another example of when you would use Object Snap is in dimensioning applications where exact endpoints and intersections are needed.

Command: **L** *(For LINE)*
Specify first point: End *(For Endpoint mode)*
of *(Pick the endpoint of the fillet illustrated in Figure 1–25)*
Specify next point or [Undo]: End *(For Endpoint mode)*
of *(Pick the endpoint of the line at "A")*
Specify next point or [Undo]: *(Press ENTER to exit this command)*

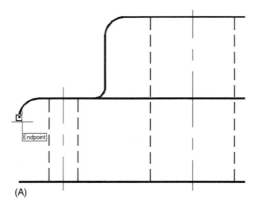

(A)

Figure 1–25

You can choose Object Snap modes in a number of different ways. Illustrated in Figure 1–26 is the standard Object Snap toolbar. Figure 1–27 shows the Object Snap modes that can activated when you hold down SHIFT and press the right mouse button. This pop-up menu will display on the screen wherever the cursor is currently positioned. Another method of activating Object Snap modes is from the keyboard. When you enter Object Snap modes from the keyboard, only the first three letters are required. The following pages give examples of the application of all Object Snap modes.

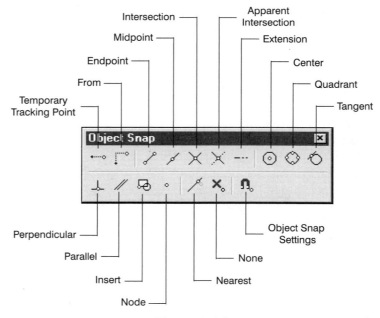

Figure 1–26

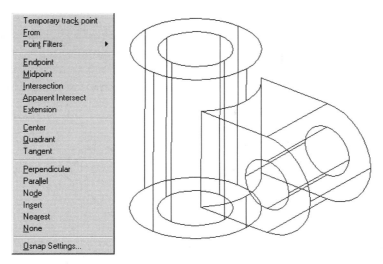

Figure 1–27

OBJECT SNAP MODES

OBJECT SNAP CENTER

Use this mode to snap to the center of a circle or arc. To accomplish this, activate the mode by clicking on the Object Snap Center button and moving the cursor along the edge of the circle or arc similar to Figure 1–28. Notice the AutoSnap symbol appearing at the center of the circle or arc.

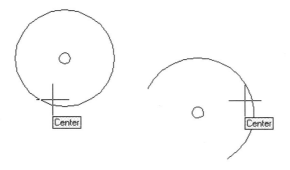

Figure 1–28

OBJECT SNAP ENDPOINT

This is one of the more popular Object Snap modes; it is helpful in snapping to the endpoints of lines or arcs. One application of Object Snap Endpoint is during the dimensioning process where exact distances are needed to produce the desired dimension. Activate this mode by clicking on the Object Snap Endpoint button, and then move the cursor along the edge of the object to snap to the endpoint. In the case of the line or arc shown in Figure 1–29, the cursor does not actually have to be positioned at the endpoint; favoring one end automatically snaps to the closest endpoint.

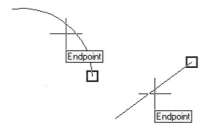

Figure 1–29

OBJECT SNAP EXTENSION

When you select a line or an arc, this Object Snap mode creates a temporary path that extends from the object. Clicking on the end of the line at "A" in Figure 1–30 acquires the angle of the line. A tooltip displays the current extension distance and angle. Clicking on the end of the arc at "B" acquires the radius of the arc and displays the current length in the tooltip.

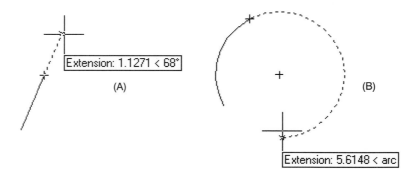

Figure 1–30

OBJECT SNAP FROM

Use this Object Snap mode along with a Secondary Object Snap mode to establish a reference point and construct an offset from that point. In Figure 1–31, the circle needed to be drawn 1.50 units in the X and Y directions from point "A." The CIRCLE command is activated and the Object Snap From mode is used in combination with the Object Snap Intersection mode. The From option requires a base point. Identify the base point at the intersection of corner "A." The next prompt asks for an offset value; enter the relative coordinate value of @1.50,1.50. This completes the use of the From option and identifies the center of the circle at "B." Study the following command sequence to accomplish this operation:

Command: **C** *(For CIRCLE)*
Specify center point for circle or [3P/2P/Ttr (tan tan radius)]: From
Base point: Int *(For Intersection Mode)*
of *(Select the intersection at "A" in Figure 1–31)*
<Offset>: @1.50,1.50
Specify radius of circle or [Diameter]: D *(For Diameter)*
Specify diameter of circle: 1.00

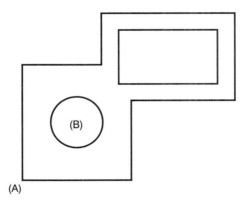

Figure 1–31

OBJECT SNAP INSERT

This Object Snap mode snaps to the insertion point of an object. In the case of the text object in Figure 1–32, activating the Object Snap Insert mode and positioning the cursor any place on the text snaps to its insertion point, in this case at the lower left corner of the text at "A." The other object illustrated in Figure 1–32 is called a block. It appears to be constructed of numerous line objects; however, all objects that make up the block are considered to be a single object. Blocks can be inserted in a drawing. Typical types of blocks are symbols such as doors, windows, bolts, and so on—anything that is used numerous times in a drawing. In order for a block to be brought into a drawing, it needs an insertion point, or a point of reference. The Object Snap Insert mode, when you position the cursor on a block, will snap to the insertion point of a block.

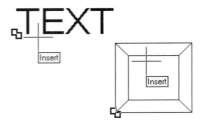

Figure 1–32

OBJECT SNAP INTERSECTION

Another popular Object Snap mode is Intersection. Use this mode to snap to the intersection of two objects. Position the cursor anywhere near the intersection of two objects and the intersection symbol appears. See Figure 1–33.

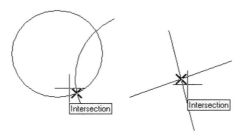

Figure 1–33

OBJECT SNAP EXTENDED INTERSECTION

Another type of intersection snap is the Object Snap Extended Intersection mode, which is used to snap to an intersection not considered obvious from the previous example. The same Object Snap Intersection button is for performing an extended intersection operation. Figure 1–34 shows two lines that do not intersect. Activate the Object Snap Extended Intersection mode and click on both lines. Notice the intersection symbol present where the two lines, if extended, would intersect.

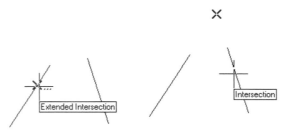

Figure 1–34

OBJECT SNAP MIDPOINT

This Object Snap mode snaps to the midpoint of objects. Line and arc examples are shown in Figure 1–35. When activating the Object Snap Midpoint mode, touch the object anywhere with some portion of the cursor; the midpoint symbol will appear at the exact midpoint of the object.

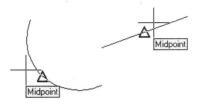

Figure 1–35

OBJECT SNAP NEAREST

This Object Snap mode snaps to the nearest point it finds on an object. Use this mode when you need to grab onto an object for the purposes of further editing. The nearest point is calculated based on the closest distance from the intersection of the crosshairs perpendicular to the object or the shortest distance from the crosshairs to the object. In Figure 1–36, the appearance of the Object Snap Nearest symbol helps to show where the point identified by this mode is actually located.

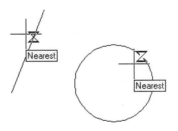

Figure 1–36

OBJECT SNAP NODE

This Object Snap mode snaps to a node or point. Touching the point in Figure 1–37 snaps to its center. The Object Snap Nearest mode shown in Figure 1–36 can also be used to snap to a point.

Figure 1–37

OBJECT SNAP PARALLEL

Use this Object Snap mode to construct a line parallel to another line. In Figure 1–38, the LINE command is started and a beginning point of the line is picked. The Object Snap Parallel mode is activated and an existing line is highlighted at "A"; the Object Snap Parallel symbol appears. Finally, move your cursor to the position that makes the new line parallel to the one just selected. The line snaps parallel along with the presence of the tracking path and tooltip giving the current distance and angle.

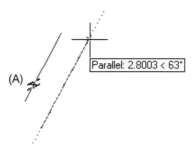

Figure 1–38

OBJECT SNAP PERPENDICULAR

This is a helpful Object Snap mode for snapping to an object normal (or perpendicular) from to a previously identified point. Figure 1–39 shows a line segment drawn perpendicular from the point at "A" to the inclined line "B." A 90° angle is formed with the perpendicular line segment and the inclined line "B." With this mode, you can also construct lines perpendicular to circles.

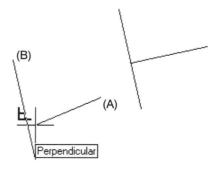

Figure 1–39

OBJECT SNAP QUADRANT

Circle quadrants are defined as points located at the 0°, 90°, 180°, and 270° positions of a circle, as in Figure 1–40. Using the Object Snap Quadrant mode will snap to one of these four positions as the edge of a circle or arc is selected. In the example of the circle in Figure 1–40, the edge of the circle is selected by the cursor location. The closest quadrant to the cursor is selected.

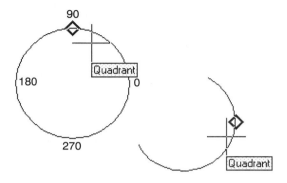

Figure 1–40

OBJECT SNAP TANGENT

This Object Snap mode is helpful in constructing lines tangent to other objects such as the circles in Figure 1–41. In this case, the Object Snap Tangent mode is being used in conjunction with the LINE command. The point at "A" is anchored to the top of the circle using the Object Snap Quadrant mode. When dragged to the next location, the line will not appear to be tangent. However, with Object Snap Tangent activated and the location at "B" picked, the line will be tangent to the large circle near "B." Follow this command sequence for constructing a line segment tangent to two circles:

Command: **L** *(For LINE)*
Specify first point: Qua *(For Quadrant mode)*
from *(Select the circle near "A")*
Specify next point or [Undo]: Tan *(For Tangent mode)*
to *(Select the circle near "B")*

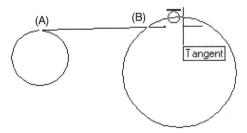

Figure 1–41

OBJECT SNAP DEFERRED TANGENT

When a line is constructed tangent from one point and tangent to another, the Deferred Tangent mode is automatically activated (see Figure 1–42). This means that instead of activating the Rubber-band Cursor mode where a line is present, the deferred tangent must have two tangent points identified before the line is constructed and displayed.

Command: **L** *(For LINE)*
Specify first point: Tan *(For Tangent mode)*
from *(Select the circle near "A")*
Specify next point or [Undo]: Tan *(For Tangent mode)*
to *(Select the circle near "B")*

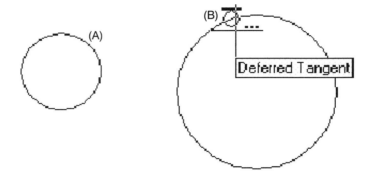

Figure 1–42

OBJECT SNAP TRACKING

In Figure 1–43A, a hole represented in the front view as a series of hidden lines needs to be displayed as a circle in the right side view. The hole will be placed in the center of the rectangle identified by "A" and "B." Tracking mode provides a quick and easy way to locate the center of the rectangle without drawing any construction geometry.

When you start tracking and pick a point on the screen, AutoCAD 2000 forces the next point to conform to an orthogonal path that could extend vertically or horizontally from the first point. The first point may retain an X or Y value and the second point may retain separate X or Y values. Use the following command sequence to get an idea of the operation of Tracking mode. For best results, be sure Ortho mode is turned off.

Command: **C** *(For CIRCLE)*
3P/2P/TTR/<Center point>: TK *(For Tracking)*

First tracking point: Mid *(For Midpoint mode)*
of *(Select the midpoint at "A" in Figure 1–107A)*
Next point *(Press* ENTER *to end tracking)*: Mid
of *(Select the midpoint at "B" in Figure 1–107A)*
Next point *(Press* ENTER *to end tracking)*: *(Press* ENTER*)*
Diameter/<Radius>: D *(For Diameter)*
Diameter: 0.50

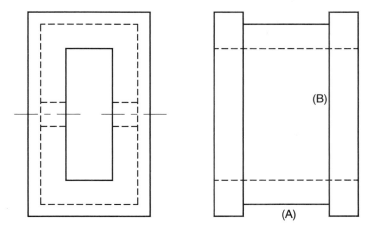

Figure 1–43A

The result is illustrated in Figure 1–43B with the hole drawn in the center of the rectangular shape.

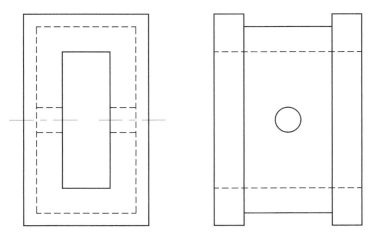

Figure 1–43B

CHOOSING RUNNING OBJECT SNAP

So far, all Object Snap modes have continuously been selected from the pop-up menu or entered at the keyboard. The problem with these methods is that if you're using a certain mode over a series of commands, you have to select the Object Snap mode every time. It is possible to make the Object Snap mode or modes continuously present through Running Osnap. Select Drafting Settings… from the Tools pull-down menu in Figure 1-44A to activate the Drafting Settings dialog box and the Object Snap tab, illustrated in Figure 1–44B. By default, the Endpoint, Center, Intersection, and Extension modes are automatically selected. Whenever your cursor lands over the top of an object supported by one of these four modes, a yellow symbol appears to alert you to the mode. It is important to know that when you make changes inside this dialog box, the changes are applied to all drawing files. Even if you quit a drawing session without saving, the changes made to the Object Snap modes are automatic. You can select other Object Snap modes by checking their appropriate boxes in the dialog box; removing the check disables the mode.

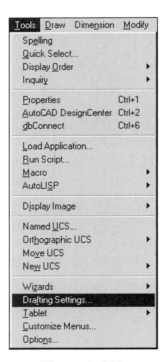

Figure 1–44A

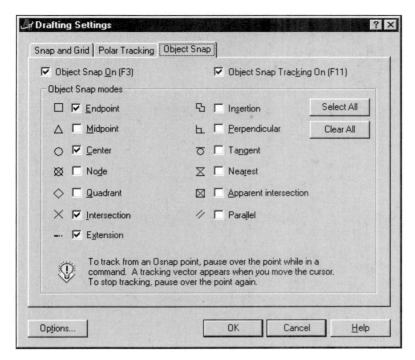

Figure 1–44B

These Object Snap modes remain in effect during drawing until you click on the OSNAP button illustrated in the Status Bar in Figure 1–45; this turns off the current running Object Snap modes. To reinstate the running Object Snap modes, click again on the OSNAP button and the previous set of Object Snap modes will be back in effect.

Figure 1–45

POLAR TRACKING

Previously in this chapter, the Direct Distance mode was highlighted as an efficient means of constructing orthogonal lines without the need for using one of the coordinate modes of entry. However it did not lend itself to lines at angles other than 0°, 90°, 180°, and 270°. Polar tracking has been designed to remedy this. This mode allows the cursor to follow a tracking path that is controlled by a preset angular increment. The POLAR button located at the bottom of the display in the Status area turns this mode On or Off. Choosing Drafting Settings... from the Tools pull-down menu in Figure 1–44A displays the Drafting Settings dialog box shown in Figure 1–46 and the Polar Tracking tab.

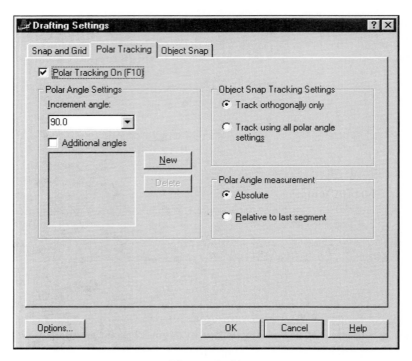

Figure 1–46

Before continuing, you need to be familiar with a few general terms:

Tracking Path – This is a temporary dotted line that can be considered a type of construction line. Your cursor will glide or track along this path (see Figure 1–47).

Tracking Point – This point is referred to as an "acquired" point. It is a point that begins the tracking path (see Figure 1–47).

Tooltip – This displays the current cursor distance and angle away from the tracking point (see Figure 1–47).

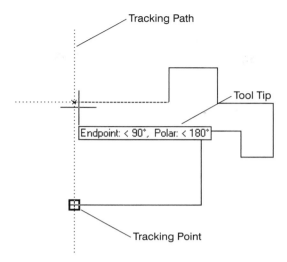

Figure 1–47

To see how the Polar Tracking mode functions, let's see how to construct an object that consists of line segments at 10° angular increments. First set the angle increment through the Polar Tracking tab of the Drafting Settings dialog box shown in Figure 1–48.

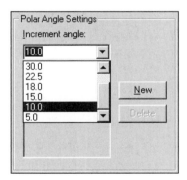

Figure 1–48

Start the LINE command, anchor a starting point at "A," move your cursor to the upper right until the tooltip reads 20°. Enter a value of 2 units for the length of the line segment (see Figure 1–49).

Move your cursor up and to the left until the tooltip reads 110° as in Figure 1–50 and enter a value of 2 units. (This will form a 90° angle with the first line.)

Figure 1–49   **Figure 1–50**

Move your cursor until the tooltip reads 20° as in Figure 1–51 and enter a value of 1 unit. Move your cursor until the tooltip reads 110° as in Figure 1–52 and enter a value of 1 unit.

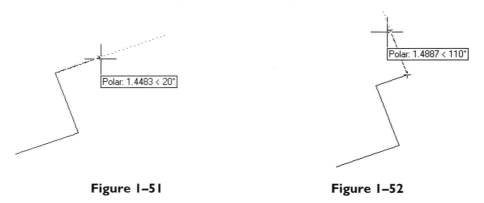

Figure 1–51   **Figure 1–52**

Move your cursor until the tooltip reads 200° as in Figure 1–53 and enter a value of 3 units.

Move your cursor to the endpoint as in Figure 1–54 or use the Close option of the LINE command to close the shape and exit the command.

Figure 1–53   **Figure 1–54**

ACQUIRING POINTS FOR POLAR TRACKING

An interesting side function of Polar Tracking is the ability to choose or acquire points to be used for construction purposes. For example, two line segments need to be added to the object in Figure 1–55 to form a rectangle. Here is how to perform this operation using Polar Tracking.

Enter the LINE command and pick a starting point for the line at "A". Then, move your cursor directly to the left until the tooltip reads 180°. The starting point for the next line segment is considered acquired (see Figure 1–56).

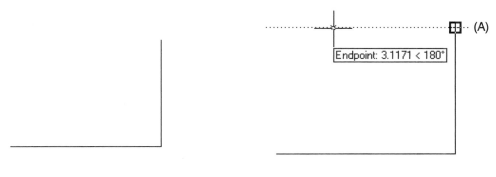

Figure 1–55	**Figure 1–56**

Rather than enter the length of this line segment, move your cursor over the top of the corner at "B" to acquire this point. Then move your cursor up until the tooltip reads 90° (see Figure 1–57).

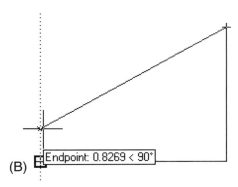

Figure 1–57

Move your cursor up until the tooltip now reads angles of 90° and 180°. Also notice the two tracking paths intersecting at the point of the two acquired points. Picking this point at "C" will construct the horizontal line segment (see Figure 1–58).

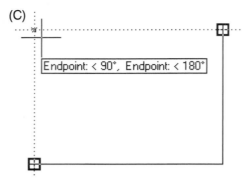

Figure 1–58

Finally, slide your cursor to the endpoint at "D" to complete the rectangle (see Figure 1–59).

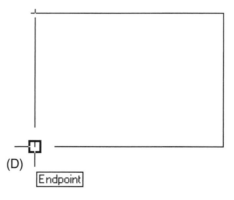

Figure 1–59

Pausing your crosshairs over an existing acquired point a second time will remove the tracking point from the object.

USING TEMPORARY TRACKING POINTS

A very powerful feature of Polar Tracking includes the ability to use an extension path along with the Temporary Tracking Point tool to construct objects under difficult situations, as illustrated in Figure 1–60.

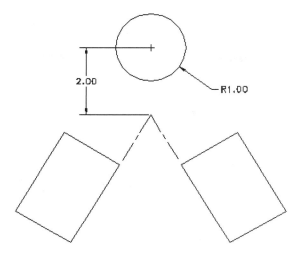

Figure 1–60

An extension path is similar to a tracking path with the exception that it is present when the Extension Object Snap mode is activated. To construct the circle in relation to the two inclined rectangles, follow the next series of steps. First, activate the CIRCLE command; this will prompt you to specify the center point for the circle. Move your cursor over the corner of the right rectangle at "A" to acquire this point. Move your cursor up and to the left, making sure the tooltip lists the Extension mode (see Figure 1–61).

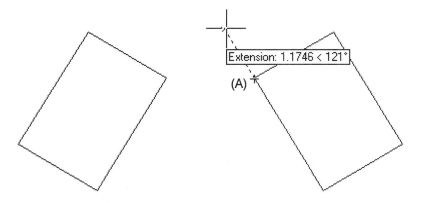

Figure 1–61

With the point acquired at "A", move your cursor over the corner of the left rectangle at "B" and acquire this point. Move your cursor up and to the right, making sure the tooltip lists the Extension mode (see Figure 1–62).

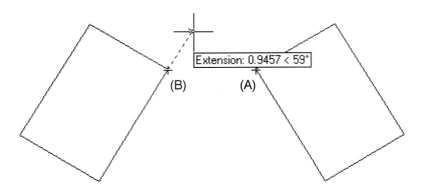

Figure 1–62

Keep moving your cursor up until both acquired points intersect as in Figure 1–63. The center of the circle is located 2 units above this intersection. Click on the Object Snap Temporary Tracking button and pick this intersection.

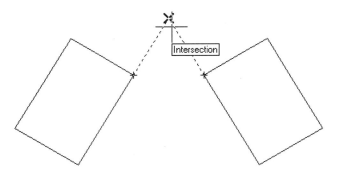

Figure 1–63

Next, move your cursor directly above the temporary tracking point as in Figure 1–64. The tooltip should read 90°. Entering a value of 2 units identifies the center of the circle.

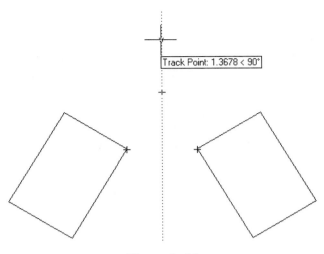

Figure 1–64

The completed construction operation is illustrated in Figure 1–65.

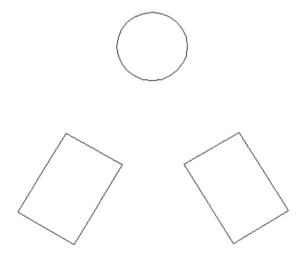

Figure 1–65

OBJECT SNAP TRACKING MODE

Object Snap Tracking works in conjunction with Object Snap. Before you can track from an Object Snap point, you must first set an Object Snap mode or modes from the Object Snap tab of the Drafting Settings dialog box. Object Snap Tracking can be toggled on or off with the OTRACK button located in the Status bar in Figure 1–66.

SNAP GRID ORTHO POLAR OSNAP OTRACK LWT MODEL

Figure 1–66

ALTERNATE METHODS USED FOR PRECISION DRAWING CARTESIAN COORDINATES

Before drawing precision geometry such as lines and circles, you must have an understanding of coordinate systems. The Cartesian or rectangular coordinate system is used to place geometry at exact distances through a series of coordinates. A coordinate is made up of an ordered pair of numbers identified as X and Y. The coordinates are then plotted on a type of graph or chart. The graph, shown in Figure 1–67, is made up of two perpendicular number lines called coordinate axes. The horizontal axis is called the X-axis. The vertical axis is called the Y-axis.

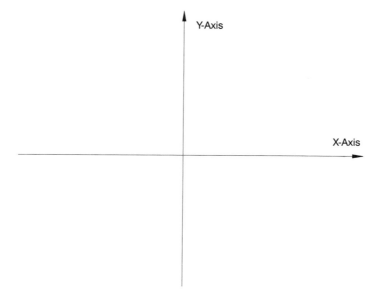

Figure 1–67

As shown in Figure 1–68, the intersection of the two coordinate axes forms a point called the origin. Coordinates used to describe the origin are 0,0. From the origin, all positive directions move up and to the right. All negative directions move down and to the left.

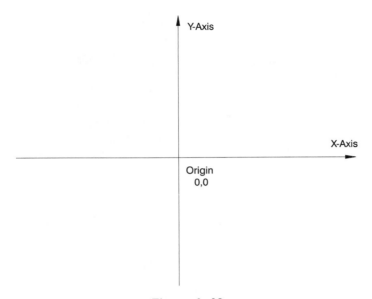

Figure 1–68

The coordinate axes are divided into four quadrants that are labeled I, II, III, and IV, as shown in Figure 1–69. In Quadrant I, all X and Y values are positive. Quadrant II has a negative X value and positive Y value. Quadrant III has negative values for X and Y. Quadrant IV has positive X values and negative Y values.

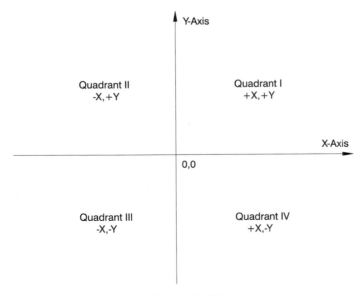

Figure 1–69

For each ordered pair of (X,Y) coordinates, X means to move from the origin to the right if positive and to the left if negative. Y means to move from the origin up if positive and down if negative. Figure 1–70 shows a series of coordinates plotted on the number lines. One coordinate is identified in each quadrant to show the positive and negative values. As an example, coordinate 3,2 in the Quadrant I means to move 3 units to the right of the origin and up 2 units. The coordinate -5,3 in Quadrant II means to move 5 units to the left of the origin and up 3 units. Coordinate -2,-2 in Quadrant III means to move 2 units to the left of the origin and down 2. Lastly, coordinate 2,-4 in Quadrant IV means to move 2 units to the right of the origin and down -4.

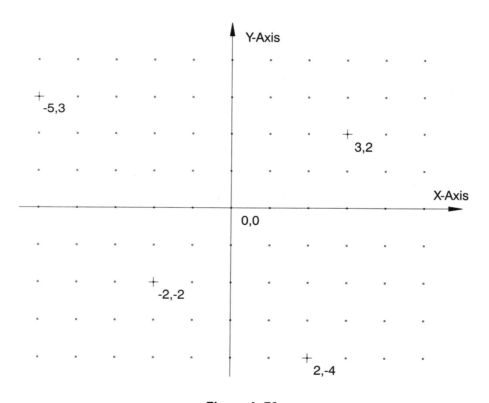

Figure 1–70

When you begin a drawing in AutoCAD 2000, the screen display reflects Quadrant I of the Cartesian coordinate system. As shown in Figure 1–71, the origin 0,0 is located in the lower left corner of the drawing screen.

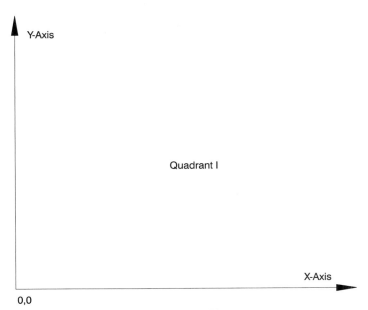

Figure 1–71

ABSOLUTE COORDINATE MODE FOR DRAWING LINES

When drawing geometry such as lines, you must use a method of entering precise distances, especially when accuracy is important. This is the main purpose of using coordinates. The simplest and most elementary form of coordinate values is absolute coordinates. Absolute coordinates conform to the following format:

$$X,Y$$

One problem with using absolute coordinates is that all coordinate values refer back to the origin 0,0. This origin on the AutoCAD 2000 screen is usually located in the lower left corner when a new drawing is created. The origin will remain in this corner unless it is altered with the LIMITS command. Study the following LINE command prompts as well as Figure 1–72.

Command: **L** *(For LINE)*
Specify first point: **2,2** *(at "A")*
Specify next point or [Undo]: **2,7** *(at "B")*
Specify next point or [Undo]: **5,7** *(at "C")*
Specify next point or [Close/Undo]: **7,4** *(at "D")*
Specify next point or [Close/Undo]: **10,4** *(at "E")*
Specify next point or [Close/Undo]: **10,2** *(at "F")*
Specify next point or [Close/Undo]: **C** *(To close the shape and exit the command)*

As you can see, all points on the object make reference to the origin at 0,0. Even though absolute coordinates are useful in starting lines, there are more efficient ways to construct lines and draw objects.

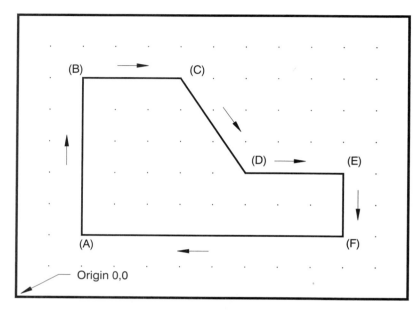

Figure 1–72

RELATIVE COORDINATE MODE FOR DRAWING LINES

In absolute coordinates, the origin at 0,0 must be kept track of at all times in order for the correct coordinate to be entered. With complicated objects, this is sometimes difficult to accomplish and as a result, the wrong coordinate is entered. It is possible to reset the last coordinate to become a new origin or 0,0 point. The new point would be relative to the previous point, and for this reason, this point is called a relative coordinate. The format is as follows:

$$@X,Y$$

In this format, we use the same X and Y values with one exception: the At symbol or @ resets the previous point to 0,0 and makes entering coordinates less confusing. Study the following LINE command prompts as well as Figure 1–73.

Command: **L** *(For LINE)*
Specify first point: 2,2 *(at "A")*
Specify next point or [Undo]: @0,4 *(to "B")*

Specify next point or [Undo]: @4,2 *(to "C")*
Specify next point or [Close/Undo]: @3,0 *(to "D")*
Specify next point or [Close/Undo]: @3,-4 *(to "E")*
Specify next point or [Close/Undo]: @-3,-2 *(to "F")*
Specify next point or [Close/Undo]: @-7,0 *(back to "A")*
Specify next point or [Close/Undo]: *(Press* ENTER *to exit this command)*

In each command, the @ symbol resets the previous point to 0,0.

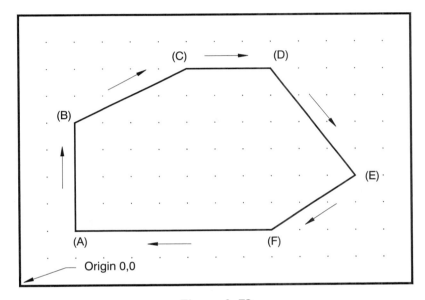

Figure 1–73

POLAR COORDINATE MODE FOR DRAWING LINES

Another popular method of entering coordinates is the Polar mode. The format is as follows:

@Distance<Direction

As the preceding format implies, the Polar Coordinate mode requires a known distance and a direction. The @ symbol resets the previous point to 0,0. The direction is preceded by the < symbol, which reads the next number as a polar or angular direction. Figure 1–74A describes the directions supported by the Polar mode. Study the following LINE command prompts as well as Figure 1–74B to better understand the Polar Coordinate mode.

Command: **L** *(For LINE)*
Specify first point: 3,2 *(at "A")*
Specify next point or [Undo]: @8<0 *(to "B")*
Specify next point or [Undo]: @5<90 *(to "C")*
Specify next point or [Close/Undo]: @5<180 *(to "D")*
Specify next point or [Close/Undo]: @4<270 *(to "E")*
Specify next point or [Close/Undo]: @2<180 *(to "F")*
Specify next point or [Close/Undo]: @2<90 *(to "G")*
Specify next point or [Close/Undo]: @1<180 *(to "H")*
Specify next point or [Close/Undo]: @3<270 *(to close back to "A")*
Specify next point or [Close/Undo]: *(Press* ENTER *to exit this command)*

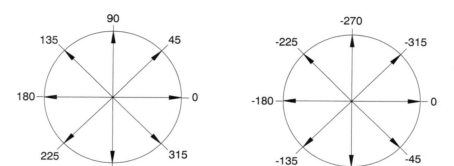

Figure 1–74A

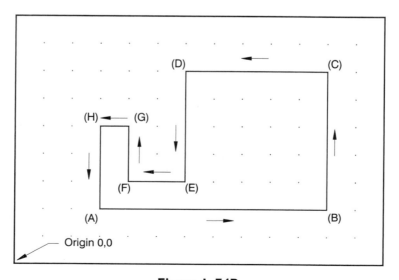

Figure 1–74B

COMBINING COORDINATE MODES FOR DRAWING

So far, the preceding pages concentrated on using each example of coordinate modes (absolute, relative, and polar) separately to create geometry. At this point, we do not want to give the impression that once you start with a particular coordinate mode, you must stay with the mode. Rather, you create drawings using one, two, or three coordinate modes in combination with each other. In Figure 1–75, the drawing starts with an absolute coordinate, changes to a polar coordinate, and changes again to a relative coordinate. It is your responsibility as a CAD operator to choose the most efficient coordinate mode to fit the drawing. Study the following LINE command prompts as well as Figure 1–75.

```
Command: L (For LINE)
Specify first point: 2,2 (at "A")
Specify next point or [Undo]: @3<90 (to "B")
Specify next point or [Undo]: @2,2 (to "C")
Specify next point or [Close/Undo]: @6<0 (to "D")
Specify next point or [Close/Undo]: @5<270 (to "E")
Specify next point or [Close/Undo]: @3<180 (to "F")
Specify next point or [Close/Undo]: @3<90 (to "G")
Specify next point or [Close/Undo]: @2<180 (to "H")
Specify next point or [Close/Undo]: @-3,-3 (back to "A")
Specify next point or [Close/Undo]: (Press ENTER to exit this command)
```

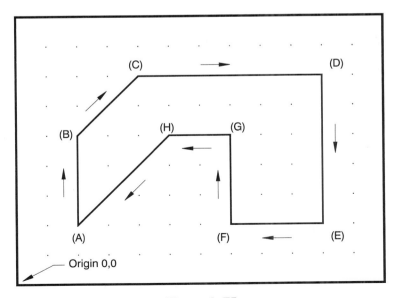

Figure 1–75

THE CIRCLE COMMAND

Choosing Circle from the Draw pull-down menu displays the cascading menu shown in Figure 1–76. All supported methods of constructing circles are displayed in the list. Circles may be constructed by radius or diameter. This command also supports circles defined by three points and constructing a circle tangent to other objects in the drawing. These last two modes will be discussed in Chapter 5, "Performing Geometric Constructions."

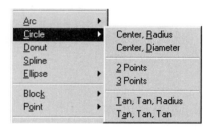

Figure 1–76

CIRCLE BY RADIUS MODE

Use the CIRCLE command and the Radius mode to construct a circle by a radius value that you specify. After selecting a center point for the circle, you are prompted to enter a radius for the desired circle. Study the following prompts and Figure 1–77A for constructing a circle using the Radius mode.

Command: **C** *(For CIRCLE)*
Specify center point for circle or [3P/2P/Ttr *(tan tan radius)*]: *(Mark the center at "A")*
Specify radius of circle or [Diameter]: 1.50

CIRCLE BY DIAMETER MODE

Use the CIRCLE command and the Diameter mode to construct a circle by a diameter value that you specify. After selecting a center point for the circle, you are prompted to enter a diameter for the desired circle. Study the following prompts and Figure 1–77B to construct a circle using the Diameter mode.

Command: **C** *(For CIRCLE)*
Specify center point for circle or [3P/2P/Ttr (tan tan radius)]: *(Mark the center at "A")*
Specify radius of circle or [Diameter]: D *(For Diameter)*
Specify diameter of circle: 3.00

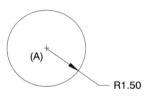

Figure 1–77A

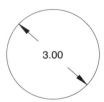

Figure 1–77B

USING THE PLINE COMMAND

Polylines are similar to individual line segments except that a polyline may consist of numerous segments and still be considered a single object. Width may also be assigned to a polyline compared to regular line segments; this makes polylines perfect for drawing borders and title blocks. Study Figures 1–78A and 1–78B and both command sequences that follow to use the PLINE command.

Command: **PL** *(For PLINE)*
From point: (Select a point at "A" in Figure 1–78A)
Current line-width is 0.0000
Arc/Close/Halfwidth/Length/Undo/Width/<Endpoint of line>: *(Mark a point at "A")*
Arc/Close/Halfwidth/Length/Undo/Width/<Endpoint of line>: W *(For Width)*
Starting width <0.0000>: 0.10
Ending width <0.1000>: *(Press ENTER to accept default)*
Arc/Close/Halfwidth/Length/Undo/Width/<Endpoint of line>: *(Mark a point at "B")*
Arc/Close/Halfwidth/Length/Undo/Width/<Endpoint of line>: *(Mark a point at "C")*
Arc/Close/Halfwidth/Length/Undo/Width/<Endpoint of line>: *(Mark a point at "D")*
Arc/Close/Halfwidth/Length/Undo/Width/<Endpoint of line>: *(Mark a point at "E")*
Arc/Close/Halfwidth/Length/Undo/Width/<Endpoint of line>: *(Press ENTER to exit this
    command)*

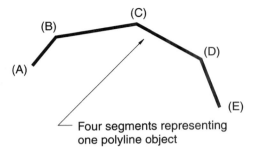

Four segments representing
one polyline object

Figure 1–78A

Command: **PL** *(For PLINE)*
From point: *(Select a point at "A" in Figure 1–78B)*
Current line-width is 0.0000
Arc/Close/Halfwidth/Length/Undo/Width/<Endpoint of line>: @1.00<0 *(To "B")*
Arc/Close/Halfwidth/Length/Undo/Width/<Endpoint of line>: @2.00<90 *(To "C")*
Arc/Close/Halfwidth/Length/Undo/Width/<Endpoint of line>: @0.50<0 *(To "D")*
Arc/Close/Halfwidth/Length/Undo/Width/<Endpoint of line>: @0.75<90 *(To "E")*
Arc/Close/Halfwidth/Length/Undo/Width/<Endpoint of line>: @0.75<180 *(To "F")*
Arc/Close/Halfwidth/Length/Undo/Width/<Endpoint of line>: @2.00<270 *(To "G")*
Arc/Close/Halfwidth/Length/Undo/Width/<Endpoint of line>: @0.50<180 *(To "H")*
Arc/Close/Halfwidth/Length/Undo/Width/<Endpoint of line>: @2.00<90 *(To "I")*
Arc/Close/Halfwidth/Length/Undo/Width/<Endpoint of line>: @0.75<180 *(To "J")*
Arc/Close/Halfwidth/Length/Undo/Width/<Endpoint of line>: @0.75<270 *(To "K")*
Arc/Close/Halfwidth/Length/Undo/Width/<Endpoint of line>: @0.50<0 *(To "L")*
Arc/Close/Halfwidth/Length/Undo/Width/<Endpoint of line>: C *(For Close)*

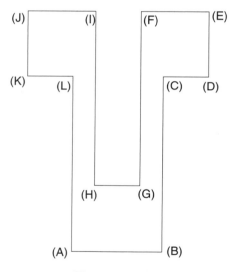

Figure 1–78B

ERASING OBJECTS

Throughout the design process, as objects such as lines and circles are placed in a drawing, changes in the design will require the removal of objects. The ERASE command will delete objects from the database. In Figure 1–79, line segments "A" and "B" need to be removed in order for a new line to be constructed closing the shape. Two ways of erasing these lines will be introduced here.

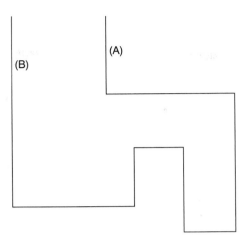

Figure 1–79

When first entering the ERASE command, you are prompted to "Select objects:" to erase. Notice your cursor changes in appearance from crosshairs to a pickbox, shown in Figure 1–80. Move the pickbox over the object to be selected and pick this item. Notice it will highlight as a dashed object to signify it is now selected. At this point, the "Select objects:" prompt appears again. If you need to pick more objects to erase, select them now. If you are finished selecting objects, press ENTER. This will perform the erase operation.

Command: **E** *(For ERASE)*
Select objects: *(Pick line "A" in Figure 1–80)*
Select objects: *(Press* ENTER *to perform the erase operation)*

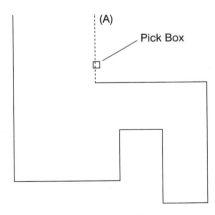

Figure 1–80

Another method of erasing is illustrated in Figure 1–81. Instead of using the ERASE command, highlight the line by selecting it with your crosshairs from the "Command:" prompt. Square boxes also appear at the endpoints and midpoints of the line. With this line segment highlighted, press DELETE on your keyboard, and this will remove the line from the drawing.

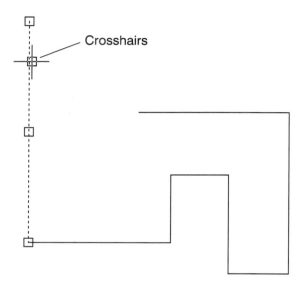

Figure 1–81

With both line segments erased, a new line is constructed from the endpoint at "A" to the endpoint at "B" as in Figure 1–82.

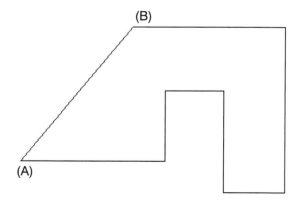

Figure 1–82

SAVING A DRAWING FILE

You can save drawings using the QSAVE, SAVE, and SAVEAS commands. Two of these three commands are on the File menu, as shown in Figure 1–83A.

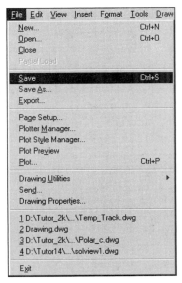

Figure 1–83A

SAVE...

This command is short for QSAVE, which stands for Quick Save. If a drawing file has never been saved and this command is selected, the dialog box shown in Figure 1–83B displays. Once a drawing file has been initially saved, selecting this command performs an automatic save and no longer displays the Save Drawing As dialog box.

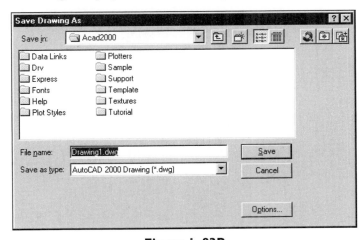

Figure 1–83B

SAVE AS...

Using the Save As... command always displays the dialog box shown in Figure 1–83B. Simply click on the Save button or press ENTER to save the drawing under the current name displayed in the edit box. This command is more popular for saving the current drawing under an entirely different name. Simply enter the new name in place of the highlighted name in the edit box. Once a drawing is given a new name through this command, it also becomes the new current drawing file.

The ability to exchange drawings with past releases of AutoCAD is still important to many industry users. When you click in the Save as type edit box in Figure 1–83B, a drop-down list appears. Use this list to save a drawing file in the following formats: AutoCAD R14/LT 98/LT 97 and R13/LT 95 (see Figure 1–83C).

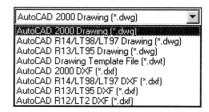

Figure 1–83C

THE HELP COMMAND

Selecting Help from the Help menu shown in Figure 1–84A, activates the AutoCAD 2000 Help System dialog box, shown in Figure 1–84B. Three tabs, Contents, Index and Find, can be used to navigate around Help. When you first enter Help, the Contents tab displays, giving various topics to switch to.

Figure 1–84A

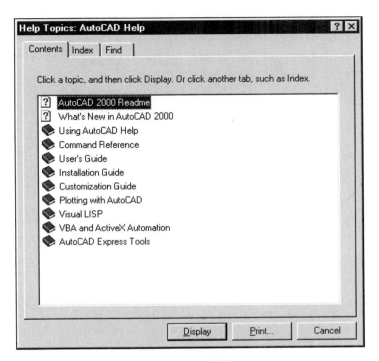

Figure 1–84B

Clicking on the Index tab displays the dialog box shown in Figure 1–84C. Use this dialog box to enter any AutoCAD 2000 command at the edit box and receive help on this command. For example, notice how the LINE command has been entered in the Index edit box. Clicking on the Display button at the bottom of the dialog box displays an additional dialog box giving help on the LINE command (see Figure 1–84D). Notice the use of text and graphical examples on the LINE command. Using the AutoCAD 2000 Help System can prove very beneficial in answering basic questions about any command.

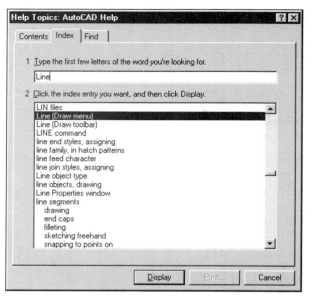

Figure 1–84C

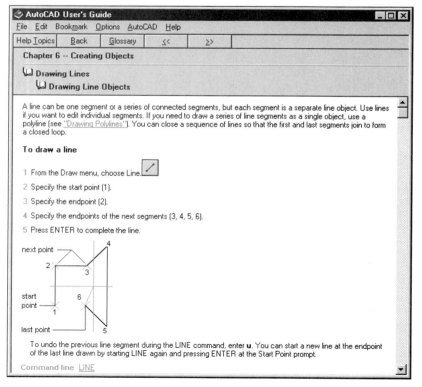

Figure 1–84D

EXITING AN AUTOCAD 2000 DRAWING SESSION

It is considered good practice to properly exit any drawing session. One way of exiting is by choosing the Exit option from the File menu (see Figure 1–85A). Another command that exits AutoCAD 2000 is QUIT. Use this command with extreme caution because QUIT exits the drawing file without saving any changes. This command is useful when you are just browsing drawing files where no changes were made. If you are a new user of AutoCAD 2000, do not use this command until you have gained sufficient experience. You can also use the CLOSE command to end an AutoCAD drawing session and still remain software.

Whichever command you use to exit AutoCAD 2000, a built-in safeguard gives you a second chance to save the drawing before exiting, especially if changes were made and a Save was not performed. You may be confronted with three options illustrated in the AutoCAD 2000 alert dialog box shown in Figure 1–85B. By default, the Yes button is highlighted.

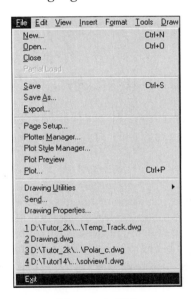

Figure I–85A

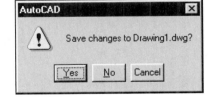

Figure I–85B

If you made changes to the drawing but did not save them, you may want to click on the Yes button before exiting the drawing. Changes to the drawing will be saved and the software exits back to the operating system.

If you made changes but do not want to save them, you may want to click on the No button. Changes to the drawing will not be saved and the software exits back to the operating system.

If you made changes to the drawing and chose the Exit option mistakenly, choosing the Cancel button cancels the Exit option and returns you to the current drawing.

TUTORIAL EXERCISE: PATTERN.DWG

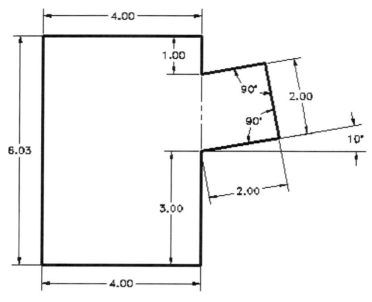

Figure 1–86

Purpose

This tutorial is designed to allow you to construct a one-view drawing of the Pattern using Polar Tracking techniques. See Figure 1–86.

System Settings

Use the current default settings for the limits of this drawing, (0,0) for the lower left corner and (12,9) for the upper right corner.

Layers

Create the following layer with the format:

Name	Color	Linetype
Object	Green	Continuous

Suggested Commands

The line command will be used entirely for this tutorial in addition to the Polar Tracking mode. Running Object Snap should already be set to the following modes; Endpoint, Center, Intersection, and Extension.

Whenever possible, substitute the appropriate command alias in place of the full AutoCAD command in each tutorial step. For example, use "CP" for the copy command, "L" for the line command, and so on. The complete listing of all command aliases is located in Table 1–2.

STEP 1

Activate the Drafting Settings dialog box, click on the Polar Tracking tab, and change the polar tracking angle setting to 10° (see Figure 1–87).

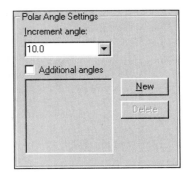

Figure 1–87

STEP 2

Activate the LINE command, select a starting point, move your cursor to the right, and enter a value of 4 units (see Figure 1–88).

Command: **L** *(For LINE)*
Specify first point: 3,1
Specify next point or [Undo]: *(Move your cursor to the right and enter 4)*

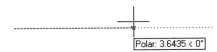

Figure 1–88

STEP 3

While still in the LINE command, move your cursor directly up, and enter a value of 3 units (see Figure 1–89).

Specify next point or [Undo]: *(Move your cursor up and enter 3)*

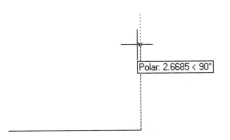

Figure 1–89

STEP 4

While still in the LINE command, move your cursor up and to the right until the tooltip reads 10°, and enter a value of 2 units (see Figure 1–90).

Specify next point or [Close/Undo]: *(Move your cursor up and to the right at a 10° angle and enter 2)*

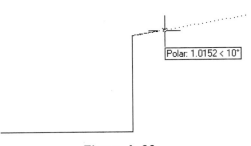

Figure 1–90

tutorial **EXERCISE**

STEP 5

While still in the LINE command, move your cursor up and to the left until the tooltip reads 100°, and enter a value of 2 units (see Figure 1–91).

Specify next point or [Close/Undo]:
    (Move your cursor up and to the left at a 100° angle and enter 2)

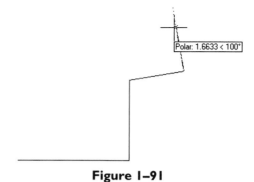

Figure 1–91

STEP 6

While still in the LINE command, first acquire the point at "A". Then move your cursor below and to the left until the tooltip reads 190°, and pick the point at "B" (see Figure 1–92).

Specify next point or [Close/Undo]:
    (Acquire the point at "A" and pick the new point at "B")

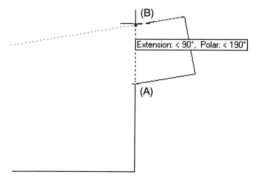

Figure 1–92

STEP 7

While still in the LINE command, move your cursor directly up, and enter a value of 1 unit (see Figure 1–93).

Specify next point or [Close/Undo]:
    (Move your cursor up and enter 1)

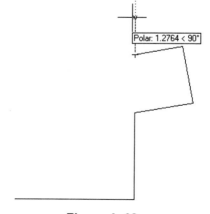

Figure 1–93

STEP 8

While still in the LINE command, first acquire the point at "C". Then move your cursor to the left until the tooltip reads Polar: < 180°, and pick the point at "D" (see Figure 1–94).

Specify next point or [Close/Undo]:
(Acquire the point at "C" and pick the new point at "D")

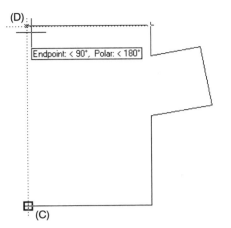

Figure 1–94

STEP 9

While still in the LINE command, complete the object by closing the shape (see Figure 1–95).

Specify next point or [Close/Undo]: C
(To close the shape and exit the LINE command)

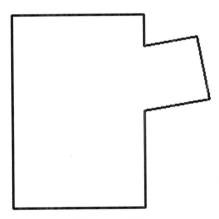

Figure 1–95

TUTORIAL EXERCISE: 01_POLYGON.DWG

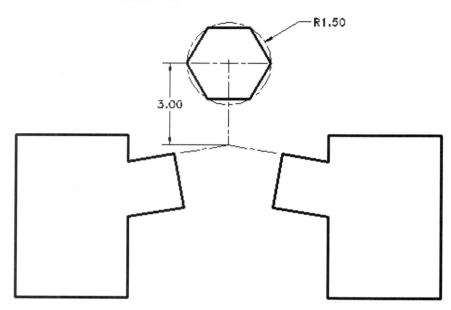

Figure 1–96

Purpose

This tutorial is designed to allow you to construct a polygon using existing geometry as a guide. Polar tracking and Object Snap tracking will be used to accomplish this task (see Figure 1–96). This drawing is available on the CD under the file name 01_Polygon.Dwg.

System Settings

No special system settings need to be made to this drawing file.

Layers

The drawing file 01_Polygon.Dwg has the following layers already created:

Name	Color	Linetype
Center	Yellow	Center
Dimension	Yellow	Continuous
Object	Green	Continuous

Suggested Commands

The polygon command is used to start this tutorial exercise. To identify the center of the polygon, Polar Tracking will be used along with the Osnap Extension mode and the Temporary Tracking Point tool.

Whenever possible, substitute the appropriate command alias in place of the full AutoCAD command in each tutorial step. For example, use "CP" for the copy command, "L" for the line command, and so on. The complete listing of all command aliases is located in Table 1–2.

STEP 1

Open the drawing file 01_Polygon.Dwg and activate the POLYGON command. Polar tracking and Object Snap tracking will be used to identify the center of the polygon (see Figure 1–97).

Command: **POL** *(For POLYGON)*
Enter number of sides <4>: 6

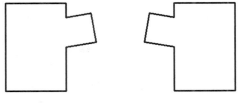

Figure 1–97

STEP 2

At the prompt "Specify center of polygon or [Edge]:", begin locating the center of the polygon by clicking on the endpoint at "A" in Figure 1–98. This creates a tracking path at the same angle as the line selected. Move your cursor above and to the right as illustrated in this figure.

Specify center of polygon or [Edge]:
(Acquire the point at "A")

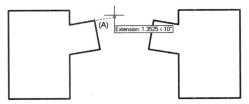

Figure 1–98

STEP 3

While still at the prompt "Specify center of polygon or [Edge]", click on the endpoint at "B" in Figure 1–99. This creates another tracking path at the same angle as the line selected. Move your cursor above and to the left as illustrated in this figure.

(Acquire the point at "B" in Figure 1–99)

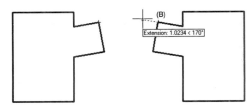

Figure 1–99

STEP 4

Now move both temporary tracking points to meet at their intersection in Figure 1–100 but do not pick a point at this location. The polygon needs to be constructed 3 units above this point. Click on the Temporary Tracking Point Objects Snap mode and then pick a point at the intersection of "C." Notice the following prompt sequence that appears:

Specify temporary OTRACK point: *(Pick a point at the intersection at "C")*

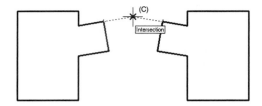

Figure 1–100

STEP 5

Move your cursor directly above this tracking point in Figure 1–101 and enter a value of 3 units. This will locate the center of the polygon 3 units directly above the last tracking point. Complete the construction by following the remaining POLYGON command prompts:

(Move your cursor up and enter 3)
Enter an option [Inscribed in circle/ Circumscribed about circle] <I>:
(Press ENTER to accept this default value)
Specify radius of circle: 1.50

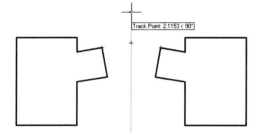

Figure 1–101

STEP 6

The completed construction task is illustrated in Figure 1–102.

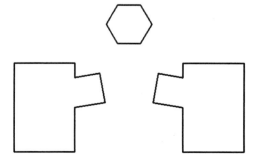

Figure 1–102

TUTORIAL EXERCISE: TEMPLATE.DWG

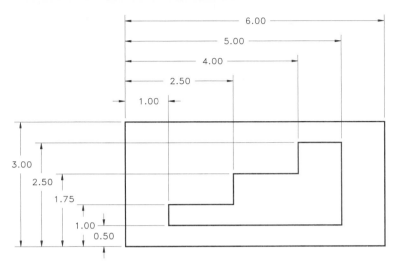

Figure 1–103

Purpose

This tutorial is designed to allow you to construct a one-view drawing of the Template using the Absolute, Relative, and Polar Coordinate modes. See Figure 1–103. The Direct Distance mode can also be used to perform this exercise.

System Settings

Use the current default settings for the limits of this drawing, (0,0) for the lower left corner and (12,9) for the upper right corner.

Use the grid command and change the grid spacing from 1.0000 to 0.25 units. The grid will be used only as a guide for constructing this object. Do not turn the Snap or Ortho modes on.

Layers

Create the following layer with the format:

Name	Color	Linetype
Object	Green	Continuous

Suggested Commands

The line command will be used entirely for this tutorial in addition to a combination of coordinate systems. The erase command could be used, although a more elaborate method of correcting a mistake would be to use the line command and the Undo option to erase a previously drawn line and still stay in the line command. The OSNAP-From mode will also be used to construct lines from a point of reference. The coordinate mode of entry and the Direct Distance mode will be used throughout this tutorial exercise.

Whenever possible, substitute the appropriate command alias in place of the full AutoCAD command in each tutorial step. For example, use "CP" for the copy command, "L" for the line command, and so on. The complete listing of all command aliases is located in Table 1–2.

STEP 1

Begin this tutorial exercise by making the Object layer current. Then use the LINE command to draw the outer perimeter of the box. One method of constructing the box is to use an absolute coordinate point followed by polar coordinates using the following command sequence and the illustration in Figure 1–104 as guides.

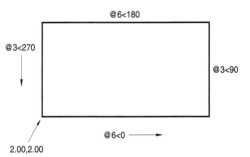

Figure 1–104

Command: **L** *(For LINE)*
Specify first point: 2,2
Specify next point or [Undo]: @6<0
Specify next point or [Undo]: @3<90
Specify next point or [Close/Undo]: @6<180
Specify next point or [Close/Undo]: @3<270
Specify next point or [Close/Undo]: *(Press ENTER to exit this command)*

An alternate method is to use the Direct Distance mode. Since the box consists of horizontal and vertical lines, Ortho mode is first turned on; this will force all movements to be in the horizontal or vertical direction. To construct a line segment, move the cursor in the direction in which the line is to be drawn and enter the exact value of the line. The line is drawn at the designated distance in the current direction of the cursor. Repeat this procedure for the other lines that make up the box, as shown in Figure 1–105.

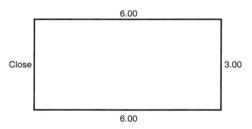

Figure 1–105

Command: **OR** *(For ORTHO)*
ON/OFF <OFF>: ON

Command: **L** *(For LINE)*
Specify first point: 2,2
Specify next point or [Undo]: *(Move the cursor to the right and enter a value of 6.00 units)*

Specify next point or [Undo]: *(Move the cursor up and enter a value of 3.00 units)*
Specify next point or [Close/Undo]: *(Move the cursor to the left and enter a value of 6.00 units)*
Specify next point or [Close/Undo]: C *(To close the shape)*

STEP 2

The next step will be to draw the stair step outline of the template using the LINE command again. However, we first need to identify the starting point of the template. Absolute coordinates could be calculated, but in more complex objects, this would be difficult. A more efficient method would be to use the OSNAP-From mode along with the OSNAP-Intersection mode to start the line relative to another point. Use the following command sequence and Figure 1–106 as guides for performing this operation.

Command: **L** *(For LINE)*
Specify first point: From
Base point: Int
of *(Pick the intersection at "A")*
<Offset>: @1.00,0.50

The relative coordinate offset value begins a new line a distance of 1.00 units in the X direction and 0.50 units in the Y direction. Continue with the LINE command to construct the stair step outline as shown in Figure 1–107.

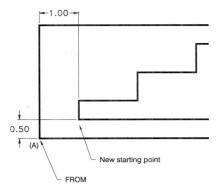

Figure 1–106

Specify next point or [Undo]: @4.00<0
Specify next point or [Undo]: @2.00<90
Specify next point or [Close/Undo]:
   @1.00<180
Specify next point or [Close/Undo]:
   @0.75<270
Specify next point or [Close/Undo]:
   @1.50<180
Specify next point or [Close/Undo]:
   @0.75<270
Specify next point or [Close/Undo]:
   @1.50<180
Specify next point or [Close/Undo]:
   @0.50<270
Specify next point or [Close/Undo]:
   (Press ENTER to exit this command)

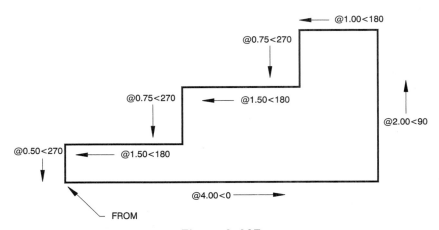

Figure 1–107

An alternate step would be to use the OSNAP-From and OSNAP-Intersect modes in combination with the Direct Distance mode to construct the inner stair step outline (see Figure 1–108). Again, Direct Distance mode is a good choice to use on this object especially since all lines are either horizontal or vertical. Use the following command sequence to construct the object with this alternate method.

Command: **L** *(For LINE)*
Specify first point: From
Base point: Int
of *(Pick the intersection at "A" in Figure 1–108)*
<Offset>: @1.00,0.50
Specify next point or [Undo]: *(Move the cursor to the right and enter a value of 4.00 units)*
Specify next point or [Undo]: *(Move the cursor up and enter a value of 2.00 units)*

Specify next point or [Close/Undo]: *(Move the cursor to the left and enter a value of 1.00 units)*
Specify next point or [Close/Undo]: *(Move the cursor down and enter a value of 0.75 units)*
Specify next point or [Close/Undo]: *(Move the cursor to the left and enter a value of 1.50 units)*
Specify next point or [Close/Undo]: *(Move the cursor down and enter a value of 0.75 units)*
Specify next point or [Close/Undo]: *(Move the cursor to the left and enter a value of 1.50 units)*
Specify next point or [Close/Undo]: C *(To close the shape)*

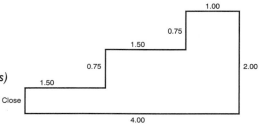

Figure 1–108

STEP 3

Dimensions may be added at a later time upon the request of your instructor. The completed problem is shown in Figure 1–109.

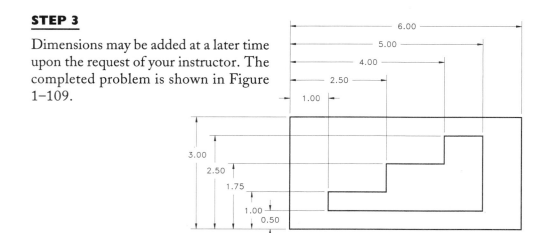

Figure 1–109

PROBLEMS FOR CHAPTER 1

Construct one-view drawings of the following figures using the line command along with coordinate or Direct Distance modes.

PROBLEM 1-1

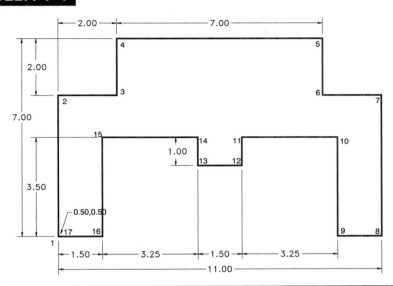

	Absolute	**Relative**	**Polar**
From Pt (1)	0.50,0.50	0.50,0.50	0.50,0.50
To Pt (2)			
To Pt (3)			
To Pt (4)			
To Pt (5)			
To Pt (6)			
To Pt (7)			
To Pt (8)			
To Pt (9)			
To Pt (10)			
To Pt (11)			
To Pt (12)			
To Pt (13)			
To Pt (14)			
To Pt (15)			
To Pt (16)			
To Pt (17)			
To Pt	Enter	Enter	Enter

PROBLEM 1-2

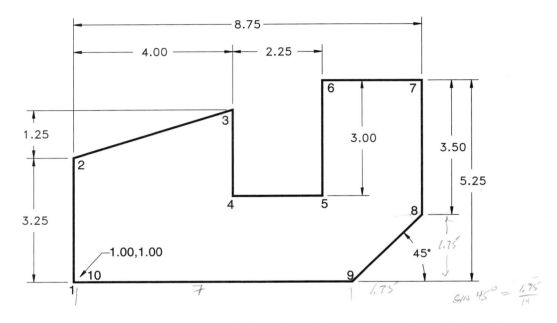

	Absolute	Relative
From Pt (1)	1.00,1.00	1.00,1.00
To Pt (2)		
To Pt (3)		
To Pt (4)		
To Pt (5)		
To Pt (6)		
To Pt (7)		
To Pt (8)		
To Pt (9)		
To Pt (10)		
To Pt	Enter	Enter

PROBLEM 1–3

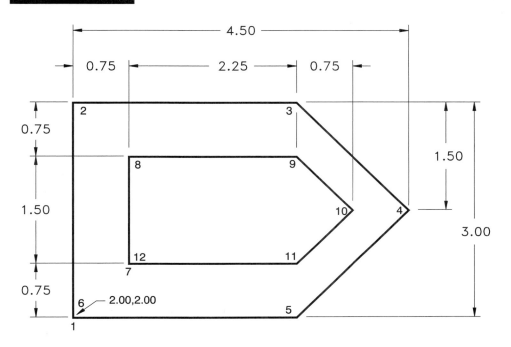

	Absolute	Relative
From Pt (1)	2.00,2.00	2.00,2.00
To Pt (2)		
To Pt (3)		
To Pt (4)		
To Pt (5)		
To Pt (6)		
To Pt	Enter	Enter
From Pt (7)		
To Pt (8)		
To Pt (9)		
To Pt (10)		
To Pt (11)		
To Pt (12)		
To Pt	Enter	Enter

PROBLEM 1–4

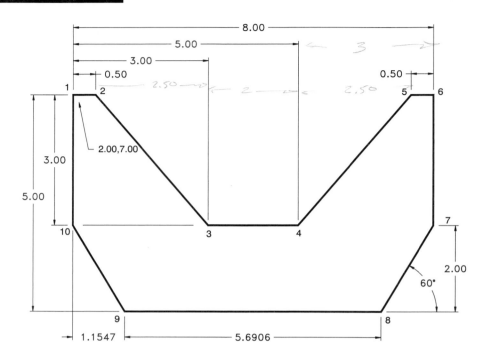

	Absolute
From Pt (1)	2.00,7.00
To Pt (2)	
To Pt (3)	
To Pt (4)	
To Pt (5)	
To Pt (6)	
To Pt (7)	
To Pt (8)	
To Pt (9)	
To Pt (10)	
To Pt	Enter

problem EXERCISE

PROBLEM 1-5

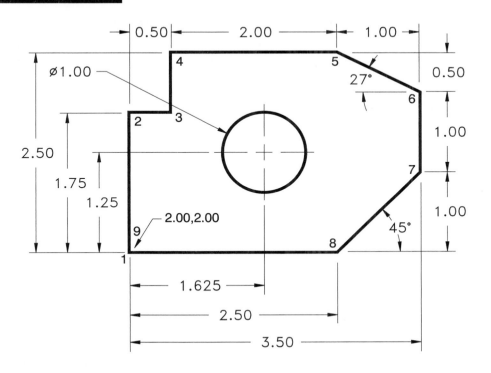

	Absolute	Relative
From Pt (1)	2.00,2.00	2.00,2.00
To Pt (2)		
To Pt (3)		
To Pt (4)		
To Pt (5)		
To Pt (6)		
To Pt (7)		
To Pt (8)		
To Pt (9)		
To Pt	Enter	Enter
Center Pt (10)		

PROBLEM 1-6

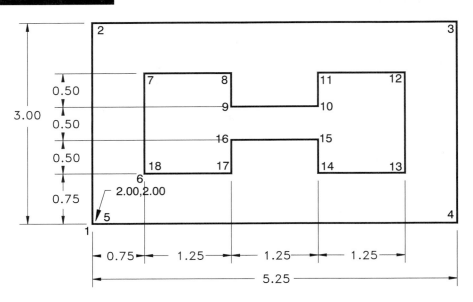

	Absolute	Relative	Polar
From Pt (1)	2.00,2.00	2.00,2.00	2.00,2.00
To Pt (2)			
To Pt (3)			
To Pt (4)			
To Pt (5)			
To Pt	Enter	Enter	Enter
From Pt (6)			
To Pt (7)			
To Pt (8)			
To Pt (9)			
To Pt (10)			
To Pt (11)			
To Pt (12)			
To Pt (13)			
To Pt (14)			
To Pt (15)			
To Pt (16)			
To Pt (17)			
To Pt (18)			
To Pt	Enter	Enter	Enter

PROBLEM 1–7

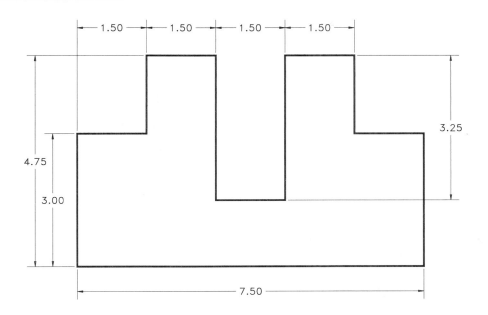

PROBLEM 1–8

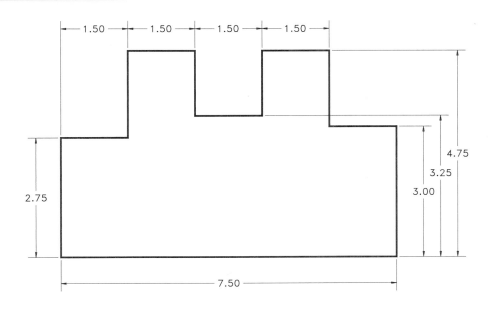

PROBLEM 1-9

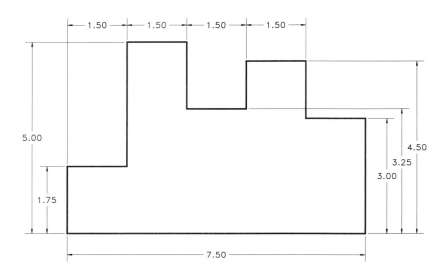

PROBLEM 1-10

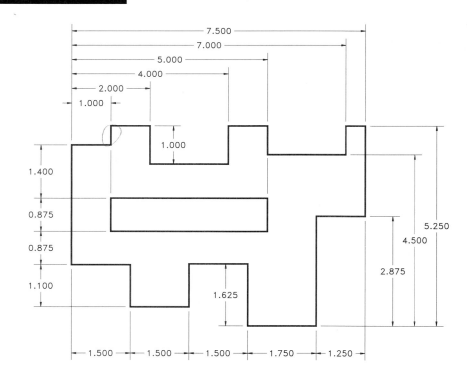

PROBLEM 1-11

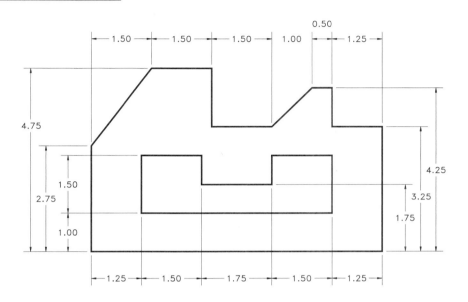

PROBLEM 1-12

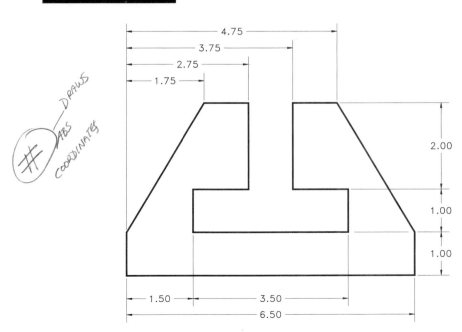

PROBLEM 1-13

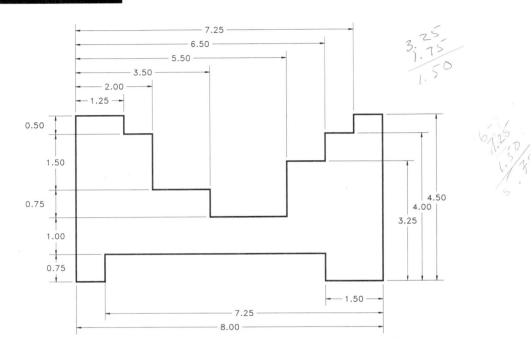

PROBLEM 1-14

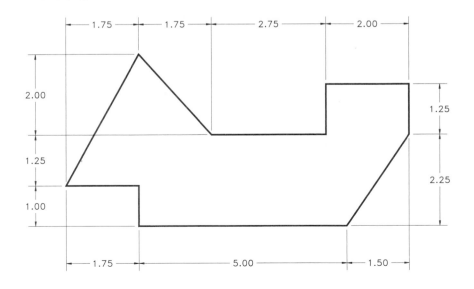

PROBLEM 1–15

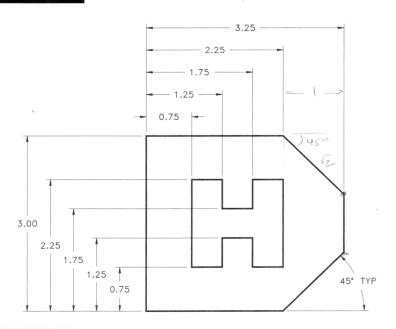

PROBLEM 1–16

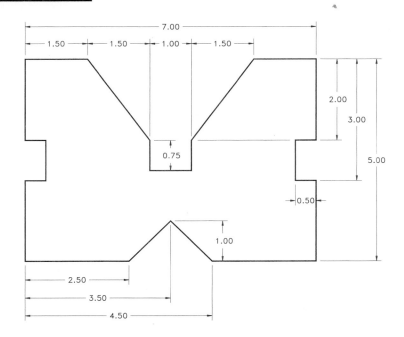

PROBLEM 1-17

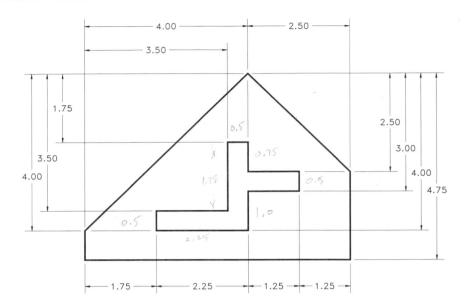

PROBLEM 1-18

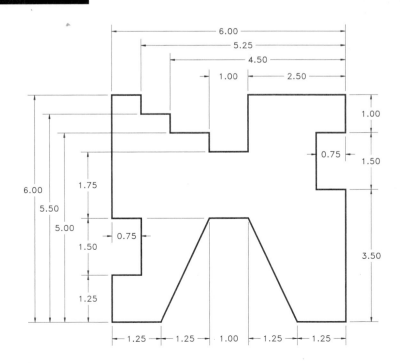

PROBLEM 1–19

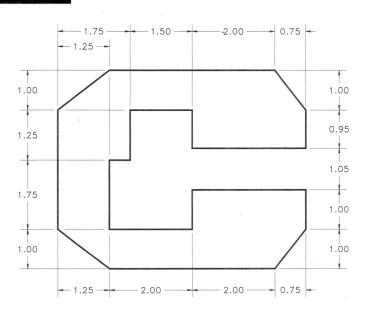

PROBLEM 1–20

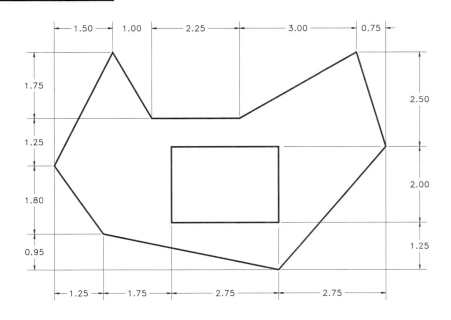

CHAPTER 2

Drawing Setup and Organization

THE DRAWING UNITS DIALOG BOX

The Drawing Units dialog box is available for interactively setting the units of a drawing. Choosing Units… from the Format pull-down menu as shown in Figure 2–1A activates the dialog box illustrated in Figure 2–1B.

Figure 2–1A

Figure 2–1B

By default, decimal units are set along with four-decimal-place precision. The following systems of units are available: Architectural, Decimal, Engineering, Fractional and Scientific (see Figure 2–2). Architectural units are displayed in feet and fractional inches. Engineering units are displayed in feet and decimal inches. Fractional units are displayed in fractional inches. Scientific units are displayed in exponential format.

Methods of measuring angles supported in the Drawing Units dialog box include Decimal Degrees, Degrees/Minutes/Seconds, Grads, Radians, and Surveyor's Units (see Figure 2–3). Accuracy of decimal degree for angles may be set between zero and eight places.

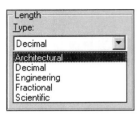

Figure 2–2

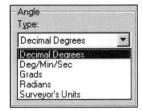

Figure 2–3

Selecting Direction… in the Drawing Units dialog box in Figure 2–1B displays the Direction Control dialog box, shown in Figure 2–4. This dialog box is used to control the direction of angle zero in addition to changing whether angles are measured in the counterclockwise or clockwise direction. By default, angles are measured in the counterclockwise direction.

Figure 2–4

THE LIMITS COMMAND

By default, the size of the drawing screen in a new drawing file measures 12 units in the X direction and 9 units in the Y direction. This size may be ideal for small objects, but larger drawings require more drawing screen area. Use the LIMITS command for increasing the size of the drawing area. You can select this command by picking Drawing Limits from the Format pull-down menu (see Figure 2–5); you can also enter this command directly at the command prompt. Illustrated in Figure 2–6 is a section view drawing. This drawing fits in a screen size of 24 units in the X direction and 18 units in the Y direction. Follow the next command sequence to change the limits of a drawing.

Figure 2–5

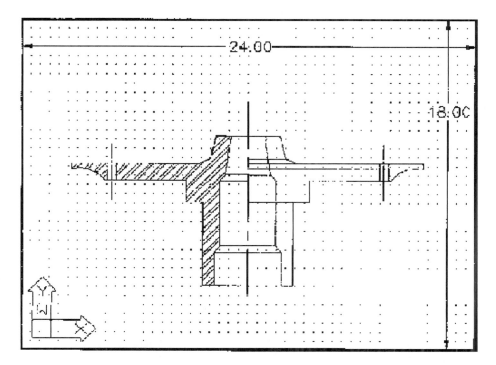

Figure 2–6

Command: **LIMITS**
ON/OFF/<Lower left corner>: <0.0000,0.0000>: *(Press* ENTER *to accept this value)*
Upper right corner <12.0000,9.0000>: **24,18**

Before continuing, perform a ZOOM-All to change the size of the display screen to reflect the changes in the limits of the drawing. You can find ZOOM-All on the View menu. It is also found on the Zoom toolbar, which you can activate by choosing Toolbars… from the View pull-down menu and placing a check in the box next to Zoom.

Command: **Z** *(For Zoom)*
All/Center/Dynamic/Extents/Left/Previous/Vmax/Window/<Scale(X/XP)>: **A** *(For All)*

CALCULATING THE LIMITS OF THE DRAWING

Before you construct any drawing, it is considered good practice to first calculate the limits or sheet size of the drawing. Two items are needed to accomplish this task: the scale of the drawing and the paper size the drawing will be plotted out on. Most operators who use CAD systems in industry already have some idea as to the scale of the drawing and whether it will fit on a certain size sheet of paper.

For the purposes of this example, the following steps will illustrate the creation of drawing limits based on a D-size sheet and at a scale of 1/4" = 1'0". This scale is commonly used to construct residential house plans. Because the following steps use this scale and sheet, the method works for calculating limits at any scale and on any sheet of paper.

STEP 1

Determine that a "D" size drawing sheet measures 36" x 24".

STEP 2

Determine the multiplication factor based on the drawing scale. In Computer-Aided Design, most drawings are created full size. The limits set a full-size drawing sheet used to match the full size drawing. The multiplication factor for the scale 1/4" = 1'0" is 48 (1' divided by 1/4). Use this multiplier to increase the size of the paper, enabling an operator to produce a full-size drawing on the computer. This is vastly different from manual drawing practices, where an object had to be reduced to 1/48th of its normal size in order to fit on the D-size paper at a scale of 1/4" = 1'0".

STEP 3

Using the multiplication factor of 48, increase the size of the D-size sheet of paper:

Paper Size from the Plot dialog box	36	24
Multiplication Factor	48	48
	1,728	1,152

Convert the previous values to feet:

$$1,728 / 12 = 144' \qquad 1,152 / 12 = 96'$$

The limits of the drawing become 144' wide and 96' high.

STEP 4

Use the Drawing Units dialog box to set the current units to Architectural (Decimal mode does not accept feet as valid units). Next, use the LIMITS command and enter the lower left corner as 0,0 and the upper right corner as 144',96'. Once the limits have been set, use the ZOOM command and the All option to display the new sheet of paper. To view the active drawing area, use the RECTANG command and construct a rectangular border around the drawing using 0,0 as the lower left corner and 144',96' as the upper right corner. Construct the drawing inside the rectangular area.

Decimal Scales

SCALE	SCALE FACTOR	ANSI "A" 11"x 8.5"	ANSI "B" 17"x11"	ANSI "C" 22"x17"	ANSI "D" 34"x22"	ANSI "E" 44"x34"
.125=1	8	88,68	136,88	176,136	272,176	352,272
.25=1	4	44,34	68,44	88,68	136,88	176,136
.50=1	2	22,17	34,22	44,34	68,44	88,68
1.00=1	1	11,8.5	17,11	22,17	34,22	44,34
2.00=1	.50	5.5,4.25	8.5,5.5	11,8.5	17,11	22,17

Architechtural Scales

SCALE	SCALE FACTOR	ANSI "A" 11"x 8.5"	ANSI "B" 17"x11"	ANSI "C" 22"x17"	ANSI "D" 34"x22"	ANSI "E" 44"x34"
1/8"=1'-0"	96	88',68'	136',88'	76',136'	272',176'	352',272'
1/4"=1'-0"	48	44',34'	68',44'	88',68'	136',88'	176',136'
1/2"=1'-0"	24	22',17'	34',22'	44',34'	68',44'	88',68'
3/4"=1'-0"	16	14.7',11.3'	22.7',14.7'	29.3',22.7'	45.3',29.3'	58.7',45.3'
1"=1'-0"	12	11',8.5'	17',11'	22',17'	34',22'	44',34'
2"=1'-0"	6	5.5',4.25'	8.5',5.5'	11',8.5'	17',11'	22',17'

Metric Scales

SCALE	SCALE FACTOR	ANSI "A" 11"x 8.5"	ANSI "B" 17"x11"	ANSI "C" 22"x17"	ANSI "D" 34"x22"	ANSI "E" 44"x34"
1=1	25.4 mm	279,216	432,279	559,432	864,559	1118,864
1=10	10 cm	110,85	170,110	220,170	340,220	440,340
1=20	20 cm	220,170	340,220	440,340	680,440	880,680
1=50	50 cm	550,425	850,550	1100,850	1700,1100	2200,1700
1=100	10 cm	1100,850	1700,1100	2200,1700	3400,2200	4400,3400

THE QUICK SETUP WIZARD

Starting a new drawing and clicking on the Use a Wizard button, shown in Figure 2–7, and then choosing the Quick Setup mode displays the Quick Setup dialog box, as shown in Figure 2–8, consisting of two categories. The Units category (see Figure 2–8) is used to graphically control the units of the drawing. Five units of measure are available through this tab, namely Decimal, Engineering, Architectural, Fractional, and Scientific. Clicking on the radio button displays a sample of the units' appearance in the drawing. When you have completed setting the units of the drawing, click on the Next> button to display the next dialog box in the Quick Setup sequence.

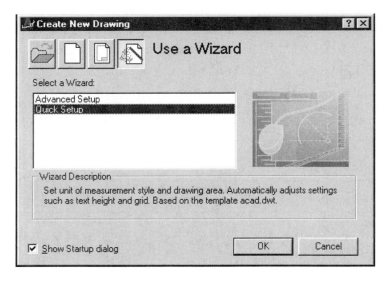

Figure 2–7

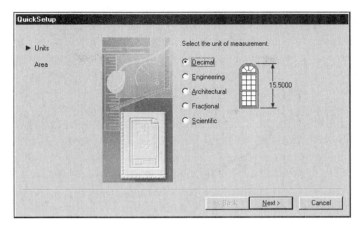

Figure 2–8

The Area category deals with the drawing area and is given as a Width and a Length value (see Figure 2–9). Setting this is comparable to using the LIMITS command. As you enter the width and length of the drawing area, the sample area image updates to reflect the changes in the values. At times, operators mistakenly substitute the width value for the length and vice versa. The image will allow you to preview the limits or drawing area before returning to the drawing screen. If the drawing area is incorrect, you can make changes, and the sample area image will update to show the latest changes. At any point, you may elect to return to the Units category by clicking on the <Back button. When you complete the setting of the drawing units and area, click on the Finish button to return to the drawing editor, where the units and area will be updated to reflect the changes you made in the Quick Setup Wizard.

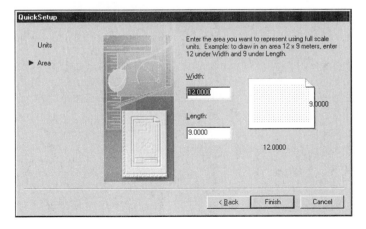

Figure 2–9

THE ADVANCED SETUP WIZARD

When you activate the Quick Setup Wizard, you are guided through two categories to make changes in the units and area of the drawing. Clicking on Advanced Setup in the Create New Drawing dialog box displays the Advanced Setup dialog box illustrated in Figure 2–10. The Advanced Wizard contains five categories for making changes to the initial drawing setup. The first category deals with the units of the drawing. This tab is almost identical to the dialog box found in the Quick Setup Wizard; it allows you to choose among Decimal, Engineering, Architectural, Fractional, and Scientific units. When you change the units, they will preview in the sample units area. An additional control for units in the Advanced Setup Wizard allows you to change the precision of the main units. In Figure 2–10, the precision for the decimal units is four decimal places. Changing the precision will update the Sample Units area, as shown in Figure 2–10.

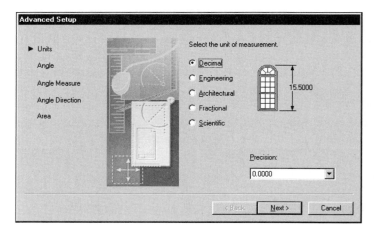

Figure 2–10

The second category of the Advanced Setup Wizard deals with how angles will be measured (see Figure 2–11). Five methods of angle measurements include the default of Decimal Degrees followed by Degrees/Minutes/Seconds, Grads, Radians, and Surveyor. A precision box for the measurement of angles allows you to change to different angular precision values. Changes to the measurement of angles and angular precision will be displayed in the image icon supplied.

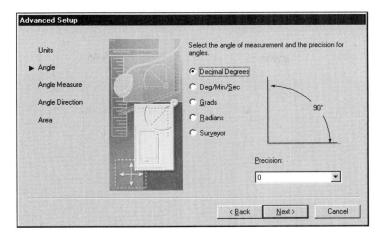

Figure 2–11

The third category controls the direction for angle measurement. By default, angles are measured starting with East for an angle of 0, North for an angle of 90°, West for 180°, and South for 270°. You have the option of changing the direction of zero; to make these changes, update the Angle Zero Direction area illustrated in Figure 2–12.

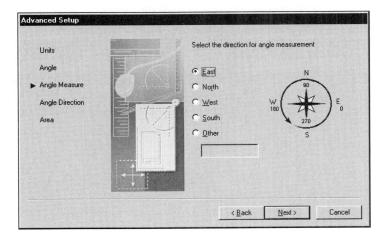

Figure 2–12

The fourth category of the Advanced Setup dialog box deals with Angle Direction. All angles are by default measured in the counterclockwise direction. Use the dialog box illustrated in Figure 2–13 to change from counterclockwise measurement of angles to clockwise angular measurements. Throughout this book, all examples dealing with angular measurements will keep the default setting of counterclockwise for angular direction.

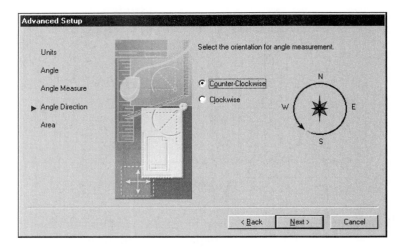

Figure 2–13

The fifth category of the Advanced Setup dialog box is used to change the drawing area (see Figure 2–14). Enter values for the Width and Length in the edit boxes provided, and watch the Sample Area image update to show the new drawing area.

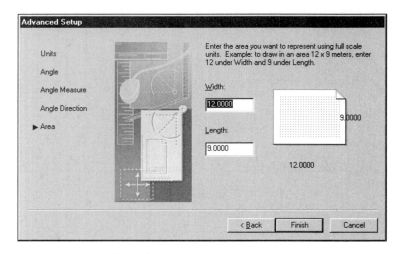

Figure 2–14

THE GRID COMMAND

Use grid to get a relative idea as to the size of objects. Grid is also used to define the size of the display screen originally set by the LIMITS command. The dots that make up the grid will never plot out on paper even if they are visible on the display screen. You can turn grid dots on or off by using the GRID command or by pressing F7, or by single-clicking on GRID, located in the status bar at the bottom of the display screen (see Figure 1–4). By default, the grid is displayed in 0.50-unit intervals similar to Figure 2–15. Use the following command sequence for the GRID command:

Command: **GRID**
Specify grid spacing(X) or [ON/OFF/Snap/Aspect] <0.5000>: *(Enter the desired grid spacing value)*

Illustrated in Figure 2–16 is a grid that has been set to a value of 0.25, or half its original size. Use the following command sequence to perform this change.

Command: **GRID**
Specify grid spacing(X) or [ON/OFF/Snap/Aspect] <0.5000>: **0.25**

While a grid is a useful aid for construction purposes, it may reduce the overall performance of the computer system. If the grid is set to a small value and it is visible on the display screen, it takes time for the grid to display. If a very small value is used for grid, a prompt will warn you that the grid value is too small to display on the screen.

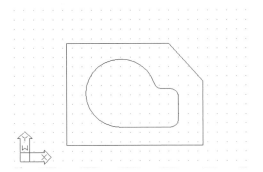
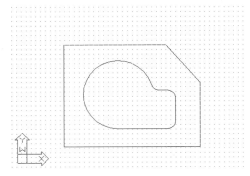

Figure 2–15   **Figure 2–16**

THE SNAP COMMAND

It is possible to have the cursor lock onto or snap to a grid dot as illustrated in Figure 2–17; this is the purpose of the SNAP command. By default, the current snap spacing is 0.50 units. Even though a value is set, the snap must be turned on for the cursor to be positioned on a grid dot. You can accomplish this by using the SNAP command as shown in the following sequence, by pressing F9, or by single-clicking on SNAP in the status bar at the bottom of the display screen (see Figure 1–4).

Command: **SN** *(For SNAP)*
Specify snap spacing or [ON/OFF/Aspect/Rotate/Style/Type] <0.5000>: *(Enter the desired snap spacing value)*

The current snap value may even affect the grid. If the current grid value is zero, the Snap value is used for the grid spacing.

Some drawing applications require that the snap be rotated at a specific angular value (see Figure 2–18). Changing the snap in this fashion also affects the cursor. Use the following command sequence for rotating the snap.

Command: **SN** *(For SNAP)*
Specify snap spacing or [ON/OFF/Aspect/Rotate/Style/Type] <0.5000>: **R** *(For Rotate)*
Specify base point <0.0000,0.0000>: *(Press ENTER to accept this value)*
Specify rotation angle <0>: **30**

One application of rotating snap is for auxiliary views where an inclined surface needs to be projected in a perpendicular direction. This will be explained and demonstrated in a later chapter.

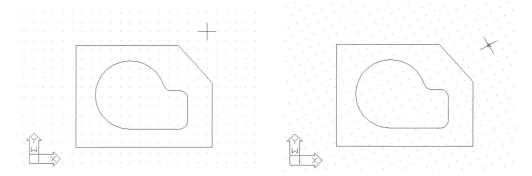

Figure 2–17   **Figure 2–18**

THE DRAFTING SETTINGS DIALOG BOX

Choosing Drafting Settings… from the Tools pull-down menu, as shown in Figure 2–19A, displays the Drafting Settings dialog box, shown in Figure 2–19B. Use this dialog box for making changes to the grid and snap settings. The Snap type & style area controls whether isometric grid is present or not.

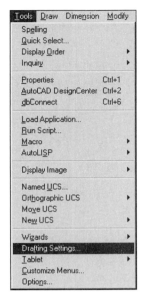

Figure 2–19A

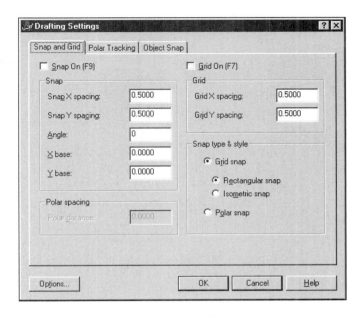

Figure 2–19B

THE ALPHABET OF LINES

Before you construct engineering drawings, the quality of the lines that make up the drawing must first be discussed. Some lines of a drawing should be made thick; others need to be made thin. This is to emphasize certain parts of the drawing and it is controlled through a line quality system. Illustrated in Figure 2–20 is a two-view drawing of an object complete with various lines that will be explained further.

The most important line of a drawing is the Object line, which outlines the basic shape of the object. Because of their importance, Object lines are made thick and continuous so they stand out among the other lines in the drawing. It does not mean that the other lines are considered unimportant; rather, the Object line takes precedence over all other lines.

The Cutting Plane line is another thick line; it is used to determine the placement in the drawing where an imaginary saw will cut into the drawing to expose interior details. It stands out by being drawn as a series of long dashes separated by spaces. Arrowheads determine how the adjacent view will be looked at. This line will be discussed in greater detail in Chapter 14, "Section Views."

The Hidden line is a medium weight line used to identify edges that become invisible in a view. It consists of a series of dashes separated by spaces. Whether an edge is visible or invisible, it still must be shown with a line.

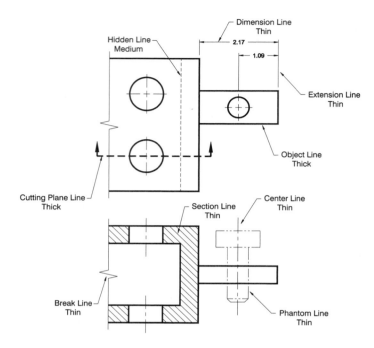

Figure 2–20

The Dimension line is a thin line used to show the numerical distance between two points. The dimension text is placed within the dimension line, and arrowheads are placed at opposite ends of the dimension line.

The Extension line is another thin continuous line, used as a part of the overall dimension. Extension lines show the edge being dimensioned. In Figure 2–20, the vertical extension lines are used to indicate the horizontal dimension distance.

When you use the Cutting Plane line to create an area to cut or slice, the surfaces in the adjacent view are section lined using the Section line, a thin continuous line.

Another important line used to identify the centers of circles is the Center line. It is a thin line consisting of a series of long and short dashes. It is a good practice to dimension to center lines; for this reason center lines, extension lines, and dimensions are made the same line thickness.

The Phantom line consists of a thin line made with a series of two dashes and one long dash. It is used to simulate the placement or movement of a part or component without actually detailing the component.

The Break line is a thin line with a "zigzag" symbol used to establish where an object is broken to simulate a continuation of the object.

ORGANIZING A DRAWING THROUGH LAYERS

As a means of organizing objects, a series of layers should be devised for every drawing. You can think of layers as a group of transparent sheets that combine to form the completed drawing. Figure 2–21 shows a drawing consisting of object lines, dimension lines, and border. Organizing these three drawing components by layers is illustrated in Figure 2–22. Only the drawing border occupies a layer that could be called "Border." The object lines occupy a layer that could be called "Object," and the dimension lines could be drawn on a layer called "Dim." At times, it may be necessary to turn off the dimension lines for a clearer view of the object. Creating all dimensions on a specific layer will allow you to turn off the dimensions while viewing all other objects on layers still turned on.

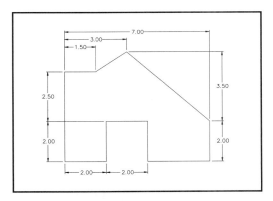

Figure 2–21

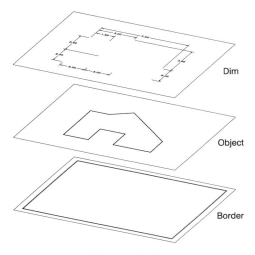

Figure 2–22

THE LAYER PROPERTIES MANAGER DIALOG BOX

An efficient way of creating and managing layers is through the Layer Properties Manager dialog box. Choose Layer...from the Format menu, as in Figure 2–23A, to display the Layer Properties Manager dialog box in Figure 2–23B.

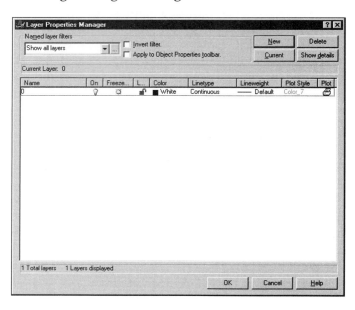

Figure 2–23A   **Figure 2–23B**

Another way of activating this dialog box is by picking the Layers button located in the Object Properties toolbar in Figure 2–24.

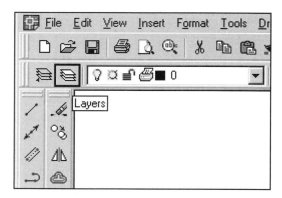

Figure 2–24

Once the dialog box displays, buttons located along the right side of this dialog box are designed for creating new layers, making an existing layer current to draw on, deleting a layer that is unused, or showing a series of details regarding a selected layer (Layer 0 cannot be deleted).

In addition to these buttons, a number of layer states exist that enable you to perform the following operations: turning layers on or off, freezing or thawing layers, locking or unlocking layers, assigning a color, linetype, lineweight, and plot style to a layer or group of layers and whether to plot a layer or group of layers (see Figure 2–23B). These functions will be discussed as follows:

On/Off – Makes all objects created on a certain layer visible or invisible on the display screen. The On state is symbolized by a yellow light bulb. The Off state has a light bulb shaded black.

Freeze – This state is similar to the Off mode; objects frozen will appear invisible on the display screen. Freeze, however, is considered a major productivity tool used to speed up the performance of a drawing. This is accomplished by not calculating any frozen layers during drawing regenerations. A snowflake symbolizes this layer state.

Thaw – This state is similar to the On mode; objects on frozen layers will reappear on the display screen when they are thawed. The sun symbolizes this layer state.

Lock – This state allows objects on a certain layer to be visible on the display screen while protecting them from accidentally being modified through an editing command. A closed padlock symbolizes this layer state.

Unlock – This state unlocks a previously locked layer and is symbolized by an open padlock.

Color – This state displays a color that is assigned to a layer and is symbolized by a square color swatch along with the name of the color. By default, the color White is assigned to a layer.

Linetype – This state displays the name of a linetype that is assigned to a layer. By default, the Continuous linetype is assigned to a layer.

Lineweight – This state sets a lineweight to a layer. An image of this lineweight value will be visible in this layer state column.

Plot Style – A plot style allows you to override the color, linetype, and lineweight settings made in the Layer Properties Manager dialog box. Notice how this area is grayed out. When working with a plot style that is color dependent, you cannot change the plot style. Plot styles will be discussed in greater detail later in this book.

Plot – This layer state controls which layers will be plotted. The presence of the printer icon symbolizes a layer that will be plotted. A printer icon with a red circle and slash signifies a layer that will not be plotted.

CREATING NEW LAYERS

Clicking on the New button of the Layer Properties Manager dialog box creates a new layer called Layer1, which displays in the layer list box as shown in Figure 2–25. Since this layer name is completely highlighted, you may elect to change its name to something more meaningful, such as a layer called Object or Hidden to hold all object or hidden lines in a drawing.

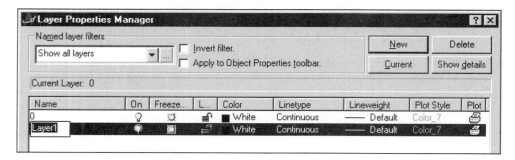

Figure 2–25

Illustrated in Figure 2–26 is the result of changing the name Layer1 to Object. You could also have entered "OBJECT" or "object" and these names would appear in the dialog box.

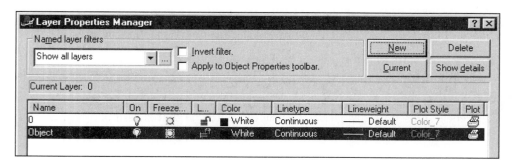

Figure 2–26

You can also be really descriptive with layer names. In Figure 2–27, a layer has been created called "SECTION (This layer is designed to hold crosshatch lines)." You are allowed to add spaces and other characters in the naming of a layer. Because of space limitations, the entire layer name will not display until you move your cursor over the top of the layer name. As a result, the layer name will appear truncated as in the layer "OBJECT (…l object lines)" in Figure 2–27.

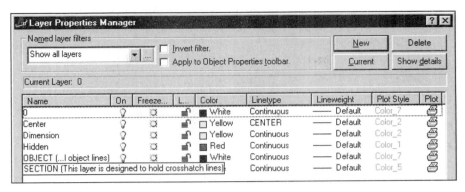

Figure 2–27

If more than one layer needs to be created, it is not necessary to continually click on the New button. Because this operation will create new layers, a more efficient method would be to perform either of the following operations: After creating a new "Layer1", change its name to Dim followed by a comma (,). This will automatically create a new Layer1 (see Figure 2–28). Change its name followed by a comma and a new layer is created, and so on. You could also have pressed ENTER twice, which would create another new layer.

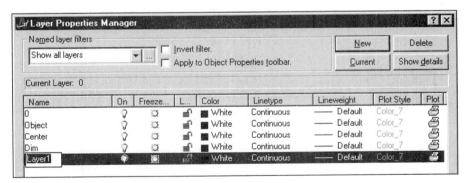

Figure 2–28

ASSIGNING COLOR TO LAYERS

Once you select a layer from the list box of the Layer Properties Manager dialog box and the color swatch is selected in the same row as the layer name, the Select Color dialog box shown in Figure 2–29 displays. Select the desired color from the Standard Colors area, Gray Shades area, or from the Full Color Palette area. The standard colors consist of the basic colors available on most color display screens. Depending on the video driver, different shades of standard colors may be selected from the Full Color Palette.

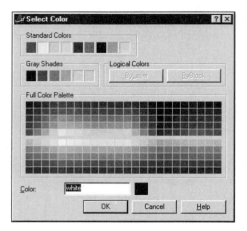

Figure 2–29

ASSIGNING LINETYPES TO LAYERS

Selecting the name Continuous next to the highlighted layer activates the Select Linetype dialog box, shown in Figure 2–30. Use this dialog box to dynamically select preloaded linetypes to be assigned to various layers. By default, the Continuous linetype is loaded for all new drawings. To load other linetypes, click on the Load button of the Select Linetype dialog box; this displays the Load or Reload Linetypes dialog box, shown in Figure 2–31.

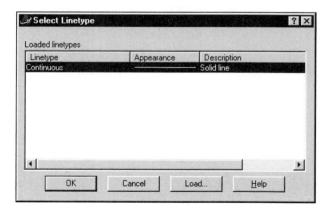

Figure 2–30

Once the Load or Reload Linetypes dialog box is displayed, as in Figure 2–31, use the scroll bars to view all linetypes contained in the file ACAD.LIN. Notice that, in addition to standard linetypes such as HIDDEN and PHANTOM, a few linetypes are provided that have text automatically embedded in the linetype. As the linetype is drawn, the text is placed depending on how it was originally designed. Notice also three variations of Hidden linetypes; namely HIDDEN, HIDDEN2, and HIDDENX2. The HIDDEN2 represents a linetype where the distances of the dashes and spaces in between each dash are half of the original HIDDEN linetype. HIDDENX2 represents a linetype where the distances of the dashes and spaces in between each dash of the original HIDDEN linetype are doubled. Click on the desired linetypes to load. When finished, click on the OK button.

The loaded linetypes now appear in the Select Linetype dialog box, as in Figure 2–32. It must be pointed out that the linetypes in this list are only loaded into the drawing and are not assigned to a particular layer. Clicking on the linetype in this dialog box will assign this linetype to the layer currently highlighted in the Layer Properties Manager dialog box.

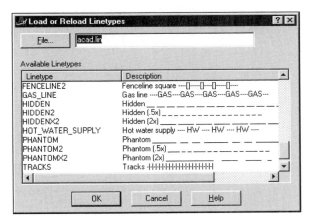

Figure 2–31

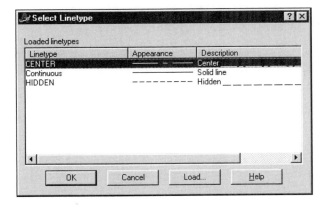

Figure 2–32

ASSIGNING LINEWEIGHT TO LAYERS

Selecting the name Default under the Lineweight heading of the Layer Properties Manager dialog box activates the Lineweight dialog box, shown in Figure 2–33. Use this to attach a lineweight to a layer. Lineweights are very important to a drawing file—they give contrast to the drawing. As stated earlier, the object lines should stand out over all other lines in the drawing. A thick lineweight would then be assigned to the object line layer.

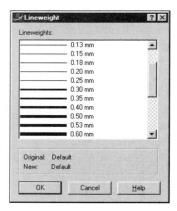

Figure 2–33

In Figure 2–34, a lineweight of 0.50 mm has been assigned to all object lines and a lineweight of 0.30 has been assigned to hidden lines. However, all lines in this figure appear to consist of the same lineweight. This is due to the lineweight feature being turned off. Notice, in this figure, the LWT button in the Status bar. Use this button to toggle on or off the display of assigned lineweights.

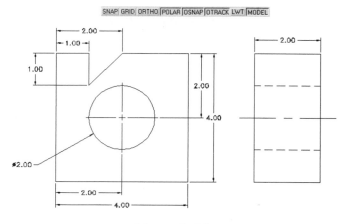

Figure 2–34

Single-clicking on the LWT button in Figure 2–35 turns on the lineweight function. The results are displayed in the illustration.

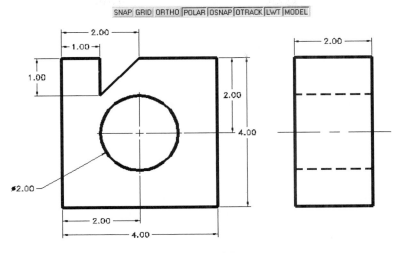

Figure 2–35

SHOWING LAYER DETAILS

Once layers have been created along with color, linetype, and lineweight assignments, the display of the Layer Properties Manager dialog box will be similar to Figure 2–36. Initially, when layers are created, they are placed in the dialog box in the exact order in which they were created. If you close the dialog box by clicking on the OK button and then revisit it at a later time, all layer names are reordered and displayed in alphabetical order (numbers preceding letters).

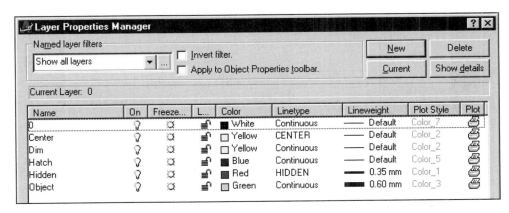

Figure 2–36

Clicking on the Show details button (see Figure 2–36) expands the bottom of the dialog box to include more detailed information about the selected layer (see Figure 2–37). Name, Color, Lineweight, and Linetype assignments are isolated in individual edit boxes. Also, the properties of the layer (On/Off, Lock/Unlock, Plot/No Plot, and Freeze/Thaw in all viewports) are displayed in a column to the right of the layer name. Clicking on the Hide details button compresses the dialog box again so that it appears like Figure 2–36.

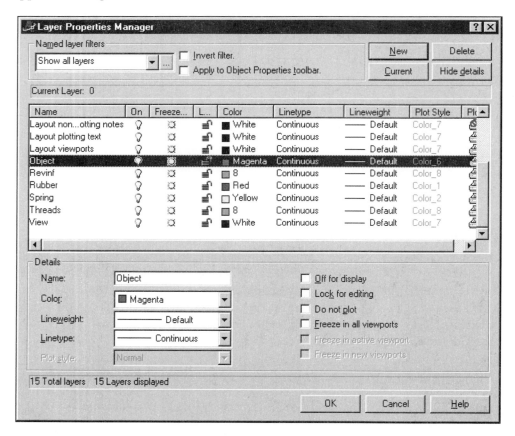

Figure 2–37

THE LINETYPE MANAGER DIALOG BOX

Choosing Linetype… from the Format pull-down menu, in Figure 2–38A, displays the Linetype Manager dialog box in Figure 2–38B. This dialog box is designed mainly to pre-load linetypes.

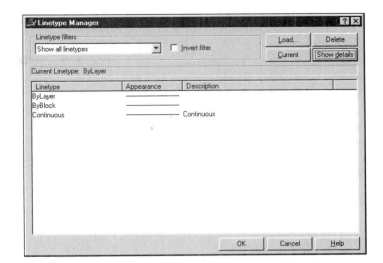

Figure 2–38A   **Figure 2–38B**

Clicking on the Load… button activates the Load or Reload Linetypes dialog box in Figure 2–39. You can select individual linetypes in this dialog box or load numerous linetypes at once by pressing CTRL and clicking on each linetype. Clicking the OK button will load the selected linetypes into the dialog box in Figure 2–40.

Figure 2–39

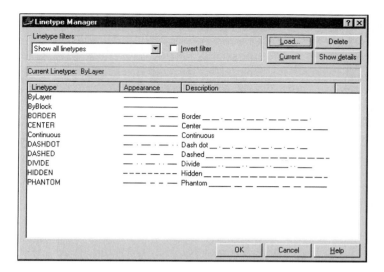

Figure 2–40

THE OBJECT PROPERTIES TOOLBAR

The Object Properties toolbar provides four areas to better access options for the control of Layers, Colors, Linetypes, and Lineweights. All four areas are illustrated in Figure 2–41. The Layer Control area will be discussed in greater detail.

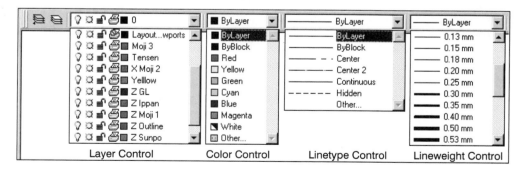

Figure 2–41

CONTROL OF LAYER PROPERTIES

Clicking in the long edit box next to the Layer button cascades all layers defined in the drawing in addition to their properties, identified by symbols (see Figure 2–42). The presence of the light bulb signifies that the layer is turned on. Clicking on the light bulb symbol turns the layer off. The sun symbol signifies that the layer is thawed. Clicking on the sun turns it into a snowflake symbol, signifying that the layer is now frozen. The padlock symbol controls whether a layer is locked or unlocked. By default, all layers are unlocked. Clicking on the padlock changes the symbol to display the image of a locked symbol, signifying that the layer is locked. The printer symbol allows you to print or not print objects on a layer.

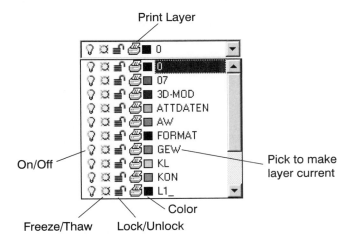

Figure 2–42

Study Figure 2–43 for a better idea of how the symbols affect the state of certain layers.

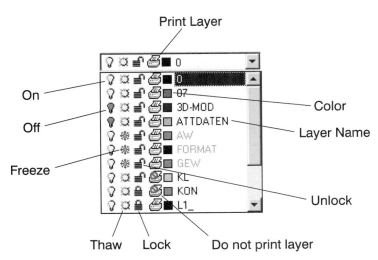

Figure 2–43

MAKING A LAYER CURRENT

Various methods can be employed to make a layer current to draw on. Select a layer in the Layer Properties Manager dialog box and then click on the Current button to make the layer current.

Picking a layer name from the Layer Control box in Figure 2–42 will also make the layer current.

The Make Object's Layer Current button, shown in Figure 2–44, allows you to make a layer the new current layer by just clicking on an object in the drawing. The layer is now made current based on the layer of the selected object.

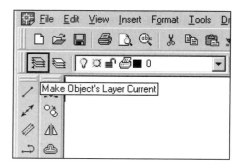

Figure 2–44

RIGHT-CLICK SUPPORT FOR LAYERS

While inside the Layer Properties Manager, right-clicking inside the layer information area displays the shortcut menu in Figure 2–45. Use this menu to make the selected layer current, to make a new layer based on the selected layer, select all layers in the dialog box or clear all layers. You can even select all layers except for the current layer. This shortcut menu provides you with easier access to commonly used layer manipulation tools.

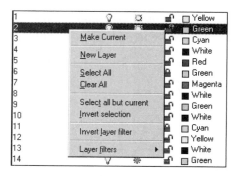

Figure 2–45

CONTROLLING THE LINETYPE SCALE

Once linetypes are associated with layers and placed in a drawing, you can control their scale with the LTSCALE command. In Figure 2–46, the default linetype scale value of 1.00 is in effect. This scale value acts as a multiplier for all linetype distances. In other words, if the hidden linetype is designed to have dashes 0.125 units long, a linetype scale value of 1.00 displays the dash of the hidden line at a value of 0.125 units. The LTSCALE command displays the following command sequence:

Command: **LTS** *(For LTSCALE)*
Enter new linetype scale factor <1.0000>: *(Press ENTER to accept the default or enter another value)*

In Figure 2–47, a linetype scale value of 0.50 units has been applied to all linetypes. As a result of the 0.50 multiplier, instead of all hidden line dashes measuring 0.125 units, they now measure 0.0625 units.

Command: **LTS** *(For LTSCALE)*
Enter new linetype scale factor <1.0000>: **0.50**

In Figure 2–48, a linetype scale value of 2.00 units has been applied to all linetypes. As a result of the 2.00 multiplier, instead of all hidden line dashes measuring 0.125 units, they now measure 0.25 units. Notice how a large multiplier displays the center lines as what appears to be a continuous linetype.

Command: **LTS** *(For LTSCALE)*
Enter new linetype scale factor <0.5000>: **2.00**

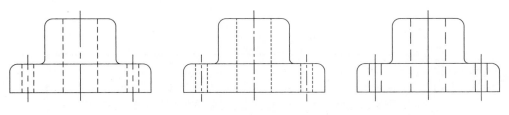

| **Figure 2–46** | **Figure 2–47** | **Figure 2–48** |

Figure 2–49 shows a facilities drawing that is designed to plot out at a scale factor of 1/4"=1'0". This creates a multiplier of 48 units (found by dividing 1' by 1/4). For all linetypes to show as hidden or center lines, this multiplier should be applied to the drawing through the LTSCALE command.

Command: **LTS** *(For LTSCALE)*
Enter new linetype scale factor <1.0000>: **48**

Since the drawing was constructed in real-world units or full size, the linetypes are converted to these units beginning with the multiplier of 48 through the LTSCALE command.

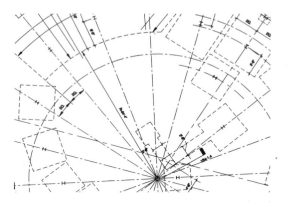

Figure 2–49

Expanding the Linetype Manager dialog box in Figure 2–50 displays a number of linetype details. The Global scale factor value has the same effect as the LTSCALE command. Also, setting a different value in the Current object scale box can scale the linetype of an individual object.

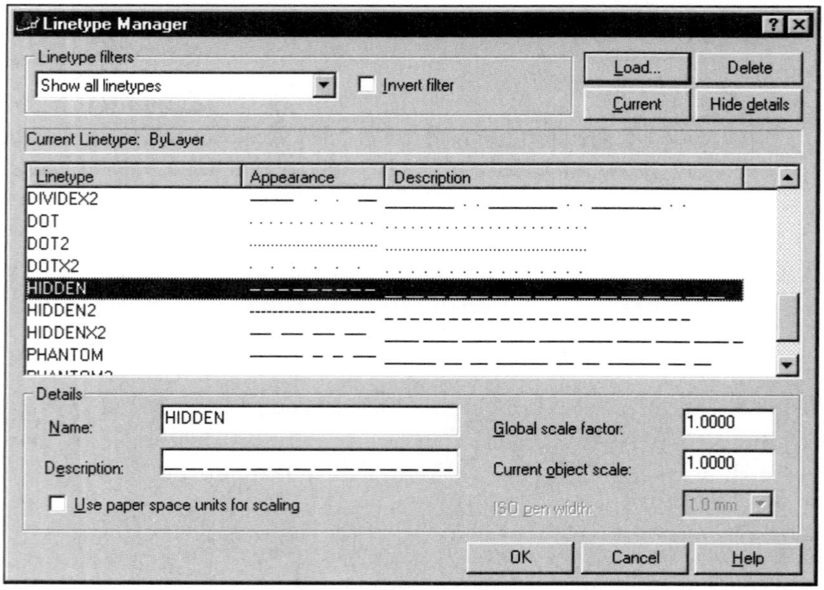

Figure 2–50

FILTERING LAYERS

Some drawings have so many layers that it is difficult to perform a search of the layers by using the scroll bars. When confronted with numerous layers in a drawing file, filter out all layers except those you want to view. For example, if you would like to list all layers that are assigned the color Red and the Hidden linetype, this technique would work. To activate layer filters, click in the outlined area of the Layer Properties Manager dialog box as illustrated in Figure 2–51. The Named Layer Filters dialog box will appear, similar to Figure 2–52. Through this dialog box, layers can be listed based on whether they are On or Off, Frozen or Thawed, Locked or Unlocked, by Color, or Linetype, or any combination of these parameters. Once a filter set has been created, you can even name this set for later use in the drawing design process.

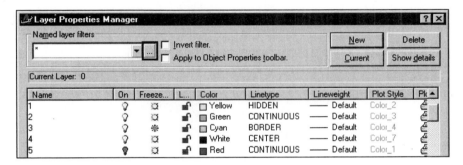

Figure 2–51

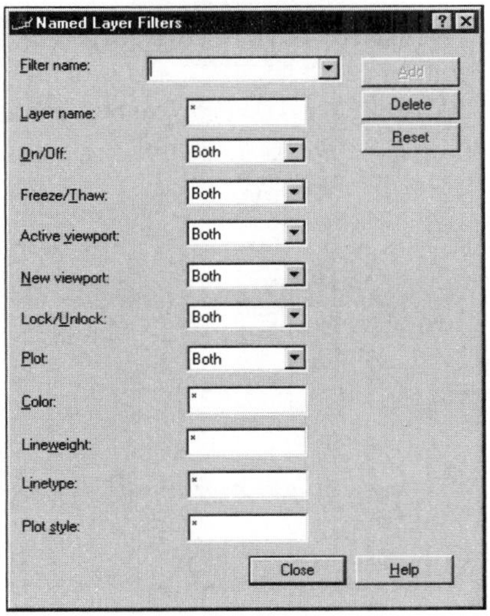

Figure 2–52

For example, to list all of the layers that are assigned to the color Cyan and are currently turned Off, click in the On/Off edit box and set the property to Off. Also click in the edit box next to Colors: and add the Cyan color to this area (see Figure 2–53). The results are displayed in Figure 2–54, showing the Layer Properties Manager dialog box. Only the layers assigned the color Cyan and turned Off appear. The following steps review the creation of a layer filter set:

1. Set the Freeze/Thaw: edit box to Off
2. Set the Color: edit box to CYAN
3. Set the Filter name: edit box to CYAN_OFF
4. Click the Add button to save this name.
5. Click the Close button to return to the Layer Properties Manager dialog box.

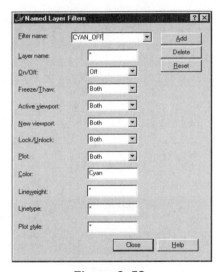

Figure 2–53

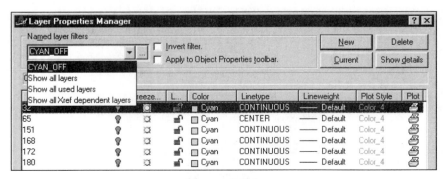

Figure 2–54

To redisplay all layers in the entire drawing, click in the Named Layer Filters edit box as shown in Figure 2–54, and change the property to "Show all layers", which will display all layers in the current drawing file. Another technique for listing all layers after performing a filter search would be to click on the Reset button, as shown in Figure 2–53. This would also list all layers in the drawing.

EXPRESS TOOLS APPLICATIONS FOR LAYERS

The Express Layer Tools toolbar shown in Figure 2–55A contains eight programs to assist in controlling layers. The Layer Manager command (LMAN) allows you to name, save, and restore the settings made in the Layer Properties Manager dialog box. These layer settings may be saved to a .LAY extension, which may be imported to other drawings. Other Express Tools commands include the ability to match an object's layer with that of another, pick an object to make the layer of that object current, and pick a layer and isolate it by turning all other layers off. Still other commands allow you to freeze, turn off, lock, or unlock an object's layer by picking only the object. The complete listing of Express Tools that pertain to layers can easily be accessed through the Express pull-down menu area illustrated in Figure 2–55B.

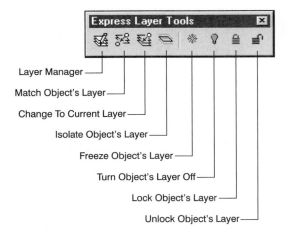

Figure 2–55A

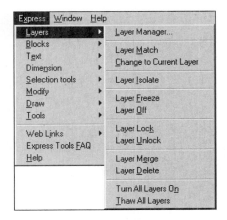

Figure 2–55B

THE LAYER MANAGER DIALOG BOX

This powerful Express tool is used to manage the settings made in the Layer Properties Manager dialog box. The Layer Manager or LMAN command allows you to save and restore layer configurations into 'layer states'. Certain layers can be turned on or off and saved in the Layer Manager dialog box in Figure 2–56. Layer states are saved in the drawing and can also be exported to or imported from a .LAY file.

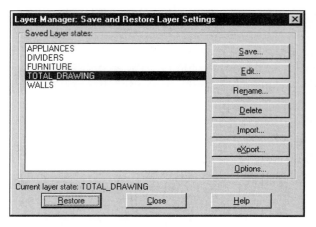

Figure 2–56

Illustrated in Figure 2–57 is an office plan drawing. The following layer states have been saved in Figure 2–56:

Appliances

Dividers

Furniture

Total_Drawing

Walls

Since the layer state called "Walls" has all layers but the building walls frozen, clicking on WALLS in the Layer Manager dialog box in Figure 2–56 displays the image in Figure 2–58.

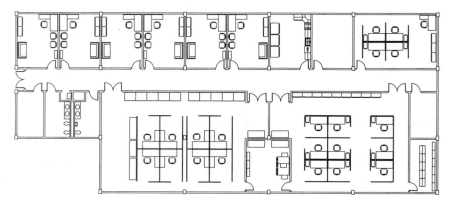

Figure 2–57

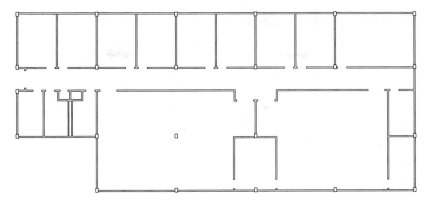

Figure 2–58

MATCH OBJECTS LAYER

This layer bonus routine allows you to change the layers of selected objects to match the layer of a selected destination object. The command is LAYMCH. First select the objects to be changed, such as the three chairs illustrated in Figure 2–59. Next select the object on the destination layer as in the shelf at "D." The three chairs now take on the same layer as the shelf.

Command: **LAYMCH**
Select objects to be changed:
Select objects: *(Select the three chairs labeled "A", "B", and "C")*
Select objects: *(Press* ENTER *to continue)*
3 found.
Type name/Select entity on destination layer: *(Pick the shelf at "D")*
3 objects changed to layer FURNITURE.

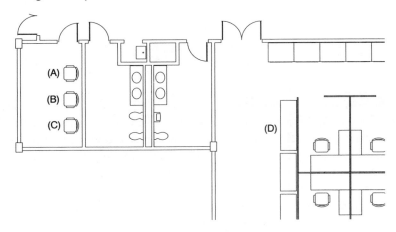

Figure 2–59

CHANGE TO CURRENT LAYER

This layer bonus routine is used to change the layer of one or more objects to the current layer. The command is LAYCUR. This is particularly helpful if objects were constructed on the wrong layer. In Figure 2–60, the three chairs labeled "A", "B", and "C" were drawn on the wrong layer. First, make the desired layer current; in Figure 2–60, the current layer is FURNITURE. Selecting the three chairs after issuing the LAYCUR command changes the three chairs to the FURNITURE layer.

Command: **LAYCUR**
Select objects to be CHANGED to the current layer:
Select objects: *(Select the three chairs "A", "B", and "C")*
Select objects: *(Press* ENTER *to continue)*
3 found.
3 objects changed to layer FURNITURE *(the current layer).*

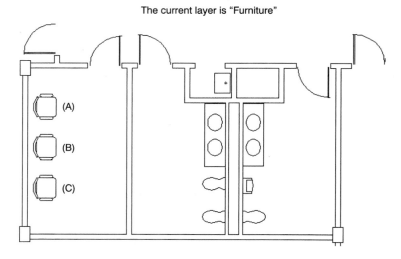

The current layer is "Furniture"

Figure 2–60

ISOLATE OBJECT'S LAYER

This layer Express tool isolates the layer or layers of one or more selected objects by turning all other layers off. The effects of this command are the same as using the Layer Properties Manager dialog box and manually picking the layer names to turn off. The LAYISO command is yet another Express tool that allows you to operate in a more efficient manner than if using conventional AutoCAD commands. After issuing the command, you prompted to select the object or objects on the layer or layers to be isolated. In Figure 2–61, pick the outer wall at "A" or any outer wall and press

ENTER to execute the command. A message in the command prompt area alerts you that layer 1-WALL was isolated. All objects on 1-WALL are illustrated in Figure 2–62; all other layers are turned off.

Command: **LAYISO**
Select object(s) on the layer(s) to be ISOLATED:
Select objects: *(Pick the wall at "A")*
Select objects: *(Press ENTER to isolate the layers based on the objects selected)*
Layer 1-WALL has been isolated.

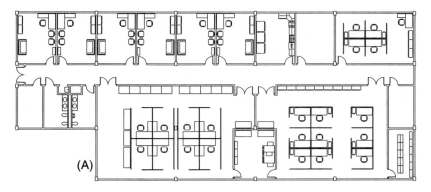

Figure 2–61

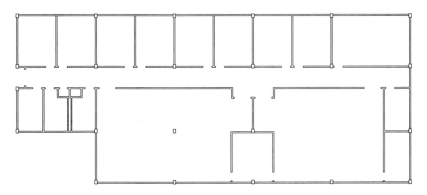

Figure 2–62

FREEZE OBJECT'S LAYER

This layer bonus routine freezes layers by picking objects to control the layers to be frozen. Again, the Layer Properties Manager dialog box is normally used to freeze and thaw layers.

Command: **LAYFRZ**

Options/Undo/<Pick an object on the layer to be FROZEN>: *(Pick one of the chairs in Figure 2–61)*

Layer CHAIRS has been frozen.

Options/Undo/<Pick an object on the layer to be FROZEN>: *(Pick one of the shelves in Figure 2–61)*

Layer FURNITURE has been frozen.

Options/Undo/<Pick an object on the layer to be FROZEN>: *(Pick one of the bathroom sinks in Figure 2–61)*

Layer FIXTURES has been frozen.

Options/Undo/<Pick an object on the layer to be FROZEN>: *(Pick one of the kitchen refrigerators in Figure 2–61)*

Layer KITCHEN has been frozen.

Options/Undo/<Pick an object on the layer to be FROZEN>: *(Pick one of the doors in Fig Figure 2–61)*

Layer DOOR has been frozen.

Options/Undo/<Pick an object on the layer to be FROZEN>: *(Press ENTER to execute this command. The results should be similar to Figure 2–63)*

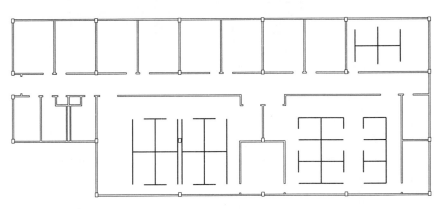

Figure 2–63

TURN OBJECT'S LAYER OFF

This layer bonus routine is similar to the LAYFRZ command, except instead of freezing layers, you can turn off a layer or group of layers by selecting an object or group of objects. The command LAYOFF is illustrated below and references Figure 2–61.

Command: **LAYOFF**

Options/Undo/<Pick an object on the layer to be turned OFF>: *(Pick one of the shelves Figure 2–61)*

Layer FURNITURE has been turned off.

Options/Undo/<Pick an object on the layer to be turned OFF>: *(Pick one of the office partitions in Figure 2–61)*
Layer I-PART has been turned off.
Options/Undo/<Pick an object on the layer to be turned OFF>: *(Pick one of the chairs in Fig Figure 2–61)*
Layer CHAIRS has been turned off.
Options/Undo/<Pick an object on the layer to be turned OFF>: *(Press* ENTER *to execute this command. The results should be similar to Figure 2–64)*

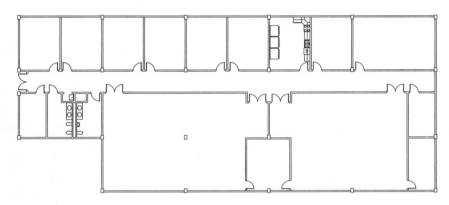

Figure 2–64

LOCK OBJECT'S LAYER

This layer bonus routine locks the layer of a selected object. A locked layer is visible on the display screen; however any object associated with a locked layer is non-selectable. The command to lock a layer is LAYLCK.

Command: **LAYLCK**
Pick an object on the layer to be LOCKED: *(Pick one of the exterior walls in Figure 2–61)*
Layer I-WALL has been locked.

UNLOCK OBJECT'S LAYER

This layer bonus routine unlocks the layer of a selected object. Objects on an unlocked layer can now be selected whenever the prompt "Select objects:" appears. The command to unlock a layer is LAYULK.

Command: **LAYULK**
Pick an object on the layer to be UNLOCKED: *(Pick one of the exterior walls in Figure 2–61)*
Layer I-WALL has been unlocked.

MERGING LAYERS INTO ONE

This interesting Express tools command, LAYMRG, will merge the contents of one layer with a target layer. The layer holding the chairs called CHAIRS, illustrated in Figure 2–65, needs to be merged with the layer holding the cabinet called FURNI-TURE. One of the chairs is selected as the layer to merge. Next, one of the cabinets at "A" is selected as the target layer. A warning message alerts you that the CHAIRS layer will be permanently merged into the FURNITURE layer. After you perform this operation, the CHAIRS layer will automatically be purged from the drawing.

Command: **LAYMRG**
Select object on layer to merge or [Type-it/Undo]: *(Select one of the chairs in Figure 2–65)*
Selected layers: CHAIRS
Select object on layer to merge or [Type-it/Undo] <done>: *(Press* ENTER *to continue)* ("CHAIRS")
Select object on target layer or [Type-it]: *(Pick cabinet "A")*
******** WARNING ********
You are about to permanently merge layer CHAIRS into layer FURNITURE.
Do you wish to continue? [Yes/No] <No>:**Y**
Merging layer CHAIRS into layer FURNITURE.
All entities which were on layer CHAIRS have been moved to layer FURNITURE.

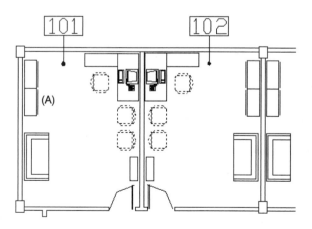

Figure 2–65

DELETING ALL OBJECTS ON A LAYER

Use this Express tools command, LAYDEL, to delete all objects on a specified layer. In Figure 2–66A, a computer terminal is selected that is identified with the CPU layer. Before all objects on this layer are deleted, a message appears asking if you really want to perform this operation. Answering yes deletes all objects that are assigned to the layer in addition to purging the layer from the drawing. Although this tool has many beneficial advantages, care must be exercised as to when and when not to use it.

Command: **LAYDEL**
Select object on layer to delete or [Type-it/Undo]: *(Select the computer monitor at "A")*
Selected layers: CPU
Select object on layer to delete or [Type-it/Undo] <done>: *(Press* ENTER *to continue)* ("CPU")
You are about to permanently delete layer CPU from this drawing.
Do you wish to continue? [Yes/No] <No>: **Y**

The results are displayed in Figure 2–66B, with all objects on the CPU layer being deleted.

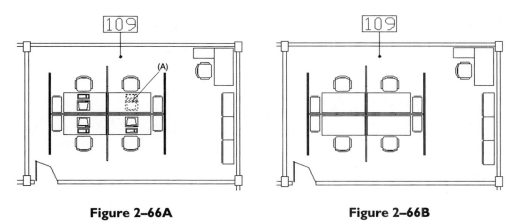

Figure 2–66A   Figure 2–66B

TURNING ON ALL LAYERS/ THAWING ALL LAYERS

Use these Express tools layer commands to turn on all layers that were previously turned off or to thaw all layers that were previously frozen.

TUTORIAL EXERCISE:
Creating Layers Using the Layer Properties Manager Dialog Box

Layer Name	Color	Linetype	Lineweight
Object	White	Continuous	0.60
Hidden	Red	Hidden	0.30
Center	Yellow	Center	Default
Dim	Yellow	Continuous	Default
Section	Blue	Continuous	Default

Purpose
Use the following steps to create layers according to the above specifications.

STEP 1

Activate the Layer Properties Manager dialog box by clicking on the Layer button located on the Object Properties toolbar as shown in Figure 2–67.

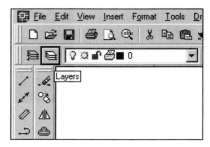

Figure 2–67

STEP 2

Once the Layer Properties Manager dialog box displays in Figure 2–68, notice on your screen that only one layer is currently listed, namely Layer 0. This happens to be the default layer, the layer that is automatically available to all new drawings. Since it is considered poor practice to construct any objects on Layer 0, new layers will be created not only for object lines but for hidden lines, center lines, dimension objects, and section lines as well. To create these layers, click on the New button and notice that a layer is automatically added to the list of layers. This layer is called Layer1 (see Figure 2–69). While this layer is highlighted, enter the first layer name, "Object."

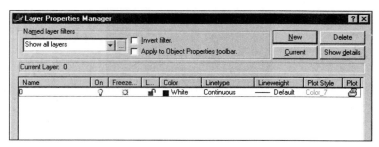

Figure 2–68

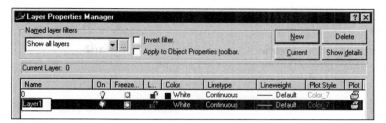

Figure 2–69

Entering a comma after the name of the layer allows more layers to be added to the listing of layers. Once you've entered the comma after the layer "Object" and the new layer appears, enter the new name of the layer as "Hidden." Repeat this pro-

cedure of using the comma to create multiple layers for "Center," "Dim," and "Section." Press ENTER after typing in "Section." The complete list of all layers created should be similar to the illustration shown in Figure 2–70.

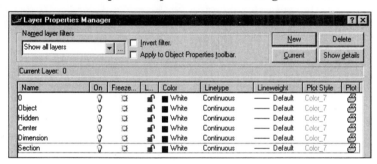

Figure 2–70

STEP 3

As all new layers are displayed, the names may be different, but they all have the same color and linetype assignments (see

Figure 2–71). At this point, the dialog box comes in handy in assigning color and linetypes to layers in a quick and easy

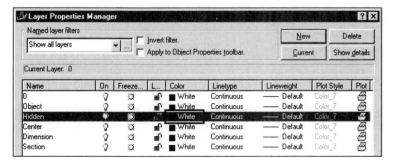

Figure 2–71

manner. First, highlight the desired layer to add color or linetype by picking the layer. A horizontal bar displays, signifying that this is the selected layer. Click on the color swatch identified by the box in Figure 2–71 and follow the next step to assign the color "Red" to the "Hidden" layer name.

STEP 4

Clicking on the color swatch in the previous step displays the Select Color dialog box shown in Figure 2–72. Select the desired color from one of the following areas: Standard Colors, Gray Shades, Full Color Palette. The standard colors represent colors 1 (Red) through 9 (Grey). On display terminals with a high-resolution graphics card, the full color palette displays different shades of the standard colors. This gives you a greater variety of colors to choose from. For the purposes of this tutorial, the color "Red" will be assigned to the "Hidden" layer. Select the box displaying the color red; a box outlines the color and echoes the color in the bottom portion of the dialog box. Click on the OK button to complete the color assignment; if you select Cancel, the color

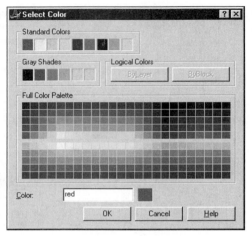

Figure 2–72

assignment will be removed, requiring this operation to be used again. Continue with this step by assigning the color Yellow to the Center and Dim layers; assign the color Blue to the Section layer.

STEP 5

Once the color has been assigned to a layer, the next step is to assign a linetype, if any, to the layer. The "Hidden" layer requires a linetype called "HIDDEN." Click on the "Continuous" linetype that is highlighted in Figure 2–71 to display the Select Linetype dialog box illustrated in Figure 2–73. By default, Continuous is the only linetype

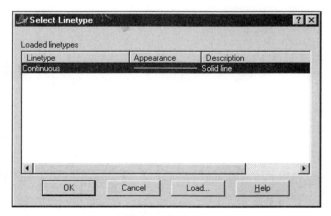

Figure 2–73

loaded. Clicking on the Load button displays the next dialog box, illustrated in Figure 2–74.

Choose the desired linetype to load from the Load or Reload the Linetype dialog box shown in Figure 2–74. Scroll through the linetypes until the "Hidden" linetype is found. Click on the OK button to return to the Select Linetype dialog box.

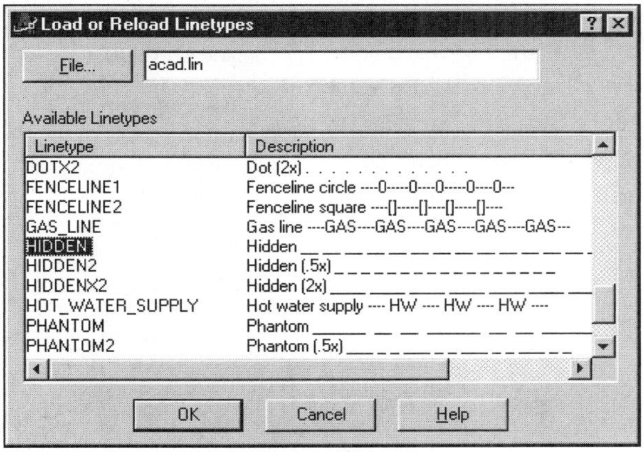

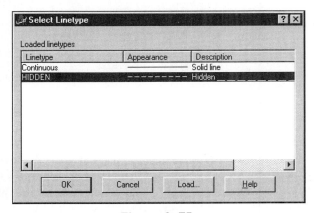

Figure 2–74

Once you're back in the Select Linetype dialog box, shown in Figure 2–75, notice the "Hidden" linetype listed along with "Continuous." Because this linetype has just been loaded, it still has not been assigned to the "Hidden" layer. Click on the Hidden linetype listed in the Select Linetype dialog box, and click on the OK button. Once the Layer Properties Manager dialog box reappears, notice that the "Hidden" linetype has been assigned to the "Hidden" layer. Repeat this procedure to assign the "Center" linetype to the "Center" layer.

Figure 2–75

STEP 6

Another requirement of the "Hidden" layer is that it carry a lineweight of 0.30 mm to have the hidden lines stand out when compared with the object in other layers. Clicking on the highlighted default Lineweight in Figure 2–71 displays the Lineweight dialog box in Figure 2–76. Click on the 0.30 mm lineweight followed by the OK button to assign this lineweight to the "Hidden" line. Use the same procedure to assign a lineweight of 0.60 mm to the Object layer.

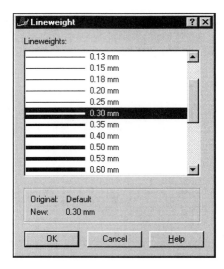

Figure 2–76

STEP 7

Once you've completed all color and linetype assignments, the Layer Properties Manager dialog box should appear similar to Figure 2–77. Click on the OK button to save all layer assignments and return to the drawing editor.

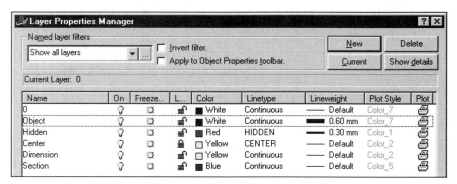

Figure 2–77

STEP 8

When layers are first created, they are listed in the order they were entered. When the layers are saved and the Layer Properties Manager dialog box is displayed again, all layers are reordered alphabetically (see Figure 2–78.)

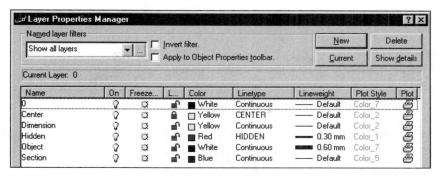

Figure 2–78

STEP 9

Clicking in the Name box at the top of the list of layers reorders the layers. Now, the layers are listed in reverse alphabetical order (see Figure 2–79). This same effect occurs when you click in the On, Freeze, Color, Linetype, Lineweight, and Plot header boxes. Figure 2–80 displays the results of clicking on the Color header where all colors are reordered starting with Red (1), Yellow (2), Blue (5), and White (7).

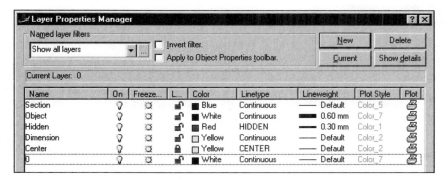

Figure 2–79

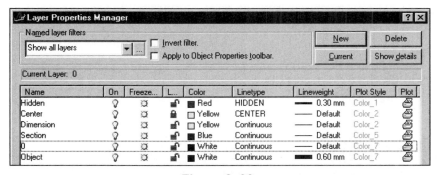

Figure 2–80

PROBLEMS FOR CHAPTER 2

PROBLEM 2–1

Open the existing drawing called 02-Storage.Dwg. Use the Layer Properties Manager along with the Named Layer Filters dialog box to answer the following layer-related questions:

1. The total number of layers frozen in the database of this drawing is _____.
2. The total number of layers locked in the database of this drawing is _____.
3. The total number of layers turned off in the database of this drawing is _____.
4. The total number of layers assigned the color red and frozen in the database of this drawing is _____.
5. The total number of layers that begin with the letter "L", assigned the color yellow, and turned off in the database of this drawing is _____.

PROBLEM 2–2

Open the existing drawing called 02_I_Pattern.Dwg. Use the Layer Properties Manager along with the Named Layer Filters dialog box to answer the following layer-related questions:

1. The total number of layers assigned the color red and turned off in the database of this drawing is _____.
2. The total number of layers assigned the color red and locked in the database of this drawing is _____.
3. The total number of layers assigned the color white and frozen in the database of this drawing is _____.
4. The total number of layers assigned the color yellow and assigned the hidden linetype in the database of this drawing is _____.
5. The total number of layers assigned the color red and assigned the phantom linetype in the database of this drawing is _____.
6. The total number of layers assigned the center linetype in the database of this drawing is _____.

CHAPTER 3

AutoCAD Display and Selection Operations

THE ZOOM COMMAND

The ability to magnify details in a drawing or demagnify the drawing to see it in its entirety is a function of the ZOOM command. It does not take much for a drawing to become very busy, complicated, or dense when displayed in the drawing editor. Therefore, use the ZOOM command to work on details or view different parts of the drawing. One way to select the options of this command is from the View pull-down menu as shown in Figure 3–1. Choosing Zoom displays the various options of the ZOOM command. These options include zooming in real time, zooming the previous display, using a window to define a boxed area to zoom to, dynamic zooming, zooming to a user-defined scale factor, zooming based on a center point and a scale factor, or performing routine operations such as zooming in or out, zooming all, or zooming the extents of the drawing. All of these modes will be discussed in the pages that follow.

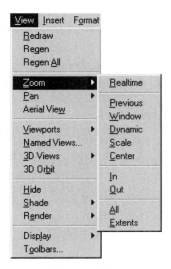

Figure 3–1

A second way to choose options of the ZOOM command is through the Standard toolbar area at the top of the display screen. Figure 3–2 shows a series of buttons that perform various options of the ZOOM command. Four main buttons initially appear in the Standard toolbar, namely Realtime PAN, Realtime ZOOM, ZOOM-Window, and ZOOM-Previous. Pressing down on the ZOOM-Window option displays a series of additional zoom option buttons that cascade down. See Figure 3–2 for the meaning of each button. Click on the button to perform the desired zoom operation.

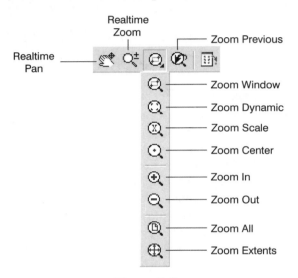

Figure 3–2

Figure 3–3 shows a third method of performing ZOOM command options using the dedicated Zoom floating toolbar. This toolbar contains the same buttons found in the Standard toolbar; it differs in the fact that the toolbar in Figure 3–3 can be moved to different positions around the display screen. You can also activate the ZOOM command from the keyboard by entering either ZOOM or the letter Z, which is its command alias.

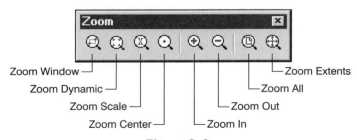

Figure 3–3

Figure 3–4 is a complex drawing of a part that consists of all required orthographic views. To work on details of this and other drawings, use the ZOOM command to magnify or demagnify the display screen. The following are the options of the ZOOM command:

Command: **Z** *(For ZOOM)*
Specify corner of window, enter a scale factor (nX or nXP), or
[All/Center/Dynamic/Extents/Previous/Scale/Window] <real time>: *(enter one of the listed options)*

Executing the ZOOM command and picking a blank part of the screen places you in automatic ZOOM-Window mode. Selecting another point zooms in to the specified area. Refer to the following command sequence to use this mode of the ZOOM command on the object illustrated in Figure 3–4.

Command: **Z** *(For ZOOM)*
Specify corner of window, enter a scale factor (nX or nXP), or
[All/Center/Dynamic/Extents/Previous/Scale/Window] <real time>: *(Mark a point at "A")*
Other corner: *(Mark a point at "B")*

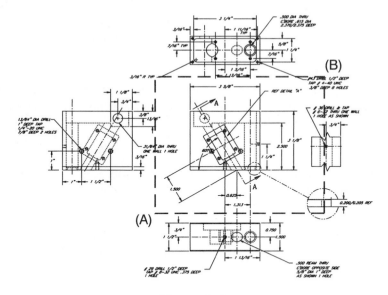

Figure 3–4

The ZOOM-Window option is automatically invoked once you select a blank part of the screen and then pick a second point. The resulting magnified portion of the screen appears in Figure 3–5.

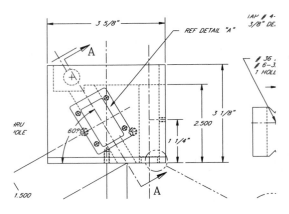

Figure 3–5

ZOOMING IN REAL TIME

A powerful option of the ZOOM command is performing screen magnifications or demagnifications in real time. This is the default option of the command. Issuing the Realtime option of the ZOOM command displays a magnifying glass icon with a positive sign and a negative sign above the magnifier icon. Identify a blank part of the drawing editor, press down the Pick button of the mouse (the left mouse button), and move in an upward direction to zoom in to the drawing in real time. Identify a blank part of the drawing editor, press down the Pick button of the mouse, and move in a downward direction to zoom out of the drawing in real time. Use the following command sequence and see also Figure 3–6.

Command: **Z** *(For ZOOM)*
Specify corner of window, enter a scale factor (nX or nXP), or
[All/Center/Dynamic/Extents/Previous/Scale/Window] <real time>: *(Press ENTER to accept Realtime as the default)*

Identify the lower portion of the drawing editor, press and hold down the Pick button of the mouse, and move the Realtime cursor up; notice the image zooming in.

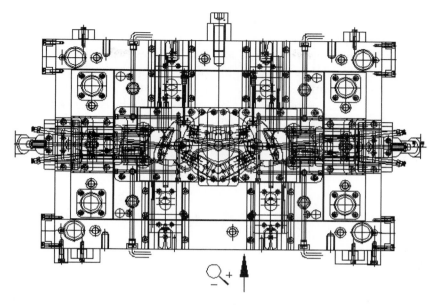

Figure 3–6

Once you are in the Realtime mode of the ZOOM command, press the right mouse button to activate the cursor menu shown in Figure 3–7. Use this menu to switch between Realtime ZOOM and Realtime PAN, which gives you the ability to pan across the screen in real time. The ZOOM Window, Original (Previous), and Extents options are also available in the cursor menu in Figure 3–7.

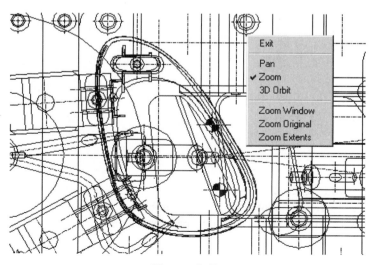

Figure 3–7

USING AERIAL VIEW ZOOMING

Another dynamic way of performing zooms is through the Aerial View option, which is selected from the View pull-down menu in Figure 3–8. Choosing Aerial View activates a dialog box in the lower right corner of the screen displaying a smaller image of the total drawing (see Figure 3–9). You create a window inside the Aerial View dialog box, which performs the zoom operation in the background of the main AutoCAD screen.

Figure 3–8

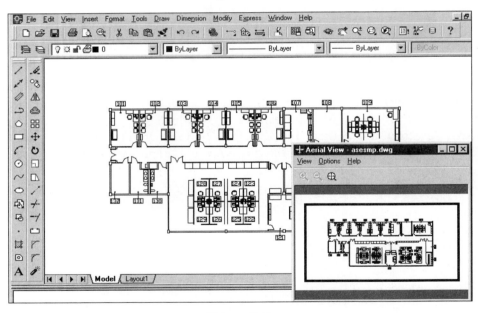

Figure 3–9

The first time you click inside the Aerial View image, a box similar to Figure 3–10A appears. This box represents the current zoom window. Notice that, as you move this box around, your drawing pans to different locations of the screen. The trick to performing zooms with this box is to make it smaller— by single-clicking on the left mouse button, your pick button. When you perform this operation, the box changes to an image of a new box as in Figure 3–10B. An arrow is present along the right edge of the rectangular box. Single-clicking the pick button of the mouse allows you to increase or decrease the size of the rectangle. Making the box smaller zooms in to the drawing. Single-clicking again with the pick button of the mouse returns the rectangle to the image illustrated in Figure 3–10C. Here the rectangle is smaller; as you move the rectangle around the screen, you can notice the magnified image panning around in the background. To complete the operation, right-click the mouse button to anchor the small rectangle to the image in the Aerial View dialog box.

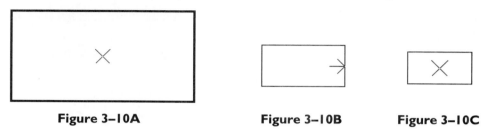

Figure 3–10A   **Figure 3–10B**   **Figure 3–10C**

In Figure 3–11, a small rectangular box has been created in the Aerial View dialog box. The current position of this rectangle in the drawing is displayed in the background of the main AutoCAD screen, where room numbers 107 and 108 are visible. Right-clicking the mouse button makes the rectangular box stationary in the Aerial View dialog box, as in Figure 3–12. This frees you up to activate the main AutoCAD screen and work on your drawing in its zoomed-in state.

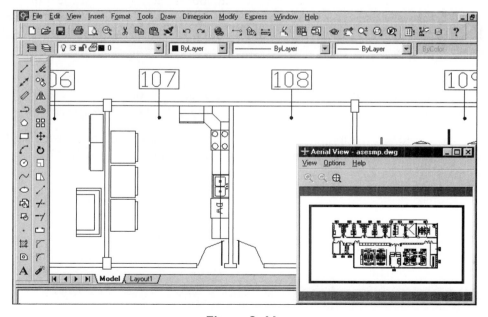

Figure 3–11

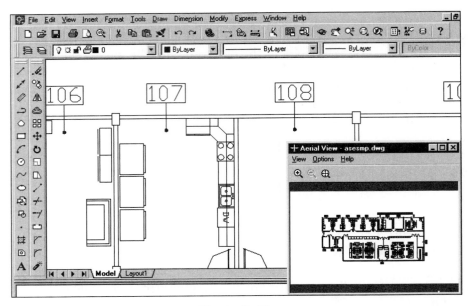

Figure 3–12

USING ZOOM-ALL

Another option of the ZOOM command is All. Use this option to zoom to the current limits of the drawing as set by the LIMITS command. In fact, right after the limits of a drawing have been changed, issuing a ZOOM-All updates the drawing file to reflect the latest screen size. To use the ZOOM-All option, refer to the following command sequence.

Command: **Z** *(For ZOOM)*
Specify corner of window, enter a scale factor (nX or nXP), or
[All/Center/Dynamic/Extents/Previous/Scale/Window] <real time>: **A** *(For All)*

In Figure 3–13A, the top illustration shows a zoomed-in portion of a part. Use the ZOOM-All option to zoom to the drawing's current limits in Figure 3–13B.

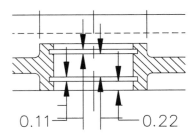

Figure 3–13A

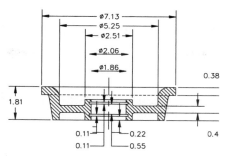

Figure 3–13B

USING ZOOM-CENTER

The ZOOM-Center option allows you to specify a new display based on a selected center point (see Figure 3–14). A window height controls whether the image on the display screen is magnified or demagnified. If a smaller value is specified for the magnification or height, the magnification of the image is increased (you zoom in to the object). If a larger value is specified for the magnification or height, the image gets smaller, or a ZOOM-Out is performed (see Figure 3–14).

Command: **Z** *(For ZOOM)*
Specify corner of window, enter a scale factor (nX or nXP), or
[All/Center/Dynamic/Extents/Previous/Scale/Window] <real time>: **C** *(For Center)*
Center point: *(Mark a point at the center of circle "A" shown in Figure 3–14)*
Magnification or Height <7.776>: **2**

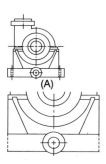

Figure 3–14

USING ZOOM-EXTENTS

The top image of the pump in Figure 3–15 reflects a ZOOM-All operation. This option displays the entire drawing area based on the drawing limits even if the objects that make up the image appear small. Instead of performing a zoom based on the drawing limits, ZOOM-Extents uses the extents of the image on the display screen to perform the zoom. The bottom image in Figure 3–15 shows the largest possible image displayed as a result of using the ZOOM command and the Extents option.

Command: **Z** *(For ZOOM)*
Specify corner of window, enter a scale factor (nX or nXP), or
[All/Center/Dynamic/Extents/Previous/Scale/Window] <real time>: **E** *(For Extents)*

Zoom-All

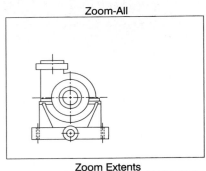

Zoom Extents

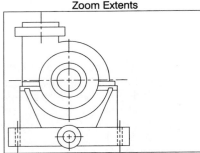

Figure 3–15

USING ZOOM-WINDOW

The ZOOM-Window option allows you to specify the area to be magnified by marking two points representing a rectangle, as in the top image of Figure 3–16. The center of the rectangle becomes the center of the new image display; the image inside the rectangle is either enlarged (see the lower image of Figure 3–16) or reduced. The following command sequence demonstrates ZOOM-Window:

Command: **Z** *(For ZOOM)*
Specify corner of window, enter a scale factor (nX or nXP), or
[All/Center/Dynamic/Extents/Previous/Scale/Window] <real time>: **W** *(For Window)*
First corner: *(Mark a point at "A")*
Other corner: *(Mark a point at "B")*

By default, the window option of zoom is automatic; in other words, without entering the Window option, the first point you pick identifies the first corner of the window box; the prompt "Other corner:" completes ZOOM-Window as indicated in the following prompts:

Command: **Z** *(For ZOOM)*
Specify corner of window, enter a scale factor (nX or nXP), or
[All/Center/Dynamic/Extents/Previous/Scale/Window] <real time>: *(Mark a point at "A")*
Other corner: *(Mark a point at "B")*

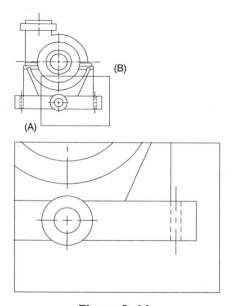

Figure 3–16

USING ZOOM-PREVIOUS

After magnifying a small area of the display screen, use the Previous option of the ZOOM command to return to the previous display. The system automatically saves up to ten views when zooming. This means you can begin with an overall display, perform two zooms, and use the ZOOM-Previous command twice to return to the original display. Zoom-Previous is also less likely to create a drawing regeneration (see Figure 3–17).

Command: **Z** *(For ZOOM)*
Specify corner of window, enter a scale factor (nX or nXP), or
[All/Center/Dynamic/Extents/Previous/Scale/Window] <real time>: **P** *(For Previous)*

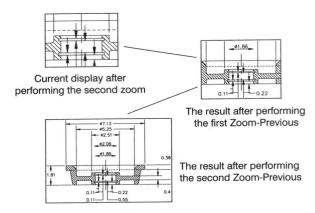

Current display after
performing the second zoom

The result after performing
the first Zoom-Previous

The result after performing
the second Zoom-Previous

Figure 3–17

USING ZOOM-SCALE

Clicking on this button prompts you to enter a zoom scale factor. If a scale factor of 0.50 is used, the zoom is performed into the drawing at a factor of 0.50, based on the original limits of the drawing.

Command: **Z** *(For ZOOM)*
Specify corner of window, enter a scale factor (nX or nXP), or
[All/Center/Dynamic/Extents/Previous/Scale/Window] <real time>: 0.50

If a scale factor of 0.50X is used, the zoom is performed into the drawing again at a factor of 0.50; however, the zoom is based on the current display screen.

Command: **Z** *(For ZOOM)*
Specify corner of window, enter a scale factor (nX or nXP), or
[All/Center/Dynamic/Extents/Previous/Scale/Window] <real time>: 0.50X

USING ZOOM-IN

Clicking on this button automatically performs a zoom-in operation at a scale factor of 0.5X; the "X" uses the current screen to perform the zoom-in operation.

USING ZOOM-OUT

Clicking on this button automatically performs a zoom-out operation at a scale factor of 2X; the "X" uses the current screen to perform the zoom-in operation.

THE PAN COMMAND

As you perform numerous ZOOM-Window and ZOOM-Previous operations, it becomes apparent that it would be nice to zoom in to a detail of a drawing and simply slide the drawing to a new area without changing the magnification; this is the purpose of the PAN command. In Figure 3–18, the Top view is magnified with ZOOM-

Window; the result is shown in Figure 3–19. Now, the Bottom view needs to be magnified to view certain dimensions. Rather than use ZOOM-Previous and then ZOOM-Window again to magnify the Bottom view, use the PAN command.

Command: P *(For PAN)*
Press ESC or ENTER to exit, or right-click to display shortcut menu.

Issuing the PAN command displays the Hand symbol. In the illustration in Figure 3–19, pressing the Pick button down at "A" and moving the hand symbol to the right at "B" pans the screen and displays a new area of the drawing in the current zoom magnification.

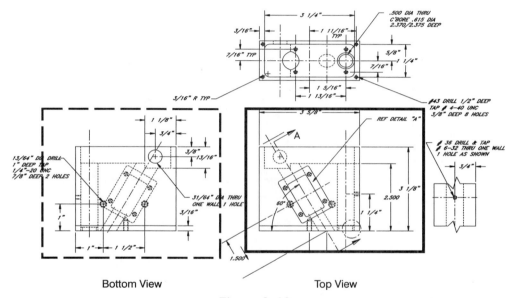

Bottom View   Top View

Figure 3–18

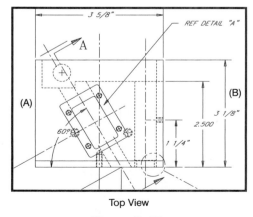

Top View

Figure 3–19

In Figure 3–20, the Bottom view is now visible after the drawing is panned from the Top view to the Bottom view, with the same display screen magnification. PAN can also be used transparently; that is, while in a current command, you can select the PAN command, which temporarily interrupts the current command, performs the pan, and restores the current command.

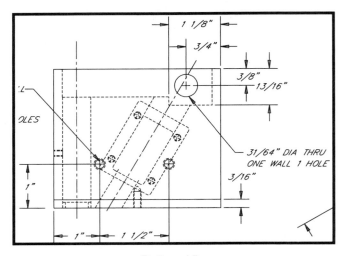

Bottom View

Figure 3–20

You can choose Pan from the View pull-down menu, shown in Figure 3–21, or you can enter P at the keyboard; P is the command alias for the PAN command. Also, don't forget about the scroll bars at the bottom and right side of the AutoCAD display screen. They also allow you to pan across the screen (see Figure 1-1).

Figure 3–21

SUPPORT FOR THE MICROSOFT® INTELLIMOUSE™

AutoCAD 2000 provides extra support for zooming and panning operations through the IntelliMouse™ by Microsoft®. It consists of the standard Microsoft two-button mouse with the addition of a wheel, as illustrated in Figure 3–22. Rolling the wheel forward zooms in to or magnifies the drawing. Rolling the wheel backward zooms out or demagnifies the drawing.

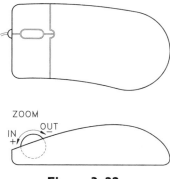

Figure 3–22

Pressing and holding the wheel down, as in Figure 3–23, places you in realtime pan mode. The familiar hand icon on the display screen identifies this mode.

Figure 3–23

Pressing CTRL while the wheel is depressed places you in Joystick Pan mode (see Figure 3–24A). This mode is identified by a pan icon similar to that shown in Figure 3–24B. This icon denotes all directions in which panning may occur. Moving the mouse in a direction with the wheel depressed displays the icon in Figure 3–24C, which shows the direction of the pan.

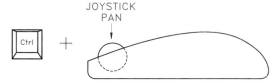

Figure 3–24A

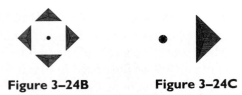

Figure 3–24B   **Figure 3–24C**

The wheel can also function like a mouse button. Double-clicking on the wheel, as in Figure 3–25, performs a ZOOM-Extents and is an extremely productive method of performing this operation. As you can see, the IntelliMouse provides for realtime zooming and panning with the addition of performing a ZOOM-Extents, all at your fingertips.

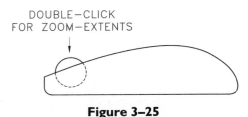

Figure 3–25

THE VIEW DIALOG BOX

An alternate method of performing numerous zooms is to create a series of views of key parts of a drawing. Then, instead of using the ZOOM command, restore the named view to perform detail work. This named view is saved in the database of the drawing for use in future editing sessions. Selecting Named Views... from the View pull-down menu as shown in Figure 3–26A activates the View dialog box, shown in Figure 3–26B. You can activate this same dialog box through the keyboard by entering the following at the command prompt:

Command: **V** *(For VIEW)*

Figure 3–26A

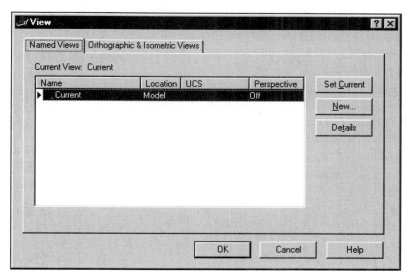

Figure 3–26B

Clicking the New… button activates the New View dialog box shown in Figure 3–27. Use this dialog box to guide you in creating a new view. By definition, a view is created from the current display screen. This is the purpose of the Current Display radio button. Many views are created with the Define Window radio button, which will create a view based on the contents of a window that you define. Choosing the Define View Window button returns you to the display screen and prompts you for the first corner and other corner required to create a new view by window.

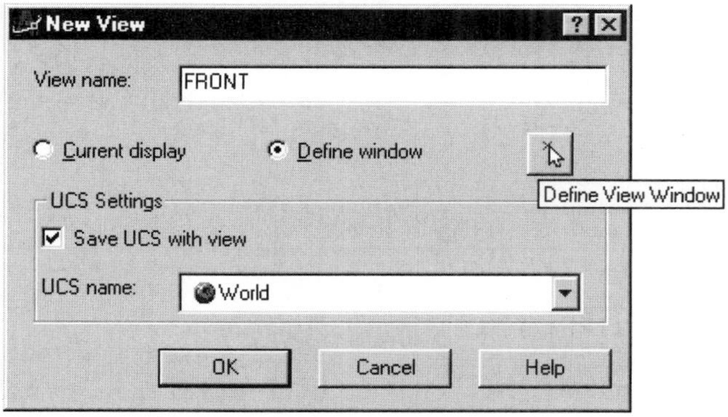

Figure 3–27

In Figure 3–28, a rectangular window is defined around the Front view using points "A" and "B" as the corners.

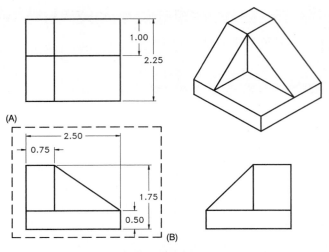

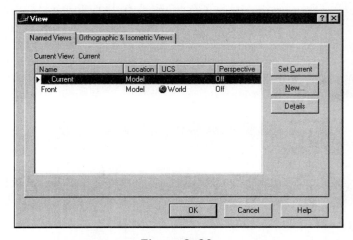

Figure 3–28

When the window is created, the Define New View dialog box redisplays. Clicking OK saves the view name in the View dialog box in Figure 3–29.

Figure 3–29

The image in Figure 3–30 illustrates numerous views created from the drawing in Figure 3–28.

Clicking on a defined view name, as in Figure 3–31, followed by right-clicking the mouse displays the shortcut menu used to set the view current. You can also rename, delete, or obtain details of the view through this shortcut menu. Clicking on Set Current displays the Front view, as in Figure 3–32.

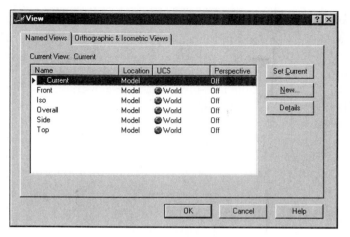

Figure 3–30

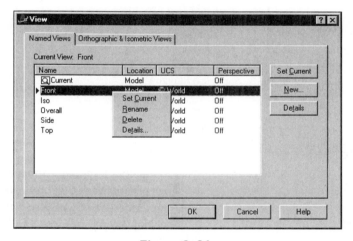

Figure 3–31

View Name = FRONT

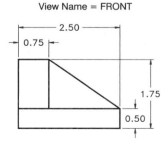

Figure 3–32

CREATING OBJECT SELECTION SETS

Selection sets are used to group a number of objects together for the purpose of editing. Applications of selection sets are covered in the following pages in addition to being illustrated in the next chapter. Once a selection set has been created, the group of objects may all be moved, copied, or mirrored. These operations supported by selection sets will be covered in Chapter 4, "Modify Commands." An object manipulation command supports the creation of selection sets if it prompts you to "Select objects." Any command displaying this prompt supports the use of selection sets. Selection set options (how a selection set is made) appear in Figure 3–33. Figures 3–34 through 3–41 present a few examples of how selection sets are used for manipulating groups of objects.

Add
All
CPolygon
Crossing
Fence
Last
Previous
Remove
Window
WPolygon
Undo

Figure 3–33

SELECTING OBJECTS BY INDIVIDUAL PICKS

When AutoCAD prompts you with "Select objects", a pickbox appears as the cursor is moved across the display screen. Any object enclosed by this box when picked will be considered selected. To show the difference between a selected and unselected object, the selected object highlights on the display screen. In Figure 3–34, the pickbox is placed over the arc segment at "A" and picked. To signify that the object is selected, the arc highlights.

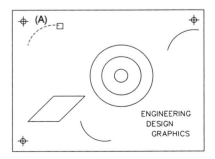

Figure 3–34

SELECTING OBJECTS BY WINDOW

The individual pick method previously outlined works fine for small numbers of objects. However, when numerous objects need to be edited, selecting each individual object could prove time-consuming. Instead, you can select all objects that you want to become part of a selection set by using the Window selection mode. This mode requires you to create a rectangular box by picking two diagonal points. In Figure 3–35, a selection window has been created with point "A" as the first corner and "B" as the other corner. When you use this selection mode, only those objects completely enclosed by the window box are selected. The window box selected four line segments, two arcs, and two points (too small to display highlighted). Even though the window touches the three circles, they are not completely enclosed by the window and therefore are not selected.

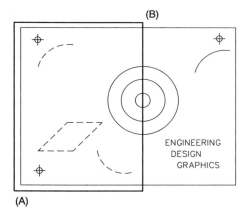

Figure 3–35

SELECTING OBJECTS BY CROSSING WINDOW

In the previous example of producing a selection set by a window, the window selected only those objects completely enclosed by it. Figure 3–36 is an example of selecting objects by a crossing window. The Crossing window option requires two points to define a rectangle as does the window selection option. In Figure 3–36, a dashed rectangle is used to select objects using "C" and "D" as corners for the rectangle; however, this time the crossing window was used. The highlighted objects illustrate the results. All objects that are touched by or enclosed by the crossing rectangle are selected. Because the crossing rectangle passes through the three circles without enclosing them, they are still selected by this object selection mode.

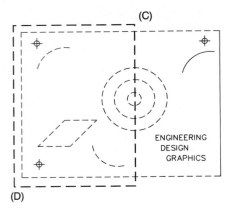

Figure 3–36

SELECTING OBJECTS BY A FENCE

Use this mode to create a selection set by drawing a line or group of line segments called a fence. Any object touched by the fence is selected. The fence does not have to end exactly where it was started. In Figure 3–37, all objects touched by the fence are selected, as represented by the dashed lines.

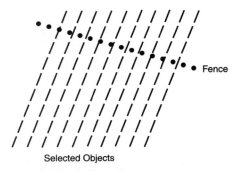

Figure 3–37

REMOVING OBJECTS FROM A SELECTION SET

All of the previous examples of creating selection sets have shown you how to create new selection sets. What if you select the wrong object or objects? Instead of canceling out of the command and trying to select the correct objects, you can use the Remove option to remove objects from an existing selection set. In Figure 3–38, a selection set has been created and made up of all of the highlighted objects. However, the large circle was mistakenly selected as part of the selection set. The Remove option allows you to remove highlighted objects from a selection set. To activate Remove, press SHIFT and pick the object you want removed; this only works if the "Select objects" prompt is present. When a highlighted object is removed from the selection set, as shown in circle "A" in Figure 3–39, it deselects and regains its original display intensity.

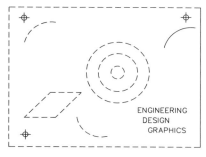

Figure 3–38

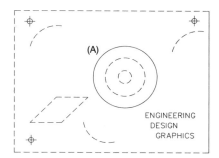

Figure 3–39

SELECTING THE PREVIOUS SELECTION SET

When you create a selection set of objects, this grouping is remembered until another selection set is made. The new selection set replaces the original set of objects. Let's say you moved a group of objects to a new location on the display screen. Now you want to rotate these same objects at a certain angle. Rather than select the same set of objects to rotate, you would pick the Previous option or type "P" at the "Select objects" prompt. This selects the previous selection set. The buffer holding the selection set is cleared whenever you use the U command to undo the previous command.

SELECTING OBJECTS BY A CROSSING POLYGON

When you use the Window or Crossing Window mode to create selection sets, two points specify a rectangular box for selecting objects. At times, it is difficult to select objects by the rectangular window or crossing box because in more cases than not, extra objects are selected and have to be removed from the selection set. Figure 3–40 shows a mechanical part with a "C"-shaped slot. Rather than use Window or Crossing Window modes, you can pick the Crossing Polygon mode (CPolygon) or type "CP" at the "Select objects" prompt. You simply pick points representing a polygon. Any object that touches or is inside the polygon is added to a selection set of objects. In Figure 3–40, the crossing polygon is constructed using points "A" through "G." A similar but different selection set mode is the Window Polygon (WPolygon). Objects are selected using this mode when they lie completely inside the Window Polygon, which is similar to the regular Window mode.

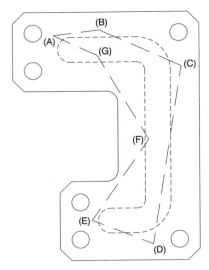

Figure 3–40

APPLICATIONS OF SELECTING OBJECTS WITH THE ERASE COMMAND

Figures 3–41 and 3–42 illustrate the use of the Window and Crossing options of deleting the circle and center marker with the ERASE command.

ERASE—WINDOW

When erasing objects by Window, be sure the objects to be erased are completely enclosed by the window, as in Figure 3–41.

Command: **E** *(For ERASE)*
Select objects: *(Pick a point at "A" and move the cursor to the right; this automatically invokes the window option)*
Other corner: *(Mark a point at "B" and notice the objects that highlight)*
Select objects: *(Press ENTER to execute the ERASE command)*

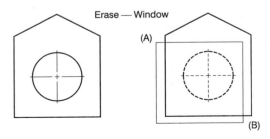

Erase — Window

Figure 3–41

ERASE—CROSSING

Erasing objects by a crossing box is similar to the Window Box mode; however, any object that touches the crossing box or is completely enclosed by the crossing box is selected. See Figure 3–42.

Command: **E** *(For ERASE)*
Select objects: *(Pick a point at "A" and move the cursor to the left; this automatically invokes the crossing option)*
Other corner: *(Mark a point at "B" and notice the objects that highlight)*
Select objects: *(Press ENTER to execute the ERASE command)*

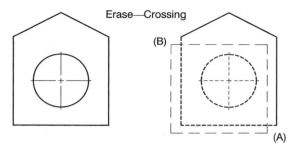

Figure 3–42

Figure 3–43 displays the objects that were selected by the Window option. However, the group of objects on the far right and left have mistakenly been selected and should not be erased. These objects need to be removed or deselected from the current selection set of objects with the Remove option of "Select objects."

Command: **E** *(For ERASE)*
Select objects: *(Pick a point at "A" in Figure 3–148 and move the cursor to the right; this automatically invokes the window option)*
Other corner: *(Mark a point at "B" in Figure 3–148 and notice the objects that highlight)*

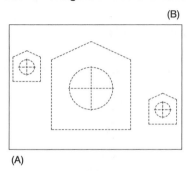

Figure 3–43

Before performing the ERASE command on the objects in Figure 3–43, issue the Re-move option to deselect the group of objects shown in Figure 3–44. You can accomplish this easily by holding down SHIFT and picking the objects you do not want to erase.

Select objects: *(While pressing the Shift key, begin picking all objects in Figures at "A" and "B" in Figure 3–44. Notice the objects deselecting)*
Select objects: *(Press ENTER to execute the ERASE command)*

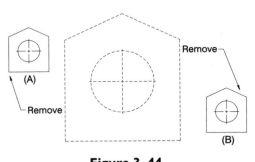

Figure 3–44

By deselecting objects through Shift-Pick, you remain in the command instead of having to cancel the command for picking the wrong objects and starting over. See Figure 3–45.

Figure 3–45

OBJECT SELECTION CYCLING

At times, the process of selecting objects can become quite tedious. Often, objects lie directly on top of each other. As you select the object to delete, the other object selects instead. To remedy this, press CTRL when prompted to "Select objects." This activates Object Selection Cycling and enables you to scroll through all objects in the vicinity of the pickbox. A message appears in the prompt area alerting you that cycling is on; you can now pick objects until the desired object is highlighted. Pressing ENTER not only accepts the highlighted object but toggles cycling off. In Figure 3–46 and with selection cycling on, the first pick selects the line segment; the second pick selects the circle. Keep picking until the desired object highlights.

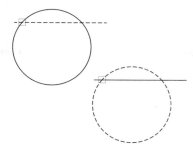

Figure 3–46

THE QSELECT COMMAND

Yet another way of creating a selection set is by matching the object type and property with objects currently in use in a drawing. This is the purpose of the QSELECT command (Quick Select). This command can be chosen from the Tools pull-down menu, as in Figure 3–47A. Choosing Quick Select... displays the dialog box in Figure 3–47B.

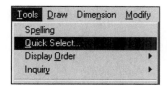

Figure 3–47A

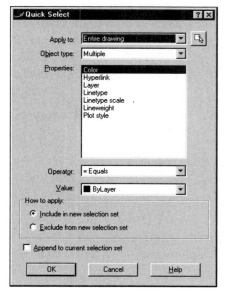

Figure 3–47B

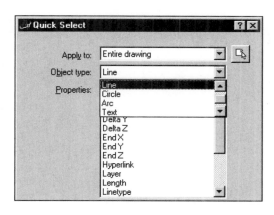

Figure 3–48

This command works only if objects are defined in a drawing; the Quick Select dialog box does not display in a drawing file without any objects drawn. Clicking in the Object type edit box displays all object types currently used in the drawing. This enables you to create a selection set by the object type. For instance, to select all line segments in the drawing file, click on Line in the Object type edit box in Figure 3–48. Clicking the OK button at the bottom of the dialog box returns you to the drawing and applies the object properties to objects in the drawing.

Notice, in Figure 3–49, that all line segments are highlighted. (The square boxes positioned around the drawing are called grips and will be discussed later in Chapter 7.) Other controls of Quick Select include the ability to select the object type from the entire drawing or from just a segment of the drawing. You can narrow the selection criteria by adding various properties to the selection mode such as Color, Layer, and Linetype, to name a few. You can also create a reverse selection set. Clicking on the radio button to exclude from the new selection set, shown in Figure 3–47B, would create a selection set of all objects not counting those identified in the Object type edit box. The Quick Select dialog box lives up to its name—it enables you to create a quick selection set.

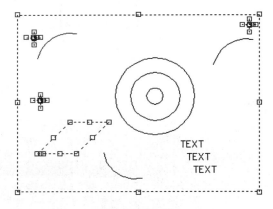

Figure 3–49

EXPRESS TOOLS APPLICATION SELECTION SETS

SELECTION SET TOOLS

A set of extra selection set tools is available to produce the opposite results of those that normal selection set options yield; they select all objects except those in the selection area. The following is a list of what are referred to as exclusionary selection set tools. These tools are available through the Express pull-down menu, shown in Figure 3–50.

EXW   Exclusionary window
EXC   Exclusionary crossing window

EXP   Exclusionary previous

EXF   Exclusionary fence

EXWP   Exclusionary Window Polygon

EXCP   Exclusionary Crossing Polygon

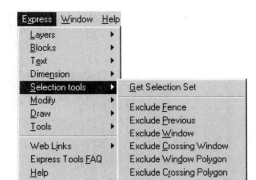

Figure 3–50

To see how to apply these exclusionary selection set tools to a drawing, refer to Figure 3–51. You need to erase the room numbers and room identifying boxes along with the box stems. Rather than concentrating on selecting the room numbers, the ERASE command is started and the Exclude Window tool is selected from the Express pull-down menu in Figure 3–50. You identify the first corner of the window at "A" and the other corner at "B." When you select objects using conventional tools under normal circumstances, the floor plan would highlight. When you use the exclusionary window tool, notice in Figure 3–51 that all objects not included in the window highlight or select. You could also accomplish the same results from the keyboard using the ERASE command along with the exclusionary window option in the following prompt sequence:

Command: **E** *(For Erase)*
Select objects: **'EXW** *(For Exclusionary Window)*
First corner: *(Pick a point at "A" in Figure 3–51)*
Other corner: *(Pick a point at "B" in Figure 3–51)*
Select objects: *(Press* ENTER *to perform the erase operation)*

When invoking the 'EXW option from the keyboard, an apostrophe is used in front of the command to designate the option as transparent. Without the apostrophe, an "invalid" message signals an unknown command or option. This same concept of excluding objects can be applied to building selection sets by fence, crossing, Window Polygon, and Crossing Polygon.

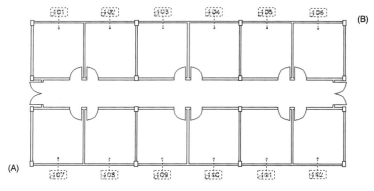

Figure 3–51

GET SELECTION SET

Yet another powerful selection set mode taken from the Express pull-down menu is the ability to create a selection set by selecting an object by layer and type. All other objects that match the selected object are also selected. In Figure 3–52, all of the walls need to be selected. Unfortunately, the doors happen to be on the same layer as the walls. Issue the GETSEL command or pick the command from the Express pull-down menu as in Figure 3–50. When prompted for an object on a source layer, select the edge of the wall at "A"). When prompted to select the type of object, again pick the edge of the wall at "A"). This not only identifies the layer of the object but selects only lines on the layer. The result is illustrated in Figure 3–52, with the walls highlighted making up the selection set. The following is the command line version:

Command: **GETSEL**
Select an object on the Source layer <*>: *(Select the edge of the wall at "A")*
Select an object of the Type you want <*>: *(Select the edge of the wall again at "A")*
Collecting all LINE objects on layer I-WALL...
218 objects have been placed in the active selection set.

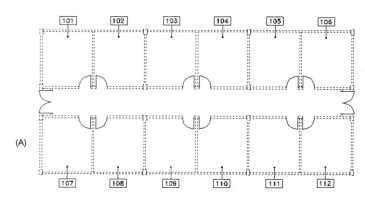

Figure 3–52

CHAPTER 4

Modify Commands

METHODS OF SELECTING MODIFY COMMANDS

The heart of any CAD system is its ability to modify and manipulate existing geometry, and AutoCAD is no exception. Many modify commands relieve the designer of drudgery and mundane tasks, and this allows more productive time for conceptualizing the design.

As with all commands, you can find the main body of modify commands on the Modify pull-down menu, shown in Figure 4–1.

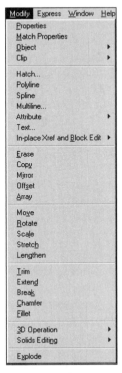

Figure 4–1

Another convenient way of selecting modify commands is through the Modify toolbar shown in Figure 4–2. You can also enter all modify commands directly from the keyboard either using their entire name or through command aliasing as in the following examples:

Enter E for the ERASE command
Enter M for the MOVE command

For a complete listing of other commands and aliases, refer to the list located in Chapter 1, Table 1–2.

These modify commands will be discussed in greater detail in the following pages:

ARRAY	MIRROR	BREAK
MOVE	CHAMFER	OFFSET
COPY	ROTATE	EXPLODE
SCALE	EXTEND	STRETCH
FILLET	TRIM	LENGTHEN

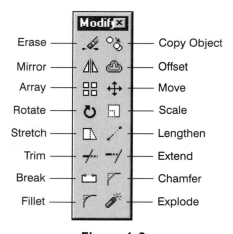

Figure 4–2

CREATING RECTANGULAR ARRAYS

The ARRAY command allows you to arrange multiple copies of an object or group of objects similar to Figure 4–4 in a rectangular pattern. Suppose a rectangular pattern of the object in Figure 4–3 needs to be made. The final result is to create three rows and three columns of the object. Finally, the spacing between rows is to be 0.50 units and the spacing between columns is 1.25 units. Study the following prompts and Figure 4–4.

Command: **AR** *(For ARRAY)*
Select objects: **All**
(This will select all objects in the original figure)
Select objects: *(Press* ENTER *to continue)*
Enter the type of array [Rectangular/Polar] <R>: **R** *(For Rectangular)*
Enter the number of rows (—) <1>: **3**
Enter the number of columns (||||) <1>: **3**
Enter the distance between rows or specify unit cell (—): **1.00**
Specify the distance between columns (||||): **2.00**

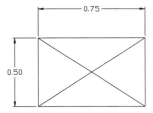

Figure 4–3

For rectangular objects to be arrayed, a reference point in the lower left corner of the figure becomes a point where you may calculate the spacing between rows and columns. Not only must the spacing distance be used, but the overall size of the object plays a role in determining the spacing distances. With the total height of the original object at 0.50 and a required spacing between rows of 0.50, both object height and spacing result in a distance of 1.00 between rows. In the same manner, with the original length of the object at 0.75 and a spacing of 1.25 units between columns, the total spacing results in a distance of 2.00.

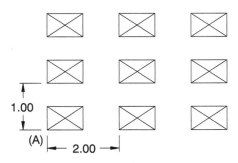

Figure 4–4

In Figure 4–4, the rectangular array illustrates an array that runs to the right and above of the original figure. At times these directions change to the left and below the original object. The only change occurs in the distances between rows and columns where negative values dictate the direction of the rectangular array. See Figure 4–5.

Command: **AR** *(For ARRAY)*
Select objects: **All**
(This will select all objects in the original figure)
Select objects: *(Press* ENTER *to continue)*
Enter the type of array [Rectangular/Polar] <R>: **R** *(For Rectangular)*
Enter the number of rows (—) <1>: **3**
Enter the number of columns (||||) <1>: **2**
Enter the distance between rows or specify unit cell (—): **-1.50**
Specify the distance between columns (||||): **-2.50**

The results for the preceding command sequence are illustrated in Figure 4–5.

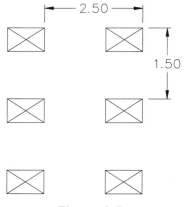

Figure 4–5

CREATING POLAR ARRAYS

Polar arrays allow you to create multiple copies of objects in a circular or polar pattern. Study the following prompts and Figure 4–6 for performing polar arrays:

Command: **AR** *(For ARRAY)*
Select objects: *(Select a point at "A")*
Other corner: *(Select a point at "B")*
Select objects: *(Press* ENTER *to continue)*
Enter the type of array [Rectangular/Polar] <R>: **P** *(For Polar)*
Specify center point of array: *(Select the edge of the large circle at "C" to identify the center)*

Enter the number of items in the array: **8**
Specify the angle to fill (+=ccw, -=cw) <360>: *(Press ENTER to accept)*
Rotate arrayed objects? [Yes/No] <Y>: *(Press ENTER to perform the array operation)*

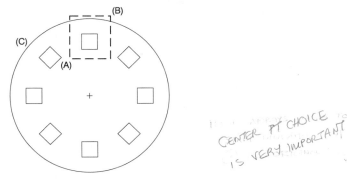

Figure 4–6

To array rectangular or square objects in a polar pattern without rotating the objects, first convert the square or rectangle to a block with an insertion point located in the center of the square. Now all squares lie an equal distance from their common center. See Figure 4–7.

Command: **AR** *(For ARRAY)*
Select objects: *(Select the square block at "A" in Figure 4–7)*
Select objects: *(Press ENTER to continue)*
Enter the type of array [Rectangular/Polar] <R>: **P** *(For Polar)*
Specify center point of array: *(Select the edge of the large circle at "B" in Figure 4–7 to identify the center)*
Enter the number of items in the array: **8**
Specify the angle to fill (+=ccw, -=cw) <360>: *(Press ENTER to accept)*
Rotate arrayed objects? [Yes/No] <Y>: **N** *(For No)*

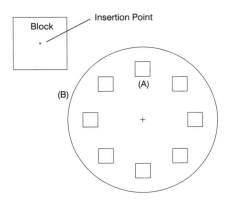

Figure 4–7

ARRAYING BOLT HOLES—METHOD #1

Multiple copies of objects such as bolt holes are easily duplicated in circular patterns through the ARRAY command with the following prompts:

Command: **AR** *(For ARRAY)*
Select objects: *(Select both small circles in Figure 4–8)*
Enter the type of array [Rectangular/Polar] <R>: **P** *(For Polar)*
Specify center point of array: *(Select the intersection at "A")*
Enter the number of items in the array: **8**
Specify the angle to fill (+=ccw, -=cw) <360>: *(Press ENTER for default)*
Rotate arrayed objects? [Yes/No] <Y>: *(Press ENTER to perform the array operation)*

Because the circles are copied in the circular pattern, centerlines need to be updated to the new bolt hole positions (see Figure 4–9). Again, the ARRAY command is used to copy and duplicate one centerline over a 45° angle to fill in the counterclockwise direction.

Command: **AR** *(For ARRAY)*
Select objects: *(Select the vertical centerline at "A" in Figure 4–9)*
Enter the type of array [Rectangular/Polar] <R>: **P** *(For Polar)*
Specify center point of array: *(Select the intersection at "B")*
Enter the number of items in the array: **2**
Specify the angle to fill (+=ccw, -=cw) <360>: **45**
Rotate arrayed objects? [Yes/No] <Y>: *(Press ENTER to perform the array operation)*

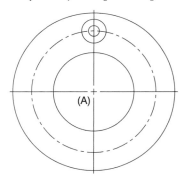

Figure 4–8

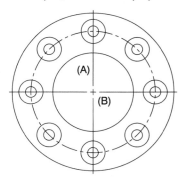

Figure 4–9

Now use the ARRAY command to copy and duplicate the last centerline and mark the remaining bolt hole circles as shown in Figure 4–10.

Command: **AR** *(For ARRAY)*
Select objects: **L** *(This should select the last line)*
Select objects: *(Press ENTER to continue)*

Enter the type of array [Rectangular/Polar] <R>: **P** *(For Polar)*
Specify center point of array: *(Select the intersection at "B" in Figure 4–10)*
Enter the number of items in the array: **4**
Specify the angle to fill (+=ccw, -=cw) <360>: *(Press* ENTER *for default)*
Rotate arrayed objects? [Yes/No] <Y>: *(Press* ENTER *to perform the array operation)*

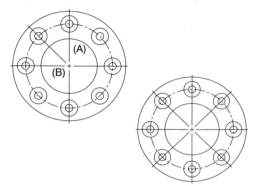

Figure 4–10

FORMING BOLT HOLES—METHOD #2

From the previous example, you have seen how easily the ARRAY command can be used for making multiple copies of objects equally spaced around an entire circle. What if the objects are copied only partially around a circle, such as the bolt holes in Figure 4–11? The ARRAY command is used here to copy the bolt holes in 40° increments. Follow the next command sequence for performing the operation shown in Figure 4–11.

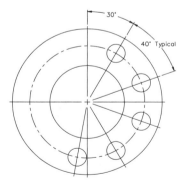

Figure 4–11

First use the ARRAY command to rotate and copy the vertical centerline at a -30° angle to fill. The -30° will copy and rotate the centerline in the clockwise direction. Next place the circle at the intersection of the centerlines using the CIRCLE command. Use the ARRAY command, select the circle and centerline, and copy the selected objects at 40° increments using the following command sequence.

Command: **AR** *(For ARRAY)*
Select objects: *(Select the small circle at "A" and line at "B" in Figure 4–12)*
Select objects: *(Press ENTER to continue)*
Enter the type of array [Rectangular/Polar] <R>: **P** *(For Polar)*
Specify center point of array: *(Select the large circle anywhere near "C" to identify the center)*
Number of items: *(Press ENTER to continue with this command)*
Specify the angle to fill (+=ccw, -=cw) <360>: **-160**
Angle between items: **-40** *(To copy 40° clockwise)*
Rotate arrayed objects? [Yes/No] <Y>: *(Press ENTER to perform the array operation)*

The result is illustrated in Figure 4–13. This method of identifying bolt holes shows how you can control the number by specifying the total angle to fill and the angle between items. In both angle specifications, a negative value is entered to force the array to be performed in the clockwise direction.

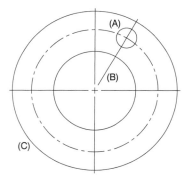

Figure 4–12

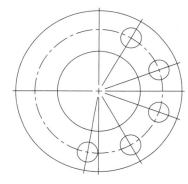

Figure 4–13

FORMING BOLT HOLES—METHOD #3

The object in Figure 4–14 is similar to the object shown in Forming Bolt Holes—Method #2. A new series of holes are to be placed 20° away from each other through the ARRAY command.

Begin placing the holes by laying out one centerline using the ARRAY command with an angle of 15° to fill. Add one circle using the CIRCLE command at the intersection of the centerlines. Use the following prompts to add the remaining holes:

Command: **AR** *(For ARRAY)*
Select objects: *(Select the small circle at "A" and line at "B" in Figure 4–15)*
Select objects: *(Press ENTER to continue)*
Enter the type of array [Rectangular/Polar] <R>: **P** *(For Polar)*

Specify center point of array: *(Select the large circle anywhere near "C" to identify the center)*
Number of items: *(Press ENTER to continue with this command)*
Specify the angle to fill (+=ccw, -=cw) <360>: **120**
Angle between items: **20**
Rotate arrayed objects? [Yes/No] <Y>: *(Press ENTER to perform the array operation)*

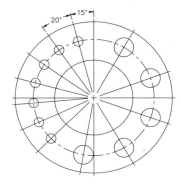

Figure 4–14

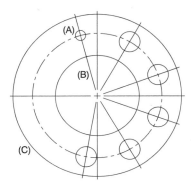

Figure 4–15

Because the direction of rotation for the array is in the counterclockwise direction, all angles are specified in positive values. See Figure 4–16.

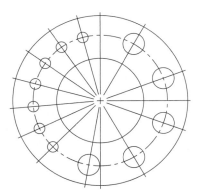

Figure 4–16

THE BREAK COMMAND

The BREAK command is used to partially delete a segment of an object. The following command sequence and Figure 4–17 refer to using the BREAK command.

Command: **BR** *(For BREAK)*
Select objects: *(Select the line at "A" in Figure 4–17)*
Specify second break point or [First point]: *(Select the line at "B" in Figure 4–17)*

To select key objects to break using Osnap options, utilize the First option of the BREAK command. This option resets the command and allows you to select an object to break followed by two different points that identify the break. The following command sequence and Figure 4–18 demonstrate using the First option of the BREAK command:

Command: **BR** *(For BREAK)*
Select objects: *(Select the line shown in Figure 4–18)*
Specify second break point or [First point]: **F** *(For First)*
   (For First)
Specify first break point: *(Select the intersection of the two lines at "A")*
Specify second break point: *(Select the endpoint of the line at "B")*

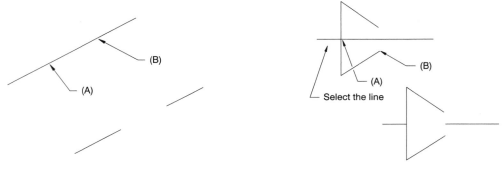

Figure 4–17   **Figure 4–18**

Breaking circles is always accomplished in the counterclockwise direction. Study the following command sequence and Figure 4–19 for breaking circles.

Command: **BR** *(For BREAK)*
Select objects: *(Select the circle in Figure 4–19)*
Specify second break point or [First point]: **F** *(For First))*
Specify first break point: *(Select "A" in either circle in Figure 4–19)*
Specify second break point: *(Select "B" in either circle in Figure 4–19)*

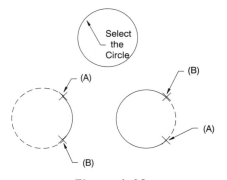

Figure 4–19

THE CHAMFER COMMAND

Chamfers represent a way to finish a sharp corner of an object. The CHAMFER command produces an inclined surface at an edge of two intersecting line segments. Distances determine how far from the corner the chamfer is made. Figure 4–20 is one example of an object that has been chamfered along its top edge.

The CHAMFER command is designed to draw an angle across a sharp corner given two chamfer distances. The most popular chamfer involves a 45° angle, which is illustrated in Figure 4–21. Even though this command does not allow you to specify an angle, you can control the angle by the distances entered. In the example in Figure 4–21, if you specify the same numeric value for both chamfer distances, a 45° chamfer will automatically be formed. As long as both distances are the same, a 45° chamfer will always be drawn. Study the illustration in Figure 4–21 and the following prompts:

Command: **CHA** *(For CHAMFER)*
(TRIM mode) Current chamfer Dist1 = 0.5000, Dist2 = 0.5000
Select first line or [Polyline/Distance/Angle/Trim/Method]: **D** *(For Distance)*
Specify first chamfer distance <0.5000>: **0.15**
Specify second chamfer distance <0.1500>: *(Press ENTER to accept the default)*
Command: **CHA** *(For CHAMFER)*
(TRIM mode) Current chamfer Dist1 = 0.1500, Dist2 = 0.1500
Select first line or [Polyline/Distance/Angle/Trim/Method]: *(Select the line at "A")*
Select second line: *(Select the line at "B")*

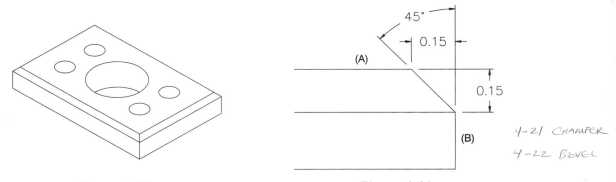

Figure 4–20   **Figure 4–21**

Figure 4–22 is similar to the 45° chamfer shown in Figure 4–21 except that the angle is different. This results from specifying two different chamfer distances, as outlined in the following prompts. This type of edge is commonly called a bevel.

Command: **CHA** *(For CHAMFER)*
(TRIM mode) Current chamfer Dist1 = 0.5000, Dist2 = 0.5000
Select first line or [Polyline/Distance/Angle/Trim/Method]: **D** *(For Distance)*
Specify first chamfer distance <0.5000>: **0.30**

Specify second chamfer distance <0.3000>: **0.15**
Command: **CHA** *(For CHAMFER)*
(TRIM mode) Current chamfer Dist1 = 0.3000, Dist2 = 0.1500
Select first line or [Polyline/Distance/Angle/Trim/Method]: *(Select the line at "A")*
Select second line: (Select the line at "B")

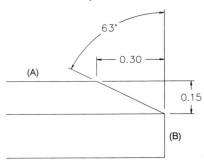

Figure 4–22

With non-intersecting corners, you could use the CHAMFER command to connect both lines. The CHAMFER command distances are both set to a value of 0 to accomplish this task. See Figure 4–23.

Command: **CHA** *(For CHAMFER)*
(TRIM mode) Current chamfer Dist1 = 0.5000, Dist2 = 0.2500
Select first line or [Polyline/Distance/Angle/Trim/Method]: **D** *(For Distance)*
Enter first chamfer distance <0.5000>: **0**
Enter second chamfer distance <0.0000>: *(Press ENTER to accept the default)*

Command: **CHA** *(For CHAMFER)*
(TRIM mode) Current chamfer Dist1 = 0.0000, Dist2 = 0.0000
Select first line or [Polyline/Distance/Angle/Trim/Method]: *(Select the line at "A")*
Select second line: *(Select the line at "B")*

Since a polyline consists of numerous segments representing a single object, using the CHAMFER command with the Polyline option produces corners throughout the entire polyline. See Figure 4–24.

Command: **CHA** *(For CHAMFER)*
(TRIM mode) Current chamfer Dist1 = 0.00, Dist2 = 0.00
Select first line or [Polyline/Distance/Angle/Trim/Method]: **D** *(For Distance)*
Enter first chamfer distance <0.00>: **0.50**
Enter second chamfer distance <0.50>: *(Press ENTER to accept the default)*

Command: **CHA** *(For CHAMFER)*
(TRIM mode) Current chamfer Dist1 = 0.5000, Dist2 = 0.5000
Select first line or [Polyline/Distance/Angle/Trim/Method]: **P** *(For Polyline)*
Select 2D Polyline: *(Select the Polyline)*

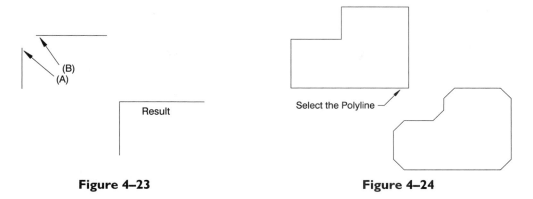

Figure 4–23   **Figure 4–24**

The CHAMFER command supports a Trim/No trim option, enabling a chamfer to be placed with lines trimmed or not trimmed as in Figure 4–25. Follow the prompt sequence carefully to set the No trim option.

Command: **CHA** *(For CHAMFER)*
(TRIM mode) Current chamfer Dist1 = 0.2000, Dist2 = 0.2000
Select first line or [Polyline/Distance/Angle/Trim/Method]: **T** *(For Trim)*
Trim/No trim <Trim>: **N** *(For No trim)*
Polyline/Distance/Angle/Trim/Method/<Select first line>: **D** *(For Distance)*
Enter first chamfer distance <0.0000>: **1.00**
Enter second chamfer distance <1.0000>: *(Press ENTER to accept the default)*

Command: **CHA** *(For CHAMFER)*
(NOTRIM mode) Current chamfer Dist1 = 1.00, Dist2 = 1.00
Select first line or [Polyline/Distance/Angle/Trim/Method]: **P** *(For Polyline)*
Select 2D Polyline: *(Select the polyline at "A")*

All edges are chamfered without being trimmed.

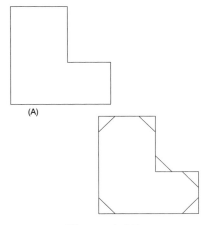

Figure 4–25

THE COPY COMMAND

The COPY command is used to duplicate an object or group of objects. In Figure 4–26, the Window mode is used to select all objects to copy. Point "C" is used as the base point of displacement, or where you want to copy the objects from. Point "D" is used as the second point of displacement, or the destination for copied objects.

Command: **CP** *(For COPY)*
Select objects: *(Pick a point at "A")*
Other corner: *(Pick a point at "B")*
Select objects: *(Press* ENTER *to continue)*
Specify base point or displacement, or [Multiple]: *(Select the endpoint of the corner at "C")*
Specify second point of displacement or <use first point as displacement>: *(Select a point near "D")*

You can also duplicate numerous objects while staying inside the COPY command. Figure 4–27 shows a group of objects copied with the Multiple option of the COPY command. See the following command sequence for using the Multiple option of the COPY command:

Command: **CP** *(For COPY)*
Select objects: *(Select all objects that make up the object at "A")*
Select objects: *(Press* ENTER *to continue)*
Specify base point or displacement, or [Multiple]: **M** *(For Multiple)*
Specify base point: *(Select the endpoint of the corner at "A")*
Specify second point of displacement or <use first point as displacement>: *(Select a point near "B")*
Specify second point of displacement or <use first point as displacement>: *(Select a point near "C")*
Specify second point of displacement or <use first point as displacement>: *(Select a point near "D")*
Specify second point of displacement or <use first point as displacement>: *(Select a point near "E")*
Specify second point of displacement or <use first point as displacement>: *(Select a point near "F")*
Specify second point of displacement or <use first point as displacement>: *(Press* ENTER *to exit this command)*

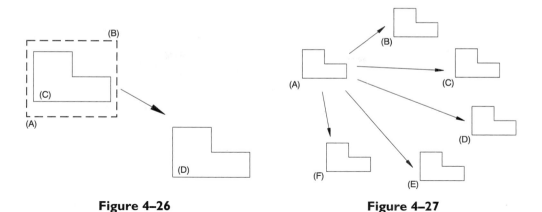

Figure 4–26	Figure 4–27

THE EXPLODE COMMAND

Using the EXPLODE command on a polyline, dimension, or block separates the single object into its individual parts. Figure 4–28 is a polyline that is considered one object. Using the EXPLODE command and selecting the polyline breaks the polyline into four individual objects.

Command: **X** *(For EXPLODE)*
Select objects: *(Select the polyline in Figure 4–28)*
Select objects: *(Press ENTER to perform the explode operation)*

Polyline
Before Explode—1 object
After Explode—4 objects

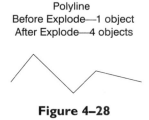

Figure 4–28

Dimensions consist of extension lines, dimension lines, dimension text, and arrowheads, all grouped into a single object. Dimensions will be covered in Chapter 11, "Dimensioning Basics." Using the EXPLODE command on a dimension breaks the dimension down into individual extension lines, dimension lines, arrowheads, and dimension text. This is not advisable, because the dimension value will not update if the dimension is stretched along with the object being dimensioned. Also the ability to manipulate dimensions with a feature called grips is lost. Grips will be discussed in Chapter 7.

Command: **X** *(For EXPLODE)*
Select objects: *(Select the dimension object shown in Figure 4–29)*
Select objects: *(Press ENTER to perform the explode operation)*

Using the EXPLODE command on a block converts the block to individual objects, lines, and circles. If the block had yet another block nested in it, you would need to perform an additional explode operation to break this object into individual objects.

Command: **X** *(For EXPLODE)*
Select objects: *(Select the block shown in Figure 4–30)*
Select objects: *(Press ENTER to perform the explode operation)*

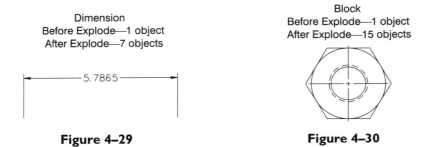

Dimension
Before Explode—1 object
After Explode—7 objects

—5.7865—

Figure 4–29

Block
Before Explode—1 object
After Explode—15 objects

Figure 4–30

You can also use the EXPLODE command to explode non-uniformly scaled block objects as shown in Figure 4–31. All blocks in this figure will be broken into individual objects without the block being redefined.

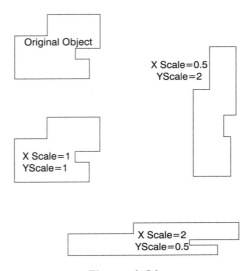

Original Object

X Scale=0.5
YScale=2

X Scale=1
YScale=1

X Scale=2
YScale=0.5

Figure 4–31

THE EXTEND COMMAND

The EXTEND command is used to extend objects to a specified boundary edge. In Figure 4–32, select the large circle "A" as the boundary edge. After pressing ENTER to continue with the command, select the arc at "B," the line at "C," and the arc at "D" to extend these objects to the circle. If you select the wrong end of an object, use the Undo feature, which is an option of the command, to undo the change and repeat the procedure at the correct end of the object.

Command: **EX** *(For EXTEND)*
Current settings: Projection=UCS Edge=None
Select boundary edges ...
Select objects: *(Select the large circle at "A")*
Select objects: *(Press ENTER to continue)*
Select object to extend or [Project/Edge/Undo]: *(Select the arc at "B")*
Select object to extend or [Project/Edge/Undo]: *(Select the line at "C")*
Select object to extend or [Project/Edge/Undo]: *(Select the arc at "D")*
Select object to extend or [Project/Edge/Undo]: *(Press ENTER to exit this command)*

An alternate method of selecting boundary edges is to press ENTER in response to the "Select objects" prompt. This automatically creates boundary edges out of all objects in the drawing. When you use this method, the boundary edges do not highlight.

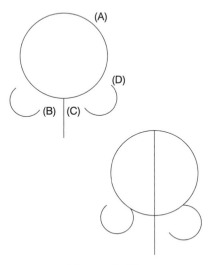

Figure 4–32

To extend multiple objects such as the five line segments shown in Figure 4–33, select the line at "A" as the boundary edge and use the Fence mode to create a crossing line from "B" to "C". This will extend all five line segments to intersect with the boundary.

Command: **EX** *(For EXTEND)*
Current settings: Projection=UCS Edge=None
Select boundary edges ...
Select objects: *(Select the line at "A" in Figure 4–33)*
Select objects: *(Press ENTER to continue)*
Select object to extend or [Project/Edge/Undo]: **F** *(For Fence)*
First fence point: *(Pick a point at "B")*
Specify endpoint of line or [Undo]: *(Pick a point at "C")*
Specify endpoint of line or [Undo]: *(Press ENTER to end the Fence and execute the extend operation)*
Select object to extend or [Project/Edge/Undo]: *(Press ENTER to exit this command)*

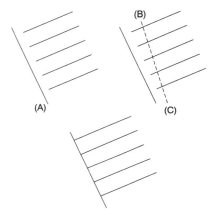

Figure 4–33

Certain conditions require the boundary edge to be extended where an imaginary edge is projected, enabling objects not in direct sight of the boundary edge to still be extended. Study Figure 4–34 and the following prompts on this special case involving the EXTEND command.

Command: **EX** *(For EXTEND)*
Current settings: Projection=UCS Edge=None
Select boundary edges ...
Select objects: *(Pick line "A")*
Select objects: *(Press ENTER to continue)*
Select object to extend or [Project/Edge/Undo]: **E** *(For Edge)*
Enter an implied edge extension mode [Extend/No extend] <No extend>: **E** *(For Extend the edge)*
Select object to extend or [Project/Edge/Undo]: *(Pick line "B")*

Continue picking the remaining lines to be extended to the extended boundary edge. You could also use the Fence mode to select all line segments to extend. After all lines are extended, press ENTER to exit the command.

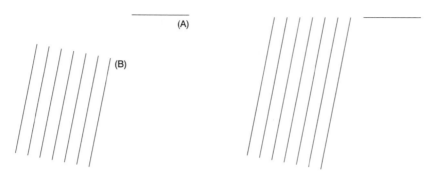

Figure 4–34

THE FILLET COMMAND

Many objects require highly finished and polished surfaces consisting of extremely sharp corners. Fillets and rounds represent the opposite case where corners are rounded off either for ornamental purposes or as required by design. Generally a fillet consists of a rounded edge formed in the corner of an object, as illustrated in Figure 4–35 at "A." A round is formed at an outside corner, similar to "B." Fillets and rounds are primarily used where objects are cast or made from poured metal. The metal will form more easily around a pattern that has rounded corners instead of sharp corners, which usually break away. Some drawings have so many fillets and rounds that a note is used to convey the size of all, similar to "All Fillets and Rounds 0.125 Radius."

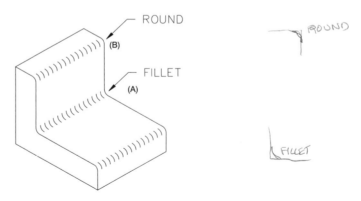

Figure 4–35

AutoCAD provides the FILLET command, which allows you to enter a radius followed by the selection of two lines. The result will be a fillet of the specified radius to the two lines selected. The two lines are also automatically trimmed, leaving the radius drawn from the endpoint of one line to the endpoint of the other line. Illustrated in Figure 4–36 are examples of the use of the FILLET command. Because sometimes the command is used over and over again, MULTIPLE automatically repeats the next command entered from the keyboard, in this case the FILLET command. Since MULTIPLE continually repeats the command, press ESC to cancel the command and return to the command prompt.

Command: **MULTIPLE**
Enter command name to repeat: **F** *(For Fillet)*
Current settings: Mode = TRIM, Radius = 0.5000
Select first object or [Polyline/Radius/Trim]: **R** *(For Radius)*
Enter fillet radius <0.0000>: **0.25**
Select first object or [Polyline/Radius/Trim]: *(Select at "A" and "B")*
Select first object or [Polyline/Radius/Trim]: *(Select at "B" and "C")*
Select first object or [Polyline/Radius/Trim]: *(Select at "C" and "D")*
Select first object or [Polyline/Radius/Trim]: *(Press the ESC key to cancel the multiple filleting operation)*

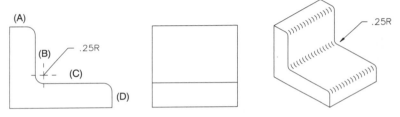

Figure 4–36

A very productive feature of the FILLET command is its use as a cornering tool. To accomplish this, set the fillet radius to a value of 0. This produces a corner out of two non-intersecting objects. See Figure 4–37.

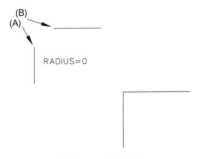

Figure 4–37

Command: **F** *(For FILLET)*
Current settings: Mode = TRIM, Radius = 0.5000
Select first object or [Polyline/Radius/Trim]: **R** *(For Radius)*
Enter fillet radius <0.5000>: **0**

Command: **F** *(For FILLET)*
Current settings: Mode = TRIM, Radius = 0.0000
Select first object or [Polyline/Radius/Trim]: *(Select line "A")*
Select second object: *(Select line "B")*

Using the FILLET command on a polyline object produces rounded edges at all corners of the polyline in a single operation (see Figure 4–38).

Command: **F** *(For FILLET)*
Current settings: Mode = TRIM, Radius = 0.0000
Select first object or [Polyline/Radius/Trim]: **R** *(For Radius)*
Enter fillet radius <0.0000>: **0.25**

Command: **F** *(For FILLET)*
Current settings: Mode = TRIM, Radius = 0.2500
Select first object or [Polyline/Radius/Trim]: **P** *(For Polyline)*
Select 2D polyline: *(Select the polyline in Figure 4–38)*

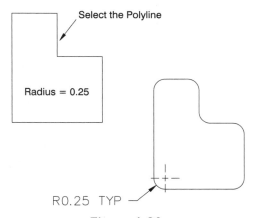

Select the Polyline

Radius = 0.25

R0.25 TYP

Figure 4–38

The FILLET command can also be used to control whether or not to trim the excess corners after a fillet is placed. Figure 4–39 shows a typical fillet operation where the polyline at "A" is selected. However, instead of the corners of the polyline automatically being trimmed, a new Trim/No trim option allows the lines to remain. The following prompts illustrate this operation.

Command: **F** *(For FILLET)*
Current settings: Mode = TRIM, Radius = 0.0000
Select first object or [Polyline/Radius/Trim]: **T** *(For Trim)*
Enter Trim mode option [Trim/No trim] <Trim>: **N** *(For No trim)*
Select first object or [Polyline/Radius/Trim]: **R** *(For Radius)*
Specify fillet radius <0.0000>: **1.00**

Command: **F** *(For FILLET)*
(NOTRIM mode) Current fillet radius = 1.00
Polyline/Radius/Trim/<Select first object>: **P** *(For Polyline)*
Select 2D polyline: *(Select polyline "A")*
6 lines were filleted

Filleting two parallel lines in Figure 4–40 automatically constructs a semicircular arc object connecting both lines at their endpoints.

Command: **F** *(For FILLET)*
Current settings: Mode = TRIM, Radius = 1.0000
Select first object or [Polyline/Radius/Trim]: *(Select line "B")*
Select second object: *(Select line "C")*

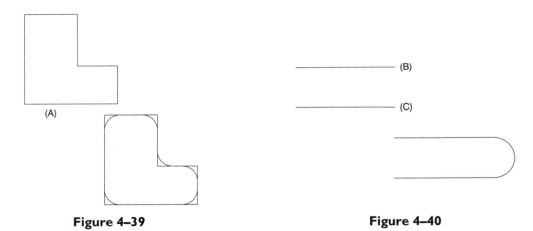

Figure 4–39   **Figure 4–40**

THE LENGTHEN COMMAND

The LENGTHEN command is used to change the length of a selected object without disturbing other object qualities such as angles of lines or radii of arcs. See Figure 4–41.

Command: **LEN** *(For LENGTHEN)*
Select an object or [DElta/Percent/Total/DYnamic]: *(Select line "A")*
Current length: 12.3649
Select an object or [DElta/Percent/Total/DYnamic]: **T** *(For Total)*
Specify total length or [Angle] (1.0000)>: **20**
Select an object to change or [Undo]: *(Select the line at "A")*
Select an object to change or [Undo]: *(Press ENTER to exit)*

When you use the LENGTHEN command on an arc segment, both the length and included angle information are displayed before you make any changes, as shown in Figure 4–42. After supplying the new total length of any object, be sure to select the desired end to lengthen when you are prompted for "Select object to change."

Command: **LEN** *(For LENGTHEN)*
Select an object or [DElta/Percent/Total/DYnamic]: *(Select arc "B")*
Current length: 1.4459, included angle: 54.9597
Select an object or [DElta/Percent/Total/DYnamic]: **T** *(For Total)*
Specify total length or [Angle] (1.0000)>: **5**
Select an object to change or [Undo]: *(Pick the arc at "B")*
Select an object to change or [Undo]: *(Press ENTER to exit)*

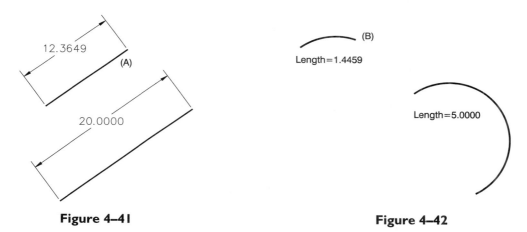

Figure 4–41

Figure 4–42

THE MIRROR COMMAND—METHOD #1

The MIRROR command is used to create a mirrored copy of an object or group of objects. When performing a mirror operation, you have the option of deleting the original object, which would be the same as flipping the object, or keeping the original object along with the mirror image, which would be the same as flipping and copying. Refer to the following prompts along with Figure 4–43 for using the MIRROR command:

Command: **MI** *(For MIRROR)*
Select objects: *(Select a point near "X")*
Specify opposite corner: *(Select a point near "Y")*
Select objects: *(Press ENTER to continue)*
Specify first point of mirror line: *(Select the endpoint of the centerline at "A")*
Specify second point of mirror line: *(Select the endpoint of the centerline at "B")*
Delete source objects? [Yes/No] <N>: *(Press ENTER for default)*

Since the original object was required to be retained by the mirror operation, the image result is shown in Figure 4–44. The MIRROR command works well when symmetry is required.

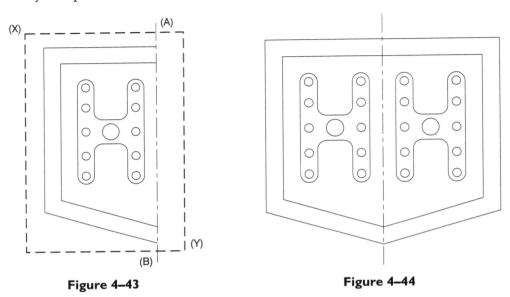

Figure 4–43   Figure 4–44

THE MIRROR COMMAND—METHOD #2

The illustration in Figure 4–45 is a different application of the MIRROR command. It is required to have all items that make up the bathroom plan flip but not copy to the other side. This is a typical process involving "what if" scenarios. Use the following command prompts to perform this type of mirror operation. The results are displayed in Figure 4–46.

Command: **MI** *(For MIRROR)*
Select objects: **All** *(This will select all objects shown in Figure 4–45)*
Select objects: *(Press ENTER to continue)*
Specify first point of mirror line: **Mid**
of *(Select the midpoint of the line at "A")*
Specify second point of mirror line: **Per**
to *(Select line "B," which is perpendicular to point "A")*
Delete source objects? [Yes/No] <N>: **Y** *(For Yes)*

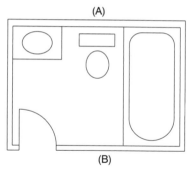

(A)

(B)

Figure 4–45

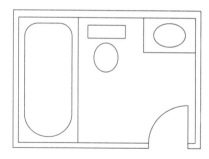

Figure 4–46

THE MIRROR COMMAND—METHOD #3

Situations sometimes involve mirroring text as in Figure 4–47. A system variable called MIRRTEXT controls this occurrence—by default it is set to a value of "1" or "On." Use the following prompts to see the results of the MIRROR command on text.

Command: **MI** *(For MIRROR)*
Select objects: *(Select a point near "X")*
Specify opposite corner: *(Select a point near "Y")*
Select objects: *(Press ENTER to continue)*
Specify first point of mirror line: *(Select the endpoint of the centerline at "A")*
Specify second point of mirror line: *(Select the endpoint of the centerline at "B")*
Delete source objects? [Yes/No] <N>: *(Press ENTER for default)*

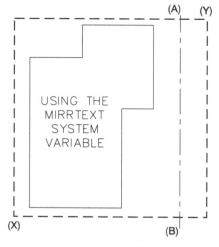

(A)   (Y)

USING THE
MIRRTEXT
SYSTEM
VARIABLE

(X)   (B)

Figure 4–47

Notice that with MIRRTEXT turned on, text is mirrored, which makes it unreadable (see Figure 4–48).

To mirror text and have it read properly, set the MIRRTEXT system variable to a value of 0 or Off as in the following command sequence. See Figure 4–49.

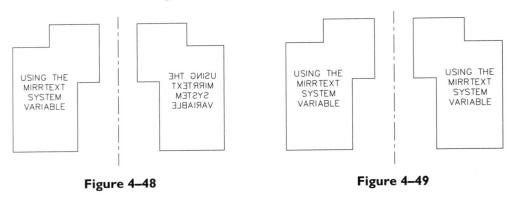

Figure 4–48   **Figure 4–49**

Command: **MIRRTEXT**
New value for MIRRTEXT <1>: **0**

Using the MIRROR command with the MIRRTEXT system variable set to a value of 0 results in text being legible, similar to Figure 4–49. Follow this command sequence to accomplish this.

Command: **MI** *(For MIRROR)*
Select objects: *(Create a selection set similar to Figure 4–47)*
Select objects: *(Press ENTER to continue)*
Specify first point of mirror line: *(Select the endpoint of the centerline at "A")*
Specify second point of mirror line: *(Select the endpoint of the centerline at "B")*
Delete source objects? [Yes/No] <N>: *(Press ENTER for default)*

THE MOVE COMMAND

The MOVE command repositions an object or group of objects at a new location. Once the objects to move are selected, a base point of displacement (where the object is to move from) is found. Next, a second point of displacement (where the object is to move to) is needed (See Figure 4–50)

Command: **M** *(For MOVE)*
Select objects: *(Select all dashed objects in Figure 4–50)*
Select objects: *(Press ENTER to continue)*
Specify base point or displacement: *(Select the endpoint of the line at "A")*
Specify second point of displacement or <use first point as displacement>: *(Mark a point at "B")*

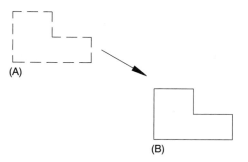

Figure 4–50

The slot shown in Figure 4–51 is incorrectly positioned; it needs to be placed 1.00 unit away from the left edge of the object. You can use the MOVE command in combination with a polar coordinate or Direct Distance mode to perform this operation.

Command: **M** *(For MOVE)*
Select objects: *(Select the slot and all centerlines in Figure 4–51)*
Select objects: *(Press ENTER to continue)*
Specify base point or displacement: *(Select the edge of arc "A")*
Specify second point of displacement or <use first point as displacement>: **@0.50<0**

As the slot is moved to a new position with the MOVE command, a new horizontal dimension must be placed to reflect the correct distance from the edge of the object to the centerline of the arc. See Figure 4–52. Another command will be explained in the following pages that affects a group of objects along with the dimension.

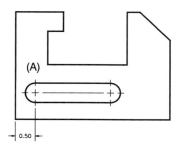

Figure 4–51

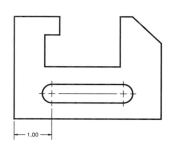

Figure 4–52

THE OFFSET COMMAND

The OFFSET command is commonly used for creating one object parallel to another. One method of offsetting is to identify a point to offset through, called a through point. Once an object is selected to offset, a through point is identified. The selected object offsets to the point shown in Figure 4–53. Refer to the following command sequence to use this method of the OFFSET command.

Command: **OF** *(For OFFSET)*
Specify offset distance or [Through] <1.00>: **T** *(For Through)*
Select object to offset or <exit>: *(Select the line at "A")*
Specify through point: **Nod**
of *(Select the point at "B")*
Select object to offset or <exit>: *(Press ENTER to exit this command)*

Another method of offsetting is by a specified offset distance. In Figure 4–54, an offset distance of 0.50 is set. The line segment "A" is identified as the object to offset. To complete the command, you must identify a side to offset to give the offset a direction in which to operate. Use the following command sequence for using this method of offsetting objects.

Command: **OF** *(For OFFSET)*
Specify offset distance or [Through] <1.00>: **0.50**
Select object to offset or <exit>: *(Select the line at "A")*
Specify point on side to offset: *(Pick a point at "B")*
Select object to offset or <exit>: *(Press ENTER to exit this command)*

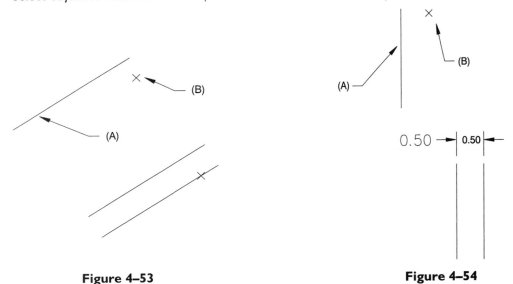

Figure 4–53   **Figure 4–54**

Another method of offsetting is shown in Figure 4–55, where the objects need to be duplicated at a set distance from existing geometry. The COPY command could be used for this operation; a better command would be OFFSET. This allows you to specify a distance and a side for the offset to occur. The result is an object parallel to the original object at a specified distance. All objects in Figure 4–55 need to be offset 0.50 toward the inside of the original object. See the command sequences that follow.

Command: **OF** *(For OFFSET)*
Specify offset distance or [Through] <1.00>: **0.50**
Select object to offset or <exit>: *(Select the horizontal line at "A")*
Specify point on side to offset: *(Pick a point anywhere on the inside near "B")*

Repeat the preceding procedure for lines C through J.

Notice that when all lines were offset, the original lengths of all line segments were maintained. Since all offsetting occurs inside, the segments overlap at their intersection points (see Figure 4–56). In one case, at "A" and "B", the lines did not even meet at all. The CHAMFER command may be used to edit all lines to form a sharp corner. You can accomplish this by assigning chamfer distances of 0.00 units.

Command: **CHA** *(For CHAMFER)*
(TRIM mode) Current chamfer Dist1 = 0.5000, Dist2 = 0.5000
Select first line or [Polyline/Distance/Angle/Trim/Method]: **D** *(For Distance)*
Specify first chamfer distance <0.5000>: **0**
Specify second chamfer distance <0.0000>: *(Press ENTER to accept the default)*

Command: **CHA** *(For CHAMFER)*
(TRIM mode) Current chamfer Dist1 = 0.0000, Dist2 = 0.0000
Select first line or [Polyline/Distance/Angle/Trim/Method]: *(Select the line at "A" in Figure 4–56)*
Select second line: *(Select the line at "B")*

Repeat the above procedure for lines "B" through "I" in Figure 4–56.

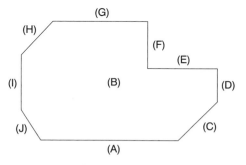

Figure 4–55

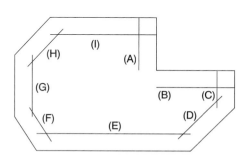

Figure 4–56

Using the OFFSET command along with the CHAMFER command produces the result shown in Figure 4–57. The chamfer distances must be set to a value of 0 for this special effect. The FILLET command produces the same result when set to a radius value of 0.

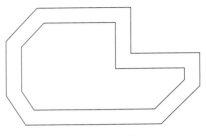

Figure 4–57

THE PEDIT COMMAND

Editing of polylines can lead to interesting results. A few of these options will be explained in the following pages. Figure 4–58 is a polyline of width 0.00. The PEDIT command was used to change the width of the polyline to 0.10 units. Refer to the following command sequence to use the PEDIT command with the Width option.

Command: **PE** *(For PEDIT)*
Select polyline: *(Select the polyline at "A")*
Enter an option [Open/Join/Width/Edit vertex/Fit/Spline/Decurve/Ltype gen/Undo]:
    W *(For Width)*
Specify new width for all segments: **0.10**
Enter an option [Open/Join/Width/Edit vertex/Fit/Spline/Decurve/Ltype gen/Undo]:
    (Press ENTER to exit this command)

It is possible to convert regular objects to polylines. In Figure 4–59, the arc segment and individual line segments may be converted to a polyline. The circle cannot be converted unless part of the circle is broken, resulting in an arc segment. Use the following prompts for converting the line segments to a polyline.

Command: **PE** *(For PEDIT)*
Select polyline: *(Select the line at "A")*
Object selected is not a polyline
Do you want to turn it into one? <Y> *(Press ENTER to accept the default)*
Enter an option [Close/Join/Width/Edit vertex/Fit/Spline/Decurve/Ltype gen/Undo]: **J**
    (For Join)
Select objects: *(Select lines "B" through "D")*
Select objects: *(Press ENTER to join the lines)*
3 segments added to polyline
Enter an option [Open/Join/Width/Edit vertex/Fit/Spline/Decurve/Ltype gen/Undo]:
    (Press ENTER to exit this command)

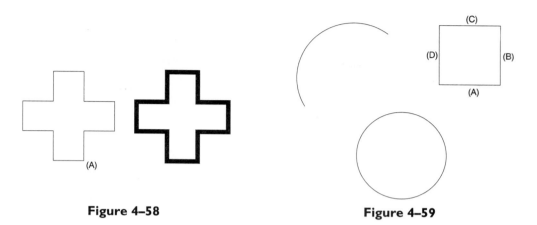

Figure 4–58   **Figure 4–59**

In the previous example, regular objects were selected individually before being converted to a polyline. For more complex objects, use the Window option to select numerous objects and perform the PEDIT-Join operation, which is faster. Refer to the following command sequence and Figure 4–60 to use this command.

Command: **PE** *(For PEDIT)*
Select polyline: *(Select the line at "A")*
Object selected is not a polyline
Do you want to turn it into one? <Y> *(Press* ENTER *for default)*
Enter an option [Close/Join/Width/Edit vertex/Fit/Spline/Decurve/Ltype gen/Undo]: **J**
   (For Join)
Select objects: *(Pick a point at "B")*
Specify opposite corner: *(Pick a point at "C")*
Select objects: *(Press* ENTER *to join the lines)*
56 segments added to polyline
Enter an option [Open/Join/Width/Edit vertex/Fit/Spline/Decurve/Ltype gen/Undo]:
   (Press ENTER *to exit this command)*

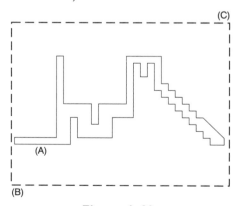

Figure 4–60

The polyline in Figure 4–61 will be used as an example of the use of the PEDIT command along with various curve-fitting utilities. In Figures 4–62A and 4–62B, the Spline option and Fit Curve option are shown. The Spline option produces a smooth fitting curve based on control points in the form of the vertices of the polyline. The Fit Curve option passes entirely through the control points, producing a less desirable curve. Study Figures 4–62B and 4–62C, which illustrate both curve options of the PEDIT command.

Figure 4–61

SPLINE CURVE GENERATION

Command: **PE** *(For PEDIT)*
Select polyline: *(Select the polyline at "A")*
Enter an option [Close/Join/Width/Edit vertex/Fit/Spline/Decurve/Ltype gen/Undo]: **S**
   (For Spline)
Enter an option [Close/Join/Width/Edit vertex/Fit/Spline/Decurve/Ltype gen/Undo]:
   (Press ENTER to exit this command)
The original polyline frame is usually not visible when a spline is created and it is shown only for illustrative purposes.

(A)

Figure 4–62A

FIT CURVE GENERATION

Command: **PE** *(For PEDIT)*
Select polyline: *(Select the polyline at "B")*
Enter an option [Close/Join/Width/Edit vertex/Fit/Spline/Decurve/Ltype gen/Undo]: **F**
   (For Fit)
Enter an option [Close/Join/Width/Edit vertex/Fit/Spline/Decurve/Ltype gen/Undo]:
   (Press ENTER to exit this command)

The Linetype Generation option of the PEDIT command controls the pattern of the linetype from polyline vertex to vertex. In the polyline at "C" in Figure 4–62C, the hidden linetype is generated from the first vertex to the second vertex. An entirely different pattern is formed from the second vertex to the third vertex, and so on. The polyline at "D" has the linetype generated throughout the entire polyline. In this way, the hidden linetype is smoothed throughout the polyline.

Command: **PE** *(For PEDIT)*
Select polyline: *(Select the polyline at "D" in Figure 4–62C)*
Enter an option [Close/Join/Width/Edit vertex/Fit/Spline/Decurve/Ltype gen/Undo]: **Lt**
    (For Ltype gen)
Enter polyline linetype generation option [ON/OFF] <Off>: **On**
Enter an option [Close/Join/Width/Edit vertex/Fit/Spline/Decurve/Ltype gen/Undo]:
    (Press ENTER to exit this command)

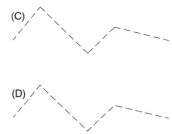

Figure 4–62B   **Figure 4–62C**

The object shown in Figure 4–63 is identical to that of Figure 4–55. Also in Figure 4–55, each individual line had to be offset to copy the lines parallel at a specified distance. Then the CHAMFER command was used to clean up the corners. There is an easier way to perform this operation. First convert all individual line segments to one polyline using the PEDIT command.

Command: **PE** *(For PEDIT)*
Select polyline: *(Select the line at "A")*
Object selected is not a polyline.
Do you want it to turn into one? <Y> *(Press ENTER for default)*
Enter an option [Close/Join/Width/Edit vertex/Fit/Spline/Decurve/Ltype gen/Undo]: **J**
    (For Join)
Select objects: *(Pick a point at "X")*
Specify opposite corner: *(Mark a point at "Y")*
Select objects: *(Press ENTER to continue)*
8 lines added to polyline.
Enter an option [Close/Join/Width/Edit vertex/Fit/Spline/Decurve/Ltype gen/Undo]:
    (Press ENTER to exit this command)

The OFFSET command is used to copy the shape 0.50 units to the inside. Because the object was converted to a polyline, all objects are offset at the same time. This procedure bypasses the need to use the CHAMFER or FILLET command to corner all intersections. See Figure 4–64.

Command: **OF** *(For OFFSET)*
Specify offset distance or [Through] <1.0000>: **0.50**
Select object to offset or <exit>: *(Select the polyline at "A")*
Specify point on side to offset: *(Select a point anywhere near "B")*
Select object to offset or <exit>: *(Press ENTER to exit this command)*

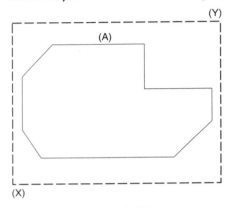

Figure 4–63

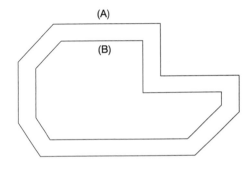

Figure 4–64

THE ROTATE COMMAND

The ROTATE command changes the orientation of an object or group of objects by identifying a base point and a rotation angle that completes the new orientation. Figure 4–65 shows an object complete with crosshatch pattern, which needs to be rotated to a 30° angle using point "A" as the base point. Use the following prompts and Figure 4–65 to perform the rotation.

Command: **RO** *(For ROTATE)*
Current positive angle in UCS: ANGDIR=counterclockwise ANGBASE=0
Select objects: **All**
Select objects: *(Press ENTER to continue)*
Specify base point: *(Select the endpoint of the line at "A")*
Specify rotation angle or [Reference]: **30**

ROTATE—REFERENCE

At times it is necessary to rotate an object to a desired angular position. However, this must be accomplished even if the current angle of the object is unknown. To maintain the accuracy of the rotation operation, use the Reference option of the ROTATE command. Figure 4–66 shows an object that needs to be rotated to the 30° angle position. Unfortunately, we do not know the angle the object currently lies in. Entering the Reference angle option and identifying two points creates a known angle of reference. Entering a new angle of 30° rotates the object to the 30° position from the reference angle. Use the following prompts and Figure 4–66 to accomplish this.

Command: **RO** *(For ROTATE)*
Current positive angle in UCS: ANGDIR=counterclockwise ANGBASE=0
Select objects: *(Select the object in Figure 4–66)*
Select objects: *(Press ENTER to continue)*
Specify base point: *(Pick either the edge of the circle or two arc segments to locate the center)*
Specify rotation angle or [Reference]: **R** *(For Reference)*
Specify the reference angle <0>: *(Pick either the edge of the circle or two arc segments to locate the center)*
Specify second point: **Mid**
of *(Select the line at "A" to establish the reference angle)*
Specify the new angle: **30**

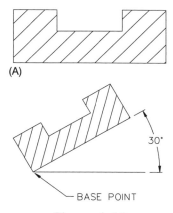

(A)

BASE POINT

Figure 4–65

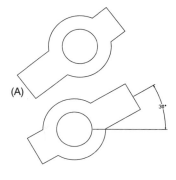

(A)

30°

Figure 4–66

THE SCALE COMMAND

Use the SCALE command to change the overall size of an object. The size may be larger or smaller in relation to the original object or group of objects. The SCALE command requires a base point and scale factor to complete the command. The object in Figure 4–67 will be scaled to different sizes in Figures 4–68 and 4–69.

With a base point at "A" and a scale factor of 0.50, the results of using the SCALE command on a group of objects are shown in Figures 4–68.

Command: **SC** *(For SCALE)*
Select objects: **All**
Select objects: *(Press* ENTER *to continue)*
Specify base point: *(Select the endpoint of the line at "A")*
Specify scale factor or [Reference]: **0.50**

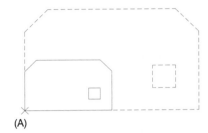

Figure 4–67   Figure 4–68

The example in Figure 4–69 shows the effects of identifying a new base point in the center of the object.

Command: **SC** *(For SCALE)*
Select objects: **All**
Select objects: *(Press* ENTER *to continue)*
Specify base point: *(Pick a point near "A")*
Specify scale factor or [Reference]: **0.40**

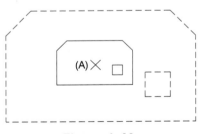

Figure 4–69

SCALE—REFERENCE

Suppose you are given a drawing that has been scaled down in size. However no one knows what scale factor was used. You do know what one of the distances should be. In this special case, you can use the Reference option of the SCALE command to identify endpoints of a line segment that act as a reference length. Entering a new length value could increase or decrease the entire object proportionally. Study Figure 4–70 and the following prompts for performing this operation.

Command: **SC** *(For SCALE)*
Select objects: *(Pick a point at "A")*
Specify opposite corner: *(Pick a point at "B")*
Select objects: *(Press ENTER to continue)*
Specify base point: *(Select the edge of the circle to identify its center)*
Specify scale factor or [Reference]: **R** *(For Reference)*
Specify reference length <1>: *(Select the endpoint of the line at "C")*
Specify second point: *(Select the endpoint of the line at "C")*
Specify new length: **2.00**

Since the length of line "CD" was not known, the endpoints were picked after the Reference option was entered. This provided the length of the line to AutoCAD. The final step to perform was to make the line 2.00 units, which increased the size of the object while also keeping its proportions.

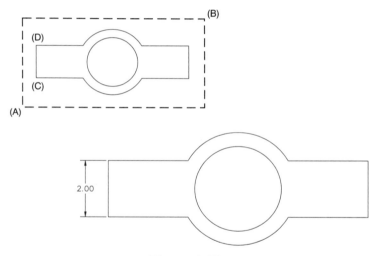

Figure 4–70

THE STRETCH COMMAND

Use the STRETCH command to move a portion of a drawing while still preserving the connections to parts of the drawing remaining in place. To perform this type of operation, you must use the Crossing option of "Select objects." In Figure 4–71, a group of objects is selected with the crossing box. Next, a base point is identified by the endpoint at "C." Finally, a second point of displacement is identified with a polar coordinate; the Direct Distance mode could also be used. Once the objects selected in the crossing box are stretched, the objects not only move to the new location but also mend themselves.

Command: **S** *(For STRETCH)*
Select objects to stretch by crossing-window or crossing-polygon...
Select objects: *(Pick a point at "A")*
Specify opposite corner: *(Pick a point at "B")*
Select objects: *(Press* ENTER *to continue)*
Specify base point or displacement: *(Select the endpoint of the line at "C")*
Specify second point of displacement: **@1.75<180**

Figure 4–72 is another example of the use of the STRETCH command. The crossing window is employed along with a base point at "C" and a polar coordinate.

Command: **S** *(For STRETCH)*
Select objects to stretch by crossing-window or crossing-polygon...
Select objects: *(Pick a point at "A")*
Specify opposite corner: *(Pick a point at "B")*
Select objects: *(Press* ENTER *to continue)*
Specify base point or displacement: *(Select the endpoint of the line at "C")*
Specify second point of displacement: **@2.75<0**

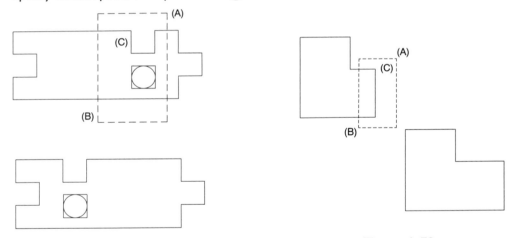

Figure 4–71   **Figure 4–72**

Applications of the STRETCH command include Figure 4–73, where a window needs to be positioned at a new location. Use the following command sequence to stretch the window at a set distance using a polar coordinate. Also see Figure 4–73.

Command: **S** *(For STRETCH)*
Select objects to stretch by crossing-window or crossing-polygon...
Select objects: *(Pick a point at "A")*
Specify opposite corner: *(Pick a point at "B")*
Select objects: *(Press* ENTER *to continue)*
Specify base point or displacement: Mid

of (Select the midpoint of the line at "C")
Specify second point of displacement: **@10'6<0**

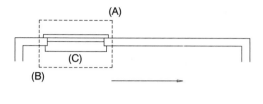
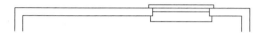
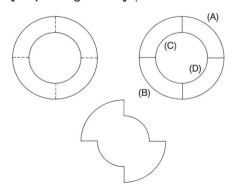

Figure 4–73

THE TRIM COMMAND

Use the TRIM command to partially delete an object or group of objects based on a cutting edge. In Figure 4–74, the four dashed lines are selected as cutting edges. Next, segments of the circles are selected to be trimmed between the cutting edges.

Command: **TR** *(For TRIM)*
Current settings: Projection=UCS Edge=None
Select cutting edges ...
Select objects: *(Select the four dashed lines in Figure 4–74)*
Select objects: *(Press ENTER to continue)*
Select object to trim or [Project/Edge/Undo]: *(Select the circle at "A")*
Select object to trim or [Project/Edge/Undo]: *(Select the circle at "B")*
Select object to trim or [Project/Edge/Undo]: *(Select the circle at "C")*
Select object to trim or [Project/Edge/Undo]: *(Select the circle at "D")*
Select object to trim or [Project/Edge/Undo]: *(Press ENTER to exit this command)*

Figure 4–74

An alternate method of selecting cutting edges is to press ENTER in response to the prompt "Select objects." This automatically creates cutting edges out of all objects in the drawing. When you use this method, the cutting edges do not highlight. In Figure 4–75, even though the four dashed lines are selected as cutting edges, the middle of the cutting edge may be trimmed out at all locations identified by "A" through "D."

Command: **TR** *(For TRIM)*
Current settings: Projection=UCS Edge=None
Select cutting edges ...
Select objects: *(Press ENTER to select all four dashed lines as cutting edges)*
Select object to trim or [Project/Edge/Undo]: *(Select the segment at "A")*
Select object to trim or [Project/Edge/Undo]: *(Select the segment at "B")*
Select object to trim or [Project/Edge/Undo]: *(Select the segment at "C")*
Select object to trim or [Project/Edge/Undo]: *(Select the segment at "D")*
Select object to trim or [Project/Edge/Undo]: *(Press ENTER to exit this command)*

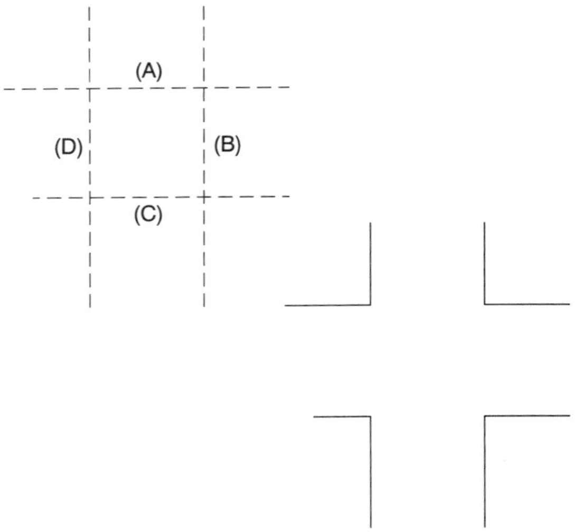

Figure 4–75

Yet another application of the TRIM command uses the Fence option of "Select objects." First, invoke the TRIM command and select the small circle as the cutting edge. Begin the prompt of "Select object to trim" with "Fence." See Figure 4–76.

Command: **TR** *(For TRIM)*
Current settings: Projection=UCS Edge=None
Select cutting edges ...
Select objects: *(Select the small circle)*
Select objects: *(Press ENTER to continue)*
Select object to trim or [Project/Edge/Undo]: **F** *(For Fence)*

Continue with the TRIM command by identifying a Fence. This consists of a series of line segments that take on a dashed appearance. This means the fence will select any object it crosses. When you have completed the construction of the desired fence shown in Figure 4–77, press ENTER.

First fence point: *(Pick a point at "A")*
Specify endpoint of line or [Undo]: *(Pick a point at "B")*
Specify endpoint of line or [Undo]: *(Pick a point at "C")*
Specify endpoint of line or [Undo]: *(Pick a point at "D")*
Specify endpoint of line or [Undo]: *(Pick a point at "E")*
Specify endpoint of line or [Undo]: *(Pick a point at "F")*
Specify endpoint of line or [Undo]: *(Pick a point at "G")*
Specify endpoint of line or [Undo]: *(Pick a point at "H")*
Specify endpoint of line or [Undo]: *(Press* ENTER *to end the Fence and execute the trim operation)*
Select object to trim or [Project/Edge/Undo]: *(Press* ENTER *to exit this command)*

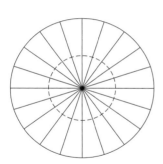

Figure 4–76

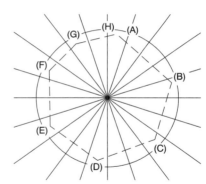

Figure 4–77

The power of the Fence option of "Select objects" is shown in Figure 4–78. Eliminating the need to select each individual line segment inside the small circle to trim, the Fence trims all objects it touches in relation to the cutting edge.

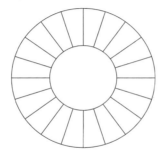

Figure 4–78

The TRIM command also allows you to trim to an extended cutting edge. In Extended Cutting Edge mode, an imaginary cutting edge is formed; all objects sliced along this cutting edge will be trimmed if selected individually or by the Fence mode. Study Figure 4–79 and the following prompts on this feature of the TRIM command.

Command: **TR** *(For TRIM)*
Current settings: Projection=UCS Edge=None
Select cutting edges ...
Select objects: *(Pick line "A")*
Select objects: *(Press ENTER to continue)*
Select object to trim or [Project/Edge/Undo]: **E** *(For Edge)*
Enter an implied edge extension mode [Extend/No extend] <No extend>: **E** *(For Extend)*
Select object to trim or [Project/Edge/Undo]: *(Pick the line at "B" along with the other segments)*
Select object to trim or [Project/Edge/Undo]: *(Press ENTER to exit this command)*

The Fence mode can also be used to select all line segments at once to trim.

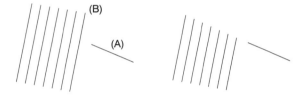

Figure 4–79

EXPRESS TOOLS APPLICATIONS FOR MODIFYING

A series of Express tools are available to provide more functionality with modifying objects. A few of these tools can be selected from the Express Standard Toolbar in Figure 4–80. Four tools are available in this toolbar, namely Multiple Entity Stretch, Move Copy Rotate, Extended Trim, and Multiple Pedit.

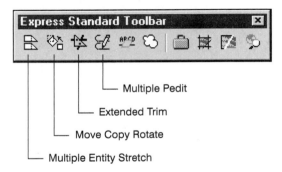

Figure 4–80

These Express tools may also be selected from the Modify option of Express pull-down menu shown in Figure 4–81. In addition to the tools listed in Figure 4–80, a tool called Polyline Join is listed here. All five of these tools will now be explained.

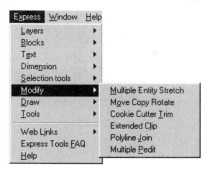

Figure 4–81

MULTIPLE ENTITY STRETCH

Normally, the STRETCH command allows for only one crossing box to be created for stretching a group of objects. Use this Express tool to create multiple crossing boxes to stretch multiple objects while inside the same STRETCH command. The MSTRETCH command is automatically activated in a Crossing Box mode. The direction of your picks does not matter—you are always in the Crossing Box mode.

Use the illustration in Figure 4–82 and the prompt sequence below as guides for using the MSTRETCH command.

Command: **MSTRETCH**
Define crossing windows or crossing polygons...
Options: Crossing Polygon or Crossing first point
Specify an option [CO/C] <Crossing first point>: *(Pick at "A")*
Specify other corner: *(Pick at "B")*
Options: Crossing Polygon, Crossing first point or Undo
Specify an option [CO/C/Undo] <Crossing first point>: *(Pick at "C")*
Specify other corner: *(Pick at "D")*
Options: Crossing Polygon, Crossing first point or Undo
Specify an option [CO/C/Undo] <Crossing first point>: *(Pick at "E")*
Specify other corner: *(Pick at "F")*
Options: Crossing Polygon, Crossing first point or Undo
Specify an option [CO/C/Undo] <Crossing first point>: *(Pick at "G")*
Specify other corner: *(Pick at "H")*
Options: Crossing Polygon, Crossing first point or Undo
Specify an option [CO/C/Undo] <Crossing first point>: *(Press* ENTER *to continue)*
Done defining windows for stretch...
Specify an option [Remove objects] <Base point>: *(Pick a point anywhere on the screen)*
Second base point: **@1<0**

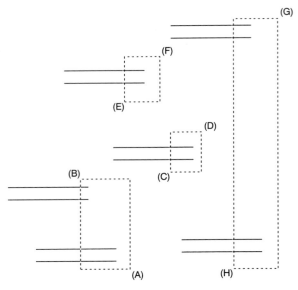

Figure 4–82

The result of the multiple stretch operation is illustrated in Figure 4–83, with all groups of lines stretched by 1 unit.

Figure 4–83

MOVE COPY ROTATE

The MOCORO or Move/Copy/Rotate command contains options to move, copy, and/ or rotate an object. A scale option is also present in this command. The object in Figure 4–84A will be used to illustrate the use of this command. When you first activate this command and select objects such as the square and hexagon shapes in Figure 4–84B, the object highlights along with a marker identifying the current base point. Follow the prompt sequence and the illustrations below to see how this command is used.

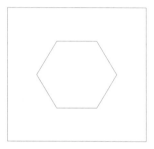

Figure 4–84A

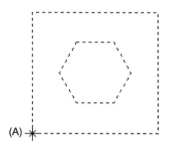

(A)

Figure 4–84B

Command: **MOCORO**
Select objects: *(Select the square and hexagon in Figure 4–84A)*
Select objects: *(Press* ENTER *to continue)*
Base point: *(Pick the endpoint at "A"; a marker appears in Figure 4–84B)*
[Move/Copy/Rotate/Scale/Base/Undo]<eXit>: **M** *(For Move)*
Second point of displacement: *(Pick a new location to move)*
[Move/Copy/Rotate/Scale/Base/Undo]<eXit>: **R** *(For Rotate)*
Second Point or Rotation angle: **45** *(See Figure 4–85A)*
[Move/Copy/Rotate/Scale/Base/Undo]<eXit>: **S** *(For Scale)*
Second Point or Scale factor: **0.75** *(See Figure 4–85B)*
[Move/Copy/Rotate/Scale/Base/Undo]<eXit>: *(Press* ENTER *to exit this command)*

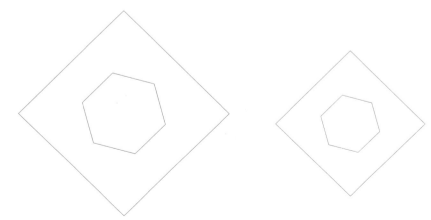

Figure 4–85A   **Figure 4–85B**

COOKIE CUTTER TRIM (EXTENDED TRIM)

Use this Express tool to select a cutting edge, pick a side to trim, and have all objects in contact with the cutting edge trim to the selected side. Use the prompt sequence below and Figure 4–86 to see how this command is used.

Command: **EXTRIM**
Pick a POLYLINE, LINE, CIRCLE, ARC, ELLIPSE, IMAGE or TEXT for cutting edge...
Select objects: *(Pick the edge of the circle at "A" in Figure 4–86)*
Specify the side to trim on: *(Pick a point outside of the circle at "B")*

The result is illustrated in Figure 4–87, with all lines outside the cutting edges are trimmed. If a point were selected inside the cutting edge, all interior lines would be trimmed to the cutting edge.

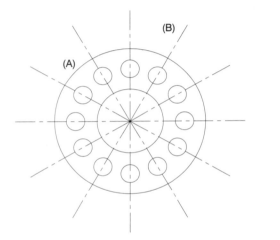

Figure 4–86

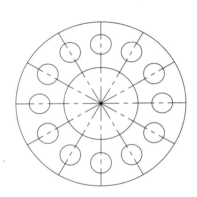

Figure 4–87

POLYLINE JOIN

As a general rule when joining polylines, you cannot have gaps present or overlapping occurring when performing this operation. If gaps or overlapping happens to exist in the polylines to join, use the Polyline Join Express tool. This command works best when joining two objects that have a gap or overlap. After selecting the two objects to join, you will be asked to enter a fuzz factor. This is the distance used by this command to bridge a gap or trim overlapping lines. You could measure the distance between two objects to determine this value. Study Figure 4–88A to better understand the concept of a fuzz factor. Also in Figure 4–88A, the thick lines represent the polylines to join. The lines to join can also consist of polylines or regular line segments. Study the command prompts and figures for performing this operation.

Command: **PLJOIN**
Select objects: *(Select lines "A" and "B" in Figure 4–88A)*
Select objects: *(Press* ENTER *to continue)*
 Join Type = Both (Fillet and Add)
Enter fuzz distance or [Jointype] <0.00>: **0.44**
Processing pline data... Done.
Command: **PLJOIN**
Select objects: *(Select lines "C" and "D" in Figure 4–88B)*
Select objects: *(Press* ENTER *to continue)*
 Join Type = Both (Fillet and Add)
Enter fuzz distance or [Jointype] <0.00>: **0.92**
Processing pline data... Done.

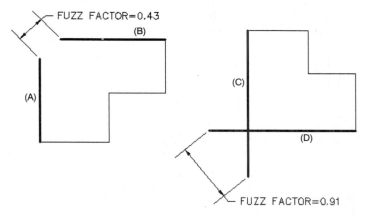

Figure 4–88A

The results are displayed in Figure 4–88B, with both examples of lines being converted to polylines and joined even though a gap or overlap was present before.

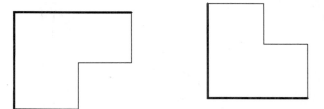

Figure 4–88B

MULTIPLE PEDIT

Multiple editing of polylines or MPEDIT allows for multiple objects to be converted to polylines. With the PEDIT command, only one object can be converted to a polyline at a time. Of course the Join option would convert other adjacent objects to polylines. However, for separate objects, PEDIT only converts one object at a time to a polyline. Use the prompt sequence below and Figure 4–89 to illustrate how the MPEDIT command is used.

Command: **MPEDIT**
Select objects: *(Select the arc and line segments in Figure 4–89)*
Select objects: *(Press Enter to continue)*
Convert Lines and Arcs to polylines? [Yes/No] <Yes>: *(Press Enter to accept the default)*
Enter an option [Open/Close/Join/Width/Fit/Spline/Decurve/Ltype gen/eXit]
<eXit>: **W** *(For Width)*
Enter new width for all segments: **0.05**
Enter an option [Open/Close/Join/Width/Fit/Spline/Decurve/Ltype gen/eXit]
<eXit>: *(Press ENTER to exit this command)*

Figure 4–89

The results of using the MPEDIT command are illustrated in Figure 4–90, with all objects converted to polylines in addition to the width being assigned to 0.05 units.

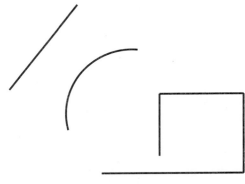

Figure 4–90

TUTORIAL EXERCISE: TILE.DWG

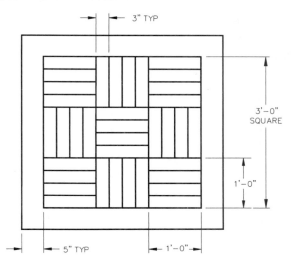

Figure 4–91

Purpose

This tutorial is designed to use the OFFSET and TRIM commands to complete the drawing of the floor tile shown in Figure 4–91.

System Settings

Use the Drawing Units dialog box and change the units of measure from decimal to architectural units. Keep the remaining default settings. Use the LIMITS command and change the limits of the drawing to (0,0) for the lower left corner and (17',11') for the upper right corner. Use the ZOOM command and the All option fit the new drawing limits to the display screen. Use the SNAP command and change the value from 1/2" to 3". (If the GRID command is set to 0, the snap setting of 3" will also change the grid spacing to 3"). Check to see that the following Object Snap modes are currently set: Endpoint, Extension, Intersection, Center.

Layers

Create the following layer with the format:

Name	Color	Linetype
Object	White	Continuous

Suggested Commands

Make the "Object" layer current. Use the LINE command to begin the inside square of the Tile. The ARRAY command is used to copy selected line segments in a rectangular pattern at a specified distance. The TRIM command is then used to form the inside tile patterns. The OFFSET command is used to copy lines parallel that form the outer square and the FILLET command is used to form the corners of the square.

Whenever possible, substitute the appropriate command alias in place of the full AutoCAD command in each tutorial step. For example, use "CP" for the COPY command, "L" for the LINE command, and so on. The complete listing of all command aliases is located in Chapter 1, Table 1–2.

STEP 1

Begin this exercise by using the LINE command and Polar Coordinate mode to draw a 3'0" square shown in Figure 4–92. The Direct Distance mode can also be used to construct the square.

Command: **L** *(For LINE)*
Specify first point: **6',3'**
Specify next point or [Undo]: *(Move your cursor to the right and enter **3'**)*
Specify next point or [Undo]: *(Move your cursor up and enter **3'**)*
Specify next point or [Close/Undo]: *(Move your cursor to the left and enter **3'**)*
Specify next point or [Close/Undo]: **C** *(To Close)*

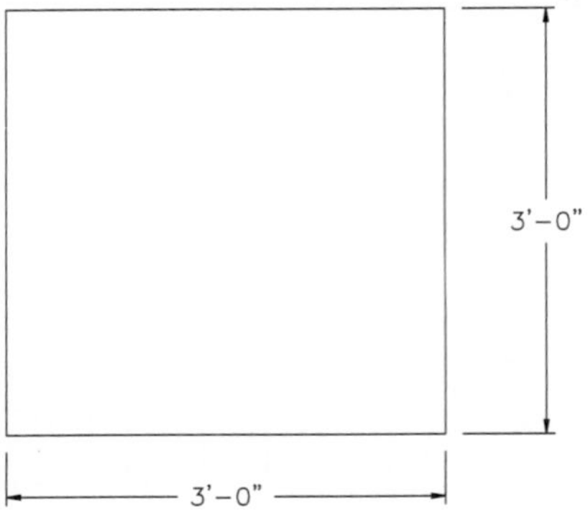

3'–0"

3'–0"

Figure 4–92

STEP 2

Use the ARRAY command to copy the top line in a rectangular pattern and have all lines spaced 3 units from each other. Select the top line as the object to array and perform a rectangular array consisting of twelve rows and one column. See Figure 4–93. Since the top line selected will be copied straight down, a negative distance must be entered to perform this operation. Another popular command that could be used here is OFFSET. However since each line must be offset separately, the ARRAY command is the more efficient command to use.

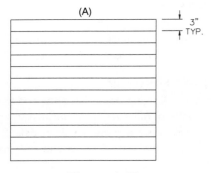

(A)

3"
TYP.

Figure 4–93

Command: **AR** *(For ARRAY)*
Select objects: *(Select the top horizontal line at "A")*
Select objects: *(Press ENTER to continue)*
Enter the type of array [Rectangular/Polar] <R>: *(Press ENTER to accept this default)*

Enter the number of rows (—) <1>: **12**
Enter the number of columns (|||) <1> **1**
Enter the distance between rows or specify unit cell (—): **-3**

STEP 3

Again use the ARRAY command but this time copy the right vertical line in a rectangular pattern and have all lines spaced 3 units away from each other. Select the right line as the object to array and perform a rectangular array consisting of one row and twelve columns (see Figure 4–94). Since the right line selected will be copied to the left, you must enter a negative distance to perform this operation.

Command: **AR** *(For ARRAY)*
Select objects: *(Select the right vertical line at "A")*
Select objects: *(Press ENTER to continue)*
Enter the type of array [Rectangular/ Polar] <R>: *(Press ENTER to accept this default)*
Enter the number of rows (—) <1>: **1**
Enter the number of columns (||||) <1> **12**
Specify the distance between columns (||||): **-3**

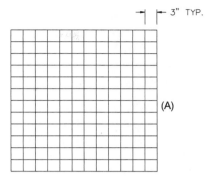

Figure 4–94

STEP 4

Magnify the image by using the ZOOM command and a rectangular window. Use the TRIM command. Select the two vertical dashed lines shown in Figure 4–95 as cutting edges; use Figure 4–96 as a guide for determining which lines to trim out.

Command: **TR** *(For TRIM)*
Current settings: Projection=UCS Edge=None
Select cutting edges ...
Select objects: *(Select the two dashed lines in Figure 4–95)*
Select objects: *(Press ENTER to continue)*
Select object to trim or [Project/Edge/ Undo]: *(Continue to Step 5)*

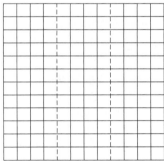

Figure 4–95

STEP 5

For the last prompt of the TRIM command in Step 4, select all horizontal lines in the areas marked "A," "B," "C," and "D" in Figure 4–96. When finished selecting the objects to trim, press ENTER to exit the command.

Select object to trim or [Project/Edge/ Undo]: *(Select all horizontal lines in area "A")*

Select object to trim or [Project/Edge/ Undo]: *(Select all horizontal lines in area "B")*

Select object to trim or [Project/Edge/ Undo]: *(Select all horizontal lines in area "C")*

Select object to trim or [Project/Edge/ Undo]: *(Select all horizontal lines in area "D")*

Select object to trim or [Project/Edge/ Undo]: *(Press ENTER to exit this command)*

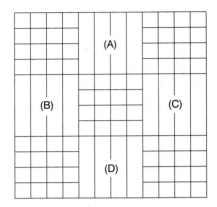

Figure 4–96

STEP 6

Use the TRIM command and select the two horizontal dashed lines as cutting edges in Figure 4–97.

  Command: **TR** *(For TRIM)*
Current settings: Projection=UCS Edge=None
Select cutting edges ...
Select objects: *(Select the two dashed lines in Figure 4–97)*
Select objects: *(Press ENTER to continue)*
Select object to trim or [Project/Edge/ Undo]: *(Continue to Step 7)*

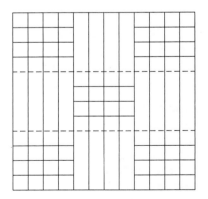

Figure 4–97

STEP 7

For the last step of the TRIM command in Step 6, select all vertical lines in the areas marked "A," "B," "C," "D," and "E" in Figure 4–98. When finished selecting the objects to trim, press ENTER to exit the command.

Select object to trim or [Project/Edge/ Undo]: *(Select all vertical lines in area "A")*

Select object to trim or [Project/Edge/ Undo]: *(Select all vertical lines in area "B")*

Select object to trim or [Project/Edge/ Undo]: *(Select all vertical lines in area "C")*

Select object to trim or [Project/Edge/ Undo]: *(Select all vertical lines in area "D")*

Select object to trim or [Project/Edge/ Undo]: *(Select all vertical lines in area "E")*

Select object to trim or [Project/Edge/ Undo]: *(Press ENTER to exit this command)*

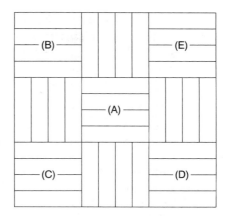

Figure 4–98

STEP 8

Use the OFFSET command to offset the lines "A," "B," "C," and "D" five units in the directions shown in Figure 4–99.

Command: **O** *(For OFFSET)*
Specify offset distance or [Through] <0'-1">: **5**
Select object to offset or <exit>: *(Select the line at "A")*
Specify point on side to offset: *(Pick a point above the line)*
Select object to offset or <exit>: *(Select the line at "B")*
Specify point on side to offset: *(Pick a point to the right of the line)*

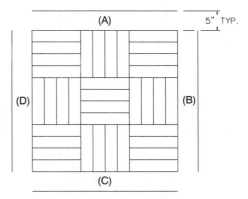

Figure 4–99

Select object to offset or <exit>: *(Select the line at "C")*
Specify point on side to offset: *(Pick a point below the line)*
Select object to offset or <exit>: *(Select the line at "D")*
Specify point on side to offset: *(Pick a point to the left of the line)*
Select object to offset or <exit>: *(Press ENTER to exit this command)*

STEP 9

Use the FILLET command set to a radius of 0 to place a corner at the intersection of lines "A" and "B." See Figure 4–100. The radius should already be set to 0 by default, so simply pick the two lines and the corner is formed.

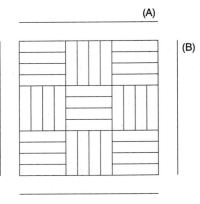

(A)

(B)

Figure 4–100

```
Command: F (For FILLET)
Current settings: Mode = TRIM, Radius =
   0'-0 1/2"
Select first object or [Polyline/Radius/
   Trim]: R (For Radius)
Specify fillet radius <0'-0 1/2">: 0
Command: F (For FILLET)
Current settings: Mode = TRIM, Radius =
   0'-0"
Select first object or [Polyline/Radius/
   Trim]: (Select line "A")
Select second object: (Select line "B")
```

Repeat this procedure for the other three corners.

STEP 10

The completed tile drawing is shown in Figure 4–101. Use the ZOOM command and the Previous option to return to the previous display image. Follow on to Step 11 to add more tiles in a rectangular pattern using the ARRAY command.

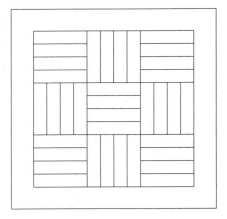

Figure 4–101

```
Command: Z (For ZOOM)
Specify corner of window, enter a scale
   factor (nX or nXP), or [All/Center/
   Dynamic/Extents/Previous/Scale/
   Window] <real time>: P (For Previous)
```

STEP II

As an alternate step, use the ARRAY command to copy the initial design in a rectangular pattern by row and column. Use the following prompts to perform this operation to construct a series of tiles as in Figure 4–102.

Command: **AR** *(For ARRAY)*
Select objects: **All**
Select objects: *(Press* ENTER *to continue)*
Enter the type of array [Rectangular/
  Polar] <R>: *(Press* ENTER *to accept this
  default)*
Enter the number of rows (—) <I>: **3**
Enter the number of columns (||||) <I> **2**
Enter the distance between rows or
  specify unit cell (—): **3'10**
Specify the distance between columns
  (||||): **3'10**

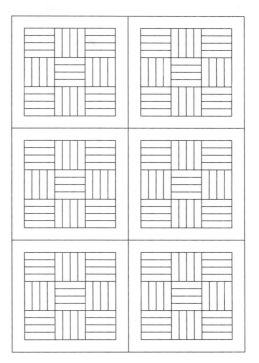

Figure 4–102

TUTORIAL EXERCISE: INLAY.DWG

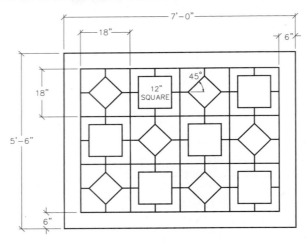

Figure 4–103

Purpose

This tutorial is designed to allow you to construct a drawing of the inlay shown in Figure 4–103 using the COPY command and the Multiple modifier.

System Settings

Use the Drawing Units dialog box and change the units of measure from decimal to architectural units. Keep the remaining default settings. Use the LIMITS command and change the limits of the drawing to (0,0) for the lower left corner and (17', 11') for the upper right corner. Use the ZOOM command and the All option to fit the new drawing limits to the display screen. Use the SNAP command and change the value from 0' 0-1/2" to 3". (If the Grid is set to 0, the snap setting of 3" will also change the grid spacing to 3".) Check to see that the following Object Snap modes are currently set: Endpoint, Extension, Intersection, Center.

Layers

Create the following layer with the format:

Name	Color	Linetypes
Object	White	Continuous

Suggested Commands

Make Layer "Object" current. Then begin this tutorial by drawing a 6'0" × 4'6" rectangle using the LINE command. Use the ARRAY command to copy designated lines in a rectangular pattern at a distance of 18". Then, using the 3" grid as a guide along with the Snap-On, draw the diamond and square shapes. Use the COPY command along with the Multiple modifier to copy the diamond and square shapes numerous times to the designated areas.

Whenever possible, substitute the appropriate command alias in place of the full AutoCAD command in each tutorial step. For example, use "CP" for the COPY command, "L" for the LINE command, and so on. The complete listing of all command aliases is located in Chapter 1, Table 1–2.

STEP 1

Begin this exercise by using the LINE command and Polar coordinate mode to draw a rectangle 6'0" by 4'6 as shown in Figure 4–104. The Direct Distance mode may also be used to construct the rectangle.

Command: **L** *(For LINE)*
Specify first point: **3',2'**
Specify next point or [Undo]: **@6'<0**
Specify next point or [Undo]: **@4'6<90**
Specify next point or [Close/Undo]:
  (Move your cursor over the intersection at "A" to acquire this point. Then move your cursor up until it locks with the horizontal snap position. Pick this point at "B").
Specify next point or [Close/Undo]: **C**
  (To Close)

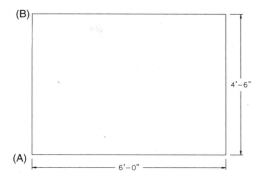

Figure 4–104

STEP 2

Use the ARRAY command to copy the top line in a rectangular pattern and have all lines spaced 18 units away from each other. Select the top horizontal line at "A" as the object to array. Since this line will be copied straight down, a value of -18 for the spacing between rows will perform this operation. Repeat this command to copy the right vertical line at "B" three times to the left at a distance of 18 units. Enter a value of -18 units for the spacing in between columns since the copying is performed in the left direction as shown in Figure 4–105.

 Command: **AR** *(For ARRAY)*
Select objects: *(Select the top horizontal line at "A")*
Select objects: *(Press ENTER to continue)*
Enter the type of array [Rectangular/ Polar] <R>: *(Press ENTER to accept this default)*
Enter the number of rows (—) <1>: **3**

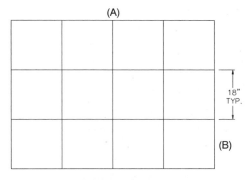

Figure 4–105

Enter the number of columns (||||) <1> **1**
Enter the distance between rows or specify unit cell (—): **-18**

Repeat the preceding rectangular ARRAY command for the vertical line at "B." Use 1 for the number of rows, 4 for the number of columns, and -18 as the distance between columns.

STEP 3

Draw one 12" × 12" diamond shape in the position shown in Figure 4–106. Use the ZOOM command with the Window option to magnify the area around the position of the diamond figure.

Command: **Z** *(For ZOOM)*
Specify corner of window, enter a scale factor (nX or nXP), or
[All/Center/Dynamic/Extents/Previous/ Scale/Window] <real time>: *(Pick a point at "A")*
Specify opposite corner: *(Pick a point at "B")*

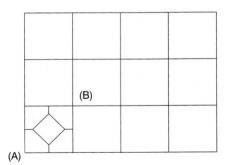

Figure 4–106

STEP 4

Turn OSNAP off by pressing F3. Be sure the Grid and Snap values are set to 3"; the Snap should already be turned on. Then use the LINE command and snap to the grid dots to draw the four lines shown in Figure 4–107. Use "A," "B," "C," and "D" as the starting points for the four lines.

Command: *(Press F3 to turn OSNAP off)*
Command: **L** *(For LINE)*
Specify first point: *(Pick a point at "A")*
Specify next point or [Undo]: *(Pick a point at "B")*
Specify next point or [Undo]: *(Press ENTER to exit this command)*

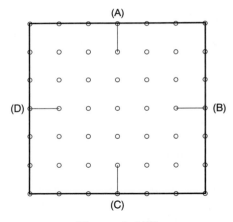

Figure 4–107

STEP 5

Turn OSNAP on by pressing F3. Use the LINE command to draw the diamond-shaped figure shown in Figure 4–108. Be sure Ortho mode is turned off before constructing the line segments. When finished, perform a **ZOOM-Previous** operation; turn the snap off by pressing F9.

Command: *(Press F3 to turn OSNAP back on)*
Command: **L** *(For LINE)*
Specify first point: *(Pick the endpoint at "A")*
Specify next point or [Undo]: *(Pick the endpoint at "B")*
Specify next point or [Undo]: *(Pick the endpoint at "C")*
Specify next point or [Close/Undo]: *(Pick the endpoint at "D")*
Specify next point or [Close/Undo]: **C** *(To Close)*
Command: **Z** *(For ZOOM)*
Specify corner of window, enter a scale factor (nX or nXP), or
[All/Center/Dynamic/Extents/Previous/Scale/Window] <real time>: **P** *(For Previous)*
Command: *(Press F9 to turn the SNAP off)*

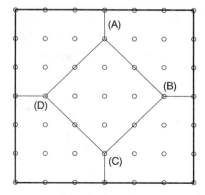

Figure 4–108

STEP 6

Use the COPY command and the Multiple option to repeat the diamond-shaped pattern in Figure 4–109. The OSNAP-Intersection and Endpoint modes should already be enabled.

Command: **CP** *(For COPY)*
Select objects: *(Select all dashed lines in Figure 4–109)*
Select objects: *(Press ENTER to continue)*
Specify base point or displacement, or [Multiple]: **M** *(For Multiple)*
Specify base point: *(Select the endpoint at "A")*
Specify second point of displacement or <use first point as displacement>: *(Select the endpoint at "B")*
Specify second point of displacement or <use first point as displacement>: *(Select the intersection at "C")*

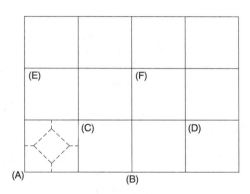

Figure 4–109

Specify second point of displacement or <use first point as displacement>: *(Select the intersection at "D")*
Specify second point of displacement or <use first point as displacement>: *(Select the endpoint at "E")*
Specify second point of displacement or <use first point as displacement>: *(Select the intersection at "F")*
Specify second point of displacement or <use first point as displacement>: *(Press ENTER to exit this command)*

STEP 7

The diamond pattern of the Inlay floor tile should be similar to Figure 4–110.

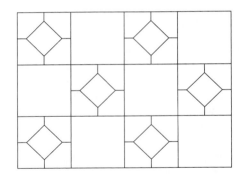

Figure 4–110

STEP 8

Draw one 12" × 12" square shape in the position shown in Figure 4–111. Use the ZOOM command along with the Window option to magnify the area around the position of the square figure.

Command: **Z** *(For ZOOM)*
Specify corner of window, enter a scale factor (nX or nXP), or
[All/Center/Dynamic/Extents/Previous/Scale/Window] <real time>: *(Pick a point at "A")*
Specify opposite corner: *(Pick a point at "B")*

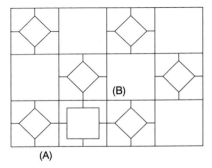

Figure 4–111

STEP 9

Be sure the Grid and Snap values are set to 3" and turn the snap back on by pressing F9. Also, turn off Running Osnap mode by either clicking on OSNAP in the status bar or pressing F3. Then use the LINE command to draw the four lines shown in Figure 4–112. Use "A," "B," "C," and "D" as the starting points for the four lines.

Command: *(Press F3 to turn OSNAP off)*
Command: *(Press F9 to turn SNAP on)*
Command: **L** *(For LINE)*
Specify first point: *(Pick a point at "A")*
Specify next point or [Undo]: *(Pick a point at "B")*
Specify next point or [Undo]: *(Press ENTER to exit this command)*

Repeat this procedure for the other three lines at "C", "D", and "E".

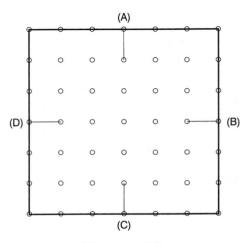

Figure 4–112

STEP 10

Use the RECTANG command to construct the square-shaped object in Figure 4–113. Be sure Snap mode is turned on.

Command: *(Press F3 to turn OSNAP back on)*
Command: **REC** *(For RECTANG)*
Specify first corner point or [Chamfer/ Elevation/Fillet/Thickness/Width]: *(Pick a point at "A")*
Specify other corner point: *(Pick a point at "B")*
Command: *(Press F7 to turn the GRID off and F9 to turn the SNAP off)*

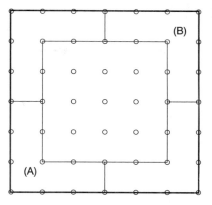

Figure 4–113

STEP 11

Return to the previous display using the ZOOM command and the Previous option. Use the COPY command and the Multiple option to repeat the square-shaped pattern in Figure 4–114.

Command: **Z** *(For ZOOM)*
Specify corner of window, enter a scale factor (nX or nXP), or
[All/Center/Dynamic/Extents/Previous/ Scale/Window] <real time>: **P** *(For Previous)*
Command: **CP** *(For COPY)*
Select objects: Specify opposite corner: *(Select all dashed lines in Figure 4–114)*
Select objects: *(Press ENTER to continue)*
Specify base point or displacement, or [Multiple]: **M** *(For Multiple)*
Specify base point: *(Select the endpoint at "A")*
Specify second point of displacement or <use first point as displacement>: *(Select the endpoint at "B")*
Specify second point of displacement or <use first point as displacement>: *(Select the endpoint at "C")*

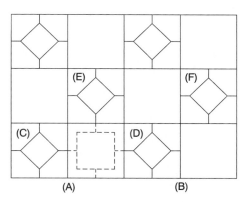

Figure 4–114

Specify second point of displacement or <use first point as displacement>: *(Select the intersection at "D")*
Specify second point of displacement or <use first point as displacement>: *(Select the intersection at "E")*
Specify second point of displacement or <use first point as displacement>: *(Select the intersection at "F")*
Specify second point of displacement or <use first point as displacement>: *(Press ENTER to exit this command)*

tutorial EXERCISE

STEP 12

The drawing of the Inlay should appear similar to Figure 4–115.

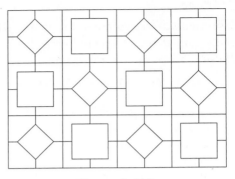

Figure 4–115

STEP 13

Use the OFFSET command to copy the outline of the Inlay outward at a distance of six units as in Figure 4–116.

Command: **O** *(For OFFSET)*
Specify offset distance or [Through] <Through>: **6**
Select object to offset or <exit>: *(Select line "A")*
Specify point on side to offset: *(Pick a point above the line)*
Select object to offset or <exit>: *(Select line "B")*
Specify point on side to offset: *(Pick a point to the left of the line)*
Select object to offset or <exit>: *(Select line "C")*
Specify point on side to offset: *(Pick a point below the line)*
Select object to offset or <exit>: *(Select line "D")*
Specify point on side to offset: *(Pick a point to the right of the line)*
Select object to offset or <exit>: *(Press ENTER to exit this command)*

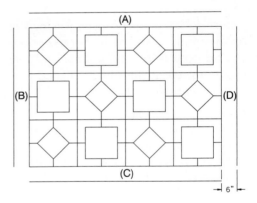

Figure 4–116

STEP 14

Use the FILLET command set to a radius of "0" to place a corner at the intersection of lines "A" and "B." See Figure 4–117.

Command: **F** *(For FILLET)*
Current settings: Mode = TRIM, Radius =
   0'-0 1/2"
Select first object or [Polyline/Radius/
   Trim]: **R** *(For Radius)*
Specify fillet radius <0'-0 1/2">: **0**
Command: **F** *(For FILLET)*
Current settings: Mode = TRIM, Radius =
   0'-0"
Select first object or [Polyline/Radius/
   Trim]: *(Select line "A" in Figure 4–117)*
Select second object: *(Select line "B")*
Repeat this procedure for the remaining three corners.

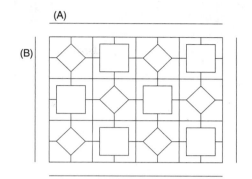

Figure 4–117

STEP 15

Perform a ZOOM-All. The completed problem is shown in Figure 4–118. Dimensions may be added upon the request of your instructor.

Command: **Z** *(For ZOOM)*
Specify corner of window, enter a scale
   factor (nX or nXP), or
[All/Center/Dynamic/Extents/Previous/
   Scale/Window] <real time>: **A** *(For All)*

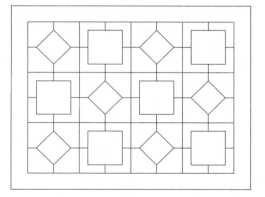

Figure 4–118

TUTORIAL EXERCISE: CLUTCH.DWG

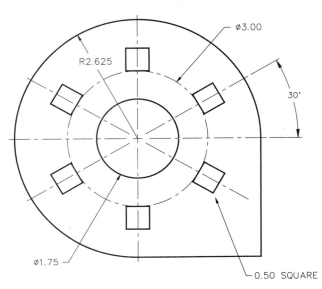

Figure 4–119

Purpose

This tutorial is designed to allow you to construct a one-view drawing of the Clutch shown in Figure 4–119 using a combination of coordinate modes and the ARRAY command.

System Settings

Use the current default settings for the units and limits of this drawing, (0,0) for the lower left corner and (12,9) for the upper right corner. Check to see that the following Object Snap modes are already set: Endpoint, Extension, Intersection, Center and Quadrant.

Layers

Create the following layers with the format:

Name	Color	Linetype
Object	White	Continuous
Center	Yellow	Center

Suggested Commands

Draw the basic shape of the object using the LINE and CIRCLE commands. Lay out a centerline circle, draw one square shape, and use ARRAY to create a multiple copy of the square in a circular pattern.

Whenever possible, substitute the appropriate command alias in place of the full AutoCAD command in each tutorial step. For example, use "CP" for the COPY command, "L" for the LINE command, and so on. The complete listing of all command aliases is located in Chapter 1, Table 1–2.

STEP 1

Check that the current layer is set to "Object". Use the Layer Control box in Figure 4–120 to accomplish this task. Begin drawing the clutch by placing a circle with the center at absolute coordinate (6.00,5.00) and radius of 2.625 units as shown in Figure 4–121. Place another circle using the same center point and a diameter of 1.75 units.

Command: **C** *(For CIRCLE)*
Specify center point for circle or [3P/2P/
   Ttr (tan tan radius)]: **6.00,5.00**
Specify radius of circle or [Diameter]:
   2.625
Command: **C** *(For CIRCLE)*
Specify center point for circle or [3P/2P/
   Ttr (tan tan radius)]: **@**
Specify radius of circle or [Diameter]
   <2.6250>: **D** *(For Diameter)*
Specify diameter of circle <5.2500>: **1.75**

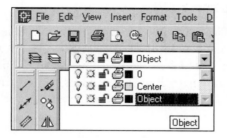

Figure 4–120

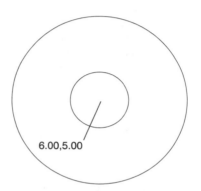

Figure 4–121

STEP 2

Use the Direct Distance mode along with Polar tracking to draw the lower right corner of the Clutch. See Figure 4–122.

 Command: **L** *(For LINE)*
Specify first point: *(Pick the quadrant of the
   circle at "A")*
Specify next point or [Undo]: *(Move your
   cursor over the quadrant at "B" to
   acquire this point. Do not pick this
   location. Slide your cursor to the down
   until it aligns with the previous quadrant
   location and pick this point at "C")*
Specify next point or [Undo]: *(Select the
   quadrant or polar intersection at "B")*
Specify next point or [Undo]: *(Press
   ENTER to exit this command)*

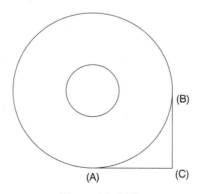

Figure 4–122

STEP 3

Use the TRIM command to partially delete one-fourth of the circle. Press ENTER to select the horizontal and vertical lines shown in Figure 4–123 as the cutting edges, and select the circle at "A" as the object to trim.

Command: **TR** *(For TRIM)*
Current settings: Projection=UCS
   Edge=None
Select cutting edges ...
Select objects: *(Press ENTER which will select all objects cutting edges)*
Select object to trim or [Project/Edge/Undo]: *(Select the circle at "A")*
Select object to trim or [Project/Edge/Undo]: *(Press ENTER to exit this command)*

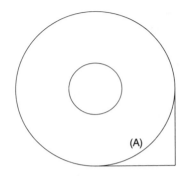

Figure 4–123

STEP 4

Make the "Center" layer current using the Layer Control box in Figure 4–124. Set the DIMCEN variable to a value of -0.09 units. Use DIMCENTER to construct a centerline. This will place a center mark at the center of the circle and extend centerlines just outside the larger circle, as shown in Figure 4–125. Erase the bottom centerline segment. This line will be placed in a later step.

Command: **DIMCEN**
Enter new value for DIMCEN <0.0900>:
   -0.09
Command: **DCE** *(For DIMCENTER)*
Select arc or circle: *(Select the edge of the arc at "A")*
Command: **E** *(For ERASE)*
Select objects: *(Select the bottom centerline at "B")*
Select objects: *(Press ENTER to perform the erase operation)*

DIMCEN
DCE/ DIMCENTER

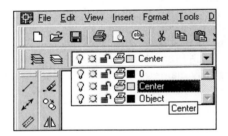

Figure 4–124

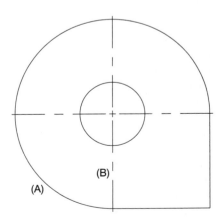

Figure 4–125

STEP 5

While in the "Center" layer, construct the 3.00 diameter centerline circle shown in Figure 4–126.

Command: **C** *(For CIRCLE)*
Specify center point for circle or [3P/2P/ Ttr (tan tan radius)]: **6.00,5.00**
Specify radius of circle or [Diameter] <0.8750>: **D** *(For Diameter)*
Specify diameter of circle <1.7500>: **3.00**

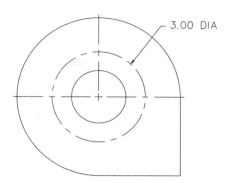

3.00 DIA

Figure 4–126

STEP 6

Set the current layer back to "Object" (see Figure 4–120). Then use the ZOOM command to magnify the upper portion of the clutch shown in Figure 4–127 for constructing a square in the next step.

Command: **Z** *(For ZOOM)*
Specify corner of window, enter a scale factor (nX or nXP), or
[All/Center/Dynamic/Extents/Previous/ Scale/Window] <real time>: *(Pick a point at "A')*
Specify opposite corner: *(Pick a point at "B")*

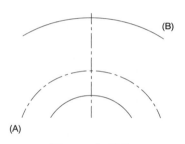

(B)

(A)

Figure 4–127

STEP 7

Construct the 0.50-unit square using the RECTANG command with the assistance of the From and Intersection OSNAP options (See Figure 4–128).

Command: **REC** *(For RECTANG)*
Specify first corner point or [Chamfer/ Elevation/Fillet/Thickness/Width]:
From
Base point: *(Select the intersection at "A" in Figure 4–128)*
<Offset>: **@0.25<180**
Specify other corner point: **@0.50,0.50**
Perform a ZOOM-Previous operation to return to the original display.

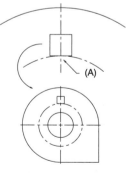

(A)

Figure 4–128

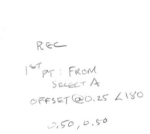

REC
1ST PT : FROM
SELECT A
OFFSET @0.25 ∠180
0.50,0.50

Command: **Z** *(For ZOOM)*
Specify corner of window, enter a scale factor (nX or nXP), or
[All/Center/Dynamic/Extents/Previous/ Scale/Window] <real time>: **P** *(For Previous)*

tutorial **EXERCISE**

STEP 8

Use the ARRAY command to copy the square and vertical centerline in a circular pattern five additional times. Use the following prompts to perform this operation and complete the clutch, as in Figure 4–129.

Command: **AR** *(For ARRAY)*
Select objects: *(Select the square and vertical centerline at "A")*
Select objects: *(Press* ENTER *to continue)*
Enter the type of array [Rectangular/ Polar] <R>: **P** *(For Polar)*
Specify center point of array: *(Select the intersection at "B")*
Enter the number of items in the array: **6**
Specify the angle to fill (+=ccw, -=cw) <360>: *(Press* ENTER *to accept this default value)*
Rotate arrayed objects? [Yes/No] <Y>: *(Press* ENTER *to perform the array operation)*

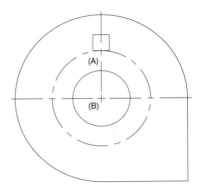

Figure 4–129

STEP 9

The completed problem is shown in Figure 4–130. Dimensions may be added at a later date.

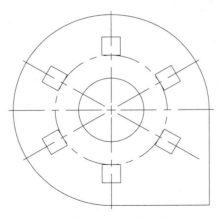

Figure 4–130

TUTORIAL EXERCISE: ANGLE.DWG

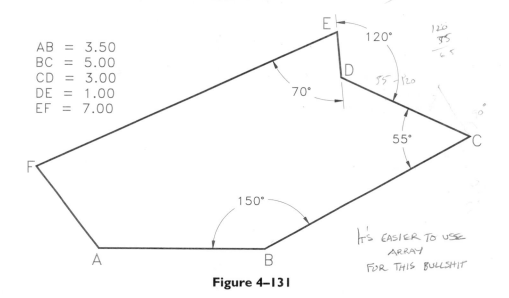

AB = 3.50
BC = 5.00
CD = 3.00
DE = 1.00
EF = 7.00

Figure 4–131

Purpose

This tutorial is designed to allow you to construct a one-view drawing of the angle shown in Figure 4–131 using the ARRAY and LENGTHEN commands.

System Settings

Use the Drawing Units dialog box and change the precision of decimal units from 4 to 2 places. Use the current default settings for the limits of this drawing, (0,0) for the lower left corner and (12,9) for the upper right corner. Check to see that the following Object Snap modes are already set: Endpoint, Extension, Intersection, Center.

Layers

Create the following layer with the format:

Name	Color	Linetype
Object	White	Continuous

Suggested Commands

Make the "Object" layer current. Begin this drawing by constructing line AB, which is horizontal. Use the ARRAY command to copy and rotate line AB at an angle of 150° in the clockwise direction. Once the line is copied, use the LENGTHEN command and modify the line to the proper length. Repeat this procedure for lines CD, DE, and EF. Complete the drawing by constructing a line segment from the endpoint at vertex "F" to the endpoint at vertex "A."

Whenever possible, substitute the appropriate command alias in place of the full AutoCAD command in each tutorial step. For example, use "CP" for the COPY command, "L" for the LINE command, and so on. The complete listing of all command aliases is located in Chapter 1, Table 1–2.

STEP 1

Draw line AB using the Polar coordinate or Direct Distance mode, as shown in Figure 4–132. (Note: Line AB is considered a horizontal line.)

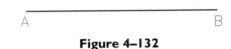

A   B

Figure 4–132

Command: **L** *(For LINE)*
Specify first point: *2,1*
Specify next point or [Undo]: *(Move your cursor directly to the right of the last point and enter a value of 3.50)*
Specify next point or [Undo]: *(Press ENTER to exit this command)*

STEP 2

One technique of constructing the adjacent line at 150° from line AB is to use the ARRAY command. Since the line needs to be arrayed in the clockwise direction, enter a value of -150° for the angle to fill.

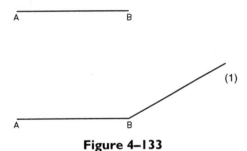

Figure 4–133

Command: **AR** *(For ARRAY)*
Select objects: *(Select line AB in Figure 4–133)*
Select objects: *(Press ENTER to continue)*
Enter the type of array [Rectangular/Polar] <R>: **P** *(For Polar)*
Specify center point of array: *(Pick the endpoint of the line at "B")*
Enter the number of items in the array: **2**
Specify the angle to fill (+=ccw, -=cw) <360>: **-150**
Rotate arrayed objects? [Yes/No] <Y>: *(Press ENTER to perform the array operation)*

The result is shown in Figure 4–133.

STEP 3

The Array operation allowed line AB to be rotated and copied at the correct angle, namely -150°. However, the new line is the same length as line AB. Use the LENGTHEN command to increase the length of the new line to a distance of 5.00 units.

Command: **LEN** *(For LENGTHEN)*
Select an object or [DElta/Percent/Total/ DYnamic]: **T** *(For Total)*
Specify total length or [Angle] <1.00)>: **5.00**
Select an object to change or [Undo]: *(Pick the end of the line at "1")*
Select an object to change or [Undo]: *(Press ENTER to exit this command)*

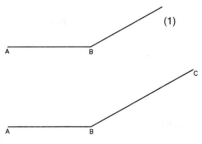

Figure 4–134

The result is shown in Figure 4–134.

STEP 4

Use the ARRAY command to rotate and copy line BC at an angle of -55°, since the direction it is being copied is in the clockwise direction. Then use the LENGTHEN command to reduce the length of the new line from 5.00 units to 3.00 units.

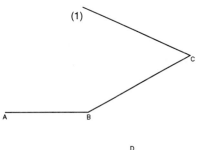

Figure 4–135

Command: **AR** *(For ARRAY)*
Select objects: *(Select line BC in Figure 4–135)*
Select objects: *(Press ENTER to continue)*
Enter the type of array [Rectangular/ Polar] <P>: *(Press ENTER to accept the Polar default setting)*
Specify center point of array: *(Pick the endpoint of the line at "C")*
Enter the number of items in the array: **2**
Specify the angle to fill (+=ccw, -=cw) <360>: **-55**
Rotate arrayed objects? [Yes/No] <Y>: *(Press ENTER to perform the array operation)*

Command: **LEN** *(For LENGTHEN)*
Select an object or [DElta/Percent/Total/ DYnamic]: **T** *(For Total)*
Specify total length or [Angle] <5.00)>: **3.00**
Select an object to change or [Undo]: *(Pick the end of the line at "1")*
Select an object to change or [Undo]: *(Press ENTER to exit this command)*

The result is shown in Figure 4–135.

STEP 5

Use the ARRAY command to rotate and copy line CD at an angle of 120°. A positive angle is entered because it is being copied in the counterclockwise direction. Then, use the LENGTHEN command to reduce the length of the new line from 3.00 units to 1.00 unit.

Command: **AR** *(For ARRAY)*
Select objects: *(Select line CD in Figure 4–136)*
Select objects: *(Press ENTER to continue)*
Enter the type of array [Rectangular/ Polar] <P>: *(Press ENTER to accept the Polar default setting)*
Specify center point of array: *(Pick the endpoint of the line at "D")*
Enter the number of items in the array: **2**
Specify the angle to fill (+=ccw, -=cw) <360>: **120**
Rotate arrayed objects? [Yes/No] <Y>: *(Press ENTER to perform the array operation)*
Command: **LEN** *(For LENGTHEN)*
Select an object or [DElta/Percent/Total/ DYnamic]: **T** *(For Total)*
Specify total length or [Angle] <3.00)>: **1.00**
Select an object to change or [Undo]: *(Pick the end of the line at "1")*
Select an object to change or [Undo]: *(Press ENTER to exit this command)*

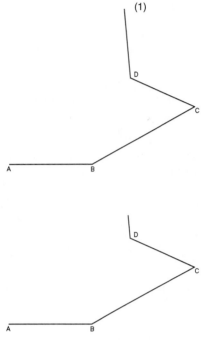

(1)

Figure 4–136

The result is shown in Figure 4–136.

STEP 6

Use the ARRAY command to rotate and copy line DE at an angle of -70° since the direction it is being copied is again in the clockwise direction. Then use the LENGTHEN command to increase the length of the new line from 1.00 unit to 7.00 units.

Command: **AR** *(For ARRAY)*
Select objects: *(Select line DE in Figure 4–137)*
Select objects: *(Press ENTER to continue)*
Enter the type of array [Rectangular/Polar] <P>: *(Press ENTER to accept the Polar default setting)*
Specify center point of array: *(Pick the endpoint of the line at "E")*
Enter the number of items in the array: **2**
Specify the angle to fill (+=ccw, -=cw) <360>: **-70**
Rotate arrayed objects? [Yes/No] <Y>: *(Press ENTER to perform the array operation)*
Command: **LEN** *(For LENGTHEN)*
Select an object or [DElta/Percent/Total/DYnamic]: **T** *(For Total)*
Specify total length or [Angle] <1.00)>: **7.00**
Select an object to change or [Undo]: *(Pick the end of the line at "1")*

Select an object to change or [Undo]: *(Press ENTER to exit this command)*

The result is shown in Figure 4–137.

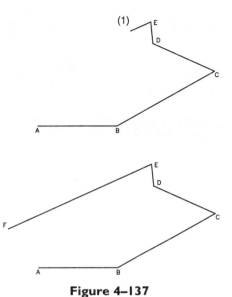

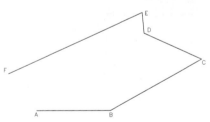

Figure 4–137

STEP 7

Connect endpoints "F" and "A" with a line as shown in Figure 4–138A.

Command: **L** *(For LINE)*
Specify first point: *(Pick the endpoint of the line at "F")*
Specify next point or [Undo]: *(Pick the endpoint of the line at "A")*
Specify next point or [Undo]: *(Press ENTER to exit this command)*

The completed drawing is illustrated in Figure 4–138B. You may add dimensions at a later date.

Figure 4–138A

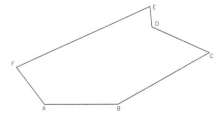

Figure 4–138B

PROBLEMS FOR CHAPTER 4

Directions for Problems 4–1 through 4–13:

Construct each one-view drawing using the appropriate coordinate mode or Direct Distance mode. Utilize advanced commands such as ARRAY and MIRROR whenever possible.

PROBLEM 4–1

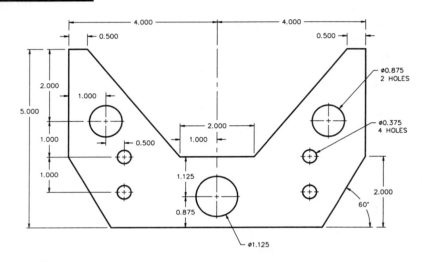

PROBLEM 4–2

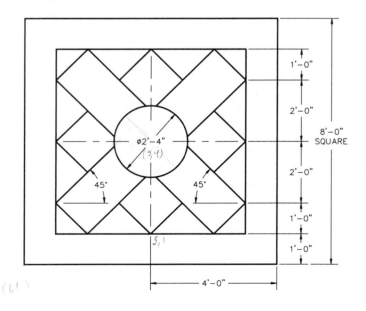

PROBLEM 4–3

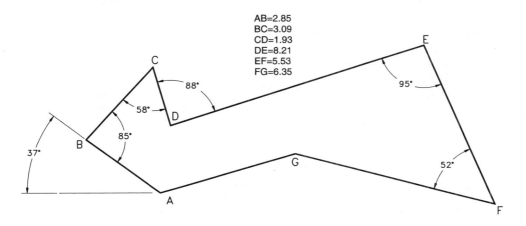

AB=2.85
BC=3.09
CD=1.93
DE=8.21
EF=5.53
FG=6.35

PROBLEM 4–4

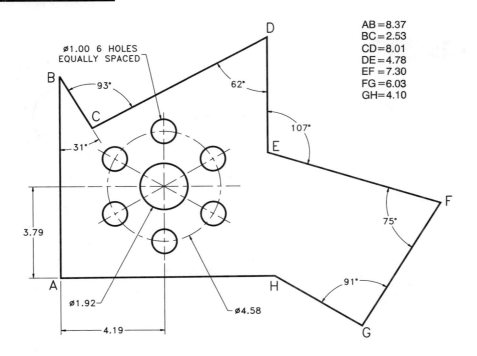

AB =8.37
BC =2.53
CD =8.01
DE =4.78
EF =7.30
FG =6.03
GH=4.10

Ø1.00 6 HOLES
EQUALLY SPACED

93°
62°
31°
107°
75°
91°

3.79
4.19

Ø1.92
Ø4.58

PROBLEM 4–5

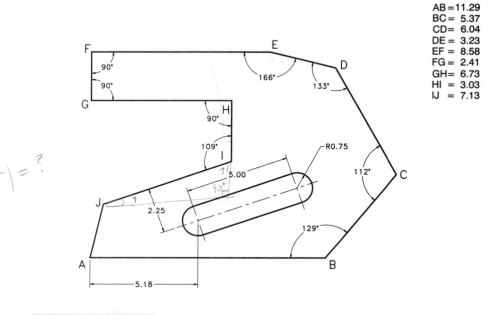

AB = 11.29
BC = 5.37
CD = 6.04
DE = 3.23
EF = 8.58
FG = 2.41
GH = 6.73
HI = 3.03
IJ = 7.13

$|\overline{AJ}| = ?$

PROBLEM 4–6

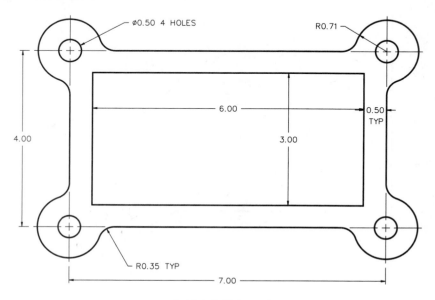

ø0.50 4 HOLES

R0.71

R0.35 TYP

6.00

3.00

0.50 TYP

4.00

7.00

0.125 GASKET THICKNESS

PROBLEM 4-7

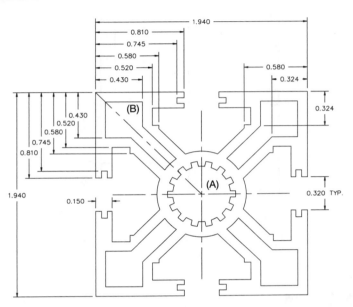

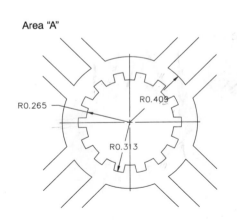

Area "A"

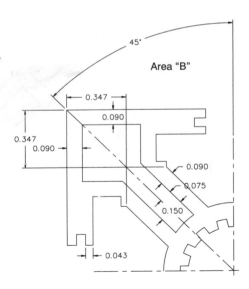

Area "B"

PROBLEM 4–8

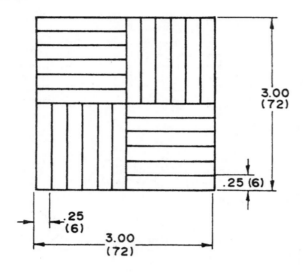

3.00
(72)

.25 (6)

.25
(6)

3.00
(72)

PROBLEM 4–9

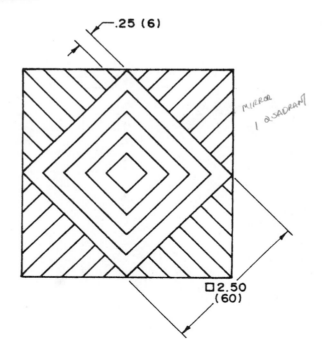

.25 (6)

MIRROR
1 QUADRANT

□2.50
(60)

PROBLEM 4–10

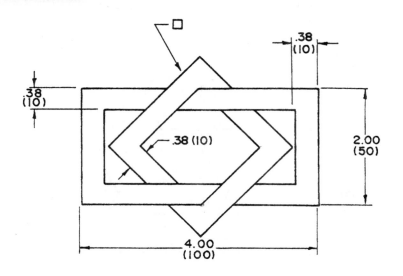

PROBLEM 4–11 *R*

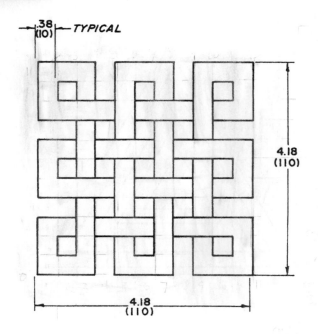

COUNT EDGES

PROBLEM 4-12

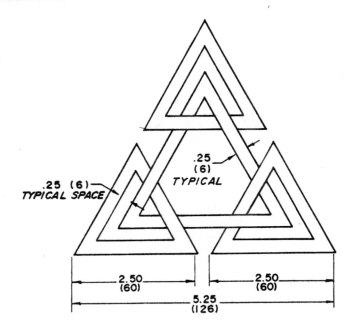

.25
(6)
TYPICAL

.25 (6)
TYPICAL SPACE

2.50
(60)

2.50
(60)

5.25
(126)

PROBLEM 4-13

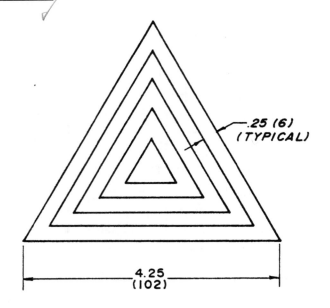

.25 (6)
(TYPICAL)

4.25
(102)

Directions for Problems 4–14 through 4–17:

Construct one-view drawings of the following figures using a grid spacing of 0.25 units. Do not dimension these drawings unless otherwise specified by your instructor.

PROBLEM 4–14

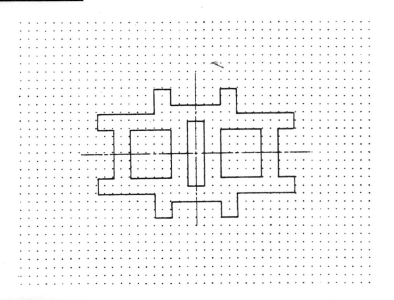

PROBLEM 4–15

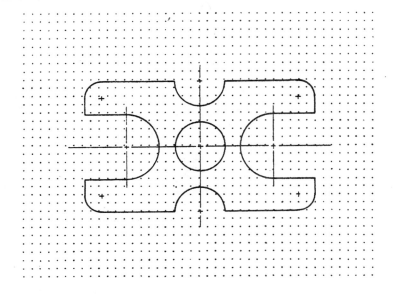

PROBLEM 4–16

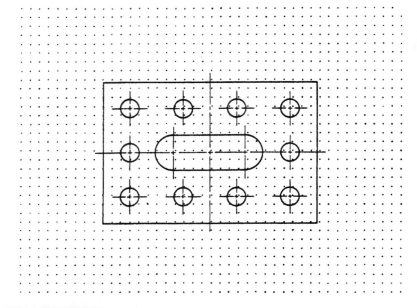

PROBLEM 4–17

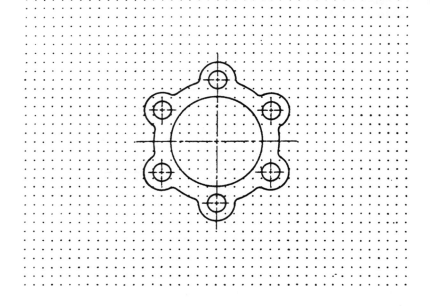

CHAPTER 5

Performing Geometric Constructions

METHODS OF SELECTING OTHER DRAW COMMANDS

This chapter introduces the commands used for object creation. The following commands will be explained in this section:

ARC	ELLIPSE	MEASURE
BOUNDARY	MLINE	POLYGON
CIRCLE-2P	MLSTYLE	RAY
CIRCLE-3P	MLEDIT	RECTANG
CIRCLE-TTR	POINT	SPLINE
DONUT	DIVIDE	XLINE

The LINE and CIRCLE commands have already been discussed in Chapter 1.

Most drawing commands can be found on the Draw menu, shown in Figure 5–1. Arrowheads displayed to the right of the command indicate a cascading menu that holds additional options of the main command.

Figure 5–2 illustrates the Draw toolbar, which holds most drawing commands.

You can also enter most drawing commands directly from the keyboard by using their entire name, such as POINT for the POINT command. The following commands may be entered with only the first letter of the command as part of AutoCAD's command aliasing feature:

> Enter A for the ARC command
>
> Enter C for the CIRCLE command
>
> Enter L for the LINE command

See Chapter 1, Table 1–2 for the complete listing of all command aliases supported in AutoCAD 2000.

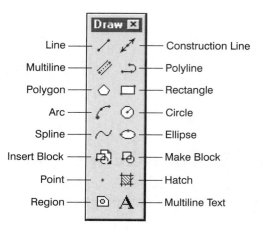

Figure 5–1

Figure 5–2

THE ARC COMMAND

Choosing Arc from the Draw pull-down menu displays the cascading menu shown in Figure 5–3. All supported methods of constructing arcs are displayed in the list. By default, the 3 Points Arc mode supports arc constructions in the clockwise as well as the counterclockwise direction. All other arc modes support the ability to construct arcs only in the counterclockwise direction. The following pages detail most of the arc modes labeled in Figure 5–3.

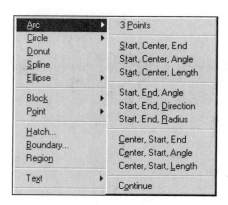

Figure 5–3

3 POINTS ARC MODE

By default, arcs are drawn with the 3 Points method. The first and third points identify the endpoints of the arc. This arc may be drawn in either the clockwise or counterclockwise direction. Use the following command sequence along with Figure 5–4 to construct a 3-point arc.

Command: **A** *(For ARC)*
Specify start point of arc or [CEnter]: *(Pick a point at "A")*
Specify second point of arc or [CEnter/ENd]: *(Pick a point at "B")*
Specify end point of arc: *(Pick a point at "C")*

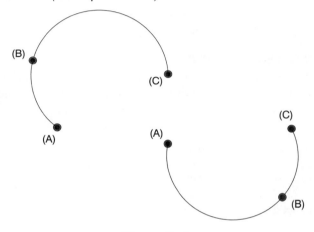

Figure 5–4

START, CENTER, END MODE

Use this ARC mode to construct an arc by defining its start point, center point, and endpoint. This arc will always be constructed in a counterclockwise direction. Use the following command sequence along with Figure 5–5 for constructing an arc by start, center, and endpoints.

Command: **A** *(For ARC)*
Specify start point of arc or [CEnter]: *(Pick a point at "A")*
Specify second point of arc or [CEnter/ENd]: **CE** *(For Center)*
Specify center point of arc: *(Pick a point at "B")*
Specify end point of arc or [Angle/chord Length]: *(Pick a point at "C")*

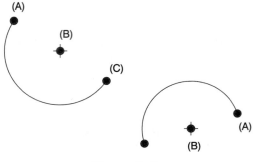

Figure 5–5

START, CENTER, ANGLE MODE

Use this mode to construct an arc by start point, center point, and included angle. If the angle is positive, the arc is drawn in the counterclockwise direction; a negative angle constructs the arc in a clockwise direction. See Figure 5–6.

Command: **A** *(For ARC)*
Specify start point of arc or [CEnter]: *(Pick a point at "A")*
Specify second point of arc or [CEnter/ENd]: **CE** *(For Center)*
Specify center point of arc: *(Pick a point at "B")*
Specify end point of arc or [Angle/chord Length]: **A** *(For Angle)*
Specify included angle: **135**

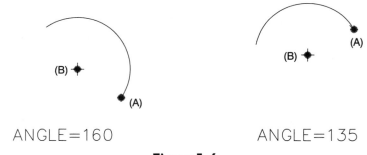

ANGLE=160   ANGLE=135

Figure 5–6

START, CENTER, LENGTH MODE

Use this mode to construct an arc by start point, center point, and length of chord. Figure 5–7 illustrates the definition of a chord.

Command: **A** *(For ARC)*
Specify start point of arc or [CEnter]: *(Pick a point at "A")*
Specify second point of arc or [CEnter/ENd]: **CE** *(For Center)*
Specify center point of arc: *(Pick a point at "B")*

Specify end point of arc or [Angle/chord Length]: **L** *(For Length)*
Specify length of chord: **2.5**

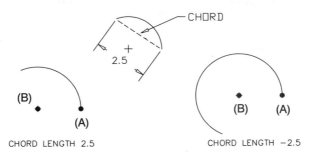

Figure 5–7

START, END, ANGLE MODE

Use this Arc mode to construct an arc by defining its starting point, endpoint, and included angle. This arc is drawn in a counterclockwise direction when a positive angle is entered; if the angle is negative, the arc is drawn clockwise. Use the following command sequence along with Figure 5–8 for constructing an arc by start point, endpoint, and included angle.

Command: **A** *(For ARC)*
Specify start point of arc or [CEnter]: *(Pick a point at "A")*
Specify second point of arc or [CEnter/ENd]: **EN** *(For End)*
Specify end point of arc: *(Pick a point at "B")*
Specify center point of arc or [Angle/Direction/Radius]: **A** *(For Angle)*
Specify included angle: **90**

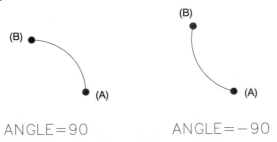

Figure 5–8

START, END, DIRECTION MODE

Use this mode to construct an arc in a specified direction. This mode is especially helpful for drawing arcs tangent to other objects. Use the following prompt sequence along with Figure 5–9 for constructing an arc by direction.

Command: **A** *(For ARC)*
Specify start point of arc or [CEnter]: *(Pick a point at "A")*
Specify second point of arc or [CEnter/ENd]: **EN** *(For End)*
Specify end point of arc: *(Pick a point at "B")*
Specify center point of arc or [Angle/Direction/Radius]: **D** *(For Direction)*
Specify tangent direction for the start point of arc: **@1<90**

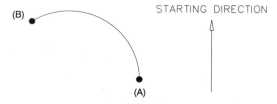

Figure 5–9

START, END, RADIUS MODE

Use this mode to construct an arc by start point, endpoint, and radius. A positive radius draws a minor arc; a negative radius draws a major arc. See Figure 5–10.

Command: **A** *(For ARC)*
Specify start point of arc or [CEnter]: *(Pick a point at "A")*
Specify second point of arc or [CEnter/ENd]: **EN** *(For End)*
Specify end point of arc: *(Pick a point at "B")*
Specify center point of arc or [Angle/Direction/Radius]: **R** *(For Radius)*
Specify radius of arc: **1.00**

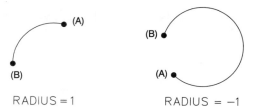

RADIUS = 1   RADIUS = −1

Figure 5–10

CENTER, START, END MODE

Use this mode to construct an arc by first locating the center point of the arc and then locating the start point and endpoint; this type of arc is constructed in the counter-clockwise direction. See Figure 5–11.

Command: **A** *(For ARC)*
Specify start point of arc or [CEnter]: **CE** *(For Center)*
Specify center point of arc: *(Pick a point at "A")*
Specify start point of arc: *(Pick a point at "B")*
Specify end point of arc or [Angle/chord Length]: *(Pick a point at "C")*

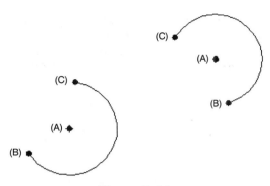

Figure 5–11

CENTER, START, ANGLE MODE

Use this mode to construct an arc by first locating the center point of the arc and then locating the start point and included angle. If the angle is positive, the arc is drawn in the counterclockwise direction; a negative angle constructs the arc in a clockwise direction. See Figure 5–12.

Command: **A** *(For ARC)*
Specify start point of arc or [CEnter]: **CE** *(For Center)*
Specify center point of arc: *(Pick a point at "A")*
Specify start point of arc: *(Pick a point at "B")*
Specify end point of arc or [Angle/chord Length]: **A** *(For Angle)*
Specify included angle: **135**

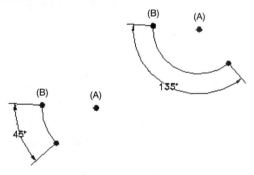

Figure 5–12

CONTINUE MODE

Use this mode to continue a previously drawn arc. All arcs drawn through Continue mode are automatically constructed tangent to the previous arc. See Figure 5–13.

(Choose Continue from the Arc menu in Figure 5–3)
Command: **A** *(For ARC)*

Specify start point of arc or [CEnter]: *(The Continue mode is automatically activated)*
Specify end point of arc: *(Pick an ending point for the continued arc)*

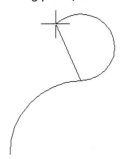

Figure 5–13

ADVANCED POLYLINE CONSTRUCTIONS
THE BOUNDARY COMMAND

The BOUNDARY command is used to create a polyline boundary around any closed shape. It has already been demonstrated that the Join option of the PEDIT command is used to join object segments into one continuous polyline. The BOUNDARY command automates this process even more. Start this command by choosing Boundary from the Draw pull-down menu, as shown in Figure 5–14A. This activates the Boundary Creation dialog box, shown in Figure 5–14B.

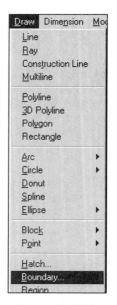

Figure 5–14A

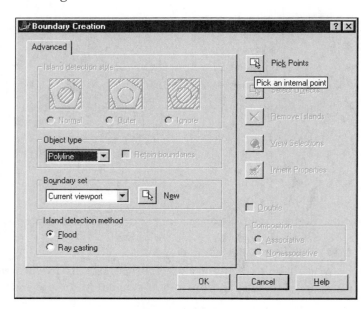

Figure 5–14B

Before you use this command, it is considered good practice to create a separate layer to hold the polyline object; this layer could be called "Boundary" or "BP" for Boundary Polyline. Unlike the Join option of the PEDIT command, which converts individual objects to polyline objects, the BOUNDARY command will trace a polyline in the current layer on top of individual objects. Click on the Pick Points button (see Figure 5–14B). Then pick a point inside the object illustrated in Figure 5–15. Notice how the entire object highlights. To complete the command, press ENTER when prompted to select another internal point, and the polyline will be traced over the top of the existing objects.

Command: **BOUNDARY**
(The dialog box in Figure 5–14B appears. Click on the Pick Points button.)
Select internal point: *(Pick a point at "A" in Figure 5–15)*
Selecting everything...
Selecting everything visible...
Analyzing the selected data...
Analyzing internal islands...
Select internal point: *(Press ENTER to construct the boundary)*
BOUNDARY created 1 polyline

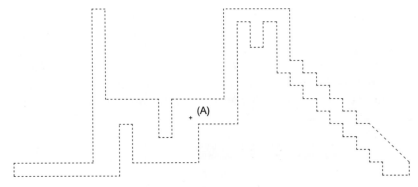

(A)

Figure 5–15

Once the boundary polyline is created, the boundary may be relocated to a different position on the screen with the MOVE command. The results are illustrated in Figure 5–16. The top object in the figure consists of the original individual objects; when the bottom object is selected, all objects highlight, signifying that it is made up of a polyline object made through the use of the BOUNDARY command.

When the BOUNDARY command is used on an object consisting of an outline and internal islands similar to the drawing in Figure 5–17, a polyline object is also traced over these internal islands.

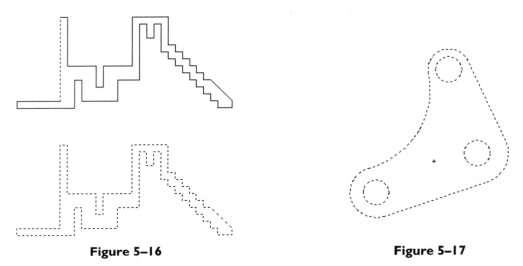

Figure 5–16

Figure 5–17

In addition to creating a special layer to hold the boundary polyline, another important rule to follow when using the BOUNDARY command is to be sure there are no gaps in the object. In Figure 5–18, it is acceptable to have lines cross at "A"; however, when the BOUNDARY command encounters the gap at "B," a dialog box displays informing you that no internal boundary could be found because the object is not completely closed. In this case, you must exit the command, close the object, and activate the BOUNDARY command again.

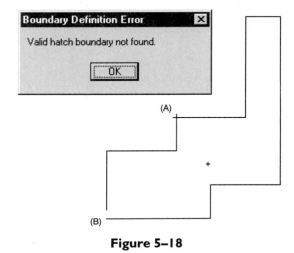

Figure 5–18

MORE OPTIONS OF THE CIRCLE COMMAND

Additional options of the CIRCLE command may be selected from the Circle cascading menu in Figure 5–19. The 2 Points, 3 Points, Tan Tan Radius, and Tan Tan Tan modes will be explained in the next series of examples.

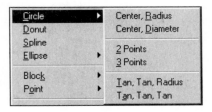

Figure 5–19

3 POINTS CIRCLE MODE

Use the CIRCLE command and the 3 Points mode to construct a circle by three points that you identify. No center point is required when you enter the 3 Points mode. Simply select three points and the circle is drawn. Study the following prompts and Figure 5–20 for constructing a circle using the 3 Points mode.

Command: **C** *(For CIRCLE)*
Specify center point for circle or [3P/2P/Ttr (tan tan radius)]: **3P**
Specify first point on circle: *(Pick a point at "A")*
Specify second point on circle: *(Pick a point at "B")*
Specify third point on circle: *(Pick a point at "C")*

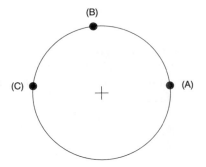

Figure 5–20

2 POINTS CIRCLE MODE

Use the CIRCLE command and the 2 Points mode to construct a circle by selecting two points. These points will form the diameter of the circle. No center point is required when you use the 2 Points mode. Study the following prompts and Figure 5–21 for constructing a circle using the 2 Points mode.

Command: **C** *(For CIRCLE)*
Specify center point for circle or [3P/2P/Ttr (tan tan radius)]: **2P**
Specify first end point of circle's diameter: *(Pick a point at "A")*
Specify second end point of circle's diameter: *(Pick a point at "B")*

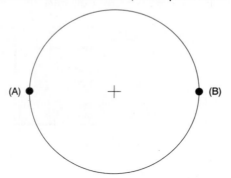

Figure 5–21

CONSTRUCTING AN ARC TANGENT TO TWO LINES USING CIRCLE-TTR

Illustrated in Figure 5–22A are two inclined lines. The purpose of this example is to connect an arc tangent to the two lines at a specified radius. The CIRCLE-TTR (Tangent-Tangent-Radius) command will be used here along with the TRIM command to clean up the excess geometry. To assist with this operation, the OSNAP-Tangent mode is automatically activated when you use the TTR option of the CIRCLE command.

First, use the CIRCLE-TTR command to construct an arc tangent to both lines, as shown in Figure 5–22B.

Command: **C** *(For CIRCLE)*
Specify center point for circle or [3P/2P/Ttr (tan tan radius)]: **TTR**
Specify point on object for first tangent of circle: *(Select the line at "A")*
Specify point on object for second tangent of circle: *(Select the line at "B")*
Specify radius of circle: *(Enter a radius value)*

Figure 5–22A

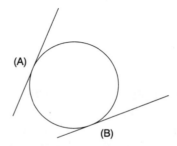

Figure 5–22B

Use the TRIM command to clean up the lines and arc. The completed result is illustrated in Figure 5–22C. The FILLET command could also have been used for this procedure. Not only will the curve be drawn, but also this command will automatically trim the lines.

The object in Figure 5–22D is an example of a typical application where this procedure might be used.

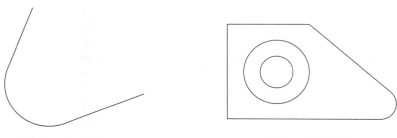

Figure 5–22C   **Figure 5–22D**

CONSTRUCTING AN ARC TANGENT TO A LINE AND ARC USING CIRCLE-TTR

Illustrated in Figure 5–23A are an arc and an inclined line. The purpose of this example is to connect an additional arc tangent to the original arc and line at a specified radius. The CIRCLE-TTR command will be used here along with the TRIM command to clean up the excess geometry.

First, use the CIRCLE-TTR command to construct an arc tangent to the arc and inclined line as shown in Figure 5–23B.

Command: **C** *(For CIRCLE)*
Specify center point for circle or [3P/2P/Ttr (tan tan radius)]: **TTR**
Specify point on object for first tangent of circle: *(Select the arc at "A")*
Specify point on object for second tangent of circle: *(Select the line at "B")*
Specify radius of circle: *(Enter a radius value)*

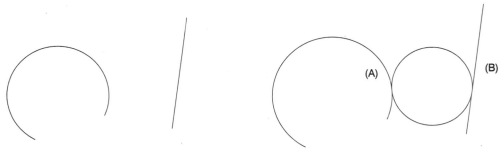

Figure 5–23A   **Figure 5–23B**

Use the TRIM command to clean up the arc and line. The completed result is illustrated in Figure 5–23C.

The object in Figure 5–23D is an example of a typical application where this procedure might be used.

Figure 5–23C   **Figure 5–23D**

CONSTRUCTING AN ARC TANGENT TO TWO ARCS USING CIRCLE-TTR

Method #1

Illustrated in Figure 5–24A are two arcs. The purpose of this example is to connect a third arc tangent to the original two at a specified radius. The CIRCLE-TTR command will be used here along with the TRIM command to clean up the excess geometry.

Use the CIRCLE-TTR command to construct an arc tangent to the two original arcs as shown in Figure 5–24B.

Command: **C** *(For CIRCLE)*
Specify center point for circle or [3P/2P/Ttr (tan tan radius)]: **TTR**
Specify point on object for first tangent of circle: *(Select the arc at "A")*
Specify point on object for second tangent of circle: *(Select the arc at "B")*
Specify radius of circle: *(Enter a radius value)*

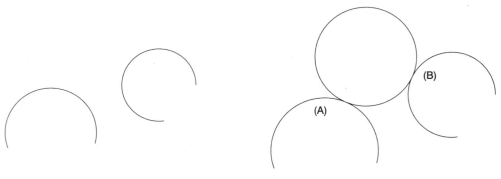

Figure 5–24A   **Figure 5–24B**

Use the TRIM command to clean up the two arcs, using the circle as a cutting edge. The completed result is illustrated in Figure 5–24C.

The object in Figure 5–24D is an example of a typical application where this procedure might be used.

Figure 5–24C   **Figure 5–24D**

CONSTRUCTING AN ARC TANGENT TO TWO ARCS USING CIRCLE-TTR

Method #2

Illustrated in Figure 5–25A are two arcs. The purpose of this example is to connect an additional arc tangent to and enclosing both arcs at a specified radius. The CIRCLE-TTR command will be used here along with the TRIM command.

First, use the CIRCLE-TTR command to construct an arc tangent to and enclosing both arcs, using the indications in Figure 5–25B.

```
Command: C (For CIRCLE)
Specify center point for circle or [3P/2P/Ttr (tan tan radius)]: TTR
Specify point on object for first tangent of circle: (Select the arc at "A")
Specify point on object for second tangent of circle: (Select the arc at "B")
Specify radius of circle: (Enter a radius value)
```

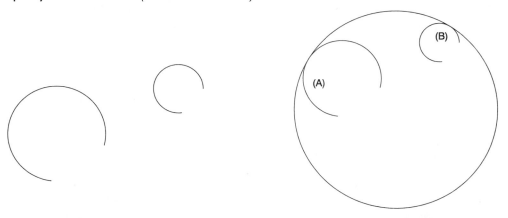

Figure 5–25A   **Figure 5–25B**

Use the TRIM command to clean up all arcs. The completed result is illustrated in Figure 5–25C.

The object in Figure 5–25D is an example of a typical application where this procedure might be used.

Figure 5–25C   Figure 5–25D

CONSTRUCTING AN ARC TANGENT TO TWO ARCS USING CIRCLE-TTR

Method #3

Illustrated in Figure 5–26A are two arcs. The purpose of this example is to connect an additional arc tangent to one arc and enclosing the other. The CIRCLE-TTR command will be used here along with the TRIM command to clean up unnecessary geometry.

First, use the CIRCLE-TTR command to construct an arc tangent to the two arcs. Study the illustration in Figure 5–26B and the following prompts to understand the proper pick points for this operation.

Command: **C** *(For CIRCLE)*
Specify center point for circle or [3P/2P/Ttr (tan tan radius)]: **TTR**
Specify point on object for first tangent of circle: *(Select the arc at "A")*
Specify point on object for second tangent of circle: *(Select the arc at "B")*
Specify radius of circle: *(Enter a radius value)*

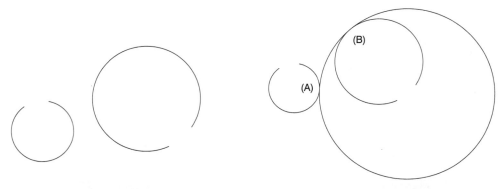

Figure 5–26A   Figure 5–26B

Use the TRIM command to clean up the arcs. The completed result is illustrated in Figure 5–26C.

The object in Figure 5–26D is an example of a typical application where this procedure might be used.

Figure 5–26C

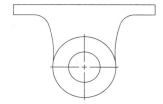

Figure 5–26D

CONSTRUCTING A LINE TANGENT TO TWO ARCS OR CIRCLES

Illustrated in Figure 5–27A are two circles. The purpose of this example is to connect the two circles with two tangent lines. This can be accomplished with the LINE command and the OSNAP-Tangent option.

Use the LINE command to connect two lines tangent to the circles as shown in Figure 5–27B. The following procedure is used for the first line. Use the same procedure for the second.

Command: **L** *(For LINE)*
Specify first point: **Tan**
to *(Select the circle near "A")*
Specify next point or [Undo]: **Tan**
to *(Select the circle near "B")*
Specify next point or [Undo]: *(Press ENTER to exit this command)*

When you use the Tangent option, the rubber-band cursor is not present when you draw the beginning of the line. This is due to calculations required when you identify the second point.

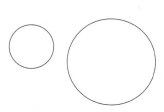

Figure 5–27A

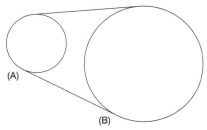

Figure 5–27B

Use the TRIM command to clean up the circles so that the appearance of the object is similar to the illustration in Figure 5–27C.

The object in Figure 5–27D is an example of a typical application where drawing lines tangent to circles might be used.

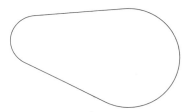

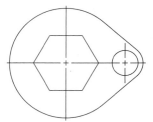

Figure 5–27C   **Figure 5–27D**

CIRCLE – TAN TAN TAN

Yet another mode of the CIRCLE command allows you to construct a circle based on three tangent points. This mode is actually a variation of the 3P mode together with using the OSNAP-Tangent mode three times. Clicking on this mode from the Circle cascading menu in Figure 5–19 automates the process, enabling you to select three objects, as in the example in Figure 5–28A. Study this figure and the following command prompt sequence:

Command: **C** *(For CIRCLE)*
Specify center point for circle or [3P/2P/Ttr (tan tan radius)]: **3p**
Specify first point on circle: **Tan**
to *(Select the line at "A")*
Specify second point on circle: **Tan**
to *(Select the line at "B")*
Specify third point on circle: **Tan**
to *(Select the line at "C")*

The result is illustrated in Figure 5–28B, with the circle being constructed tangent to the edges of all three line segments.

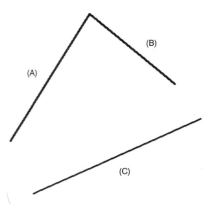

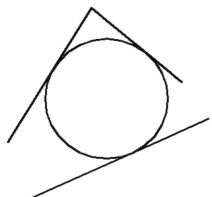

Figure 5–28A   **Figure 5–28B**

QUADRANT VERSUS TANGENT OSNAP OPTION

Various examples have been given on previous pages concerning drawing lines tangent to two circles, two arcs, or any combination of the two. The object in Figure 5–29 illustrates the use of the OSNAP-Tangent option when used along with the LINE command.

Command: **L** *(For LINE)*
Specify first point: **Tan**
to *(Select the arc near "A")*
Specify next point or [Undo]: **Tan**
to *(Select the arc near "B")*
Specify next point or [Undo]: *(Press* ENTER *to exit this command)*

Note that the angle of the line formed by points "A" and "B" is neither horizontal nor vertical. The object in Figure 5–29 is a typical example of the capabilities of the OSNAP-Tangent option.

The object illustrated in Figure 5–30 is a modification of the drawing in Figure 5–29 with the inclined tangent lines changing to horizontal and vertical tangent lines. This example is to inform you that two OSNAP options are available for performing tangencies, namely OSNAP-Tangent and OSNAP-Quadrant. However, it is up to you to evaluate under what conditions to use these OSNAP options. In Figure 5–30, you can use the OSNAP-Tangent or OSNAP-Quadrant option to draw the lines tangent to the arcs. The Quadrant option could be used only because the lines to be drawn are perfectly horizontal or vertical. Usually it is impossible to know this ahead of time, and in this case, the OSNAP-Tangent option should be used whenever possible.

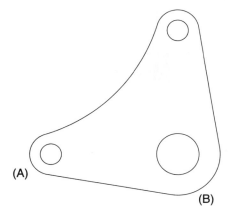

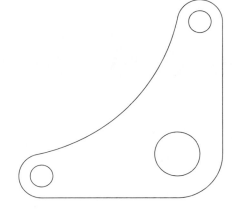

Figure 5–29   **Figure 5–30**

THE DONUT COMMAND

Use the DONUT command to construct a filled-in circle. This object belongs to the polyline family. Figure 5–31 is an example of a donut with an inside diameter of 0.50 units and an outside diameter of 1.00 units. When you place donuts in a drawing, the multiple option is automatically invoked. This means you can place as many donuts as you like until you choose another command from the menu or issue a Cancel by pressing ESC.

Command: **DO** *(For DONUT)*
Specify inside diameter of donut <0.50>: *(Press ENTER to accept the default)*
Specify outside diameter of donut <1.00>: *(Press ENTER to accept the default)*
Specify center of donut or <exit>: *(Pick a point to place the donut)*
Specify center of donut or <exit>: *(Pick a point to place another donut or press ENTER to exit this command)*

Setting the inside diameter of a donut to a value of zero (0) and an outside diameter to any other value constructs a donut representing a dot. See Figure 5–32.

Command: **DO** *(For DONUT)*
Specify inside diameter of donut <0.50>: **0**
Specify outside diameter of donut <1.00>: **0.25**
Specify center of donut or <exit>: *(Pick a point to place the donut)*
Specify center of donut or <exit>: *(Pick a point to place another donut or press ENTER to exit this command)*

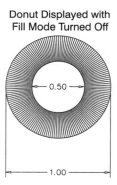

Donut Displayed with
Fill Mode Turned Off

0.50

1.00

Figure 5–31

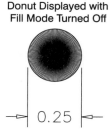

Donut Displayed with
Fill Mode Turned Off

0.25

Figure 5–32

Figures 5–33 and 5–34 show two examples where donuts could be useful; donuts are sometimes used in place of arrows to act as terminators for dimension lines, as shown in Figure 5–33. Figure 5–34 illustrates how donuts might be used to act as connection points in an electrical schematic.

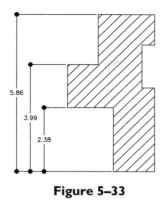

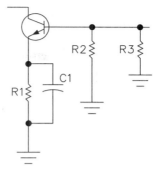

Figure 5–33	**Figure 5–34**

CONSTRUCTING ELLIPTICAL SHAPES
THE ELLIPSE COMMAND

Use the ELLIPSE command to construct a true elliptical shape. Choose Ellipse from the Draw pull-down menu, shown in Figure 5–35.

Figure 5–35

Before studying the three examples for ellipse construction, see Figure 5–36 to view two important parts of any ellipse, namely its major and minor diameters.

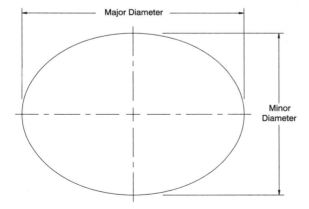

Figure 5–36

You can construct an ellipse by marking two points, which specify one of its axes (see Figure 5–37). These first two points also identify the angle with which the ellipse will be drawn. Responding to the prompt "Specify distance to other axis or [Rotation]" with another point identifies half of the other axis. The rubber-banded line is added to assist you in this ellipse construction method.

Command: **EL** *(For ELLIPSE)*
Specify axis endpoint of ellipse or [Arc/Center]: *(Pick a point at "A")*
Specify other endpoint of axis: *(Pick a point at "B")*
Specify distance to other axis or [Rotation]: *(Pick a point at "C")*

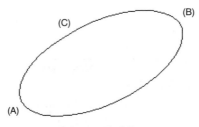

Figure 5–37

You can also construct an ellipse by first identifying its center. You can pick points to identify its axes or use polar coordinates to accurately define the major and minor diameters of the ellipse. See Figure 5–38 and the following command sequence to construct this type of ellipse. Use the Polar coordinate or Direct Distance modes for locating the two axis endpoints of the ellipse.

Command: **EL** *(For ELLIPSE)*
Specify axis endpoint of ellipse or [Arc/Center]: **C** *(For Center)*
Specify center of ellipse: *(Pick a point at "A")*
Specify endpoint of axis: **@2.50<0** *(To point "B")*
Specify distance to other axis or [Rotation]: **@1.50<90** *(To point "C")*

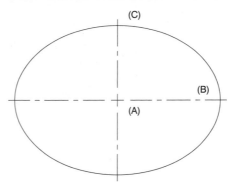

Figure 5–38

This last method, in Figure 5–39, illustrates constructing an ellipse by way of rotation. Identify the first two points for the first axis. Reply to the prompt "Specify distance to other axis or [Rotation]" with Rotation. The first axis defined is now used as an axis of rotation that rotates the ellipse into a third dimension.

Command: **EL** *(For ELLIPSE)*
Specify axis endpoint of ellipse or [Arc/Center]: *(Pick a point at "A")*
Specify other endpoint of axis: *(Pick a point at "B")*
Specify distance to other axis or [Rotation]: **R** *(For Rotation)*
Specify rotation around major axis: **80**

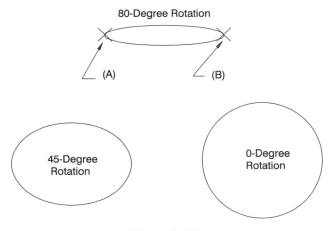

Figure 5–39

MULTILINES

A multiline consists of a series of parallel line segments. You can group as many as 16 multilines together to form a single multiline object. Multilines may take on color, may be spaced apart at different increments, and may take on different linetypes. In Figures 5–40A through 5–40C, the floor plan layout was made with the default multiline style of two parallel line segments. The spacing was changed to an offset distance of 4 units. Once the multilines are laid out as in Figure 5–40A, the Multiline Edit (MLEDIT) command cleans up all corners and intersections, as displayed in Figure 5–40B. Because multilines cannot be partially deleted with the BREAK command, the MLEDIT command allows you to cut between two points that you identify, as in Figure 5–40C. Also, you have the option of breaking all or only one multiline segment depending on your desired results. Once a multiline segment has been cut, it can be welded back together again with the MLEDIT command. The following pages outline the creation of a multiline style with the Multiline Style (MLSTYLE) command. Two additional dialog boxes are contained inside the Multiline Style dialog box: The Element Properties dialog box sets the spacing, color, and linetype of the individual multiline segments. The Multiline Properties dialog box provides other information on multilines, including capping modes and the ability to fill the entire multiline in a selected color.

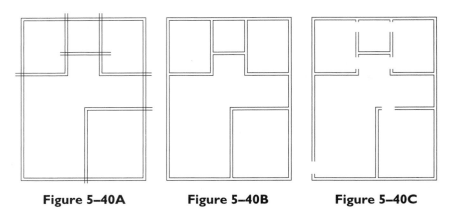

Figure 5–40A   Figure 5–40B   Figure 5–40C

CONSTRUCTING MULTIPLE PARALLEL LINES
THE MLINE COMMAND

Once the element and multiline properties have been set and saved to a multiline style, loading a new current style and clicking the OK button of the Multiline Styles dialog box draws the multiline to the current multiline style configuration. Figure 5–41 is the result of setting the offset distances with the Multiline Style command. This configuration may be used for an architectural application where a block wall 8 units thick and a footing of 16 units thick is needed. Multilines are drawn with the MLINE command and the following command sequence.

Command: **ML** *(For MLINE)*
Current settings: Justification = Top, Scale = 1.00, Style = STANDARD
Specify start point or [Justification/Scale/STyle]: **J** *(For Justification)*
Enter justification type [Top/Zero/Bottom] <top>: **Z** *(For Zero)*
Current settings: Justification = Zero, Scale = 1.00, Style = STANDARD
Specify start point or [Justification/Scale/STyle]: *(Pick a point at "A")*
Specify next point: *(Pick a point at "B")*
Specify next point or [Undo]: *(Pick a point at "C")*
Specify next point or [Close/Undo]: *(Pick a point at "D")*
Specify next point or [Close/Undo]: *(Press ENTER to exit this command)*

See also Figure 5–42 for an example of changing the justification of multilines. Justification modes reference the top of the multiline, the bottom of the multiline, or the zero location of the multiline, which is the multiline's center.

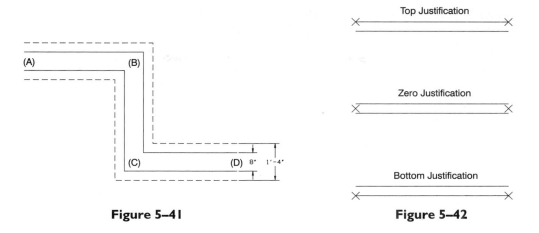

Figure 5–41   **Figure 5–42**

CREATING MULTILINE STYLES
THE MULTILINE STYLE COMMAND

The Multiline Styles dialog box is used to create a new multiline style or to make an existing style current. This command is located on the Format pull-down menu, shown in Figure 5–43A. Choose Multiline Style… to display the dialog box in Figure 5–43B. The dialog box lists the current multiline style, which by default is named "STANDARD." To create a new multiline style, click on the Element Properties… button or Multiline Properties… button, make the desired changes, and when finished, return to the main Multiline Styles dialog box. Change the name STANDARD to a name depicting the purpose of the new multiline style and click the ADD button to add the new multiline style to the database of the current drawing.

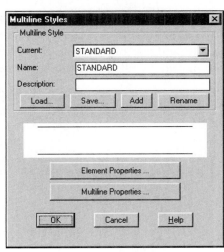

Figure 5–43A   **Figure 5–43B**

MULTILINE STYLE—ELEMENT PROPERTIES

Clicking on the Element Properties... button displays the Element Properties dialog box in Figure 5–44. Use this dialog box to make offset, color, and linetype assignments to the various multiline segments. Clicking the Add button adds a new multiline listing to the Elements box. Offset distances can now be changed along with Color and Linetype assignments to the selected multiline object. If a linetype is not present, a special Load button is available at the bottom of the Linetype dialog box. This in turn displays all supported linetypes. Clicking the OK button in the Element Properties dialog box returns you to the main Multiline Styles dialog box where a new multiline style name may be entered and the changes added to the new multiline style.

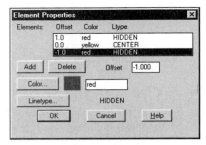

Figure 5–44

MULTILINE STYLE—MULTILINE PROPERTIES

Clicking on the Multiline Properties... button displays the Multiline Properties dialog box in Figure 5–45. This dialog box allows additional control of multilines: whether line segments are drawn at each multiline joint, whether the beginning and/or end of the multiline is capped with a line segment, whether the beginning and/or end of the outer parts of a multiline are capped by arcs, whether the beginning and/or end of the inner parts of a multiline are capped by arcs, or whether a user-defined angle is used to cap the start and end of the multiline. With Multiline Fill mode turned on, the entire multiline will be filled in with the selected color.

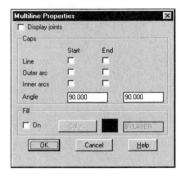

Figure 5–45

EDITING MULTILINES
THE MLEDIT COMMAND

The MLEDIT command is a special feature designed for editing or cleaning up intersections of multiline objects. Choose Multiline from the Modify pull-down menu area, shown in Figure 5–46A. This activates the Multiline Edit Tools dialog box in Figure 5–46B. Clicking on one of the buttons in the dialog box displays the title of the tool in the lower left corner.

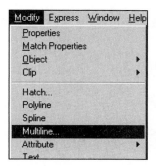

Figure 5–46A

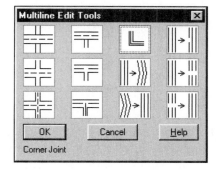

Figure 5–46B

Various cleanup modes are available, such as creating a closed cross, creating an open cross, or creating a merged cross. Options are available for creating "Tee" sections in multilines. Corner joints may be created along with the adding or deleting of a vertex of a multiline. Figures 5–47A, 5–47B and 5–47C depict the creation of open intersections and tees or the corner operation. In these figures, the identifying letters "A" and "B" show where to pick the multiline when using the Multiline Edit Tools dialog box.

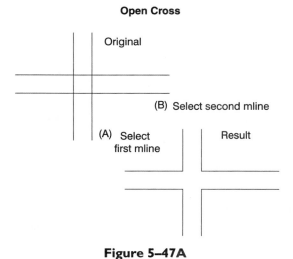

Figure 5–47A

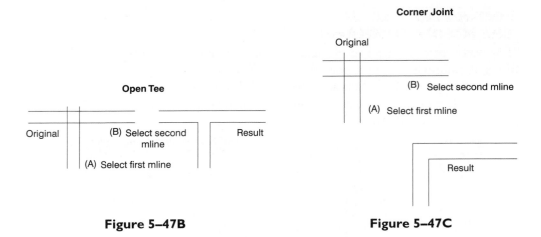

Figure 5–47B

Figure 5–47C

In the vertex example in Figure 5–47D, deleting a vertex forces the multiline segment to straighten out. When you add a vertex as in Figure 5–47E, the results may not be as evident. After the vertex was added to the straight multiline segment, the STRETCH command was used to create the shape in Figure 5–47E. Multiline segments cannot be broken with the BREAK command. They must be cut with the Cut option of the Multiline Edit Tools dialog box, shown in Figure 5–46B. To mend the cut, the Multiline Weld option is used. Both of these tools are located in the far right column of the Multiline Edit Tools dialog box.

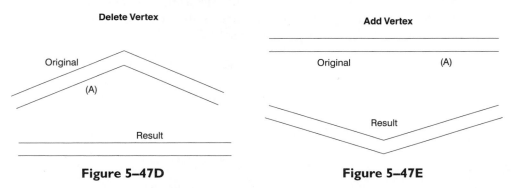

Figure 5–47D

Figure 5–47E

THE POINT COMMAND

Use the POINT command to identify the location of a point on a drawing, which may be used for reference purposes. Choose Point from the Draw pull-down menu shown in Figure 5–48. The OSNAP-Node or Nearest option is used to snap to points. By default, a point is displayed as a dot on the screen.

Command: **PO** *(For POINT)*
Current point modes: PDMODE=0 PDSIZE=0.0000
Specify a point: *(Pick the new position of a point)*
Specify a point: *(Either pick another point location or press* ESC *to exit this command)*

Figure 5–48

Since the appearance of the default point as a dot may be confused with the existing grid dots already on the screen, a mechanism is available to change the appearance of the point. Choosing Point Style from the Format pull-down menu, shown in Figure 5–49A, displays the Point Style dialog box in Figure 5–49B. Use this icon menu to set a different point mode and point size. Only one point style may be current in a drawing. Once a point is changed to a current style, the next drawing regeneration updates all points to this style.

Figure 5–49A

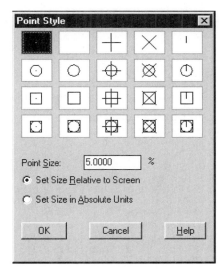

Figure 5–49B

DIVIDING AN OBJECT INTO EQUAL PARTS
THE DIVIDE COMMAND

Illustrated in Figure 5–50 is an inclined line. The purpose of this example is to divide the line into an equal number of parts. This was a tedious task with manual drafting methods, but thanks to the DIVIDE command, this operation is much easier to perform. The DIVIDE command instructs you to supply the number of divisions and then performs the division by placing a point along the object to be divided. The Point Style dialog box shown in Figure-5–49B controls the point size and shape. Be sure the point style appearance is set to produce a visible point. Otherwise, the results of the DIVIDE command will not be obvious.

Next, use the DIVIDE command. Select the inclined line as the object to divide, enter a value for the number of segments, and the command divides the object by a series of points as shown in Figure 5–51. This command is located on the Point cascading menu (from the Draw pull-down menu).

Command: **DIV** (For DIVIDE)
Select object to divide: (Select the inclined line)
Enter the number of segments or [Block]: **9**

Figure 5–50   **Figure 5–51**

One practical application of the DIVIDE command would be in drawing screw threads where a number of threads per inch are needed to form the profile of the thread. See Figure 5–52.

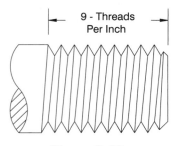

Figure 5–52

SETTING OFF EQUAL DISTANCES
THE MEASURE COMMAND

The MEASURE command takes an object such as a line or arc and measures along it depending on the length of the segment. The MEASURE command, similar to the DI-

VIDE command, places a point on the object at a specified distance given in the MEA-SURE command. It is important to note that as a point is placed along an object during the measuring process, the object is not automatically broken at the location of the point. Rather, the point is commonly used to construct from, along with the OSNAP-Node option. Also, the measuring starts at the endpoint closest to the point you used to select the object (See Figure 5–53).

Choose Point from the Draw pull-down menu, and then enter the MEASURE command.

Command: **ME** *(For MEASURE)*
Select object to measure: *(Select the end of the line in Figure 5–53)*
Specify length of segment or [Block]: **1.00**

The results in Figure 5–54 show various points placed at 1.00 increments. As with the DIVIDE command, the appearance of the points placed along the line is controlled through the Point Style dialog box.

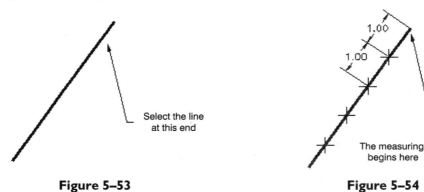

Figure 5–53   Figure 5–54

CONSTRUCTING MULTI-SIDED SHAPES
THE POLYGON COMMAND

The POLYGON command is used to construct a regular polygon. You create polygons by identifying the number of sides for the polygon, locating a point on the screen as the center of the polygon, specifying whether the polygon is inscribed or circumscribed, and specifying a circle radius for the size of the polygon. Polygons consist of a closed polyline object with width set to zero. It is possible to create a polygon that consists of a maximum number of 1028 sides. Use the following command sequence to construct the inscribed polygon in Figure 5–55.

Command: **POL** *(For POLYGON)*
Enter number of sides <4>: **6**
Specify center of polygon or [Edge]: *(Mark a point at "A")*
Enter an option [Inscribed in circle/Circumscribed about circle] <I>: **I** *(For Inscribed)*
Specify radius of circle: **1.00**

Use the following command sequence to construct the circumscribed polygon in Figure 5–56.

Command: **POL** *(For POLYGON)*
Enter number of sides <4>: **7**
Specify center of polygon or [Edge]: *(Pick a point at "A")*
Enter an option [Inscribed in circle/Circumscribed about circle] <I>: **C** *(For Circumscribed)*
Specify radius of circle: **1.00**

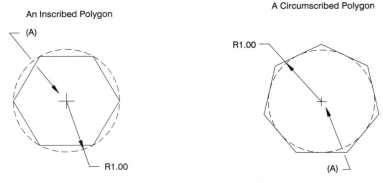

An Inscribed Polygon

A Circumscribed Polygon

Figure 5–55   **Figure 5–56**

Locating the endpoints of one of its edges may specify polygons. The polygon is then drawn in a counterclockwise direction. Study Figure 5–57 and the following command sequence to construct a polygon by edge.

Command: **POL** *(For POLYGON)*
Enter number of sides <4>: **5**
Specify center of polygon or [Edge]: **E** *(For Edge)*
Specify first endpoint of edge: *(Mark a point at "A")*
Specify second endpoint of edge: *(Mark a point at "B")*

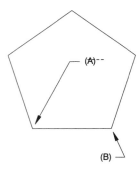

A Polygon by Edge

Figure 5–57

THE RAY COMMAND

A ray is a type of construction line object that begins at a user-defined point and extends to infinity only in one direction. In Figure 5–58, the quadrants of the circles identify all points where the ray objects begin and are drawn to infinity to the right. You should organize ray objects on specific layers. You should also exercise care in the editing of rays, and take special care not to leave segments of objects in the drawing database as a result of breaking ray objects. Breaking the ray object at "A" in Figure 5–58 converts one object to an individual line segment, and the other object remains as a ray. Study the following command sequence for constructing a ray.

Command: **RAY**
Specify start point: *(Pick a point on an object)*
Specify through point: *(Pick an additional point to construct the ray object)*
Specify through point: *(Pick another point to construct the ray object or press* ENTER *to exit this command)*

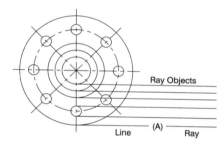

Figure 5–58

THE RECTANG COMMAND

Use the RECTANG command to construct a rectangle by defining two points. In Figure 5–59, two diagonal points are picked to define the rectangle. The rectangle is drawn as a single polyline object.

Command: **REC** *(For RECTANG)*
Specify first corner point or [Chamfer/Elevation/Fillet/Thickness/Width]: *(Pick a point at "A")*
Specify other corner point: *(Pick a point at "B")*

Figure 5–59

Other options of the RECTANG command enable you to construct a chamfer or fillet at all corners of the rectangle, to assign a width to the rectangle, and to have the rectangle drawn at a specific elevation and at a thickness for 3D purposes. In Figure 5–60 and the following command sequence, a rectangle is constructed with a chamfer distance of 0.20 units; the width of the rectangle is also set at 0.05 units. A relative coordinate value of 1.00,2.00 will be used to construct the rectangle 1 unit in the "X" direction and 2 units in the "Y" direction. The @ symbol resets the previous point at "A" to zero.

Command: **REC** *(For RECTANGLE)*
Specify first corner point or [Chamfer/Elevation/Fillet/Thickness/Width]: **C** *(For Chamfer)*
Specify first chamfer distance for rectangles <0.0000>: **0.20**
Specify second chamfer distance for rectangles <0.2000>: *(Press ENTER to accept this default value)*
Specify first corner point or [Chamfer/Elevation/Fillet/Thickness/Width]: **W** *(For Width)*
Specify line width for rectangles <0.0000>: **0.05**
Specify first corner point or [Chamfer/Elevation/Fillet/Thickness/Width]: *(Pick a point at "A")*
Specify other corner point: **@1.00,2.00** *(To identify the other corner at "B")*

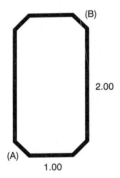

Figure 5–60

THE SPLINE COMMAND

Use the SPLINE command to construct a smooth curve given a sequence of points. You have the option of changing the accuracy of the curve given a tolerance range. The basic command sequence follows, which constructs the spline segment shown in Figure 5–61.

Command: **SPL** *(For SPLINE)*
Specify first point or [Object]: *(Pick a first point)*
Specify next point: *(Pick another point)*
Specify next point or [Close/Fit tolerance] <start tangent>: *(Pick another point)*

Specify next point or [Close/Fit tolerance] <start tangent>: *(Pick another point)*
Specify next point or [Close/Fit tolerance] <start tangent>: *(Press ENTER to continue)*
Specify start tangent: *(Press ENTER to accept)*
Specify end tangent: *(Press ENTER to accept the end tangent position, which exits the command and places the spline)*

The spline may be closed to display a continuous segment, as shown in Figure 5–62. Entering a different tangent point at the end of the command changes the shape of the curve connecting the beginning and end of the spline.

Command: **SPL** *(For SPLINE)*
Specify first point or [Object]: *(Pick a first point)*
Specify next point: *(Pick another point)*
Specify next point or [Close/Fit tolerance] <start tangent>: *(Pick another point)*
Specify next point or [Close/Fit tolerance] <start tangent>: *(Pick another point)*
Specify next point or [Close/Fit tolerance] <start tangent>: **C** *(To Close)*
Specify tangent: *(Press ENTER to exit the command and place the spline)*

Figure 5–61   **Figure 5–62**

Figure 5–63 illustrates an application of using splines in the creation of a plastic bottle to hold such household products as dishwashing liquid. The bottle illustrated is in the form of a 3D wireframe model.

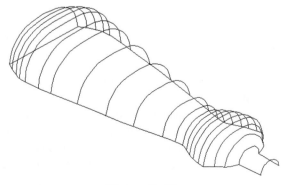

Figure 5–63

Once the bottle is defined with a series of splines, surfaces are applied to the spline frame forming a more realistic 3D model of the bottle in Figure 5–64. This image was created with the surface module of the Mechanical Desktop by Autodesk.

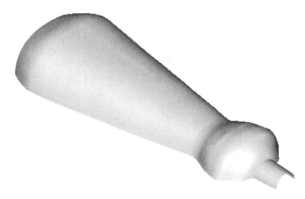

Figure 5–64

THE XLINE COMMAND

Xlines are construction lines drawn from a user-defined point. You are not prompted for any length information because the Xline extends an unlimited length, beginning at the user-defined point and going off to infinity in opposite directions from the point. Xlines can be drawn horizontal, vertical, and angular. You can bisect an angle using an Xline or offset the Xline at a specific distance. Figure 5–65, the circular view represents the Front view of a flange. To begin the creation of the Side views, lines are usually projected from key features on the adjacent view. In the case of the Front view in Figure 5–65, the key features are the top of the plate in addition to the other circular features. In this case, the Xlines were drawn with the Horizontal mode from the Quadrant of all circles. The following prompts outline the XLINE command sequence:

Command: **XL** *(For XLINE)*
Specify a point or [Hor/Ver/Ang/Bisect/Offset]: **H** *(For Horizontal)*
Specify through point: *(Pick a point on the display screen to place the first Xline)*
Specify through point: *(Pick a point on the display screen to place the second Xline)*

Since the Xlines continue to be drawn in both directions, care must be taken to manage these objects. Construction management techniques of Xlines could take the form of placing all Xlines on a specific layer to be turned off or frozen when not needed. When editing Xlines (especially with the BREAK command), you need to take special care to remove all excess objects that will still remain on the drawing screen. In Figure 5–66, breaking the Xline converts the object to a ray object. Use the ERASE command to remove any access Xlines.

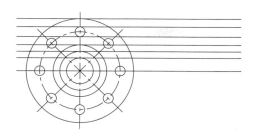

Figure 5–65   **Figure 5–66**

Another application of Xlines is illustrated in Figure 5–67. Three horizontal and vertical Xlines are constructed. You must corner the Xlines to create the object illustrated in Figure 5–68. Follow the command prompt sequence for the FILLET command to accomplish this task. Because numerous fillet operations need to be performed, the MULTIPLE command followed by FILLET will keep you in the command. When finished, cancel the command by pressing ESC.

Command: **MULTIPLE**
Enter command name to repeat: **FILLET**
Current settings: Mode = TRIM, Radius = 0.5000
Select first object or [Polyline/Radius/Trim]: **R** *(For Radius)*
Specify fillet radius <0.5000>: **0**
FILLET
Current settings: Mode = TRIM, Radius = 0.0000
Select first object or [Polyline/Radius/Trim]: *(Select the line at "A")*
Select second object: *(Select the line at "B")*
FILLET
Current settings: Mode = TRIM, Radius = 0.0000
Select first object or [Polyline/Radius/Trim]: *(Select the line at "B")*
Select second object: *(Select the line at "C")*
FILLET
Current settings: Mode = TRIM, Radius = 0.0000
Select first object or [Polyline/Radius/Trim]: *(Select the line at "C")*
Select second object: *(Select the line at "D")*
FILLET
Current settings: Mode = TRIM, Radius = 0.0000
Select first object or [Polyline/Radius/Trim]: *(Select the line at "D")*
Select second object: *(Select the line at "E")*
FILLET
Current settings: Mode = TRIM, Radius = 0.0000
Select first object or [Polyline/Radius/Trim]: *(Select the line at "E")*
Select second object: *(Select the line at "F")*

FILLET
Current settings: Mode = TRIM, Radius = 0.0000
Select first object or [Polyline/Radius/Trim]: *(Select the line at "F")*
Select second object: *(Select the line at "A")*
FILLET
Current settings: Mode = TRIM, Radius = 0.0000
Select first object or [Polyline/Radius/Trim]: *(Press the ESC key to exit this command)*

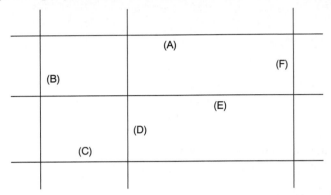

Figure 5–67

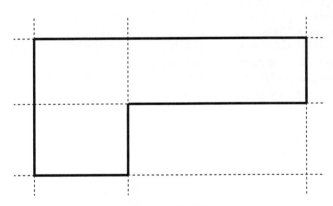

Figure 5–68

OGEE OR REVERSE CURVE CONSTRUCTION

An ogee curve connects two parallel lines with a smooth flowing curve that reverses itself in symmetrical form. To begin constructing an ogee curve to line segments "AB" and "CD," first draw line "BC," which connects both parallel line segments. See Figure 5–69.

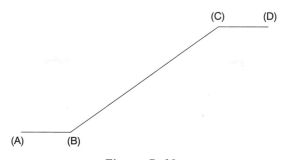

Figure 5–69

Use the DIVIDE command to divide line segment "BC" into four equal parts. Be sure to set a new point mode by picking a new point from the Point Style dialog box. Construct vertical lines from "B" and "C." Complete this step by constructing line segment "XY," which is perpendicular to line "BC" as shown in Figure 5–70. Do not worry about where line "XY" is located at this time.

Move line "XY" to the location identified by the point in Figure 5–71. Complete this step by copying line "XY" to the location identified by point "Z" illustrated in Figure 5–71.

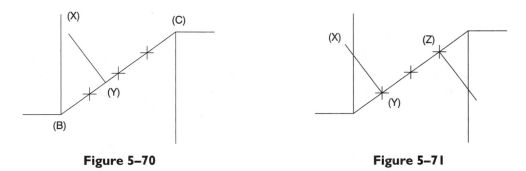

Figure 5–70   **Figure 5–71**

Construct two circles with centers located at points "X" and "Y" in Figure 5–72. Use the OSNAP-Intersection mode to accurately locate the centers. Note: If an intersection is not found from the previous step, use the EXTEND command to find the intersection and continue with this step. The radii of both circles are equivalent to distances "XB" and "YC."

Use the TRIM command to trim away any excess arc segments to form the ogee curve as shown in Figure 5–73. This forms the frame of the ogee for the construction of objects such as the wrench illustrated in Figure 5–74.

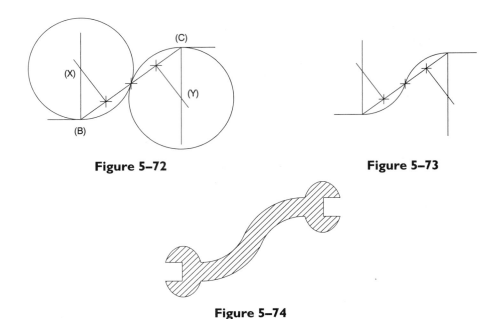

| Figure 5–72 | Figure 5–73 |

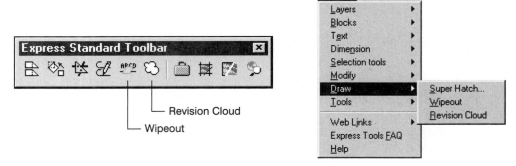

Figure 5–74

EXPRESS TOOLS APPLICATIONS FOR DRAWING

Two Express tools that relate to drawing commands are available through the Express Standard Toolbar in Figure 5–75: the WIPEOUT and REVCLOUD commands. These commands are also found on the Draw cascading menu on the Express pull-down menu, as shown in Figure 5–76.

| Figure 5–75 | Figure 5–76 |

THE WIPEOUT COMMAND

This Express tool covers a selected area with a mask of the current background color. One method of using this tool is to pick a series of points that define a closed polyline.

Objects that lie inside this closed area will be masked. Another method involves first creating a polyline and then selecting this shape, which will mask all objects inside. After you select the polyline, an additional prompt gives you the choice of erasing or saving the polyline.

Command: **WIPEOUT**
Select first point or [Frame/New from Polyline] <New>: *(Press* ENTER *to accept the default value)*
Select a polyline: *(Pick the polyline at "A" in Figure 5–77A)*
Erase polyline? [Yes/No] <No>: **Y** *(For Yes)*
Wipeout created.

The results of using WIPEOUT are illustrated in Figure 5–77B, with the island in the center of the kitchen being wiped out or masked over. In addition to the island being masked, also notice the centerline struck from the window and the hidden line at the right side of the countertop are also masked. If it appears the polyline is not deleted, it is because the frame is still turned on. Reissue the WIPEOUT command, use the Frame option, and turn it off.

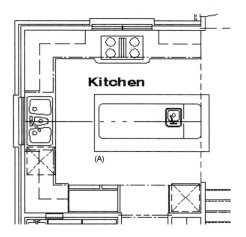

Figure 5–77A

Figure 5–77B

THE REVCLOUD COMMAND

Use this Express tool to create a revision cloud. The components of the cloud consist of a series of arcs in sequence that form a closed polyline object. You can adjust an arc length to reflect a cloud that has larger or smaller arcs. First pick a starting point for the arc and then move the crosshairs in the counterclockwise direction (See Figure 5–78A). This automatically creates the cloud in the current direction. Continuing to move the cursor back to the original starting position of the revision cloud closes the shape and exits the command.

Command: **REVCLOUD**
Arc length = 0.5000, Arc style = Normal
Specify cloud starting point or [eXit/Options] <eXit>: *(Pick the start of the cloud)*
Guide crosshairs along cloud path...
(Move your cursor in a counterclockwise direction. Continue to move back to the original start of the cloud to close the shape)
Cloud finished.

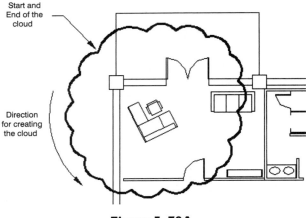

Start and End of the cloud

Direction for creating the cloud

Figure 5–78A

Figure 5–78B shows a 3D assembly model that has a revision cloud drawn around one of the components. It is recommended that the REVCLOUD command be used with Snap turned off. If you get caught in an endless loop because snap was turned on, press ESC to cancel the command.

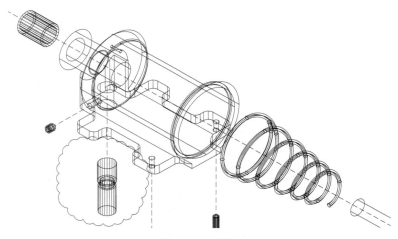

Figure 5–78B

TUTORIAL EXERCISE: PATTERN1.DWG

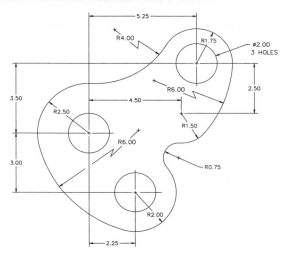

Figure 5–79

Purpose:

This tutorial is designed to use various Draw commands to construct a one-view drawing of Pattern1, as shown in Figure 5–79. Refer to the following for special system settings and suggested command sequences.

System Settings:

Use the Drawing Units dialog box to change the number of decimal places past the zero from four to two. Keep the remaining default unit values. Using the LIMITS command, keep (0,0) for the lower left corner and change the upper right corner from (12,9) to (21.00,16.00). Perform a ZOOM-All after changing the drawing limits. Check to see that the following Object Snap modes are already set: Endpoint, Extension, Intersection, Center.

Layers:

Create the following layers with the format:

Name	Color	Linetype
Object	White	Continuous
Center	Yellow	Center
Dimension	Yellow	Continuous

Suggested Commands:

Begin a new drawing called "Pattern1." Begin constructing this object by first laying out four points, which will be used as centers for circles. Use the CIRCLE-TTR command to construct tangent arcs to the circles already drawn. Use the TRIM command to clean up and partially delete circles to obtain the outline of the pattern. Then add the 2.00-diameter holes followed by the center markers, using the DIMCENTER command.

Whenever possible, substitute the appropriate command alias in place of the full AutoCAD command in each tutorial step. For example, use "CP" for the COPY command, "L" for the LINE command, and so on. The complete listing of all command aliases is located in Chapter 1, Table 1–2.

STEP I

Check that the current layer is set to "Object." Use the Layer Control box in Figure 5–80A to accomplish this task. Change the default point appearance from a "dot" to the "plus" through the Point Style dialog box (see Figure 5–80B), which you can open by choosing Point Style... from the Format pull-down menu.

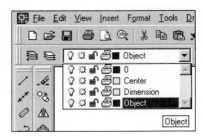

Figure 5–80A

STEP 2

Locate one point at absolute coordinate 7.50,7.50. Then use the COPY command and the dimensions in Figure 5–81 as a guide for duplicating the remaining points.

Command: **PO** *(For Point)*
Current point modes: PDMODE=2
   PDSIZE=0.00
Specify a point: **7.50,7.50** *(Locates the point at "A" in Figure 5–81)*
Command: **CP** *(For COPY)*
Select objects: **L** *(This should select the point)*
Select objects: *(Press ENTER to continue)*
Specify base point or displacement, or [Multiple]: **M** *(For Multiple)*
Specify base point: **Nod**
of *(Select the point at "A" in Figure 5–81)*
Specify second point of displacement or
   <use first point as displacement>:
   @2.25,-3.00 *(To locate point "B")*
Specify second point of displacement or
   <use first point as displacement>:
   @4.5,1.00 *(To locate point "C")*
Specify second point of displacement or
   <use first point as displacement>:
   @5.25,3.50 *(To locate point "D")*
Specify second point of displacement or
   <use first point as displacement>:
   (Press ENTER to exit this command)

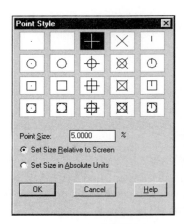

Figure 5–80B

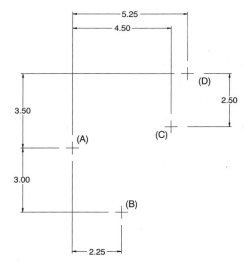

Figure 5–81

STEP 3

Use the CIRCLE command to place four circles of different sizes from points located at "A," "B," "C," and "D" as shown in Figure 5–82. When you have completed drawing the four circles, use the ERASE command to erase points "A," "B," "C," and "D."

Command: **C** *(For CIRCLE)*
Specify center point for circle or [3P/2P/
   Ttr (tan tan radius)]: **Nod**
of *(Select the point at "A")*
Specify radius of circle or [Diameter]:
   2.50
Command: **C** *(For CIRCLE)*
Specify center point for circle or [3P/2P/
   Ttr (tan tan radius)]: **Nod**
of *(Select the point at "B")*
Specify radius of circle or [Diameter]
   <2.50>: **2.00**
Command: **C** *(For CIRCLE)*
Specify center point for circle or [3P/2P/
   Ttr (tan tan radius)]: **Nod**
of *(Select the point at "C")*
Specify radius of circle or [Diameter]
   <2.00>: **1.50**
Command: **C** *(For CIRCLE)*
Specify center point for circle or [3P/2P/
   Ttr (tan tan radius)]: **Nod**
of *(Select the point at "D")*
Specify radius of circle or [Diameter]
   <1.50>: **1.75**
Command: **E** *(For ERASE)*
Select objects: *(Pick the four points labeled
   "A", "B", "C", and "D")*
Select objects: *(Press ENTER to execute this
   command)*

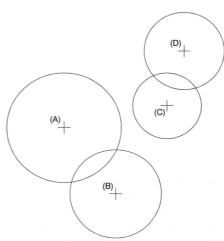

Figure 5–82

STEP 4

Use the CIRCLE-TTR command to construct a 4.00-radius circle tangent to the two dashed circles in Figure 5–83. Then use the TRIM command to trim away part of circle "C."

Command: **C** *(For CIRCLE)*
Specify center point for circle or [3P/2P/ Ttr (tan tan radius)]: **T** *(For TTR)*
Specify point on object for first tangent of circle: *(Select the dashed circle at "A")*
Specify point on object for second tangent of circle: *(Select the dashed circle at "B")*
Specify radius of circle <1.75>: **4.00**
Command: **TR** *(For TRIM)*
Current settings: Projection=UCS Edge=None
Select cutting edges ...
Select objects: *(Select the two dashed circles shown in Figure 5–83)*
Select objects: *(Press ENTER to continue)*
Select object to trim or [Project/Edge/ Undo]: *(Select the large circle at "C")*
Select object to trim or [Project/Edge/ Undo]: *(Press ENTER to exit this command)*

If, during the trimming process, you trim the wrong segment by mistake, you can use a built-in Undo to trim the correct item.

STEP 5

Use the CIRCLE-TTR command to construct a 6.00-radius circle tangent to the two dashed circles in Figure 5–84. Then use the TRIM command to trim away part of circle "C."

Command: **C** *(For CIRCLE)*
Specify center point for circle or [3P/2P/ Ttr (tan tan radius)]: **T** *(For TTR)*
Specify point on object for first tangent of circle: *(Select the dashed circle at "A")*

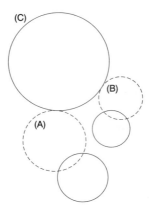

Figure 5–83

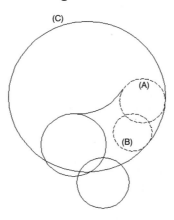

Figure 5–84

Specify point on object for second tangent of circle: *(Select the dashed circle at "B")*
Specify radius of circle <4.00>: **6.00**
Command: **TR** *(For TRIM)*
Current settings: Projection=UCS Edge=None
Select cutting edges ...
Select objects: *(Select the two dashed circles "A" and "B" in Figure 5–84)*
Select objects: *(Press ENTER to continue)*
Select object to trim or [Project/Edge/ Undo]: *(Select the large circle at "C")*
Select object to trim or [Project/Edge/ Undo]: *(Press ENTER to exit this command)*

STEP 6

Use the CIRCLE-TTR command to construct a 6.00-radius circle tangent to the two dashed circles in Figure 5–85. Then use the TRIM command to trim away part of circle "C."

Command: **C** *(For CIRCLE)*
Specify center point for circle or [3P/2P/ Ttr (tan tan radius)]: **T** *(For TTR)*
Specify point on object for first tangent of circle: *(Select the dashed circle at "A")*
Specify point on object for second tangent of circle: *(Select the dashed circle at "B")*
Specify radius of circle <6.00>: *(Press ENTER to accept this default value)*
Command: **TR** *(For TRIM)*
Current settings: Projection=UCS Edge=None
Select cutting edges ...
Select objects: *(Select the two dashed circles "A" and "B" in Figure 5–85)*
Select objects: *(Press ENTER to continue)*
Select object to trim or [Project/Edge/ Undo]: *(Select the large circle at "C")*
Select object to trim or [Project/Edge/ Undo]: *(Press ENTER to exit this command)*

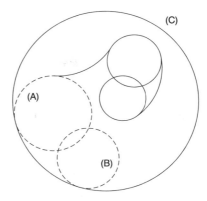

Figure 5–85

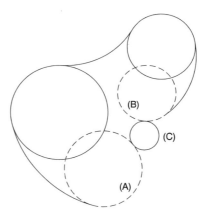

Figure 5–86

STEP 7

Use the CIRCLE-TTR command to construct a 0.75-radius circle tangent to the two dashed circles in Figure 5–86. Then use the TRIM command to trim away part of circle "C."

Command: **C** *(For CIRCLE)*
Specify center point for circle or [3P/2P/ Ttr (tan tan radius)]: **T** *(For TTR)*
Specify point on object for first tangent of circle: *(Select the dashed circle at "A")*
Specify point on object for second tangent of circle: *(Select the dashed circle at "B")*

Specify radius of circle <6.00>: **0.75**
Command: **TR** *(For TRIM)*
Current settings: Projection=UCS Edge=None
Select cutting edges ...
Select objects: *(Select the two dashed circles "A" and "B" in Figure 5–86)*
Select objects: *(Press ENTER to continue)*
Select object to trim or [Project/Edge/ Undo]: *(Select the circle at "C")*
Select object to trim or [Project/Edge/ Undo]: *(Press ENTER to exit this command)*

STEP 8

Use the TRIM command, select all dashed arcs in Figure 5–87 as cutting edges, and trim away the circular segments to form the outline of the Pattern1 drawing.

Command: **TR** *(For TRIM)*
Current settings: Projection=UCS
    Edge=None
Select cutting edges ...
Select objects: *(Select the four dashed arcs in Figure 5–87)*
Select objects: *(Press ENTER to continue)*
Select object to trim or [Project/Edge/
    Undo]: *(Select the circle at "A")*
Select object to trim or [Project/Edge/
    Undo]: *(Select the circle at "B")*
Select object to trim or [Project/Edge/
    Undo]: *(Select the circle at "C")*
Select object to trim or [Project/Edge/
    Undo]: *(Select the circle at "D")*
Select object to trim or [Project/Edge/
    Undo]: *(Press ENTER to exit this command)*

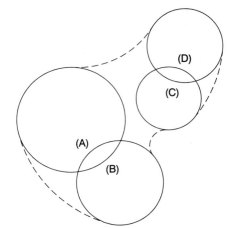

Figure 5–87

STEP 9

Your drawing should be similar to the illustration in Figure 5–88.

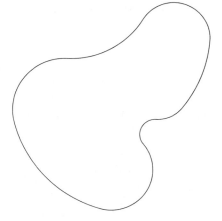

Figure 5–88

STEP 10

Use the CIRCLE command to place a circle
of 2.00-unit diameter at the center of arc
"A." Then use the COPY command to duplicate the circle at the center of arcs "B"
and "C" (the OSNAP-Center mode should
already be running). See Figure 5–89.

Command: **C** *(For CIRCLE)*
Specify center point for circle or [3P/2P/
    Ttr (tan tan radius)]: *(Select the edge of
    arc "A")*
Specify radius of circle or [Diameter]
    <0.7500>: **D** *(For Diameter)*
Specify diameter of circle <1.50>: **2.00**
Command: **CP** *(For COPY)*
Select objects: **L** *(For Last)*
Select objects: *(Press ENTER to continue)*
Specify base point or displacement, or
    [Multiple]: **M** *(For Multiple)*
Specify base point: *(Select the edge of the
    arc at "A")*
Specify second point of displacement or
    <use first point as displacement>:
    (Select the edge of the arc at "B")
Specify second point of displacement or
    <use first point as displacement>:
    (Select the edge of the arc at "C")
Specify second point of displacement or
    <use first point as displacement>:
    (Press ENTER to exit this command)

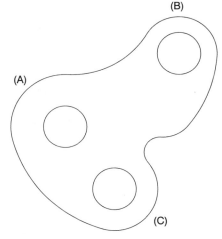

Figure 5–89

STEP 11

Use the Layer Control box to set the current layer to "Center" as in Figure 5–90.

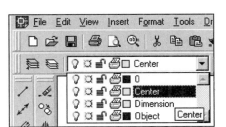

Figure 5–90

STEP 12

Change the DIMCEN variable from a value of 0.09 to -0.12. This will place the center marker identifying the centers of circles and arcs when the DIMCENTER command is used. See Figure 5–91.

Command: **DIMCEN**
Enter new value for DIMCEN <0.09>: **-0.12**
Command: **DCE** *(For DIMCENTER)*
Select arc or circle: *(Select the arc at "A")*
Command: **DCE** *(For DIMCENTER)*
Select arc or circle: *(Select the arc at "B")*
Command: **DCE** *(For DIMCENTER)*
Select arc or circle: *(Select the arc at "C")*
Command: **DCE** *(For DIMCENTER)*
Select arc or circle: *(Select the arc at "D")*

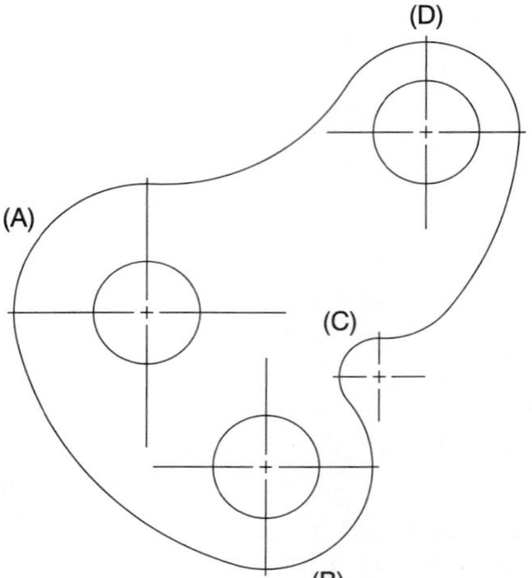

Figure 5–91

STEP 13

Dimensions may be added to this drawing at a later time. Place them on the Dimension layer as shown in Figure 5–92.

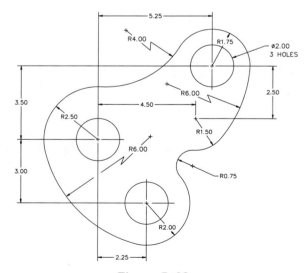

Figure 5–92

TUTORIAL EXERCISE: GEAR-ARM.DWG

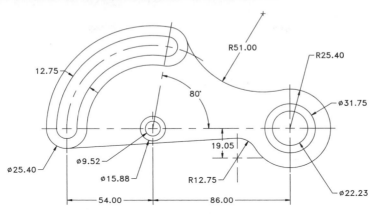

Figure 5–93

Purpose:

This tutorial is designed to use geometric commands to construct a one-view drawing of the Gear-arm in metric format as illustrated in Figure 5–93.

System Settings:

Use the Drawing Units dialog box to change the number of decimal places past the zero from four to two. Keep the remaining default unit values. Using the LIMITS command, keep (0,0) for the lower left corner and change the upper right corner from (12,9) to (265.00,200.00). Perform a ZOOM-All after changing the drawing limits. Since a layer called "Center" must be created to display centerlines, use the LTSCALE command and change the default value of 1.00 to 25.40. This will make the long and short dashes of the centerlines appear on the display screen. Check to see that the following Object Snap modes are already set: Endpoint, Extension, Intersection, Center.

Layers:

Create the following layers with the format:

Name	Color	Linetype
Object	White	Continuous
Center	Yellow	Continuous
Dimension	Yellow	Continuous

Suggested Commands:

Begin a new drawing called "Gear-arm." The object consists of a combination of circles and arcs along with tangent lines and arcs. Before beginning, be sure to set "Object" as the new current layer. Use the POINT command to identify and lay out the centers of all circles for construction purposes. Use the ARC command to construct a series of arcs for the left side of the Gear-arm. The TRIM command will be used to trim circles, lines, and arcs to form the basic shape. Also, use the CIRCLE-TTR command for tangent arcs to existing geometry.

Whenever possible, substitute the appropriate command alias in place of the full AutoCAD command in each tutorial step. For example, use "CP" for the COPY command, "L" for the LINE command, and so on. The complete listing of all command aliases is located in Chapter 1, Table 1–2.

STEP I

Check that the current layer is set to "Object." Use the Layer Control box in Figure 5–94A to accomplish this task.

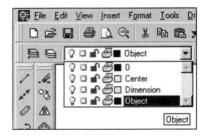

Figure 5–94A

Begin the gear-arm by drawing two circles of diameters 9.52 and 15.88 using the CIRCLE command and coordinate 112.00,90.00 as the center of both circles. See Figure 5–94B.

Command: **C** *(For CIRCLE)*
Specify center point for circle or [3P/2P/
    Ttr (tan tan radius)]: **112.00,90.00** *(At
    Point "A")*
Specify radius of circle or [Diameter]: **D**
    (For Diameter)
Specify diameter of circle: **9.52**
Command: **C** *(For CIRCLE)*
Specify center point for circle or [3P/2P/
    Ttr (tan tan radius)]: **@** *(To identify the
    last known point as the center of the
    circle)*
Specify radius of circle or [Diameter]
    <4.76>: **D** *(For Diameter)*
Specify diameter of circle <9.52>: **15.88**

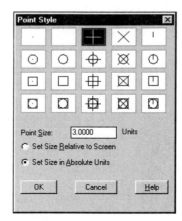

Figure 5–94B

STEP 2

Change the default point appearance from a "dot" to the "plus" through the Point Style dialog box (see Figure 5–95A), which you can open by choosing Point Style... from the Format pull-down menu. While inside this dialog box, change the Point Size: to 3.00 units because this is a Metric drawing. Then use the POINT command to place a point at the center of the two circles. See Figure 5–95B.

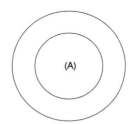

Figure 5–95A

Command: **PO** *(For POINT)*
Current point modes: PDMODE=2
    PDSIZE=3.00
Specify a point: *(Select the edge of the
    circle at "A" in Figure 5–95B)*

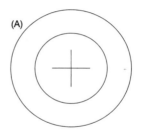

Figure 5–95B

STEP 3

Use the COPY command to duplicate the point using a polar coordinate distance of 86 units in the 0-degree direction. The Direct Distance mode may also be used for this operation. See Figure 5–96.

Command: **CP** *(For COPY)*
Select objects: **L** *(For Last to select the last point)*

Select objects: *(Press ENTER to continue)*
Specify base point or displacement, or [Multiple]: *(Select the edge of the circle at "A" to identify its center)*
Specify second point of displacement or <use first point as displacement>: **@86<0**

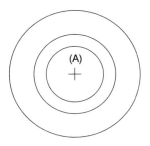

Figure 5–96

STEP 4

Use the CIRCLE command to place three circles of different sizes from the same point at "A." See Figure 5–97.

Command: **C** *(For CIRCLE)*
Specify center point for circle or [3P/2P/Ttr (tan tan radius)]: **Nod**
of *(Select the point at "A")*
Specify radius of circle or [Diameter] <7.94>: **25.40**
Command: **C** *(For CIRCLE)*
Specify center point for circle or [3P/2P/Ttr (tan tan radius)]: **@** *(To identify the last known point as the center of the circle)*

Specify radius of circle or [Diameter] <25.40>: **D** *(For Diameter)*
Specify diameter of circle <50.80>: **31.75**
Command: **C** *(For CIRCLE)*
Specify center point for circle or [3P/2P/Ttr (tan tan radius)]: **@** *(To identify the last known point as the center of the circle)*
Specify radius of circle or [Diameter] <15.88>: **D** *(For Diameter)*

Specify diameter of circle <31.75>: **22.23**

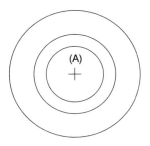

Figure 5–97

STEP 5

Use the COPY command to duplicate point "A" using a polar coordinate distance of 54 units in the 180-degree direction. The Direct Distance mode could also be used for this application. See Figure 5–98.

Command: **CP** *(For COPY)*
Select objects: *(Select the point at "A")*

Select objects: *(Press ENTER to continue)*
Specify base point or displacement, or
    [Multiple]: *(Select the edge of the large*
    circle at "A" to identify its center)
Specify second point of displacement or
    <use first point as displacement>:
    @54<180

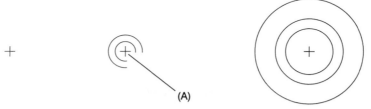

(A)

Figure 5–98

STEP 6

Use the CIRCLE command to place two circles of diameters 25.40 and 12.75 using point "A" in Figure 5–99 as the center. These circles will be converted to arcs in later steps.

Command: **C** *(For CIRCLE)*
Specify center point for circle or [3P/2P/
    Ttr (tan tan radius)]: **Nod**
of *(Select the point at "A")*
Specify radius of circle or [Diameter]
    <11.12>: **D** *(For Diameter)*

Specify diameter of circle <22.23>: **25.40**
Command: **C** *(For CIRCLE)*
Specify center point for circle or [3P/2P/
    Ttr (tan tan radius)]: **@** *(To identify the*
    last known point as the center of the
    circle)
Specify radius of circle or [Diameter]
    <12.70>: **D** *(For Diameter)*
Specify diameter of circle <25.40>: **12.75**

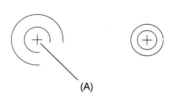

(A)

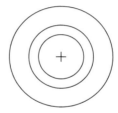

Figure 5–99

STEP 7

Use the COPY command to duplicate point "A" using a polar coordinate distance of 54 units in the 80-degree direction. See Figure 5–100.

Command: **CP** *(For COPY)*
Select objects: *(Select the point at "A")*

Select objects: *(Press* ENTER *to continue)*
Specify base point or displacement, or [Multiple]: *(Select the edge of the large circle at "A" to identify its center)*
Specify second point of displacement or <use first point as displacement>: **@54<80**

Figure 5–100

STEP 8

Use the LINE command to draw a line using a polar coordinate distance of 70 and an 80-degree direction. Start the line at point "A." Then use the CIRCLE command to place two circles of diameters 25.40 and 12.75 at point "B" as in Figure 5–101. These circles will be converted to arcs in later steps.

Command: **L** *(For LINE)*
Specify first point: *(Select the edge of the large circle at "A" to identify its center)*
Specify next point or [Undo]: **@70<80**
Specify next point or [Undo]: *(Press* ENTER *to exit this command)*

Command: **C** (For CIRCLE)
Specify center point for circle or [3P/2P/ Ttr (tan tan radius)]: **Nod**

of *(Select the point at "B")*
Specify radius of circle or [Diameter] <6.38>: **D** *(For Diameter)*
Specify diameter of circle <12.75>: **25.40**
Command: **C** *(For CIRCLE)*
Specify center point for circle or [3P/2P/ Ttr (tan tan radius)]: **@** *(To identify the last known point as the center of the circle)*
Specify radius of circle or [Diameter] <12.70>: **D** *(For Diameter)*
Specify diameter of circle <25.40>: **12.75**

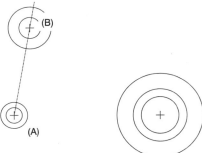

Figure 5–101

STEP 9

Use the ARC command and draw an arc using point "A" as the center, point "B" as the start point, and point "C" as the endpoint, as shown in Figure 5–102.

Command: **A** *(For ARC)*
Specify start point of arc or [CEnter]: **C** *(For Center)*

Specify center point of arc: *(Select the edge of the circle at "A" to identify its center)*
Specify start point of arc: *(Select the intersection of the line and circle at "B")*
Specify end point of arc or [Angle/chord Length]: *(Using polar tracking, select the polar intersection at "C")*

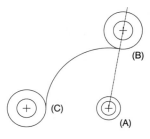

Figure 5–102

STEP 10

Use the ARC command and draw an arc using point "A" as the center, point "B" as the start point, and point "C" as the endpoint, as shown in Figure 5–103.

Command: **A** *(For ARC)*
Specify start point of arc or [CEnter]: **C** *(For Center)*

Specify center point of arc: *(Select the edge of the circle at "A")*
Specify start point of arc: *(Select the intersection of the line and circle at "B")*
Specify end point of arc or [Angle/chord Length]: *(Using polar tracking, select the polar intersection at "C")*

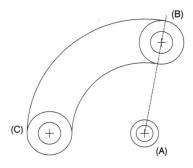
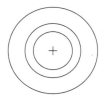

Figure 5–103

STEP 11

Use the TRIM command, select the two dashed arcs in Figure 5–104 as cutting edges, and trim the two circles at points "C" and "D."

Command: **TR** *(For TRIM)*
Current settings: Projection=UCS Edge=None
Select cutting edges ...
Select objects: *(Select the two dashed arcs "A" and "B")*
Select objects: *(Press ENTER to continue)*

Select object to trim or [Project/Edge/Undo]: *(Select the circle at "C")*
Select object to trim or [Project/Edge/Undo]: *(Select the circle at "D")*
Select object to trim or [Project/Edge/Undo]: *(Press ENTER to exit this command)*

If, during the trimming process, you trim the wrong segment, you can use a built-in Undo to trim the correct item.

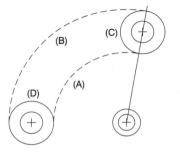
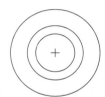

Figure 5–104

STEP 12

Your drawing should be similar to the illustration in Figure 5–105.

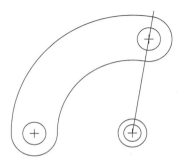
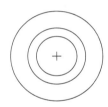

Figure 5–105

STEP 13

Use the ARC command and draw an arc using point "A" as the center, point "B" as the start point, and point "C" as the endpoint, as shown in Figure 5–106.

Command: **A** *(For ARC)*
Specify start point of arc or [CEnter]: **C** *(For Center)*
Specify center point of arc: *(Select the edge of the circle at "A")*
Specify start point of arc: *(Select the intersection of the line and circle at "B")*
Specify end point of arc or [Angle/chord Length]: *(Using polar tracking, select the polar intersection at "C")*

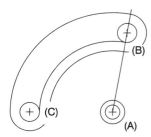

Figure 5–106

STEP 14

Use the ARC command and draw an arc using point "A" as the center, point "B" as the start point, and point "C" as the endpoint as shown in Figure 5–107.

Command: **A** *(For ARC)*
Specify start point of arc or [CEnter]: **C** *(For Center)*
Specify center point of arc: *(Select the edge of the circle at "A")*
Specify start point of arc: *(Select the intersection of the line and circle at "B")*
Specify end point of arc or [Angle/chord Length]: *(Using polar tracking, select the polar intersection at "C")*

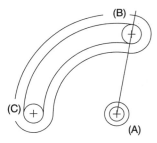

Figure 5–107

STEP 15

Use the TRIM command, select the two dashed arcs at the right as cutting edges, and trim the two circles at points "C" and "D," as shown in Figure 5–108.

Command: **TR** *(For TRIM)*
Current settings: Projection=UCS
  Edge=None
Select cutting edges ...
Select objects: *(Select the two dashed arcs "A" and "B")*
Select objects: *(Press ENTER to continue)*
Select object to trim or [Project/Edge/ Undo]: *(Select the circle at "C")*
Select object to trim or [Project/Edge/ Undo]: *(Select the circle at "D")*

Select object to trim or [Project/Edge/ Undo]: *(Press ENTER to exit this command)*

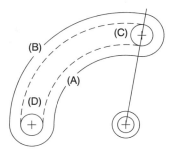

Figure 5–108

STEP 16

Your drawing should be similar to the illustration in Figure 5–109. Always perform periodic screen redraws using the REDRAW command to clean up the display screen if necessary.

Command: **R** *(For REDRAW)*

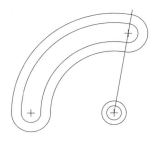

Figure 5–109

STEP 17

Use the LINE command and draw a line from the quadrant of the small circle to the quadrant of the large circle in Figure 5–110. This line will be used only for construction purposes.

Command: **L** *(For LINE)*
Specify first point: **Qua**
of *(Select the circle at "A")*
Specify next point or [Undo]: *(Use polar tracking to select the polar intersection at "B")*
Specify next point or [Undo]: *(Press ENTER to exit this command)*

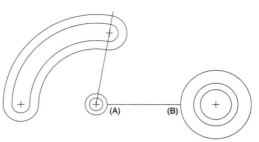

Figure 5–110

STEP 18

Use the MOVE command to move the dashed line down the distance of 19.05 units using the Polar Coordinate or Direct Distance mode. Then use the OFFSET command to offset the dashed circle the distance 12.75 units. The intersection of these two objects will be used to draw a 12.75-radius circle, as shown in Figure 5–111.

Command: **M** *(For MOVE)*
Select objects: *(Select the dashed line in Figure 5–111)*
Select objects: *(Press ENTER to continue)*

Specify base point or displacement: *(Select the endpoint of the line at "A")*
Specify second point of displacement or <use first point as displacement>: *(Move your cursor straight down and enter a value of 19.05)*
Command: **O** *(For OFFSET)*
Specify offset distance or [Through] <1.00>: **12.75**
Select object to offset or <exit>: *(Select the dashed circle at "B")*
Specify point on side to offset: *(Pick a blank part of the screen at "C")*
Select object to offset or <exit>: *(Press ENTER to exit this command)*

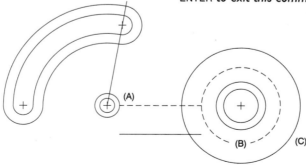

Figure 5–111

STEP 19

Use the CIRCLE command to draw a circle with a radius of 12.75. Use the center of the circle as the intersection of the large dashed circle and the dashed horizontal line illustrated in Figure 5–112. Use the ERASE command to delete the dashed circle and dashed line.

Command: **C** *(For CIRCLE)*

Specify center point for circle or [3P/2P/ Ttr (tan tan radius)]: *(Select the intersection of the line and circle at "A")*
Specify radius of circle or [Diameter] <6.38>: **12.75**
Command: **E** *(For ERASE)*
Select objects: *(Select the dashed circle and dashed line in Figure 5–112)*
Select objects: *(Press ENTER to execute this command)*

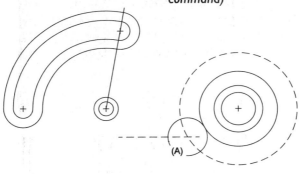

Figure 5–112

STEP 20

Use the LINE command to draw a line from a point tangent to the arc at "A" to a point tangent to the circle at "B," as shown in Figure 5–113. Use the OSNAP-Tangent option to accomplish this.

Command: **L** *(For LINE)*

Specify first point: **Tan**
to *(Select the arc at "A")*
Specify next point or [Undo]: **Tan**
to *(Select the arc at "B")*
Specify next point or [Undo]: *(Press ENTER to exit this command)*

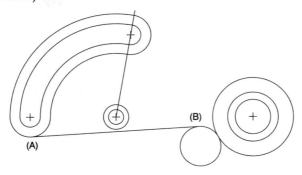

Figure 5–113

STEP 21

Use the TRIM command, select the dashed line and dashed circle illustrated in Figure 5–114 as cutting edges, and trim the circle at "A."

Command: **TR** *(For TRIM)*
Current settings: Projection=UCS
  Edge=None
Select cutting edges ...

Select objects: *(Select the dashed line and the dashed circle in Figure 5–114)*
Select objects: *(Press ENTER to continue)*
Select object to trim or [Project/Edge/Undo]: *(Select the circle at "A")*
Select object to trim or [Project/Edge/Undo]: *(Press ENTER to exit this command)*

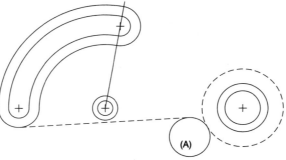

Figure 5–114

STEP 22

Use the CIRCLE-TTR command to draw a circle tangent to the arc at "A" and tangent to the circle at "B" with a radius of 51 as in Figure 5–115. When you are using the TTR option of the CIRCLE command, the OSNAP-Tangent option is automatically invoked.

Command: **C** *(For CIRCLE)*
Specify center point for circle or [3P/2P/Ttr (tan tan radius)]: **T** *(For TTR)*
Specify point on object for first tangent of circle: *(Select the arc at "A")*
Specify point on object for second tangent of circle: *(Select the arc at "B")*
Specify radius of circle <12.75>: **51**

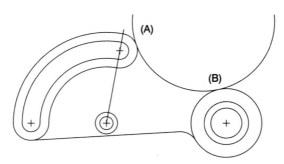

Figure 5–115

STEP 23

Use the TRIM command, select the dashed arc and dashed circle illustrated in Figure 5–116 as cutting edges, and trim the large 51-radius circle at "A."

Command: **TR** *(For TRIM)*
Current settings: Projection=UCS
   Edge=None
Select cutting edges ...

Select objects: *(Select the dashed arc and dashed circle in Figure 5–116)*
Select objects: *(Press ENTER to continue)*
Select object to trim or [Project/Edge/ Undo]: *(Select the circle at "A")*
Select object to trim or [Project/Edge/ Undo]: *(Press ENTER to exit this command)*

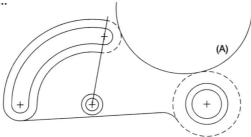

Figure 5–116

STEP 24

Use the TRIM command, select the two dashed arcs illustrated in Figure 5–117 as cutting edges, and trim the circle at "C." Use the ERASE command to delete all four points used to construct the circles.

Command: **TR** *(For TRIM)*
Current settings: Projection=UCS
   Edge=None
Select cutting edges ...
Select objects: *(Select the dashed arcs "A" and "B" in Figure 5–117)*

Select objects: *(Press ENTER to continue)*
Select object to trim or [Project/Edge/ Undo]: *(Select the circle at "C")*
Select object to trim or [Project/Edge/ Undo]: *(Press ENTER to exit this command)*
Command: **E** *(For ERASE)*
Select objects: *(Select the four points in Figure 5–117)*
Select objects: *(Press ENTER to execute this command)*

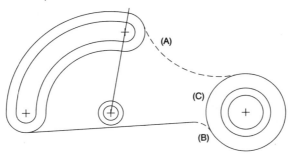

Figure 5–117

STEP 25

Your display should appear similar to the illustration in Figure 5–118A. Notice the absence of the points that were erased in the previous step. Standard centerlines will be placed to mark the center of all circles in the next series of steps. Use the Layer Control box to set the current layer to "Center" in Figure 5–118B. If you have not already done so, use the LTSCALE command to change the linetype scale from 1.00 to a new value of 10.00.

Command: **LTS** *(For LTSCALE)*
New scale factor <1.00>: **10.00**

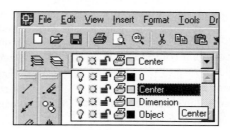

Figure 5–118A

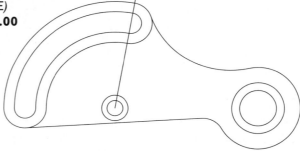

Figure 5–118B

STEP 26

To place centerlines for all circles and arcs, two dimension commands or variables need to be changed. The dimension variable DIMSCALE is set to a value of 1. Since this is a metric drawing, the DIMSCALE value needs to be changed to 25.4. This will increase all variables by this value, which is necessary because you are drawing in metric units. Also, the DIMCEN variable needs to be changed from a value of 0.09 to -0.09. The negative value will extend the centerlines past the edge of the circle when you use the DIMCENTER command. After all centerlines are placed, erase the top of the centerline at "E." See Figure 5–119.

Command: **DIMSCALE**
Enter new value for DIMSCALE <1.00>:
  25.4

Command: **DIMCEN**
Enter new value for DIMCEN <0.09>: **-0.09**
Command: **DCE** *(For DIMCENTER)*
Select arc or circle: *(Select the arc at "A")*
Repeat the preceding command to place a center marker at arcs "B," "C," and "D."

Command: **E** *(For ERASE)*
Select objects: *(Select the top of the line at "E")*
Select objects: *(Press ENTER to execute this command)*

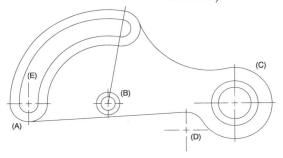

Figure 5–119

STEP 27

Use the OFFSET command to offset the inside arc at "A" the distance 6.375 units. Indicate a point in the vicinity of "B" for the side to perform the offset as shown in Figure 5–120.

Command: **O** *(For OFFSET)*
Specify offset distance or [Through] <12.75>: **6.375**

Select object to offset or <exit>: *(Select the arc at "A")*
Specify point on side to offset: *(Pick a point in the vicinity of "B")*
Select object to offset or <exit>: *(Press ENTER to exit this command)*

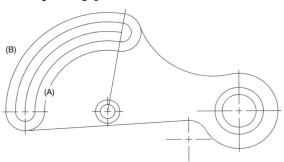

Figure 5–120

STEP 28

Change the middle arc at "A" and the line at "B" in Figure 5–121A to the "Center" layer. First select the lines. Then activate Layer Control in Figure 5–121B and pick the "Center" layer.

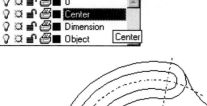

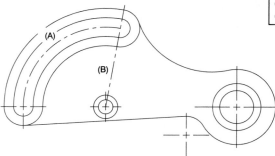

Figure 5–121A   **Figure 5–121B**

STEP 29

Use the LENGTHEN command to lengthen the centerline arc "A" in Figure 5–122. Then use the DIMCEN variable and change the default value from -0.09 to 0.09. This will change the center point to a "plus" for arc "B" without the centerlines extending beyond the arc when you use the DIMCENTER command.

Command: **LEN** *(For LENGTHEN)*
Select an object or [DElta/Percent/Total/ DYnamic]: *(Select the arc at "A")*
Current length: 94.25, included angle: 100
Select an object or [DElta/Percent/Total/ DYnamic]: **T** *(For Total)*
Specify total length or [Angle] <1.00)>: **120**

Select an object to change or [Undo]:
  (Press ENTER *to exit this command)*
Select an object to change or [Undo]:
  (Press ENTER *to exit this command)*

Command: **DIMCEN**
Enter new value for DIMCEN <-0.09>:
  0.09
Command: **DCE** *(For DIMCENTER)*
Select arc or circle: *(Select the arc at "B")*

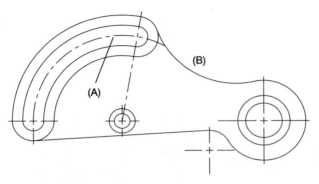

Figure 5–122

The finished object may be dimensioned
as an optional step as illustrated in Figure
5–123.

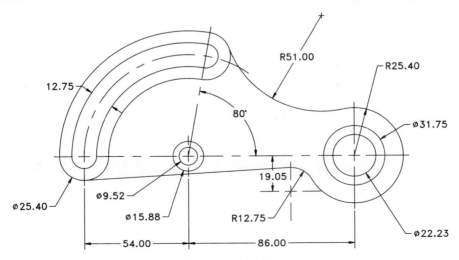

Figure 5–123

PROBLEMS FOR CHAPTER 5

Directions for Problems 5–1 through 5–18:

Construct these geometric construction figures using existing AutoCAD commands.

PROBLEM 5–1

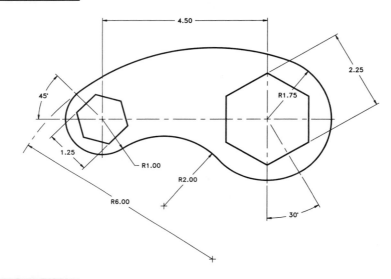

PROBLEM 5–2

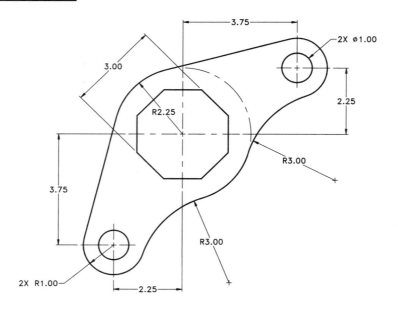

PROBLEM 5-3

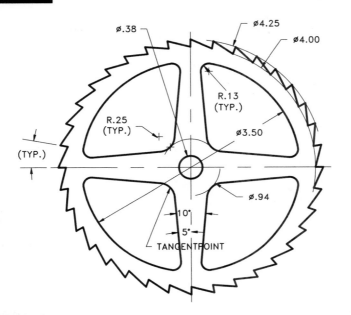

PROBLEM 5-4

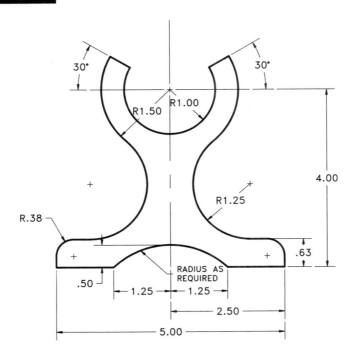

PROBLEM 5-5

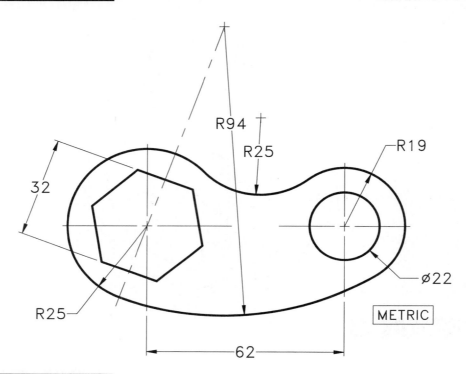

PROBLEM 5-6

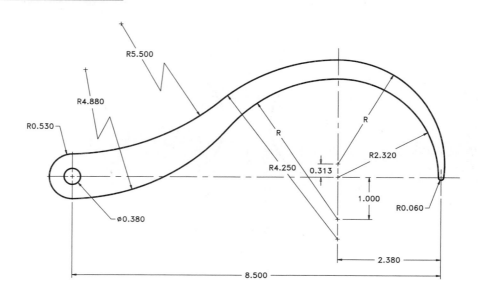

PROBLEM 5–7

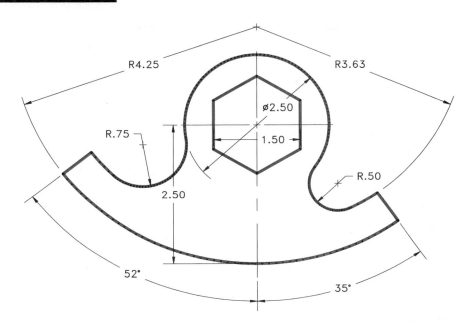

R4.25

R3.63

R.75

ø2.50

1.50

R.50

2.50

52°

35°

PROBLEM 5–8

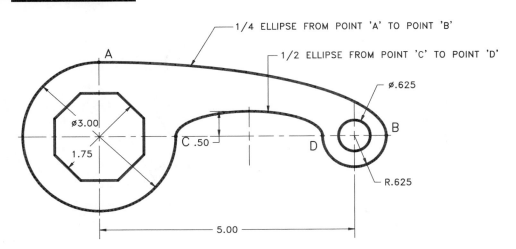

1/4 ELLIPSE FROM POINT 'A' TO POINT 'B'

1/2 ELLIPSE FROM POINT 'C' TO POINT 'D'

A

ø.625

ø3.00

C .50

D

B

1.75

R.625

5.00

PROBLEM 5-9

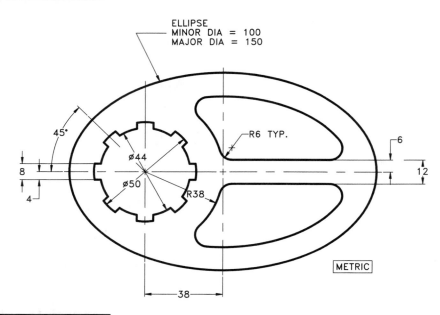

ELLIPSE
MINOR DIA = 100
MAJOR DIA = 150

R6 TYP.

45°

Ø44

Ø50

R38

8

4

6

12

38

METRIC

PROBLEM 5-10

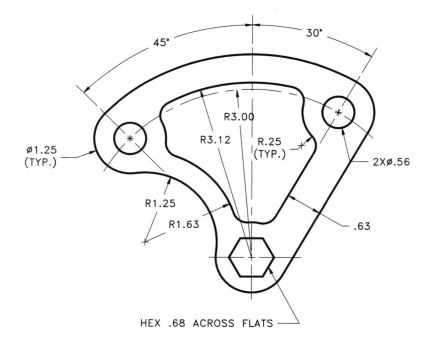

45°

30°

R3.00

Ø1.25
(TYP.)

R3.12

R.25
(TYP.)

2XØ.56

R1.25

R1.63

.63

HEX .68 ACROSS FLATS

PROBLEM 5-11

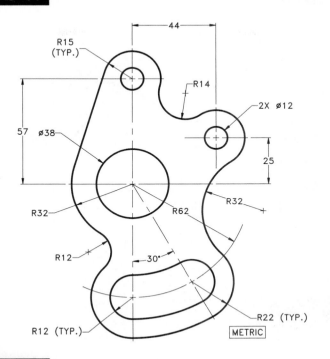

R15
(TYP.)

44

R14

2X ⌀12

57   ⌀38

25

R32

R62

R32

R12

30°

R22 (TYP.)

R12 (TYP.)

METRIC

PROBLEM 5-12

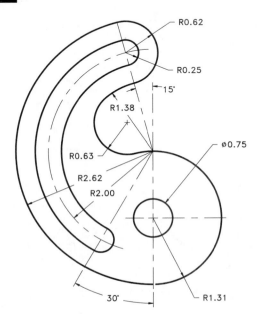

R0.62

R0.25

15°

R1.38

⌀0.75

R0.63

R2.62

R2.00

30°

R1.31

problem EXERCISE

PROBLEM 5-13

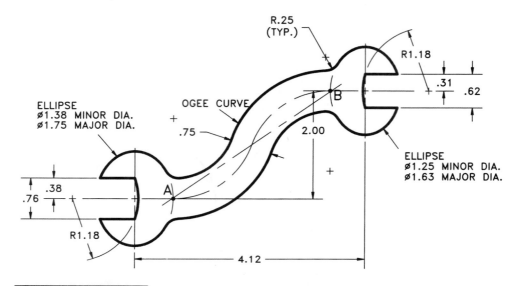

R.25
(TYP.)

R1.18

.31

.62

ELLIPSE
ø1.38 MINOR DIA.
ø1.75 MAJOR DIA.

OGEE CURVE

B

.75

2.00

ELLIPSE
ø1.25 MINOR DIA.
ø1.63 MAJOR DIA.

.38

A

.76

R1.18

4.12

PROBLEM 5-14

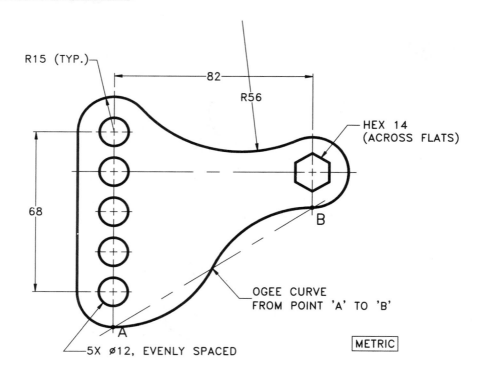

R15 (TYP.)

82

R56

HEX 14
(ACROSS FLATS)

68

B

OGEE CURVE
FROM POINT 'A' TO 'B'

A

5X ø12, EVENLY SPACED

METRIC

PROBLEM 5-15

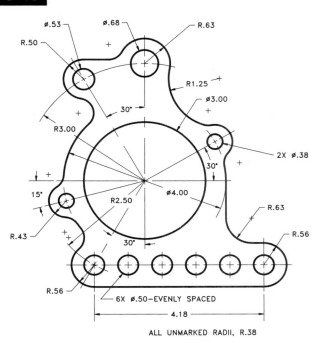

ALL UNMARKED RADII, R.38

PROBLEM 5-16

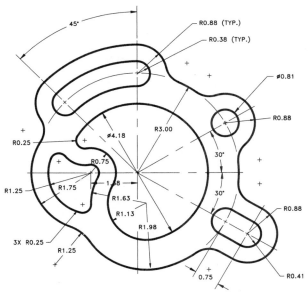

ALL UNMARKED RADII R .63

320

problem EXERCISE

PROBLEM 5-17

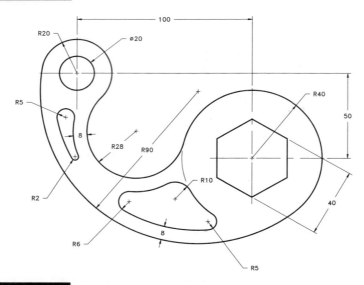

PROBLEM 5-18

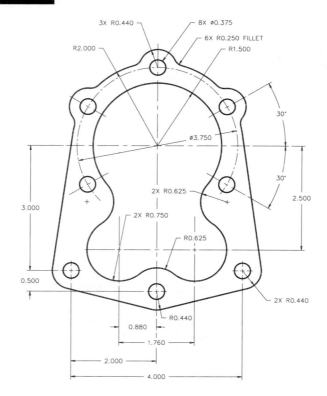

CHAPTER 6

Adding Text to Your Drawing

AUTOCAD TEXT COMMANDS

Text can be added to your AutoCAD drawing file through two methods: Multiline Text (the MTEXT command) and Single Line Text (the DTEXT command). You can find both these commands by choosing the Text cascading menu from the Draw pull-down menu, as shown in Figure 6–1.

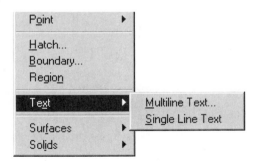

Figure 6–1

THE DTEXT COMMAND

The DTEXT command stands for Dynamic Text mode and allows you to place text in a drawing and view the text as you type it. Choose the Draw menu and then choose Single Line Text. All of the following text examples are displayed in a font called RomanS even though the default text font is called Txt. More information about text fonts will be discussed with the STYLE command.

When using the DTEXT command, you are prompted to specify a start point, height, and rotation angle. You are then prompted to enter the actual text. As you do this, each letter displays on the screen. When you are finished with one line of text, pressing ENTER drops the Insert bar to the next line where you can enter more text. Again pressing ENTER drops the Insert bar down to yet another line of text (see Figure 6–2A). Pressing ENTER at the "Enter text" prompt exits the DTEXT command and permanently adds the text to the database of the drawing (see Figure 6–2B).

Command: **DT** *(For DTEXT)*
Current text style: "Standard" Text height: 0.2000
Specify start point of text or [Justify/Style]: *(Pick a point at "A")*
Specify height <0.2000>: **0.50**
Specify rotation angle of text <0>: *(Press ENTER to accept this default value)*
Enter text: **AutoCAD** *(After this text is entered, press ENTER to drop to the next line of text)*
Enter text: **2000** *(After this text is entered, press ENTER to drop to the next line of text)*
Enter text: *(Either add more text or press ENTER to exit this command and place the text)*

(A)

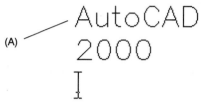

Figure 6–2A    **Figure 6–2B**

By default, the justification mode used by the DTEXT command is left-justified. Study Figure 6–3 and the following command sequence to place the text string "MECHANICAL."

Command: **DT** *(For DTEXT)*
Current text style: "Standard" Text height: 0.2000
Specify start point of text or [Justify/Style]: *(Pick a point at "A")*
Specify height <0.2000>: **0.50**
Specify rotation angle of text <0>: *(Press ENTER to accept this default value)*
Enter text: **MECHANICAL**
Enter text: *(Press ENTER to place the text and exit this command)*

Figure 6–4 and the following command sequence demonstrate justifying text by a center point.

Command: **DT** *(For DTEXT)*
Current text style: "Standard" Text height: 0.2000
Specify start point of text or [Justify/Style]: **C** *(For Center)*
Specify center point of text: *(Pick a point at "A")*
Specify height <0.2000>: **0.50**
Specify rotation angle of text <0>: *(Press ENTER to accept this default value)*
Enter text: **CIVIL ENGINEERING**
Enter text: *(Press ENTER to place the text and exit this command)*

MECHANICAL   CIVIL ENGINEERING
(A)   (A)

Figure 6–3    **Figure 6–4**

Figure 6–5 and the following command sequence demonstrate justifying text by a middle point.

Command: **DT** *(For DTEXT)*
Current text style: "Standard" Text height: 0.2000
Specify start point of text or [Justify/Style]: **M** *(For Middle)*
Specify middle point of text: *(Pick a point at "A")*
Specify height <0.2000>: **0.50**
Specify rotation angle of text <0>: *(Press* ENTER *to accept this default value)*
Enter text: **CIVIL ENGINEERING**
Enter text: *(Press* ENTER *to place the text and exit this command)*

Figure 6–6 and the following command sequence demonstrate justifying text by aligning the text between two points. The text height is automatically scaled depending on the length of the points and the number of letters that make up the text.

Command: **DT** *(For DTEXT)*
Current text style: "Standard" Text height: 0.5000
Specify start point of text or [Justify/Style]: **A** *(For Aligned)*
Specify first endpoint of text baseline: *(Pick a point at "A")*
Specify second endpoint of text baseline: *(Pick a point at "B")*
Enter text: **MECHANICAL**
Enter text: *(Press* ENTER *to place the text and exit this command)*

CIVIL ENGINEERING
(A)

Figure 6–5

MECHANICAL
(A)   (B)

Figure 6–6

Figure 6–7 and the following command sequence demonstrate justifying text by fitting the text in between two points and specifying the text height. Notice how the text appears compressed due to the large text height and short distance of the text line.

Command: **DT** *(For DTEXT)*
Current text style: "Standard" Text height: 0.2000
Specify start point of text or [Justify/Style]: **F** *(For Fit)*
Specify first endpoint of text baseline: *(Pick a point at "A")*
Specify second endpoint of text baseline: *(Pick a point at "B")*
Specify height <0.2000>: **0.50**
Enter text: **MECHANICAL**
Enter text: *(Press* ENTER *to place the text and exit this command)*

Figure 6–8 and the following command sequence demonstrate justifying text by a point at the right.

Command: **DT** *(For DTEXT)*
Current text style: "Standard" Text height: 0.2000
Specify start point of text or [Justify/Style]: **R** *(For Right)*
Specify right endpoint of text baseline: *(Pick a point at "A")*
Specify height <0.2000>: **0.50**
Specify rotation angle of text <0>: *(Press ENTER to accept this default value)*
Enter text: **MECHANICAL**
Enter text: *(Press ENTER to place the text and exit this command)*

(A)   (B)

Figure 6–7

MECHANICAL

(A)

Figure 6–8

Figure 6–9 and the following command sequence demonstrate rotating text along a user-specified angle.

Command: **DT** *(For DTEXT)*
Current text style: "Standard" Text height: 0.2000
Specify start point of text or [Justify/Style]: *(Pick a point at "A")*
Specify height <0.2000>: **0.50**
Specify rotation angle of text <0>: **45**
Enter text: **ENGINEERING**
Enter text: **DESIGN**
Enter text: **GRAPHICS**
Enter text: *(Press ENTER to place the text and exit this command)*

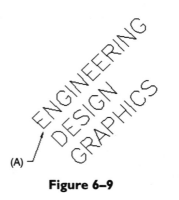

(A)

Figure 6–9

TEXT JUSTIFICATION MODES

Figure 6–10A illustrates a sample text item and the various locations by which the text can be justified. By default, when you place text using the DTEXT command, it is left-justified. Enter one of the sets of letters shown in Figure 6–10B to justify text in a different location when prompted to "Specify start point of text or [Justify/Style]."

Figure 6–10A

LEFT	Align Left (Default)
C	Center
M	Middle
R	Right
TL	Top/Left
TC	Top/Center
TR	Top/Right
ML	Middle/Left
MC	Middle/Center
MR	Middle/Right
BL	Bottom/Left
BC	Bottom/Center
BR	Bottom/Right
A	Align
F	Fit

Figure 6–10B

SPECIAL TEXT CHARACTERS USED WITH DTEXT

Special text characters called control codes enable you to apply certain symbols and styles to text objects. All control codes begin with the double percent sign (%%) followed by the special character that invokes the symbol. These special text characters are explained in the following paragraphs.

UNDERSCORE (%%U)

Use the double percent signs followed by the letter "U" to underscore a particular text item. Figure 6–11 shows the word "MECHANICAL," which is underscored. When you are prompted to enter the text, type the following:

Enter text: **%%UMECHANICAL**

MECHANICAL

Figure 6–11

DIAMETER SYMBOL (%%C)

The double percent signs followed by the letter "C" create the diameter symbol shown in Figure 6–12. When you are prompted to enter text, the diameter symbol is displayed if you type the following:

Enter text: **%%C0.375**

Ø0.375

Figure 6–12

PLUS/MINUS SYMBOL (%%P)

The double percent signs followed by the letter "P" create the plus/minus symbol shown in Figure 6–13. When you are prompted to enter text, the plus/minus symbol is displayed if you type the following:

Enter text: **%%P0.005**

±0.005

Figure 6–13

DEGREE SYMBOL (%%D)

The double percent signs followed by the letter "D" create the degree symbol shown in Figure 6–14. When you are prompted to enter text, the degree symbol is displayed if you type the following:

Enter text: **37%%D**

37°

Figure 6–14

OVERSCORE (%%O)

Similar to the underscore, the double percent signs followed by the letter "O" overscore a text object as shown in Figure 6–15. When you are prompted to enter text, the overscore is displayed if you type the following:

Enter text: **%%OMECHANICAL**

MECHANICAL

Figure 6–15

COMBINING SPECIAL TEXT CHARACTER MODES

Figure 6–16 shows how the control codes are used to toggle on or off the special text characters. Enter the following at the text prompt:

Enter text: **%%UTEMPERATURE%%U 29%%D F**

The first "%%U" toggles underscore mode on and the second "%%U" turns under-score off.

TEMPERATURE 29° F

Figure 6–16

ADDITIONAL DYNAMIC TEXT APPLICATIONS

You place a line of text using the DTEXT command and press ENTER. This drops the Insert bar down to the next line of text. This continues until you press ENTER at the "Enter text" prompt, which exits the command and places the text permanently in the drawing. You can also control the placement of the Insert bar by clicking a new location in response to the "Enter text" prompt. In Figure 6–17, various labels need to be placed in the pulley assembly. The first label, "BASE PLATE," is placed with the DTEXT command. After pressing ENTER, pick a new location at "B" and place the text "SUPPORT." Continue this process with the other labels. The Insert bar at "E" denotes the last label that needs to be placed. When you perform this operation, pressing ENTER one last time at the "Enter text" prompt places the text and exits the command. Follow the command sequence below for a better idea on how to perform this operation.

Command: **DT** *(For DTEXT)*
Current text style: "STANDARD" Text height: 0.3000
Specify start point of text or [Justify/Style]: *(Pick a point at "A")*
Specify height <0.3000>: **0.25**
Specify rotation angle of text <0>: *(Press* ENTER *to accept this default value)*
Enter text: **BASE PLATE** *(After this text is entered, pick approximately at "B")*
Enter text: **SUPPORT** *(After this text is entered, pick approximately at "C")*
Enter text: **SHAFT** *(After this text is entered, pick approximately at "D")*
Enter text: **PULLEY** *(After this text is entered, pick approximately at "E")*
Enter text: *(You would enter* **BUSHING** *for this part and then press* ENTER*)*
Enter text: *(Press* ENTER *to exit this command)*

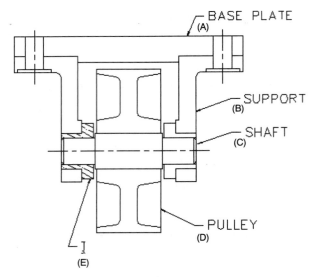

Figure 6–17

CREATING DIFFERENT TEXT STYLES

A text style is a collection of settings that are applied to text placed with the DTEXT or MTEXT command. These settings could include presetting the text height and font in addition to providing special affects, such as an oblique angle for inclined text. Choose Text Style... from the Format pull-down menu shown in Figure 6–18A to create different text styles through the Text Style dialog box, shown in Figure 6–18B.

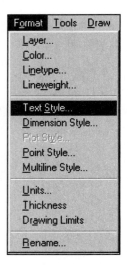

Figure 6–18A

Use the Text Style dialog box in Figure 6–18B to create new text styles. As numerous styles are created, this dialog can also be used to make an existing style current in the drawing. By default, when you first begin a drawing, the current Style Name is STAN-DARD. It is considered good practice to create your own text style and not rely on STANDARD. Once a new style is created, a font name is matched with the style. Clicking in the edit box for Font Name displays a list of all text fonts supported by the operating system. These fonts have different extensions such as SHX and TTF. TTF or TrueType Fonts are especially helpful in AutoCAD because these fonts display in the drawing in their true form. If the font is bold and filled-in, the font in the drawing file displays as bold and filled-in. When a Font Name is selected, it displays in the Preview area located in the lower right corner of the dialog box. The Effects area allows you to display the text upside down, backwards, or vertical. Other effects include a width factor, explained later, and the oblique angle for text displayed at a slant.

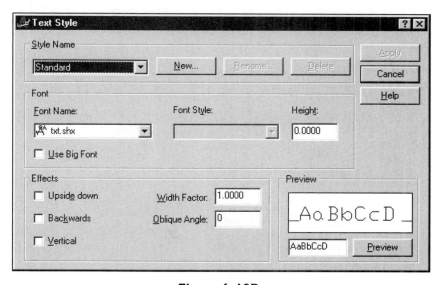

Figure 6–18B

Clicking on the New... button of the Text Style dialog box displays the New Text Style dialog box shown in Figure 6–19. A new style is created called GENERAL NOTES. Clicking on the OK button returns you to the Text Style dialog box shown in Figure 6–20. Clicking in the Font Name edit box displays all supported fonts. Clicking on Romand.shx assigns the font to the style name GENERAL NOTES. Clicking on the Apply button saves the font to the database of the current drawing file. In the Text Style dialog box shown in Figure 6–20, an edit box is present that deals with the text height. By default, the value is set to 0.0000 units. This allows the DTEXT or MTEXT command to control the text height. Entering a value for Height in the edit box places all text under this style at that height. When you use the DTEXT or MTEXT commands, the text height is already set to a value defined in the Text Style dialog box.

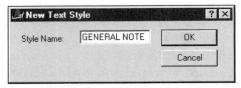

Figure 6–19

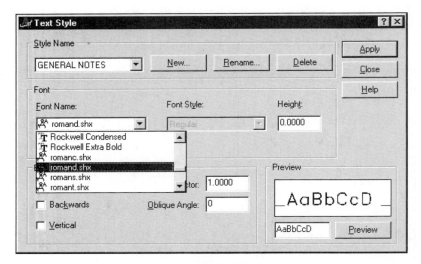

Figure 6–20

The Effects area of the Text Style dialog box, shown in Figure 6–21A, allows for other settings to be applied to the text style being created. A few of the more unusual settings are "Upside down" and "Backwards," illustrated in Figure 6–21B. Other more useful settings include the ability to have text displayed vertically, adding a Width Factor to condense or expand text, and an Obliquing Angle used to provide a slant or incline in the text (see Figure 6–21B).

Figure 6–21A

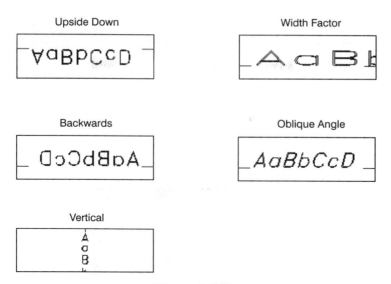

Figure 6–21B

All of the images in Figure 6–21B were taken from the Preview area located in the lower right corner of the Text Style dialog box. All changes to fonts and effects will be updated in this area. Changing the text height will not be reflected in the Preview area. By default, the letters "AaBbCcDd" preview. You can enter different letters and have them preview to have an idea of how these letters or numbers will appear based on the selected font. In Figure 6–22, the characters "1 Q X 9" are entered in the box provided. Clicking on the Preview button displays these characters in the Preview area.

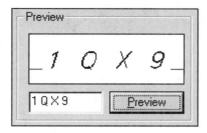

Figure 6–22

TEXT FONTS SUPPORTED IN AUTOCAD 2000

Two types of text fonts can be used inside AutoCAD drawings. One type of font has the extension .SHX. This is a native AutoCAD font, which has been compiled after the font was defined through a series of pen motions that created the shape of a letter. The other type of font has an extension .TTF, which stands for true type font. This font consists of high quality text. Also, a variety of true type fonts are available to add better contrast to your drawing. Figure 6–23 shows the names of a few of these native AutoCAD and true type fonts. The AutoCAD-defined fonts have a Caliper icon adjacent to the name of the font. The overlapping letters TT, also adjacent to the name of the font, identify the true type font.

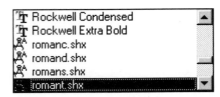

Figure 6–23

To have an idea of how the various fonts would appear using the same text, see Figure 6–24. The text "AutoCAD 2000" is displayed in the Txt, Swiss Block Outline Bold, City Blueprint, Roman Simplex, and Roman Duplex fonts. By default, a new drawing is automatically assigned to construct text in the Txt font. Use the Text Style dialog box to assign new text fonts. While the true type fonts appear inviting, care needs to be taken with using these fonts in your drawing. Having a large number of text items and notes in a true type font will cause your drawing to slow down considerably compared to native AutoCAD font use. In Figure 6–24, two very popular native AutoCAD fonts are displayed. The Roman Simplex font is similar to the standard block lettering style used in early engineering drawings. To make the text stand out, Roman Duplex is also available. This is similar to the Roman Simplex font, with the exception that two sets of lines are present to describe each letter and number.

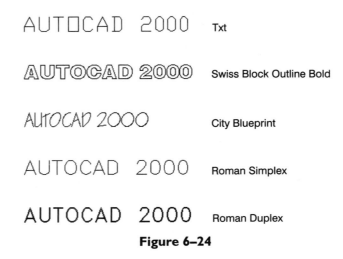

Figure 6–24

CHARACTER MAPPING FOR SYMBOL FONTS

A few of the native AutoCAD fonts actually consist of symbols, which can be displayed in your drawing through the DTEXT command. These symbols include mathematical, astronomical, music, and even mapping symbols. In order for these symbols to display, each symbol must be mapped to a particular letter of the alphabet. The image in Figure 6–25 displays all symbols and the corresponding letter used to bring up the symbol.

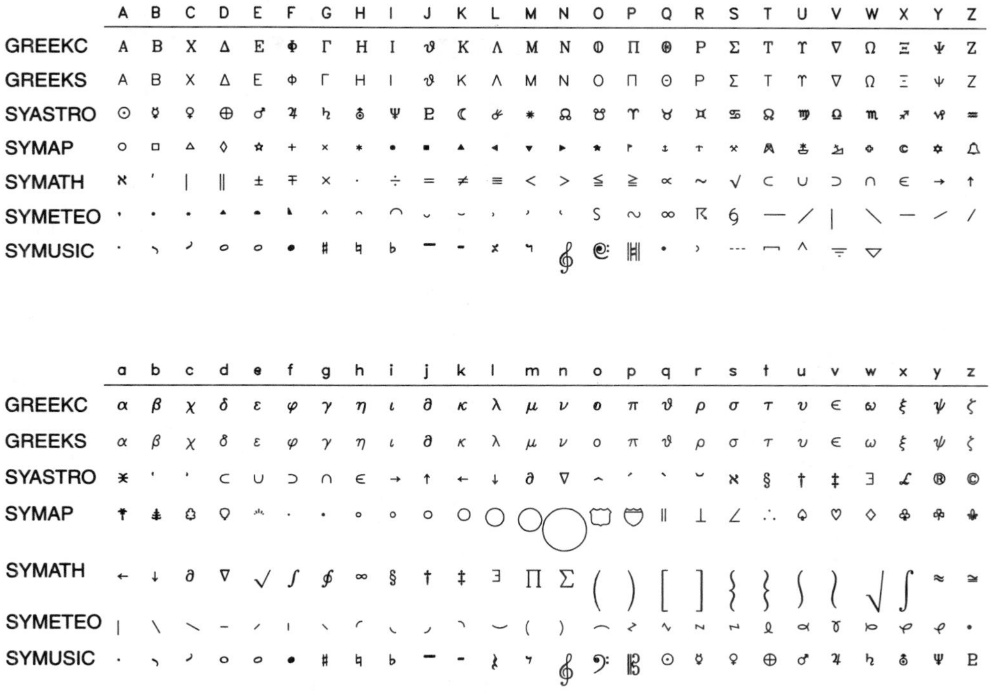

	A	B	C	D	E	F	G	H	I	J	K	L	M	N	O	P	Q	R	S	T	U	V	W	X	Y	Z
GREEKC	Α	Β	Χ	Δ	Ε	Φ	Γ	Η	Ι	ϑ	Κ	Λ	Μ	Ν	Θ	Π	Θ	Ρ	Σ	Τ	Υ	∇	Ω	Ξ	Ψ	Ζ
GREEKS	Α	Β	Χ	Δ	Ε	Φ	Γ	Η	Ι	ϑ	Κ	Λ	Μ	Ν	Ο	Π	Θ	Ρ	Σ	Τ	Υ	∇	Ω	Ξ	Ψ	Ζ
SYASTRO	☉	☿	♀	⊕	♂	♃	♄	♅	Ψ	♇	☽	☌	✱	♈	♉	♊	♋	♌	♍	♎	♏	♐	♑	♒	♓	☷
SYMAP	○	□	△	◇	☆	+	×	*	•	■	▲	◀	▼	▶	★	⌐	⌙	⊤	×	☗	♨	♺	◆	●	✿	△
SYMATH	א	′	|	‖	±	∓	×	·	÷	=	≠	≡	<	>	≤	≥	∝	~	√	⊂	∪	⊃	∩	∈	→	↑
SYMETEO	·	·	·	▲	■	╲	^	⌒	∩	⌣	⌣	′	′	′	S	∽	∞	Γ	ϛ	—	∕	|	╲	−	⁄	/
SYMUSIC	·	♩	♪	○	○	●	♯	♮	♭	—	-	✗	⌐	𝄞	𝄢	𝄃	·	♪	⋯	⌐	∧	≂	▽

	a	b	c	d	e	f	g	h	i	j	k	l	m	n	o	p	q	r	s	t	u	v	w	x	y	z
GREEKC	α	β	χ	δ	ε	φ	γ	η	ι	∂	κ	λ	μ	ν	ο	π	ϑ	ρ	σ	τ	υ	∈	ω	ξ	ψ	ζ
GREEKS	α	β	χ	δ	ε	φ	γ	η	ι	∂	κ	λ	μ	ν	ο	π	ϑ	ρ	σ	τ	υ	∈	ω	ξ	ψ	ζ
SYASTRO	✳	′	′	⊂	∪	⊃	∩	∈	→	↑	←	↓	∂	∇	⌢	′	`	˘	א	§	†	‡	∃	ℒ	®	©
SYMAP	♁	♆	⚘	♀	ˌ	·	·	○	○	○	○	◯	◯	◯	⬡	⬢	‖	⊥	∠	∴	◊	♡	◇	♣	♠	♣
SYMATH	←	↓	∂	∇	√	∫	∮	∞	§	†	‡	∃	Π	Σ	(	)	[	]	}	{	}	{	√	∫	≈	≅
SYMETEO	|	╲	╲	−	⁄	|	╲	⌐	⌣	⌣	⌣	⌣	(	)	⌐	⌐	∧	∼	⌐	♌	♈	♉	♀	ϕ	ϱ	·
SYMUSIC	·	♩	♪	○	○	●	♯	♮	♭	—	-	♪	⌐	𝄞	𝄢	𝄡	☉	☿	♀	⊕	♂	♃	♄	♅	Ψ	♇

Figure 6–25

The first step before displaying these symbols is to create a new text style using the Text Style dialog box in Figure 6–26. Here, the style name Map_Symbols is assigned the font called "Symap.SHX," which contains all mapping symbols. Notice the appearance of a few of these symbols in the Preview area located in the lower right corner of the dialog box.

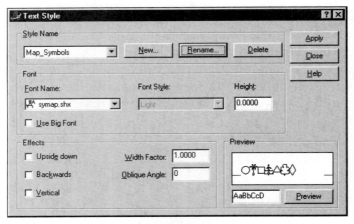

Figure 6–26

Next, use the chart in Figure 6–25 along with the DTEXT command to place a few of the symbols in the drawing. Figure 6–27 shows the effects of the DTEXT command while the letters "o," "p," "Q," "R," and "S" are entered. It is important to note that a different symbol is mapped to each uppercase and lowercase letter.

Command: **DT** *(For DTEXT)*
Current text style: "Map_Symbols" Text height: 0.5000
Specify start point of text or [Justify/Style]:
Specify height <0.2000>: **0.50**
Specify rotation angle of text <0>: *(Press ENTER to accept this default value)*
Enter text: **o**
Enter text: **p**
Enter text: **Q**
Enter text: **R**
Enter text: **S**
Enter text: *(Press ENTER to place the map symbols and exit this command)*

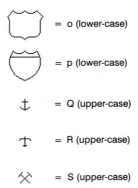

Figure 6–27

THE MTEXT COMMAND

The MTEXT command allows for the placement of text in multiple lines. Entering the MTEXT command from the command prompt displays the following prompts:

Command: **MT** *(For MTEXT)*
Current text style: "GENERAL NOTES" Text height: 0.2000
Specify first corner: *(Pick a point at "A" to identify one corner of the Mtext box)*
Specify opposite corner or [Height/Justify/Line spacing/Rotation/Style/Width]: *(Pick another corner forming a box).*

Picking a first corner displays a user-defined box with an arrow at the bottom similar to that shown in Figure 6–28. This box will define the area where the multiline text will be placed. If the text cannot fit on one line, it will wrap to the next line automatically. After you click on a second point marking the other corner of the insertion box, the Multiline Text Editor dialog box appears, similar to Figure 6–29. As you begin typing text, it appears in this box.

(A)

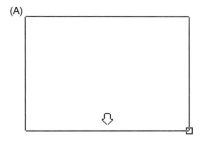

Figure 6–28

Figure 6–29

In the dialog box shown in Figure 6–29, as the multiline text is entered, it automatically wraps to the next line. The current text font is displayed in addition to the text height. You can make the text bold, italicized, or underlined by using the B, I, and U buttons present at the top of the dialog box. When you have entered the text, click on the OK button to dismiss the dialog box and place the text in the drawing, as shown in Figure 6–30.

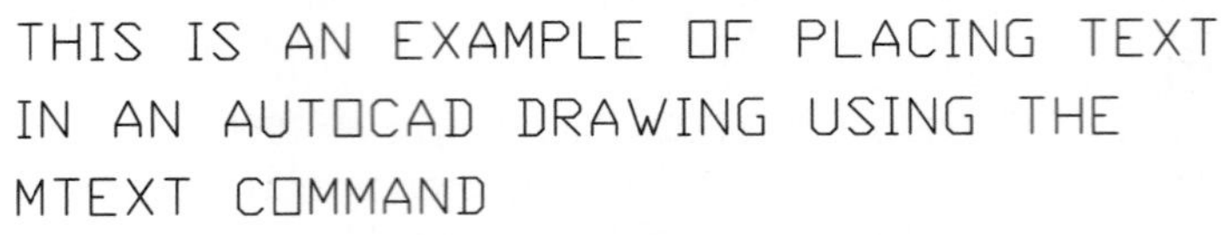

THIS IS AN EXAMPLE OF PLACING TEXT
IN AN AUTOCAD DRAWING USING THE
MTEXT COMMAND

Figure 6–30

Another feature of the Character tab is the ability to assign a symbol in the form of a degree, diameter, or plus/minus symbol. Click on the Symbol button as shown in Figure 6–31A. In the case of the diameter symbol, notice the appearance of the special code "%%c." This is used to place the diameter symbol in the Mtext object. Clicking on the OK button exits the Multiline Text Editor dialog box and displays the text in Figure 6–31B.

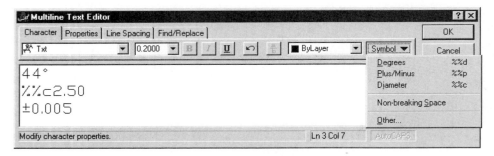

Figure 6–31A

44°
Ø2.50
±0.005

Figure 6–31B

Other capabilities of the character tab include the reformatting of fractional text. Entering a space after the fraction value in Figure 6–32A displays the AutoStack Properties dialog box in Figure 6–32B. Use this dialog box to enable the AutoStacking of fractions. You can also remove the leading space in between the whole number and the fraction. Dismissing this dialog box and completing the fractional text expression displays the results in Figure 6–32C.

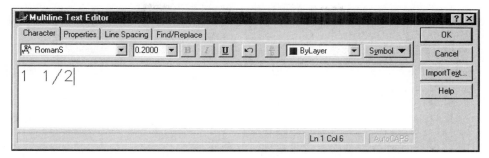

Figure 6–32A

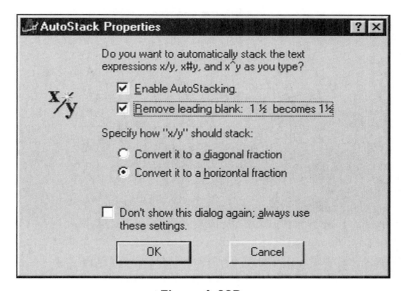

Figure 6–32B

$1\frac{1}{2}$ O.D. PIPE DIAMETER

Figure 6–32C

Clicking on the Properties tab of the Multiline Text Editor displays information illustrated in Figure 6–33A. Under this tab, you can change the text style, choose a different justification mode, change the width of the original box created when the MTEXT command was started, and change the rotation angle of the text. Making any of these changes will affect new text entered in the dialog box. If you want to make changes to existing text displayed in the Multiline Text Editor dialog box, you must first highlight the text for the changes to take place. Figure 6–33B is an example of left-justified versus center-justified text placed with the MTEXT command.

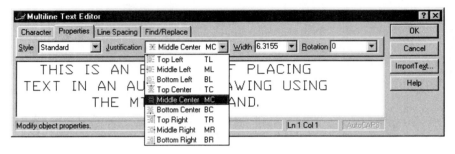

Figure 6–33A

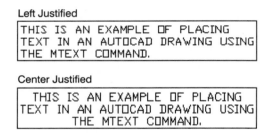

Figure 6–33B

Clicking on the Line Spacing tab displays the dialog box in Figure 6–34A. Use this area to control the spacing between lines of text. Notice the three settings to control this operation. Lines of text can be spaced Single (or Normal), 1.5 Lines, or Double spaced. Figure 6–34B illustrates all three of these settings and how they affect the Mtext object.

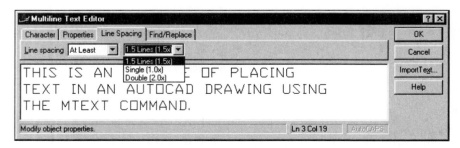

Figure 6–34A

Single

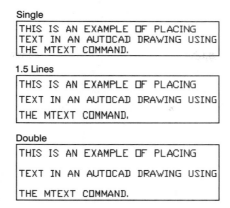

Figure 6–34B

Clicking on the Find/Replace tab of the Multiline Text Editor displays information illustrated in Figure 6–35. Under this tab, you can accomplish the following: you can locate text using the Find edit box; you can replace text using the Replace With edit box; and you can specify the exact case (whether upper- or lowercase or a combination of both) by placing a check in the Match Case box. A button is available at the right of the dialog box designed to import text in ASCII or RTF format. This is especially helpful if the text was generated through a word processor or from such popular Windows Accessories such as Notepad. Figure 6–36 displays the results of replacing "AutoCAD" with "AutoCAD 2000."

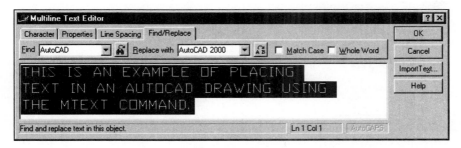

Figure 6–35

Figure 6–36

THE DDEDIT COMMAND

Text constructed with the MTEXT and DTEXT commands is easily modified with the DDEDIT command. Start this command by choosing Text... from the Modify pull-down menu, shown in Figure 6–37A. Selecting the text object in Figure 6–37B displays the Multiline Text Editor in Figure 6–37C. This is the same dialog box as that used to initially create text through the MTEXT command. Use this dialog box to change the text height, font, color, and justification.

Command: **ED** *(For DDEDIT)*
Select an annotation object or [Undo]: *(Select the Mtext object and the Multiline Text Editor dialog box will appear)*
Select an annotation object or [Undo]: *(Pick another Mtext object to edit or press ENTER to exit this command)*

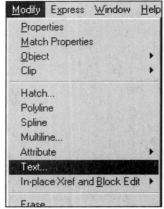

THIS IS AN EXAMPLE OF PLACING TEXT IN AN AUTOCAD DRAWING USING THE MTEXT COMMAND

Figure 6–37A   **Figure 6–37B**

Figure 6–37C

Examples of editing Mtext objects while inside the Multiline Text Editor dialog box are illustrated in Figure 6–38A. In this dialog box, the text currently drawn in the Txt font needs to be changed to RomanD. To accomplish this, first highlight all of the text. Next, change to the desired font. Since all text was highlighted, changing the font to RomanD updates all text in Figure 6–38B. Clicking on the OK button dismisses the dialog box and changes the text font in the drawing shown in Figure 6–38C.

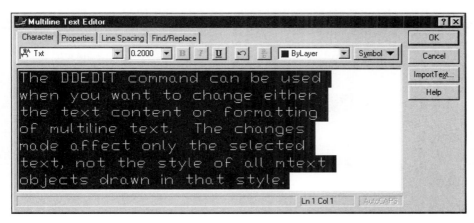

Figure 6–38A

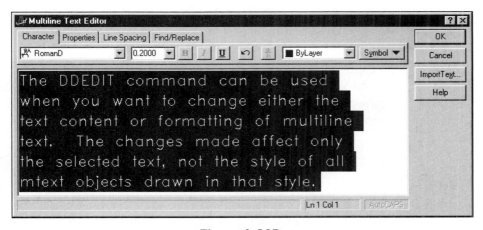

Figure 6–38B

THE DDEDIT COMMAND CAN BE USED
WHEN YOU WANT TO CHANGE EITHER THE
TEXT CONTENT OR FORMATTING OF A
MULTILINE TEXT. THE CHANGES MADE
AFFECT ONLY THE SELECTED TEXT, NOT
THE STYLE THE MTEXT WAS DRAWN IN.

Figure 6–38C

When editing Mtext objects, you can selectively edit only certain words that are part of a multiline text string. In Figure 6–39A, the text "PLACING" is underscored; the text "AutoCAD" is changed to a Swiss font; the text "MTEXT" is increased to a text height of 0.30 units. When performing this type of operation, you only need to highlight the text object you want to change. Clicking on the OK button dismisses the dialog box and changes the individual text objects in the drawing, as in Figure 6–39B.

Figure 6–39A

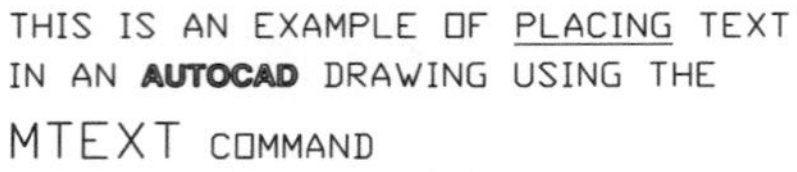

Figure 6–39B

Using the DDEDIT command on a text object created with the DTEXT command, as shown in Figure 6–40A, displays the Edit Text dialog box shown in Figure 6–40B. Use this to change text in the edit box provided. Font, justification, and text height are not supported in this dialog box.

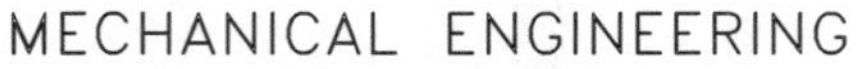

MECHANICAL ENGINEERING

Figure 6–40A

Figure 6–40B

THE SPELL COMMAND

Issuing the SPELL command and selecting the multiline text object displays the Check Spelling dialog box shown in Figure 6–41A, with the following components: the current dictionary and the Current word identified by the spell checker as being misspelled. The presence of the word does not necessarily mean that the word is misspelled, such as "AUTOCAD." The word is not part of the current dictionary.

The Suggestions area displays all possible alternatives to the word identified as being misspelled. The Ignore button allows you to skip the current word; this would be applicable especially in the case of acronyms such as CAD and GDT. Clicking on Ignore All skips all remaining words that match the current word. If the word "AUTOCAD" keeps coming up as a misspelled word, instead of constantly choosing the Ignore button, use Ignore All to skip the word "AUTOCAD" in any future instances. The Change button replaces the word in the Current word box with a word in the Suggestions box. The Add button adds the current word to the current dictionary. The Lookup button checks the spelling of a selected word found in the Suggestions box. The Change Dictionaries button allows you to change to a different dictionary containing other types of words. The Context area displays a phrase showing how the current word is being used in a sentence. Use this area to check proper sentence structure. In Figure 6–41A, the word "AUTOCAD" was identified as being misspelled. Clicking on the Ignore button continues with the spell checking operation until completed. After undergoing spell checking, the Mtext object is displayed in Figure 6–41B.

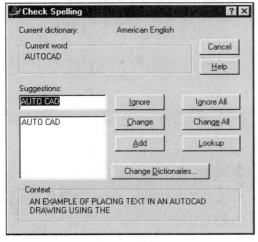

THIS IS AN EXAMPLE OF PLACING TEXT
IN AN AUTOCAD DRAWING USING THE
MTEXT COMMAND

Figure 6–41A   **Figure 6–41B**

THE FIND COMMAND

AutoCAD 2000 provides more capabilities for editing text with a Find and Replace tool. In the multiline text object in Figure 6–42, all words that contain "entity" must be replaced with the new word "object."

Begin this process by starting the Find and Replace tool, which can be selected from the Standard toolbar shown in Figure 6–43A. This activates the Find and Replace dialog box in Figure 6–43B. While inside the dialog box, enter the text to find (entity) and the text to replace with (object). Clicking on the Find button performs a search through the entire drawing and displays the text objects in the Context edit box. AutoCAD finds the search text and highlights the item. Click on the Replace button to replace this text item. Click on the Replace All box to automatically replace all text with the new text "object."

Part of an entity can be removed using the BREAK command. You can break lines, circles, arcs, polylines, ellipses, splines, xlines, and rays. When breaking an entity, you can either select the entity at the first break point and then specify a second break point, or you can select the entire entity and then specify the two break points.

Figure 6–42   **Figure 6–43A**

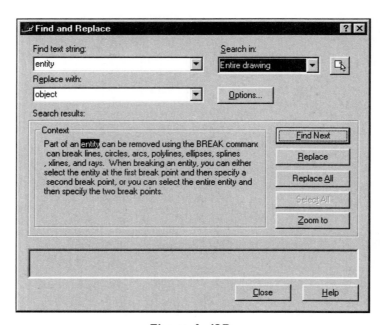

Figure 6–43B

Clicking on the Options… button displays the Find and Replace Options dialog box in Figure 6–43C. You can control the type of text AutoCAD will search for through the check boxes in the dialog box. By default, all text is searched for. To concentrate your search on all text and Mtext objects, be sure to the box adjacent to Text (Mtext, Dtext, Text) is the only one checked.

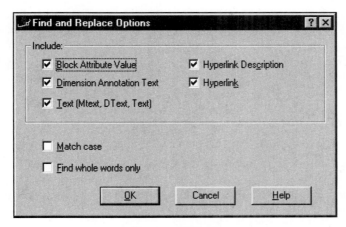

Figure 6–43C

Once the text "entity" is replaced with "object," click the Close button of the Find and Replace text dialog box in Figure 6–43B. This will display the text paragraph in Figure 6–44.

```
Part of an object can be removed using the
BREAK command.  You can break lines,
circles, arcs, polylines, ellipses, splines,
xlines, and rays.  When breaking an object,
you can either select the object at the first
break point and then specify a second
break point, or you can select the entire
object and then specify the two break
points.
```

Figure 6–44

EXPRESS TOOLS APPLICATIONS FOR TEXT

Four Express tools that relate to text are available through the Express Text Toolbar shown in Figure 6–45: the Text Fit (TEXTFIT), Text Mask (TEXTMASK), Explode Text and Arc Aligned Text (ARCTEXT) commands. These and other Express tool commands can also found on the Text cascading menu under the Express pull-down menu, shown in Figure 6–46.

346

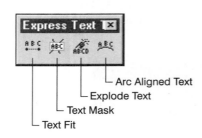

Figure 6–45

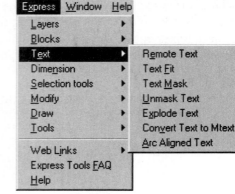

Figure 6–46

REMOTE TEXT

The RTEXT command (Remote Text) allows for an ASCII Text file to be imported into the current drawing. When you identify an Rtext object by File, a box appears similar to Figure 6–47. This is the area that will be occupied by the ASCII text. Picking a location for the text displays the image in Figure 6–48A, where the Rtext object is placed in the drawing but is highlighted because you remain in the command. Entering H for height allows you to change the height of the Rtext object. Entering a new height of 0.40 displays the Rtext object in Figure 6–48B.

Command: **RTEXT**
Current settings: Style=Standard Height=0.2000 Rotation=0
Enter an option [Style/Height/Rotation/File/Diesel] <File>: *(Press* ENTER *to accept this default value)*
Specify start point of RText: *(Pick a location for the Rtext object)*
Current values: Style=Standard Height=0.2000 Rotation=0
Enter an option [Style/Height/Rotation/Edit]: **H** *(For Height)*
Specify height <0.2000>: **0.40**
Current values: Style=Standard Height=0.4000 Rotation=0
Enter an option [Style/Height/Rotation/Edit]: *(Press* ENTER *to exit this command)*

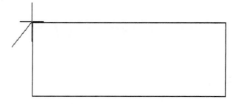

Figure 6–47

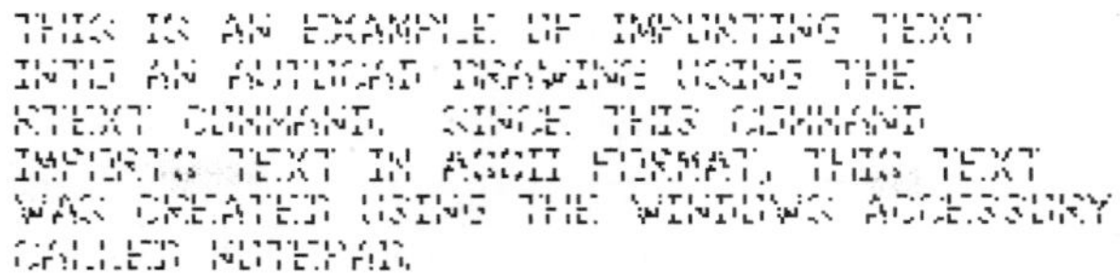

THIS IS AN EXAMPLE OF IMPORTING TEXT
INTO AN AUTOCAD DRAWING USING THE
RTEXT COMMAND. SINCE THIS COMMAND
IMPORTS TEXT IN ASCII FORMAT, THIS TEXT
WAS CREATED USING THE WINDOWS ACCESSORY
CALLED NOTEPAD.

Figure 6–48A

THIS IS AN EXAMPLE OF IMPORTING TEXT
INTO AN AUTOCAD DRAWING USING THE
RTEXT COMMAND. SINCE THIS COMMAND
IMPORTS TEXT IN ASCII FORMAT, THIS TEXT
WAS CREATED USING THE WINDOWS ACCESSORY
CALLED NOTEPAD.

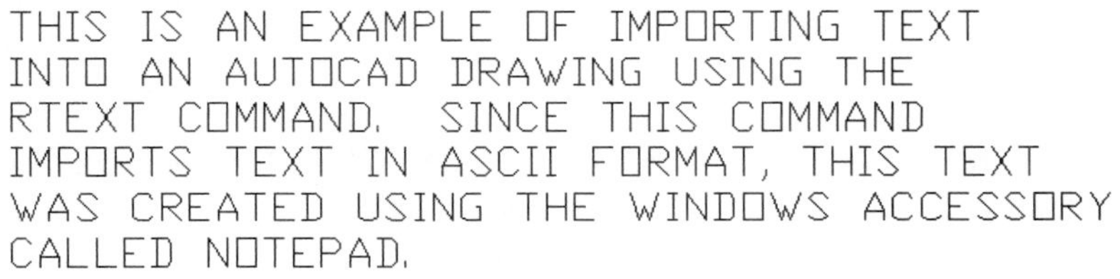

Figure 6–48B

Because this object is considered remote text, conventional text editing commands such as DDEDIT will not function on this object. Use the RTEDIT command to edit the Rtext object. When you enter RTEXT, the application used to originally create the Rtext object launches. If Figure 6–49, the Notepad application launches to edit the Rtext object. Using the EXPLODE command on the Rtext object will convert this object to multiline text.

Command: **RTEDIT**
Select objects: *(Select an Rtext object)*
Current values: Style=Standard Height=0.4000 Rotation=0
Enter an option [Style/Height/Rotation/Edit]: **E** *(For edit, which displays the application the Rtext was created in. When finished, save and exit the application)*
Updates may not be apparent until next regen.
Current values: Style=Standard Height=0.4000 Rotation=0
Enter an option [Style/Height/Rotation/Edit]: *(Pick another option or press* ENTER *to exit this command)*

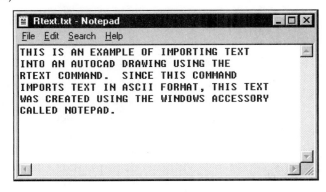

Figure 6–49

TEXT FIT

Use this Express tool to compress or stretch text objects by picking new start and end points. At this time, the command will only operate on text objects made through the DTEXT command. When you select the text to stretch or compress, the operation begins at the original insertion point of the text object. You identify a new start point and the distance of the ending point will determine whether the text is expanded or compressed. To compress text, use the prompt sequence below and the illustration in Figure 6–50.

Command: **TEXTFIT**
Select Text to stretch or shrink: *(Pick the text "INDUSTRIAL TECHNOLOGY")*
Specify end point or [Start point]: **S** *(For starting point)*
Specify new starting point: *(Pick a point at "A" in Figure 6–50)*
ending point: *(Pick a point at "B")*

To expand text, use the prompt sequence below and the illustration in Figure 6–51.

Command: **TEXTFIT**
Select Text to stretch or shrink: *(Pick the text "INDUSTRIAL")*
Specify end point or [Start point]: **S** *(For starting point)*
Specify new starting point: *(Pick a point at "A" in Figure 6–51)*
ending point: *(Pick a point at "B")*

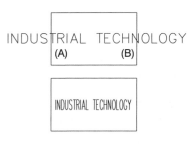

Figure 6–50

Figure 6–51

MASKING AND UNMASKING TEXT

Use the TEXTMASK command to hide all objects that come in contact with a selected text object. This command works on both text and Mtext objects. It uses a rectangular frame and an offset distance from the text. This distance may be specified at the command prompt. TEXTMASK actually works in conjunction with the WIPEOUT command to hide objects crossing over the top of text. Use the prompt sequence below and the illustration in Figure 6–52 to utilize this command.

Command: **TEXTMASK**
Current settings: Offset factor = 0.3500, Mask type = Wipeout
Select text objects to mask or [Masktype/Offset]: **M** *(For Masktype)*
Mask type currently set to Wipeout
Specify entity type to use for mask [Wipeout/3dface/Solid] <Wipeout>: *(Press* ENTER
   to accept this default)
Current settings: Offset factor = 0.3500, Mask type = Wipeout
Select text objects to mask or [Masktype/Offset]: **O** *(For Offset)*
Mask offset currently set to 0.3500
Enter offset factor relative to text height <0.3500>: *(Press* ENTER *to accept this default)*
Current settings: Offset factor = 0.3500, Mask type = Wipeout
Select text objects to mask or [Masktype/Offset]: *(Pick the text objects at "A", "B", and
   "C" in Figure 6–52)*
Current settings: Offset factor = 0.3500, Mask type = Wipeout
Select text objects to mask or [Masktype/Offset]: *(Press* ENTER *to perform the text mask
   operation)*
Masking text with a Wipeout
Wipeout created.
Wipeout created.
Wipeout created.
3 text items have been masked with a Wipeout.

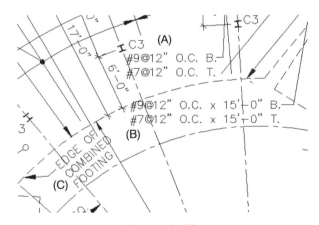

Figure 6–52

The result of using the TEXTMASK command is illustrated at the right in Figure 6–53.
Before, center and object lines crossed through the text objects. Now, the text has a
mask underneath, preventing the lines from crossing through the text object.

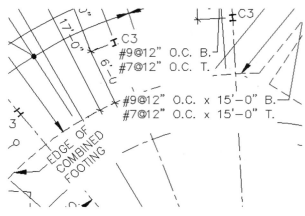

Figure 6–53

Use the TEXTUNMASK command to remove the text mask, using the following command prompt sequence.

Command: **TEXTUNMASK**
Select text or MText object from which mask is to be removed.
Select objects: *(Pick a text object that has been masked)*
Select objects: *(Press ENTER to perform the unmasking operation)*
Removed mask from one text object.

EXPLODING TEXT

Use this Express tool to explode text or Mtext objects into geometry consisting of polylines.

The illustration at "A" in Figure 6–54 shows the word "NORMAL," which consists of a regular Mtext object. When this object is selected, all lines and arcs that define the text highlight.

The illustration at "B" in Figure 6–54 shows the word "EXPLODED." When this text object at "C" is selected, only the line highlights. This is the result of exploding a text or Mtext object; it is broken down into the individual polyline segments. In some cases, the polylines of an exploded text object may be used to machine text into an object using a CAM (Computer-Aided Manufacturing) package. The text polylines act as contours, which guide the milling cutter.

Follow the prompt sequence below for exploding text.

Command: **TXTEXP**
Select text to be EXPLODED:
Select objects: *(Select the word "EXPLODED" in Figure 6–54)*
Select objects: *(Press ENTER to perform the text explode operation)*

Deleting block "WMF0".
I block deleted.
I text object(s) have been exploded to lines.
The line objects have been placed on layer 0.

(A)

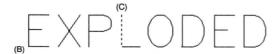
(C)
(B)

Figure 6–54

CONVERTING TEXT TO MTEXT

The TXT2MTXT Express tool is designed to take text that was originally created with the DTEXT command and convert it to a single Mtext object. In Figure 6–55A, five individual lines of text were created with the DTEXT command. From earlier discussions of this command, the editing capabilities are very limited when compared to editing text created with the MTEXT command. Follow the command prompt sequence below to perform this text conversion operation.

Command: **TXT2MTXT**
Select text objects, or [Options]<Options>:
Select objects: *(Press* ENTER *to display the Text to MText Options dialog box in Figure 6–55B)*
Select objects: *(Pick the 5 lines of text in Figure 6–55A)*
Select objects: *(Press* ENTER *to perform the text conversion operation)*
5 Text objects removed, I MText object added.

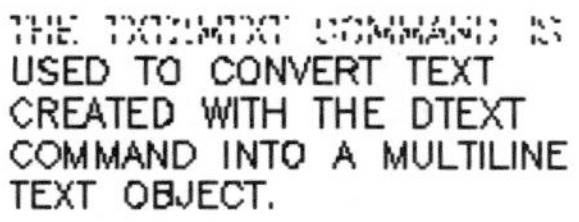

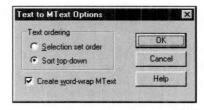

Figure 6–55A

Figure 6–55B

When a single line of text is highlighted in Figure 6–55C, all text highlights, signifying the individual lines of text converted to a single Mtext object.

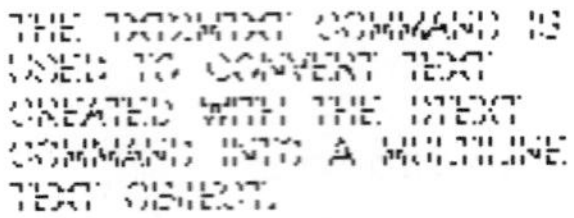

THE TXT2MTXT COMMAND IS
USED TO CONVERT TEXT
CREATED WITH THE DTEXT
COMMAND INTO A MULTILINE
TEXT OBJECT.

Figure 6–55C

ARC ALIGNED TEXT

The ARCTEXT command is used to place text along an arc. Activating the command and selecting an arc displays the dialog box in Figure 6–56. Numerous settings such as text height, width factor, offset distance from the arc, and text font and style information can be changed through this dialog box. Also, controls at the top of the dialog box allow you to place text above or below the arc, to have the text left, right, or center justified, or bold and/or underline the text. Arc text is placed with the following command sequence.

Command: **ARCTEXT**
Select an Arc or an ArcAlignedText: *(Select an arc or an existing arc aligned text object)*

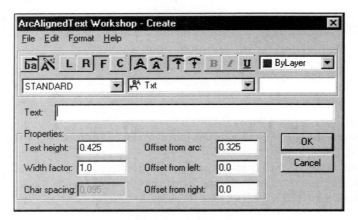

Figure 6–56

To edit arc text, use the same command (ARCTEXT) and select the arc text already placed. The dialog box will again appear similar to Figure 6–56.

Figure 6–57A displays arc text placed above an arc; Figure 6–57B displays arc text placed below an arc; Figure 6–57C displays arc text that has its font changed to RomanD.

Figure 6–57A

Figure 6–57B

Figure 6–57C

TUTORIAL EXERCISE: 06_PUMP_ASSEMBLY.DWG

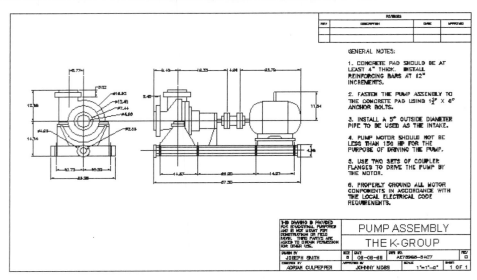

Figure 6–58

Purpose:

This tutorial is designed to create numerous text styles and add different types of text objects to the title block illustrated in Figure 6–58.

System Settings:

Since this drawing is already provided on the CD, open an existing drawing file called "06_Pump_Assembly." Follow the steps in this tutorial for creating a number of text styles and then placing text in the title block area. Check to see that the following Object Snap modes are already set: Endpoint, Extension, Intersection, Center.

Layers:

Various layers have already been created for this drawing. Since this tutorial covers the topic of text, the current layer is "Text."

Suggested Commands:

Open the drawing called "06_Pump_Assembly." Create four text styles called "Company_Drawing," "Title Block Text," "Disclaimer," and "General Notes." The text style "Company_Drawing" will be applied to one part of the title block. "Title Block Text" will be applied to a majority of the title block where questions such as Drawn By, Date, and Scale are asked. A disclaimer will be imported into the title block on the "Disclaimer" text style. Finally a series of notes will be imported in the "General Notes" text style. Once these general notes are imported, a spell check operation will be performed on the notes.

Whenever possible, substitute the appropriate command alias in place of the full AutoCAD command in each tutorial step. For example, use "CP" for the COPY command, "L" for the LINE command, and so on. The complete listing of all command aliases is located in Chapter 1, Table 1–2.

STEP I

Opening the drawing file "06_Pump _Assembly" displays the image similar to Figure 6–59. The purpose of this tutorial is to fill in the title block area with a series of text and Mtext objects. Also, a series of general notes will be placed to the right of the pump assembly. Before you place the text, four text styles will be created to assist with the text creation. The four text styles are also identified in Figure 6–59. The text style "Company_Drawing" will be used to place the name of the company and drawing title. Various information in the form of scale, date, and who performed the drawing will be handled by the text style "Title Block Text." A disclaimer will be imported from an existing .TXT file available on the CD; this will be accomplished in the "Disclaimer" text style. A listing of six general notes is also available on the CD. It is in .RTF format, originally created in Microsoft Word. It will be imported into the drawing and placed in the "General Notes" text style.

Since you will be locating various justification points for the placement of text, turn OSNAP off by single-clicking on OSNAP in the Status bar at the bottom of the display screen.

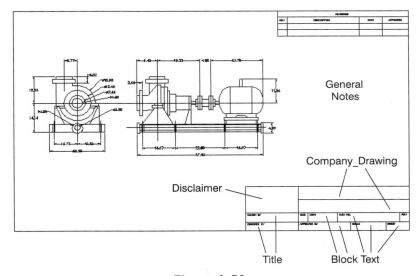

Figure 6–59

STEP 2

Before creating the first text style, first see what styles are already defined in the drawing by choosing Text Style... from the Format pull-down menu in Figure 6–60A. This displays the Text Style dialog box in Figure 6–60B. Click in the Style Name edit box to display the current text styles defined in the drawing. Every AutoCAD drawing contains the STANDARD text style. This is created by default and cannot be deleted. Also, the "Title_Block_Headings" text style is present. This text style is used to create the headings in each title block box such as "Scale" and "Date."

Figure 6–60A

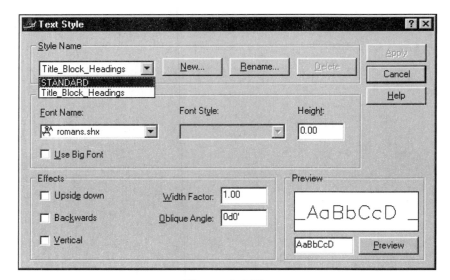

Figure 6–60B

STEP 3

Create the first text style by clicking on the New... button in Figure 6–60B. This will activate the New Text Style dialog box in Figure 6–61A. For Style Name, enter "Company_Drawing." When finished, click the OK button. This will take you back to the Text Style dialog box in Figure 6–61B. In the Font area, change the name of the font to Arial. Notice the font appearing in the Preview area. All company and drawing names will be drawn to a height of 0.25 units. Rather than enter the text height in the DTEXT command, change the text height in the Height edit box from 0.00 to 0.25. This locks the text at 0.25 units whenever you place text in the "Company_Drawing" text style. When finished, click the Apply button to complete the text style creation process.

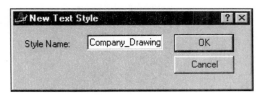

Figure 6–61A

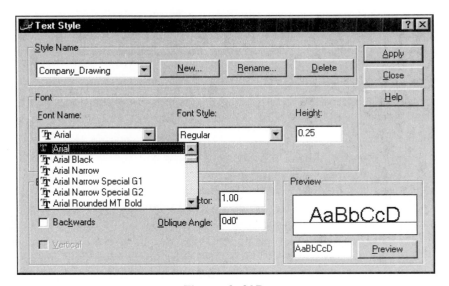

Figure 6–61B

STEP 4

While in the Text Style dialog box, create the next text style by clicking on the New... button and entering "General Notes" as the new text style name. When finished, click the OK button. In the Font area of the Text Style dialog box, change the name of the font to RomanD. This font will appear in the Preview area. All general drawing notes will be drawn to a height of 0.12 units. Change the text height in the Height: edit box from 0.25 to 0.12. This locks the text at 0.12 units whenever you place text in the "General Notes" text style. Your dialog box should appear similar to Figure 6–62. When finished, click the Apply button to complete the text style creation process.

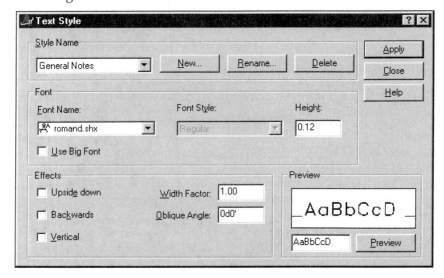

Figure 6–62

STEP 5

While in the Text Style dialog box, create the next text style by clicking on the New... button and entering "Title Block Text" as the new text style name. When finished, click the OK button. In the Font area of the Text Style dialog box, change the name of the font to RomanS. This font will appear in the Preview area. All title block text information such as date and scale will be drawn to a height of 0.10 units. Change the text height in the Height edit box from 0.12 to 0.10. This locks the text at 0.10 units whenever you place text in the "Title Block Text" text style. Your dialog box should appear similar to Figure 6–63. When finished, click the Apply button to complete the text style creation process.

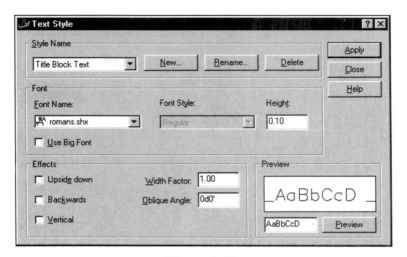

Figure 6–63

STEP 6

While in the Text Style dialog box, create the final text style by clicking on the New... button and entering "Disclaimer" as the new text style name. When finished, click the OK button. In the Font area of the Text Style dialog box, ensure that Romans is still the font. This font will appear in the Preview area. A disclaimer will be imported into an area of the title block and will be drawn to a height of 0.08 units. Change the text height in the Height edit box from 0.10 to 0.08. This locks the text at 0.08 units whenever you place text in the "Disclaimer" text style. Your dialog box should appear similar to Figure 6–64. When finished, click the Apply button to complete the text style creation process.

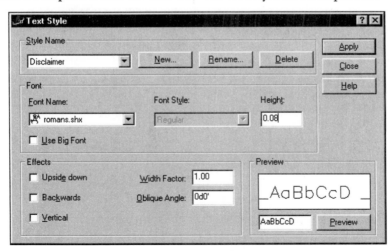

Figure 6–64

STEP 7

Double check that all text styles have been created by clicking in the dropdown box in the Style Name area of the Text Style dialog box. All text styles should be created and appear similar to Figure 6–65.

STEP 8

It is now time to begin placing the first series of text objects. We will start with the drawing and company titles. First, while in the Text Style dialog box, click on the "Company_Drawing" text style in Figure 6–66A. Click the Close button when finished to return to the drawing. Zoom in to the title block in Figure 6–66B. To better assist in picking the text location, turn OSNAP off by clicking in the appropriate button located in the Status bar at the bottom of the screen. Use the DTEXT command to place the two text objects in Figure 6–66B. Both these text objects will be middle justified. The height of the text is already set through the current text style.

Command: **DT** *(For DTEXT)*
Current text style: "Company_Drawing"
   Text height: 0.25
Specify start point of text or [Justify/
   Style]: **M** *(For Middle)*
Specify middle point of text: *(Pick a point
   approximately at "A" in Figure 6–66B)*

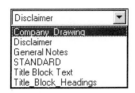

Figure 6–65

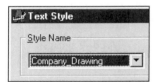

Figure 6–66A

Specify rotation angle of text <0d0'>:
   (Press ENTER *to accept this default value)*
Enter text: **PUMP ASSEMBLY**
Enter text: *(Press* ENTER *to exit this
   command)*
Command: **DT** *(For DTEXT)*
Current text style: "Company_Drawing"
   Text height: 0.25
Specify start point of text or [Justify/
   Style]: **M** *(For Middle)*
Specify middle point of text: *(Pick a point
   approximately at "B" in Figure 6–66B)*
Specify rotation angle of text <0d0'>:
   (Press ENTER *to accept this default value)*
Enter text: **THE K-GROUP**
Enter text: *(Press* ENTER *to exit this
   command)*

(A)

	PUMP ASSEMBLY	
	THE K-GROUP	
DRAWN BY	SIZE	DATE	DWG NO. (B)	REV
CHECKED BY	APPROVED BY	SCALE	SHEET	

Figure 6–66B

STEP 9

Open the Text Style dialog box and click on the "Title Block Text" text style in Figure 6–67A. Click the Close button when finished to return to the drawing. While still zoomed in to the title block, use the DTEXT command to place the nine text objects in Figure 6–67B. All of these text objects will be left justified. The height of the text is already set through the current text style.

Command: **DT** *(For DTEXT)*
Current text style: "Title Block Text"
    Text height: 0.12
Specify start point of text or [Justify/
    Style]: *(Pick a point approximately at "A")*
Specify rotation angle of text <0d0'>:
    (Press ENTER to accept this default value)
Enter text: **JOSEPH SMITH** *(After this text is entered, pick approximately at "B")*
Enter text: **ADRIAN CULPEPPER** *(After this text is entered, pick approximately at "C")*

Enter text: **B** *(After this text is entered, pick approximately at "D")*
Enter text: **JOHNNY MOSS** *(After this text is entered, pick approximately at "E")*
Enter text: **09-08-99** *(After this text is entered, pick approximately at "F")*
Enter text: **AE76998-54C7** *(After this text is entered, pick approximately at "G")*
Enter text: **1" = 1'-0"** *(After this text is entered, pick approximately at "H")*
Enter text: **0** *(After this text is entered, pick approximately at "I")*
Enter text: **1 OF 1** *(After this text is entered, press ENTER)*
Enter text: *(Press ENTER to exit this command)*

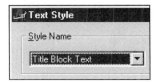

Figure 6–67A

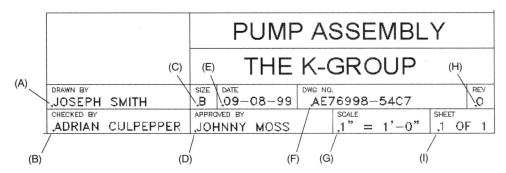

Figure 6–67B

STEP 10

Open the Text Style dialog box and click on the "Disclaimer" text style in Figure 6–68A. Click the Close button when finished to return to the drawing. While still zoomed in to the title block, use the MTEXT command to create a rectangle, as in Figure 6–68B. This will be used to hold the multiline text.

Command: **MT** *(For MTEXT)*
Current text style: "Disclaimer" Text height: 0.08

Specify first corner: *(Pick a point at "A")*
Specify opposite corner or [Height/ Justify/Line spacing/Rotation/Style/ Width]: *(Pick a point at "B")*

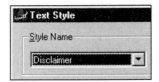

Figure 6–68A

PUMP ASSEMBLY

THE K-GROUP

DRAWN BY	SIZE	DATE	DWG NO.		REV
JOSEPH SMITH	B	09–08–99	AE76998–54C7		0
CHECKED BY	APPROVED BY		SCALE	SHEET	
ADRIAN CULPEPPER	JOHNNY MOSS		1" = 1'–0"	1 OF 1	

Figure 6–68B

When the Mutliline Text Editor dialog box appears, click on the Import Text… button. Then click on the location of your CD and pick the file Disclaimer.txt, as in Figure 6–68C. Since this information was already created in an application outside AutoCAD, it will be imported into the Multiline Text Editor dialog box in Figure 6–68D. The text height and font are already set through the current text style. Click the OK button to place the text in the title block.

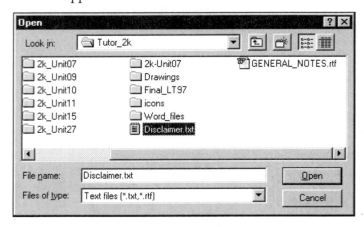

Figure 6–68C

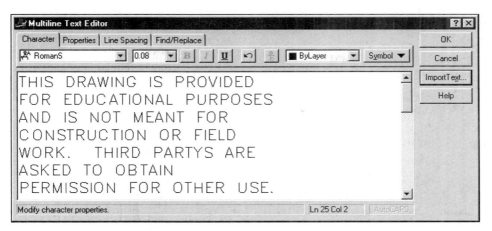

Figure 6–68D

The results of this operation are displayed
in Figure 6–68E.

THIS DRAWING IS PROVIDED FOR EDUCATIONAL PURPOSES AND IS NOT MEANT FOR CONSTRUCTION OR FIELD WORK. THIRD PARTYS ARE ASKED TO OBTAIN PERMISSION FOR OTHER USE.			PUMP ASSEMBLY			
			THE K-GROUP			
DRAWN BY JOSEPH SMITH	SIZE B	DATE 09–08–99	DWG NO. AE76998–54C7			REV O
CHECKED BY ADRIAN CULPEPPER	APPROVED BY JOHNNY MOSS			SCALE 1" = 1'–0"	SHEET 1 OF 1	

Figure 6–68E

STEP 11

Before placing the last text object, zoom to the extents of the drawing. Then open the Text Style dialog box and click on the "General Notes" text style in Figure 6–69A. Click the Close button when finished to return to the drawing. Use the MTEXT command to create a rectangle to the right of the Pump Assembly in Figure 6–69B. This will be used to hold a series of general notes in multiline text format.

Command: **MT** *(For MTEXT)*

Current text style: "General Notes"
   Text height: 0.12
Specify first corner: *(Pick a point at "A")*
Specify opposite corner or [Height/
   Justify/Line spacing/Rotation/Style/
   Width]: *(Pick a point at "B")*

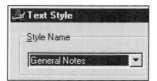

Figure 6–69A

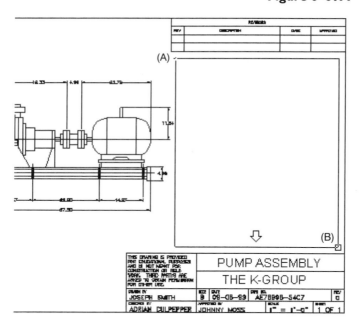

Figure 6–69B

When the Mutliline Text Editor dialog box appears, click on the Import Text… button. Then click on the location of your CD and select the file General_Notes.rtf, as in Figure 6–69C. This rtf file (Rich Text File) was created outside AutoCAD in Microsoft Word. Notice that this file is imported into the Multiline Text Editor dialog box, as in Figure 6–69D. Click the OK button to place the text in the title block.

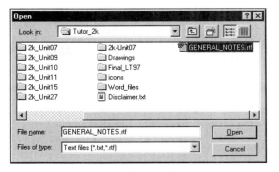

Figure 6–69C

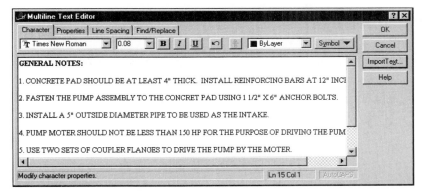

Figure 6–69D

The results of this operation are displayed in Figure 6–69E. However, the text appears smaller in height than 0.12. This is due to the fact that when files are imported in Rtf format, the original format of the text is kept. This means that the Times New Roman font at a height of 0.08 was used even though the current text style has a height of 0.12 and a font of Romand.shx.

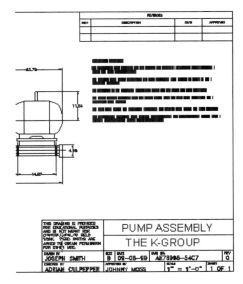

Figure 6–69E

To remedy this, use the DDEDIT command and select one of the general notes. When the Multiline Text Editor dialog box displays, highlight all of the general notes. Change the font to RomanD and the text height to 0.12 units, as in Figure 6–69F.

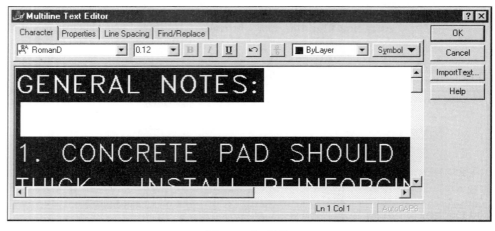

Figure 6–69F

Clicking the OK button in the Multiline Text Editor dialog box displays the text that has been edited. Notice the text height is back to 0.12 (see Figure 6–69G).

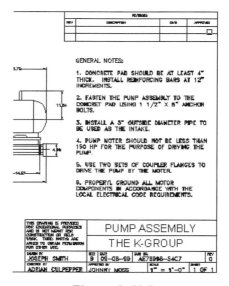

Figure 6–69G

STEP 12

Use the SPELL command and click on the multiline text object that holds the six general notes. When the Check Spelling dialog box appears in Figure 6–70, make the corrections to the words CONCRETE, MOTOR, and PROPERLY.

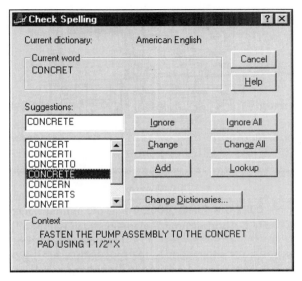

Figure 6–70

STEP 13

Use the DDEDIT command one more time on the general notes. Selecting the general notes displays the Multiline Text Editor dialog box in Figure 6–71A. In the second general note, highlight the fraction "1/2" in Figure 6–71A. Then click on the Stack/Unstack button.

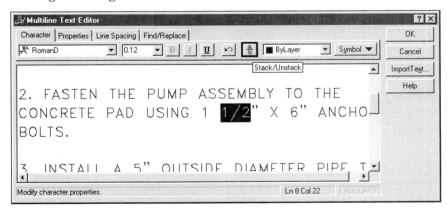

Figure 6–71A

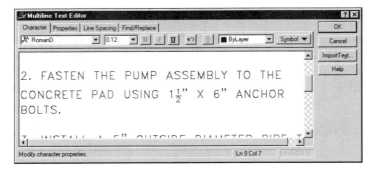

Figure 6–71B

This stacks the fraction in Figure 6–71B.

Clicking the OK button returns to the drawing editor with all changes made to the general notes in Figure 6–71C.

STEP 14

The completed title block is displayed in Figure 6–72.

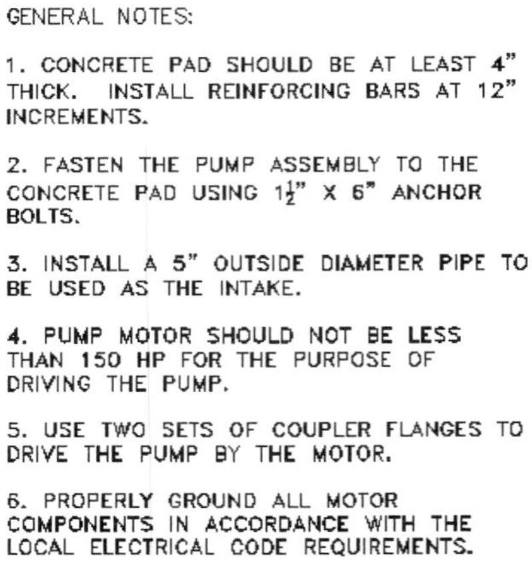

GENERAL NOTES:

1. CONCRETE PAD SHOULD BE AT LEAST 4" THICK. INSTALL REINFORCING BARS AT 12" INCREMENTS.

2. FASTEN THE PUMP ASSEMBLY TO THE CONCRETE PAD USING 1½" X 6" ANCHOR BOLTS.

3. INSTALL A 5" OUTSIDE DIAMETER PIPE TO BE USED AS THE INTAKE.

4. PUMP MOTOR SHOULD NOT BE LESS THAN 150 HP FOR THE PURPOSE OF DRIVING THE PUMP.

5. USE TWO SETS OF COUPLER FLANGES TO DRIVE THE PUMP BY THE MOTOR.

6. PROPERLY GROUND ALL MOTOR COMPONENTS IN ACCORDANCE WITH THE LOCAL ELECTRICAL CODE REQUIREMENTS.

Figure 6–71C

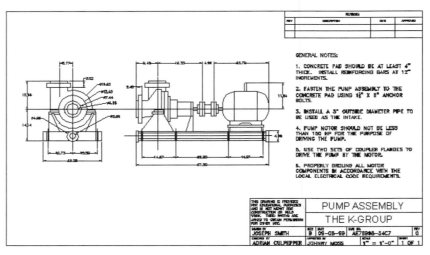

Figure 6–72

CHAPTER 7

Object Grips and Changing the Properties of Objects

USING OBJECT GRIPS

An alternate method of editing is to use object grips. A grip is a small box appearing at key object locations such as the endpoints and midpoints of lines and arcs or the center and quadrants of circles. Once grips are selected, the object may be stretched, moved, rotated, scaled, or mirrored. Grips are at times referred to as visual Osnaps because your cursor automatically snaps to all grips displayed along an object. Displayed in Figure 7–1 are other examples of grip locations on various objects.

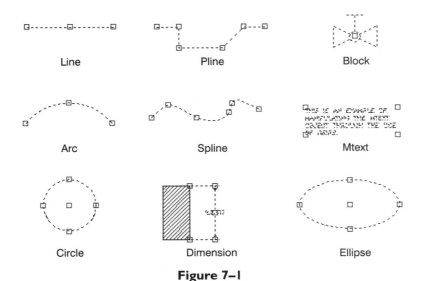

Line	Pline	Block
Arc	Spline	Mtext
Circle	Dimension	Ellipse

Figure 7–1

The Grips dialog box is available for changing settings, color, and grip size. Choose Options… from the Tools pull-down menu, as shown in Figure 7–2A, which will display the main Options dialog box. Click on the Selection tab to display the grip settings (see Figure 7–2B). By default, grips are enabled; a check in the Enable grips box means that grips will display when you select an object. Also by default, a grip is placed at the insertion point when a block is selected. Check the Enable grips within blocks box if you want grips to be displayed along with all individual objects that makeup the block. Color is applied to selected and unselected grips. Selecting the arrow in the Unselected grip color box or Selected grip color box displays a color dialog box used to change the color of selected or unselected grips. It is considered good practice to change the Unselected grip color to a light color if you are using a black screen background. Use the Grip Size area to move a slider bar left to make the grip smaller or to the right to make the grip larger.

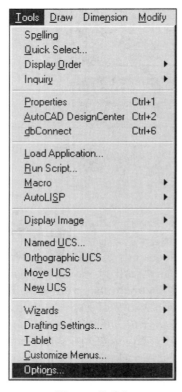

Figure 7–2A

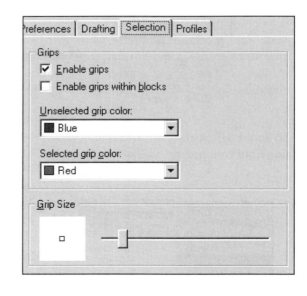

Figure 7–2B

OBJECT GRIP MODES

Figure 7–3 shows three types of grips. When an object is first selected with the grip pickbox located at the intersection of the crosshairs, the object highlights and the square grips are displayed; this type of grip is called a warm grip. The entire object is subject to the many grip edit commands. When one of the grips is selected, it fills in with the current selected grip color; this type of grip is called a hot grip. Once a hot grip is selected, the following prompts appear in the command prompt area:

** STRETCH **
Specify stretch point or [Base point/Copy/Undo/eXit]: *(Press the Spacebar)*
** MOVE **
Specify move point or [Base point/Copy/Undo/eXit]: *(Press the Spacebar)*
** ROTATE **
Specify rotation angle or [Base point/Copy/Undo/Reference/eXit]: *(Press the Spacebar)*
** SCALE **
Specify scale factor or [Base point/Copy/Undo/Reference/eXit]: *(Press the Spacebar)*
** MIRROR **
Specify second point or [Base point/Copy/Undo/eXit]: *(Press the Spacebar to begin STRETCH mode again or enter X to exit grip mode)*

Figure 7–3

To move from one edit command mode to another, press the Spacebar. Once an editing operation is completed, pressing ESC once removes the highlight from the object but leaves the grip; this type of grip is considered cold. Pressing ESC again removes the grips from the object. Figure 7–4 shows various examples of each editing mode.

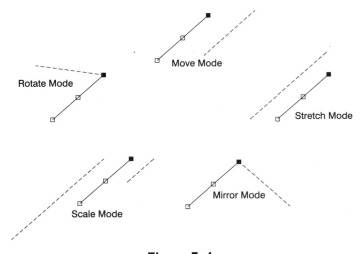

Figure 7–4

ACTIVATING THE GRIP CURSOR MENU

In Figure 7–5, a horizontal line has been selected and grips appear. The rightmost endpoint grip has been selected. Rather than use the Spacebar to scroll through the various grip modes, click the right mouse button. Notice that a cursor menu on grips appears. This provides an easier way of navigating from one grip mode to another.

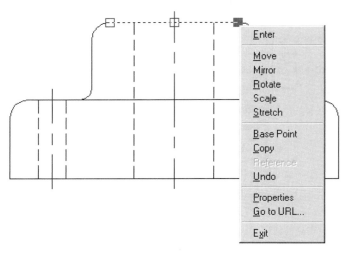

Figure 7–5

USING THE GRIP—STRETCH MODE

The STRETCH mode of grips operates similarly to the normal STRETCH command. Use STRETCH mode to move an object or group of objects and have the results mend themselves similar to Figure 7–6. The line segments "A" and "B" are both too long by one unit. To decrease these line segments by one unit, use the STRETCH mode by selecting lines "A," "B," and "C" with the grip pickbox at the command prompt. Next, while holding down SHIFT, select the warm grips "D," "E," and "F." This will make the grips hot and ready for the stretch operation. Release SHIFT and pick the grip at "E" again. The STRETCH mode appears in the command prompt area. Since these hot grips are considered base points, entering a polar coordinate value of @1.00<180 will stretch the three highlighted grip objects to the left a distance of one unit. The Direct Distance mode could also be used with grips. To remove the object highlight and grips, press ESC twice at the command prompt.

Command: *(Select the three dashed lines shown in Figure 7–6. Then, while holding down SHIFT, select the warm grips at "D," "E," and "F" to make them hot. Release the SHIFT key and pick the grip at "E" again)*
STRETCH
Specify stretch point or [Base point/Copy/Undo/eXit]: **@1<180**

Command: *(Press ESC to remove the object highlight)*
Command: *(Press ESC to remove the grips)*

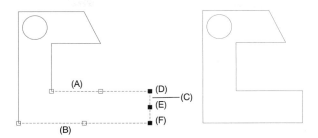

Figure 7–6

USING THE GRIP—SCALE MODE

Using the SCALE mode of object grips allows an object to be uniformly scaled in both the X and Y directions. This means that a circle, such as the one shown in Figure 7–7, cannot be converted to an ellipse through different X and Y values. As the warm grip is converted to a hot grip, any cursor movement will drag the scale factor until a point is marked where the object will be scaled to that factor. Figure 7–7 and the following prompt show the use of an absolute value to perform the scaling operation of half the circle's normal size.

Command: *(Select the circle to enable grips, then select the warm grip at the center of the circle to make it hot. Press the Spacebar until the SCALE mode appears at the bottom of the prompt line or press the right mouse button to activate the grip cursor menu to choose Scale)*

SCALE
Specify scale factor or [Base point/Copy/Undo/Reference/eXit]: **0.50**
Command: *(Press ESC to remove the object highlight)*
Command: *(Press ESC to remove the grips)*

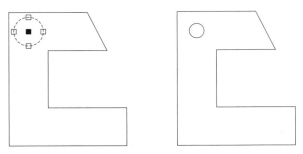

Figure 7–7

USING THE GRIP—MOVE/COPY MODE

The Multiple Copy option of the MOVE mode is demonstrated with the circle shown in Figure 7–8 being copied with polar coordinates at distances 2.50 and 5.00, both in the 270° direction. The Direct Distance mode could also be used to accomplish this task. This Multiple Copy option is actually disguised under the command options of object grips.

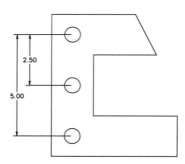

Figure 7–8

Select the circle, and then select the center grip of the circle to convert the warm grip to a hot grip. Use the Spacebar to scroll past the STRETCH mode to the MOVE mode. Issue a Copy within the MOVE mode to be placed in Multiple MOVE mode. See Figures 7–9A and 7–9B.

Command: *(Select the circle to activate the grips at the center and quadrants; select the warm grip at the center of the circle to make it hot. Then press the Spacebar until the MOVE mode appears at the bottom of the prompt line or press the right mouse button to activate the grip cursor menu to choose Move)*

MOVE
Specify move point or [Base point/Copy/Undo/eXit]: **C** *(For Copy)*
MOVE (multiple)
Specify move point or [Base point/Copy/Undo/eXit]: **@2.50<270**
MOVE (multiple)
Specify move point or [Base point/Copy/Undo/eXit]: **@5.00<270**
MOVE (multiple)
Specify move point or [Base point/Copy/Undo/eXit]: **X** *(For exit)*
Command: *(Press ESC to remove the object highlight)*
Command: *(Press ESC to remove the grips)*

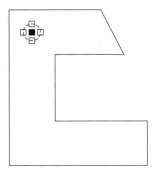

Figure 7–9A

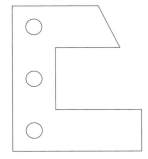

Figure 7–9B

USING THE GRIP—MIRROR MODE

Use the grip MIRROR mode to flip an object along a mirror line similar to the one used in the regular MIRROR command. Follow the prompts for the MIRROR option if an object needs to be mirrored but the original does not need to be saved. This performs the mirror operation but does not produce a copy of the original. If the original object needs to be saved during the mirror operation, use the Copy option of MIRROR mode. This places you in multiple MIRROR mode. Locate a base point and a second point to perform the mirror operation. See Figure 7–10.

Command: *(Select the circle in Figure 7–10 to enable grips, then select the warm grip at the center of the circle to make it hot. Press the Spacebar until the MIRROR mode appears at the bottom of the prompt line or press the right mouse button to activate the grip cursor menu to choose Mirror)*

MIRROR
Specify second point or [Base point/Copy/Undo/eXit]: **C** *(For Copy)*
MIRROR (multiple)
Specify second point or [Base point/Copy/Undo/eXit]: **B** *(For Base Point)*
Base point: **Mid**
of *(Pick the midpoint at "A")*
MIRROR (multiple)
Specify second point or [Base point/Copy/Undo/eXit]: **@1<90**
MIRROR (multiple)
Specify second point or [Base point/Copy/Undo/eXit]: **X** *(For Exit)*
Command: *(Press ESC to remove the object highlight)*
Command: *(Press ESC to remove the grips)*

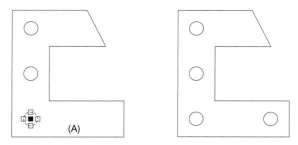

Figure 7–10

USING THE GRIP—ROTATE MODE

Numerous grips may be selected with window or crossing boxes. At the command prompt, pick a blank part of the screen; this should place you in Window/Crossing selection mode. Picking up or below and to the right of the previous point places you in Window selection mode; picking up or below and to the left of the previous point places you in Crossing selection mode. This method is used on all objects shown in Figure 7–11. Selecting the lower left grip and using the Spacebar to advance to the ROTATE option allows all objects to be rotated at a defined angle in relation to the previously selected grip.

Command: *(Pick near "X," then near "Y" to create a window selection set and enable all grips in all objects. Select the warm grip at the lower left corner of the object to make it hot. Then press the Spacebar until the ROTATE mode appears at the bottom of the prompt line or click the right mouse button to activate the grip cursor menu to choose Rotate)*

ROTATE
Specify rotation angle or [Base point/Copy/Undo/Reference/eXit]: **30**
Command: *(Press ESC to remove the object highlight)*
Command: *(Press ESC to remove the grips)*

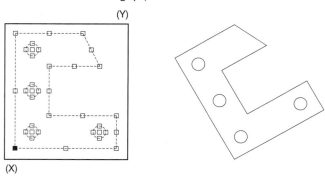

Figure 7–11

USING THE GRIP—MULTIPLE ROTATE MODE

One problem of the regular ROTATE command is that although an object can be rotated, the original location cannot be saved. You have to use points to mark the original location of an object before it is rotated. The ARRAY command has been used as a substitute to rotate and copy an object. Now with object grips, you may use ROTATE mode to rotate and copy an object without the use of reference points or the ARRAY command. Figure 7–12A is a line that needs to be rotated and copied at a 40° angle. With a positive angle, the direction of the rotation will be counterclockwise.

Selecting the line in Figure 7–12A enables grips located at the endpoints and midpoint of the line. Selecting the warm grip makes it hot. This hot grip also locates the vertex of the required angle. Press the Spacebar until the ROTATE mode is reached. Enter Multiple ROTATE mode by entering C for Copy when you are prompted in the following command sequence. Finally, enter a rotation angle of 40 to produce a copy of the original line segment at a 40° angle in the counterclockwise direction. (See Figure 7–12B.)

Command: *(Select line segment "A"; then select the warm grip at "B" to make it hot. Press the Spacebar until the ROTATE mode appears at the bottom of the prompt line or click the right mouse button to activate the grip cursor menu to choose Rotate)*
ROTATE
Specify rotation angle or [Base point/Copy/Undo/Reference/eXit]: **C** *(For Copy)*
ROTATE (multiple)
Specify rotation angle or [Base point/Copy/Undo/Reference/eXit]: **40**
Specify rotation angle or [Base point/Copy/Undo/Reference/eXit]: **X** *(For Exit)*
Command: *(Press ESC to remove the object highlight)*
Command: *(Press ESC to remove the grips)*

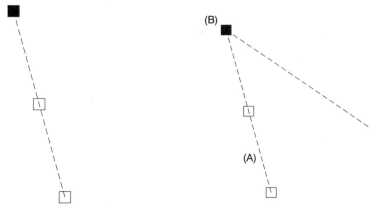

Figure 7–12A   **Figure 7–12B**

The result is illustrated in Figure 7–12C.

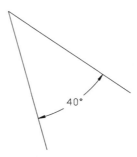

Figure 7–12C

MULTIPLE COPY MODE AND OFFSET SNAP LOCATIONS FOR ROTATIONS

All Multiple Copy modes within grips may be operated in a snap location mode while you hold down SHIFT. Here's how it works. In Figure 7–13A, the vertical centerline and circle are selected with the grip pickbox. The objects highlight and the grips appear. A multiple copy of the selected objects needs to be made at an angle of 45°. The grip ROTATE option is used in Multiple Copy mode.

Command: (Select centerline segment "A" and circle "B"; and then select the warm grip at the center of the circle to make it hot. Press the Spacebar until the ROTATE mode appears at the bottom of the prompt line)
ROTATE
Specify rotation angle or [Base point/Copy/Undo/Reference/eXit]: **C** (For Copy)
ROTATE (multiple)
Specify rotation angle or [Base point/Copy/Undo/Reference/eXit]: **B** (For Base Point)
Base point: **Cen**
of (Select the circle at "C" to snap to the center of the circle)
ROTATE (multiple)
Specify rotation angle or [Base point/Copy/Undo/Reference/eXit]: **45**

Rather than enter another angle to rotate and copy the same objects, you hold down SHIFT, which places you in offset snap location mode. Moving the cursor snaps the selected objects. This keeps the value of the original angle of rotation, namely 45° (see Figure 7–13B).

ROTATE (multiple)
Specify rotation angle or [Base point/Copy/Undo/Reference/eXit]: (Hold down the SHIFT key and move the circle and centerline until it snaps to the next 45° position shown in Figure
7–13B)
ROTATE (multiple)
Specify rotation angle or [Base point/Copy/Undo/Reference/eXit]: **X** (For Exit)
Command: (Press ESC to remove the object highlight)
Command: (Press ECS to remove the grips)

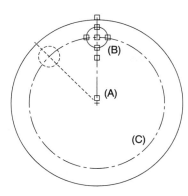

Figure 7–13A

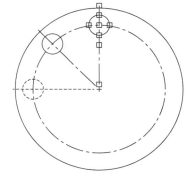

Figure 7–13B

The Rotate-Copy-Snap Location mode could allow you to create Figure 7–13C without the aid of the ARRAY command. Since all angle values are 45°, continue holding down SHIFT to snap to the next 45° location, and mark a point to place the next group of selected objects.

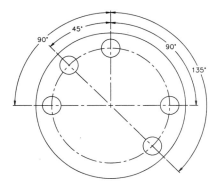

Figure 7–13C

MULTIPLE COPY MODE AND OFFSET SNAP LOCATIONS FOR MOVING

As with the previous example of using Offset Snap Locations for ROTATE Mode, these same snap locations apply to MOVE mode. Figure 7–14A shows two circles along with a common centerline. The circles and centerline are selected with the grip pickbox, which highlights these three objects and activates the grips. The intent is to move and copy the selected objects at 2-unit increments.

Command: *(Select the two circles and centerline to activate the grips; select the warm grip at the midpoint of the centerline to make it hot. Then press the Spacebar until the MOVE mode appears at the bottom of the prompt line or click the right mouse button to activate the grip cursor menu to choose Move)*

MOVE
Specify move point or [Base point/Copy/Undo/eXit]: **C** *(For Copy)*
MOVE (multiple)
Specify move point or [Base point/Copy/Undo/eXit]: **@2.00<270**

In Figure 7–14B, instead of remembering the previous distance and entering it to create another copy of the circles and centerline, hold down SHIFT and move the cursor down to see the selected objects snap to the distance already specified.

MOVE (multiple)
Specify move point or [Base point/Copy/Undo/eXit]: *(Hold down the SHIFT key and move the cursor down to have the selected entities snap to another 2.00-unit distance)*
MOVE (multiple)
Specify move point or [Base point/Copy/Undo/eXit]: **X** *(To exit)*
Command: *(Press ESC to remove the object highlight)*
Command: *(Press ESC to remove the grips)*

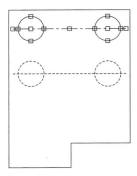

Figure 7–14A

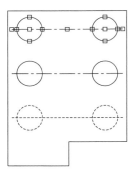

Figure 7–14B

Figure 7–14C shows the completed hole layout, the result of using the Offset Snap Location method of object grips.

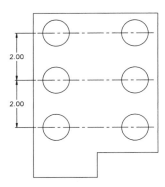

Figure 7–14C

MODIFYING THE PROPERTIES OF OBJECTS

At times, objects are drawn on the wrong layer, color, or even in the wrong linetype. The length of line segments are incorrect, or the radius values of circles and arcs are incorrect. Eliminating the need to erase these objects and reconstruct them to their correct specifications, a series of tools are available to modify the properties of these objects. Illustrated in Figure 7–15 are three line segments. One of the line segments has been pre-selected, as shown by the highlighted appearance and the presence of grips.

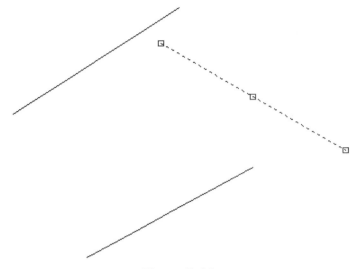

Figure 7–15

Clicking on the Properties button on the Standard toolbar in Figure 7–16A displays the Properties dialog box in Figure 7–16B. This dialog box displays information about the object already selected; in this case the information is about the line segment, which is identified at the top of the dialog box. Two tabs are present in the dialog box. They allow you to easily view the information about the line. The first tab lists properties of the line alphabetically (see Figure 7–16B).

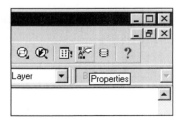

Figure 7–16A

The second tab displays information in a series of categories, as in Figure 7–16C. Notice, in the Categorized tab, that two sets of information are provided about the selected object. One area deals with General properties to change. This area allows you to change the color, layer, linetype, linetype scale, etc. of the selected object. The second area gives information about the selected object. Since a single line was selected, the starting and ending X, Y, Z values are given. As modifications are made to these values, other information displayed automatically updates, such as length and angle of the line. You cannot make modifications to values that appear grayed out.

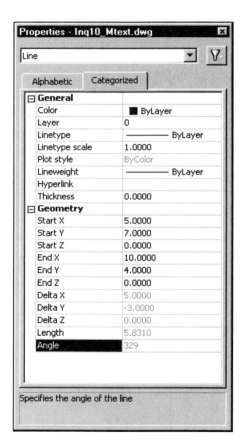

Figure 7–16B	Figure 7–16C

When changing a selected object to a different layer, click in the layer field in Figure 7–16D and the current layer displays (in this example, Layer 0). Clicking the down arrow displays all layers defined in the drawing. Clicking on one of these layers changes the selected object to a different layer.

End Z	0.0000
Hyperlink	
Layer	0
Length	0
Linetype	Center
Linetype scale	Hidden
Lineweight	Object
Plot style	ByColor
Start Y	5.0000

Figure 7–16D

Illustrated in Figure 7–17 are the same three line segments. This time, all three segments have been pre-selected, as shown by the highlighted appearance and the presence of grips.

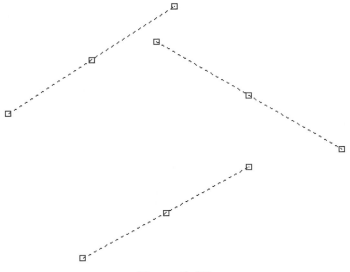

Figure 7–17

Clicking on the Properties button on the Standard toolbar displays the Properties dialog box, as in Figure 7–18. At the top of the dialog box, the three lines are identified. In the Alphabetic tab, you can change the color, layer, linetype, and other general properties of all three lines. However, you are unable to enter the Start and End X, Y, and Z values, and there is no length or angle information. Whenever you select more than one of the same object, you can change the general properties of the object but not individual values that deal with the object's geometry.

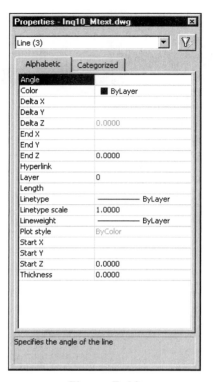

Figure 7–18

In Figure 7–19, an arc, circle, and line are pre-selected along with the display of grips. Activating the Properties dialog box in Figure 7–20 displays a number of object types at the top of the dialog box. You can click which object or group of objects to modify. With "All (3)" highlighted, you can change the general properties, such as layer and linetype, but not any geometry settings.

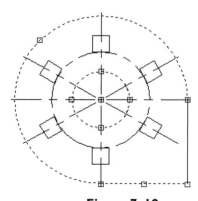

Figure 7–19

Figure 7–20

What if you need to increase the radius of the circle to 1.25 units? Click on the Circle object type at the top of the Properties dialog box. Since only one object was selected, the full complement of general and geometry settings is present for you to modify. Click in the Radius field and change the current value to 1.25 units. Pressing ENTER automatically updates the other geometry settings in addition to the actual object in the drawing (see Figure 7–20). When finished, dismiss this dialog box to return to the drawing.

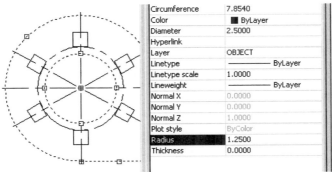

Figure 7–21

You can also open the Properties dialog box by choosing Properties from the Modify pull-down menu in Figure 7–22A. This displays the Properties dialog box in Figure 7–22B. Notice that "No selection" is listed at the top of the dialog box, meaning that no objects have been selected to modify. Even though nothing is selected, you can still change the current color, linetype, and even make a layer current.

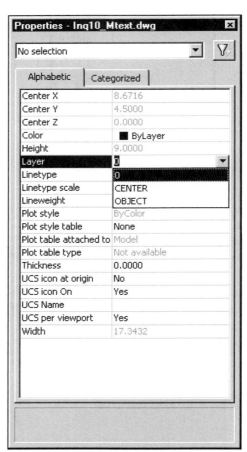

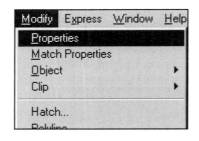

Figure 7–22A   **Figure 7–22B**

Clicking on the Quick Select button in Figure 7–22B displays the Quick Select dialog box in Figure 7–22C. Use this dialog box to build a selection set to modify its object properties.

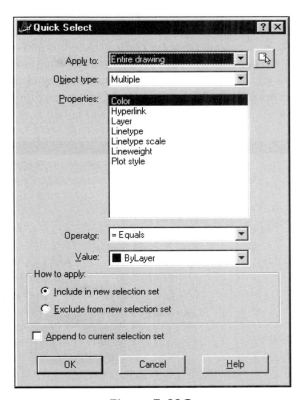

Figure 7–22C

MODIFYING THE PROPERTIES OF INDIVIDUAL OBJECTS

This segment of the chapter begins a study of the Properties dialog box and all of the properties displayed on the particular object selected. In each example, the object is already pre-selected. Grips are also present at key locations of each object. You can modify the following General properties of all objects through this dialog box: Color, Layer, Linetype, Linetype scale, Plot style, Lineweight, Hyperlink, and Thickness. The remainder of this segment will concentrate on the individual geometric properties that can be modified through the Properties dialog box.

ARCS

The Properties dialog box for an arc provides you with the following information, which is displayed in Figure 7–23. Values in bold can be modified; items grayed out cannot.

- Name of the object being listed (Arc) at the top of the dialog box

- Starting and Ending X, Y, and Z values of the arc (grayed out)

- X, Y, and Z values of the center of the arc

- Radius of the arc
- Start and End angles of the arc
- Total angle (grayed out)
- Total arc length (grayed out)
- Length of the arc (grayed out)
- Normal X, Y, and Z values (grayed out)

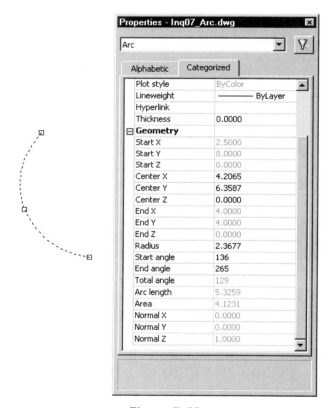

Figure 7–23

BLOCKS

The Properties dialog box for a block provides you with the following information, which is displayed in Figure 7–24. Values in bold can be modified; items grayed out cannot.

- Name of the object being listed (Block Reference) at the top of the dialog box
- Position X, Y, and Z values (otherwise known as the insertion point of the block)
- X, Y, and Z scale factors for the block

- Name of the block (grayed out)
- Rotation angle of the block

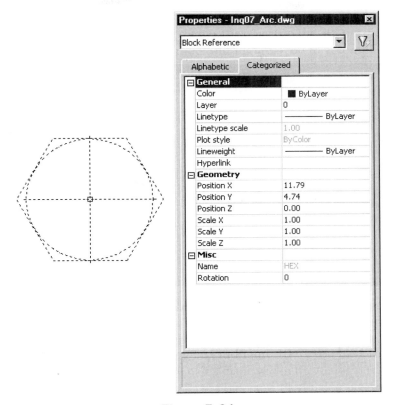

Figure 7–24

CIRCLES

The Properties dialog box for a circle provides you with the following information, which is displayed in Figure 7–25. Values in bold can be modified; items grayed out cannot.

- Name of the object being listed (Circle) at the top of the dialog box
- X, Y, and Z center point values of the circle
- Radius of the circle
- Diameter of the circle
- Circumference of the circle
- Area of the circle
- Normal X, Y, and Z values (grayed out)

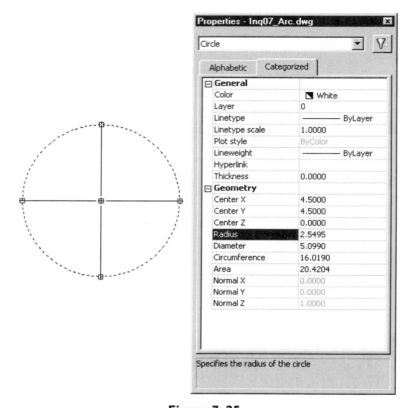

Figure 7–25

DIMENSIONS

The Properties dialog box for a dimension provides you with an incredible amount of information, which can be used to modify the dimension. This information is displayed in Figure 7–26. As an example, clicking on the "+" symbol next to the Fit category displays a number of settings that can be modified. All categories identified with the "+" symbol contain these types of settings. Values in bold can be modified; items grayed out cannot.

- Name of the object being listed (Rotated Dimension) at the top of the dialog box
- Miscellaneous information dealing with the dimension (Dimension Style)
- Dimension settings that deal with dimension and extension lines in addition to arrowheads
- Dimension text information
- Dimension fit information

- Primary and alternate unit information

- Settings that deal with tolerances

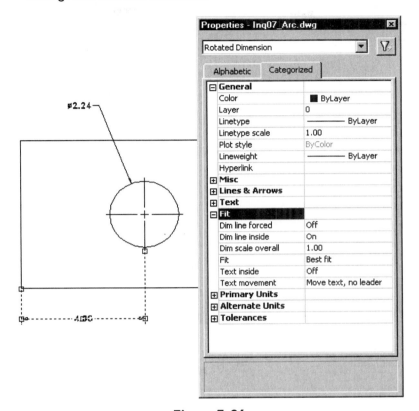

Figure 7–26

ELLIPSES

The Properties dialog box for an ellipse provides you with the following information, which is displayed in Figure 7–27. Values in bold can be modified; items grayed out cannot.

- Name of the object being listed (Ellipse) at the top of the dialog box

- Starting and Ending X, Y, and Z values of the arc (grayed out)

- X, Y, and Z values of the center of the arc

- Major and minor radius values

- Radius ratio values

- Start and end angle values

- X, Y, and Z major and minor axis vectors (grayed out)

Notice, at the top of the Properties dialog box, that an ellipse can also be classified as a polyline. This type of ellipse consists of a series of polyarcs joined together in one single polyline object. You accomplish this by first setting the system variable PELLIPSE to a value of 1, which turns the variable on. This is what generates the ellipse as the polyline object.

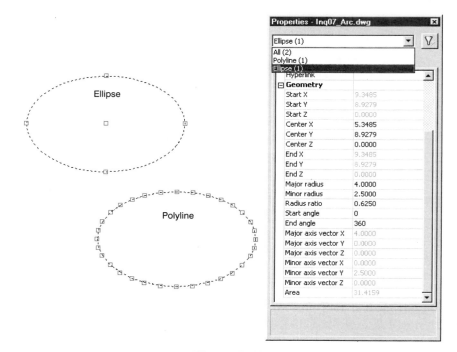

Figure 7–27

HATCHES

The Properties dialog box for a hatch pattern provides you with the following information, which is displayed in Figure 7–28. Values in bold can be modified; items grayed out cannot.

- Name of the object being listed (Hatch) at the top of the dialog box
- Pattern type
- Name, angle, and scale of the pattern
- Pattern spacing (grayed out)
- Whether the pattern is doubled or not (grayed out)

Other miscellaneous information that affects hatch patterns such as elevation, whether the pattern is associative, and the method of island detection

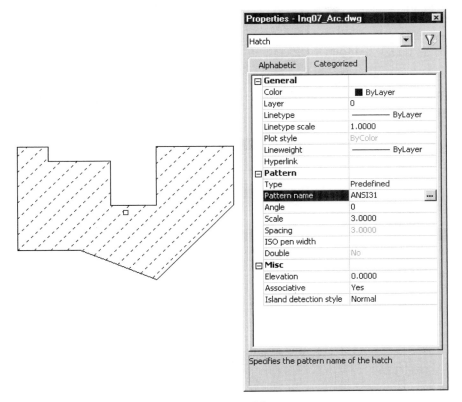

Figure 7–28

LINES

The Properties dialog box for a line provides you with the following information, which is displayed in Figure 7–29 (Lines were used as an example earlier in this chapter to introduce the Properties dialog box). Values in bold can be modified; items grayed out cannot.

- Name of the object being listed (Line) at the top of the dialog box
- Starting and Ending X, Y, and Z values of the line
- Delta (Relative) X, Y, and Z values of the line (grayed out)
- Length and angle of the line (grayed out)

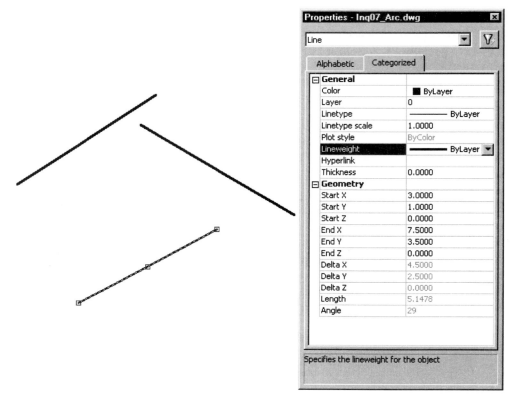

Figure 7–29

MULTILINES

In Figure 7–30, only general information about a multiline such as color, layer, linetype, linetype scale, and lineweight may be changed. Because the current multiline style is listed as STANDARD, it is grayed out and cannot be modified in the Properties dialog box.

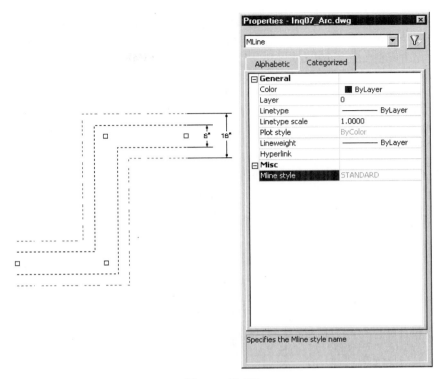

Figure 7–30

MULTILINE TEXT

The Properties dialog box for a multiline text object provides you with the following information, which is displayed in Figure 7–31. In addition to the General and Geometry categories to modify, multiline text has a separate Text category to assist in the modification. Values in bold can be modified; items grayed out cannot.

- Name of the object being listed (MText) at the top of the dialog box

In the Text category in Figure 7–31 the contents of the multiline may be modified. You can also click on the button represented by the three "..." which will take you to the Multiline Text Editor to make additional modifications to the text. You can also make the following additional modification in the Text category:

- Text style, justification, direction, width, height, rotation, line space factor, and the line space style

Information to modify in the Geometry area includes the X, Y, and Z positions of the multiline text.

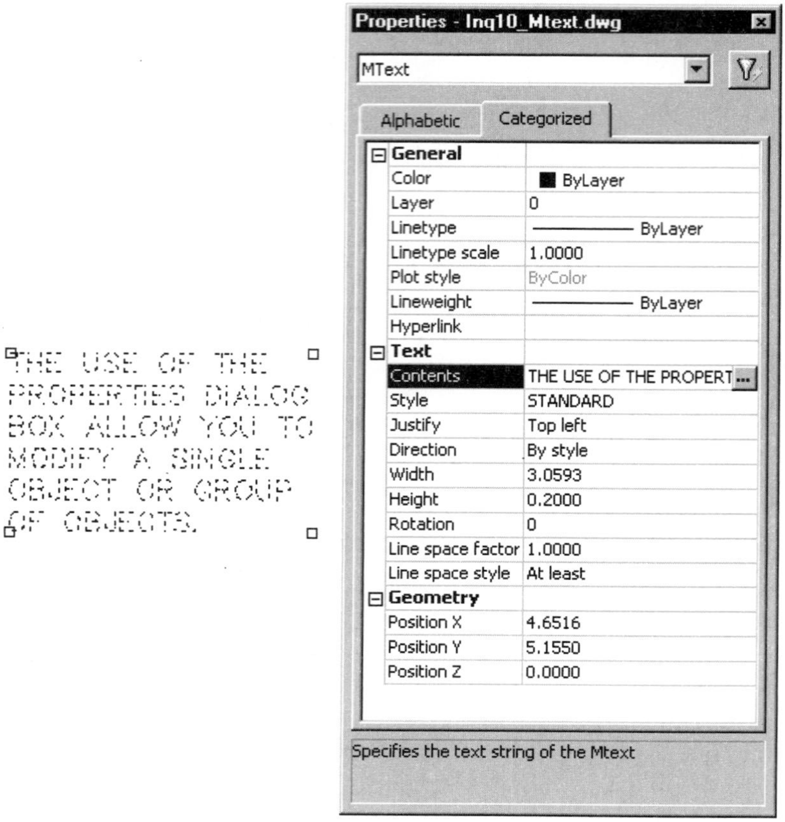

Figure 7–31

POLYLINES

Displayed in Figure 7–32 is a hexagon created through the POLYGON command, a large dot created through the DONUT command (the fill has been turned off), and an irregular shape created with the PLINE command. Even though three different commands were used to create these shapes, the three objects are all classified as polylines.

The Properties dialog box for the irregular polyline shape provides you with the following information. Values in bold can be modified; items grayed out cannot.

- Name of the object being listed (Hatch) at the top of the dialog box

- Vertex number, which is identified by an "X" on the object and can be changed in the dialog box

- X and Y values of the current vertex

- Starting and Ending widths of each segment

- Global width of the polyline
- Elevation of the polyline

Other information supplied in the Miscellaneous category includes whether the polyline is closed and the current linetype generation mode.

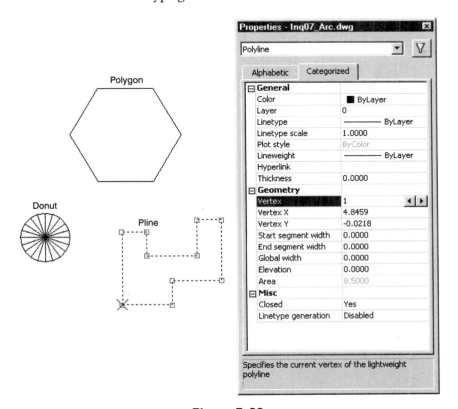

Figure 7–32

POINTS

The Properties dialog box for a point object provides you with the following information, which is displayed in Figure 7–33.

- Name of the object being listed (Point) at the top of the dialog box
- Location of the point in X, Y, and Z coordinate values, which can be modified

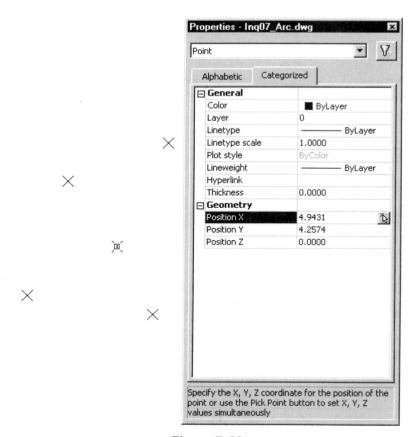

Figure 7–33

RAYS

The Properties dialog box for a ray provides you with the following information, which is displayed in Figure 7–34. Values in bold can be modified; items grayed out cannot.

- Name of the object being listed (Ray) at the top of the dialog box
- Basepoint of the ray in X, Y, and Z coordinates
- Second point of the ray in X, Y, and Z coordinates
- Notice in the message at the bottom of the dialog box that you can change the X, Y, and Z values simultaneously by clicking on the Pick Point button.
- The X, Y, and Z coordinates are also given in the direction vector, although they cannot be modified here.

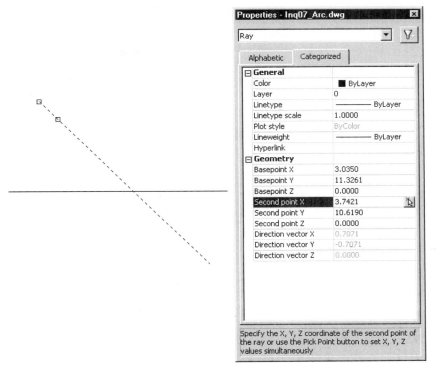

Figure 7–34

SPLINES

The Properties dialog box for a spline provides you with the following information, which is displayed in Figure 7–35. A majority of the values in the Data Points category in bold can be modified; a majority of the values in the Miscellaneous category that are grayed out cannot.

- Name of the object being listed (Spline) at the top of the dialog box
- Number of control points (grayed out)
- Control point number (1), which is identified by a "X" on the drawing (click in this field to change to a different control point number)
- X, Y, and Z coordinate values of the control point
- The weight of this control point
- The number of fit points
- The fit point number, which can be changed in a way similar to that of control points

- The X, Y, and Z coordinate value of the fit point
- Most of the information in the Miscellaneous category reacts to changes in information in the Data Points category and cannot be modified individually. This information includes:
- Degree of the spline
- Whether the spline is closed or open
- Whether the spline is planar
- Starting and Ending tangent vectors of the spline in X, Y, and Z values
- Fit tolerance of the spline (this value may be modified)
- Area of the spline

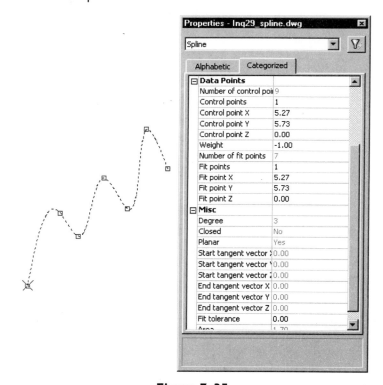

Figure 7–35

TEXT

The Properties dialog box for a text object provides you with the following information, which is displayed in Figure 7–36. Text, Geometry, and Miscellaneous categories are available to assist in the modification of the text object. Values in bold can be modified; items grayed out cannot.

- Name of the object being listed (Text) at the top of the dialog box
- The Text category consists of the following items:
- Text contents
- Text style
- Justification of the text object
- Height, rotation angle, width factor, and obliquing angle of the text object
- Text alignment position in absolute X, Y, and Z coordinates
- The Geometry category holds the position of the text object in absolute X, Y, and Z coordinates, which cannot be modified. The Miscellaneous category allows you to modify whether the text is placed upside down or even backwards.

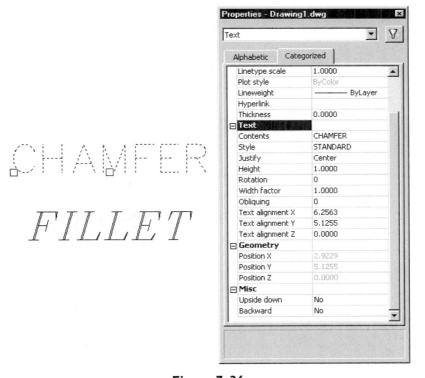

Figure 7–36

XLINES

The Properties dialog box For an Xline object provides you with the following information, which is displayed in Figure 7–37. Values in bold can be modified; items grayed out cannot.

- Name of the object being listed (XLine) at the top of the dialog box
- Basepoint location of the Xline in absolute X, Y, and Z coordinates
- Second point location of the Xline in Absolute X, Y, and Z coordinates
- Direction vector of the Xline in absolute X, Y, and Z coordinates (grayed out)

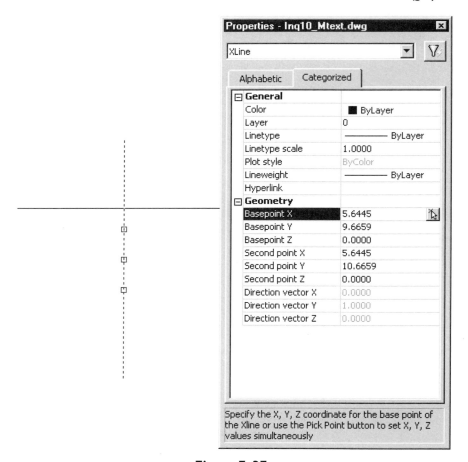

Figure 7–37

USING THE LAYER CONTROL BOX TO MODIFY OBJECT PROPERTIES

If all you need to do is to change an object or a group of objects from one layer to another, the Layer Control box can easily perform this operation. In Figure 7–38A, an arc and two line segments have been pre-selected. Notice that the current layer is 0 in the Layer Control box. These objects need to be on the OBJECT layer.

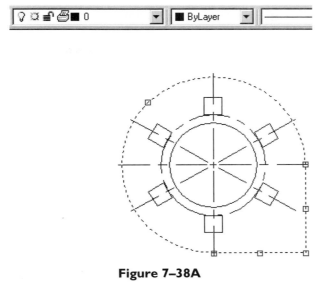

Figure 7–38A

Click in the Layer Control box to display all layers defined in the drawing. Then click on the desired layer for all highlighted objects (in this case, the OBJECT layer in Figure 7–38B).

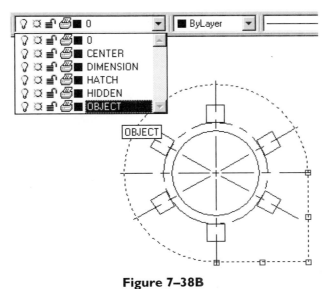

Figure 7–38B

Notice that in Figure 7–38C, with the objects still highlighted, the layer listed is OBJECT. This is one of the quickest and most productive ways of changing an object from one layer to another.

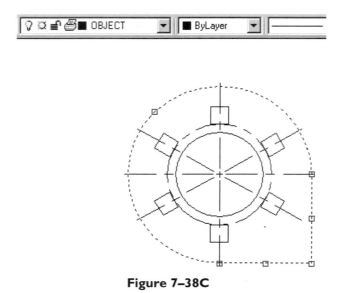

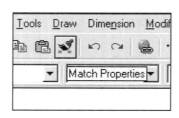

Figure 7–38C

MATCHING THE PROPERTIES OF OBJECTS

At times objects are drawn on the wrong layers or the wrong color scheme is applied to a group of objects. Text objects are sometimes drawn with an incorrect text style. You have just seen how the Properties dialog box and the Layer Control box provide quick ways to fix such problems. Yet another tool is available for changing the properties of objects—the MATCHPROP command. Choose this command from the Standard toolbar in Figure 7–39A. You could also select this command from the Modify pull-down menu shown in Figure 7–39B.

Figure 7–39A   **Figure 7–39B**

When you start the command, a source object is required. This source object transfers all of its current properties to other objects designated as "Destination Objects." In Figure 7–40A, the flange requires the object lines located at "B," "C," "D," and "E" to be converted to hidden lines. Using the MATCHPROP command, select the existing hidden line "A" as the source object. Notice the appearance of the Match Properties icon. Select lines "B" through "E" as the destination objects using this icon.

Command: **MATCHPROP**
Select source object: *(Select the hidden line at "A")*
Current active settings: Color Layer Ltype Ltscale Lineweight Thickness PlotStyle Text Dim Hatch
Select destination object(s) or [Settings]: *(Select line "B")*
Select destination object(s) or [Settings]: *(Select line "C")*
Select destination object(s) or [Settings]: *(Select line "D")*
Select destination object(s) or [Settings]: *(Select line "E")*
Select destination object(s) or [Settings]: *(Press* ENTER *to exit this command)*

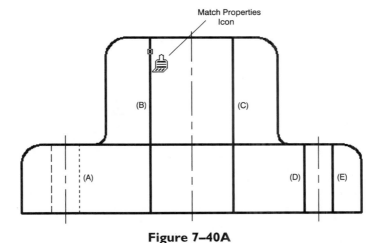

Figure 7–40A

The results appear in the flange illustrated in Figure 7–40B, where the continuous object lines were converted to hidden lines. Not only did the linetype property get transferred, but so did the color, layer, lineweight, and linetype scale information.

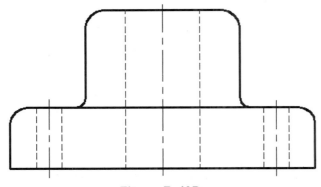

Figure 7–40B

To get a better idea of what object properties are affected by the MATCHPROP command, re-enter the command, pick a source object, and instead of picking a destination object immediately, enter s for settings. This will display the Property Settings dialog box in Figure 7–41.

Command: **MATCHPROP**
Select source object: *(Select the hidden line at "A" in Figure 7–40A)*
Current active settings: Color Layer Ltype Ltscale Lineweight Thickness
PlotStyle Text Dim Hatch
Select destination object(s) or [Settings]: **S** *(For Settings; this displays the Property Settings dialog box in Figure 7–41)*

Any box with a check displayed in it will transfer that property from the source object to all destination objects. If you need to transfer only the layer information and not the color and linetype properties of the source object, remove the checks from the Color and Linetype properties before you select the destination objects, which prevents these properties from being transferred to any destination objects.

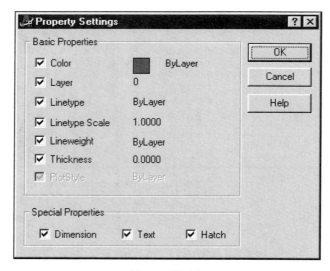

Figure 7–41

The MATCHPROP command can control special properties of dimensions, text, and hatch patterns. The Dimension Special Property will be featured next (see Figure 7–42).

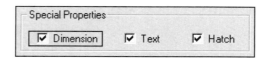

Figure 7–42

Figure 7–43A shows two blocks: the block assigned a dimension value of 46.6084 was dimensioned with the METRIC dimension style with the ROMAND font applied. The block assigned a dimension value of 2.3872 was dimensioned with the STANDARD dimension style with the TXT font applied. Both blocks need to be dimensioned with the METRIC dimension style. Issue the MATCHPROP command and select the 46.6084 dimension as the source object and then select the 2.3872 dimension as the destination object.

Command: **MATCHPROP**
Select source object: *(Select the dimension at "A" in Figure 7–43A)*
Current active settings: Color Layer Ltype Ltscale Lineweight Thickness
PlotStyle Text Dim Hatch
Select destination object(s) or [Settings]: *(Select the dimension at "B")*
Select destination object(s) or [Settings]: *(Press* ENTER *to exit this command)*

The results are shown in Figure 7–43B, with the METRIC dimension style applied to the STANDARD dimension style through the use of the MATCHPROP command. Because the text font was associated with the dimension style, it also changed in the destination object.

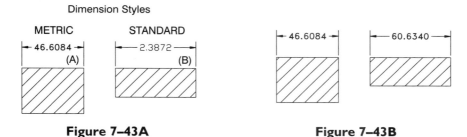

Figure 7–43A   **Figure 7–43B**

The following is an example of how the MATCHPROP command affects a text object with the Text Special Property shown in Figure 7–44.

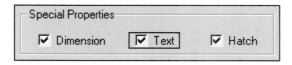

Figure 7–44

Figure 7–45A shows two text items displayed in different fonts. The text "Coarse Knurl" at "A" was constructed with a text style called ROMAND. The text "Medium Knurl" at "B" was constructed with the default text style called STANDARD. Use the following command sequence to match the STANDARD text style with the ROMAND text style using the MATCHPROP command.

Command: **MATCHPROP**
Select source object: *(Select the text at "A")*
Current active settings: Color Layer Ltype Ltscale Lineweight Thickness
PlotStyle Text Dim Hatch
Select destination object(s) or [Settings]: *(Select the text at "B")*
Select destination object(s) or [Settings]: *(Press ENTER to exit this command)*

The result is shown in Figure 7–45B. Both text items now share the same text style. Notice that the text string stays intact when text properties are matched. Only the text style of the source object is applied to the destination object.

Figure 7–45A   **Figure 7–45B**

A source hatch object can also be matched to a destination pattern with the MATCHPROP command and the Hatch Special Property (see Figure 7–46).

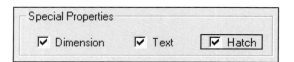

Figure 7–46

In Figure 7–47A, the crosshatch patterns at "B" and "C" are at the wrong angle and scale. They should reflect the pattern at "A" because it is the same part. Use the MATCHPROP command, select the hatch pattern at "A" as the source object, and select the patterns at "B" and "C" as the destination objects.

Command: **MATCHPROP**
Select source object: *(Select the hatch pattern at "A")*
Current active settings: Color Layer Ltype Ltscale Lineweight Thickness
PlotStyle Text Dim Hatch
Select destination object(s) or [Settings]: *(Select the hatch pattern at "B")*
Select destination object(s) or [Settings]: *(Select the hatch pattern at "C")*
Select destination object(s) or [Settings]: *(Press ENTER to exit this command*

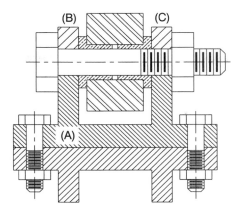

Figure 7–47A

The results appear in Figure 7–47B, where the source hatch pattern property was applied to all destination hatch patterns.

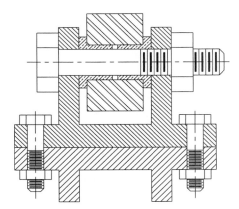

Figure 7–47B

TUTORIAL EXERCISE: 07_LUG.DWG

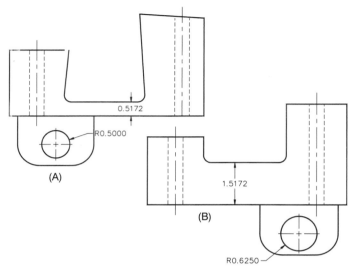

Figure 7–48

Purpose

This tutorial is designed to use object grips to edit the drawing of the lug shown in Figure 7–48 at "A" until it appears like the illustration at "B."

System Settings

Since this drawing is provided on the CD, open an existing drawing file called "07_Lug." Follow the steps in this tutorial for using object grips to edit and make changes to the Lug.

Layers

The following layers have already been created with the format:

Name	Color	Linetype
Object	White	Continuous
Center	Yellow	Center
Hidden	Red	Hidden
Dim	Yellow	Continuous
Defpoints	White	Continuous

Suggested Commands

Be sure grip mode is enabled through the Selection tab of the Options dialog box. Then use the STRETCH, MOVE, ROTATE, SCALE, and MIRROR modes of object grips to make changes to the existing drawing file.

Whenever possible, substitute the appropriate command alias in place of the full AutoCAD command in each tutorial step. For example, use "CP" for the COPY command, "L" for the LINE command, and so on. The complete listing of all command aliases is located in Chapter 1, Table 1–2.

STEP 1

Be sure that Object Grip mode is enabled by choosing Options... from the Tools pull-down menu, which activates the Options dialog box. Click on the Selection tab to determine that grips are turned on (the default setting). Your dialog box should be similar to Figure 7–49.

STEP 2

Begin by turning ORTHO off by clicking on ORTHO in the Status bar. Next, at the command prompt, use the grip cursor to select the inclined line "A" shown in Figure 7–50. Notice the appearance of the grips at the endpoints and midpoint of the line.

Command: **ORTHO**
Enter mode [ON/OFF] <ON>: **OFF**
Command: *(Select the inclined line "A")*

STEP 3

While still at the command prompt, select the grip shown in Figure 7–51 at "A." This grip becomes the current base point for the following editing options. Use the STRETCH option to reposition the endpoint of the highlighted line at the endpoint of the horizontal line shown in Figure 7–51 (OSNAP-Endpoint should already be enabled). When this operation is complete, press ESC twice at the command prompt to remove the object highlight and grips from the display screen.

Command: *(Select the warm grip at the endpoint of the line at "A" to make it hot)*
STRETCH
Specify stretch point or [Base point/ Copy/Undo/eXit]: *(Select the endpoint of the horizontal line at "B")*

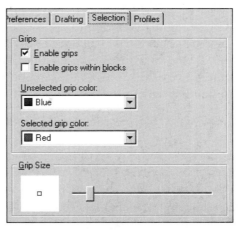

Figure 7–49

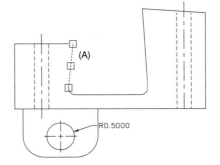

Figure 7–50

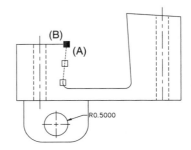

Figure 7–51

Command: *(Press ESC to remove the object highlight)*
Command: *(Press ESC to remove the grips)*

STEP 4

At the command prompt, use the grip cursor to select the two inclined lines "A" and "B" shown in Figure 7–52. Notice the appearance of the grips at the endpoints and midpoints of the lines.

Command: (Select the inclined lines "A" and "B")

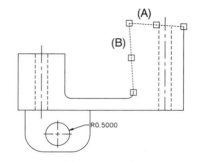

Figure 7–52

STEP 5

While still at the command prompt, select the grip shown in Figure 7–53 at "A." This grip becomes the current base point for the following editing options. Use the STRETCH option in combination with OSNAP-Tracking mode to reposition the corner of the highlighted lines to form a 90° corner. When this operation is complete, press ESC twice at the command prompt to remove the object highlight and grips from the display screen.

Command: (Select the warm grip at the endpoint of the line at "A" to make it hot)
** STRETCH **
Specify stretch point or [Base point/Copy/Undo/eXit]: **TK** (To enable Tracking)
First tracking point: (Select the grip at "B" in Figure 7–217)
Next point (Press ENTER to end tracking): (Select the grip at "C" in Figure 7–53)
Next point (Press ENTER to end tracking): (Press ENTER to end tracking and perform the operation)
Command: (Press ESC to remove the object highlight)
Command: (Press ESC to remove the grips)

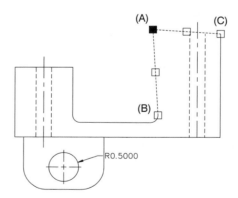

Figure 7–53

STEP 6

At the command prompt, use the grip cursor to select the two vertical lines, horizontal line, and both filleted corners shown in Figure 7–54. Notice the appearance of the grips at the endpoints and midpoints of the lines and arcs.

Command: *(Select the two vertical lines, two arcs representing fillets, and the horizontal line)*

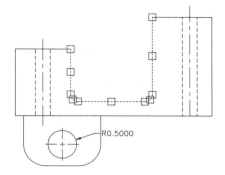

Figure 7–54

STEP 7

While still at the command prompt, select all the grips in Figure 7–55 by holding down SHIFT as you select the grips. Release SHIFT and pick the selected grip at "A" again. This grip becomes the current base point for the following editing options. Use the STRETCH option in combination with a polar coordinate to stretch the objects a distance of one unit in the 90° direction (you could also use the Direct Distance mode to accomplish this operation). When this operation is complete, press ESC at the command prompt to remove the object highlight and grips from the display screen.

Command: *(Hold down the SHIFT key and select all warm grips in Figure 7–55 to make them hot. Release the SHIFT key and pick the grip at "A" again.)*
STRETCH
Specify stretch point or [Base point/ Copy/Undo/eXit]: **@1.00<90**
Command: *(Press ESC to remove the object highlight)*
Command: *(Press ESC to remove the grips)*

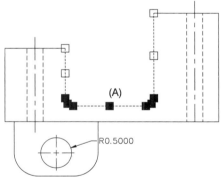

Figure 7–55

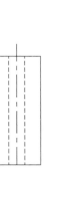

STEP 8

At the command prompt, use the grip cursor to select the circle and radius dimension in Figure 7–56. Notice the appearance of the grips at the quadrants and center of the circle and center, starting point, and text location of the radius dimension.

Command: *(Select the circle and the radius dimension)*

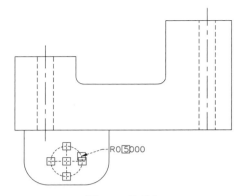

Figure 7–56

STEP 9

While still at the command prompt, select the grip at the center of the circle shown in Figure 7–57. This grip becomes the current base point for the following editing options. Use the SCALE option to increase the size of the circle using the Reference option. Notice that this will also affect the value of the dimension. When this operation is complete, press ESC twice at the command prompt to remove the object highlight and grips from the display screen.

Command: *(Select the warm grip at the center of the circle to make it hot. Press the Spacebar until the SCALE mode appears at the bottom of the prompt line or press the right mouse button to activate the grip cursor menu to select Scale)*
** SCALE **
Specify scale factor or [Base point/Copy/ Undo/Reference/eXit]: **R** *(For Reference)*
Specify reference length <1.0000>: **0.50**
** SCALE **
Specify new length or [Base point/Copy/ Undo/Reference/eXit]: **0.625**
Command: *(Press ESC to remove the object highlight)*
Command: *(Press ESC to remove the grips)*

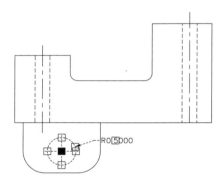

Figure 7–57

STEP 10

At the command prompt, use the grip cursor to select the 0.625 radius dimension in Figure 7–58. Notice the appearance of the grips at the center, starting point, and text location of the radius dimension. The dimension text will be relocated with grips.

Command: (*Select the dimension in Figure 7–58*)

STEP 11

Still at the command prompt, select the text grip shown in Figure 7–59. This grip becomes the current base point for the following editing options. Use the STRETCH option to reposition the dimension text at a better location. When this operation is complete, press ESC twice at the command prompt to remove the object highlight and grips from the display screen.

Command: (*Select the warm grip at the text location to make it hot*).
** STRETCH **
Specify stretch point or [Base point/ Copy/Undo/eXit]: (*Pick a convenient location on the display screen to relocate the radius dimension*)
Command: (*Press ESC to remove the object highlight*)
Command: (*Press ESC to remove the grips*)

Note: The grip operation detailed in the preceding step is only possible when you use the Dimension Style Manager dialog box and the Fit tab to change the Fit Options to Text and Arrows and the Text Placement to Beside the dimension line. This will be covered in greater detail in the Chapter 12, Dimension Style Manager dialog box.

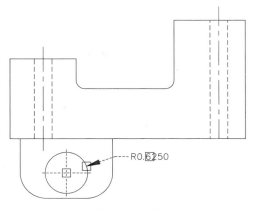

Figure 7–58

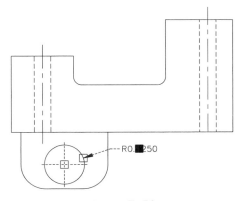

Figure 7–59

STEP 12

At the command prompt, use the grip cursor to select the centerline in Figure 7–60. Notice the appearance of the grips at the endpoints and midpoint of the centerline.

Command: (Select the centerline shown in Figure 7–60)

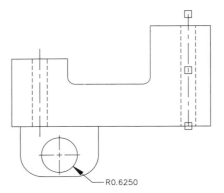

Figure 7–60

STEP 13

While still at the command prompt, select the grip shown in Figure 7–225 at "A." This grip becomes the current base point for the following editing options. Use the STRETCH mode and Base option to extend the centerline from a new base point at "B" to the intersection at "C." When this operation is complete, press ESC twice at the command prompt to remove the object highlight and grips from the display screen.

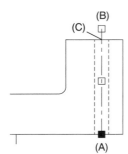

Figure 7–61

Command: (Select the warm grip at the bottom of the centerline at "A" to make it hot).
** STRETCH **
Specify stretch point or [Base point/ Copy/Undo/eXit]: **B** (For Base)
Specify base point: (Select the grip at "B")
** STRETCH **
Specify stretch point or [Base point/ Copy/Undo/eXit]: (Select the intersection of the centerline and horizontal line at "C")
Command: (Press ESC to remove the object highlight)
Command: (Press ESC to remove the grips)

STEP 14

At the command prompt, use the grip cursor to select the two hidden lines and centerline shown in Figure 7–62. Notice the appearance of the cold grips at the endpoints and midpoints of the lines.

Command: (Select the two hidden lines and the centerline)

STEP 15

While still at the command prompt, select the three grips at the midpoints of the center and hidden lines by holding down SHIFT as they are selected, as in Figure 7–63. These grips become the current base point for the following editing options. Release SHIFT and pick the selected middle grip again. Use the MOVE option to center the hidden and centerlines to the middle of the horizontal line. When this operation is complete, press ESC twice at the command prompt to remove the object highlight and grips from the display screen.

Command: (Use SHIFT to individually select the three warm grips at the middle of the three lines to make them hot. Release SHIFT and pick the middle grip again. Press the Spacebar until MOVE mode appears at the bottom of the prompt line)
** MOVE **
Specify move point or [Base point/Copy/Undo/eXit]: **B** (For Base)
Specify base point: (Pick the intersection of the center and horizontal lines at "A")
** MOVE **
Specify move point or [Base point/Copy/Undo/eXit]: **Mid**
of (Select the horizontal line at "B")
Command: (Press ESC to remove the object highlight)
Command: (Press ESC to remove the grips)

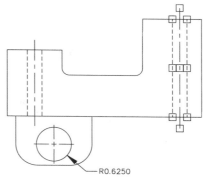

Figure 7–62

R0.6250

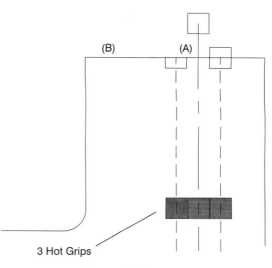

(B) (A)

3 Hot Grips

Figure 7–63

STEP 16

At the command prompt, use the grip cursor to select all of the objects shown in Figure 7–64. Notice the appearance of the grips at the various key points of these objects.

Command: *(Select all objects shown in Figure 7–64. You can accomplish this easily by using the Automatic Window mode and marking points at "X" and "Y")*

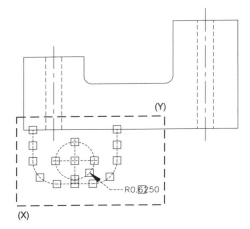

Figure 7–64

STEP 17

While still at the command prompt, pick the grip at "A" and press the Spacebar until MIRROR mode appears at the bottom of the prompt line. Mirror the selected objects using the midpoint of the line at "B" in Figure 7–65.

Command: *(Pick the warm grip at "A" to make it hot. Press the Spacebar until MIRROR mode appears at the bottom of the prompt line)*
** MIRROR**
Specify second point or [Base point/ Copy/Undo/eXit]: **B** *(For Base)*
New base point: **Mid**
of *(Select the midpoint of the horizontal line at "B")*
** MIRROR**
Specify second point or [Base point/ Copy/Undo/eXit]: **@1<90**
Command: *(Press ESC to remove the object highlight)*
Command: *(Press ESC to remove the grips)*

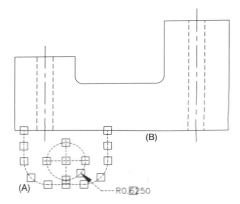

Figure 7–65

The completed object is illustrated in Figure 7–66.

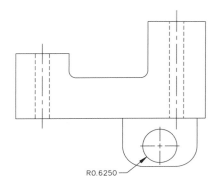

Figure 7–66

TUTORIAL EXERCISE: 07_MODIFY-EX.DWG

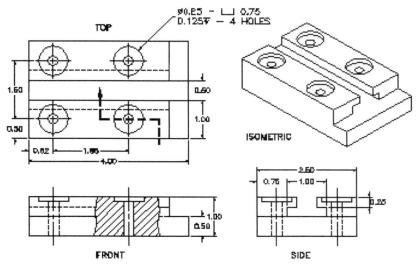

Figure 7–67

Purpose

This tutorial exercise is designed to change the properties of existing objects displayed in Figure 7–67.

System Settings

Since this drawing is provided on the CD, open an existing drawing file called "07_Modify-Ex." Follow the steps in this tutorial for changing various objects to the correct layer, text style, and dimension style.

Layers

Layers have already been created in this drawing.

Suggested Commands

Begin this tutorial by choosing the using the Properties dialog box to change the isomet-ric object to a different layer. Continue by changing the text height and layer of the view identifiers (FRONT, TOP, SIDE, ISOMET-RIC). The MATCHPROP command will be used to transfer the properties from one dimension to another, one text style to another, and one hatch pattern to another. The Layer Control box will be used to change the layer of various objects located in the Front and Top views.

Whenever possible, substitute the appropriate command alias in place of the full AutoCAD command in each tutorial step. For example, use "CP" for the COPY command, "L" for the LINE command, and so on. The complete listing of all command aliases is located in Chapter 1, Table 1–2.

STEP I

Loading this drawing displays the objects in a page layout called "Orthographic Views." Page layout is where the drawing will be plotted out in the future. This name is present next to the Model tab in the bottom portion of the drawing screen. Since a majority of changes will be made in Model mode, click on the Model tab. Your image will appear similar to Figure 7–68.

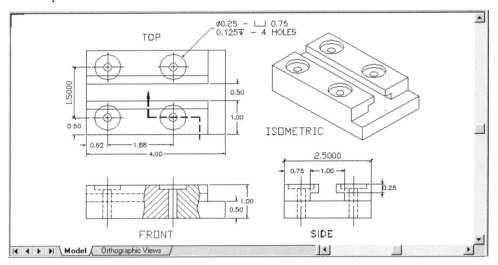

Figure 7–68

STEP 2

While in the Model environment, select all lines that make up the isometric view in Figure 7–69. You can accomplish this by using the Window mode from the Command prompt. If you accidentally select the word ISOMETRIC, de-select this word. Activate the Properties dialog box and click in the Layer field. This will display the current layer the objects are drawn on (DIM) in Figure 7–69. Click the down arrow to display the other layers and pick the OBJECT layer. This will change all selected objects that make up the isometric view to the OBJECT layer. When finished, dismiss the Properties dialog box. Press ESC twice to remove the object highlight and the grips from the drawing.

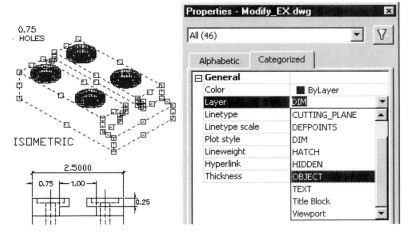

Figure 7–69

STEP 3

Pre-select the view titles (FRONT, TOP, SIDE, ISOMETRIC). These text items need to be changed to a height of 0.15 and the TEXT layer. With all four text objects highlighted, activate the Proper-

ties dialog box, click in the Layer field, and click the down arrow to change the selected objects to the TEXT layer in Figure 7–70.

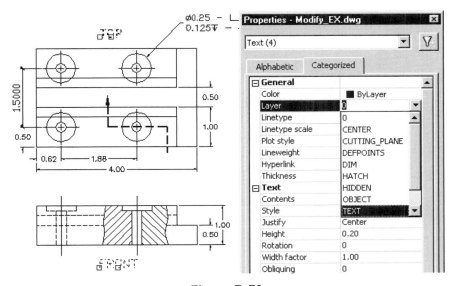

Figure 7–70

In the Properties dialog box, change the text height from 0.20 to a new value of 0.15. Pressing ENTER after making this change automatically updates all selected text objects to this new height in Figure 7–71. Dismiss this dialog box when finished. Press ESC twice to remove the object highlight and the grips from the drawing.

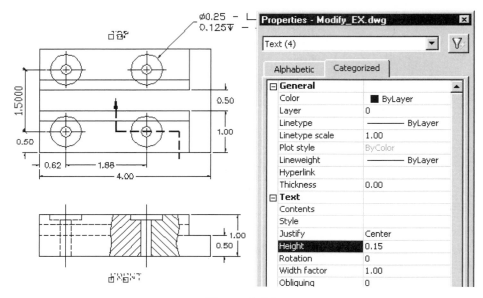

Figure 7–71

STEP 4

When you examine the view titles, TOP and SIDE are in one text style while FRONT and ISOMETRIC are in another. To remain consistent in the design process, you should make sure all text identifying the view titles has the same text style as TOP. Activate the MATCHPROP command by clicking on the button in Figure 7–72. Click the text object TOP as the source object. When the Match Property icon appears in Figure 7–73, select ISOMETRIC and FRONT as the destination objects. All properties associated with the TOP text object will be transferred to ISOMETRIC and FRONT, including the text style.

Command: **MATCHPROP**
Select source object: *(Select the text object "TOP" which should highlight in Figure 7–73)*
Current active settings: Color Layer Ltype Ltscale Lineweight Thickness PlotStyle Text Dim Hatch
Select destination object(s) or [Settings]: *(Select the text object "ISOMETRIC")*
Select destination object(s) or [Settings]: *(Select the text object "FRONT")*
Select destination object(s) or [Settings]: *(Press ENTER to exit this command)*

Figure 7–72

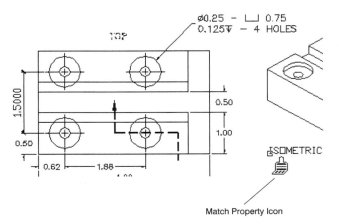

Figure 7–73

STEP 5

Notice the dimensions in this drawing. Two dimensions stand out above the rest (the 1.5000 vertical dimension in the Top view and the 2.5000 horizontal dimension in the Side view). Again to remain consistent in the design process, you should make sure all dimensions have the same appearance (dimension text height, number of decimal places, broken inside instead of placed above the dimension line). Activate the MATCHPROP command again by clicking on the button in the Standard Toolbar. Click the 4.00 horizontal dimension in the Top view as the source object. When the Match Property icon appears as in Figure 7–74, select the 1.5000 and 2.5000 vertical dimensions as

the destination objects. All dimension properties associated with the 4.00 dimension will be transferred to the 1.5000 and 2.5000 dimensions.

Command: **MATCHPROP**
Select source object: *(Select the 4.00 dimension, which should highlight)*
Current active settings: Color Layer Ltype Ltscale Lineweight Thickness PlotStyle Text Dim Hatch
Select destination object(s) or [Settings]: *(Select the 2.5000 dimension at "A")*
Select destination object(s) or [Settings]: *(Select the 1.5000 dimension at "B")*
Select destination object(s) or [Settings]: *(Press ENTER to exit this command)*

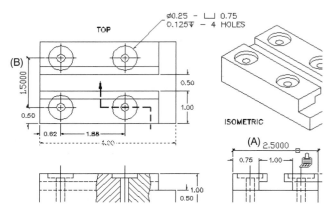

Figure 7–74

STEP 6

In the Front view, an area is crosshatched. However, both sets of crosshatching lines need to be drawn in the same direction, rather than opposing each other. Activate the MATCHPROP command again by clicking on the button in the Standard Toolbar. Click the left hatch pattern as the source object. When the Match Property icon appears, as in Figure 7–75, select the right hatch pattern as the destination object. All hatch properties asso-

ciated with the left hatch pattern will be transferred to the right hatch pattern.

Command: **MATCHPROP**
Select source object: *(Select the left hatch pattern, which should highlight)*
Current active settings: Color Layer
    Ltype Ltscale Lineweight Thickness
PlotStyle Text Dim Hatch
Select destination object(s) or [Settings]:
    (Pick the right hatch pattern)
Select destination object(s) or [Settings]:
    (Press ENTER *to exit this command.)*

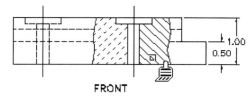

FRONT

Figure 7–75

STEP 7

Your display should appear similar to Figure 7–76. All view titles (FRONT, TOP, SIDE, ISOMETRIC) share the same text style and are at the same height. All

dimensions share the same parameters and text orientation. Both crosshatch patterns are drawn in the same direction.

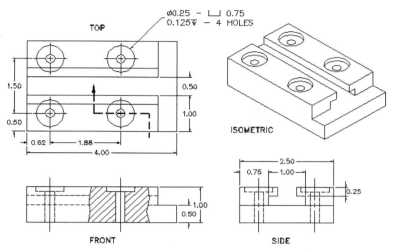

Figure 7–76

STEP 8

A different method will now be used to change the specific layer properties of objects. The two highlighted lines in the Top view in Figure 7–77 were accidentally drawn on Layer OBJECT and need to be transferred to the HIDDEN layer.

With the lines highlighted, click in the Layer Control box to display all layers. Click on the HIDDEN layer to change the highlighted lines to the HIDDEN layer. Press ESC twice to remove the object highlight and the grips from the drawing.

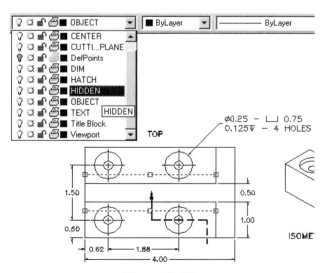

Figure 7–77

STEP 9

The two highlighted lines in the Front view in Figure 7–78 were accidentally drawn on the TEXT Layer and need to be transferred to the CENTER layer. With the lines highlighted, click in the Layer Control box to display all layers. Click on the CENTER layer to change the highlighted lines to the CENTER layer. Press ESC twice to remove the object highlight and the grips from the drawing.

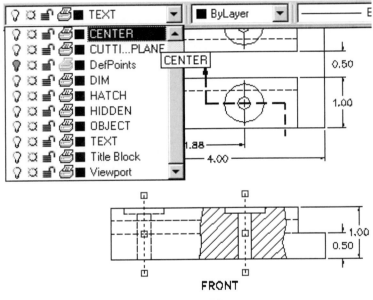

FRONT

Figure 7–78

STEP 10

Your display should appear similar to Figure 7–79. Notice the hidden lines in the Top view and the centerlines in the Front view.

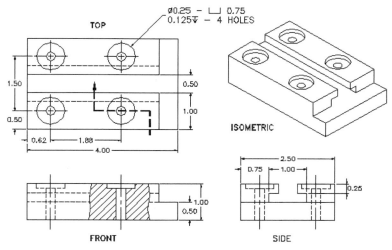

Figure 7–79

STEP 11

Click on the Orthographic Views Tab. Select the rectangular viewport and change this object's layer to "Viewport" in Figure 7–80.

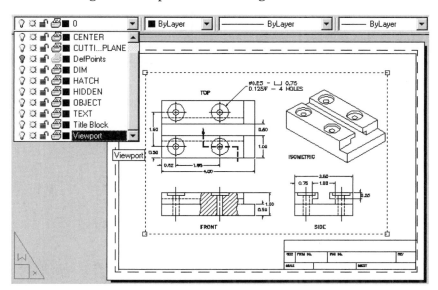

Figure 7–80

STEP 12

Turn off the Viewport layer. Your display should appear similar to Figure 7–81. This completes this tutorial exercise.

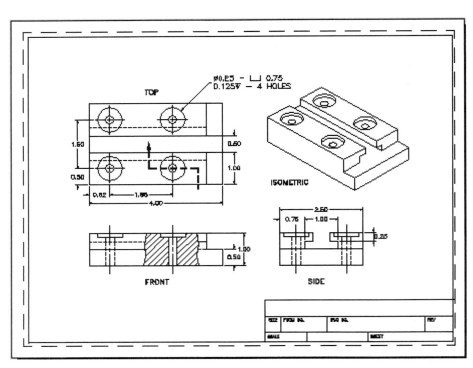

Figure 7–81

CHAPTER 8

Shape Description/ Multiview Projection

SHAPE DESCRIPTION

Before any object is made in production, some type of drawing needs to be created. This is not just any drawing, but rather an engineering drawing consisting of overall object sizes with various views of the object organized on the computer screen. This chapter introduces the topic of shape description or how many views are really needed to describe an object. The art of multiview projection includes methods of constructing one-view, two-view, and three-view drawings using AutoCAD commands.

Begin constructing an engineering drawing by first analyzing the object being drawn. To accomplish this, describe the object by views or how an observer looks at the object. Figure 8–1 shows a simple wedge; this object can be viewed at almost any angle to get a better idea of its basic shape. However, to standardize how all objects are to be viewed and to limit the confusion usually associated with complex multiview drawings, some standard method of determining how and where to view the object must be exercised.

Even though the simple wedge is easy to understand because it is currently being displayed in picture or isometric form, it would be difficult to produce this object because it is unclear what the sizes of the front and top faces are. Illustrated in Figure 8–2 are six primary ways or directions in which to view an object. The Front view begins the shape description and is followed by the Top view and Right Side view. Continuing on, the Left Side view, Back view, and Bottom view complete the primary ways to view an object.

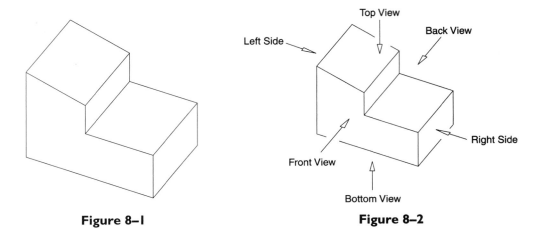

Figure 8–1

Figure 8–2

Now that the primary ways of viewing an object have been established, the views need to be organized to promote clarity and have the views reference one another. Imagine the simple wedge positioned in a transparent glass box similar to the illustration in Figure 8–3. With the entire object at the center of the box, the sides of the box represent the ways to view the object. Images of the simple wedge are projected onto the glass surfaces of the box.

With the views projected onto the sides of the glass box, we must now prepare the views to be placed on a two-dimensional (2D) drawing screen. For this to be accomplished, the glass box, which is hinged, is unfolded as in Figure 8–4. All folds occur from the Front view, which remains stationary.

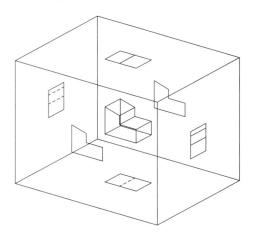

Figure 8–3

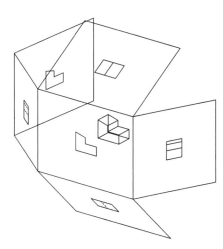

Figure 8–4

Figure 8–5 shows the views in their proper alignment to one another. However, this illustration is still in a pictorial view. These views need to be placed flat before continuing.

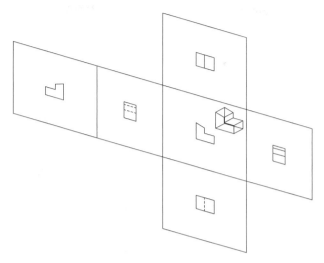

Figure 8–5

As the glass box is completely unfolded and laid flat, the result is illustrated in Figure 8–6. The Front view becomes the main view, with other views placed in relation to it. Above the Front view is the Top view. To the right of the Front view is the Right Side view. To the left of the Front view is the Left Side view followed by the Back view. Underneath the Front view is the Bottom view. This becomes the standard method of laying out the views needed to describe an object. But are all the views necessary? Upon closer inspection we find that, except for being a mirror image, the Front and Back views are identical. The Top and Bottom views appear similar, as do the Right and Left Side views. One very important rule to follow in multiview objects is to select only those views that accurately describe the object and discard the remaining views.

The complete multiview drawing of the simple wedge is illustrated in Figure 8–7. Only the Front, Top, and Right Side views are needed to describe this object. When laying out views, remember that the Front view is usually the most important—it holds the basic shape of the object being described. Directly above the Front view is the Top view, and to the right of the Front view is the Right Side view. All three views are separated by a space of varying size. This space is commonly called a dimension space because it is a good area in which to place dimensions describing the size of the object. The space also acts as a separator between views; without it, the views would touch on one another, making them difficult to read and interpret. The minimum distance of this space is usually 1.00 unit.

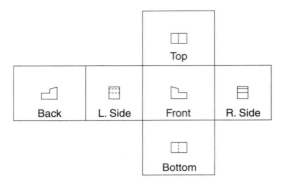

Figure 8–6

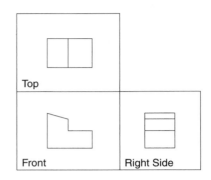

Figure 8–7

RELATIONSHIPS BETWEEN VIEWS

Some very interesting and important relationships are set up when the three views are placed in the configuration illustrated in Figure 8–8. Notice that because the Top view is directly above the Front view, both share the same width dimension. The Front and Right Side views share the same height. The relationship of depth on the Top and Right Side views can be explained with the construction of a 45°-projection line at "A" and the projection of the lines over and down or vice versa to get the depth. Another principle illustrated by this example is that of projecting lines up, over, across, and down to create views. Editing commands such as ERASE and TRIM are then used to clean up unnecessary lines.

With the three views identified in Figure 8–9, the only other step, and it is an important one, is to annotate the drawing, or add dimensions to the views. With dimensions, the object drawn in multiview projection can now be produced. Even though this has been a simple example, the methods of multiview projection work even for the most difficult and complex of objects.

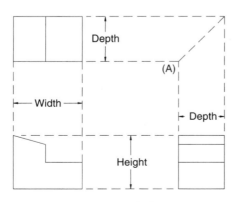

Figure 8–8

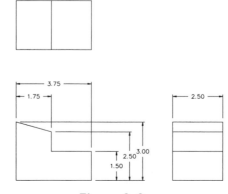

Figure 8–9

REVIEW OF LINETYPES AND CONVENTIONS

At the heart of any engineering drawing is the ability to assign different types of lines to convey meaning to the drawing. When plotted out, all lines of a drawing are dark; border and title block lines are the thickest. Object lines outline visible features of a drawing and are made thick and dark, but not as thick as a border line. Features that are invisible in an adjacent view are identified by a hidden line. This line is a series of dashes 0.12 units in length with a spacing of 0.06 units. Centerlines identify the centers of circular features such as holes or show that a hidden feature in one view is circular in another. The centerline consists of a series of long and short dashes. The short dash measures approximately 0.12 units, whereas the long dash may vary from 0.75 to 1.50 units. A gap of 0.06 separates dashes. Study the examples of these lines in Figure 8–10.

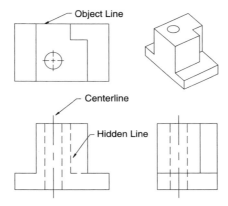

Figure 8–10

The object shown in Figure 8–11 is a good illustration of the use of phantom lines in a drawing. Phantom lines are especially useful where motion is applied. The arm on the right is shown by standard object lines consisting of the continuous linetype. To show that the arm rotates about a center pivot, the identical arm is duplicated through the ARRAY command, and all lines of this new element are converted to phantom lines through the Properties dialog box, MATCHPROP command or Layer Control box. Notice that the smaller segments are not shown as phantom lines due to their short lengths.

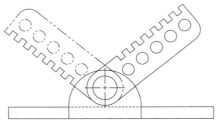

Figure 8–11

Use of linetypes in a drawing is crucial to the interpretation of the views and the final design before the object is actually made. Sometimes the linetype appears too long; in other cases, the linetype does not appear at all, even though using the LIST command on the object will show the proper layer and linetype. The LTSCALE or Linetype Scale command is used to manipulate the size of all linetypes loaded into a drawing. By default, all linetypes are assigned a scale factor of 1.00. This means that the actual dashes and/or spaces of the linetype are multiplied by this factor. The views illustrated in Figure 8–12 show linetypes that use the default value of 1.00 from the LTSCALE command.

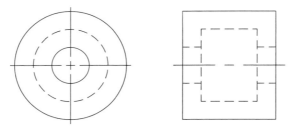

Figure 8–12

If a linetype dash appears too long, use the LTSCALE command and set a new value to less than 1.00. If a linetype appears too short, use the LTSCALE command and set a new value to greater than 1.00. The same views illustrated in Figure 8–13 show the effects of the LTSCALE command set to a new value of 0.75. Notice that the center in the Right Side view has one more series of dashes than the same object illustrated previously. The 0.75-multiplier affects all dashes and spaces defined in the linetype.

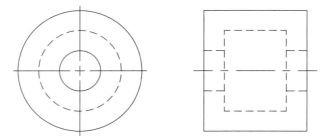

Figure 8–13

Figure 8–14 illustrates how using the LTSCALE command with a new value of 0.50 shortens the linetype dashes and spaces even more. Now even the center marks identifying the circles have been changed to centerlines. When you use the LTSCALE command, the new value, whether larger or smaller than 1.00, affects all linetypes visible on the display screen.

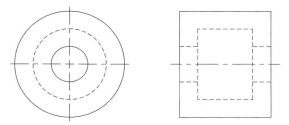

Figure 8–14

ONE-VIEW DRAWINGS

An important rule to remember concerning multiview drawings is to draw only enough views to accurately describe the object. In the drawing of the gasket in Figure 8–15, Front and Side views are shown. However, the Side view is so narrow that it is diffi-cult to interpret the hidden lines drawn inside. A better approach would be to leave out the Side view and construct a one-view drawing consisting of just the Front view.

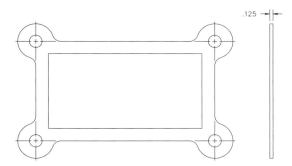

Figure 8–15

Begin the one-view drawing of the gasket by first laying out centerlines marking the centers of all circles and arcs, as in Figure 8–16. A layer containing centerlines could be used to show all lines as centerlines.

Figure 8–16

Use the CIRCLE command to lay out all circles representing the bolt holes of the gasket shown in Figure 8–17. You could use the OFFSET command to form the large rectangle on the inside of the gasket. If lines of the rectangle extend past each other, use the FILLET command set to a value of 0. Selecting two lines of the rectangle will form a corner. Repeat this procedure for any other lines that do not form exact corners.

Use the TRIM command to begin forming the outside arcs of the gasket shown in Figure 8–18.

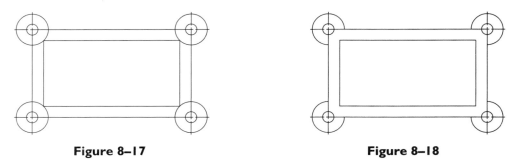

Figure 8–17   **Figure 8–18**

Use the FILLET command set to the desired radius to form a smooth transition from the arcs to the outer rectangle shown in Figure 8–19.

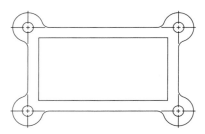

Figure 8–19

TWO-VIEW DRAWINGS

Before attempting any drawing, determine how many views need to be drawn. A minimum number of views are needed to describe an object. Drawing extra views is not only time-consuming, but it may result in two identical views with mistakes in each view. You must interpret which is the correct set of views. The illustration in Figure 8–20 is a three-view multiview drawing of a coupler. The circles and circular hidden circle identify the Front view. Except for their rotation angles, the Top and Right Side views are identical. In this example or for other symmetrical objects, only two views are needed to accurately describe the object being drawn. The Top view has been deleted to leave the Front and Right Side views. The Side view could have easily been deleted in favor of leaving the Front and Top views. This decision is up to the designer, depending on sheet size and which views are best suited for the particular application.

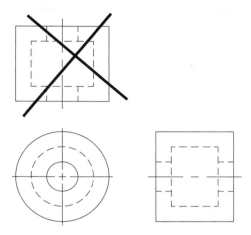

Figure 8–20

To illustrate how AutoCAD is used as the vehicle for creating a two-view engineering drawing, study the pictorial drawing illustrated in Figure 8–21 to get an idea of how the drawing will appear. Begin the two-view drawing by using the LINE command to lay out the Front and Side views. You can find the width of the Top view by projecting lines up from the front because both views share the same width. Provide a space of 1.50 units between views to act as a separator and allow for dimensions at a later time.

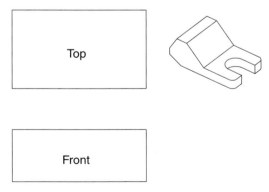

Figure 8–21

Begin adding visible details to the views, such as circles, filleted corners, and angles, as shown in Figure 8–22. Use various editing commands such as TRIM, EXTEND, and OFFSET to clean up unnecessary geometry.

From the Front view, project corners up to the Top view. These corners will form visible edges in the Top view. Use the same projection technique to project features from the Top view to the Front view (see Figure 8–23).

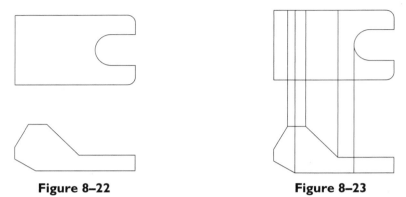

Figure 8–22   **Figure 8–23**

Use the TRIM command to delete any geometry that appears in the 1.50 dimension space. The views now must conform to engineering standards by showing which lines are visible and which are invisible, as shown in Figure 8–24. The corner at "A" represents an area hidden in the Top view. Use the Property dialog box or MATCHPROP command to convert the line in the Top view from the continuous linetype to the hidden linetype. In the same manner, the slot visible in the Top view is hidden in the Front view. Again convert the continuous line in the Front view to the hidden linetype, using one of the methods already described. Since the slot in the Top view represents a circular feature, use the DIMCENTER command to place a center marker at the center of the semicircle. To show in the Front view that the hidden line represents a circular feature, add one centerline consisting of one short dash and two short dashes. If the slot in the Top view were square instead of circular, centerlines would not be necessary.

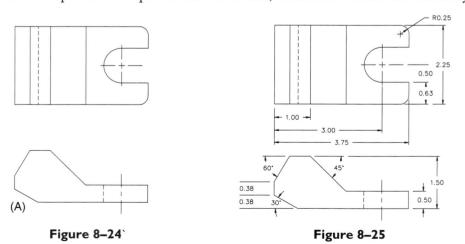

(A)

Figure 8–24`   **Figure 8–25**

Use the spaces provided to properly add dimensions to the drawing, as shown in Figure 8–25. Once the dimension spaces are filled with numbers, use outside areas to call out distances.

THREE-VIEW DRAWINGS

If two views are not enough to describe an object, draw three views. This consists of Front, Top, and Right Side views. A three-view drawing of the guide block, as illustrated in pictorial format in Figure 8–26, will be the focus of this segment. Notice the broken section exposing the Spotface operation above a drill hole. Begin this drawing by laying out all views using overall dimensions of width, depth, and height. The LINE and OFFSET commands are popular commands used to accomplish this. Provide a space between views to accommodate dimensions at a later time.

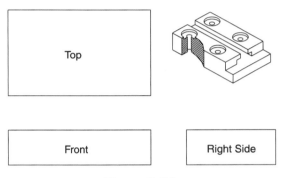

Figure 8–26

Begin drawing features in the views where they are visible, as illustrated in Figure 8–27. Since the Spotface holes appear above, draw these in the Top view. The notch appears in the Front view; draw it there. A slot is visible in the Right Side view and is drawn there.

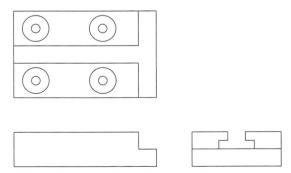

Figure 8–27

As in two-view drawings, all features are projected down from the Top to the Front view. To project depth measurements from the Top to the Right Side view, construct a 45° line at "A." See Figure 8–28.

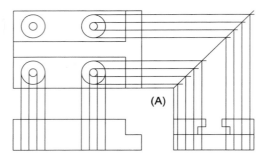

Figure 8–28

Use the 45° line to project the slot from the Right Side view to the Top view as shown in Figure 8–29. Project the height of the slot from the Right Side view to the Front view. Convert the continuous lines to hidden lines where features appear invisible, such as the holes in the Front and Right Side views.

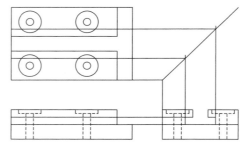

Figure 8–29

Change the remaining lines from continuous to hidden. Erase any construction lines, including the 45°-projection line (see Figure 8–30).

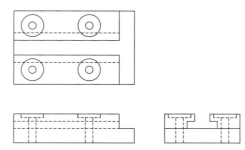

Figure 8–30

Begin adding centerlines to label circular features as shown in Figure 8–31. The DIMCENTER command is used where the circles are visible. Where features are hidden but represent circular features, the single centerline consisting of one short dash and two long dashes is used. In Figure 8–32, dimensions remain the final step in completing the engineering drawing before it is checked and shipped off for production.

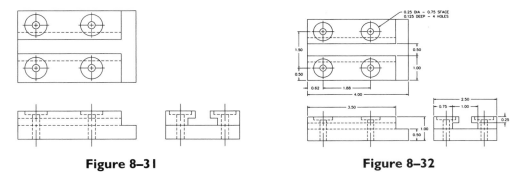

| Figure 8–31 | Figure 8–32 |

RUNOUTS

Where flat surfaces become tangent to cylinders, there must be some method of accurately representing this using fillets. In the object illustrated in Figure 8–33, the Front view shows two cylinders connected to each other by a tangent slab. The Top view is complete; the Front view has all geometry necessary to describe the object with the exception of the exact intersection of the slab with the cylinder.

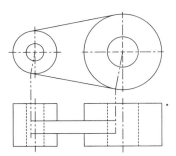

Figure 8–33

Figure 8–34 displays the correct method for finding intersections or runouts—areas where surfaces intersect others and blend in, disappear, or simply run out. A point of intersection is found at "A" in the Top view with the intersecting slab and the cylinder. This actually forms a 90° angle with the line projected from the center of the cylinder and the angle made by the slab. A line is projected from "A" in the Top view to intersect with the slab found in the Front view.

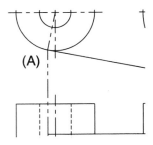

Figure 8–34

In Figure 8–35, fillets are drawn to represent the slab and cylinder intersections. This forms the runout.

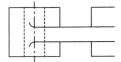

Figure 8–35

The resulting two-view drawing, complete with runouts, is illustrated in Figure 8–36.

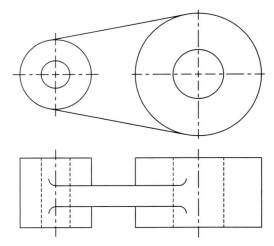

Figure 8–36

TUTORIAL EXERCISE: SHIFTER.DWG

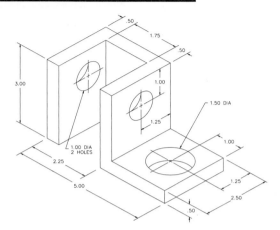

Figure 8–37

Purpose

This tutorial is designed to allow the user to construct a three-view drawing of the Shifter as shown in Figure 8–37.

System Settings

Use the Drawing Units dialog box and change the number of decimal places from four to two. Set linetype scale to 0.50 using the LTSCALE (LTS) command. Keep the remaining default settings. Use the default settings for the screen limits: (0,0) for the lower left corner and (12.00,9.00) for the upper right corner. The grid and snap do not need to be set to any certain values. Check to see that the following Object Snap modes are already set: Endpoint, Extension, Intersection, Center.

Layers

Create the following layers with the format:

Name	Color	Linetype
Object	Green	Continuous
Hidden	Red	Hidden
Center	Yellow	Center
Dimension	Yellow	Continuous
Projection	Cyan	Continuous

Suggested Commands

The primary commands used during this tutorial are OFFSET and TRIM. The OFFSET command is used for laying out all views before the TRIM command is used to clean up excess lines. Since different linetypes represent certain features of a drawing, the Layer Control box is used to convert to the desired linetype needed as set in the Layer Properties Manager dialog box. Once all visible details are identified in the primary views, project the visible features to the other views using the LINE command. A 45° inclined line is constructed to project lines from the Top view to the Right Side view and vice versa.

Whenever possible, substitute the appropriate command alias in place of the full AutoCAD command in each tutorial step; for example, use "CP" for the COPY command, "L" for the LINE command, and so on. The complete listing of all command aliases is located in Chapter 1, Table 1–2.

STEP 1

Make the "Object" layer current in Figure 8–38A. Begin the orthographic drawing of the Shifter by constructing a right angle consisting of one horizontal and one vertical line. The corner formed by the two lines will be used to orient the Front view (see Figure 8–38B).

Command: **L** *(For LINE)*
Specify first point: **1,1**
Specify next point or [Undo]: **@11,0**
Specify next point or [Undo]: *(Press ENTER to exit this command)*
Command: **L** *(For LINE)*
Specify first point: **1,1**
Specify next point or [Undo]: **@8<90**
Specify next point or [Undo]: *(Press ENTER to exit this command)*

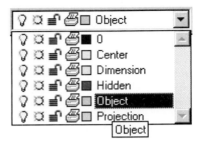

Figure 8–38A

Figure 8–38B

STEP 2

Begin the layout of the primary views by using the OFFSET command to offset the vertical line at "A" a distance of 5.00 units, which represents the length of the Shifter (see Figure 8–39).

Command: **O** *(For OFFSET)*
Specify offset distance or [Through] <1.0000>: **5.00**
Select object to offset or <exit>: *(Select the vertical line at "A")*
Specify point on side to offset: *(Pick a point anywhere near "B")*
Select object to offset or <exit>: *(Press ENTER to exit this command)*

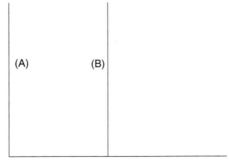

Figure 8–39

STEP 3

Use the OFFSET command to offset the horizontal line at "A" a distance of 3.00 units, which represents the height of the Shifter (see Figure 8–40).

Command: **O** *(For OFFSET)*
Specify offset distance or [Through] <5.00>: **3.00**
Select object to offset or <exit>: *(Select the horizontal line at "A")*
Specify point on side to offset: *(Pick a point anywhere near "B")*
Select object to offset or <exit>: *(Press ENTER to exit this command)*

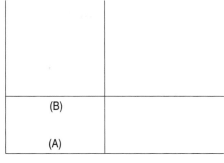

Figure 8–40

STEP 4

Begin laying out dimension spaces that will act as separators between views and allow for the placement of dimensions once the Shifter is completed (see Figure 8–41). A spacing of 1.50 units will be more than adequate for this purpose. Again, use the OFFSET command to accomplish this.

Command: **O** *(For OFFSET)*
Specify offset distance or [Through] <3.00>: **1.50**
Select object to offset or <exit>: *(Select the vertical line at "A")*
Specify point on side to offset: *(Pick a point anywhere near "B")*
Select object to offset or <exit>: *(Select the horizontal line at "C")*
Specify point on side to offset: *(Pick a point anywhere near "D")*
Select object to offset or <exit>: *(Press ENTER to exit this command)*

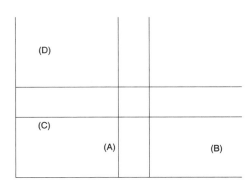

Figure 8–41

STEP 5

Use the OFFSET command to lay out the depth of the Shifter at a distance of 2.50 units (see Figure 8–42).

Command: **O** *(For OFFSET)*
Specify offset distance or [Through] <1.50>: **2.50**
Select object to offset or <exit>: *(Select the vertical line at "A")*
Specify point on side to offset: *(Pick a point anywhere near "B")*
Select object to offset or <exit>: *(Select the horizontal line at "C")*
Specify point on side to offset: *(Pick a point anywhere near "D")*
Select object to offset or <exit>: *(Press ENTER to exit this command)*

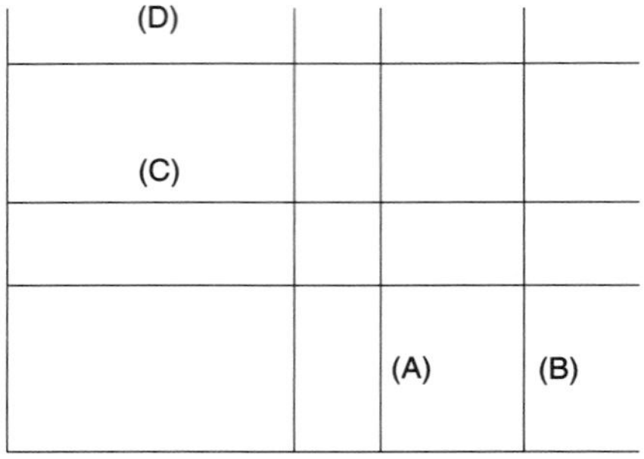

Figure 8–42

STEP 6

Use the TRIM command to trim away excess construction lines you used when laying out the primary views of the Shifter (see Figure 8–43).

Command: **TR** *(For TRIM)*
Current settings: Projection=UCS Edge=None
Select cutting edges ...
Select objects: *(Select lines "A" and "B")*
Select objects: *(Press ENTER to continue)*
Select object to trim or [Project/Edge/ Undo]: *(Select the line at "C")*
Select object to trim or [Project/Edge/ Undo]: *(Select the line at "D")*
Select object to trim or [Project/Edge/ Undo]: *(Select the line at "E")*
Select object to trim or [Project/Edge/ Undo]: *(Select the line at "F")*
Select object to trim or [Project/Edge/ Undo]: *(Press ENTER to exit this command)*

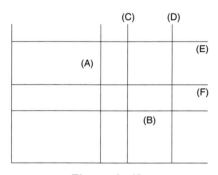

Figure 8–43

STEP 7

Use the TRIM command again to complete trimming away excess construction lines you used when laying out the primary views of the Shifter (see Figure 8–44).

Command: **TR** *(For TRIM)*
Current settings: Projection=UCS
    Edge=None
Select cutting edges ...
Select objects: *(Press ENTER to accept all objects as cutting edges)*
Select object to trim or [Project/Edge/ Undo]: *(Select the lines at "A" through "D" to trim)*
Select object to trim or [Project/Edge/ Undo]: *(Press ENTER to exit this command)*

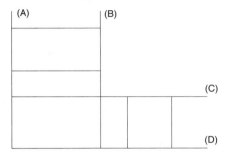

Figure 8–44

STEP 8

Your display should appear similar to the illustration in Figure 8–45 with the layout of the Front, Top, and Right Side views. Begin adding details to all views through methods of projection. Use the OFFSET command to offset lines "A," "C," and "E" a distance of 0.50.

Command: **O** *(For OFFSET)*
Specify offset distance or [Through] <2.50>: **0.50**
Select object to offset or <exit>: *(Select the vertical line at "A")*
Specify point on side to offset: *(Pick a point anywhere near "B")*
Select object to offset or <exit>: *(Select the horizontal line at "C")*
Specify point on side to offset: *(Pick a point anywhere near "D")*
Select object to offset or <exit>: *(Select the horizontal line at "E")*
Specify point on side to offset: *(Pick a point anywhere near "F")*
Select object to offset or <exit>: *(Press ENTER to exit this command)*

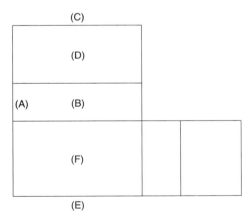

Figure 8–45

STEP 9

Use the OFFSET command and offset the vertical line at "A" a distance of 1.75 (see Figure 8–46).

Command: **O** *(For OFFSET)*
Specify offset distance or [Through] <0.50>: **1.75**
Select object to offset or <exit>: *(Select the vertical line at "A")*
Specify point on side to offset: *(Pick a point anywhere near "B")*
Select object to offset or <exit>: *(Press ENTER to exit this command)*

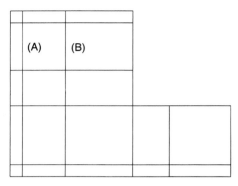

Figure 8–46

STEP 10

Use the OFFSET command to offset the vertical line at "A" a distance of 0.50 (see Figure 8–47).

Command: **OF** *(For OFFSET)*
Specify offset distance or [Through] <1.75>: **0.50**
Select object to offset or <exit>: *(Select the vertical line at "A")*
Specify point on side to offset: *(Pick a point anywhere near "B")*
Select object to offset or <exit>: *(Press ENTER to exit this command)*

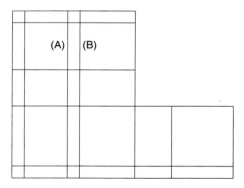

Figure 8–47

STEP 11

Your display should appear similar to the illustration in Figure 8–48. Next, use the TRIM command to partially delete the line segments located in the spaces between views.

Command: **TR** *(For TRIM)*
Current settings: Projection=UCS
   Edge=None
Select cutting edges ...
Select objects: *(Press* ENTER *to accept all objects as cutting edges)*
Select object to trim or [Project/Edge/Undo]: *(Select the lines at "A" through "H" to trim)*
Select object to trim or [Project/Edge/Undo]: *(Press* ENTER *to exit this command)*

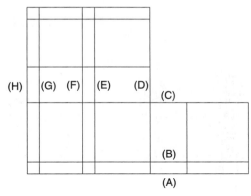

Figure 8–48

STEP 12

Use the ZOOM command and the Window option to magnify the Top view similar to the illustration in Figure 8–49. Then use the TRIM command to trim the line labeled "B."

Command: **TR** *(For TRIM)*
Current settings: Projection=UCS
   Edge=None
Select cutting edges ...
Select objects: *(Select the vertical line at "A")*
Select objects: *(Press* ENTER *to continue)*
Select object to trim or [Project/Edge/Undo]: *(Select the line at "B")*
Select object to trim or [Project/Edge/Undo]: *(Press* ENTER *to exit this command)*

Figure 8–49

STEP 13

Zoom back to the original display using the ZOOM command and the Previous option. Use the ZOOM command and the Window option to magnify the display to show the Front view illustrated in Figure 8–50. Use the TRIM command to clean up the excess lines in the Front view, using the illustration in Figure 8–50 as a guide.

Command: **TR** *(For TRIM)*
Current settings: Projection=UCS
    Edge=None
Select cutting edges ...
Select objects: *(Select lines "A" and "B")*
Select objects: *(Press ENTER to continue)*
Select object to trim or [Project/Edge/
    Undo]: *(Select lines "C" through "F" to
    trim)*
Select object to trim or [Project/Edge/
    Undo]: *(Press ENTER to exit this
    command)*

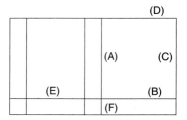

Figure 8–50

STEP 14

Use the ZOOM command and the Previous option to zoom back to the previous screen display containing the three views. Use the TRIM command to partially delete the lines at "A" through "D" in Figure 8–51.

Command: **TR** *(For TRIM)*
Current settings: Projection=UCS
    Edge=None
Select cutting edges ...
Select objects: *(Press ENTER to accept all
    objects as cutting edges)*
Select object to trim or [Project/Edge/
    Undo]: *(Select the lines at "A" through
    "D" to trim)*
Select object to trim or [Project/Edge/
    Undo]: *(Press ENTER to exit this
    command)*

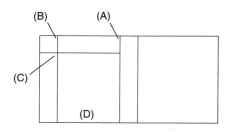

Figure 8–51

STEP 15

Begin placing a circle in the Top view representing the 1.50 diameter drill hole in Figure 8–52. Use the CIRCLE command and the OSNAP-From and OSNAP-Intersect modes to set up a temporary point of reference as illustrated in Figure 8–52.

Command: **C** *(For CIRCLE)*
Specify center point for circle or [3P/2P/
    Ttr (tan tan radius)]: **From**
Base point: *(Pick the corner intersection at
    "A")*
<Offset>: **@-1,-1.25**
Specify radius of circle or [Diameter]: **D**
    (For Diameter)
Specify diameter of circle: **1.50**

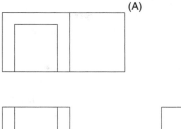
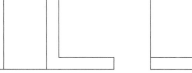

Figure 8–52

STEP 16

Place a circle in the Right Side view representing the 1.00-diameter drill hole. Use the CIRCLE command and the OSNAP-From and OSNAP-Intersect modes to set up a temporary point of reference as illustrated in Figure 8–53 at "A."

Command: **C** *(For CIRCLE)*
Specify center point for circle or [3P/2P/
    Ttr (tan tan radius)]: **From**
Base point: *(Pick the corner intersection at
    "A")*
<Offset>: **@-1.25,-1.00**
Specify radius of circle or [Diameter]
    <0.75>: **D** *(For Diameter)*
Specify diameter of circle <1.50>: **1.00**

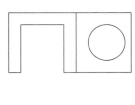
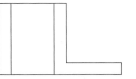

Figure 8–53

STEP 17

Before continuing with this step, first make the "Projection" layer current in Figure 8–54A. Use the LINE command to draw projection lines from both circles to the Front view as shown in Figure 8–54B. Use the OSNAP-Quadrant mode along with the Polar Intersection mode to accomplish this.

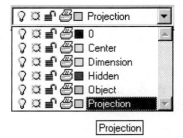

Figure 8–54A

Command: **L** *(For LINE)*
Specify first point: **Qua**
of *(Select the quadrant of the circle at "A")*
Specify next point or [Undo]: *(Select the polar intersection at "B")*
Specify next point or [Undo]: *(Press ENTER to exit this command)*

Repeat the above procedure for the circle locations at "C," "D," and "E" using quadrants and polar intersections.

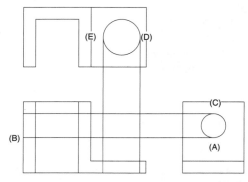

Figure 8–54B

STEP 18

Before continuing, first make the "Center" layer current in Figure 8–55A. Then place center marks at the centers of both circles as illustrated in Figure 8–55B. Before you proceed with this operation, you need to set a system variable to a certain value to achieve the desired results. Set the DIMCEN variable to a value of -0.12. This will not only place the center mark when the DIMCENTER command is used, but will also extend the centerline a short distance outside both circles.

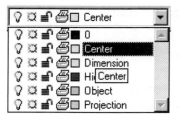

Figure 8–55A

Command: **DIMCEN**

Enter new value for DIMCEN <0.09>: **- 0.12**

Command: **DCE** *(For DIMCENTER)*

Select arc or circle: *(Select the circle at "A")*

Command: **DCE** *(For DIMCENTER)*

Select arc or circle: *(Select the circle at "B")*

When finished with these operations, make the "Projection" layer current again.

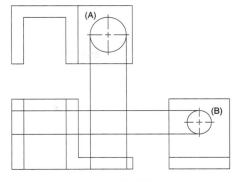

Figure 8–55B

STEP 19

With the "Projection" layer current, construct two lines from the endpoints of both center marks using the OSNAP-Endpoint mode. Turn Ortho mode on to assist with this operation. The lines will be converted to centerlines at a later step. Draw the lines 0.50 units past the Front view as illustrated in Figure 8–56.

Command: **L** *(For LINE)*

Specify first point: *(Select the endpoint of the center mark in the Top view at "A")*

Specify next point or [Undo]: *(Pick a point just below the Front view at "B")*

Specify next point or [Undo]: *(Press ENTER to exit this command)*

Command: **L** *(For LINE)*

Specify first point: *(Select the endpoint of the center mark in the Side view at "C")*

Specify next point or [Undo]: *(Pick a point to the left of the Front view at "D")*

Specify next point or [Undo]: *(Press ENTER to exit this command)*

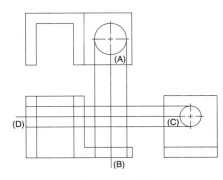

Figure 8–56

STEP 20

Use the TRIM command and the illustration in Figure 8–57 to trim away excess lines.

Command: **TR** *(For TRIM)*
Current settings: Projection=UCS
    Edge=None
Select cutting edges ...
Select objects: *(Select the lines at "A," "B," "C," and "D")*
Select objects: *(Press ENTR to continue)*
Select object to trim or [Project/Edge/
    Undo]: *(Select the lines at "E" through "J" to trim)*
Select object to trim or [Project/Edge/
    Undo]: *(Press ENTER to exit this command)*

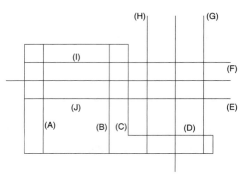

Figure 8–57

STEP 21

At the Command prompt, select the six lines in Figure 8–58. Use the Layer Control box and change all highlighted lines to the "Hidden" layer. When finished, press ESC twice to remove the object highlight and the grips.

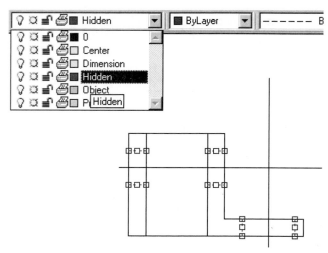

Figure 8–58

STEP 22

Before continuing with this step, turn off OSNAP mode. Then use the BREAK command to partially delete the lines illustrated in Figure 8–59 before converting them to center lines. Remember, centerlines extend past the object lines when identifying hidden drill holes; it would be inappropriate to use the TRIM command for this step.

Command: <Osnap off>
Command: **BR** *(For BREAK)*
Select object: *(Select the horizontal line approximately at "A")*
Specify second break point or [First point]: *(Select the line approximately at "B")*
Command: **BR** *(For BREAK)*
Select object: *(Select the vertical line approximately at "C")*
Specify second break point or [First point]: *(Select the line approximately at "D")*

Command: **BR** *(For BREAK)*
Select object: *(Select the horizontal line approximately at "E")*
Specify second break point or [First point]: **@** *(To break the line at the last point)*

Using the @ symbol breaks the object at the exact location identified by the first point selected. Remember, the @ symbol means "last point."

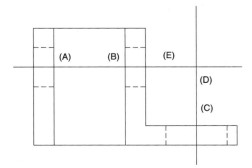

Figure 8–59

STEP 23

The purpose of the @ symbol in the previous step is to break a line into two segments without showing the break. The @ symbol means "the last known point," which completes the BREAK command by satisfying the "Second point" prompt. To prove this, use the ERASE command to delete the segments no longer needed (see Figure 8–60).

Command: **E** *(For ERASE)*
Select objects: *(Carefully select the lines at "A" and "B")*
Select objects: *(Press ENTER to perform the erase operation)*

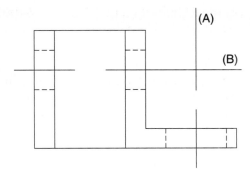

Figure 8–60

STEP 24

At the Command prompt, select the three lines in Figure 8–61. Then use the Layer Control box change all highlighted lines to the "Center" layer. When finished, press ESC twice to remove the object highlight and the grips.

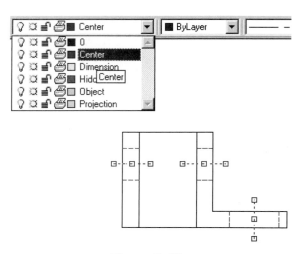

Figure 8–61

STEP 25

Turn OSNAP back on. You need to construct a 45° angle in order to begin projecting features from the Top view to the Right Side view and then back again. You form this angle by extending the bottom edge of the Top view to intersect with the left edge of the Side view, as shown in Figure 8–62. Use the FILLET command with the Radius set to 0 to accomplish this. Then draw the 45° line; the length of this line is not important. Turn Ortho mode off for this step.

Command: **F** *(For FILLET)*
Current settings: Mode = TRIM, Radius = 0.50

Select first object or [Polyline/Radius/ Trim]: **R** *(For Radius)*
Specify fillet radius <0.50>: **0**

Command: **F** *(For FILLET)*
Current settings: Mode = TRIM, Radius = 0.00
Select first object or [Polyline/Radius/ Trim]: *(Select line "A")*
Select second object: *(Select line "B")*
Command: **L** *(For LINE)*
Specify first point: *(Select the intersection at "A")*
Specify next point or [Undo]: **@4<45**
Specify next point or [Undo]: *(Press ENTER to exit this command)*

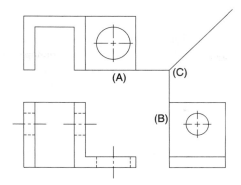

Figure 8–62

STEP 26

The "Projector" layer should still be current. Draw lines from points "A," "B," and "C" to intersect with the 45°-projection line as shown in Figure 8–63. Be sure Ortho mode is on to draw horizontal lines. Use the OSNAP-Intersect, Endpoint, and Quadrant modes to assist in this operation.

Command: **L** *(For LINE)*
Specify first point: *(Select the intersection of the corner at "A")*
Specify next point or [Undo]: *(Draw a horizontal line just past the 45° angle)*
Specify next point or [Undo]: *(Press*
ENTER *to exit this command)*
Command: **L** *(For LINE)*
Specify first point: *(Select the endpoint of the centerline at "B")*
Specify next point or [Undo]: *(Draw a horizontal line just past the 45° angle)*
Specify next point or [Undo]: *(Press*
ENTER *to exit this command)*
Command: **L** *(For LINE)*
Specify first point: **QUA**
of *(Select the quadrant of the circle at "C")*
Specify next point or [Undo]: *(Draw a horizontal line just past the 45° angle)*
Specify next point or [Undo]: *(Press*
ENTER *to exit this command)*

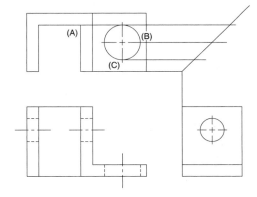

Figure 8–63

STEP 27

Draw lines from the intersection of the 45°-projection line to points in the Right Side view as shown in Figure 8–64. Use the following commands to accomplish this.

Command: **L** *(For LINE)*
Specify first point: *(Select the intersection of the angle at "A")*
Specify next point or [Undo]: *(Select the polar intersection at "B")*
Specify next point or [Undo]: *(Press ENTER to exit this command)*
Command: **L** *(For LINE)*
Specify first point: *(Select the intersection of the angle at "C")*
Specify next point or [Undo]: *(Select the polar intersection at "D")*
Specify next point or [Undo]: *(Press ENTER to exit this command)*
Command: **L** *(For LINE)*
Specify first point: *(Select the intersection of the angle at "E")*

Specify next point or [Undo]: *(Pick a point below the bottom of the Side view at "F")*
Specify next point or [Undo]: *(Press ENTER to exit this command)*

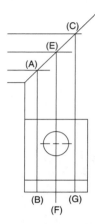

Figure 8–64

STEP 28

Delete the three projection lines from the Top view using the ERASE command as shown in Figure 8–65.

Command: **E** *(For ERASE)*
Select objects: *(Select lines "A," "B," and "C")*
Select objects: *(Press ENTER to perform this command)*

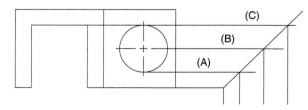

Figure 8–65

STEP 29

Use the TRIM command to trim away unnecessary geometry using the illustration
in Figure 8–66 as a guide. The hidden hole
and slot will be formed in the Side view
through this operation.

Command: **TR** *(For TRIM)*
Current settings: Projection=UCS
    Edge=None
Select cutting edges ...
Select objects: *(Select the horizontal line at
    "A")*
Select objects: *(Press ENTER to continue)*
Select object to trim or [Project/Edge/
    Undo]: *(Select the line at "B")*
Select object to trim or [Project/Edge/
    Undo]: *(Press ENTER to exit this
    command)*

Repeat this procedure for the line illustrated in Figure 8–66 using the horizontal line "C" as the cutting edge and the
vertical line "D" as the line to trim.

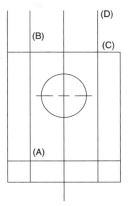

Figure 8–66

STEP 30

Use the BREAK command to split the vertical line at "A" into two separate objects
as shown in Figure 8–67. Accomplish this
by typing @ in response to the prompt "Enter second point." This will split the line in
two without the break being noticeable.

Command: **BR** *(For BREAK)*
Select object: *(Select the vertical line at
    "A")*
Specify second break point or [First
    point]: **@** *(To select the last point)*

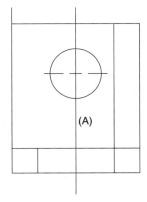

Figure 8–67

STEP 31

Use the ERASE command to delete the top half of the line broken in the previous step. This will leave a short line segment that will be converted to a centerline marking the center of the hidden hole (see Figure 8–68).

Command: **E** *(For ERASE)*
Select objects: *(Select the vertical line at "A")*

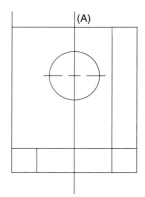

Figure 8–68

STEP 32

Change the two vertical lines labeled "A" and "B" in Figure 8–69A to the "Hidden" layer. First select the lines, activate the Layer Control box in Figure 8–69A, and pick the "Hidden" layer. When finished, press ESC twice to remove the object highlight and the grips.

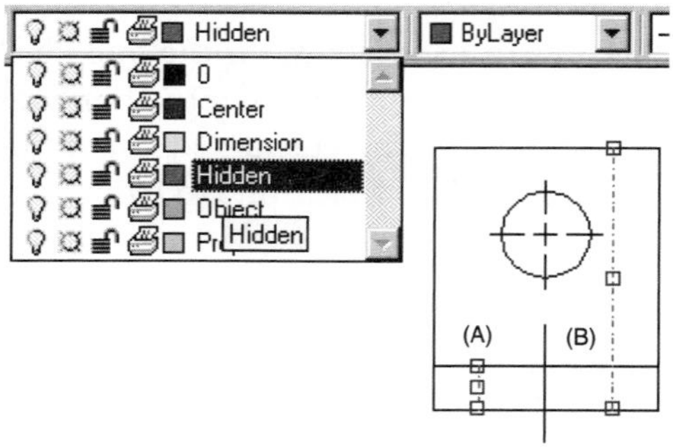

Figure 8–69A

Change vertical line "C" in Figure 8–69B to the "Center" layer. First select the line, activate the Layer dropdown list in Figure 8–69B, and pick the "Center" layer. When finished, press ESC twice to remove the object highlight and the grips.

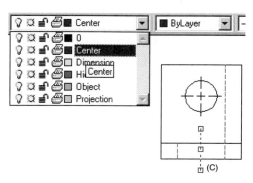

Figure 8–69B

STEP 33

Draw three lines from key features on the Side view to intersect with the 45°- projection line as shown in Figure 8–70. Use Object Snap options whenever possible. Ortho mode must be on.

Command: **L** *(For LINE)*
Specify first point: **Qua**
of *(Select the quadrant of the circle at "A")*
Specify next point or [Undo]: *(Pick a point just past the 45° angle)*
Specify next point or [Undo]: *(Press* ENTER *to exit this command)*

Repeat the same procedure for the other two projection lines. Use the OSNAP-Endpoint option and begin the second projection line from the endpoint of the center marker at "B." Begin the third projection line from the quadrant of the circle at "C."

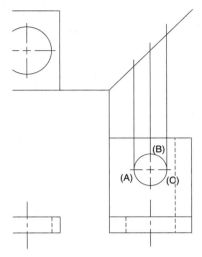

Figure 8–70

STEP 34

Draw three lines from the intersection at the 45°-projection line across to The top view as shown in Figure 8–71. Draw the middle line longer because it will be converted to a centerline at a later step.

Command: **L** *(For LINE)*
Specify first point: *(Select the intersection at "A")*
Specify next point or [Undo]: *(Snap to the polar intersection at "B")*
Specify next point or [Undo]: *(Press* ENTER *to exit this command)*

Repeat the same procedure for the other two lines. Do not use an Object Snap option for the opposite end of the middle line but rather identify a point just to the left of the vertical line at "B."

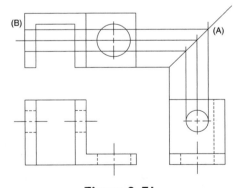

Figure 8–71

STEP 35

Use the ERASE command to erase the three vertical projection lines from the Side view as shown in Figure 8–72.

Command: **E** *(For ERASE)*
Select objects: *(Select the vertical lines labeled "A," "B," and "C")*
Select objects: *(Press ENTER to execute this command)*

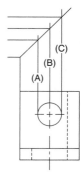

Figure 8–72

STEP 36

Use the TRIM command to trim away excess lines in the Top view as illustrated in Figure 8–73.

Command: **TR** *(For TRIM)*
Current settings: Projection=UCS Edge=None
Select cutting edges ...
Select objects: *(Select the three vertical lines labeled "A," "B," and "C")*
Select objects: *(Press ENTER to continue)*
Select object to trim or [Project/Edge/ Undo]: *(Select lines "D" through "G" to trim)*
Select object to trim or [Project/Edge/ Undo]: *(Press ENTER to exit this command)*

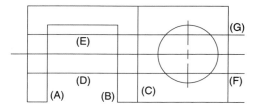

Figure 8–73

STEP 37

Before continuing, first turn OSNAP off. Then use the BREAK command to partially delete the horizontal line segment from "A" to "B" as shown in Figure 8–74. Use the BREAK command with the @ option to break the line into two segments at "C." Use the ERASE command to delete the trailing line segment at "D."

Command: **BR** *(For BREAK)*
Select object: *(Select the horizontal line at "A")*
Specify second break point or [First point]: *(Select the line at "B")*
Command: **BR** *(For BREAK)*
Select object: *(Select the horizontal line at "C")*

Specify second break point or [First point]: **@** *(To perform the break at the last point)*

Command: **E** *(For ERASE)*

Select objects: *(Select the horizontal line segment at "D")*

Select objects: *(Press* ENTER *to execute this command)*

Turn OSNAP back on.

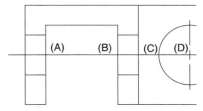

Figure 8–74

STEP 38

Change the four horizontal lines labeled "A" through "D" in Figure 8–75A to the "Hidden" layer. First select the lines, activate the Layer Control box in Figure 8–75A, and pick the "Hidden" layer. When finished, press ESC twice to remove the object highlight and the grips.

Change the two horizontal lines labeled "E" and "F" in Figure 8–75B to the "Center" layer. First select the lines and then activate the Layer Control box in Figure 8–75B and pick the "Center" layer. When finished, press ESC twice to remove the object highlight and the grips.

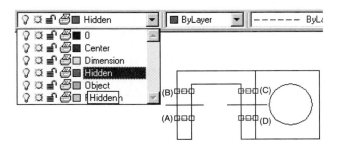

Figure 8–75A

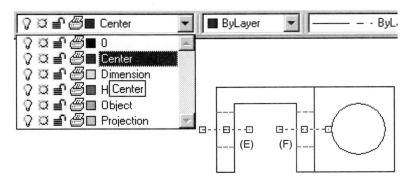

Figure 8–75B

STEP 39

Use the ERASE command to delete the 45°
line. Use the FILLET command to create
corners in the Top view and Right Side
view as shown in Figure 8–76. Check that
the fillet radius is currently set to 0.00.

Command: **E** *(For ERASE)*
Select objects: *(Select the inclined line at
"A")*
Select objects: *(Press ENTER to execute
this command)*
Command: **F** *(For FILLET)*
Current settings: Mode = TRIM, Radius =
0.00
Select first object or [Polyline/Radius/
Trim]: *(Select line "B")*
Select second object: *(Select line "C")*

Command: **F** *(For FILLET)*
Current settings: Mode = TRIM, Radius =
0.00
Select first object or [Polyline/Radius/
Trim]: *(Select line "D")*
Select second object: *(Select line "E")*

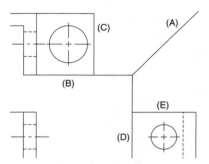

Figure 8–76

STEP 40

Perform a ZOOM-Extents to view the
entire drawing (see Figure 8–77). The final process in completing a multiview
drawing is to place dimensions to define
size and locate features. This topic will be
discussed in a later chapter.

Command: **Z** *(For ZOOM)*
Specify corner of window, enter a scale
factor (nX or nXP), or [All/Center/
Dynamic/Extents/Previous/Scale/
Window] <real time>: **E** *(For Extents)*

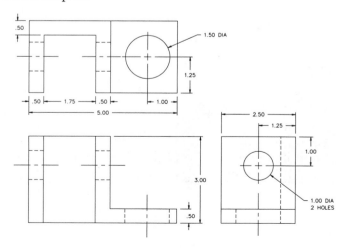

Figure 8–77

TUTORIAL EXERCISE: GAGE.DWG

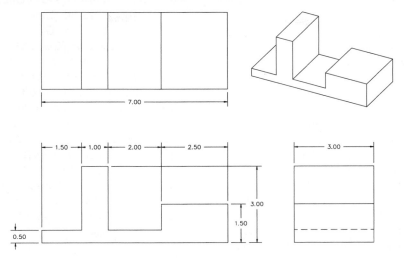

Figure 8–78

Purpose

This tutorial is designed to allow the user to construct a three-view drawing of the Gage with the aid of XYZ filters. See Figure 8–78.

System Settings

Begin a new drawing called "Gage." Use the Drawing Units dialog box to change the number of decimal places past the zero from four to two. Keep the remaining default unit values. Using the LIMITS command, keep (0,0) for the lower left corner and change the upper right corner from (12,9) to (15.50,9.50). Use the GRID command and change the grid spacing from 1.00 to 0.25 units. Do not turn the Snap or Ortho modes on. It is very important to check and see that the following Object Snap modes are already set: Endpoint, Extension, Intersection, Center.

Layers

Create the following layers with the format:

Name	Color	Linetype
Object	Green	Continuous
Hidden	Red	Hidden

Suggested Commands

Begin this tutorial by laying out the three primary views using the LINE and OFFSET commands. Use the TRIM command to clean up any excess line segments. As an alternate method used for projection, use XYZ filters in combination with Object Snap options to add features in other views.

Whenever possible, substitute the appropriate command alias in place of the full AutoCAD command in each tutorial step; for example, use "CP" for the COPY command, "L" for the LINE command, and so on. The complete listing of all command aliases is located in Chapter 1, Table 1–2.

STEP I

Begin by first making the Object layer current. Then construct the Front view of the Gage by using absolute coordinates and the Direct Distance mode in Figure 8–79. Start the Front view at coordinate 1.50,1.00.

Command: **L** *(For LINE)*
Specify first point: **1.50,1.00** *(At "A")*
Specify next point or [Undo]: *(Move your cursor to the right of the last point and enter **7.00**...for "B")*
Specify next point or [Undo]: *(Move your cursor up from the last point and enter **1.50** ... for "C")*
Specify next point or [Close/Undo]: *(Move your cursor to the left of the last point and enter **2.50** ... for "D")*
Specify next point or [Close/Undo]: *(Move your cursor down from the last point and enter **1.00** ... for "E")*
Specify next point or [Close/Undo]: *(Move your cursor to the left of the last point and enter **2.00** ... for "F")*

Specify next point or [Close/Undo]: *(Move your cursor up from the last point and enter **2.50** ... for "G")*
Specify next point or [Close/Undo]: *(Move your cursor to the left of the last point and enter **1.00** ... for "H")*
Specify next point or [Close/Undo]: *(Move your cursor to down from the last point and enter **2.50** ... for "I")*
Specify next point or [Close/Undo]: *(Move your cursor to the left of the last point and enter **1.50** ... for "J")*
Specify next point or [Close/Undo]: **C** *(To close the shape back to "A" and exit the command)*

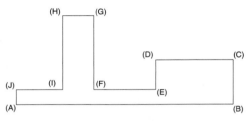

Figure 8–79

STEP 2

Begin the construction of the Top view by locating the lower left corner at coordinate 1.50,5.50, as shown in Figure 8–80.

Command: **L** *(For LINE)*
Specify first point: **1.50,5.50** *(At "A")*
Specify next point or [Undo]: *(Move your cursor to the right of the last point and enter **7.00** ...for "B")*
Specify next point or [Undo]: *(Move your cursor up from the last point and enter **3.00** ... for "C")*
Specify next point or [Close/Undo]: *(Move your cursor to the left of the last point and enter **7.00** ... for "D")*

Specify next point or [Close/Undo]: **C** *(To close the shape back to "A" and exit the command)*

Figure 8–80

STEP 3

Begin the construction of the Right Side view by locating the lower left corner at coordinate 10.50,1.00, as shown in Figure 8–81.

Command: **L** *(For LINE)*
Specify first point: **10.50,1.00** *(At "A")*
Specify next point or [Undo]: *(Move your cursor to the right of the last point and enter **3.00** ... for "B")*
Specify next point or [Undo]: *(Move your cursor to up from the last point and enter **3.00** ... for "C")*
Specify next point or [Close/Undo]: *(Move your cursor to the left of the last point and enter **3.00** ... for "D")*
Specify next point or [Close/Undo]: **C** *(To close the shape back to "A" and exit the command)*

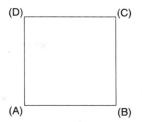

Figure 8–81

STEP 4

Your display should appear similar to the illustration in Figure 8–82, with the placement of the Top view above the Front view and the Right Side view directly to the right of the Front view. A new method of using a temporary snap point will now be used to complete the Top and Right Side views.

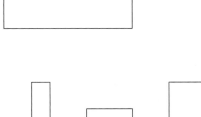

Figure 8–82

STEP 5

Zoom in to the drawing so the Front and Top views appear similar to the illustration in Figure 8–83B. Use the following steps for using a temporary snap point in conjunction with the LINE command for completing the Top view. With POLAR, OSNAP, and OTRACK (Object Tracking) all turned on, begin to draw a line in the Top view using the intersection at "A" as a temporary tracking point. The temporary tracking point allows you to jump the gap between views and begin drawing the line in the Top view.

Command: **L** *(For LINE)*
Specify first point: **TT** *(For Temporary Tracking point; Pick the Temporary Tracking Point button in the Standard toolbar in Figure 8–83A)*
Specify temporary OTRACK point: *(Pick the intersection at "A" as the temporary tracking point in Figure 8–83B)*
Specify first point: *(Pick the intersection at "B")*
Specify next point or [Undo]: *(Pick the Polar intersection at "C")*
Specify next point or [Undo]: *(Press ENTER to exit this command)*

Figure 8–83A

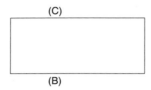

Figure 8–83B

STEP 6

Your display should appear similar to the illustration in Figure 8–84. Rather than use a temporary tracking point for the next line, simply duplicate the last line drawn using the COPY command.

Command: **CP** *(For COPY)*
Select objects: **L** *(For Last; this should select the last line drawn)*
Select objects: *(Press ENTER to continue)*
Specify base point or displacement, or [Multiple]: *(Pick the endpoint of the line at "A")*
Specify second point of displacement or <use first point as displacement>: *(Pick the endpoint of the line at "B")*

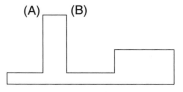

Figure 8–84

STEP 7

After you perform the COPY command, your display should appear similar to the illustration in Figure 8–85. Use a temporary tracking point to project the last object line in the Top view from the Front view. Use the procedure from Step 5 as a guide for performing this operation.

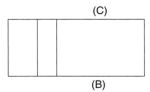

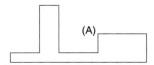

Figure 8–85

Command: **L** *(For LINE)*
Specify first point: **TT** *(For Temporary Tracking point; Pick the Temporary Tracking Point button in the Standard toolbar in Figure 8–83A)*
Specify temporary OTRACK point: *(Pick the intersection at "A" as the temporary tracking point)*
Specify first point: *(Pick the intersection at "B")*
Specify next point or [Undo]: *(Pick the Polar intersection at "C")*
Specify next point or [Undo]: *(Press ENTER to exit this command)*

STEP 8

The complete Front and Top views are illustrated in Figure 8–86. Use the same procedure with the temporary tracking point for completing the lines located in the Right Side view.

Figure 8–86

STEP 9

Before adding the missing lines to the Right Side view, first use the ZOOM command to magnify the area illustrated in Figure 8–87.

Command: **Z** *(For ZOOM)*
Specify corner of window, enter a scale factor (nX or nXP), or [All/Center/Dynamic/Extents/Previous/Scale/Window] <real time>: *(Pick a point approximately at "A")*
Specify opposite corner: *(Pick a point approximately at "B")*

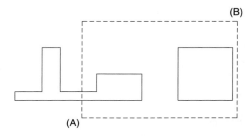

Figure 8–87

STEP 10

Use the procedure from Step 7 to construct the line in the Side view using a Temporary Tracking point. See Figure 8–88.

Command: **L** *(For LINE)*
Specify first point: **TT** *(For Temporary Tracking point; click the Temporary Tracking Point button in the Standard toolbar in Figure 8–83A)*
Specify temporary OTRACK point: *(Pick the intersection at "A" as the temporary tracking point)*
Specify first point: *(Pick the intersection at "B")*
Specify next point or [Undo]: *(Pick the Polar intersection at "C")*
Specify next point or [Undo]: *(Press ENTER to exit this command)*

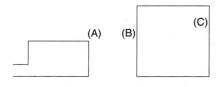

Figure 8–88

tutorial EXERCISE

STEP 11

Use the procedure from Step 10 to construct the line in the Side view using a Temporary Tracking point in Figure 8–89.

Command: **L** *(For LINE)*
Specify first point: **TT** *(For Temporary Tracking point; Pick the Temporary Tracking Point button in the Standard toolbar in Figure 8–83A)*
Specify temporary OTRACK point: *(Pick the intersection at "A" as the temporary tracking point)*
Specify first point: *(Pick the intersection at "B")*

Specify next point or [Undo]: *(Pick the Polar intersection at "C")*
Specify next point or [Undo]: *(Press* ENTER *to exit this command)*

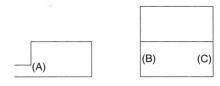

Figure 8–89

STEP 12

Use the Layer Control box in Figure 8–90 to change the layer of the last line drawn from "Object" to "Hidden." When finished, press ESC twice to remove the object highlight and the grips.

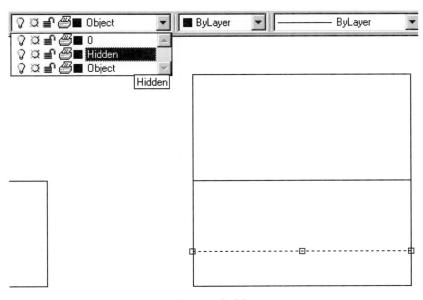

Figure 8–90

tutorial EXERCISE

STEP 13

The completed Right Side view is illustrated in Figure 8–91 complete with visible lines and hidden lines that represent invisible surfaces. Use the ZOOM-Previous or ZOOM-All options to demagnify your display and show all three views of the Gage.

Figure 8–91

With the completed three-view drawing illustrated in Figure 8–92, the next step would be to add dimensions to the views for manufacturing purposes. This topic will be discussed in a later chapter.

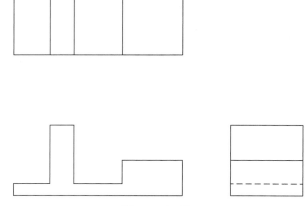

Figure 8–92

PROBLEMS FOR CHAPTER 8

Directions for Problems 8–1 and 8–2:

Find the missing lines in these problems and sketch the correct solution.

PROBLEM 8–1

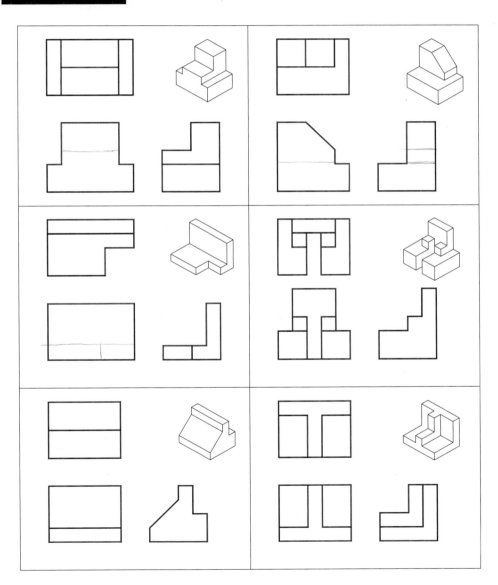

PROBLEM 8–2

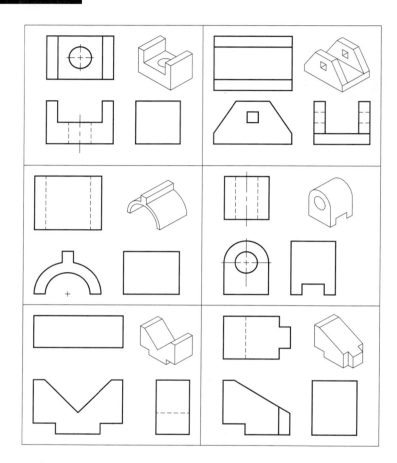

Directions for Problems 8–3 through 8–28:

Construct a multiview drawing by sketching the Front, Top, and Right Side views. Use a grid of 0.25 units to assist in the construction of the sketches. These drawings may also be constructed directly in the CAD system through AutoCAD commands. Problem 8–3 and the illustrations that follow may be used as a guide for the completion of the other sketching problems.

STEP 1

Begin Problem 8–3 by laying out the Front, Top, and Right side views using only the overall dimensions. Do not be concerned about certain details such as holes or slots; these will be added to the views at a later step. The length of the object in Figure P8–3A should measure 8 grid units. The height of the object is 5 grid units; the depth of the object is 4 units.

PROBLEM 8-3

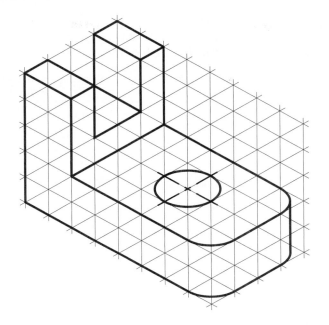

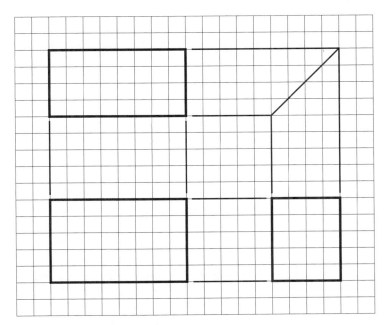

Figure P8–3A

STEP 2

Once the overall dimensions have been used to lay out the Front, Top, and Right Side views, begin adding visible details to the views. The "L" shape is added to the Front view; the hole and corner fillets are added to the Top view; the rectangular slot is added to the Right Side view (see Figure P8–3B).

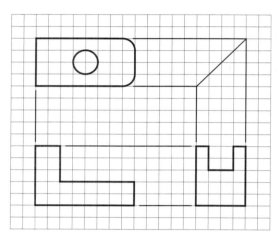

Figure P8–3B

STEP 3

Visible edges from hole and slot features are projected to other views. Slot information is added to the Top view; height information is projected to the Right Side view from the Front view. At this point, only add visible information to other views where they apply (see Figure P8–3C).

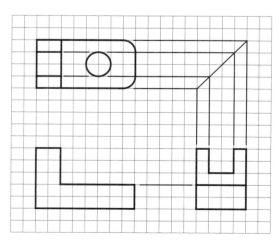

Figure P8–3C

STEP 4

Now project hidden features to the views. The hole is projected hidden in the Front view along with the slot visible in the Right Side view. The hole is also hidden in the Right Side view (see Figure P8–3D).

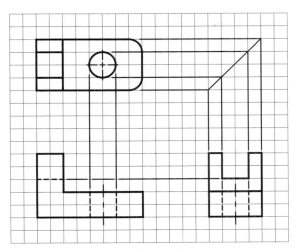

Figure P8–3D

STEP 5

The completed multiview drawing solution is illustrated in Figure P8–3E. Dimensions are added in Figure P8–3F when the object is to be constructed. Proper placement of dimensions will be discussed in a later chapter.

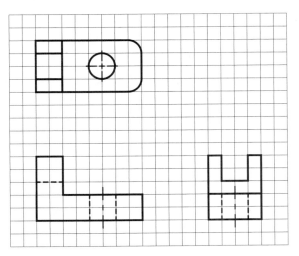

Figure P8–3E

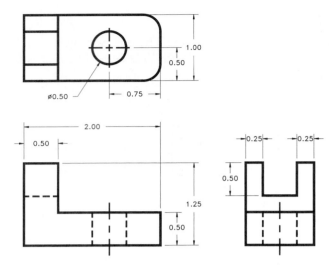

Figure P8–3F

PROBLEM 8–4

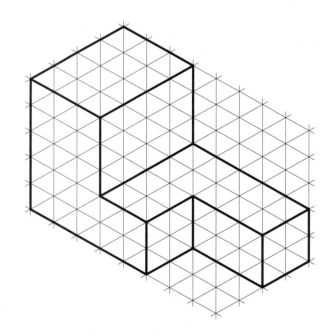

PROBLEM 8–5

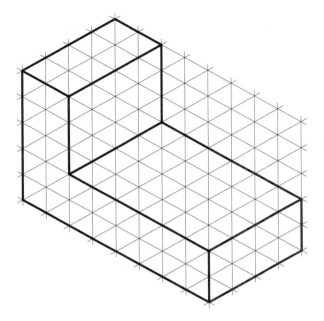

PROBLEM 8–6

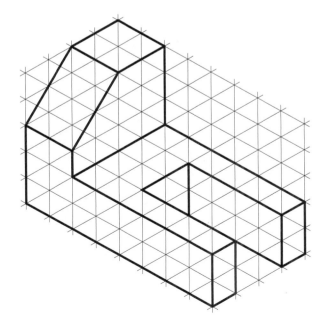

480

problem **EXERCISE**

PROBLEM 8-7

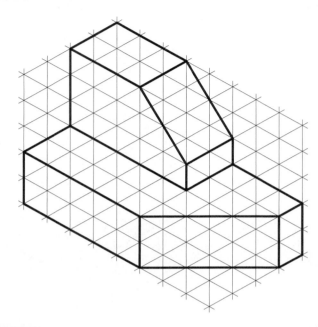

PROBLEM 8-8

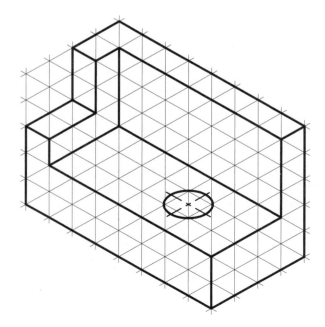

PROBLEM 8-9

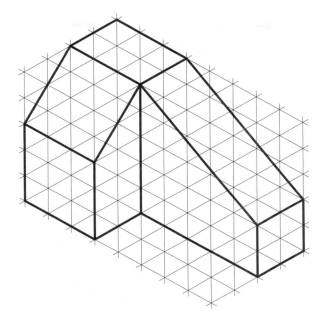

PROBLEM 8-10

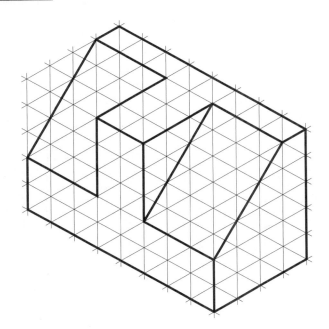

PROBLEM 8–11

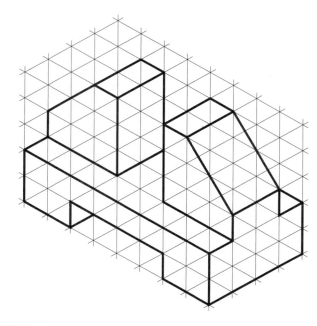

PROBLEM 8–12

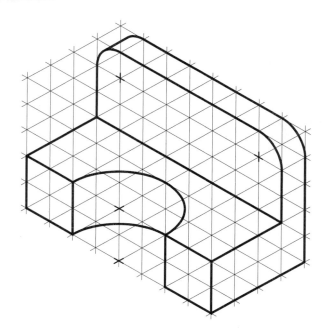

PROBLEM 8–13

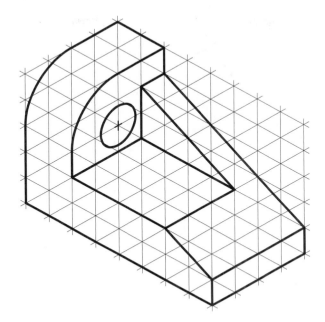

PROBLEM 8–14

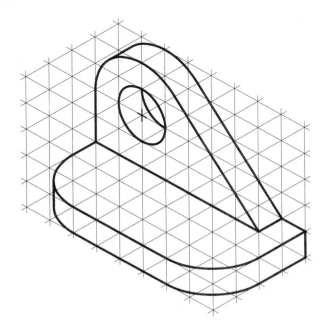

PROBLEM 8–15

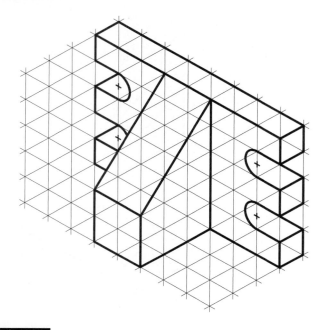

PROBLEM 8–16

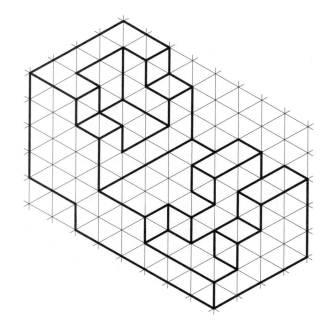

PROBLEM 8–17

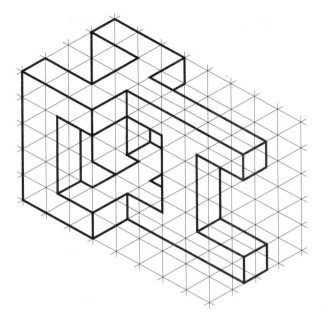

PROBLEM 8–18

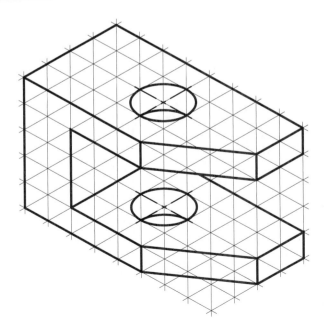

PROBLEM 8–19

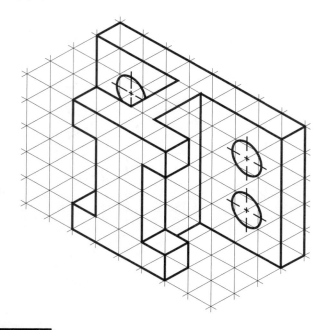

PROBLEM 8–20

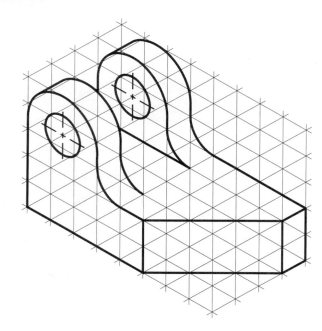

PROBLEM 8–21

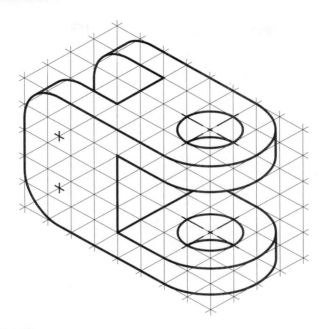

PROBLEM 8–22

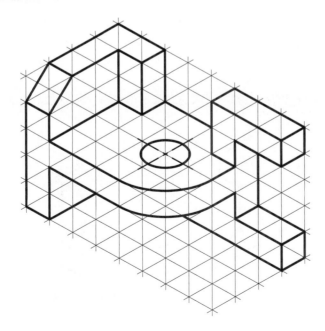

PROBLEM 8–23

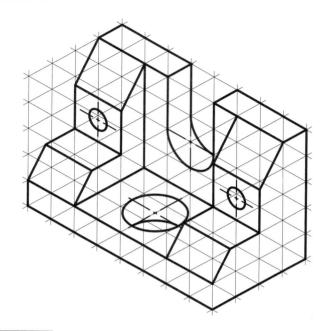

PROBLEM 8–24

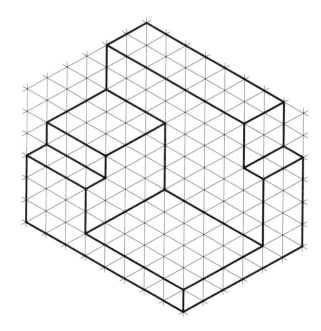

PROBLEM 8–25

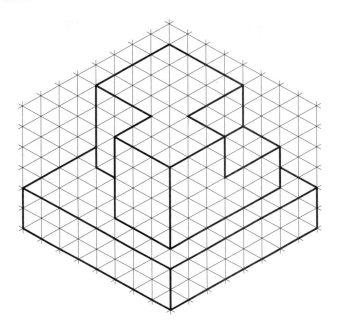

PROBLEM 8–26

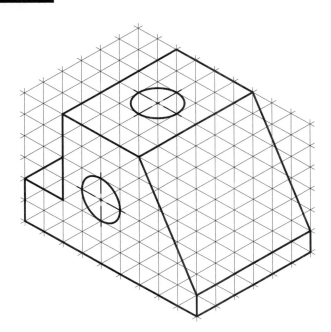

PROBLEM 8–27

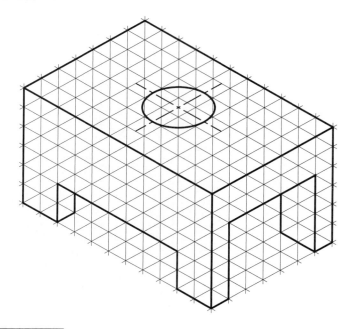

PROBLEM 8–28

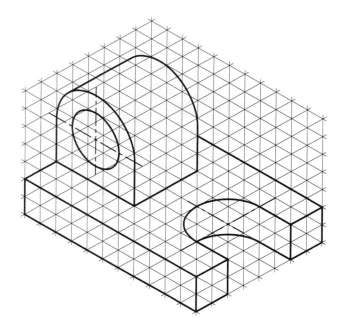

Directions for Problems 8–29 through 8–38:

Construct a multiview drawing of each object. Be sure to construct only those views that accurately describe the object.

PROBLEM 8–29

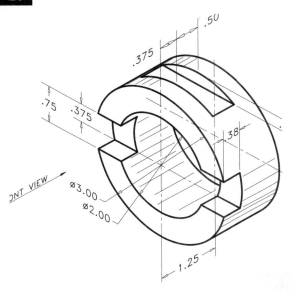

PROBLEM 8–30

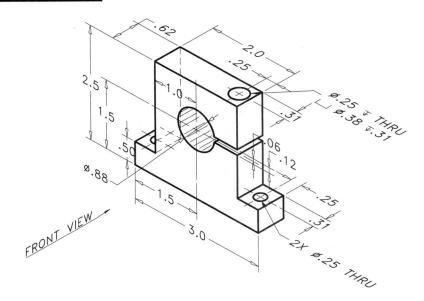

PROBLEM 8–31

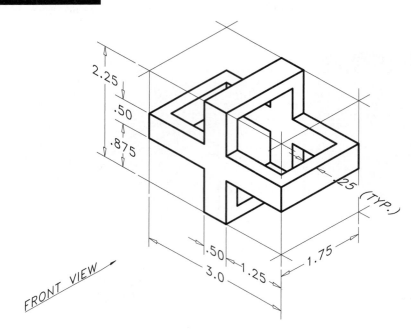

2.25

.50

.875

.125 (TYP.)

.50

3.0

1.25

1.75

FRONT VIEW

PROBLEM 8–32

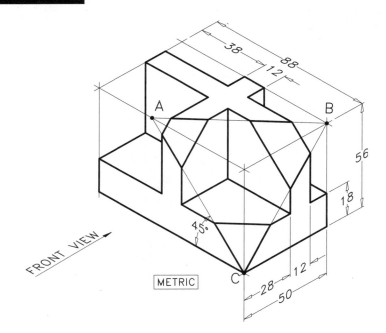

38

88

12

A

B

56

18

45°

C

FRONT VIEW

METRIC

28

12

50

PROBLEM 8–33

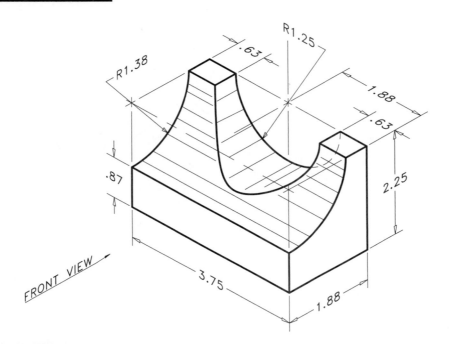

FRONT VIEW

PROBLEM 8–34

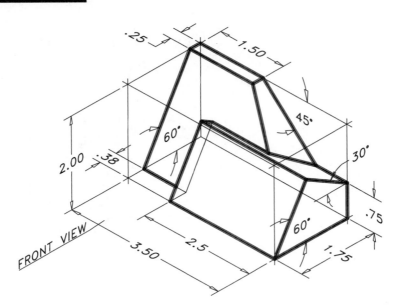

FRONT VIEW

PROBLEM 8–35

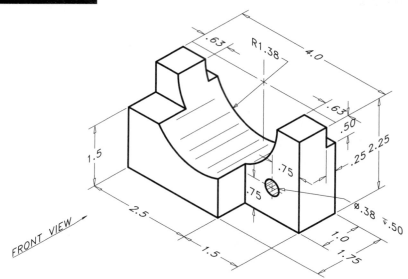

FRONT VIEW

PROBLEM 8–36

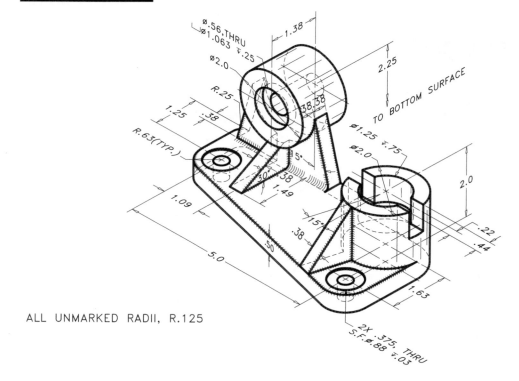

ALL UNMARKED RADII, R.125

PROBLEM 8–37

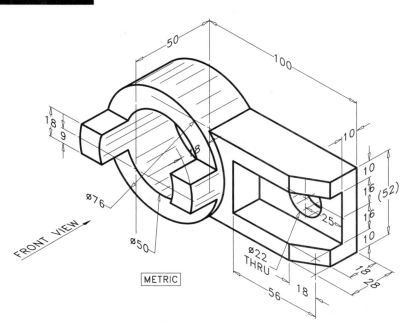

50
100
18
8
9
ø76
ø50
FRONT VIEW
METRIC
18
10
10
16
(52)
16
10
25
ø22
THRU
18
28
56
18

PROBLEM 8–38

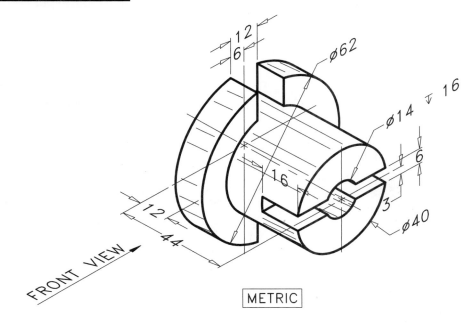

12
6
ø62
ø14 ⌵ 16
6
12
16
3
ø40
44
FRONT VIEW
METRIC

Directions for Problems 8–39 through 8–48:

Construct a three-view drawing of each object.

PROBLEM 8–39

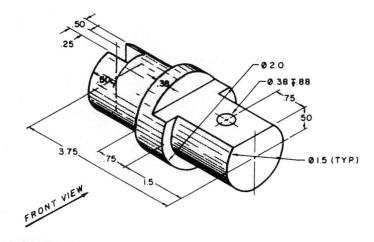

FRONT VIEW

PROBLEM 8–40

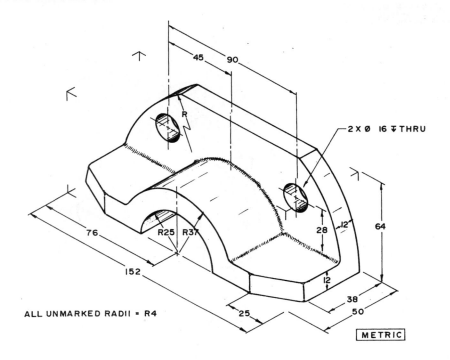

ALL UNMARKED RADII = R4

METRIC

PROBLEM 8–41

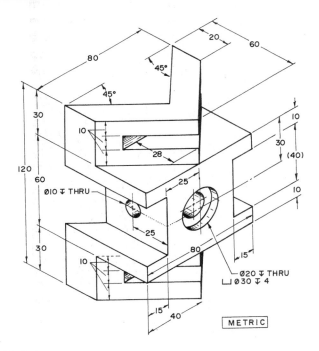

METRIC

PROBLEM 8–42

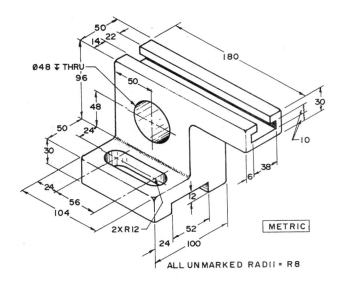

METRIC

ALL UNMARKED RADII = R8

PROBLEM 8-43

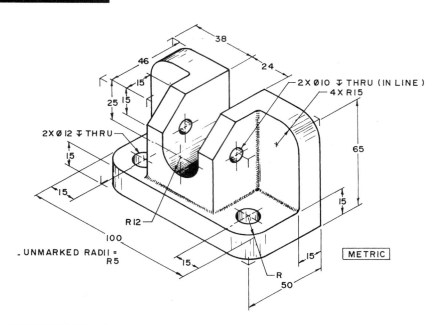

38
46
24
2 X Ø 10 ⌄ THRU (IN LINE)
4 X R15
15
25 15
2 X Ø 12 ⌄ THRU
65
15
15
R 12
15
100
UNMARKED RADII = R5
15
15
R
50
METRIC

PROBLEM 8-44

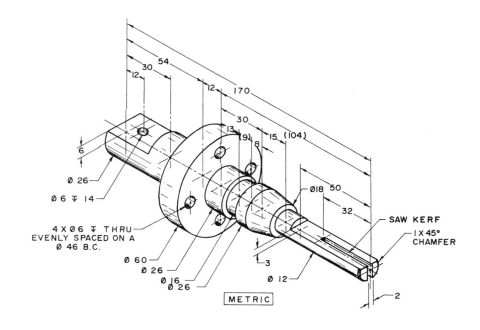

54
30
12
12
170
30
13
(9)
15 (104)
8
6
Ø 26
Ø18
50
Ø 6 ⌄ 14
32
4 X Ø6 ⌄ THRU
EVENLY SPACED ON A
Ø 46 B.C.
SAW KERF
1 X 45°
CHAMFER
Ø 60
Ø 26
3
Ø 16
Ø 26
Ø 12
2
METRIC

PROBLEM 8–45

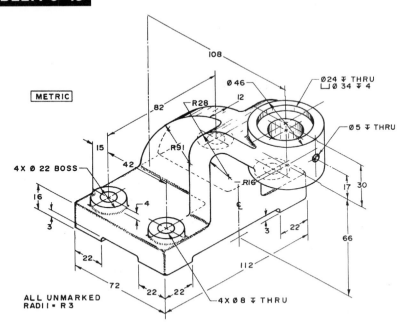

METRIC

108

Ø 46

Ø24 ⊥ THRU
⊔ Ø 34 ⊥ 4

82

R28

12

Ø5 ⊥ THRU

15

R91

42

R16

17 30

4X Ø 22 BOSS

16

4

3

3 22

66

22

112

72

22 22

4X Ø 8 ⊥ THRU

ALL UNMARKED
RADII ▪ R 3

PROBLEM 8–46

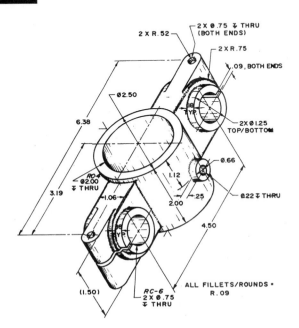

2 X Ø.75 ⊥ THRU
(BOTH ENDS)

2 X R.52

2 X R.75

.09, BOTH ENDS

Ø2.50

6.38

2X Ø1.25
TOP/BOTTOM

RC-4
Ø2.00
⊥ THRU

Ø.66

3.19

1.12

1.06

.25

Ø.22 ⊥ THRU

2.00

4.50

(1.50)

RC-6
2 X Ø.75
⊥ THRU

ALL FILLETS/ROUNDS ▪
R.09

PROBLEM 8-47

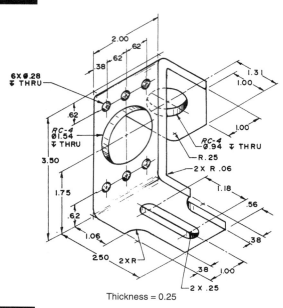

6X Ø.28
⊥ THRU

2.00
.38
62
62
62

1.31
1.00

1.00

RC-4
Ø1.54
⊥ THRU

RC-4
Ø.94 ⊥ THRU

R.25

2 X R.06

.62

3.50

1.75

1.18

.56

.62

1.06

2.50   2 X R

.38   1.00

.38   1.00

2 X .25

Thickness = 0.25

PROBLEM 8-48

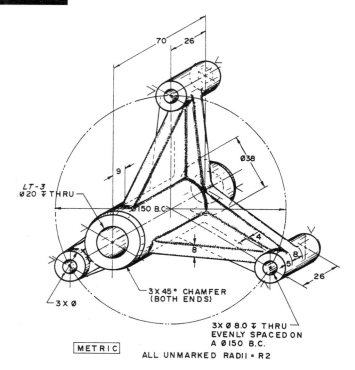

70   26

Ø38

9

LT-3
.Ø20 ⊥ THRU

Ø150 B.C.

4

8

8

5

26

3 X 45° CHAMFER
(BOTH ENDS)

3 X Ø

3X Ø 8.0 ⊥ THRU
EVENLY SPACED ON
A Ø150 B.C.

METRIC

ALL UNMARKED RADII = R2

CHAPTER 9

An Introduction to Drawing Layouts

INTRODUCING DRAWING LAYOUTS

Up to this point in the text, all drawings have been constructed in Model Space. This consisted of part geometry such as arcs, circles, lines, and even dimensions. You can even plot your drawing from Model Space; this gets a little tricky, especially when you attempt to plot the drawing at a known scale, or even trickier if multiple details need to be arranged on a single sheet at different scales.

You can elect to work in two separate spaces on your drawing; namely Model Space and Paper Space. Part geometry and dimensions are still drawn in Model Space. However, items such as notes, annotations and title blocks are laid out separately in Paper Space, which is designed to simulate an actual sheet of paper. To arrange a single view of a drawing or multiple views of different drawings, you arrange a series of viewports in Paper Space to view the images. These viewports are mainly rectangular in shape and can be made to any size. You can even create circular and polygonal viewports. An option of the ZOOM command allows you to scale the image inside the viewport. In this way, a series of images may be scaled differently in order for the drawing to be plotted out at a scale of 1=1.

This chapter introduces you to the controls used in Paper Space to manage information contained in a viewport. A series of tutorial exercises is provided to help you practice using the Paper Space environment. Problems that follow enable you to apply the tutorials in order to build a layout sheet to arrange a drawing in Paper Space.

MODEL SPACE

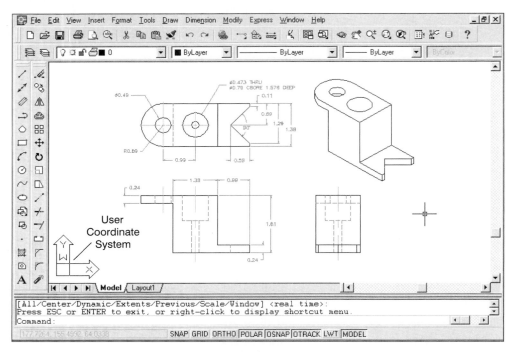

Figure 9–1

Before starting on the topic of Paper Space, you must first understand the current environment in which all drawings are originally constructed, namely Model Space. It is here that the drawing is drawn full size or in real world units. Model Space is easily identified by the appearance of the User Coordinate System icon located in the lower left corner of the active drawing area in Figure 9–1. This icon identifies itself with the Model Space environment. Past and present users of AutoCAD have felt that it clutters the display screen and gets in the way of geometry, and they have used the UCSICON command to turn the icon off. While these points may seem true, the icon should be present at all times to identify the space you are working in. Another indicator that you are in Model Space is in the presence of the Model tab located just below the User Coordinate System icon, also in Figure 9–1.

TILED VIEWPORTS IN MODEL SPACE

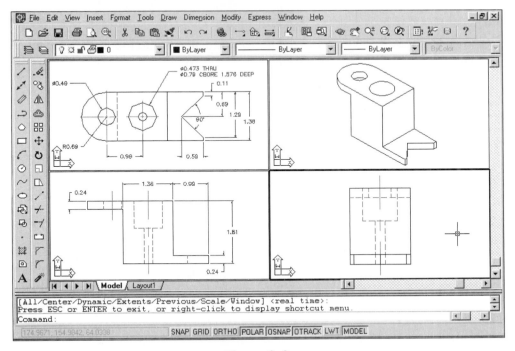

Figure 9–2

An advantage of constructing in the Model Space environment is the ability to divide the display screen into a number of smaller screens, as illustrated in Figure 9–2. In the illustration, the display screen is divided into four smaller screens or viewports. These viewports are further arranged in a certain pattern or configuration. You do not have much control over this viewport arrangement. For this reason these are called tiled viewports, because they resemble the tile pattern commonly used in floors. Do not be confused that you have four separate files open in the illustration. Rather, it is the same drawing file, which enables you to view separate parts of the drawing in each viewport. Again, notice the appearance of the familiar User Coordinate System icon present in the lower left corner of each viewport. This signifies you are in Model Space in each viewport. Also in the illustration, notice how the lower right viewport has a thick border around it along with the presence of the AutoCAD cursor. This happens to be the active viewport. Only one viewport can be active at one time. If a viewport is inactive, the appearance of an arrow instead of a cursor is present. To activate another viewport, simply move the arrow to the viewport and click inside the viewport; this activates a new current viewport. Yet another way to cycle through the viewports is to use the combination keys of CTRL+V.

THE VPORTS COMMAND

Viewports are created through the VPORTS command. You can access this command by choosing Viewports from the View pull-down menu. Choosing New Viewports… as shown in Figure 9–3A activates the Viewports dialog box in Figure 9–3B. Numerous layout examples and configurations for viewports are available. Clicking on the desired viewport in the Standard viewports area previews the viewport layout at the right of the dialog box. These changes will be made to the display after you click the OK button at the bottom of the dialog box.

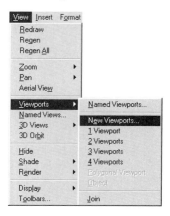

Figure 9–3A

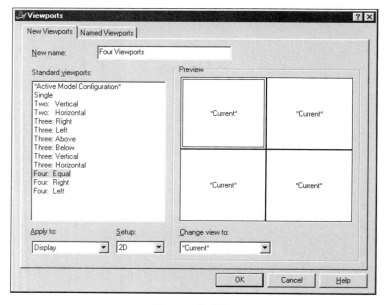

Figure 9–3B

MODEL SPACE AND PAPER SPACE

You have already seen that Model Space is the environment set aside for constructing your drawing in real world units or full size. Paper Space is considered an area used to lay out your drawing before it is plotted. Paper Space is also used to place title block information and notes associated with the drawing. The drawing illustrated in Figure 9–4 has been laid out in Paper Space and shows a sheet of paper with the drawing surrounded by a viewport. The dashed lines along the outer perimeter of the sheet are referred to as margins. Anything inside the margins will plot out; for this reason, this is called the printable area. Notice also at the bottom of the screen the Layout1 tab is activated. Both tabs at the bottom of the screen can be used to easily display your drawing in either Model Space or Paper Space. Also notice the icon in the lower left corner of the illustration; this icon is in the form of a triangle and is used to quickly identify the Paper Space environment.

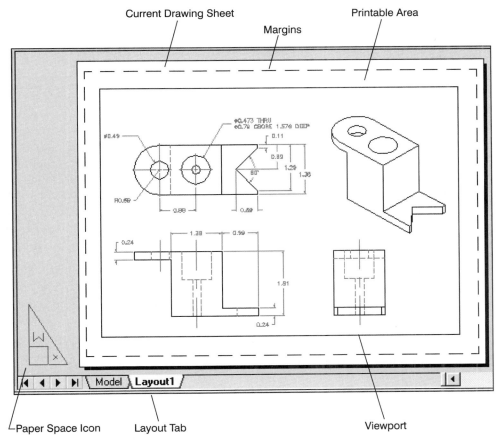

Figure 9–4

PAPER SPACE FEATURES

When you draw and plot out from Model Space, the limits of the drawing have to be properly set depending on the scale of the drawing and the sheet of paper the drawing is planned for. Also, title block information has to be scaled depending on the scale of the drawing. The text height that made up the notes in a drawing also has to be scaled depending on the scale of the drawing. Finally, the scale of the drawing is applied to Plot Settings tab of the Plot dialog box, which scales the drawing for placement on a certain sheet of paper. As this is still currently the method of plotting drawings, problems still exist when you use this method. For instance, a certain design office assigned an individual to plot out all drawings produced by the engineers and designers. It was this individual's responsibility to plot the drawings out at the proper scale. However, with numerous drawings to plot, it was easy to get the plot scales confused. As a result, drawings were plotted out at the wrong scale. This not only cost paper but valuable time. This scenario is the reason for Paper Space; All drawings, no matter how many details are arranged, can be plotted out at 1=1 with little or no error encountered, through the use of Paper Space. Plotting becomes the fundamental reason for using the Paper Space environment. Here are a few other reasons for arranging a drawing in the Paper Space:

- Paper Space is based on the actual sheet size. If you are plotting out a drawing on a "D" size sheet of paper, you use the actual size (36 x 24) in Paper Space.

- Title blocks and the text used for notes do not have to be scaled as in the past with Model Space.

- Viewports created in Paper Space are user-defined. Viewports created in Model Space are dependent on a configuration set by AutoCAD. In Paper Space, the viewports can be different sizes and shapes, depending on the information contained in the viewport.

- Multiple viewports can be created in Paper Space, as in Model space. However, the images assigned to Paper Space viewports can be at different scales. Also, the control of layers in Paper Space is viewport-dependent. In other words, layers turned on in one viewport can be frozen in another viewport.

- All drawings, no matter how many viewports are created or details arranged, are plotted out at a scale of 1=1.

CREATING A PAPER SPACE LAYOUT

Figure 9–5 shows a drawing originally created in Model Space. You decide to lay this drawing out in Paper Space.

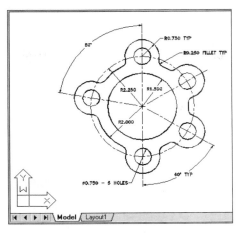

Figure 9–5

As you click on the Layout1 tab at the bottom of the display screen, you activate the Page Setup dialog box in Figure 9–6A. While in the Plot Device tab, click on the plotter configuration desired. Plotter configurations list all plotters that can be used by your drawing. The HP DesignJet 750C Plus will be used throughout this discussion of layouts in Paper Space. You can configure this plotter using the first tutorial exercise in Chapter 10, "Plotting Your Drawings." These concepts do not change for different plotting devices. Notice, at the top of this dialog box, the Layout name of One View Drawing. This is user-defined; it is considered good practice to assign a meaningful name to your layout.

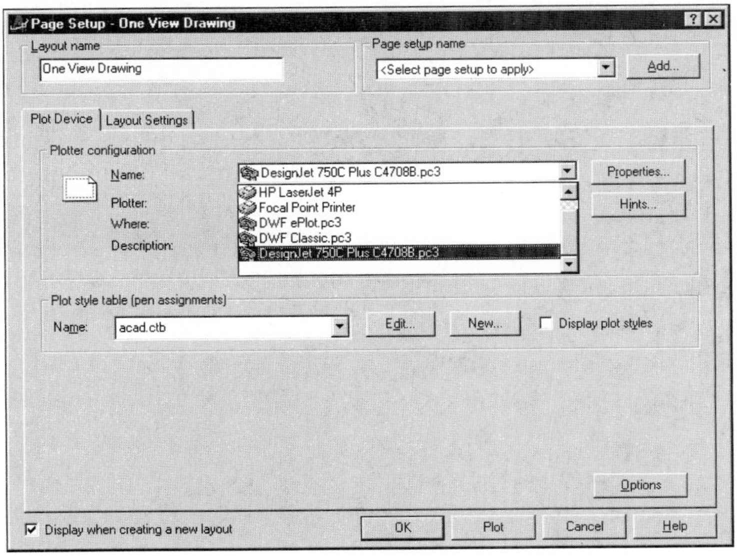

Figure 9–6A

Clicking on the Layout Settings tab of the Page Setup dialog box displays additional settings, shown in Figure 9–6B, for controlling a plot. Since the drawing will be laid out in Paper Space, the plot scale reflects 1:1. You can also control the Plot offset (where the plot begins), the Plot area, and additional Plot options. Clicking in the Paper size edit box will display all supported paper sizes for performing the plot. With the plot device currently set to the DesignJet 750C Plus, numerous paper sizes can be used, as shown in Figure 9–6C. For this example, the expanded B (17.00 x 11.00) is being used as the layout sheet.

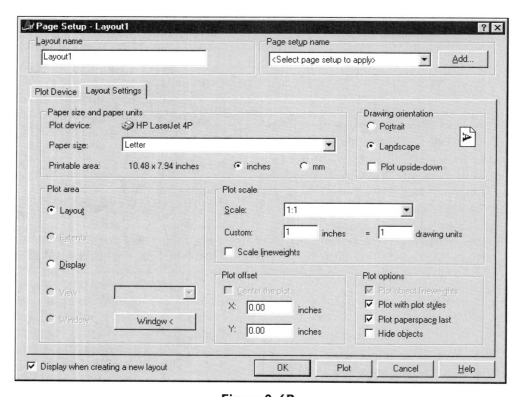

Figure 9–6B

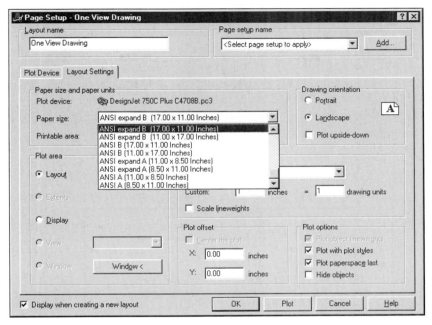

Figure 9–6C

Clicking the OK button displays the image in Figure 9–7. The expanded B size drawing sheet is displayed in this figure along with a viewport that contains the drawing image brought in from Model Space. Notice also the title of the layout is One View Drawing, located at the bottom of the screen next to the Model tab.

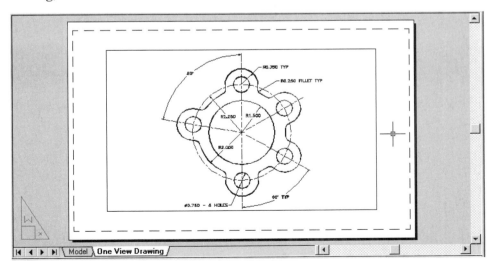

Figure 9–7

In the previous figure, the viewport was automatically created and the image displayed inside the viewport. However, a scale factor has not yet been applied to the image. Before applying a scale factor, you must first be placed in floating Model Space. This is signified by the sheet layout in Figure 9–8 along with the presence of the User Coordinate System icon inside the viewport. To accomplish this task, double-click anywhere inside the viewport. This places you in floating Model Space. Notice also that the default layout name "Layout1" has been renamed to "One View Drawing" next to the Model tab.

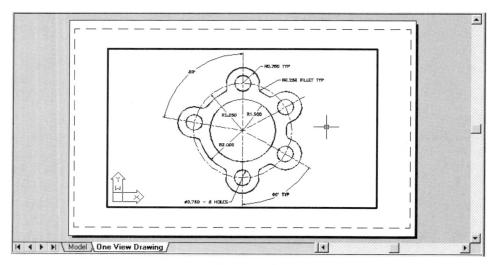

Figure 9–8

From inside floating Model Space, the image needs to be scaled to the Paper Space viewport. A special toolbar is available to control viewports. To display it, first right-click any button on your display screen. This displays the toolbar pop-up menu in Figure 9–9A. Clicking on Viewports displays the Viewports toolbar in Figure 9–9B. The number inside the edit box is the current scale of the image inside the viewport. Whenever you create a drawing layout, the image is automatically zoomed to the extents of the viewport. For this reason, the desired scale will never be correctly displayed in this area.

Figure 9–9A

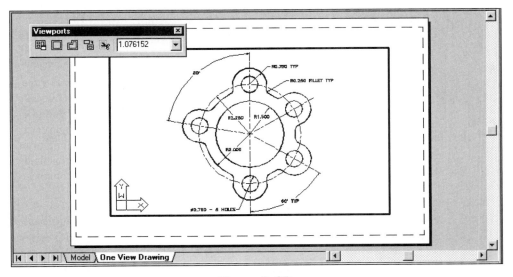

Figure 9–9B

To have the image properly scaled inside the viewport, click the down arrow in Figure 9–10. This displays the drop-down list of all standard scales. These include standard engineering and architectural scales. The image is assigned a scale of 1:1, which is applied to the viewport. You can see in the figure that the image has shrunk inside the viewport. However, it is properly scaled.

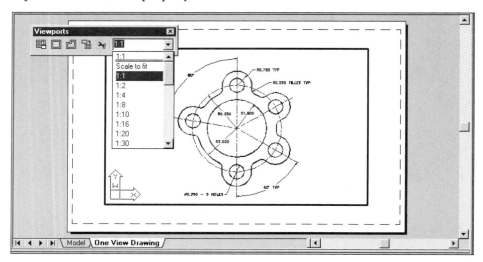

Figure 9–10

With the image scaled, double-click anywhere outside the viewport; this returns you to the Paper Space environment. Clicking once on the viewport displays grips located at the four corners as in Figure 9–11. Use these grips to size the viewport to the image. Accomplish this by clicking on one of the grips in the corner and stretching the viewport to a new location.

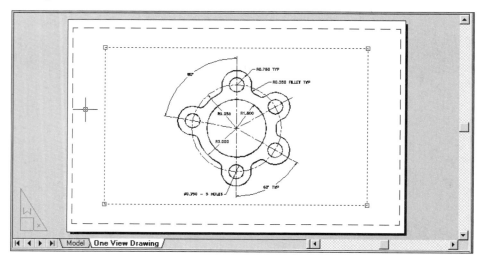

Figure 9–11

The results of this operation are displayed in Figure 9–12. Press ESC twice to remove the object highlight and the grips.

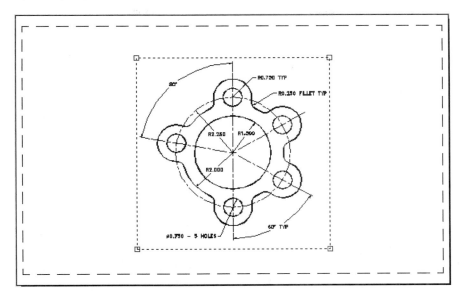

Figure 9–12

Before the drawing is plotted out, a title block containing information such as drawing scale, date, title, company name, and designer is inserted in the Paper Space environment. Accomplish this by first choosing Block... from the Insert pull-down menu as shown in Figure 9–13A. This activates the Insert dialog box shown in Figure 9–13B. This feature of AutoCAD 2000 will be covered in greater detail in Chapter 17.

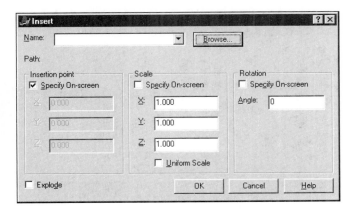

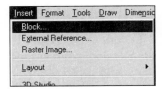

Figure 9–13A   **Figure 9–13B**

Click the Browse button to display the various folders as shown in Figure 9–13C. Double-click on the Template folder to display all title blocks that are automatically available in AutoCAD 2000.

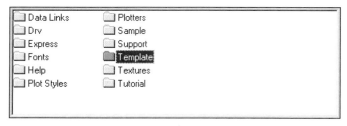

Figure 9–13C

Since a drawing layout based on the expanded B size drawing sheet was used, click on the ANSI B title block in Figure 9–13D. This displays the title block in the Preview window. If this is the correct title block to use, click the Open button.

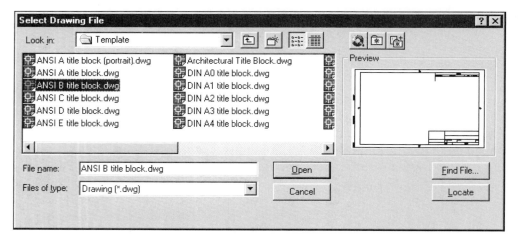

Figure 9–13D

This returns you to the Insert dialog box in Figure 9–13E. Notice the name ANSI B title block in this figure.

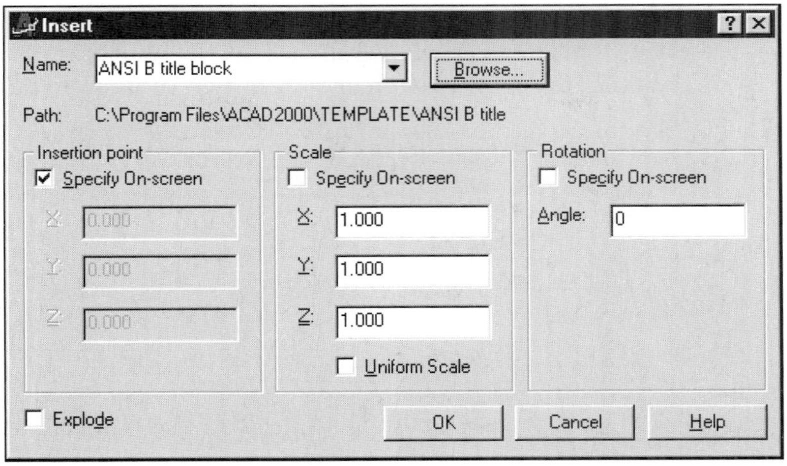

Figure 9–13E

Click the OK button to place the title block in Figure 9–14. Be sure the title block displays completely inside the paper margins; otherwise part of it will not plot out.

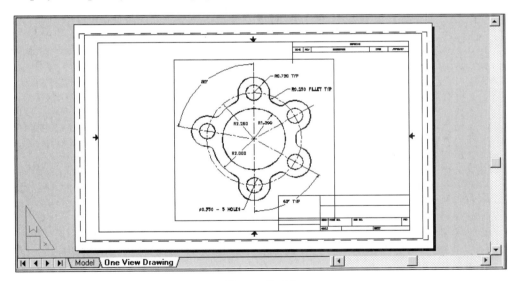

Figure 9–14

At this point, the MOVE command is used to move the viewport containing the drawing image to a better location based on the title block. Unfortunately, if the drawing were to be plotted out at this point, the viewport would also plot. It is considered good practice to assign a layer to the viewport and then turn it off. In Figure 9–15, the viewport is first selected and grips appear. Click in the Layer Control box and click in the Viewport layer. Then click the light bulb icon in the Viewport layer row to turn it off.

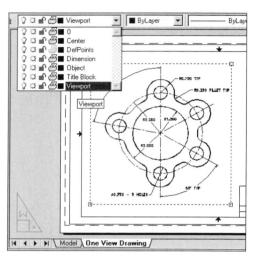

Figure 9–15

The drawing display will appear similar to Figure 9–16, with the drawing laid out and properly scaled in Paper Space.

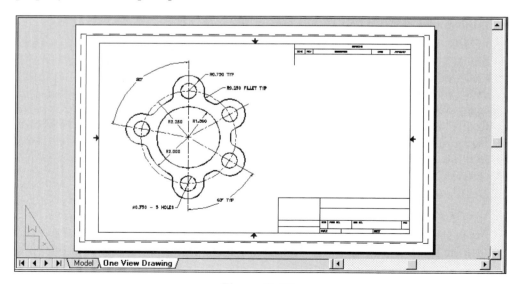

Figure 9–16

CREATING VIEWPORTS WITH THE MVIEW COMMAND

In the unlikely event that you delete a Paper Space viewport, the image tied to that viewport is also deleted (see Figure 9–17). Not to worry—you still have all drawing information contained in the Model tab. You need to re-create the viewport in Paper Space, which will retrieve the information from the Model tab.

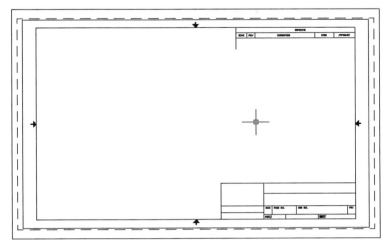

Figure 9–17

To do this, choose Viewports from the View pull-down menu. Numerous viewport options are displayed in Figure 9–18.

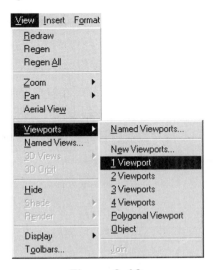

Figure 9–18

Clicking on 1 Viewport will prompt you to pick a first corner and other corner, which will create the viewport in Figure 9–19. Whenever a new viewport is created, the image automatically zooms to the extents of the viewport no matter how large or small the viewport is. Notice also in Figure 9–19 the appearance of the Status bar. Clicking on the Paper switches you to floating Model Space; the button changes to MODEL.

Remember that you scale the image inside floating Model Space. The Viewports toolbar was used earlier to accomplish this task. You could also use the ZOOM command along with the XP option; it is this special XP option that scales the image inside the viewport to Paper Space units. Follow the command prompt sequence below to perform this task:

Command: **Z** *(For ZOOM)*
Specify corner of window, enter a scale factor (nX or nXP), or
[All/Center/Dynamic/Extents/Previous/Scale/Window] <real time>: **1XP**

Complete the layout by returning to Paper Space, changing the viewport to the proper layer and then turning it off, and plotting the drawing.

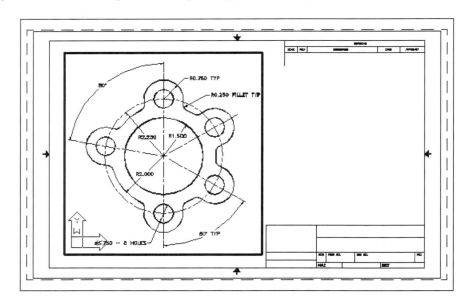

Figure 9–19

USING A WIZARD TO CREATE A LAYOUT

The AutoCAD Create Layout Wizard can be especially helpful in laying out a drawing in the Paper Space environment. Choosing Wizards from the Tools pull-down menu displays additional wizards, as shown in Figure 9–20.

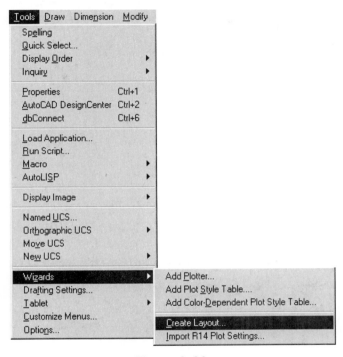

Figure 9–20

Clicking on Create Layout… displays the Create Layout-Begin dialog box in Figure 9–21. You cycle through the different categories and when finished, the drawing layout is displayed, complete with title block and viewport. Figure 9–21 displays the first category, namely Begin. You have the opportunity to name the layout of your choice in this dialog box. Notice the name "Three Views" will be used for this segment. Click the Next> button to continue to the Printer category.

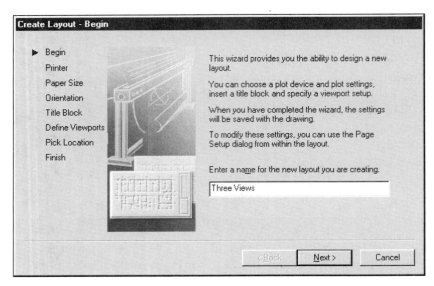

Figure 9–21

Use the Printer category to select a configured plotter from the list provided as in Figure 9–22. The HP DesignJet 750C Plus will be selected for use during this segment. Click the Next> button to continue to the Paper Size category.

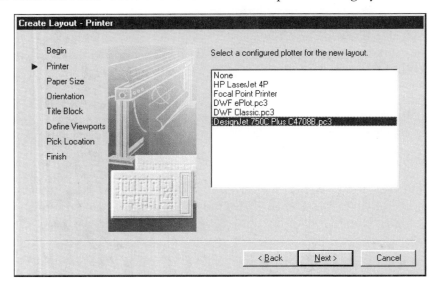

Figure 9–22

The Paper Size category displays all available paper sizes supported by the currently configured plotter. Selecting a different printer or plotter could change the number of selections. You can also select whether your paper size is in Millimeter (Metric) or Inches (English) units (see Figure 9–23). Click the Next> button to continue to the Orientation category.

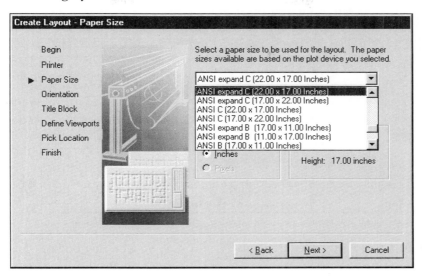

Figure 9–23

Use the Orientation category to designate whether to plot the drawing in Landscape or Portrait mode in Figure 9–24. Click the Next> button to continue to the Title Block category.

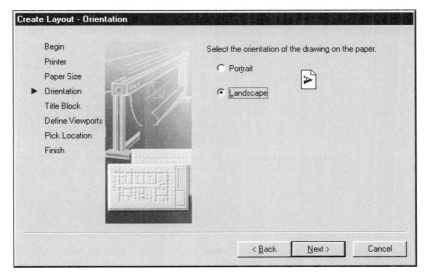

Figure 9–24

Depending on the paper size, choose a corresponding title block to automatically be inserted in the layout sheet. When you pick the title block as in Figure 9–25, a preview of the title is displayed at the right of the dialog box. Click the Next> button to continue to the Define Viewports category.

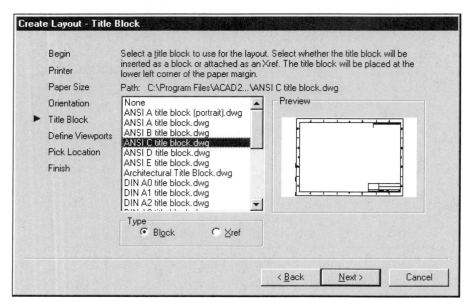

Figure 9–25

The Define Viewports category is used to either create a viewport or leave the drawing layout empty of viewports. If you select the None option, you will have to use the MVIEW command to create a user-defined viewport. Other choices include creating a single viewport, a series of standard engineering views (Front, Top, Side, and Isometric), or an array of viewports in a certain pattern (see Figure 9–26). You can also set the scale for the image inside the viewport. The scale of 1:1 shown in Figure 9–26 will automatically be applied to the image; you will not have to use the Viewports toolbar to set this value. Click the Next> button to continue to the Pick Location category.

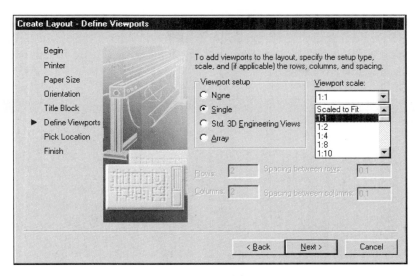

Figure 9–26

The Pick Location category creates the viewport to hold the image in Paper Space. If you click the Next> button, the viewport will be constructed to match the margins of the paper size. If you click the Select location< button in Figure 9–27, you return to the drawing and pick two diagonal points to define the viewport. In this way, you control the size of the viewport. Click the Next> button to continue to the Finish category.

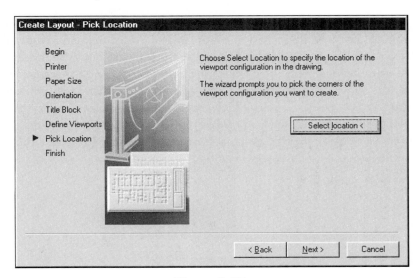

Figure 9–27

The Finish category alerts you to a new layout named "Three Views" that will be created. Once it is created, modifications can be made through the Page Setup dialog box. Click the Finish button in Figure 9–28 to dismiss the Create Layout wizard and create the drawing layout.

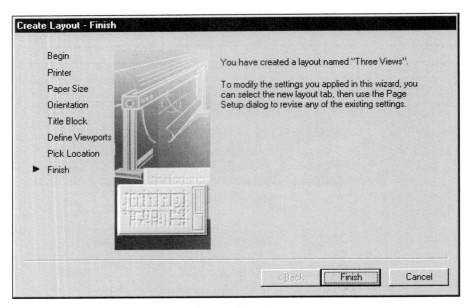

Figure 9–28

The completed layout is illustrated in Figure 9–29. Notice the new layout tab called Three Views at the bottom of the screen next to the Layout1 tab. Also, notice the lack of dashed margins to identify the edges of the plotted area. This is because a viewport the size of the paper margins was created as a result of the Pick Location category of the Create Layout wizard. Since this viewport is quite large compared to the drawing image, picking the edge of the viewport highlights the object and displays grips at its four corners (see Figure 9–29). Click on the corners of the grips to size the viewport to match the image.

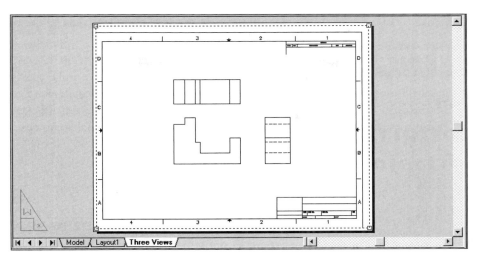

Figure 9–29

Figure 9–30 displays an image of a viewport that has been sized from the previous operation. Double-clicking inside the viewport switches you to floating Model Space. Here, the Viewports toolbar verifies the scale of the image that occupies the viewport as 1:1 (set in the Define Viewports category of the Create Layout wizard).

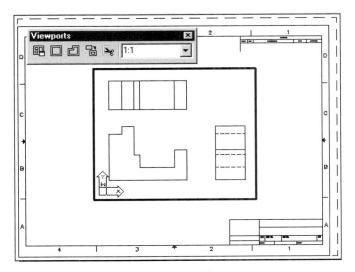

Figure 9–30

Double-clicking outside the viewport returns you to Paper Space. Selecting the viewport and then right-clicking displays the cursor menu in Figure 9–31. The upper part of this cursor menu holds special controls for viewports. One such control is the ability to lock a viewport. This will prevent any accidental zoom operations from being performed in floating Model Space.

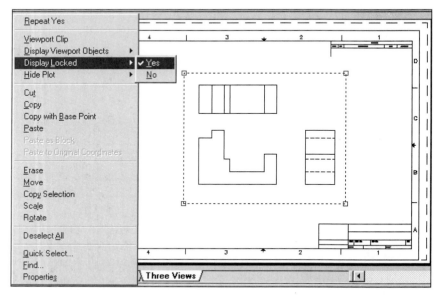

Figure 9–31

Changing the viewport to a layer and then turning the layer off displays the image as in Figure 9–32.

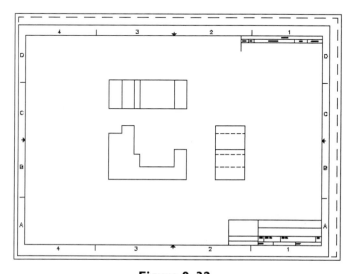

Figure 9–32

One more operation needs to be performed before you save the drawing. When you use the Create Layout wizard, the new layout tab (Three Views) was added to an existing layout (Layout1). Right-clicking on the Layout1 tab displays the cursor menu in Figure 9–33. Clicking on Delete removes the tab. (Note: You cannot remove the Model tab.) The completed drawing layout, ready for plotting, is displayed in Figure 9–34.

Figure 9–33

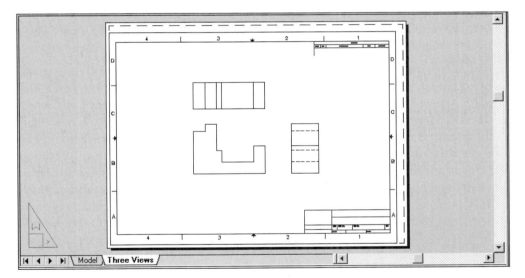

Figure 9–34

ARRANGING ARCHITECTURAL DRAWINGS IN PAPER SPACE

Architectural drawings pose certain challenges when you lay them out in Paper Space based on the scale of the drawing. For example, Figure 9–35 shows a drawing of a stair detail originally created in Model Space. The scale of this drawing is 3/8? = 1'-0?. This scale will be referred to later on in this segment on laying out architectural drawings in Paper Space.

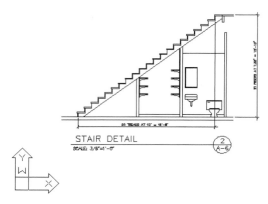

Figure 9–35

Click on the Layout1 tab or use the Create Layout wizard to arrange the stair detail in Paper Space so that your drawing appears similar to Figure 9–36. In the figure, we see the drawing sheet complete with viewport that holds the image and a title block that has been inserted. Because the image in Figure 9–36 is now visible in Paper Space, it is not necessarily scaled to 3/8? = 1'-0?. Double-clicking inside the viewport places you in floating Model Space (notice the appearance of the User Coordinate System icon only in the viewport). Activate the Viewports toolbar and scale the image in floating Model Space to the proper scale. Notice, in the figure, a complete listing of the more commonly used Architectural scales. You could also have used the ZOOM command and entered a XP value of 1/32 (found by dividing 1' or 12 by 3/8").

Command: **Z** *(For ZOOM)*
Specify corner of window, enter a scale factor (nX or nXP), or
[All/Center/Dynamic/Extents/Previous/Scale/Window] <real time>: 1/**32XP**

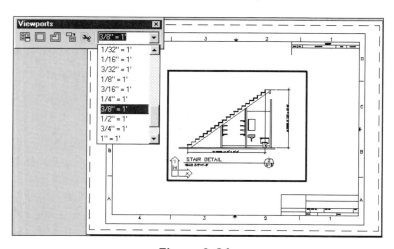

Figure 9–36

Changing the viewport to a different layer and then turning that layer off displays the completed layout of the stair detail in Figure 9–37. Since the purpose of Paper Space is to lay out a drawing and scale the image inside the viewport, this drawing will be plotted out at a scale of 1:1.

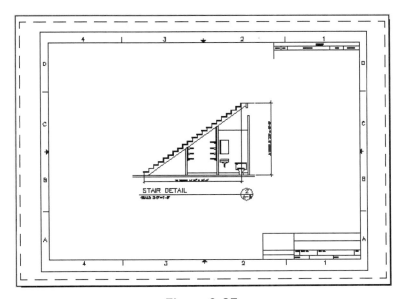

Figure 9–37

TYPICAL ARCHITECTURAL DRAWING SCALES

When the ZOOM command is used along with the XP option to scale architectural drawings drawn full size in Model Space, below are a few typical architectural scales and the XP scaling factors to use inside the floating Model Space viewport:

1/8? = 1'-0?	Zoom 1/96XP
1/4? = 1'-0?	Zoom 1/48XP
3/8? = 1'-0?	Zoom 1/32XP
1/2? = 1'-0?	Zoom 1/24XP
3/4? = 1'-0?	Zoom 1/16XP
1' = 1'-0?	Zoom 1/12XP
1 1/2? = 1'-0?	Zoom 1/8XP
3? = 1'-0?	Zoom 1/4XP

All of the XP values are arrived at by the division of the value located to the right of the "=" by the value located to the left of the "=." In other words, to arrive at the XP value for the scale 1/2? = 1'-0?, divide 1' or 12 by 1/2 or 0.50; the result is 24. The value to enter when using the ZOOM command is 1/24XP.

ARRANGING METRIC DRAWINGS IN PAPER SPACE

The methods used for arranging metric drawings in Paper Space are no different from what has been already discussed for architectural drawings. For example, Figure 9–38 shows a drawing of an object in decimal units, originally created in Model Space. It just so happens that decimal units can also support metric drawings.

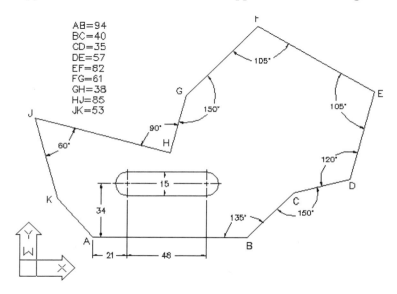

Figure 9–38

Click on the Layout1 tab or use the Create Layout wizard to arrange the metric drawing in Paper Space so that your drawing appears similar to the illustration in Figure 9–39. In the figure, we see the drawing sheet complete with viewport that holds the image and a metric title block that has been inserted. Because the image in Figure 9–39 is now visible in Paper Space, it is not necessarily scaled at 1:1. Double-clicking inside the viewport places you in floating Model Space (notice the appearance of the User Coordinate System icon only in the viewport). Activate the Viewports toolbar and scale the metric image in floating Model Space to the proper scale.

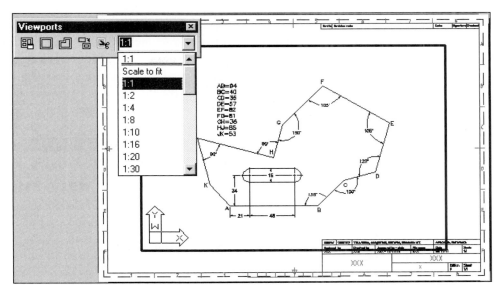

Figure 9–39

Changing the viewport to a different layer and then turning that layer off displays the completed layout of the object in Figure 9–40. Since the purpose of Paper Space is to lay out a drawing and scale the image inside the viewport, this drawing will be plotted out at a scale of 1:1.

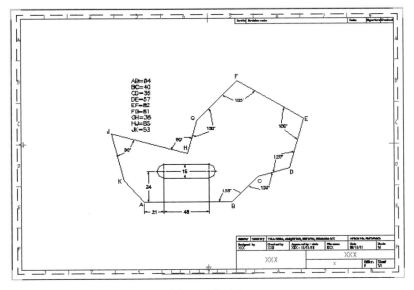

Figure 9–40

CREATING MULTIPLE DRAWING LAYOUTS

The methods explained so far have dealt with the arrangement of a single layout in Paper Space. AutoCAD 2000 provides for greater flexibility with working in Paper Space by enabling you to create multiple layouts of the same drawing. This will be explained through the illustration in Figure 9–41, which depicts a floor plan complete with dimensions. The drawing appears very busy, with a foundation plan with its dimensions and an electrical plan with its notes all overlaid on the floor plan. Because it is difficult to read each plan, individual layouts will be created to display separate images of the floor, foundation, and electrical plans. Because multiple layouts are used, a special feature of layers will allow certain layers frozen in one layout to be visible in another layout.

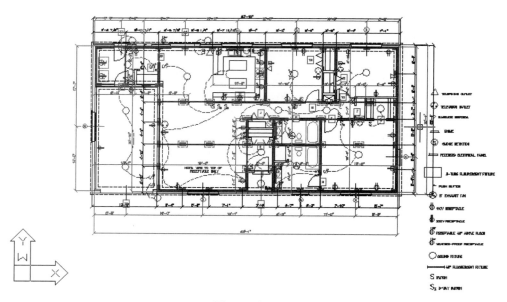

Figure 9–41

Begin the multiple layout process by clicking on the Layout1 tab. Make the proper plotter and sheet size settings to arrange the floor plan in Paper Space so that your drawing appears similar to the illustration in Figure 9–42. In the figure, we see the drawing sheet complete with viewport that holds the image and an architectural title block that has been inserted. Because the image in Figure 9–42 is now visible in Paper Space, it is not necessarily scaled to 1/4? = 1'-0?. Double-clicking inside the viewport places you in floating Model Space (notice the appearance of the User Coordinate System icon only in the viewport). Activate the Viewports toolbar and scale the image in floating Model Space to the proper scale.

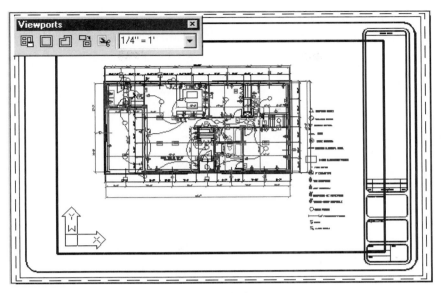

Figure 9–42

You want to see only the floor plan and its dimensions inside this viewport. Activating the Layer Properties Manager dialog box displays the layer information in Figure 9–43. When you create a new drawing layout and use this dialog box, two additional layer modes are added. These are the ability to freeze layers only in the active viewport and the ability to freeze layers in new viewports. Freezing layers in all viewports is not an effective means of controlling layers when you create multiple layouts. You need to be very familiar with the layers created in order to perform this task. To display only the floor plan information, notice that the following layers have been frozen under the Active... heading in Figure 9–43:

Construction

Electrical_Symbols

Electrical_Wiring

Foundation

Foundation_Centers

Foundation_Dimensions

Piers

All other layers such as Appliances, First_Floor, Floor_Dimensions, Title Block, Title Text, Viewport, and Windows remain visible in this viewport.

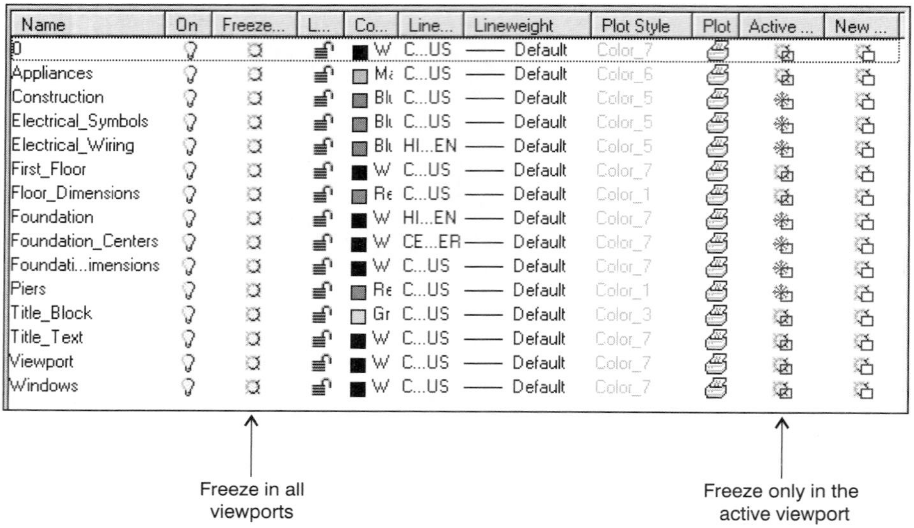

Name	On	Freeze...	L...	Co...	Line...	Lineweight	Plot Style	Plot	Active ...	New ...
0	♀	☼	⬛	■ W	C...US	—— Default	Color_7	🖨	⊠	⊠
Appliances	♀	☼	⬛	☐ Ma	C...US	—— Default	Color_6	🖨	⊠	⊠
Construction	♀	☼	⬛	■ Blu	C...US	—— Default	Color_5	🖨	❋	⊠
Electrical_Symbols	♀	☼	⬛	■ Blu	C...US	—— Default	Color_5	🖨	❋	⊠
Electrical_Wiring	♀	☼	⬛	■ Blu	HI...EN	—— Default	Color_5	🖨	❋	⊠
First_Floor	♀	☼	⬛	■ W	C...US	—— Default	Color_7	🖨	⊠	⊠
Floor_Dimensions	♀	☼	⬛	■ Re	C...US	—— Default	Color_1	🖨	⊠	⊠
Foundation	♀	☼	⬛	■ W	HI...EN	—— Default	Color_7	🖨	❋	⊠
Foundation_Centers	♀	☼	⬛	■ W	CE...ER	—— Default	Color_7	🖨	❋	⊠
Foundati...imensions	♀	☼	⬛	■ W	C...US	—— Default	Color_7	🖨	❋	⊠
Piers	♀	☼	⬛	■ Re	C...US	—— Default	Color_1	🖨	❋	⊠
Title_Block	♀	☼	⬛	☐ Gr	C...US	—— Default	Color_3	🖨	⊠	⊠
Title_Text	♀	☼	⬛	■ W	C...US	—— Default	Color_7	🖨	⊠	⊠
Viewport	♀	☼	⬛	■ W	C...US	—— Default	Color_7	🖨	⊠	⊠
Windows	♀	☼	⬛	■ W	C...US	—— Default	Color_7	🖨	⊠	⊠

Freeze in all viewports ↑   ↑ Freeze only in the active viewport

Figure 9–43

The resulting image is displayed in Figure 9–44 with only the floor plan information visible in the viewport. The next series of steps involve the creation of additional layouts to display the foundation and electrical plans.

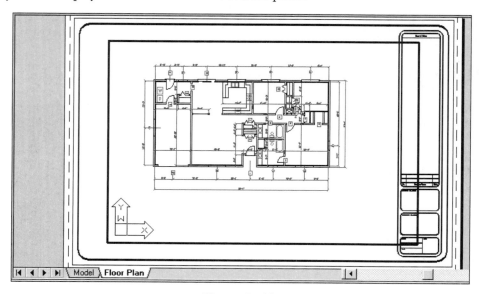

Figure 9–44

To create an additional layout, right-click on the Floor Plan tab at the bottom of the display screen as in Figure 9–44. This displays the cursor menu in Figure 9–45A. Clicking on Move or Copy… displays the Move or Copy dialog box in Figure 9–45B. You have the opportunity to shuffle drawing layouts around to change their order. You can also click in the check box to create a copy of the existing layout based on the floor plan. This new layout will be moved to the end; it will be placed after the floor plan.

Figure 9–45A

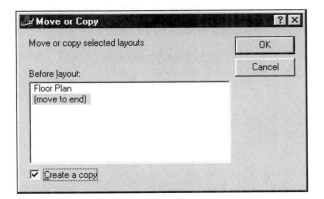

Figure 9–45B

Clicking the OK button displays the new layout in Figure 9–46A. When you copy a layout, all information such as title block, viewport, and information inside the viewport is copied. Begin the conversion of the new layout to the foundation plan by right-clicking on the Floor Plan (2) tab at the bottom of the display screen. This displays the cursor menu in Figure 9–46B. Click on the Rename option to activate the Rename Layout dialog box in Figure 9–46C. Rename this layout to Foundation Plan and click the OK button.

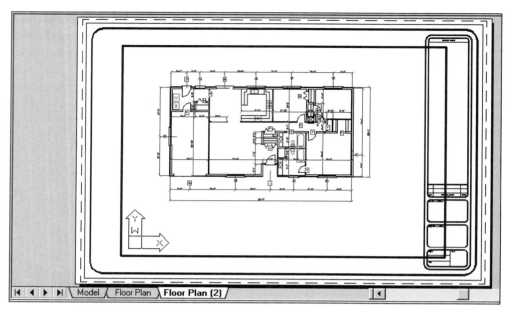

Figure 9–46A

Figure 9–46B

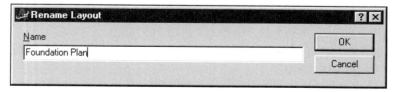

Figure 9–46C

Be sure you are in floating Model Space by double-clicking inside the viewport. Then activate the Layer Properties Manager dialog box in Figure 9–47A. Because this layout will hold all information that pertains to the Foundation Plan, the following layers are frozen in the active viewport:

> Appliances
>
> Construction
>
> Electrical_Symbols
>
> Electrical_Wiring
>
> Floor_Dimensions

The remaining layers, thawed in the active viewport, will display the foundation plan while freezing all other layers in this viewport.

The resulting image is displayed in Figure 9–47B, which shows the information needed to describe the foundation plan. Notice also at the bottom of the display screen that the new layout tab is renamed to reflect the Foundation Plan.

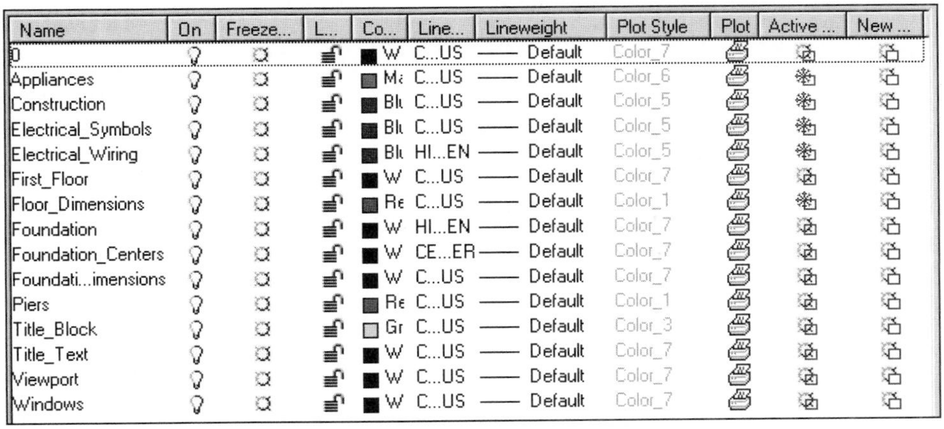

Name	On	Freeze...	L...	Co...	Line...	Lineweight	Plot Style	Plot	Active ...	New ...
0	♀	☼	⊆	■ W	C...US	Default	Color_7	🖶	✿	✿
Appliances	♀	☼	⊆	■ Ma	C...US	Default	Color_6	🖶	✿	✿
Construction	♀	☼	⊆	■ Blt	C...US	Default	Color_5	🖶	✿	✿
Electrical_Symbols	♀	☼	⊆	■ Blt	C...US	Default	Color_5	🖶	✿	✿
Electrical_Wiring	♀	☼	⊆	■ Blt	HI...EN	Default	Color_5	🖶	✿	✿
First_Floor	♀	☼	⊆	■ W	C...US	Default	Color_7	🖶	✿	✿
Floor_Dimensions	♀	☼	⊆	■ Re	C...US	Default	Color_1	🖶	✿	✿
Foundation	♀	☼	⊆	■ W	HI...EN	Default	Color_7	🖶	✿	✿
Foundation_Centers	♀	☼	⊆	■ W	CE...ER	Default	Color_7	🖶	✿	✿
Foundati...imensions	♀	☼	⊆	■ W	C...US	Default	Color_7	🖶	✿	✿
Piers	♀	☼	⊆	■ Re	C...US	Default	Color_1	🖶	✿	✿
Title_Block	♀	☼	⊆	□ Gr	C...US	Default	Color_3	🖶	✿	✿
Title_Text	♀	☼	⊆	■ W	C...US	Default	Color_7	🖶	✿	✿
Viewport	♀	☼	⊆	■ W	C...US	Default	Color_7	🖶	✿	✿
Windows	♀	☼	⊆	■ W	C...US	Default	Color_7	🖶	✿	✿

Figure 9–47A

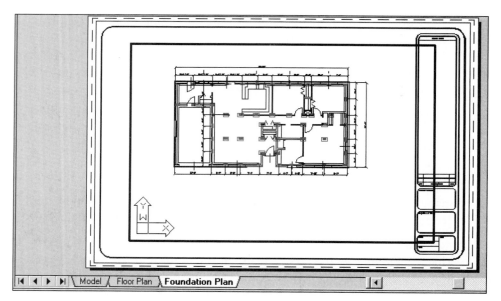

Figure 9–47B

Follow the previous steps to create a new layout and rename the layout to Electrical Plan. Then activate the Layer Properties Manager dialog box in Figure 9–48A and freeze the following layers in the active viewport:

Construction

Floor_Dimensions

Foundation

Foundation_Centers

Foundation_Dimensions

Piers

The remaining layers, thawed in the active viewport, will display the electrical plan in Figure 9–48B.

Name	On	Freeze...	L...	Co...	Line...	Lineweight	Plot Style	Plot	Active ...	New ...
0	♀	☼	🔓	■ W	C...US	—— Default	Color_7	🖨	🖫	🖫
Appliances	♀	☼	🔓	■ Mɑ	C...US	—— Default	Color_6	🖨	🖫	🖫
Construction	♀	☼	🔓	■ Blɩ	C...US	—— Default	Color_5	🖨	❋	🖫
Electrical_Symbols	♀	☼	🔓	■ Blɩ	C...US	—— Default	Color_5	🖨	🖫	🖫
Electrical_Wiring	♀	☼	🔓	■ Blɩ	HI...EN	—— Default	Color_5	🖨	🖫	🖫
First_Floor	♀	☼	🔓	■ W	C...US	—— Default	Color_7	🖨	🖫	🖫
Floor_Dimensions	♀	☼	🔓	■ Re	C...US	—— Default	Color_1	🖨	❋	🖫
Foundation	♀	☼	🔓	■ W	HI...EN	—— Default	Color_7	🖨	❋	🖫
Foundation_Centers	♀	☼	🔓	■ W	CE...ER	—— Default	Color_7	🖨	❋	🖫
Foundati...imensions	♀	☼	🔓	■ W	C...US	—— Default	Color_7	🖨	❋	🖫
Piers	♀	☼	🔓	■ Re	C...US	—— Default	Color_1	🖨	❋	🖫
Title_Block	♀	☼	🔓	□ Gr	C...US	—— Default	Color_3	🖨	🖫	🖫
Title_Text	♀	☼	🔓	■ W	C...US	—— Default	Color_7	🖨	🖫	🖫
Viewport	♀	☼	🔓	■ W	C...US	—— Default	Color_7	🖨	🖫	🖫
Windows	♀	☼	🔓	■ W	C...US	—— Default	Color_7	🖨	🖫	🖫

Figure 9–48A

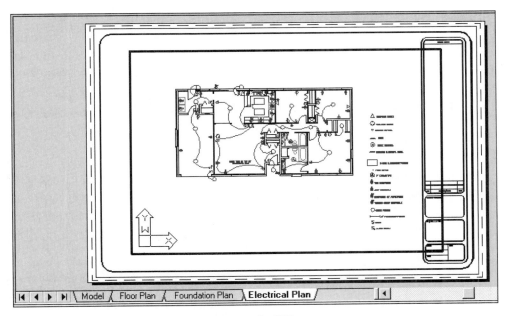

Figure 9–48B

Complete the multiple drawing layout in Figure 9–48C by assigning the viewports to a layer (called Viewports) and turn this layer off. Add text to each layout to identify the purpose and scale of each layout (Floor Plan, Foundation Plan, and Electrical Plan).

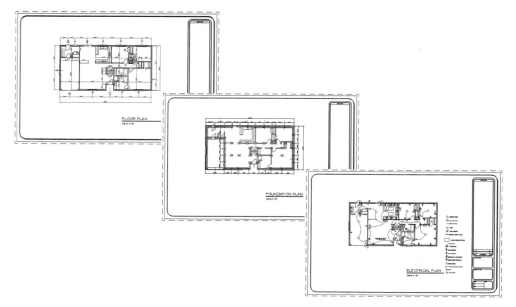

Figure 9–48C

USING THE OPTIONS DIALOG BOX TO CONTROL LAYOUTS

Moving your cursor into the command prompt area at the bottom of your screen and right-clicking the mouse button activates the cursor menu in Figure 9–49A. Click on Options to activate the Options dialog box. (You can also activate this dialog box by choosing Options from the Tools pull-down menus.) Clicking on the Display tab displays various controls that affect the display of AutoCAD. One of these controls focuses on Layout Elements in Figure 9–49B. By default, all boxes are checked, which means the individual layout settings are all turned on. Removing the check turns the setting off.

Figure 9–49A

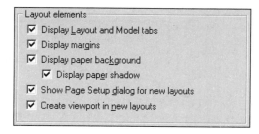

Figure 9–49B

All of these settings are labeled in Figure 9–49C. Turning off the Display Layout and Model tabs setting removes the Model and Layout1 tabs from the screen. Previously created layouts are still present; you just can't access them by clicking on the tab. Your display screen would also resemble an AutoCAD Release 14 screen. Turning off the Display margins setting turns off the dashed margin that identifies the printable area of the paper. Turning off the Display paper background setting turns the background from gray to white. At the right and bottom edges of the paper, two shadows are created to give the appearance of a 3D paper sheet. Turning off the Display paper shadow setting displays a paper border similar to the top and left edges. If the Create viewport in new layouts setting is turned off, a blank sheet empty of any viewports is displayed only for new layouts. You would have to use the MVIEW command to create your own viewport.

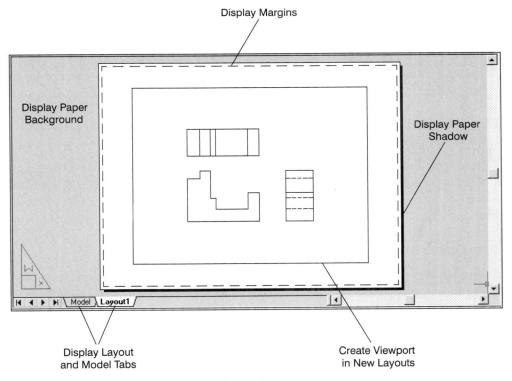

Figure 9–49C

RETAINING THE VISIBILITY OF LAYERS IN PAPER SPACE

One more command needs mention: the VISRETAIN command, the ability to retain the visibility of layers set in the Layer Properties Manager dialog box. By default, this command is turned on.

Command: **VISRETAIN**
New value for VISRETAIN <1>: *(1 for On, 0 for Off)*

If this command is turned off, none of the layer states specific to Paper Space will be saved with the drawing. As a result, all images in all viewports reappear, requiring that the images again be frozen in the active viewports where they do not appear. AutoCAD has this command turned on by default so the reappearance of images in viewports should not happen. This problem might occur if you load a drawing from a previous version where this command was turned off. Simply activate the VISRETAIN command and be sure it is turned on.

NOTES TO THE STUDENT AND INSTRUCTOR

Two tutorial exercises have been designed around this introductory topic of laying out drawings in Paper Space. As with all tutorials, you will tend to follow the steps very closely, taking care not to make a mistake. This is to be expected. However, most individuals rush through the tutorials to get the correct solution, only to forget the steps used to complete the tutorial. This is also to be expected.

Of all tutorial exercises in this book, the Drawing Layout tutorials will probably pose the greatest challenge. Certain operations have to be performed in Paper Space while other operations require floating Model Space. For example, it was already illustrated that user-defined viewports are constructed in Paper Space; scaling an image to Paper Space units through the Viewports toolbar or through the XP option of the ZOOM command must be accomplished in floating Model Space, and so on.

It is recommended to both student and instructor that all tutorial exercises, and especially those tutorials that deal with Paper Space, be performed two or even three times. Completing the tutorial the first time will give you the confidence that it can be done; however, you may not understand all of the steps involved. Completing the tutorial a second time will allow you to focus on where certain operations are performed and why things behave the way they do. This still may not be enough. Completing the tutorial exercise a third time will allow you to anticipate each step and have a better idea of which operation to perform in each step.

This recommendation should be exercised with all tutorials; however, it should especially be practiced when you work on the following Paper Space tutorials. For applying the concepts of Paper Space to other drawing files, problems have been set up at the end of this chapter that take advantage of previously completed drawings from other units. Use Paper Space as the layout mechanism for each of the problems.

TUTORIAL EXERCISE: 09_CENTER_GUIDE.DWG

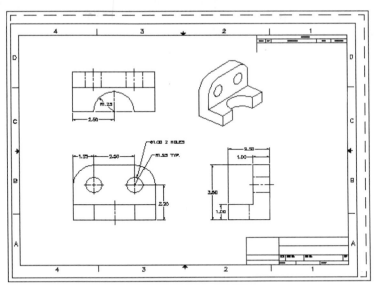

Figure 9–50

Purpose

This tutorial is designed to create a layout of the four view drawing in Figure 9–50 in Paper Space.

System Settings

All unit and limit values have been pre-set. In order for this tutorial to function properly, the HP DesignJet 750C Plus C4708B must be configured. If this not the case, please perform the first tutorial exercise in Chapter 10 to configure this plotter.

Layers

In addition to the current layers of the drawing, create the following layer:

Name	Color	Linetype
Viewport	Green	Continuous

Suggested Commands

Begin by opening the drawing file 09_Center_Guide.Dwg. Use the Create Layout wizard to perform the initial work of creating the layout of the Center Guide. When the layout is created and properly scaled, change the viewport object to the Viewport layer. Then turn off this layer to prepare the drawing to be plotted out at a scale of 1:1.

Whenever possible, substitute the appropriate command alias in place of the full AutoCAD command in each tutorial step; for example, use "CP" for the COPY command, "L" for the LINE command, and so on. The complete listing of all command aliases is located in Chapter 1, Table 1–2.

STEP 1

Illustrated in Figure 9–51 is a three-view orthographic drawing of a Center Guide. An isometric view is also provided to better interpret the three views. The drawing was created in Model Space. This is easily identified by the appearance of the User Coordinate System icon in the lower left corner. Use this as the starting point to arrange the model in Paper Space for the drawing to be plotted out at a scale of 1=1.

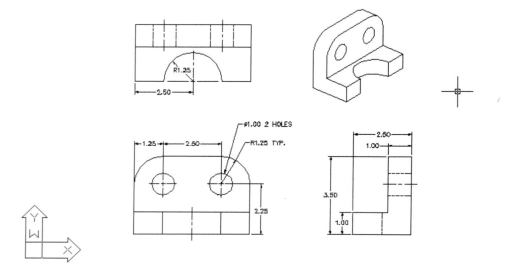

Figure 9–51

STEP 2

Choose Wizards from the Tools pull-down menu. Choosing the Create Layout... option in Figure 9–52A displays the Create Layout-Begin dialog box in Figure 9–52B. Begin the process of creating this new layout by entering "Four Views" as the name of the layout. When finished, click on the Next> button.

Figure 9–52A

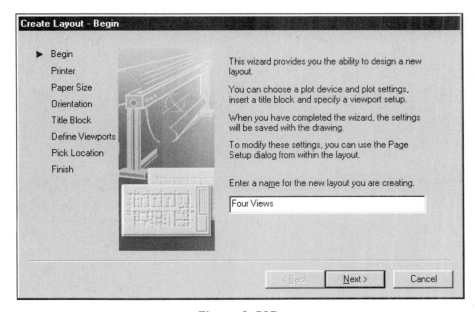

Figure 9–52B

STEP 3

In the Printer category, you must now select a printer that has already been configured. From the list of printers in Figure 9–53, click on the DesignJet 750C Plus C4708B.pc3 as the device. (Note: If this printer device is not listed, stop this tutorial, go to the first tutorial exercise in Chapter 10 and follow the steps for configuring this device; begin this exercise again.) When finished, click the Next> button.

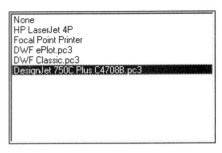

Figure 9–53

STEP 4

In the Paper category of the Create Layout dialog box, assign the paper size to be used to hold the drawing information. In the edit box in Figure 9–54, choose the ANSI expand C (22.00 x 17.00 Inches) sheet size. Keep all other default settings. When finished, click on the Next> button.

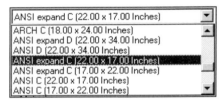

Figure 9–54

STEP 5

For orientating the drawing on the paper, be sure the Landscape radio button in 9–55 is active. When finished, click the Next> button.

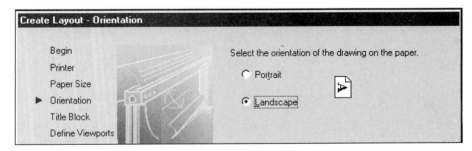

Figure 9–55

STEP 6

In the Title Block category, choose the "ANSI C title block.dwg" in Figure 9–56. This is the title block that will automatically be inserted in the new layout created. When finished, click the Next> button.

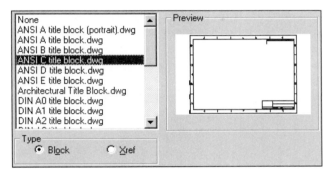

Figure 9–56

STEP 7

In the Define Viewports area, be sure the Viewport setup is set to single as in Figure 9–57. You also have the opportunity to pre-scale the drawing to paper space units. The drawing will be scaled inside the viewport at 1:1; click on this value in Figure 9–57. When finished, click the Next> button.

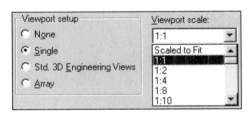

Figure 9–57

STEP 8

In the Pick Location category, you can define the size of the viewport that will be created in the drawing layout. If you choose not to click the Select location < button, a viewport is created to match the printable area—the area inside the dashed margins in the layout. Click on the Select location< button to create a smaller viewport as in 9–58A. This returns you to the drawing, where the sheet size and title block are displayed as in Figure 9–58B. Pick a point at "A" and "B" to define the new viewport. After identifying the viewport, you are returned to the Create Layout wizard. In this last dialog box, click on the Finish button.

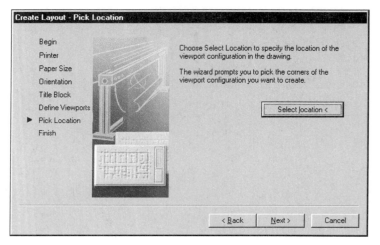

Figure 9–58A

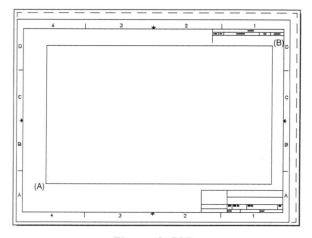

Figure 9–58B

STEP 9

The completed layout is displayed in Figure 9–59. Notice that the following items are already created in the layout: Layout title (Four Views), Viewport containing the four-view drawing, and C-size title block already inserted. The Paper Space icon is present, alerting you to the layout mode. If your image doesn't exact fit inside of the viewport or is not centered in relation to the title block, use grips to stretch the title block to a better size and then move the viewport containing the image to a better location inside of the title block.

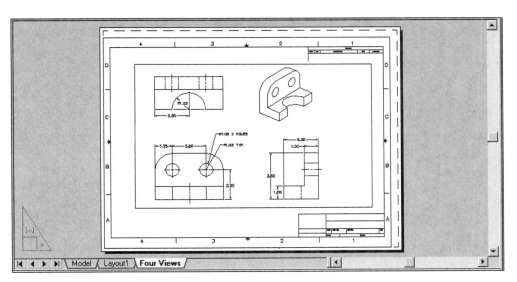

Figure 9–59

STEP 10

When the Layout Wizard created the viewport to hold the drawing information, it did not place the viewport on the correct layer. To accomplish this, first click on the viewport; it highlights and grips appear. Click in the Layer Control box and change the viewport to the Viewport layer as in Figure 9–60. When finished, press ESC twice to remove the object highlight and remove the grips.

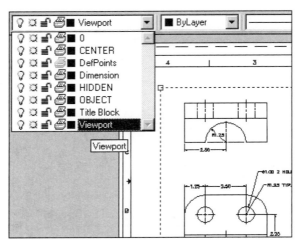

Figure 9–60

STEP 11

Verify the correct Paper Space scale by double-clicking in the viewport at "A." This places you in floating Model Space, with the appearance of the User Coordinate System icon as in Figure 9–61. Activate the Viewports toolbar and notice that the correct scale of 1:1 is reflected in the edit box. When finished verifying, dismiss the Viewports toolbar and double-click at "B" to return to Paper Space.

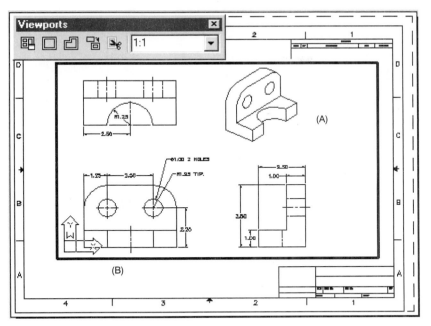

Figure 9–61

STEP 12

Your display should appear similar to Figure 9–62.

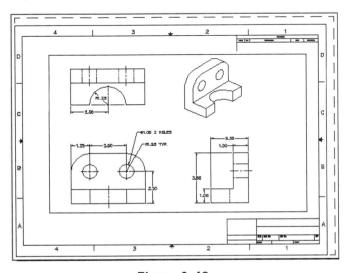

Figure 9–62

STEP 13

Notice that the Layout1 tab is still present at the bottom of your display screen. Since the Layout Wizard was used to create the Four Views tab, Layout1 still remains. Right-clicking on Layout1 displays the cursor menu in Figure 9–63A. Click on the Delete option to remove Layout1 from the drawing database. An AutoCAD alert box appears, as in Figure 9–63B, telling you the layout will be permanently deleted. Also, it reminds you that the Model tab cannot be deleted. Click the OK button to delete Layout1.

Figure 9–63A

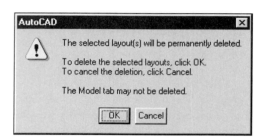

Figure 9–63B

STEP 14

Before plotting, turn off the Viewport layer in the Layer Control box in Figure 9–64.

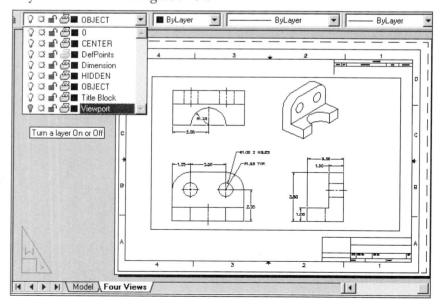

Figure 9–64

STEP 15

If at any point you need to change the drawing layout or plotter device, you can accomplish this by choosing Page Setup from the File pull-down menu as shown in Figure 9–65A, which displays the Page Setup dialog box in Figure 9–65B. Changes made here will be reflected in the current layout.

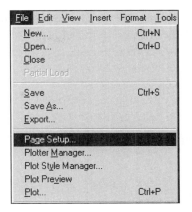

Figure 9–65A

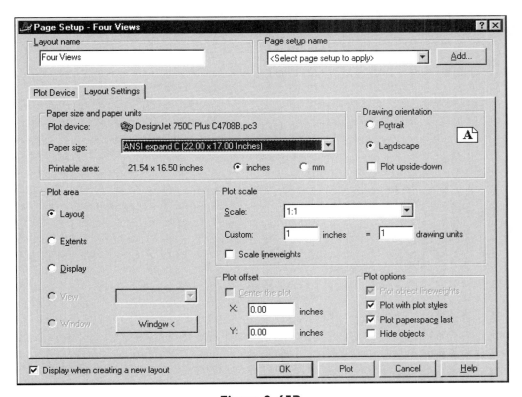

Figure 9–65B

STEP 16

The completed layout is displayed in Figure 9–66. In this final step, save this drawing file. It will be used for the purposes of plotting, which will be covered in Chapter 10. The layout will also be saved since it is part of the drawing.

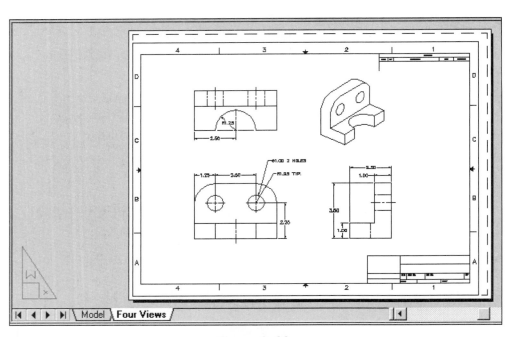

Figure 9–66

TUTORIAL EXERCISE: 09_HVAC.DWG

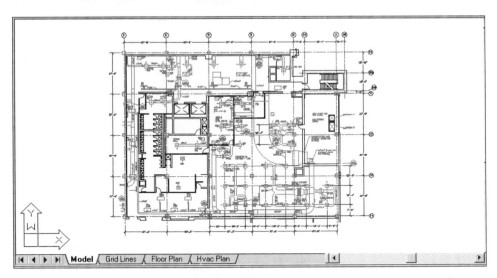

Figure 9–67

Purpose

This tutorial is designed to create multiple layouts of the HVAC drawing in Figure 9–67 in Paper Space.

System Settings

All unit and limit values have been pre-set. In order for this tutorial to function properly, the HP DesignJet 750C Plus C4708B must be configured. If this not the case, please perform the first tutorial exercise in Chapter 10 to configure this plotter.

Layers

Layers have already been created for this exercise.

Suggested Commands

Begin by opening the drawing file 09_Hvac.Dwg. Click on Layout1 to begin the first drawing layout that consists of the grid lines. When the layout is created, enter floating Model space and scale the image to

1/8?=1'-0?. Also, freeze the layers that pertain to the floor and HVAC plans only in the current viewport. Press ENTER to return to Paper Space and insert an architectural title block and change the viewport to the Viewports layer. From this layout, create another layout called Floor Plan. While in floating Model Space, freeze the layers that pertain to the grid lines and HVAC plans. From this layout, create another layout called Hvac Plan. While in floating Model Space, freeze the layers that pertain to the grid lines. Turn off all viewports and add drawing titles to each layout.

Whenever possible, substitute the appropriate command alias in place of the full AutoCAD command in each tutorial step; for example, use "CP" for the COPY command, "L" for the LINE command, and so on. The complete listing of all command aliases is located in Chapter 1, Table 1–2.

STEP I

Illustrated in Figure 9–68 is a drawing of an HVAC plan, which includes grid lines used for the layout of columns, a floor plan, and the actual HVAC plan, which consists of ductwork. The drawing was created in Model Space. This is easily identified by the appearance of the User Coordinate System icon displayed in the lower left corner. Three separate layouts will be created. The first will consist of the grid lines; the second will be the floor plan information; the third will consist of the floor plan along with HVAC plan. First, click on the Layout1 tab at the bottom of the screen.

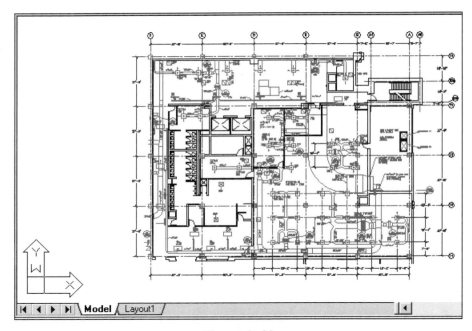

Figure 9–68

STEP 2

When the Page Setup dialog box appears in Figure 9–69A, click on the Plot Device tab and change the plotter to the DesignJet 750C Plus C4708B.pc3. (Note: If this printer device is not listed, stop this tutorial, go to the first tutorial exercise in Chapter 10 and follow the steps for configuring this device; begin this exercise again.) Also, in the upper left corner of the dialog box, change the Layout name from "Layout1" to "Grid Lines." Then click on the Layout Settings tab in Figure 9–69B and change the Paper size to ARCH expand D (22.00 x 34.00 Inches). Click the OK button to create the layout.

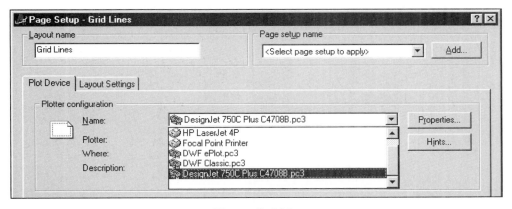

Figure 9–69A

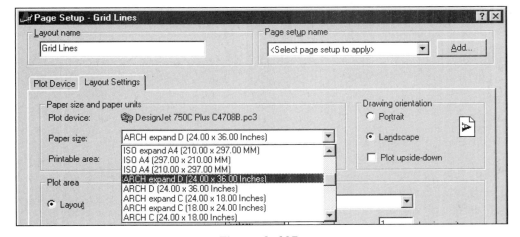

Figure 9–69B

STEP 3

Your display should be similar to Figure 9–70. Notice that the name of the layout next to the Model tab is "Grid Lines." Click on the rectangular viewport at "A." The viewport should highlight and grips will appear at the corners. Then click in the Layer Control box to change the highlighted viewport to the VIEWPORTS layer. Press ESC twice to remove the object highlight and grips.

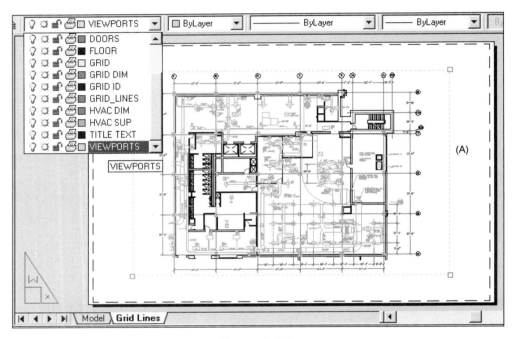

Figure 9–70

STEP 4

Double-click anywhere inside the viewport to switch to floating Model Space. Notice the re-appearance of the User Coordinate System icon in the lower left corner of the viewport. Activate the Viewports dialog box in Figure 9–71 and change the scale of the image inside the viewport to 1/8?=1'-0?. The image may get larger.

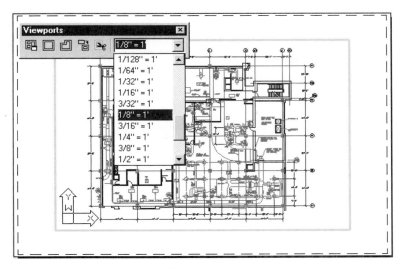

Figure 9–71

STEP 5

Double-click anywhere outside the viewport to return to Paper Space in Figure 9–72. Click on the edge of the viewport and use the corner grips to size the viewport to the image. When finished, press ESC twice to remove the object highlight and grips.

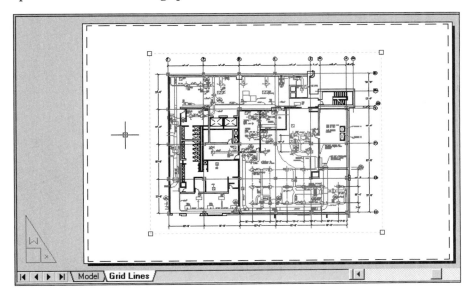

Figure 9–72

STEP 6

Choosing Block... from the Insert pull-down menu activates the Insert dialog box in Figure 9–73. Search in the Name box for a block called Architectural Title Block (this block has been pre-defined in the drawing). Click the OK button to return to the drawing.

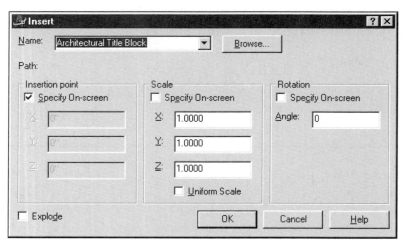

Figure 9–73

STEP 7

Position the title block in Figure 9–74. Then click on the title block to have the highlight appear along with the grips. Activate the Layer Control box and change the highlighted title block to the BORDER layer. When finished, press ESC twice to remove the object highlight and grips.

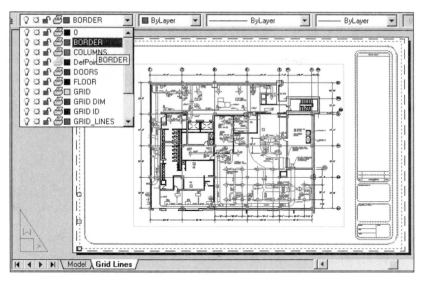

Figure 9–74

STEP 8

It will be necessary to center the viewport containing the image, moving it to a better location. Use the Insert dialog box again to insert a block called "Title." This block contains the title and scale of the drawing and is also predefined in the drawing. After inserting this in your drawing, use the EXPLODE command to break this block up into individual objects. When finished, your drawing should appear similar to Figure 9–75.

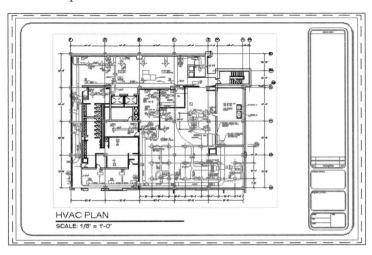

Figure 9–75

STEP 9

Click on the drawing title (HVAC PLAN); the title should highlight and grips will appear. Then activate the Properties dialog box. Change HVAC PLAN to GRID LINES PLAN in Figure 9–76; pressing ENTER automatically updates the text to the new value. Dismiss this dialog box when finished. Press ESC twice to remove the object highlight and grips.

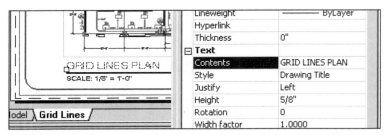

Figure 9–76

STEP 10

Prepare to create the second layout by right-clicking on the Grid Lines tab to display the cursor menu in Figure 9–77A. Click on the Move or Copy... option to make a copy of the current layout. With the Move or Copy dialog box displayed in Figure 9–77B, place a check in the box to make a copy. Also highlight the area to move the new layout to the end. When finished, click the OK button to create the new layout.

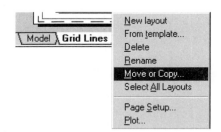

Figure 9–77A

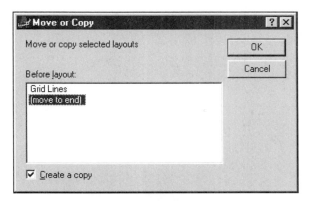

Figure 9–77B

STEP 11

Before continuing, right-click on the new layout "Grid Lines (2)" and click on the Rename option in the cursor menu in Figure 9–78A. In the Rename Layout dialog box shown in Figure 9–78B, change the current layout name "Grid Lines (2)" to the new name of "Floor Plan." Click the OK button to dismiss the dialog box and make the change.

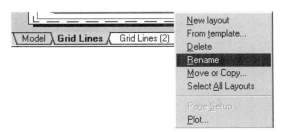

Figure 9–78A

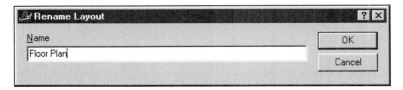

Figure 9–78B

STEP 12

Your display should appear similar to Figure 9–79. Not only has the image been copied in the new layout, but the viewport, border, and drawing title have been copied as well.

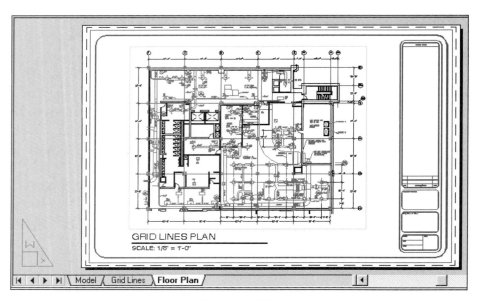

GRID LINES PLAN
SCALE: 1/8" = 1'-0"

Model | Grid Lines | **Floor Plan**

Figure 9–79

STEP 13

While in this new layout, right-click on the Floor Plan tab to display the cursor menu in Figure 9–80A. Click on the Move or Copy... mode to make one more copy of the current layout. With the "Move or Copy" dialog box displayed in Figure 9–80B, place a check in the box to create a copy. Also, highlight the area to move the new layout to the end. When finished, click the OK button to create the new layout.

Model | Grid Lines | **Floor Plan**

New layout
From template...
Delete
Rename
Move or Copy...
Select All Layouts

Page Setup...
Plot...

Figure 9–80A

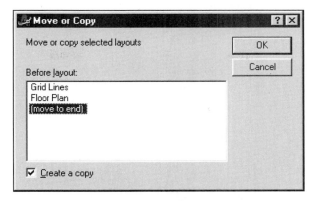

Figure 9–80B

STEP 14

Before continuing, right-click on the new layout "Floor Plan (2)" and click on the Rename option in the cursor menu in Figure 9–81A. Change the current layout name "Floor Plan (2)" to the new name of "Hvac Plan" in Figure 9–81B. Click the OK button to dismiss the dialog box and make the change.

Figure 9–81A

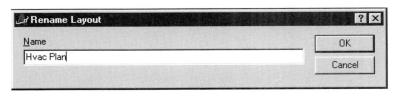

Figure 9–81B

tutorial EXERCISE

STEP 15

Your display should appear similar to Figure 9–82.

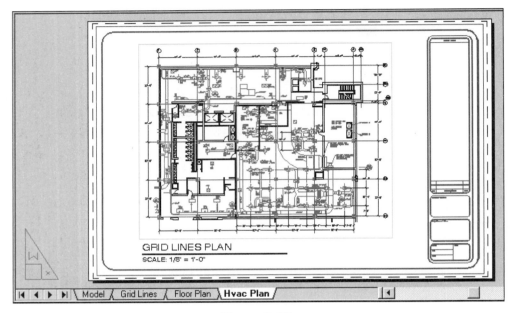

Figure 9–82

STEP 16

Click on the Grid Lines tab. Double-click anywhere inside the viewport in Figure 9–83A; this places you in floating Model Space. You will need to turn off all layers that deal with the floor plan and HVAC, which leave only the layers with the grid lines visible.

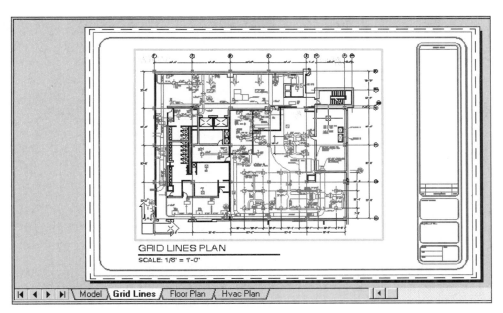

Figure 9–83A

Activate the Layer Properties Manager dialog box in Figure 9–83B and freeze the highlighted layers in the active viewport that pertain to the floor plan and HVAC plan.

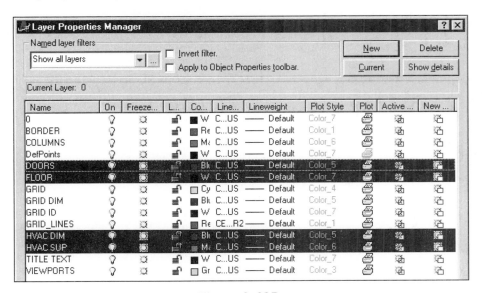

Figure 9–83B

Clicking the OK button returns to the drawing editor. Double-click anywhere outside the viewport to return to Paper Space. Your drawing should appear similar to Figure 9–83C.

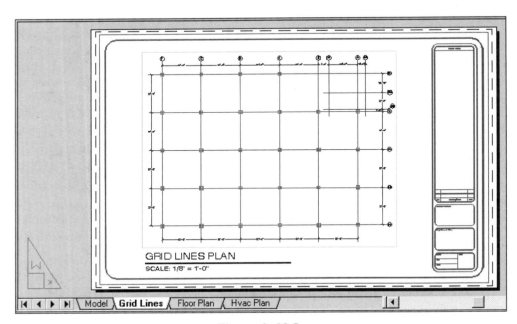

GRID LINES PLAN
SCALE: 1/8" = 1'-0"

Model Grid Lines Floor Plan Hvac Plan

Figure 9–83C

STEP 17

Click on the Floor Plan tab. Double-click anywhere inside the viewport in Figure 9–84A; this places you in floating Model Space. You will need to turn off all layers that deal with the grid lines and HVAC.

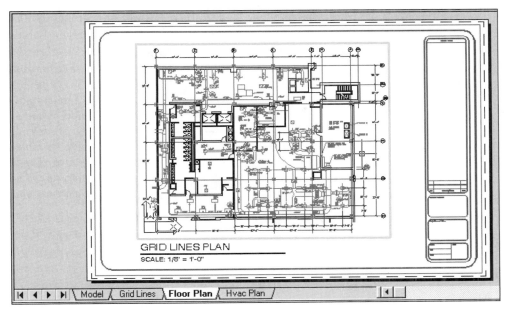

Figure 9–84A

Activate the Layer Properties Manager dialog box in Figure 9–84B and freeze the layers in the active viewport that pertain to the grid lines plan and HVAC plan.

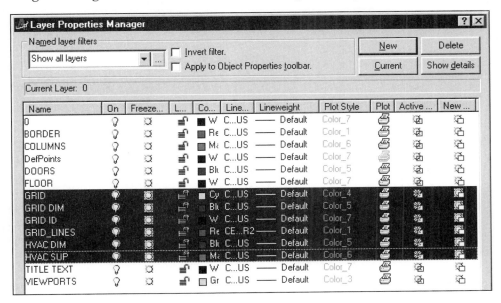

Figure 9–84B

Clicking the OK button returns you to the drawing editor. Double-click anywhere outside the viewport to return to Paper Space. Use the Properties dialog box to change the drawing title from GRID LINES PLAN in Figure 9–84C, to FLOOR PLAN. Do this by first selecting GRID LINES PLAN. With this object highlighted, activate the Properties dialog box and change the text. When finished, press ESC twice to remove the object highlight and grips.

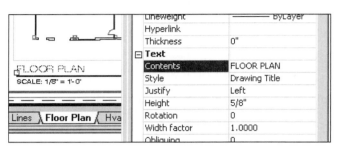

Figure 9–84C

Your drawing should appear similar to Figure 9–84D.

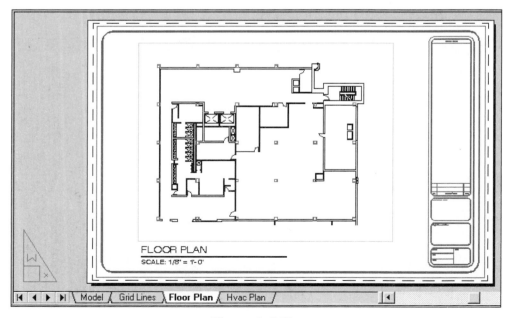

Figure 9–84D

STEP 18

Click on the Hvac Plan tab. Double-click anywhere inside the viewport in Figure 9–85A; this places you in floating Model Space. You will need to turn off all layers that deal with the grid lines plan.

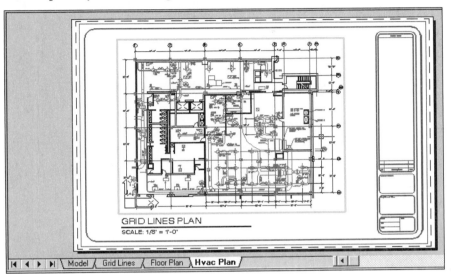

Figure 9–85A

Activate the Layer Properties Manager dialog box in Figure 9–85B and freeze the layers in the active viewport that pertain to the grid lines plan.

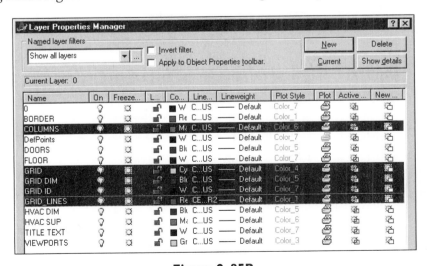

Figure 9–85B

Clicking the OK button returns to the drawing editor. Double-click anywhere outside the viewport to return to Paper Space. Use the Properties dialog box to change the drawing title from GRID LINES PLAN in Figure 9–85C to FLOOR PLAN. Do this by first selecting GRID LINES PLAN. With this object highlighted, activate the Properties dialog box and change the text. When finished, press ESC twice to remove the object highlight and grips.

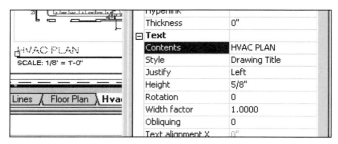

Figure 9–85C

Your drawing should appear similar to Figure 9–85D.

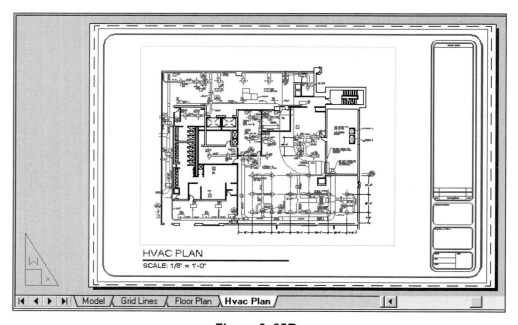

Figure 9–85D

STEP 19

Turn off the Viewports layer. The completed drawing is displayed in Figure 9–86.

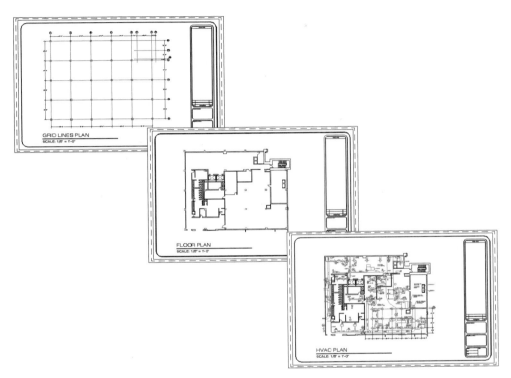

Figure 9–86

PROBLEMS FOR CHAPTER 9

PROBLEM 9–1

Arrange the following drawings from Chapter 5, "Performing Geometric Constructions." Use Figure 9–87 (showing Problem 5–1) as an example of how the layout will appear when finished.

Problems 5–1, 5–3, 5–4, 5–6, 5–7, 5–8, 5–10, 5–12, 5–13, 5–15

Use ANSI_B.Dwg as the title block inside Paper Space. This title block can be found in the Template folder of AutoCAD 2000. Scale the image inside floating Model Space at a scale of 1:1 or 1XP if using the ZOOM command.

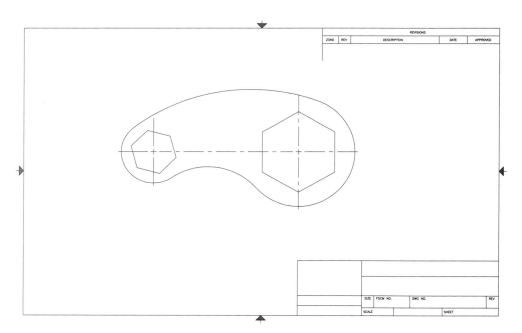

Figure 9–87

PROBLEM 9–2

Arrange the following drawings from Chapter 5, "Performing Geometric Constructions." Use Figure 9–88 (showing Problem 5–2) as an example of how the layout will appear when finished.

Problems 5–2, 5–18

Use ANSI_C.Dwg as the title block inside Paper Space. This title block can be found in the Template folder of AutoCAD 2000. Scale the image inside floating Model Space at a scale of 1:1 or 1XP if using the ZOOM command.

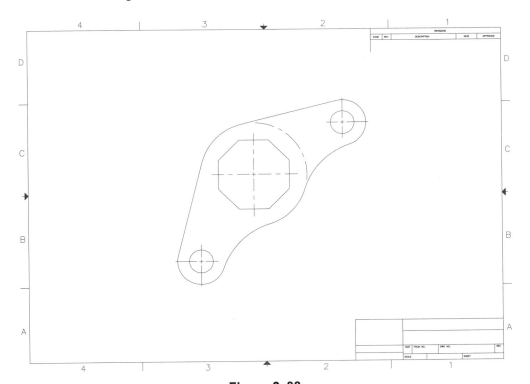

Figure 9–88

PROBLEM 9–3

Arrange the following metric drawings from Chapter 5, "Performing Geometric Construc-tions." Use Figure 9–89 (showing Problem 5–8) as an example of how the layout will appear when finished.

Problems 5–5, 5–9, 5–11, 5–14, 5–17

Use ISO_A3.Dwg as the title block inside Paper Space. This title block can be found in the Template folder of AutoCAD 2000. Scale the image inside floating Model Space at a scale of 1:1 or 1XP if using the ZOOM command.

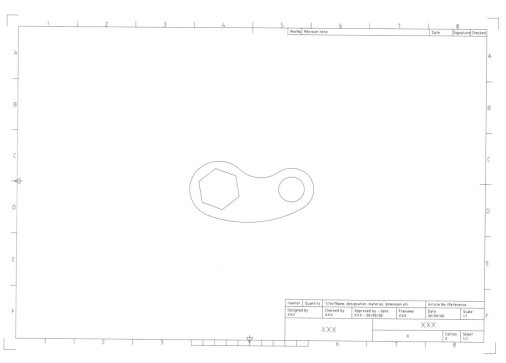

Figure 9–89

CHAPTER 10

Plotting Your Drawings

Plotting has been greatly enhanced in AutoCAD 2000 to allow you to output your drawing more efficiently. This chapter discusses plotting through a series of tutorial exercises designed to configure a new plotter, create color tables, create plot style tables, and finally execute the plot. Plotting a drawing for viewing over the Internet will also be discussed.

CONFIGURING A PLOTTER

Before plotting, you must first establish communication between AutoCAD and the plotter. This is called configuring. From a list of supported plotting devices, you choose the device that matches the model of plotter you own. This plotter becomes part of the software database, which allows you to choose this plotter many times. If you have more than one output device, each device must be configured before being used. This section discusses the configuration process used in AutoCAD 2000.

STEP I

Begin the plotter configuration process by choosing Plotter Manager from the File pull-down menu as in Figure 10–1A. This activates the Plotters program group in Figure 10–1B, which lists all valid plotters that are currently configured. The listing in Figure 10–1B displays the default plotters configured after the software is loaded. Except for the DWF devices, which allow you to publish a drawing for viewing over the Internet, a plotter has not yet been configured. Double-click on the Add-A-Plotter Wizard icon to continue with the configuration process.

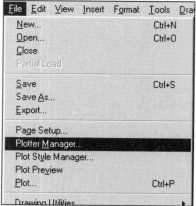

Figure 10–1A

Figure 10–1B

STEP 2

Double-clicking on the Add-A-Plotter Wizard icon displays the Add Plotter-Introduction Page in Figure 10–2. This dialog box states you are about to configure a Windows or non-Windows system plotter. This configuration information will be saved in a file with the extension .PC3. Click the Next> button to continue on to the next dialog box.

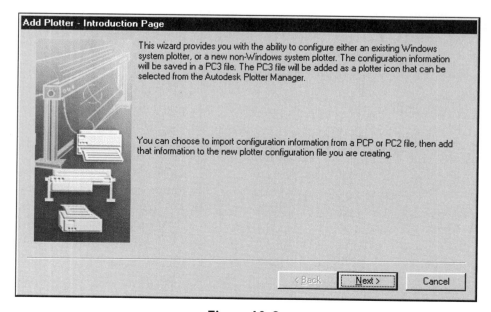

Figure 10–2

STEP 3

In the Add Plotter-Begin dialog box in Figure 10–3, decide how the plotter will be controlled: by the computer you are currently using, by a network plot server, or by an existing system printer where changes can be made specifically for AutoCAD 2000. Click on the radio button next to My Computer. Then click on the Next> button to continue on to the next dialog box.

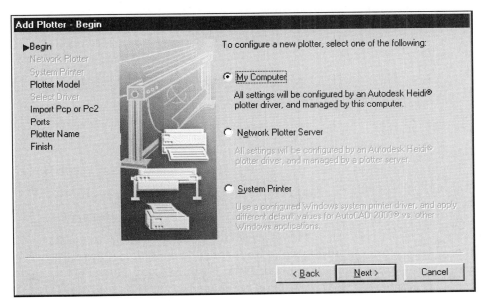

Figure 10–3

STEP 4

Use the Add Plotter-Plotter Model dialog box in Figure 10–4 to associate your model plotter with AutoCAD. You would first choose the appropriate plotter manufacturer from the list provided. Once this is done, all models supported by the manufacturer appear to the right. If your plotter model is not listed, you are told to consult the plotter documentation for a compatible plotter. If you have an installation disk containing an HDI driver, you could click on the Have Disk... button to copy this driver from disk to your computer. For the purposes of this tutorial, click on Hewlett-Packard in the list of Manufacturers. Click on the DesignJet 750C Plus C4708B for the plotter model. A Driver Info dialog box may appear giving you more directions regarding the type of HP DeskJet plotter selected. Click the Continue button to move on to the next dialog box used in the plotter configuration process.

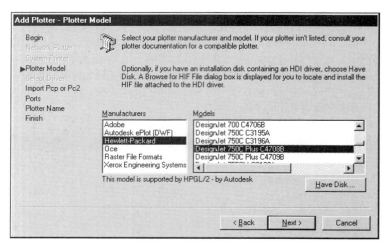

Figure 10–4

STEP 5

PCP and PC2 files have been in existence for many years. They are designed to hold plotting information such as pen assignments. In this way, you use the PCP or PC2 files to control pen settings instead of constantly making pen assignments every time you perform a plot. The Add Plotter-Import Pcp or Pc2 dialog box in Figure 10–5 allows you to import those files for use in AutoCAD 2000 in a PC3 format. If you will not be using any Pcp or Pc2 files from previous versions of AutoCAD, click on the Next> button to move on to the next dialog box.

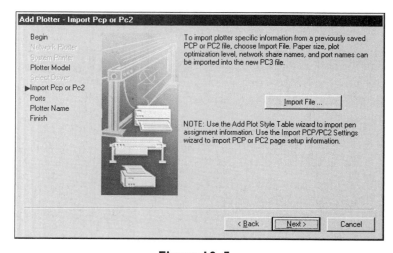

Figure 10–5

STEP 6

In the Add Plotter-Ports dialog box, shown in Figure 10–6, click on the port used for communication between your computer and the plotter. The LPT1 port will be used for the purposes of this tutorial. Place a check in its box and then click on the Next> button to continue on to the next dialog box.

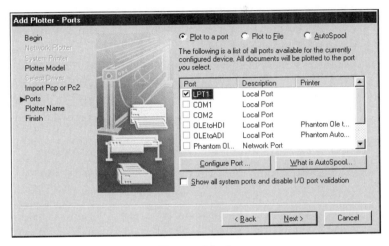

Figure 10–6

STEP 7

In the Add Plotter-Plotter Name dialog box in Figure 10–7, you have the option of giving the plotter a name other than the name displayed in the dialog box. This name will be displayed whenever you use the Page Setup and Plot dialog boxes.

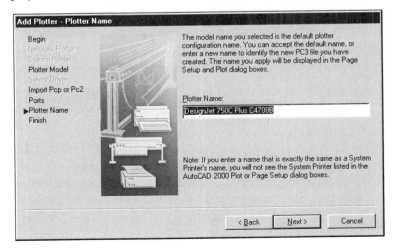

Figure 10–7

STEP 8

The last dialog box is displayed in Figure 10–8. In the Add Plotter-Finish dialog box, you can modify the default settings of the plotter you just configured. You can also test and calibrate the plotter if desired. Click the Finish button to dismiss the Add Plotter dialog box.

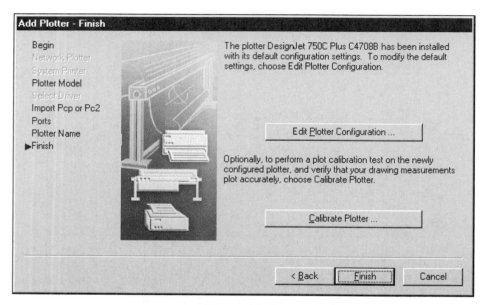

Figure 10–8

STEP 9

Exiting the Add Plotters dialog box returns you to the Plotters program group shown in Figure 10–9A. Notice that the additional icon for the DesignJet 750C Plus plotter has been added to this list. In the View pull-down menu, you have additional controls used to display the plotters. By default, the Large Icons mode is selected. Clicking on Small Icons displays the plotter list in Figure 10–9B. Clicking on Details displays additional information on your plotters in Figure 10–9C. This completes the steps used to configure the DesignJet 750C Plus plotter. Follow these same steps if you need to configure another plotter.

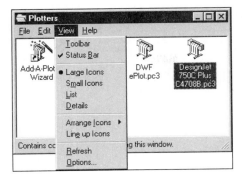

Figure 10–9A

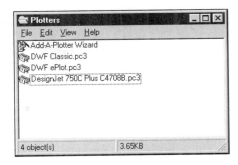

Figure 10–9B

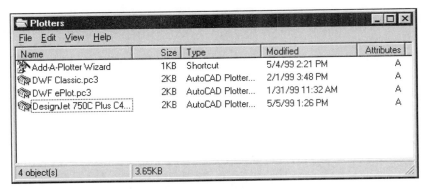

Figure 10–9C

PLOTTING IN PAPER SPACE

In the previous chapter, "An Introduction to Drawing Layouts," you were instructed to save the drawing 09_Center_Guide.Dwg. Open this drawing now to step through the process of plotting the drawing out of layout mode, or Paper Space.

STEP 1

When you open the drawing 09_Center_Guide.Dwg, your drawing should appear similar to Figure 10–10. This drawing should be laid out in Paper Space. A layout called "Four Views" should be present at the bottom of the screen next to the Model tab.

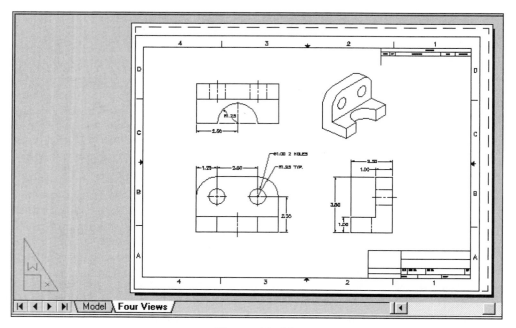

Figure 10–10

STEP 2

Begin the process of plotting this drawing by choosing Plot... from the File pull-down menu as in Figure 10–11A. This activates the Plot dialog box in Figure 10–11B. In the Plot Device tab, verify the correct plotter configuration. You could click the down arrow for the current list of all configured plotters and then pick the desired plotter from this list. For the purposes of this tutorial, the DesignJet 750C Plus is being used as the output device. Although your plotter may be different, the steps used in performing the plot are the same for any device. (The Properties... button adjacent to the plotter name will be explained in greater detail later in this chapter, as will the Plot style table area located in the middle of the dialog box.) By default, the Current tab radio button is clicked in the "What to plot area." If numerous layouts were created, you could plot all layouts at once by choosing this radio button. Also, you can opt to create additional copies of your plot; by default this value is set to 1. In the Plot to file area in the lower right corner of the dialog box, you could output your plot to a file instead of to paper (hard copy). AutoCAD drawings can be merged into other Windows applications; however, the file must be converted to a different format. Plotting the drawing to a file creates a file with the PLT extension. This will always be the extension if you configured any type of plotter manufactured by Hewlett-Packard.

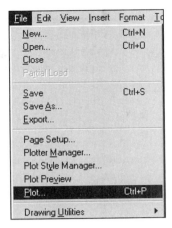

Figure 10–11A

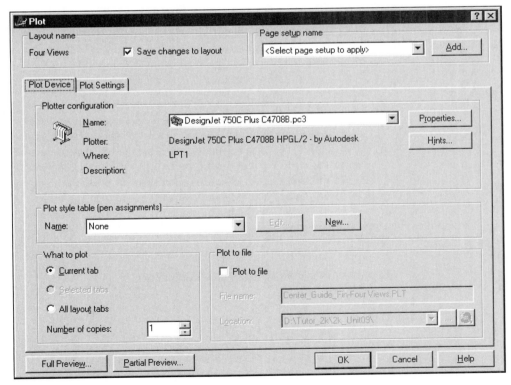

Figure 10–11B

Click on the Plot Settings tab of the Plot dialog box in Figure 10–11C and verify the following areas are properly set. Make sure the paper size is currently set to ANSI expand C (22.00 x 17.00 Inches). In the Drawing orientation area, make sure that the radio button adjacent to Landscape is selected. You could also plot the drawing out in Portrait mode, where the short edge of the paper is the top of the page. For special plots, you could plot the drawing upside-down.

In the Plot area, the Layout radio button is selected. Since you created a layout, this is the more obvious choice. The Extents mode allows you to plot the drawing based on all objects that make up the drawing. Plotting the Display will plot your current drawing view, but be careful. If you are currently zoomed in to your drawing, plotting the Display will only plot this view. In this case, it would be more predictable to use Layout or Extents to plot. When you plot a Layout in Paper Space, the Plot Scale will be set to 1:1. Since you pre-scaled the drawing to the Paper Space viewport using the Viewports toolbar, all drawings in Paper Space are designed to be plotted out at this scale. The Plot Offset is designed to move or shift the location of your plot on the paper if it appears off center. In the Plot options area, you have more control over plots by applying lineweights, using existing plot styles, plotting paperspace last (Model Space first), or even hiding objects, which is performed on a 3D solid model.

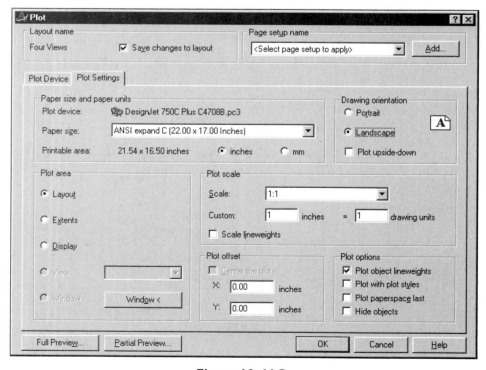

Figure 10–11C

STEP 3

One of the more efficient features of plotting is that you can preview your plot before sending the plot information to the plotter. In this way, you can determine if the entire drawing will plot based on the sheet size (this includes the border and title block). Clicking on the Full Preview... button activates the image in Figure 10–12. The sheet size is present along with the border and four-view drawing. Right-clicking anywhere on this preview image displays the cursor menu, allowing you perform various display functions such as ZOOM and PAN to assist with the verification process. If everything appears satisfactory, click on the Plot option to send the drawing information to the plotter. Clicking on the Exit option returns you to the Plot dialog box, where you can make changes in the Plot Device or Plot Settings tabs.

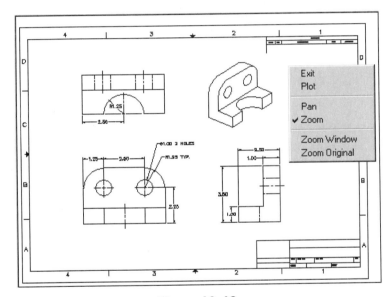

Figure 10–12

STEP 4

A faster way of previewing a plot is through the Partial Preview... button. Clicking on this button displays the Partial Plot Preview dialog box in Figure 10–13. This method of previewing a plot shows a quick representation of the effective plot area relative to the paper size and printable area. You will also be given warnings in the event that the drawing is outside the printable area. This usually affects the border and title block. This series of steps has provided an introduction to plotting your drawing. This dialog box will be used in later sections of this chapter to demonstrate additional controls for enhancing your plots.

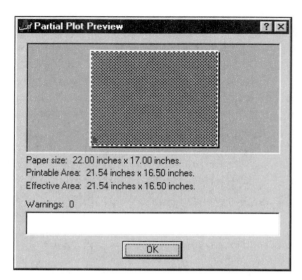

Figure 10–13

PLOTTING IN MODEL SPACE

You can also plot out of Model Space. However, there is more involved in bringing in such basic items as borders, title blocks or notes into the drawing. Open the drawing called 10_Roof_Plan.Dwg and follow the next series of steps, designed to plot out of Model Space.

STEP 1

As with Paper Space, you must determine the scale at which you wish to plot. The drawing in Figure 10–14 illustrates a Roof Plan. Verify the area occupied by the drawing using the LIMITS command.

Command: **LIMITS**
Reset Model space limits:
Specify lower left corner or [ON/OFF] <0'-0"?,0'-0?>: (Press ENTER)
Specify upper right corner <276'-0?,184'-0?>: (Press ENTER)

From the information in the LIMITS command, this drawing is currently occupying a sheet of paper 276' by 184'. This makes sense because Model Space is where the drawing is constructed full size or in real world units. So you need a full size sheet of paper to draw on. Notice also the scale of the drawing under the text "ROOF PLAN"; it reads 1/8? = 1'-0?. This becomes the plot scale, which you will use in a later step.

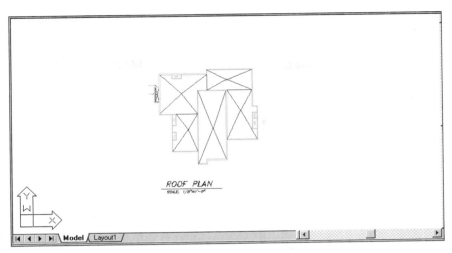

Figure 10–14

STEP 2

Since you will not be using the Layout1 tab, you must insert a title block in this drawing. The Architectural Title Block will be used in this example. This title block normally measures 34.5? x 23?. Remember, however, that the limits of the drawing used to define the sheet measure 276' x 184'. This means the title block must be blown up or enlarged when you bring it into the drawing of the Roof Plan. Also, you must enlarge it based on the scale of the drawing. Since the drawing will be plotted out at 1/8? = 1'-0?, you need to enlarge the title block 96 times its normal size. The value 96 is found by the division of 1' or 12 by 1/8 or 0.125 units. Activate the Insert dialog box in Figure 10–15 and make the changes in the dialog box such as the insertion point, scale, and rotation angle. The title block will be inserted at 0,0 and have a scale value of 96, as previously explained. A rotation angle of 0 will be used.

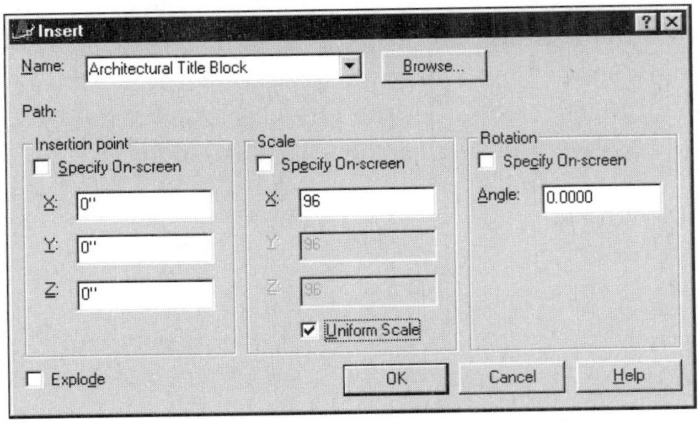

Figure 10–15

STEP 3

Your drawing should appear similar to Figure 10–16. Since the title block had to be enlarged, adding additional support items to the drawing such as dimensions and notes must also include the enlargement factor of 96 when the note height and overall dimension scale factor are calculated. This tends to make this process more involved and time consuming.

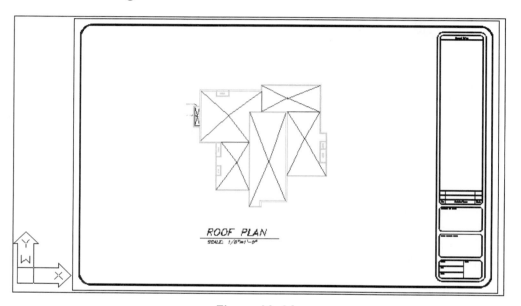

ROOF PLAN
SCALE: 1/8"=1'-0"

Figure 10–16

STEP 4

Activate the Plot dialog box and make sure you are in the Plot Settings tab in Figure 10–17. Verify that the current plot device is the HP DesignJet 750C Plus. If this is not the device, click on the Plot Device tab to activate this plotter. In addition, Landscape mode should be selected under the Drawing orientation area. Continue on to the next step for making additional changes in this dialog box.

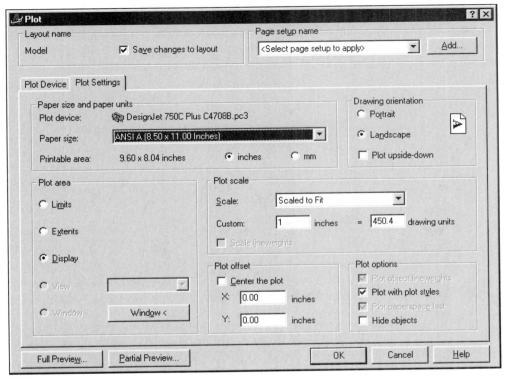

Figure 10–17

STEP 5

In the Plot area portion of the Plot dialog box in Figure 10–18A, you are required to click on the radio button of your choice to determine the type of area you desire to plot the drawing by. These modes will now be explained so that you can see their results. The image of the Roof Plan without the title block will be used to explain each mode. Clicking on the Limits radio button plots the Roof Plan based on the original sheet size created through the LIMITS command. The results are illustrated in Figure 10–18B (The title block has been temporarily turned off in this figure.) Clicking on the Extents radio button calculates the plot based on the objects placed in the drawing. This results in a plot similar to Figure 10–18C (Again, the title block has been temporality turned off in this figure.) Clicking on the Display radio button could result in a plot based on the image in Figure 10–18B. It could also result in the plot appearing similar to Figure 10–18D. Here, the drawing was magnified through the ZOOM command; however, the entire drawing needs to be plotted out. If you forget to perform a ZOOM-ALL when using the plot Display mode, the plot displays as in

Figure 10–18D, which is a common occurrence. The other two plot area modes are View and Window. It you have previously created a named view of your drawing, you can retrieve this view and plot it out. You can also define a plot area by a window by clicking on the Window< button. This returns you to your drawing, where you define a rectangular box as the plot area. Continue on to the next step for making additional changes in this dialog box.

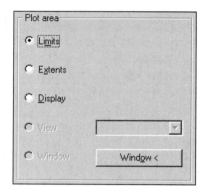

Figure 10–18A

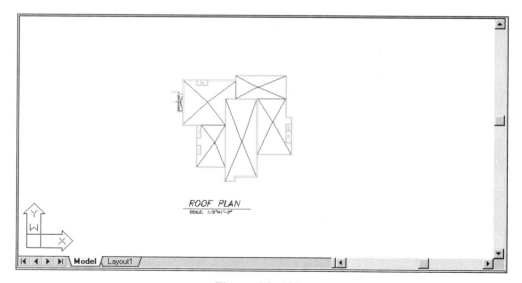

Figure 10–18B

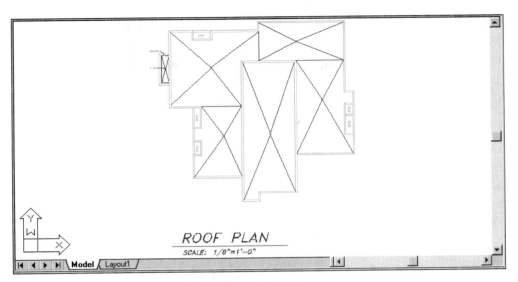

Figure 10–18C

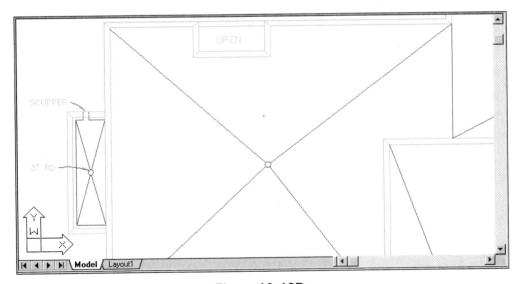

Figure 10–18D

STEP 6

Next, click in the Paper size and paper units area of the Plot dialog box. For this drawing of the Roof Plan, a D-size sheet of paper will be used. Clicking on the ARCH expand D (24.00 x 36.00 Inches) sheet size in Figure 10–19 performs this task. Continue on to the next step for making additional changes in this dialog box.

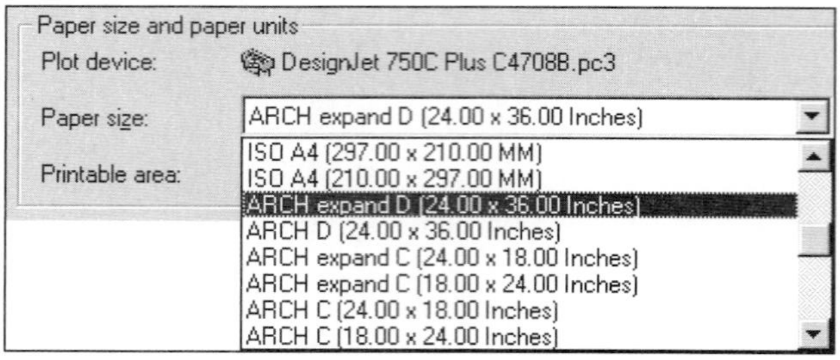

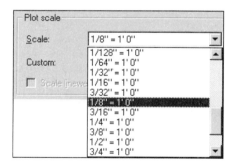

Figure 10–19

STEP 7

In the previous topic on plotting using a layout, a plot scale of 1:1 is used for all plotting from Paper Space. When plotting from Model Space, you must enter the scale of the drawing as the plot scale. In Figure 10–20, click on the scale 1/8?=1'0? in the Plot scale area of the Plot dialog box. Continue on to the next step for making additional changes in this dialog box.

Figure 10–20

STEP 8

The Plot dialog box should appear similar to Figure 10–21. Notice, in the Plot scale area, the Custom mode displays 1 inches = 96 drawing units. This means that for every 96 units in the drawing (since it is full size), the plotter plots out 1 unit. Also, in the lower right corner under Plot options, you could check the box labeled Plot object lineweights to control the line quality of your drawing. This will be discussed in the next section of this chapter. Before plotting your drawing, it is considered good practice to click on the Full Preview... button to preview the appearance of your plot.

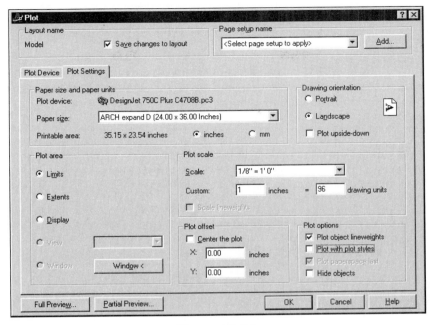

Figure 10–21

STEP 9

The previewed plot is displayed in Figure 10–22. If the results are favorable, click on Plot in the cursor menu to plot the drawing out or click on Exit to return to the Plot dialog box.

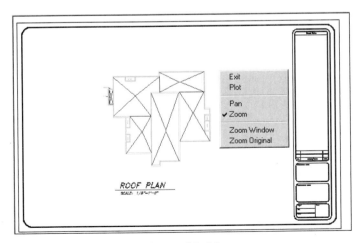

Figure 10–22

ENHANCING YOUR PLOTS WITH LINEWEIGHTS

This section on plotting will describe the process of assigning lineweights to objects and then having the lineweights appear in the finished plot. Open the drawing called 10_V_Step.Dwg in Figure 10–23 and notice you are currently in Model Space (the Model tab is current at the bottom of the screen). Follow the next series of steps to assign lineweights to a drawing before it is plotted out.

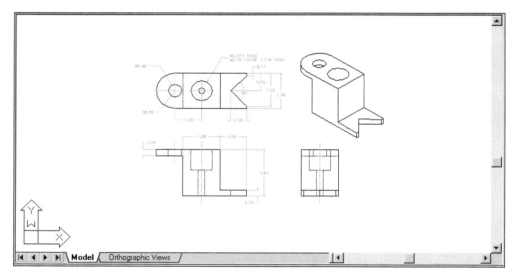

Figure 10–23

STEP 1

From the image of the drawing in Figure 10–23, all lines on the Object layer need to be assigned a lineweight of 0.60 mm. All objects on the Hidden layer need to be assigned a lineweight of 0.30 mm. There is also a border and title block that will be used with this drawing. The Title Block layer needs a lineweight assignment of 0.80 mm. Click on the Layer Properties Manager dialog box in Figure 10–24 and make these lineweight assignments.

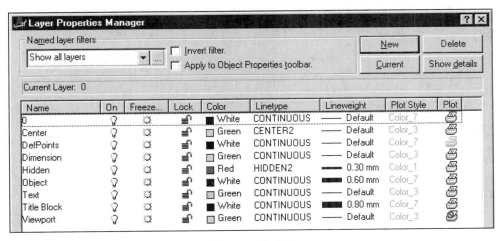

Figure 10–24

STEP 2

Click on the LWT button in the Status bar to display the lineweights in Figure 10–25A. Notice however that all lineweights appear extremely thick. You can control the lineweight scale to have the linetypes give off a more pleasing appearance inside of your drawing.

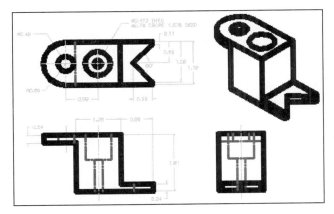

Figure 10–25A

Clicking on Lineweight in the Format pulldown menu in Figure 10–25B displays the Lineweight Settings dialog box in Figure 10–25C. Notice the slider bar in the area called Adjust Display Scale. Use this slider bar to reduce or increase the scale of the lineweights when being viewed in your drawing. In Figure 10–25C, the slider bar has been adjusted towards the minimum side of the scale. This should make the lineweights better to read in the drawing. Click the OK button to return to the drawing and observe the results.

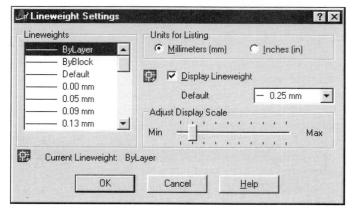

Figure 10–25B   **Figure 10–25C**

All hidden and objects lines have been reduced in scale in Figure 10–25D. This action adds to the clarity and appearance of the drawing. If after making changes to the lineweights you decide to increase the scale, return to the Lineweight Settings dialog box and readjust the slider bar until you achieve the desired results.

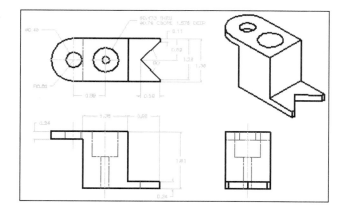

Figure 10–25D

STEP 3

Clicking on the Orthographic Views tab switches you to Paper Space in Figure 10–26. The lineweights will not appear in Paper Space at first glance. Zooming in to the drawing will display all lineweights at their proper widths.

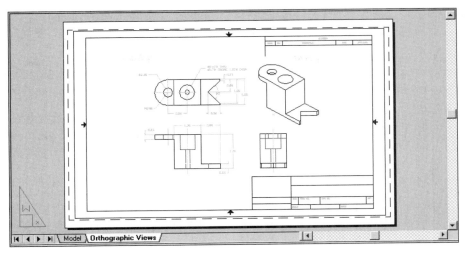

Figure 10–26

STEP 4

Activate the Plot dialog box and make sure you are in the Plot Settings tab as shown in Figure 10–27. Verify that the current plot device is the HP DesignJet 750C Plus. If this is not the device, click on the Plot Device tab to activate this plotter. In the Plot options area of the main Plot dialog box, be sure to check the Plot object lineweights box. This will ensure that the linetypes will plot out. If this box appears to be inactive, remove the check from the Plot with plot styles area.

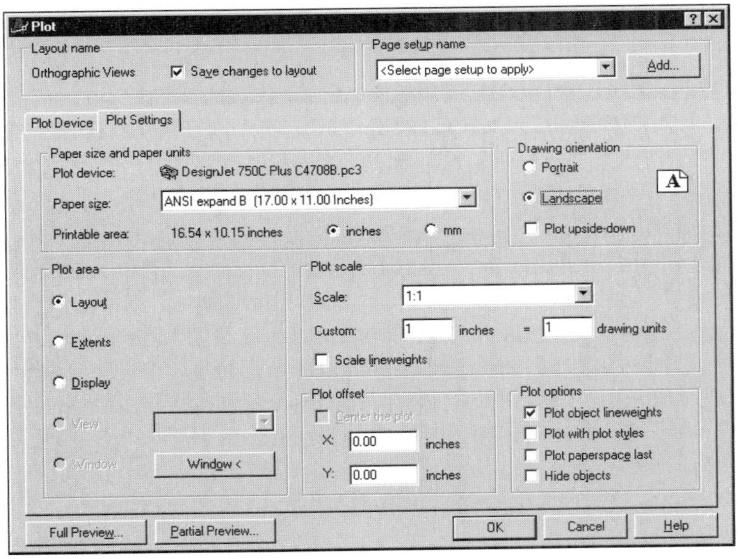

Figure 10–27

STEP 5

Click on the Full Preview… button; your display should appear similar to Figure 10–28. Notice that the viewport is not present in the plot preview. The Viewports layer was either turned off or set to a non-plot state inside the Layer Properties Manager dialog box. To view the lineweights in Preview mode, zoom in to segments of your drawing.

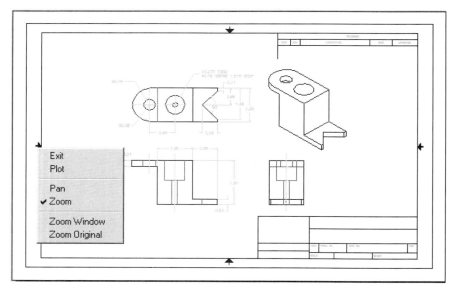

Figure 10–28

MODIFYING THE PROPERTIES OF A PLOTTER CONFIGURATION–CHANGING THE PAPER MARGINS

The illustration in Figure 10–29A shows a layout designed for the ANSI A paper dimensions of 11 x 8.5 units. The dashed margins indicate the printable area. The standard ANSI A title block is also inserted around an existing viewport. However, when the plot of the drawing is previewed, a different image is illustrated in Figure 10–29B. Here we see the vertical sides of the title block intact; however, the horizontal top and bottom edges seem to have been clipped off. After inspecting the image back in Figure 10–29A, we see that the standard title block is outside the top and bottom paper margins. This is the reason for the clipping. Unfortunately, if you follow through with the plot, the results will be similar to Figure 10–29B. It is possible to change these margins by modifying the properties of the current plotter configuration. Follow the next series of steps for performing this operation.

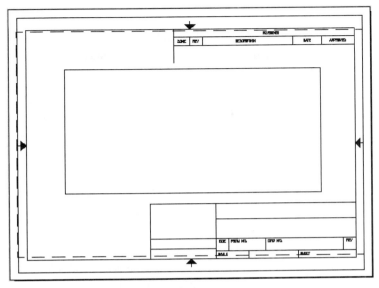

Figure 10–29A

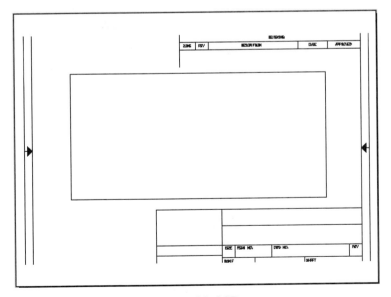

Figure 10–29B

STEP 1

First activate the Plot dialog box in Figure 10–30 and verify that the plotter is the DesignJet 750C Plus plotter (this plotter has been used in all plotting tutorials throughout this chapter). Then click on the Properties button.

Figure 10–30

STEP 2

Clicking on the Properties button activates the Plotter Configuration Editor for the current plot device in Figure 10–31. There are General and Ports tabs that deal with changing the plotter communication port. The Device and Document Settings tab will be the focus of this section of the chapter. Find the item in the display box that deals with the User-defined Paper Sizes & Calibration. Then find the section on Modify Standard Paper Sizes (Printable Area). Clicking on this section displays all standard paper sizes. Scroll down, as in Figure 10–31, until you find the ANSI A (11 x 8.50 Inches) paper size and click on it. Then click the Modify button.

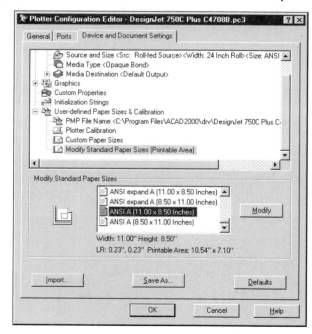

Figure 10–31

STEP 3

Clicking the Modify button takes you to the Custom Paper Size – Printable Area dialog box in Figure 10–32A. Notice the numbers that define the margins of the paper in addition to a thumbnail sketch of the paper size in the Preview box. Change the Top and Bottom margins to the values illustrated in Figure 10–32B. Notice also the sketch in the Preview box updating as these changes are made. When this operation is completed, click on the Next> button.

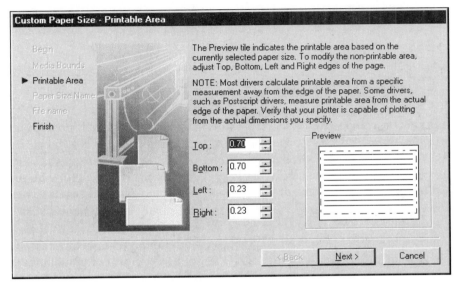

Figure 10–32A

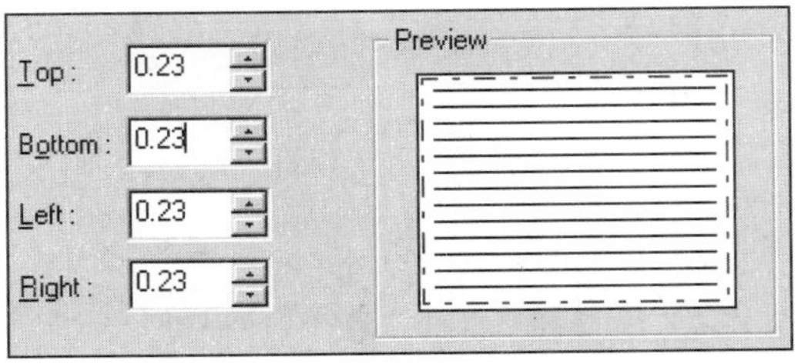

Figure 10–32B

STEP 4

Clicking on the Next> button in the previous step takes you to the Custom Paper Size – Finish dialog box in Figure 10–33. Here you select the source of paper: Roll-fed or Sheet-fed. Click on the Sheet-fed option and then click the Finish button.

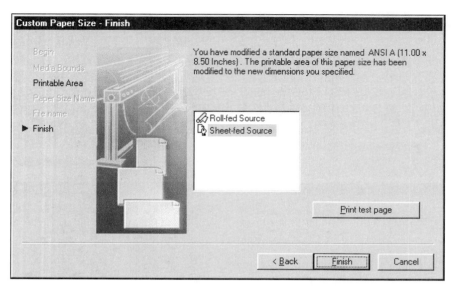

Figure 10–33

STEP 5

Once back in the main Plotter Configuration Editor for the DesignJet 750C Plus printer, click on the Save As... button. This takes you to the Save As dialog box, shown in Figure 10–34, which is designed to save the changes in the printable area to a PC3 file. Click on the Save button and, when the alert box appears telling you this file exists and asking you if you want to replace it, click the OK button. When you lay out the ANSI A sheet size 11 x 8.50 units and the standard ANSI A title block is inserted, the results are more favorable—they are illustrated in Figure 10–35.

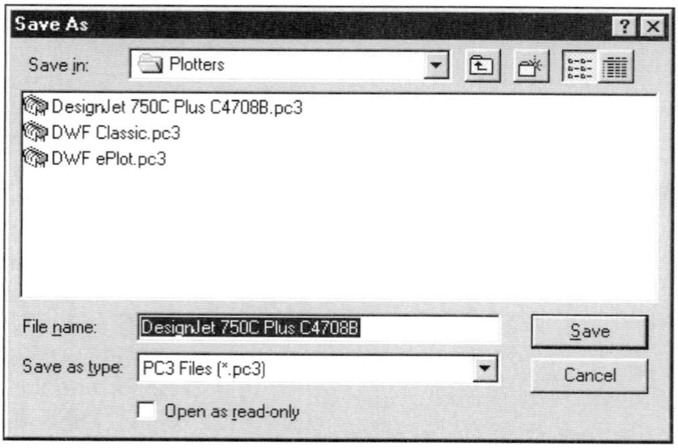

Figure 10–34

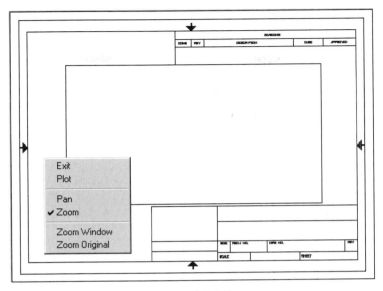

Figure 10–35

THE OPTIONS DIALOG BOX–USING THE PLOTTING TAB

Additional plotting controls can be set through the Plotting tab of the Options dialog box. To call up this dialog box, first move your cursor into the Command line area and right-click your mouse button. This displays the cursor menu in Figure 10–36A. Clicking on Options… displays the Options dialog box in Figure 10–36B. Click on the Plotting tab because we will be concentrating on this area at this time. (You can also launch the Options dialog box by choosing Options from the Tools pull-down menu.)

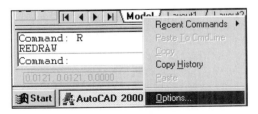

Figure 10–36A

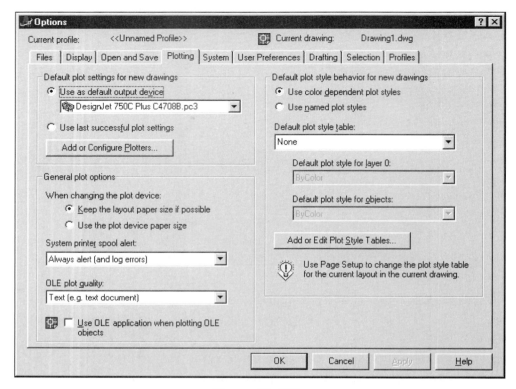

Figure 10–36B

The Default plot settings for new drawings area of the Plotting tab, shown in Figure 10–37, allows you to choose a plotter and have it serve as the default output device whenever you set up a page layout or activating the Plot dialog box. A button is also available to allow you to conveniently add or configure additional plotters.

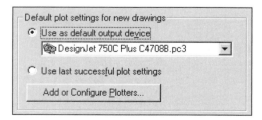

Figure 10–37

The Default plot style behavior for new drawings area of the Plotting tab, shown in Figure 10–38A, displays two types of plot styles to use with your drawings and lists available plot style table, which are device-independent files that control the appearance of color, linetypes, and lineweights. You can also apply additional controls such as screening, grayscale, dithering, end and joint styles, and fill patterns to your plots. Plot styles are attached to your drawing. You can also apply a different plot style to individual page layouts. In this way, you can produce different plots of the same drawing.

Two types of plot styles are listed in Figure 10–38A; the first is a Color-Dependent Plot Style. This type of plot style determines how your drawing plots based on color. You have 255 colors to work with when using this type of table. Plot characteristics such as linetype and lineweight are assigned to color, which determines how the drawing will plot. Color-dependent plot styles have the file extension .CTB. The plot styles listed in the figure are the default plot style tables that come with AutoCAD 2000.

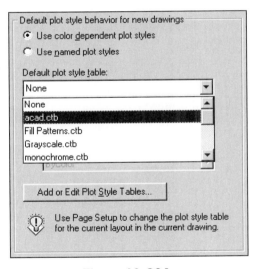

Figure 10–38A

The second type of plot style table is illustrated in Figure 10–38B; this is the named plot style table. This type of plot style does not depend on color. Rather, it is a list of names that you create. Inside each name is a list of plot characteristics such as color, linetype, and lineweight. The name is commonly referred to as a plot style. What makes this type of plot style different from the color-dependent variety is your ability to assign different plot style names to objects with the same colors. This provides a very powerful way of plotting your drawings. Named plot styles have the file extension .STB. Both plot style types will be demonstrated later in this chapter.

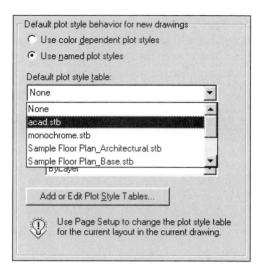

Figure 10–38B

Figure 10–39A displays the upper corner of your display screen. This is the Plot Style Control box used for changing named plot styles. However, if you are currently set to a color-dependent plot style, this area is grayed out in the figure.

Figure 10–39A

Clicking on the Use named plot styles radio button back in the Plotting tab of the Options dialog box in Figure 10–38B activates the Plot Style Control in Figure 10–39B. This area will activate only for new drawing files.

Figure 10–39B

Displaying the Layer Properties Manager dialog box in Figure 10–39C also shows the Plot Style area active when you use named plot styles. Color-dependent plot styles will have this layer area grayed out.

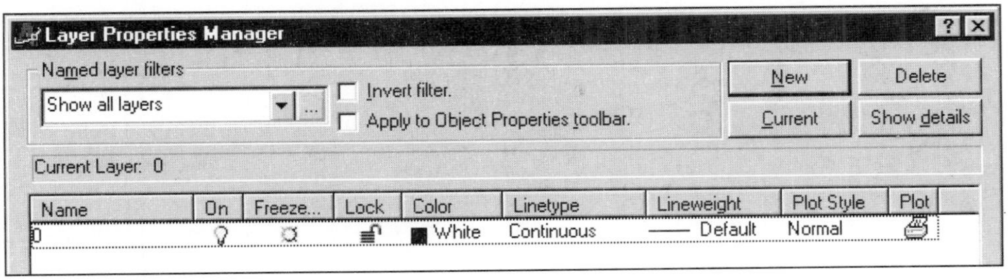

Figure 10–39C

CREATING A COLOR-DEPENDENT PLOT STYLE TABLE

This section of the chapter is devoted to the creation of a color-dependent plot style table. Once the table is created, it will be applied to a drawing. From there, the drawing will be previewed to see how this type of plot style table affects the final plot. Open the drawing 10_Color_R-Guide.Dwg. Your display should appear similar to the image in Figure 10–40. A two-view drawing together with an isometric view is arranged in a layout called "Orthographic Views." The drawing is also organized by layer names and color assignments. The object is to create a color-dependent plot style table where all layers will plot out black. Also, through the color-dependent plot style table, the hidden lines will be assigned a lineweight of 0.30 mm, object lines 0.70 mm, and the title block 0.80 mm. Follow the next series of steps to perform this task

Color	Layer
Red	Hidden
Yellow	Center
Green	Viewports
Cyan	Text
Blue	Title Block
Magenta	Dimensions
Black	Object

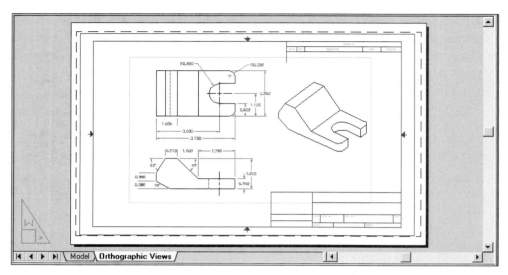

Figure 10–40

STEP 1

Begin the process of creating a color-dependent plot style table by choosing Plot Style Manager from the File pull-down menu as in Figure 10–41A. This activates the Plot Styles dialog box in Figure 10–41B. Various color-dependent and named plot styles already exist in this dialog box. To create a new plot style, double-click on the Add-A-Plot Style Table Wizard.

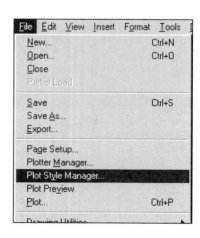

Figure 10–41A

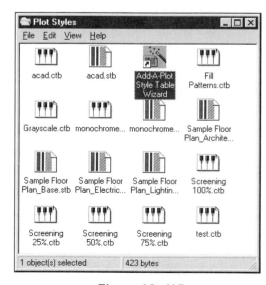

Figure 10–41B

STEP 2

The Add Plot Style Table dialog box appears as in Figure 10–42 and introduces you to the process of creating plot style tables. Plot styles contain plot definitions for color, lineweight, linetype, end capping, fill patterns, and screening. You will be presented with various choices in creating a plot style from scratch, using the parameters in an existing plot style, or importing pen assignments information from a PCP, PC2, or CFG file. You will also have the choice of saving this plot style information in a CTB (color-dependent) or STB (named) plot style. Click the Next> button to display the next dialog box.

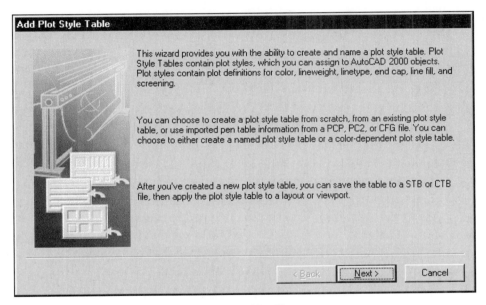

Figure 10–42

STEP 3

In the Add Plot Style Table – Begin dialog box in Figure 10–43, four options are available for you to choose, depending how you want to create the plot style table. Click on the Start from scratch radio button to create this plot style from scratch. If you have made pen assignments from previous releases of AutoCAD, you can import them through this dialog box. Click the Next> button to display the next dialog box.

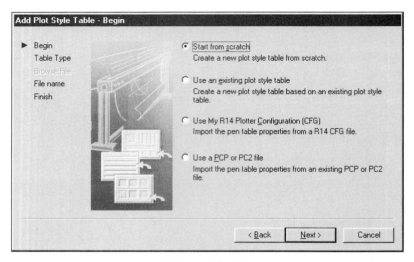

Figure 10–43

STEP 4

In the Add Plot Style Table – Pick Plot Style Table dialog box in Figure 10–44, click on the Color-Dependent Plot Style Table radio button to make this the type of plot style you will create. (The Named Plot Style Table will be discussed in the next tutorial exercise.) Click the Next> button to display the next dialog box.

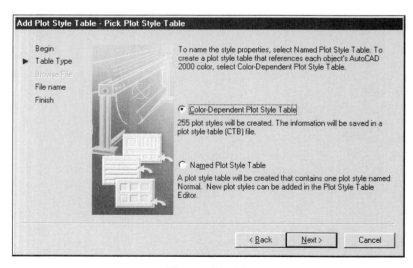

Figure 10–44

STEP 5

Use the Add Plot Style Table – File name dialog box in Figure 10–45 to assign a name to the plot style table. Enter the name Ortho_Drawings in the File name area. The extension CTB will automatically be added to this file name. Click the Next> button to display the next dialog box.

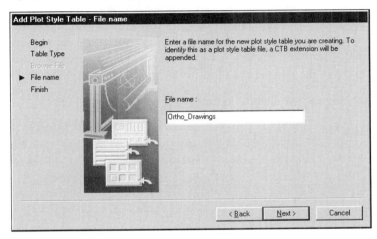

Figure 10–45

STEP 6

The Finish dialog box in Figure 10–46 alerts you that a plot style called Ortho_Drawings.ctb has been created. However, you want to have all colors plot out black and you need to assign different lineweights to a few of the layers. To accomplish this, click on the Plot Style Table Editor button.

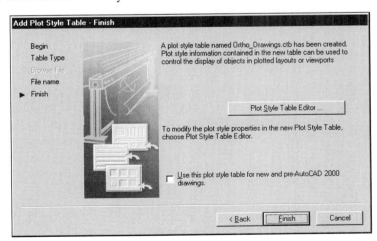

Figure 10–46

STEP 7

Clicking on the Plot Style Table Editor button displays the Plot Style Table Editor dialog box in Figure 10–47. Notice the name of the plot style table present at the top of the dialog box. Also, three tabs are available for making changes to the current plot style table (Ortho-Drawings.ctb). The first tab is General and displays file information about the current plot style table being edited. It is considered good practice to add a description to further document the purpose of this plot style table. It must be pointed out at this time that this plot style table is not going to be used only on the current drawing. Rather, if layers are standard across projects, the same plot style dialog box can be used. This is typical information that can be entered in the Description area.

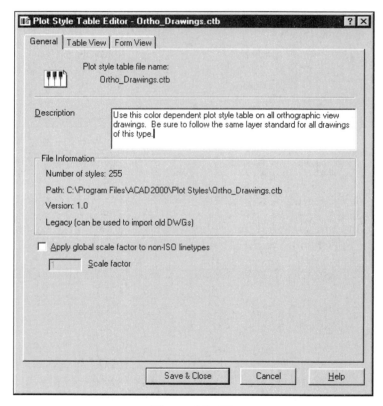

Figure 10–47

STEP 8

Clicking on the Table View tab activates the image in Figure 10–48. You can use the horizontal scroll bar to get a listing of all 255 colors along with special properties that can be changed. This information is presented in a spreadsheet format. You can click in any of the categories under a specific color and make changes, which will be applied to the current plot style table. The color and lineweight changes will be made through the next tab.

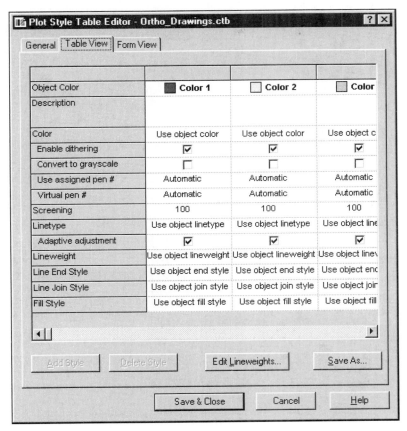

Figure 10–48

STEP 9

Clicking on the Form View tab displays the image in Figure 10–49. Here the colors are arranged vertically with the properties displayed on the right. Click on Color 1 (Red) and change the Color property to Black. Whatever is red in your drawing will plot out in the color black. Since red is used to identify hidden lines, click in the Lineweight area and set the lineweight for all red lines to 0.30 mm.

For Color 2 (Yellow), Color 3 (Green), Color 4 (Cyan), and Color 6 (Magenta), change the color to Black in the Properties area. All colors can be selected at one time by holding down CTRL while picking each of them. Changes can then be made to all colors simultaneously. Whatever is yellow, green, cyan, and magenta will plot out in the color black. No other changes need to be made in the dialog box for these colors.

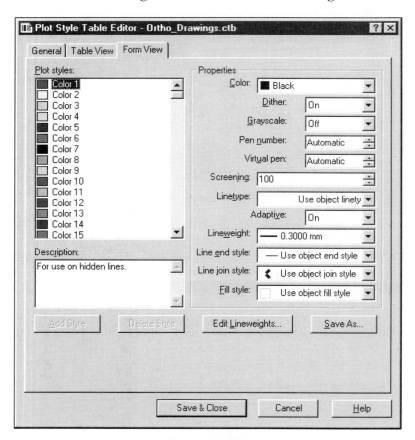

Figure 10–49

STEP 10

Click on Color 5 (Blue) in Figure 10–50 and change the Color property to Black. Whatever is blue in your drawing will plot out in the color black. Since blue is used to identify the title block lines, click in the Lineweight area and set the lineweight for all blue lines to 0.80 mm.

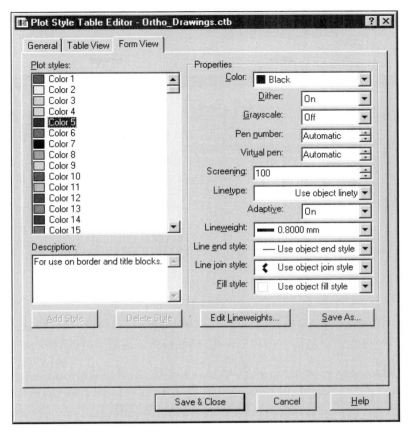

Figure 10–50

STEP 11

Click on Color 7 (Black) in Figure 10–51 and change the Color property to Black. Since black is used to identify the object lines, click in the Lineweight area and set the lineweight for all object lines to 0.70 mm. This completes the editing process of the current plot style table. Click on the Save & Close button; this returns you to the Finish dialog box. Clicking on the Finish button will return you to the Plot Styles dialog box. Close this box to display your drawing.

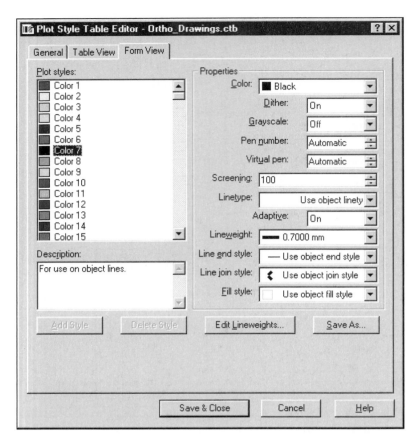

Figure 10–51

STEP 12

Activate the Plot dialog box in Figure 10–52 and, in the Plot device tab, verify that the plotter is the DesignJet 750C Plus plotter (this plotter has been used in all plotting tutorials throughout this chapter). In the Plot style table (pen assignments) area, make the current plot style Ortho_Drawings.ctb. Notice at the top of the dialog box that the plot style table will be saved to this layout. This means if you need to plot this drawing again in the future, you will not have to look for the desired plot style table.

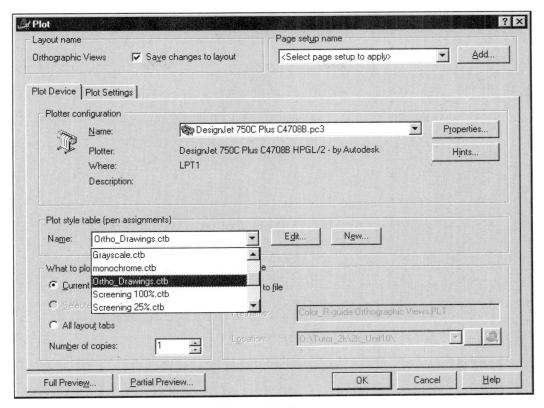

Figure 10–52

STEP 13

Click on the Plot Settings tab in Figure 10–53 and verify in the lower right corner of the dialog box under Plot options that you will be plotting with plot styles. Click on the Full Preview button to display the results. Unless a plotter is attached to your computer, it is not necessary to click the OK button.

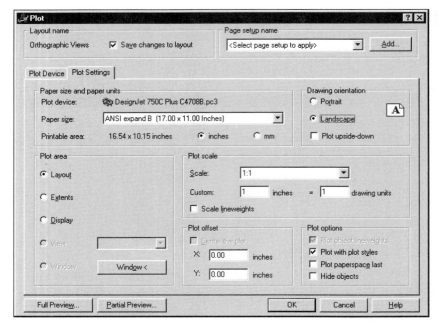

Figure 10–53

STEP 14

The results of performing a full preview are illustrated in Figure 10–54. Notice that all lines are black even though they appear in color in the drawing file. This is the result of using a color-dependent plot style table on this drawing. This file can also be attached to other drawings that share the same layer names.

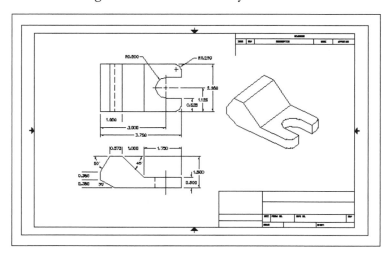

Figure 10–54

CREATING A NAMED PLOT STYLE TABLE

In the previous exercise, a color-dependent plot style table was created. This table is driven entirely by color defined in your drawing. What if you want to assign different lineweights to the same color in a drawing, and what if these colors are assigned to different layers? This is where you would create a named plot style table and attach the plot style to an individual layer. To see how this can be accomplished, open the drawing called 10_Named_R-Guide.Dwg in Figure 10–55. Notice that the entire drawing is red; yet if you examine the Layer Properties Manager dialog box, you will notice the drawing is organized in layers. We want all red objects to plot out black. We also want to assign thin, medium, thick, and heavy lineweights to certain layers as follows:

Layer	Plot Style
Hidden	Medium
Center	Thin
Viewports	Thin
Text	Medium
Title Block	Heavy
Dimensions	Thin
Object	Thick

Follow the next series of steps for creating a named plot style with the extension of STB.

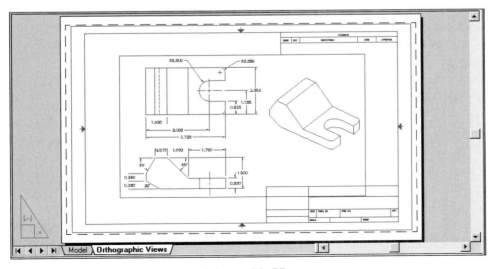

Figure 10–55

STEP 1

As with the creation of a color-dependent plot style, begin the process of creating a named plot style table by choosing Plot Style Manager from the File pull-down menu. When the Plot Styles dialog box displays, double-click on the Add-A-Plot Style Table Wizard (see Figure 10–41B). Click the Next> button in the Add Plot Style Table dialog box (see Figure 10–42), and click on the Start from scratch radio button in the Add Plot Style Table-Begin dialog box (see Figure 10–43). Clicking the Next> button takes you to the Pick Plot Style Table dialog box in Figure 10–56. Click on the Named Plot Style Table radio button and then click the Next> button to display the next dialog box.

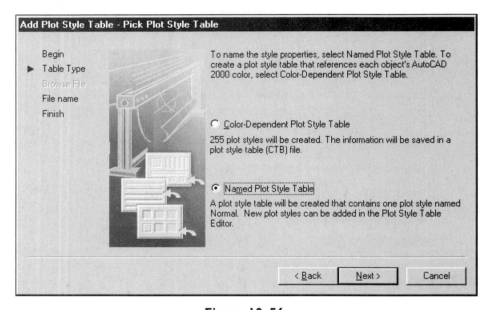

Figure 10–56

STEP 2

In the File name dialog box in Figure 10–57, enter the name "Ortho_Drawings" as the name of this plot style. Do not be concerned that this name is identical to the one just used for the color-dependent table. The named plot style extension is STB, which is quite different from the color-dependent extension of CTB. Click the Next> button to display the next dialog box.

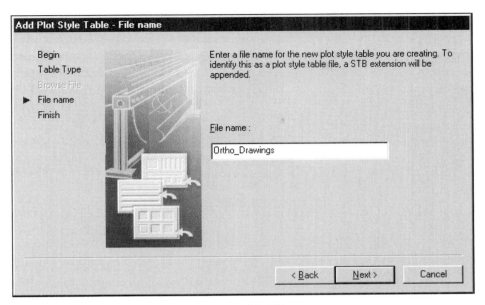

Figure 10–57

STEP 3

The Finish dialog box appears. Click on the Plot Style Table Editor button in Figure 10–58 to make changes to the named plot style properties.

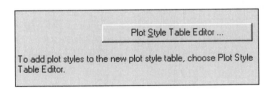

Figure 10–58

STEP 4

In the Plot Style Table Editor dialog box in Figure 10–59, notice the name of the plot style table you are editing (Ortho_Drawings.stb). Click on the Form View tab and notice the plot style "Normal" is present. By default, when you use a named plot style, "Normal" is always present. Four new plot styles need to be added to the Plot Styles list to handle thin, medium, thick, and heavy lines. We will use these same names when creating each style. Click on the Add Style button near the bottom of the dialog box.

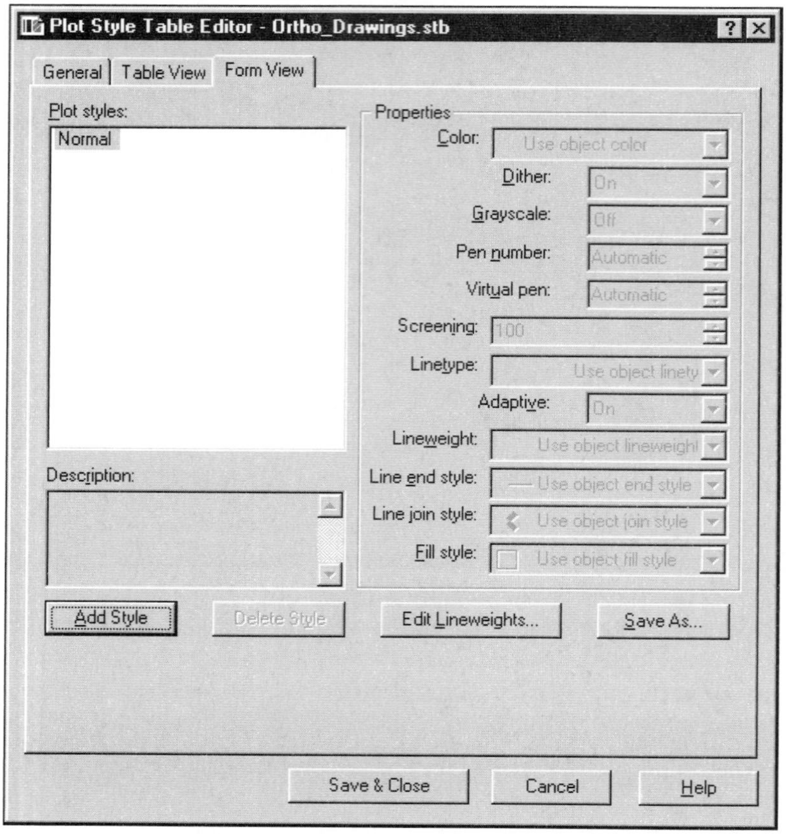

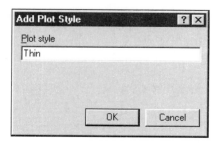

Figure 10–59

STEP 5

In the Add Plot Style dialog box in Figure 10–60, enter the name "Thin" as the new plot style name. Click the OK button to return to the Plot Style Table Editor.

Figure 10–60

STEP 6

Back in the Form View tab of the Plot Style Table Editor dialog box in Figure 10–61, change the color property to Black and the Lineweight to 0.10 mm. Notice that at the bottom of the dialog box in the Description area, you can add a note for the purpose of the "Thin" style. This will remind you that objects on the center, viewports, and dimensions layer will plot out with thin, black lines. Click on the Add Style button to create another plot style called "Medium."

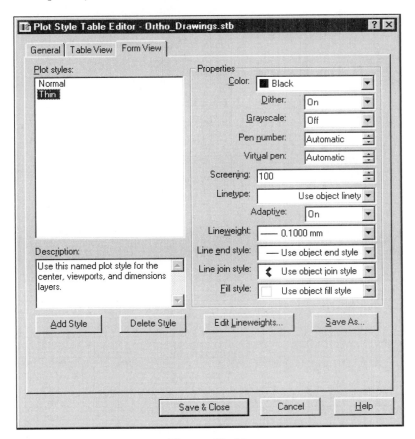

Figure 10–61

STEP 7

Back in the Form View tab in Figure 10–62, change the color property to Black and the Lineweight to 0.30 mm for the Medium plot style. Add a note in the Description area to remind you that objects on the hidden and text layers will plot out with medium, black lines. Click on the Add Style button to create another plot style called "Thick."

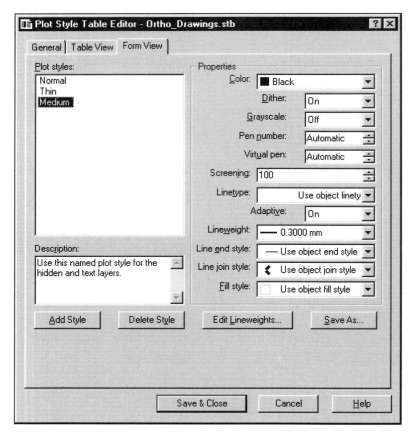

Figure 10–62

STEP 8

Back in the Form View tab in Figure 10–63, change the color property to Black and the Lineweight to 0.70 mm for the Thick plot style. Add a note in the Description area to remind you that the object layer will plot out with thick, black lines. Click on the Add Style button to create the final plot style called "Heavy."

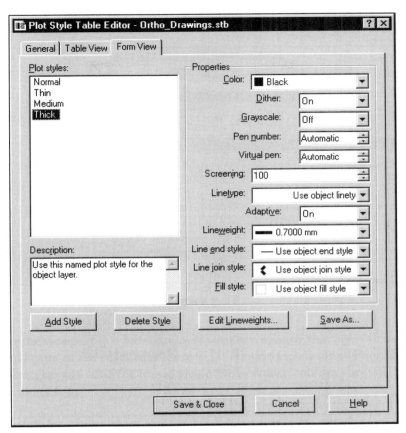

Figure 10–63

STEP 9

Back in the Form View tab in Figure 10–64, change the color property to Black and the Lineweight to 0.80 mm for the Heavy plot style. Add a note in the Description area to remind you that the title block layer will plot out with heavy, black lines. Click on the Save & Close button at the bottom of the dialog box to save these changes to the Ortho_Drawings.stb file. When the Finish dialog box displays, click on the Finish button, close the Plot Styles dialog box and return to your drawing.

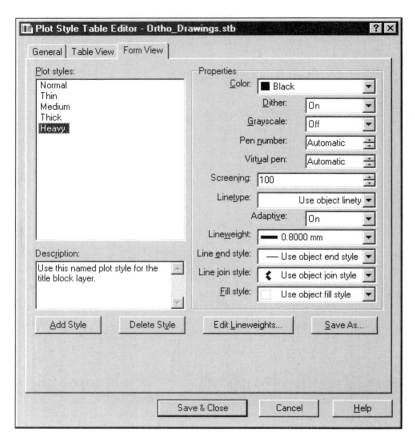

Figure 10–64

STEP 10

Before you plot out this drawing, a few more actions need to be performed in preparation for the plot. First, activate the Page Setup dialog box in Figure 10–65 and change the new current plot style to Ortho_Drawings.stb in the Plot style table (pen assignments) area. Also, in this same area, be sure a check is placed in the Display plot styles to view the results when previewing the plot. When finished, click the OK button, which will return you to your drawing.

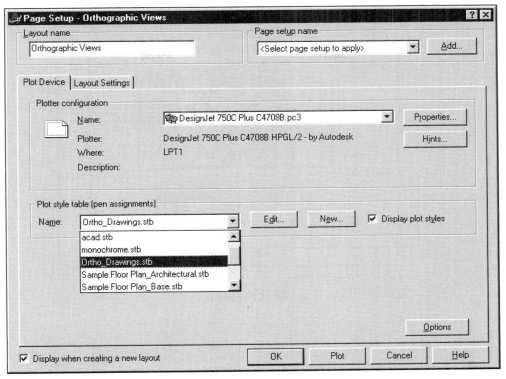

Figure 10–65

STEP 11

The purpose of the named plot style is to assign the plot style to objects and not to rely on colors. To illustrate this, click on the Layer Properties Manager dialog box. As shown in Figure 10–66, click on all three layers identified as Center, Dimension, and Viewports. Click on the name "Normal" in the Plot Style heading of any of these three layers. The Select Plot Style dialog box appears with all defined plot styles that belong to the plot style table "Ortho_Drawings.stb". Click on "Thin" and then the OK button. This will change the "Normal" plot style of these three layers to "Thin."

630

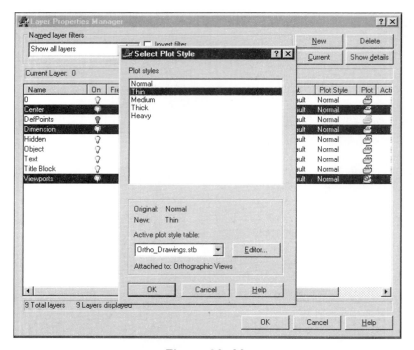

Figure 10–66

STEP 12

Repeat this procedure to change the plot style of the Hidden and Text layers to "Medium," the Object layer to "Thick" and the Title Block layer to "Heavy." You could click on the Viewports and pick the printer icon under Plot. This will make the viewport layer visible but not included in the final plot of the drawing. Your display should appear similar to Figure 10–67.

Figure 10–67

STEP 13

Activate the Plot dialog box in Figure 10–68. Verify that the current plot device is the DesignJet 750C Plus and that the current plot style table name is Ortho_Drawings.stb. Click on the Full Preview Button.

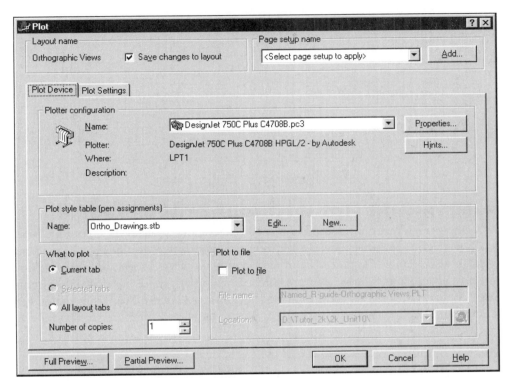

Figure 10–68

STEP 14

The results of performing a full preview are illustrated in Figure 10–69. Notice that all lines are black even though they appear in color in the drawing file. This is the result of using a named plot style table on this drawing. This file can also be attached to other drawings that share the same layer names.

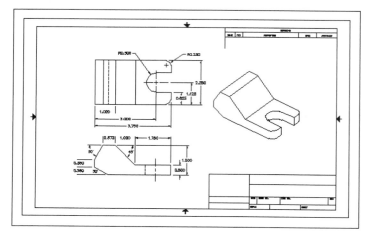

Figure 10–69

MODIFYING THE PLOT STYLE PROPERTIES OF OBJECTS

Named plot styles are considered an object property along with layers, colors, linetypes, and lineweights, to name a few. In addition to assigning a named plot style to a layer, you can also change the plot style of an individual object. In Figure 10–70, various line and arc segments have been pre-selected. As usual, these objects highlight and grips appear. Clicking on the Properties button activates the Properties dialog box. Clicking in the Plot style area displays the plot styles currently defined in the drawing. Click on "Other…" in Figure 10–70.

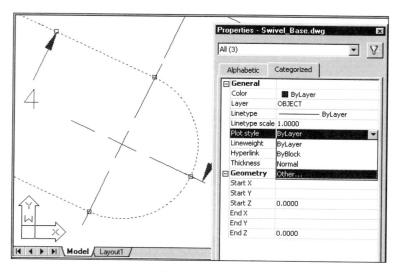

Figure 10–70

This displays the Select Plot Style dialog box in Figure 10–71. Clicking on the desired plot style name will change all highlighted objects to that plot style. In the figure, the plot style "Thick" is selected. Clicking the OK button in both this dialog box and the Properties dialog box will perform this operation.

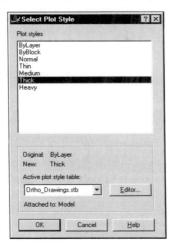

Figure 10–71

Yet another method of changing the plot style of objects is illustrated in Figure 10–72, where a series of dimensions has been pre-selected. Clicking in the Plot Style Control box in the figure displays all valid plot styles. Click on "Thin" to apply this plot style to the highlighted objects. If the desired style is not present in this list, click on "Other..." to display the Select Plot Style dialog box and pick the plot style from this list.

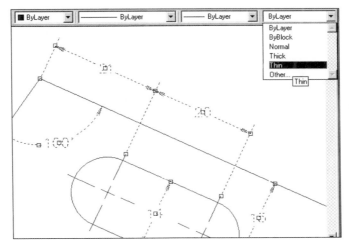

Figure 10–72

ELECTRONIC PLOTTING TO A DWF FILE

AutoCAD 2000 drawings can be viewed over the Internet after an electronic plot of the drawing is performed. You must first select the proper plotter configuration, which is in the form of an ePlot device, as shown in Figure 10–73. This also activates the Plot to file area, where the name of the file is listed with the DWF extension. You can designate the exact location for this file. You also can attach a plot style table and even click on the Plot Settings tab to change other plotter settings.

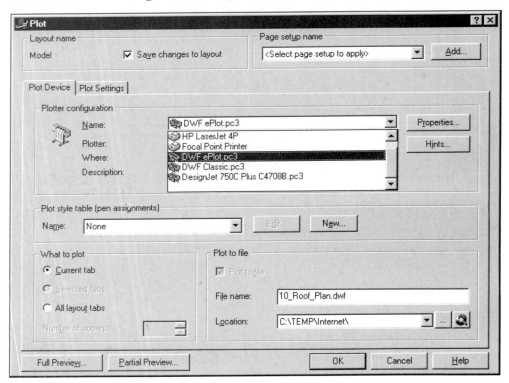

Figure 10–73

When the electronic plotting utility creates the file, it can now be viewed over the Internet through either the Internet Explorer or Netscape Navigator. The next example illustrates viewing the DWF file using Internet Explorer. First choose Open from the File pull-down menu as in Figure 10–74A. Search for the file in the Open dialog box in Figure 10–74B.

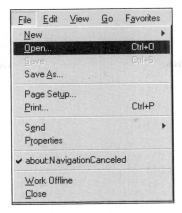

Figure 10–74A

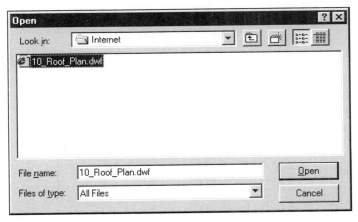

Figure 10–74B

Once the file displays in the Internet Explorer, the display screen appears similar to Figure 10–75.

636

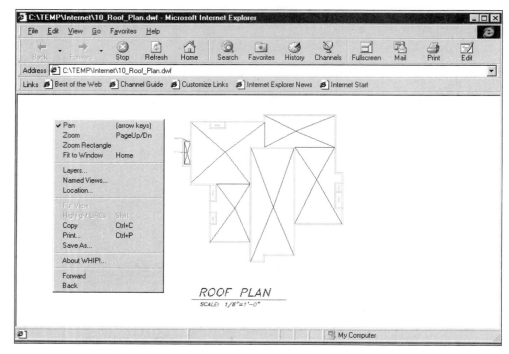

Figure 10–75

CHAPTER 11

Dimensioning Basics

DIMENSIONING BASICS

Once orthographic views have been laid out, a design is not ready for the production line until numbers describing the width, height and depth of the object are added to the drawing. However, these numbers must be added in a certain organized fashion; otherwise, the drawing becomes difficult to read. That may lead to confusion and the possible production of a part that is incorrect according to the original design. Before discussing the components of a dimension, remember that object lines (at "A") continue to be the thickest lines of a drawing, with the exception of border or title blocks (see Figure 11–1). To promote contrasting lines, dimensions become visible, yet thin, lines. The heart of a dimension is the dimension line (at "B"), which is easily identified by the location of arrow terminators at both ends (at "D"). In mechanical cases, the dimension line is broken in the middle, which provides an excellent location for the dimension text (at "E"). For architectural applications, dimension text is usually placed above an unbroken dimension line. The extremities of the dimension lines are limited by the placement of lines that act as stops for the arrow terminators. These lines, called extension lines (at "C"), begin close to the object without touching the object. Extension lines will be highlighted in greater detail in the pages that follow. For placing diameter and radius dimensions, a leader line consisting of an inclined line with a short horizontal shoulder is used (at "F"). Other applications of leader lines are for adding notes to drawings.

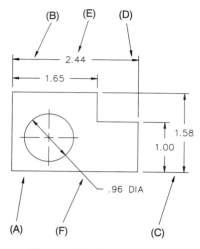

Figure 11–1

When you place dimensions in a drawing, provide a spacing of at least 0.38 units between the first dimension line and the object being dimensioned (at "A" in Figure 11–2). If you're placing stacked or baseline dimensions, provide a minimum spacing of at least 0.25 units between the first and second dimension line at "B" or any other dimension lines placed thereafter. This will prevent dimensions from being placed too close to each other.

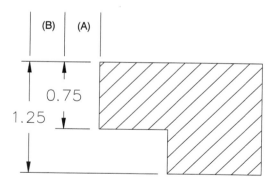

Figure 11–2

It is recommended that extensions never touch the object being dimensioned and that they begin approximately between 0.03 and 0.06 units away from the object at "A" (see Figure 11–3). As dimension lines are added, extension lines should extend no further than 0.12 beyond the arrow or any other terminator (at "B"). The height of dimension text is usually 0.125 units (at "C"). This value also applies to notes placed on objects with leader lines. Certain standards may require a taller lettering height. Become familiar with your office's standard practices, which may deviate from these recommended values.

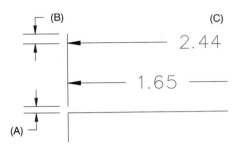

Figure 11–3

PLACEMENT OF DIMENSIONS

When placing multiple dimensions on one side of an object, place the shorter dimension closest to the object followed by the next larger dimension, as shown in Figure 11–4. When you are placing multiple horizontal and vertical dimensions involving extension lines that cross other extension lines, do not place gaps in the extension lines at their intersection points.

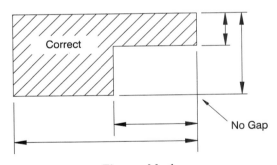

Figure 11–4

It is acceptable for extension lines to cross each other, but it is considered unacceptable practice for extension lines to cross dimension lines, as in the example in Figure 11–5. The shorter dimension is placed closest to the object followed by the next larger dimension.

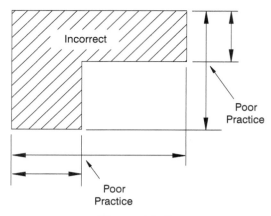

Figure 11–5

It is considered poor practice to dimension inside an object when there is sufficient room to place dimensions on the outside. There may be exceptions to this rule, however. It is also considered poor practice to cross dimension lines, because this may render the drawing confusing and possibly result in the inaccurate interpretation of the drawing. See Figure 11–6.

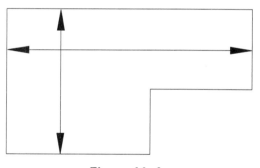

Figure 11–6

PLACING EXTENSION LINES

Because two extension lines may intersect without providing a gap, so also may extension lines and object lines intersect with each other without the need for a gap in between them. This is the same rule practiced when centerlines are used that extend beyond the object without gapping, as illustrated in Figure 11–7.

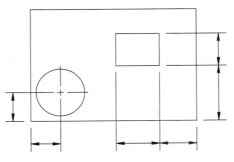

Figure 11–7

In Figure 11–8, the gaps in the extension lines may appear acceptable; however, in a very complex drawing, gaps in extension lines would render a drawing confusing. Draw extension lines as continuous lines without providing breaks in the lines.

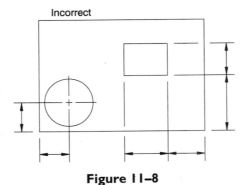

Figure 11–8

In the same manner, when centerlines are used as extension lines for dimension purposes, no gap is provided at the intersection of the centerline and the object (as in Figure 11–9). As with extension lines, the centerline should extend no further than 0.125 units past the arrow terminator.

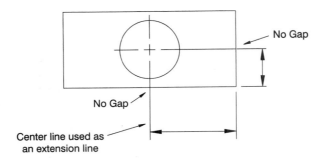

Figure 11–9

GROUPING DIMENSIONS

To promote ease of reading and interpretation, group dimensions whenever possible, as shown in Figure 11–10. In addition to making the drawing and dimensions easier to read, this promotes good organizational skills and techniques. As in previous examples, always place the shorter dimensions closest to the drawing followed by any larger or overall dimensions.

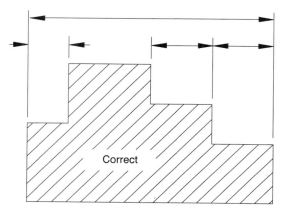

Figure 11–10

Avoid placing dimensions to an object line substituting for an extension line, as in Figure 11–11. The drawing is more difficult to follow, with the dimensions placed at different levels instead of being grouped. In some cases, however, even this practice of dimensioning is unavoidable.

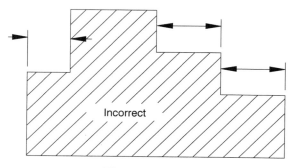

Figure 11–11

For tight spaces, arrange dimensions as in the illustration in Figure 11–12. Take extra care to follow proper dimension rules without sacrificing clarity.

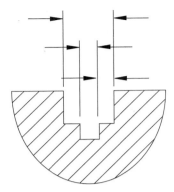

Figure 11–12

DIMENSIONING TO VISIBLE FEATURES

In Figure 11–13, the object is dimensioned correctly; however, the problem is that hidden lines are used to dimension to. Although there are always exceptions, try to avoid dimensioning to any hidden surfaces or features.

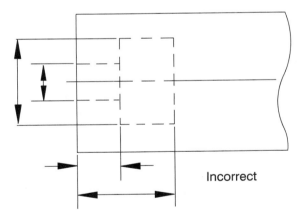

Incorrect

Figure 11–13

The object illustrated in Figure 11–14 is almost identical to that in Figure 11–13 except that it has been converted to a full section. Surfaces that were previously hidden are now exposed. This example illustrates a better way to dimension details that were previously invisible.

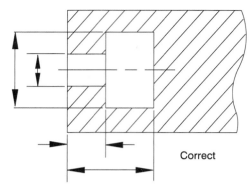

Figure 11–14

DIMENSIONING TO CENTERLINES

Centerlines are used to identify the center of circular features as in "A" in Figure 11–15. You can also use centerlines to indicate an axis of symmetry as in "B." Here, the centerline consisting of a short dash flanked by two long dashes signifies that the feature is circular in shape and form. Centerlines may take the place of extension lines when dimensions are placed in drawings.

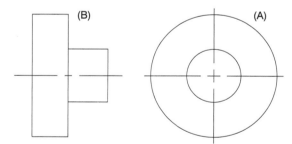

Figure 11–15

Figure 11–16 represents the top view of a U-shaped object with two holes placed at the base of the U. It also represents the correct way of utilizing centerlines as extension lines when dimensioning to holes. What makes this example correct is the rule of always dimensioning to visible features. The example in Figure 11–16 uses centerlines to dimension to holes that appear as circles. This is in direct contrast to Figure 11–17.

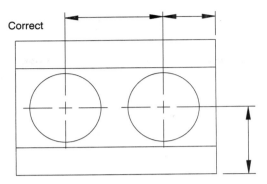

Correct

Figure 11–16

Figure 11–17 represents the front view of the U-shaped object. The hidden lines display the circular holes passing through the object along with centerlines. Centerlines are being used as extension lines for dimensioning purposes; however, it is considered poor practice to dimension to hidden features or surfaces. Always attempt to dimension to a view where the features are visible before dimensioning to hidden areas.

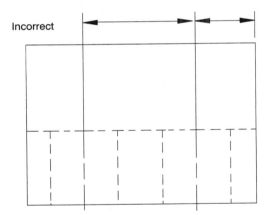

Incorrect

Figure 11–17

ARROWHEADS

Arrowheads are generally made three times as long as they are wide, or very long and narrow, as shown in Figure 11–18. The actual size of an arrowhead would measure approximately 0.125 units in length.

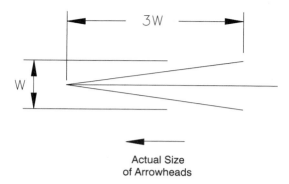

Figure 11–18

Dimension line terminators may take the form of shapes other than filled-in arrow-heads (see Figure 11–19). The 45° slash or "tick" is a favorite dimension line termina-tor used by architects, although they are sometimes seen in mechanical applications.

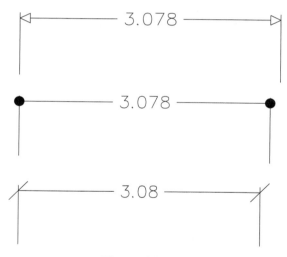

Figure 11–19

DIMENSIONING SYSTEMS

THE UNIDIRECTIONAL SYSTEM

A typical example of placing dimensions using AutoCAD is illustrated in Figure 11–20. Here, all text items are right-reading or horizontal. This applies to all vertical, aligned, angular, and diameter dimensions. When dimension text can be read from one direc-tion, this is the Unidirectional Dimensioning System. By default, all AutoCAD set-tings are set to dimension in the Unidirectional System.

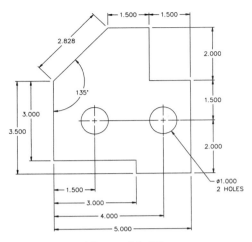

Figure 11–20

THE ALIGNED SYSTEM

The same object from the previous example is illustrated in Figure 11–21. Notice that all horizontal dimensions have the text positioned in the horizontal direction as in Figure 11–20. However, vertical and aligned dimension text is rotated or aligned with the direction being dimensioned. This is the most notable feature of the Aligned Dimensioning System. Text along vertical dimensions is rotated in such a way that the drawing must be read from the right. Angular dimensions remain unaffected in the Aligned System; however, aligned dimension text is rotated parallel with the feature being dimensioned.

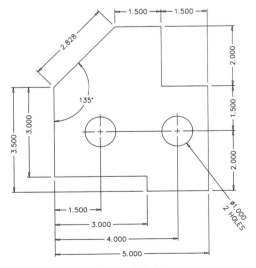

Figure 11–21

REPETITIVE DIMENSIONS

Throughout this chapter, numerous methods of dimensioning are discussed, supported by many examples. Just as it was important in multiview projection to draw only those views that accurately describe the object, so also is it important to dimension these views. In this manner, the actual production of the part may start. However, take care when placing dimensions; it takes planning to better place a dimension. The problem with the illustration in Figure 11–22 is that, even though the views are correct and the dimensions call out the overall sizes of the object, there are too many cases where dimensions are duplicated. Once a feature has been dimensioned, such as the overall width of 3.75 units, this number does not need to be placed in the Top view. This is the purpose of understanding the relationship between views or what dimensions the views have in common with each other. Adding unnecessary dimensions also makes the drawing very busy and cluttered in addition to being very confusing to read. Compare Figure 11–22 with Figure 11–23, which shows only those dimensions needed to describe the size of the object. Do not be concerned that the Top view contains no dimensions; the designer should interpret the width of the top as 3.75 units from the Front view and the depth as 2.50 from the Right Side view.

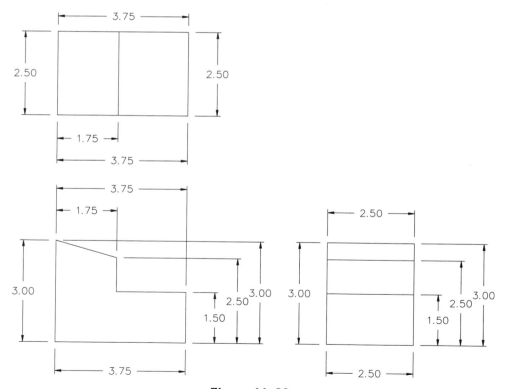

Figure 11–22

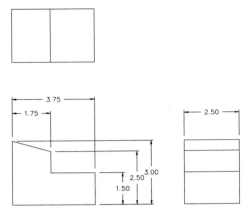

Figure 11–23

METHODS OF CHOOSING DIMENSION COMMANDS

THE DIMENSION TOOLBAR

The Dimension toolbar, illustrated in Figure 11–24A, makes the selection of dimensioning commands easier. Study this image and the corresponding commands associated with each button.

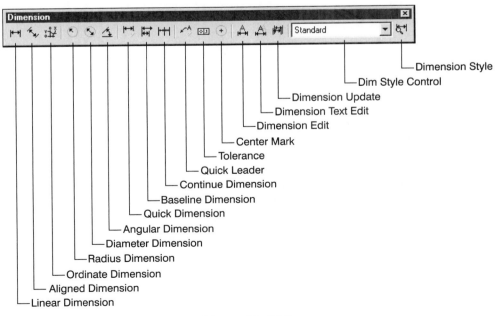

Figure 11–24A

THE DIMENSION PULL-DOWN MENU

The Dimension pull-down menu, illustrated in Figure 11–24B, is an additional area used for selecting the more commonly used dimension commands. Another way of activating dimension commands is through the keyboard. These commands tend to get long; the DIMLINEAR command is one example. To spare you the effort of entering the entire command, all dimension commands have been abbreviated to three letters. These commands are all displayed in the list of command aliases in Table 1–2. DLI is the shortcut for the DIMLINEAR command.

Figure 11–24B

BASIC DIMENSION COMMANDS

LINEAR DIMENSIONS

The Linear Dimensioning mode generates either a horizontal or vertical dimension depending on the location of the dimension. The following prompts illustrate the generation of a horizontal dimension with the DIMLINEAR command. Notice that identifying the dimension line location at "C" in Figure 11–25A automatically generates a horizontal dimension.

Command: **DLI** *(For DIMLINEAR)*
Specify first extension line origin or <select object>: *(Select the endpoint of the line at "A")*
Specify second extension line origin: *(Select the other endpoint of the line at "B")*

Specify dimension line location or [Mtext/Text/Angle/Horizontal/Vertical/Rotated]:
   (Pick a point near "C" to locate the dimension)
Dimension text = 2.00

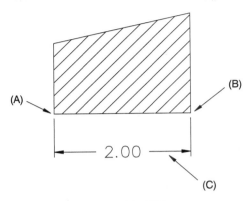

Figure 11–25A

The linear dimensioning command is also used to generate vertical dimensions. The following prompts illustrate the generation of a vertical dimension with the DIMLINEAR command. Notice that identifying the dimension line location at "C" in Figure 11–25B automatically generates a vertical dimension.

Command: **DLI** *(For DIMLINEAR)*
Specify first extension line origin or <select object>: *(Select the endpoint of the line at "A")*
Specify second extension line origin: *(Select the endpoint of the line at "B")*
Specify dimension line location or [Mtext/Text/Angle/Horizontal/Vertical/Rotated]:
   (Pick a point near "C" to locate the dimension)
Dimension text = 2.10

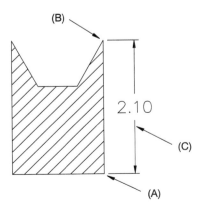

Figure 11–25B

Rather than select two separate endpoints to dimension to, certain situations allow you to press ENTER and select the object (in this case, the line). This selects the two endpoints and prompts you for the dimension location. The completed dimension is illustrated in Figure 11–26.

Command: **DLI** *(For DIMLINEAR)*
Specify first extension line origin or <select object>: *(Press* ENTER *to select an object)*
Select object to dimension: *(Select the line at "A")*
Specify dimension line location or [Mtext/Text/Angle/Horizontal/Vertical/Rotated]:
  (Pick a point near "B" to locate the dimension)
Dimension text = 5.5532

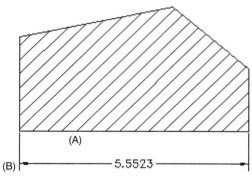

Figure 11–26

ALIGNED DIMENSIONS

The Aligned Dimensioning mode generates a dimension line parallel to the distance specified by the location of two extension line origins as shown in Figure 11–27. The following prompts illustrate the generation of an aligned dimension with the DIMALIGNED command:

Command: **DAL** *(For DIMALIGNED)*
Specify first extension line origin or <select object>: *(Select the endpoint of the line at "A")*
Specify second extension line origin: *(Select the endpoint of the line at "B")*
Specify dimension line location or [Mtext/Text/Angle]: *(Pick a point at "C" to locate the dimension)*
Dimension text = 5.6569

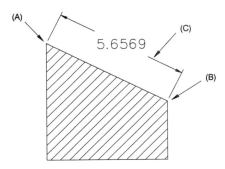

5.6569

Figure 11–27

ROTATED LINEAR DIMENSIONS

This Linear Dimensioning mode generates a dimension line, which is rotated at a specific angle (see Figure 11–28). The following prompts illustrate the generation of a rotated dimension with the DIMLINEAR command and a known angle of 45°. If you do not know the angle, you could easily establish the angle by clicking on the endpoints at "A" and "D" in Figure 11–28 when prompted to "Specify angle of dimension line <0>."

Command: **DLI** *(For DIMLINEAR)*
Specify first extension line origin or <select object>: *(Select the endpoint of the line at "A")*
Specify second extension line origin: *(Select the endpoint of the line at "B")*
Specify dimension line location or [Mtext/Text/Angle/Horizontal/Vertical/Rotated]: **R** *(For Rotated)*
Specify angle of dimension line <0>: **45**
Specify dimension line location or [Mtext/Text/Angle/Horizontal/Vertical/Rotated]: *(Pick a point at "C" to locate the dimension)*
Dimension text = 2.8284

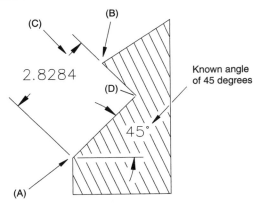

Figure 11–28

CONTINUE DIMENSIONS

The power of grouping dimensions for ease of reading has already been explained. Figure 11–29 shows yet another feature of dimensioning in AutoCAD: the practice of using Continue dimensions. With one dimension already placed, you issue the DIMCONTINUE command, which prompts you for the second extension line location. Picking the second extension line location strings the dimensions next to each other or continues the dimension.

Command: **DLI** *(For DIMLINEAR)*
Specify first extension line origin or <select object>: *(Select the endpoint of the line at "A")*
Specify second extension line origin: *(Select the endpoint of the line at "B")*
Specify dimension line location or [Mtext/Text/Angle/Horizontal/Vertical/Rotated]: *(Locate the 1.75 horizontal dimension)*
Dimension text = 1.75
Command: **DCO** *(For DIMCONTINUE)*
Specify a second extension line origin or [Undo/Select] <Select>: *(Select the endpoint of the line at "C")*
Dimension text = 1.25
Specify a second extension line origin or [Undo/Select] <Select>: *(Select the endpoint of the line at "D")*
Dimension text = 1.50
Specify a second extension line origin or [Undo/Select] <Select>: *(Select the endpoint of the line at "E")*
Dimension text = 1.00
Specify a second extension line origin or [Undo/Select] <Select>: *(Press ENTER when finished)*
Select continued dimension: *(Press ENTER to exit this command)*

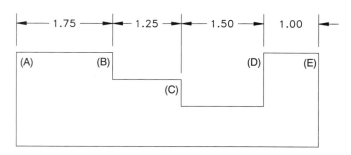

Figure 11–29

BASELINE DIMENSIONS

Yet another aid in grouping dimensions is the DIMBASELINE command. Continue dimensions place dimensions next to each other; Baseline dimensions establish a base or starting point for the first dimension, as shown in Figure 11–30. Any dimensions that follow in the DIMBASELINE command are calculated from the common base point already established. This is a very popular mode to use when one end of an object acts as a reference edge. When you place dimensions using the DIMBASELINE command, a dimension setting of 0.38 units controls the spacing of the dimensions from each other.

Command: **DLI** *(For DIMLINEAR)*
Specify first extension line origin or <select object>: *(Select the endpoint of the line at "A")*
Specify second extension line origin: *(Select the endpoint of the line at "B")*
Specify dimension line location or [Mtext/Text/Angle/Horizontal/Vertical/Rotated]: *(Locate the 1.75 horizontal dimension)*
Dimension text = 1.75
Command: **DBA** *(For DIMBASELINE)*
Specify a second extension line origin or [Undo/Select] <Select>: *(Select the endpoint of the line at "C")*
Dimension text = 3.00
Specify a second extension line origin or [Undo/Select] <Select>: *(Select the endpoint of the line at "D")*
Dimension text = 4.50
Specify a second extension line origin or [Undo/Select] <Select>: *(Select the endpoint of the line at "E")*
Dimension text = 5.50
Specify a second extension line origin or [Undo/Select] <Select>: *(Press ENTER when finished)*
Select base dimension: *(Press ENTER to exit this command)*

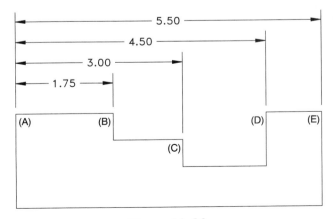

Figure 11–30

THE QDIM COMMAND

A more efficient means of placing a series of continued or baseline dimensions is the QDIM or Quick Dimension command. You identify a number of valid corners representing intersections or endpoints of an object and all dimensions are placed. In Figure 11–31A, a crossing box is used to identify all corners to dimension to. When the dimension line is identified in Figure 11–31B, all continued dimensions are placed. Follow the next series of prompts and illustrations for quickly placing a series of continued dimensions:

Command: **QDIM**
Select geometry to dimension: *(Pick a point at "A" in Figure 11-31A)*
Specify opposite corner: *(Pick a point at "B")*
Select geometry to dimension: *(Press* ENTER *to continue)*
Specify dimension line position, or
[Continuous/Staggered/Baseline/Ordinate/Radius/Diameter/datumPoint/Edit]
<Continuous>: *(Locate the dimension line at "C" in Figure 11-31B)*

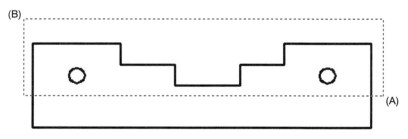

Figure 11–31A

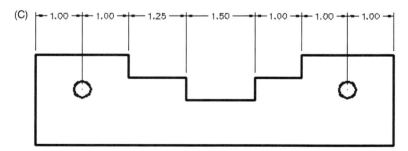

Figure 11–31B

By default, the QDIM command places continued dimensions. Before locating the dimension line, you have the option of placing the Staggered dimensions in Figure

11–32A. The process with this style begins with adding dimensions to inside details and continuing outward until all features are dimensioned.

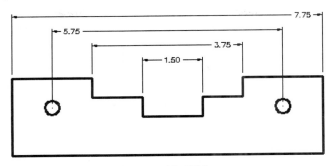

Figure 11–32A

Another option of the QDIM command is the ability to place Baseline dimensions. As with all Baseline dimensions, an edge is used as the baseline or datum. All dimensions are calculated from the left edge as in Figure 11–32B. The datumPoint option of the QDIM command would allow you to select the right edge of the object in Figure 11–32B as the new datum or baseline.

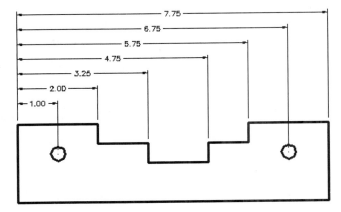

Figure 11–32B

The Ordinate option of the QDIM command calculates dimensions from a known 0,0 corner. The main Ordinate dimensioning topic will be discussed in greater detail later in this chapter. A new User Coordinate System must first be created in the corner of the object. Then the QDIM command is activated along with the Ordinate options to display the results in Figure 11–32C. Dimension lines are not used in Ordinate dimensions, in order to simplify the reading of the drawing.

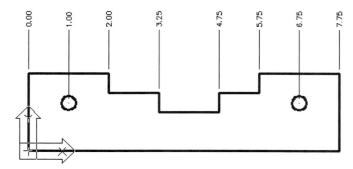

Figure 11-32C

The QDIM command can also be used to edit an existing group of dimensions. Activate the command, select all existing dimensions with a crossing box, and enter an option to change the style of all dimensions.

RADIUS AND DIAMETER DIMENSIONING

Arcs and circles are to be dimensioned in the view where their true shape is visible. The mark in the center of the circle or arc indicates its center point, as shown in Figure 11-33. You may place the dimension text either inside or outside the circle; you may also use grips to aid in the dimension text location of a diameter or radius dimension. The prompts for Diameter and Radius dimensions are as follows:

Command: **DDI** *(For DIMDIAMETER)*
Select arc or circle: *(Select the edge of the circle)*
Dimension text = 2.50
Specify dimension line location or [Mtext/Text/Angle]: *(Pick a point to locate the diameter dimension)*
Command: **DRA** *(For DIMRADIUS)*
Select arc or circle: *(Select the edge of the arc)*
Dimension text = 1.50
Specify dimension line location or [Mtext/Text/Angle]: *(Pick a point to locate the radius dimension)*

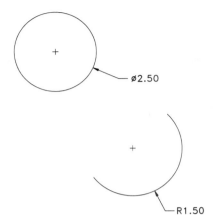

Figure 11–33

USING QDIM FOR RADIUS AND DIAMETER DIMENSIONS

As with linear dimensions, the QDIM command has options that can be applied to radius and diameter dimensions. In Figures 11–34A and 11–34B, all circles are selected. Activating either the Radius or Diameter option displays the results in the illustrations. When locating the radius or diameter dimensions, you can specify the angle of the leader. A predefined leader length is applied to all dimensions. Grips could be used to relocate the dimensions to better places.

Command: **QDIM**
Select geometry to dimension: *(Select the six circles in Figure 11–34A)*
Select geometry to dimension: *(Press* ENTER *to continue)*
Specify dimension line position, or
[Continuous/Staggered/Baseline/Ordinate/Radius/Diameter/datumPoint/Edit]
<Continuous>: **R** *(For Radius)*
Specify dimension line position, or
[Continuous/Staggered/Baseline/Ordinate/Radius/Diameter/datumPoint/Edit]
<Radius>: *(Locate the dimension line at "A")*

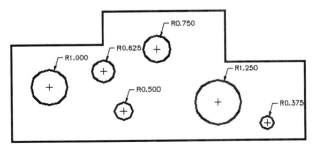

Figure 11–34A

Command: **QDIM**
Select geometry to dimension: *(Select the six circles in Figure 11–34B)*
Select geometry to dimension: *(Press* ENTER *to continue)*
Specify dimension line position, or
[Continuous/Staggered/Baseline/Ordinate/Radius/Diameter/datumPoint/Edit]
<Continuous>: **D** *(For Diameter)*
Specify dimension line position, or
[Continuous/Staggered/Baseline/Ordinate/Radius/Diameter/datumPoint/Edit]
<Diameter>: *(Locate the dimension line at "A")*

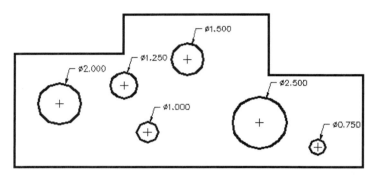

Figure 11–34B

LEADER LINES

A leader line is a thin, solid line leading from a note or dimension ending with an arrowhead, illustrated at "A" in Figure 11–35. The arrowhead should always terminate at an object line such as the edge of a hole or arc. A leader to a circle or arc should be radial; this means it is drawn so that if extended, it would pass through the center of the circle illustrated at "B." Leaders should cross as few object lines as possible and should never cross each other. The short horizontal shoulder of a leader should meet the dimension illustrated at "A." It is poor practice to underline the dimension with the horizontal shoulder illustrated at "C." Example "C" also illustrates a leader not lined up with the center or radial. This may affect the appearance of the leader. Again, check your office's standard practices to ensure that this example is acceptable.

Yet another function of a leader is to attach notes to a drawing, illustrated at "D." Notice that the two notes attached to the view have different terminators, arrows, and dots. It is considered good practice to adopt only one terminator for the duration of the drawing. The sequence for the LEADER command is as follows:

Command: **LEADER**
Specify leader start point: **Nea**
to *(Select the edge of the arc at "A")*
Specify next point: *(Pick a point to locate the end of the leader)*

Specify next point or [Annotation/Format/Undo] <Annotation>: *(Press* ENTER *to accept this default)*

Enter first line of annotation text or <options>: *(Press* ENTER *to accept this default)*

Enter an annotation option [Tolerance/Copy/Block/None/Mtext] <Mtext>: *(Press* ENTER *to display the Multiline Text Editor dialog box. Enter the desired text to make up the leader note and click the OK button to place the leader)*

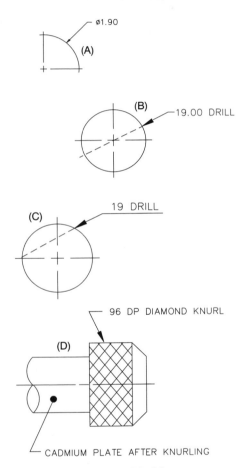

Figure 11–35

THE QLEADER COMMAND

The QLEADER command, or Quick Leader, provides numerous controls for placing leaders in your drawing. This command may appear to operate similarly to the LEADER command previously explained. Study the prompt sequence below and Figure 11–36 for this command.

Command: **LE** *(For QLEADER)*
Specify first leader point, or [Settings]<Settings>: **Nea**
to *(Pick the point nearest at "A")*
Specify next point: *(Pick a point at "B")*
Specify next point: *(Press ENTER to continue)*
Specify text width <0.00>: *(Press ENTER to accept this default)*
Enter first line of annotation text <Mtext>: *(Press ENTER to display the Multiline Text
    Editor dialog box. Enter "R0.20" and "2 PLACES" in two separate text lines. Click the OK
    button to place the leader)*

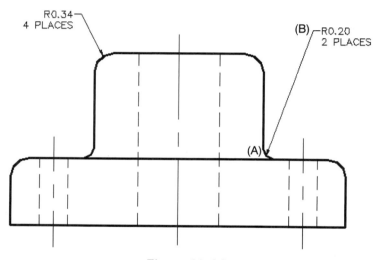

Figure 11–36

Pressing ENTER at the first Quick Leader prompt displays the Leader Settings dialog box in Figure 11–37A. Various controls are available to add greater functionality to your leaders through three tabs: Annotation, Leader Line & Arrow, and Attachment.

The Annotation tab deals with the object placed at the end of the leader. By default, the MText radio button is selected, allowing you to add a note through the Multiline Text Editor dialog box. You could also copy an object at the end of the leader, have a geometric tolerancing symbol placed in the leader, have a pre-defined block placed in the leader, or leave the leader blank.

Command: **LE** *(For QLEADER)*
Specify first leader point, or [Settings]<Settings>: *(Press ENTER to accept the default
    value and display the Leader Settings dialog box in Figure 11–37A)*

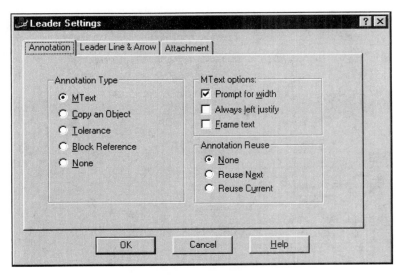

Figure 11–37A

The Leader Line & Arrow tab in Figure 11–37B allows you to draw a leader line consisting of straight segments or in the form of a spline object. You can control the number of points used to define the leader; a maximum of three points will be more than enough to create your leader. You can even change the arrowhead through the list provided in Figure 11–37C.

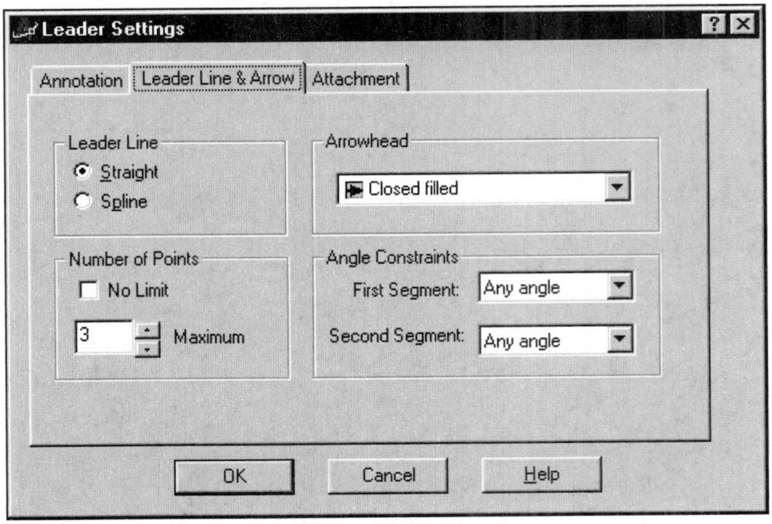

Figure 11–37B

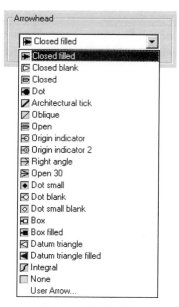

Figure 11–37C

The Attachment tab in Figure 11–37D allows you to control how text is attached to the end of the leader through various justification modes. The settings in the figure are the default values. For example, back in Figure 11–36, a Text Attachment mode of Middle of top line is used since the text is placed on the left side of the leader.

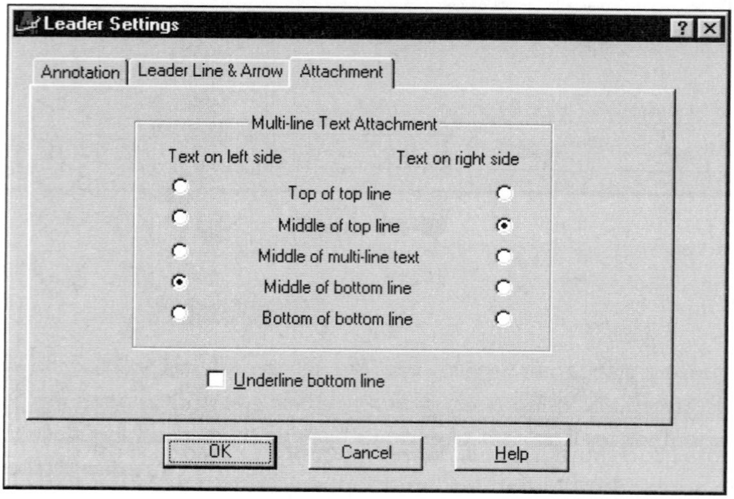

Figure 11–37D

DIMENSIONING ANGLES

Dimensioning angles requires two lines forming the angle in addition to the location of the vertex of the angle along with the dimension arc location. Before going any further, understand where the curved arc for the angular dimension is derived from. At "A" in Figure 11–38, the dimension arc is struck from an imaginary center or vertex of the arc. The following prompts are taken from the DIMANGULAR command:

Command: **DAN** *(For DIMANGULAR)*
Select arc, circle, line, or <specify vertex>: *(Select line "A")*
Select second line: *(Select line "B")*
Specify dimension arc line location or [Mtext/Text/Angle]: *(Pick a point at "C" to locate the dimension)*
Dimension text = 53

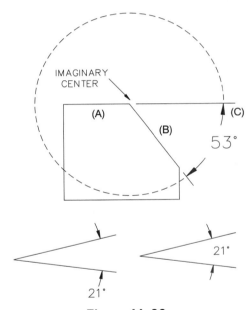

Figure 11–38

DIMENSIONING SLOTS

For slots, first select the view where the slot is visible. Two methods of dimensioning the slot are illustrated in Figure 11–39. You call out a slot by locating the center-to-center distance of the two semicircles, followed by a radius dimension to one of the semicircles; which radius dimension is selected depends on the available room to dimension. A second method involves the same center-to-center distance followed by an overall distance designating the width of the slot. This dimension happens to be the same as the diameter of the semicircles. It is also considered good practice to place this dimension inside the slot.

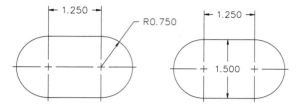

Figure 11–39

A more complex example in Figure 11–40 involves slots formed by curves and angles. Here, the radius of the circular center arc is called out. Angles reference each other for accuracy. As in Figure 11–39, the overall width of the slot is dimensioned, which happens to be the diameter of the semicircles at opposite ends of the slot. Use the DIMALIGN command to place the 0.48 dimension. Use OSNAP-Nearest to identify the location at "A" and OSNAP-Perpendicular to identify the location at "B."

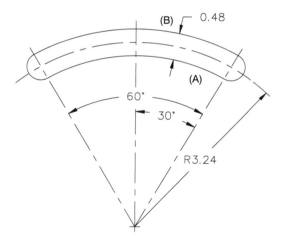

Figure 11–40

ORDINATE DIMENSIONING

The plate in Figure 11–41 consists of numerous drill holes with a few slots, in addition to numerous 90°-angle cuts along the perimeter. This object is not considered difficult to draw or make because it mainly consists of drill holes. However, conventional dimensioning techniques make the plate appear complex because a dimension is required for the location of every hole and slot in both the X and Y directions. Add standard dimension components such as extension lines, dimension lines, and arrow-heads, and it is easy to get lost in the complexity of the dimensions even on this simple object.

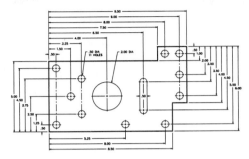

Figure 11–41

A better dimensioning method is illustrated in Figure 11–42, called Ordinate or Datum dimensioning. Here, dimension lines or arrowheads are not drawn; instead, one extension line is constructed from the selected feature to a location specified by you. A dimension is added to identify this feature in either the X or Y directions. It is important to understand that all dimension calculations occur in relation to the current User Coordinate System (UCS), or the current 0,0 origin. In Figure 11–42, with the 0,0 origin located in the lower left corner of the plate, all dimensions in the horizontal and vertical directions are calculated in relation to this 0,0 location. Holes and slots are called out with the DIMDIAMETER command. The following illustrates a typical ordinate dimensioning command sequence:

Command: **DOR** *(For DIMORDINATE)*
Specify feature location: *(Select a feature using an Osnap option)*
Specify leader endpoint or [Xdatum/Ydatum/Mtext/Text/Angle]: *(Locate a point outside of the object)*
Dimension text = Calculated value

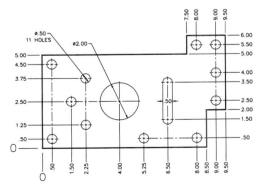

Figure 11–42

To understand how to place ordinate dimensions, see the example in Figure 11–43 and the prompt sequence below. Before you place any dimensions, a new User Coordinate System must be moved to a convenient location on the object with the UCS command and the Origin option. All ordinate dimensions will reference this new origin because it is located at coordinate 0,0. At the command prompt, enter DOR (for DIMORDINATE) to begin ordinate dimensioning. Select the quadrant of the arc at "A" as the feature. For the leader endpoint, pick a point at "B." Be sure Ortho mode is on. It is also helpful to snap to a convenient snap point for this and other dimensions along the direction. This helps in keeping all ordinate dimensions in line with each other. Follow the prompt sequence below:

Command: **DOR** *(For DIMORDINATE)*
Specify feature location: *(Select the quadrant of the slot at "A")*
Specify leader endpoint or [Xdatum/Ydatum/Mtext/Text/Angle]: *(Locate a point at "B")*
Dimension text = 1.50

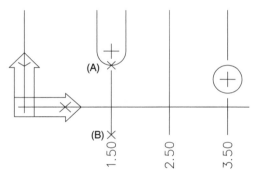

Figure 11–43

With the previous example in Figure 11–43 highlighting horizontal ordinate dimensions, placing vertical ordinate dimensions is identical (see Figure 11–44). With the UCS still located in the lower left corner of the object, select the feature at "A" using either the Endpoint or Quadrant modes. Pick a point at "B" in a convenient location on the drawing. Again, it is helpful if Ortho is on and you snap to a grid dot.

Command: **DOR** *(For DIMORDINATE)*
Specify feature location: *(Select the quadrant of the slot at "A")*
Specify leader endpoint or [Xdatum/Ydatum/Mtext/Text/Angle]: *(Locate a point at "B")*
Dimension text = 3.00

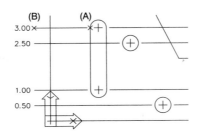

Figure 11–44

When spaces are tight to dimension to, two points not parallel to the X or Y axis will result in a "jog" being drawn (see Figure 11–45). It is still helpful to snap to a grid dot when performing this operation; however, be sure Ortho is turned off.

Command: **DOR** *(For DIMORDINATE)*
Specify feature location: *(Select the line at "A")*
Specify leader endpoint or [Xdatum/Ydatum/Mtext/Text/Angle]: *(Locate a point at "B")*
Dimension text = 2.00

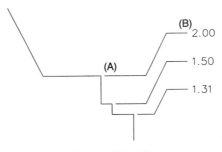

Figure 11–45

Ordinate dimensioning provides a neat and easy way of organizing dimensions for drawings where geometry leads to applications involving numerical control. Only two points are required to place the dimension that references the current location of the UCS. See Figure 11–46.

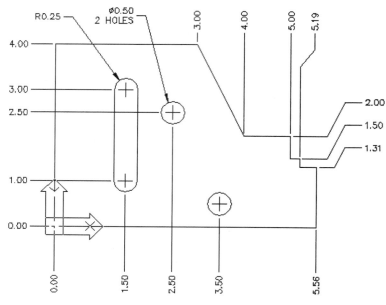

Figure 11–46

EDITING DIMENSIONS

Use the DIMEDIT command to add text to the dimension value, rotate the dimension text, rotate the extension lines for an oblique effect, or return the dimension text to its home position. Figure 11–47 shows the effects of adding text to a dimension and rotating the dimension to a user-specified angle. Follow the command prompt sequences below for accomplishing this task.

Command: **DED** *(For DIMEDIT)*
Enter type of dimension editing [Home/New/Rotate/Oblique] <Home>: **N** *(For New. This will display the Multiline Text Editor dialog box. Add the text "TYPICAL" on the other side of the <>. When finished, click the OK button)*
Select objects: *(Select the 5.00 dimension)*
Select objects: *(Press ENTER to perform the dimension edit operation)*
Command: **DED** *(For DIMEDIT)*
Enter type of dimension editing [Home/New/Rotate/Oblique] <Home>: **R** *(For Rotate)*
Specify angle for dimension text: **10**
Select objects: *(Select the 5.00 dimension)*
Select objects: *(Press ENTER to perform the dimension edit operation)*

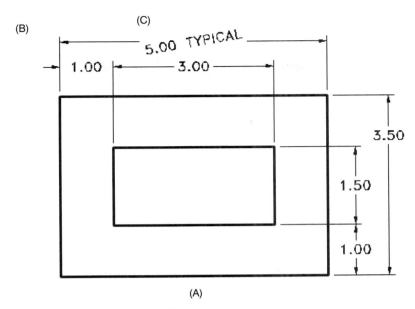

Figure 11–47

Regular AutoCAD Modify commands can also affect dimensions. When you use the STRETCH command on the object in Figure 11–47, points "A" and "B" identify a crossing window. Point "C" is the base point of displacement. The results are displayed in Figure 11–48. Not only did the object lines stretch to the new position but the dimensions all updated themselves to new values.

Command: **S** *(For STRETCH)*
Select objects to stretch by crossing-window or crossing-polygon...
Select objects: *(Pick a point at "A" in Figure 11–47)*
Specify opposite corner: *(Pick a point at "B" to activate the crossing window)*
Select objects: *(Press ENTER to continue)*
Specify base point or displacement: *(Pick a point at "C" in Figure 11–47...it could also be anywhere on the screen)*
Specify second point of displacement: **@0.75<180**

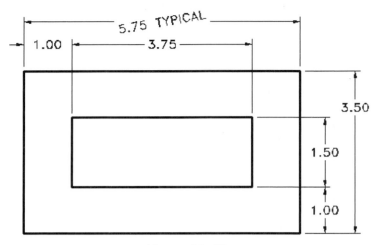

Figure 11–48

One of the other options of the DIMEDIT command is the ability for you to move and rotate the dimension text and still have the text return to its original or home location. This is the purpose of the Home option. Selecting the 5.00 dimension in Figure 11–49 will return it to its original position. However, you would have to use the New option of the DIMEDIT command to remove the text "TYPICAL."

Command: **DED** *(For DIMEDIT)*
Enter type of dimension editing [Home/New/Rotate/Oblique] <Home>: **H** *(For Home)*
Select objects: *(Select the 5.00 dimension)*
Select objects: *(Press ENTER to perform the dimension edit operation)*

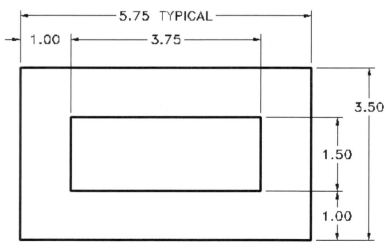

Figure 11–49

The DIMEDIT command also has an Oblique option that allows you to enter an obliquing angle, which will rotate the extension lines and reposition the dimension line. This option is useful if you are interested in placing dimensions in isometric drawings.

Command: DED *(For DIMEDIT)*
Enter type of dimension editing [Home/New/Rotate/Oblique] <Home>: O *(For Oblique)*
Select objects: *(Select the 2.50 and 1.25 dimensions in Figure 11–50)*
Select objects: *(Press* ENTER *to continue)*
Enter obliquing angle (press ENTER **for none): 150**

Figure 11–50

The results are illustrated in Figure 11–51, with both dimensions being repositioned with the Oblique option of the DIMEDIT command. Notice that the extension lines and dimension lines were affected; however, the dimension text stayed the same.

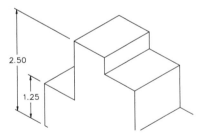

Figure 11–51

The Oblique option of the DIMEDIT command was used to rotate the dimension at "A" at an obliquing angle of 210° (see Figure 11–52). An obliquing angle of -30° was used to rotate the dimension at "B," and the dimensions at "C" required an obliquing angle of 90°. This represents proper isometric dimensions, except for the orientation of the text.

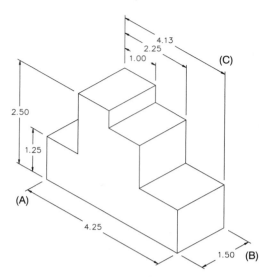

Figure 11–52

TOLERANCES

Interchangeability of parts requires replacement parts to fit in an assembly of an object no matter where the replacement part comes from. In Figure 11–53, the U-shaped channel of 2.000 units in width is to accept a mating part of 1.995 units. Under normal situations, there is no problem with this drawing or callout. However, what if the production person cannot make the channel piece exactly at 2.000? What if he or she is close and the final product measures 1.997? Again, the mating part will have no problems fitting in the 1.997 slot. What if the mating part is not made exactly 1.995 units but is instead 1.997? You see the problem. As easy as it is to attach a dimension to a drawing, some thought needs to go into the possibility that based on the numbers, maybe the part cannot be easily made. Instead of locking dimensions in using one number, a range of numbers would allow the production person the flexibility to vary in any direction and still have the parts fit together. This is the purpose of converting some basic dimensions to tolerances.

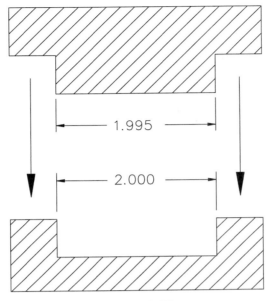

Figure 11–53

The example in Figure 11–54 shows the same two mating parts; this time, two sets of numbers for each part are assigned. For the lower part, the machinist must make the part anywhere between 2.002 and 1.998, which creates a range of 0.004 units of variance. The upper mating part must be made within 1.997 and 1.993 units in order for the two to fit correctly. The range for the upper part is also 0.004 units. As you will soon see, whether upper or lower numbers are used, the parts will always fit together. If the bottom part is made to 2.002 and the top part is made to 1.993, the parts will fit. If the bottom part is made to 1.998 and the upper part is made to 1.997, the parts will fit. In any case or combination, if the dimensions are followed exactly as stated by the tolerances, the pieces will always fit. If the bottom part is made to 1.998 and the upper part is made to 1.999, the upper piece is rejected. The method of assigning upper and lower values to dimensions is called limit dimensioning. Here, the larger value in all cases is placed above the smaller value. This is also called a clearance fit since any combinations of numbers may be used and the parts will still fit together.

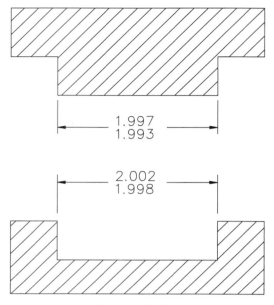

Figure 11–54

The object in Figure 11–55 has a different tolerance value assigned to it. The basic dimension is 2.000; in order for this part to be accepted, the width of this object may go as high as 2.002 units or as low as 1.999 units, giving a range of 0.003 units by which the part may vary. This type of tolerance is called a plus/minus dimension with an upper limit of 0.002 difference from the lower limit of -0.001.

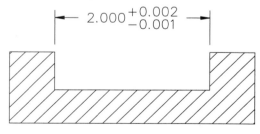

Figure 11–55

Illustrated in Figure 11–56 is yet another way to display tolerances. It is very similar to the plus/minus method except that both upper and lower limit values are the same; for this reason, this method is called plus and minus dimensions. The basic dimension is still 2.000 with upper and lower tolerance limits of 0.002 units.

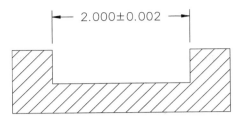

Figure 11–56

GEOMETRIC DIMENSIONING AND TOLERANCING (GD&T)

In the object in Figure 11–57, approximately 1000 items need to be manufactured based on the dimensions of the drawing. Also, once constructed, all objects needed to be tested to be sure the 90° surface did not deviate over 0.005 units from its base. Unfortunately, the wrong base was selected as the reference for all dimensions. Which should have been the correct base feature? As all items were delivered, they were quickly returned because the long 8.80 surface drastically deviated from required 0.005 unit deviation or zone. This is one simple example that demonstrates the need for using geometric dimension and tolerancing techniques. First, this method deals with setting tolerances to critical characteristics of a part. Of course, the function of the part must be totally understood by the designer in order for tolerances to be assigned. The problem with the object in Figure 11–57 was the note, which did not really specify which base feature to choose as a reference for dimensioning. Figure 11–58 shows the same object complete with dimensioning and tolerancing symbols. The letter "A" inside the rectangle identifies the datum or reference surface. The tolerance symbol at the end of the long edge tells the individual making this part that the long edge cannot deviate more than 0.005 units using surface "A" as a reference. The datum triangles that touch the part are a form of dimension arrowhead.

Using geometric tolerancing symbols ensures more accurate parts with less error in interpreting the dimensions.

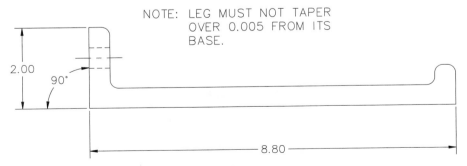

Figure 11–57

Figure 11–58

GD&T SYMBOLS

Entering TOL at the command prompt brings up the main Geometric Tolerance dialog box in Figure 11–59A. This box contains all tolerance zone boxes and datum identifier areas. Clicking in one of the dark boxes under the Sym area displays the Symbol dialog box illustrated in Figure 11–59B. This dialog box contains all major geometric dimensioning and tolerancing symbols. Choose the desired symbol by clicking on the specific symbol; this will return you to the main Geometric Tolerance dialog box.

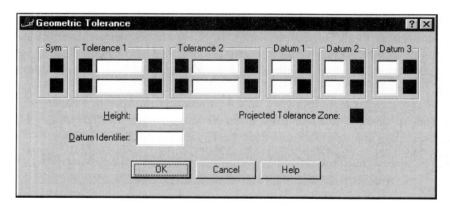

Figure 11–59A

Figure 11–59B

Figure 11–60 shows a chart outlining all geometric tolerancing symbols supported in the Symbol dialog box. Alongside each symbol is the characteristic controlled by the symbol in addition to a tolerance type. Tolerances of Form such as Flatness and Straightness can be applied to surfaces without a datum being referenced. On the other hand, tolerances of Orientation such as Angularity and Perpendicularity require datums as reference.

Symbol	Characteristic	Type of Tolerance
▱	Flatness	Form
—	Straightness	
◯	Roundness	
⌭	Cylindricity	
⌒	Profile of a Line	Profile
⌓	Profile of a Surface	
∠	Angularity	Orientation
⊥	Perpendicularity	
//	Parallelism	
⊕	Position	Location
◎	Concentricity	
⌯	Symmetry	
↗	Circular Runout	Runout
↗↗	Total Runout	

Figure 11–60

With the symbol placed inside this dialog box as in Figure 11–61, you now assign such items as tolerance values, maximum material condition modifiers, and datums. In Figure 11–61, the tolerance of Parallelism is to be applied at a tolerance value of 0.005 units to Datum "A."

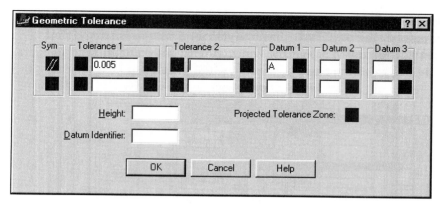

Figure 11–61

ADDITIONAL GD&T SYMBOLS

In the main Geometric Tolerance dialog box, shown in Figure 11–61, clicking in the black box next to the 0.005 tolerance value activates the Material Condition dialog box in Figure 11–62. This dialog box is devoted to the concept of Material Condition and acts as a modifier to the main tolerance value.

Figure 11–62

Figure 11–63 shows the meaning of the symbols used in this dialog box. In brief, Maximum Material Condition refers to the condition of a characteristic such as a hole when the most material exists. Least Material Condition refers to the condition of a characteristic where the least material exists. Regardless of Feature Size indicates that the characteristic tolerance such as Flatness or Straightness must be maintained regardless of the actual produced size of the object.

Datums are represented by surfaces, points, lines, or planes and are used for referencing the features of an object. The symbols illustrated in Figure 11–63 are "flags" used to identify a datum. The next series of figures illustrates how datums are used in combination with tolerances of Orientation.

Ⓜ	Maximum Material Condition
Ⓛ	Least Material Condition
Ⓢ	Regardless of Feature Size
–A–	Primary Datum
–B–	Secondary Datum
–C–	Tertiary Datum

Figure 11–63

THE FEATURE CONTROL SYMBOL

Once you have completed the desired information in the Geometric Tolerance dialog box, the result is illustrated in Figure 11–64, in the form of a Feature Control Symbol. This is the actual order in which all symbols, values, modifiers, and datums are placed. Follow the graphic and information in Figure 11–64 to view the contents of each area of the Feature Control Symbol box.

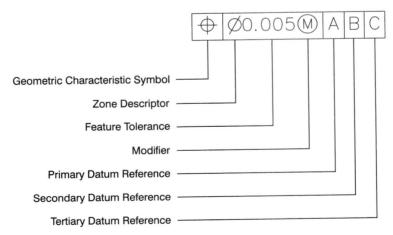

Geometric Characteristic Symbol
Zone Descriptor
Feature Tolerance
Modifier
Primary Datum Reference
Secondary Datum Reference
Tertiary Datum Reference

Figure 11–64

TOLERANCE OF ANGULARITY EXAMPLE

Illustrated in the Geometric Tolerance dialog box in Figure 11–65A is an example of applying the tolerance of Angularity to a feature. First the symbol is identified along with the tolerance value. Since this tolerance requires a datum, the letter "A" identifies primary datum "A." The results are illustrated in Figure 11–65B. Datum "A" identifies the surface of reference as the base of the object. The feature control symbol is applied to the angle and reads: "The surface dimensioned at 60° cannot deviate more than 0.005 units from datum 'A'." The graphic in Figure 11–65B shows an exaggerated tolerance zone and is used for illustrative purposes only.

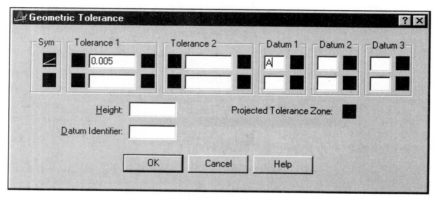

Figure 11–65A

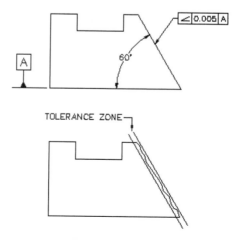

Figure 11–65B

TOLERANCE OF FLATNESS EXAMPLE

The next example applies a tolerance of Flatness to a surface. First the Geometric Tolerance dialog box has the Flatness symbol assigned in addition to a tolerance value of 0.003 units in Figure 11–66A. The results are illustrated in Figure 11–66B, with the feature control box being applied to the top surface. The tolerance box reads: "The surface must be flat with a 0.003 unit tolerance zone." The tolerance range is displayed in the second graphic and is exaggerated. The tolerance of Flatness does not usually require a datum for reference.

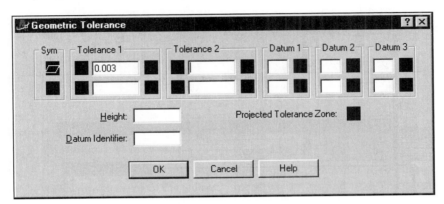

Figure 11–66A

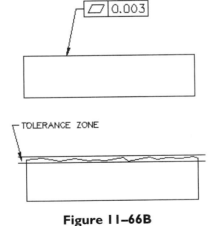

Figure 11–66B

TOLERANCE OF PERPENDICULARITY EXAMPLE

This next example illustrates a tolerance of Perpendicularity. The Geometric Tolerance dialog box reflects the perpendicularity symbol along with a tolerance value of 0.003 units. See Figure 11–67A. This tolerance characteristic requires a datum to be most effective. In Figure 11–67B, the feature control box reads: "This surface must be perpendicular to datum 'A' within a tolerance zone of 0.003 units." The second graphic shows the tolerance zone and the amount of deviation that is acceptable. It is meant to be exaggerated and is used for illustrative purposes only.

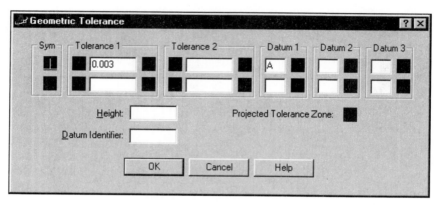

Figure 11–67A

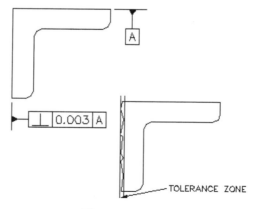

Figure 11–67B

TOLERANCE OF POSITION EXAMPLE

This last example displays a tolerance of Position that is applied to a hole feature. Since a hole is centered from two edges, two datums are identified in the Geometric

Tolerance dialog box illustrated in Figure 11–68A. This is sometimes called a circular tolerance. The feature control box illustrated in Figure 11–68B reads: "The center of the hole must lie within a circular tolerance zone of 0.050 units in relation to datums 'A' and 'B'."

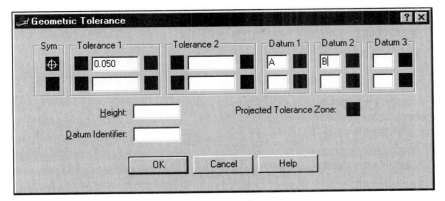

Figure 11–68A

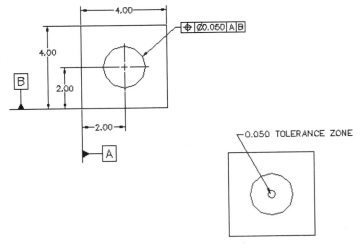

Figure 11–68B

GENERAL DIMENSIONING SYMBOLS

In today's global economy, the transfer of documents in the form of drawings is becoming a standard way of doing business. As drawings are shared with a subsidiary of an overseas company, two different forms of language may be needed to interpret the dimensions of the drawing. In the past, this has led to confusion in interpreting drawings, and as a result, a system of dimensioning symbols has been developed. It is hoped that a symbol will be easier to recognize and interpret than a note in a different language about the particular feature being dimensioned.

Figure 11–69 shows some of the more popular dimensioning symbols in use today on drawings. Notice how the symbols are designed to make as clear and consistent an interpretation of the dimension as possible. As an example, the Deep or Depth symbol displays an arrow pointing down. This symbol is used to identify how far into a part a drill hole goes. The Arc Length symbol identifies the length of an arc, and so on.

⌒	Arc Length
X.XX	Basic Dimension
▷	Conical Taper
⊔	Counterbore or Spotface
∨	Countersink
⊽	Deep or Depth
∅	Diameter
X.XX	Dimension Not to Scale
2X	Number of Times—Places
R	Radius
(X.XX)	Reference Dimension
S∅	Spherical Diameter
SR	Spherical Radius
◁	Slope
□	Square

Figure 11–69

The size of all general dimensioning symbols is illustrated in Figure 11–70A. All values are in relation to "h" or the relative height of the dimension numerals. As an example, with a dimension numeral height of 0.18, study the illustration in

Figure 11–70B for finding the size of the counterbore dimension symbol. Since the height of the counterbore symbol is defined by "h," simply substitute the dimension numeral height of 0.18 for the height of the counterbore symbol. Since the width of the symbol is "2h," multiply 0.18 by 2 to obtain a value of 0.36. These symbols may be created and saved as blocks for insertion in drawings. They may also be part of an existing text font mapped to a particular key on the keyboard. Pressing that key produces the dimension symbol.

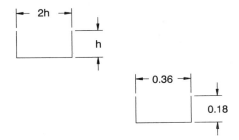

Figure 11–70A

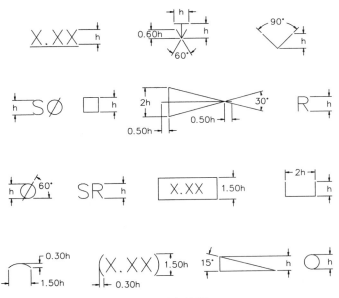

Figure 11–70B

APPLYING THE BASIC DIMENSION SYMBOL

Drawing a rectangular box around the dimension numeral (see Figure 11–71) identifies a basic dimension. The basic dimension represents the theoretical distance of the feature, such as the distance from the center hole to the left hole as 4.00 units with the distance from the left hole to the right hole as 8.00 units. It is a known fact that it is impossible to maintain tolerances to obtain an exact value of 4.00 or 8.00 units. You can construct a rectangle; however, this may be a tedious task depending on the number of decimal places past the zero, which determines the size of the rectangle. To constrain the rectangle around dimension text, choose Dimension Style… from the Format pull-down menu and then Modify. Click on the Tolerances tab and in the Method area, change None to Basic to draw the rectangle around dimension text.

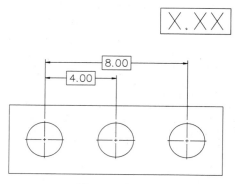

Figure 11–71

APPLYING THE REFERENCE DIMENSION SYMBOL

Illustrated in Figure 11–72 is a series of Continue dimensions strung together in one line. The dimension numeral 3.00 is enclosed in parentheses and is referred to as a reference dimension. Reference dimensions are not required drawing dimensions; they are placed for information purposes. The parentheses on the keyboard are used along with the dimension numeral to create a reference dimension.

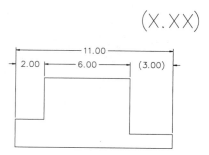

Figure 11–72

APPLYING THE NOT TO SCALE DIMENSION SYMBOL

At times it is necessary to place a dimension that is not to scale, as illustrated in Figure 11–73. To distinguish regular dimensions from a dimension that is not to scale, this dimension is identified in the drawing by a line drawn under the dimension. To do this in an AutoCAD drawing, enter the following text string for the dimension value: %%u5.00%%u. The "%%u" toggles on Underline Text mode followed by the dimension text of 5.00 units. To toggle underline off, complete the text string with another "%%u." When you use the Multiline Text Editor dialog box, an underline button is provided to simplify this task.

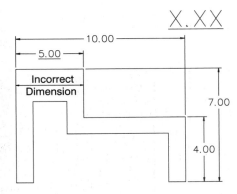

Figure 11–73

APPLYING THE SQUARE DIMENSION SYMBOL

Use this dimension symbol to identify the cross section of an object as a square. This will limit the need for two dimensions of the same value to be placed identifying a square. The square dimension symbol should be placed before the dimension numeral, as in the illustration in Figure 11–74.

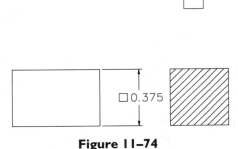

Figure 11–74

APPLYING THE SPHERICAL DIAMETER DIMENSION SYMBOL

The spherical illustrated in Figure 11–75 is in the form of a dome-shaped feature. Use the diameter symbol preceded by the letter "S" for spherical.

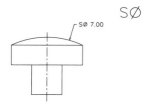

Figure 11–75

APPLYING THE SPHERICAL RADIUS DIMENSION SYMBOL

The illustration in Figure 11–76 is similar to the previous spherical example except that a radius identifies the spherical. Place the letter "S" in front of the "R" symbol.

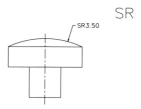

Figure 11–76

APPLYING THE DIAMETER DIMENSION SYMBOL

A circle with a diagonal line through it represents this symbol. This diameter symbol always precedes the dimension numeral with no space in between. This symbol is automatically placed before the dimension numeral when you use the DIMDIAMETER command. All holes represented by the diameter symbol are understood to pass completely through a part, as in Figure 11–77.

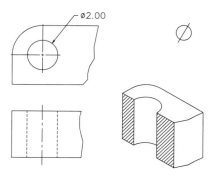

Figure 11–77

APPLYING THE DEEP DIMENSION SYMBOL

When a hole does not pass completely through a part, as in Figure 11–78, the depth or deep dimension symbol is used. The diameter of the hole is placed first, followed by the depth symbol and the distance the hole is to be drilled. This figure also shows yet another method of identifying holes, although it is considered dated.

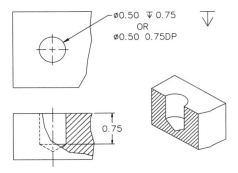

Figure 11–78

APPLYING THE DEEP DIMENSION SYMBOL

Illustrated in Figure 11–79 is yet another application of the depth dimension symbol. First the diameter of the hole is identified. Next, a counterdrill diameter with angle of the counterdrill is given. The depth of the counterdrill completes the dimension. This operation is used when a flat-head screw needs to be recessed below the surface of a part.

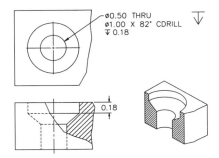

Figure 11–79

APPLYING THE COUNTERBORE DIMENSION SYMBOL

A counterbore is an enlarged portion of a previously drilled hole. The purpose of a counterbore is to receive the head of such screws as socket-head or fillister-head screws. The counterbore example in Figure 11–80 first identifies the diameter of the thru hole. On the next line come the counterbore symbol and its diameter. The final specification is the depth of the counterbore identified by the Depth dimension symbol. Figure 11–80 shows another method of identifying counterbores, although it is considered dated.

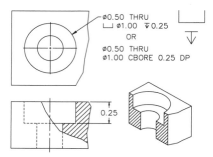

Figure 11–80

APPLYING THE SPOTFACE DIMENSION SYMBOL

A spotface is similar to the counterbore except that it is usually made quite shallower than the counterbore. The purpose of the spotface is to seat a washer, preventing it from moving around along the surface of a part. The counterbore symbol is used to identify a spotface, as in Figure 11–81. The diameter of the thru hole is first given. On the next line, the counterbore symbol followed by diameter is given. Finally the depth of the counterbore is identified by the distance and the Depth dimension symbol. Figure 11–81 also shows another method of identifying spotfaces, although it is considered dated.

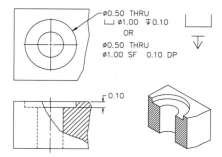

Figure 11–81

APPLYING THE SLOPE DIMENSION SYMBOL

The slope dimension symbol applies to an inclined surface and the amount of rise in the surface given by two dimensions, as in Figure 11–82. This rise is indicated by the change in height per unit distance along a baseline. The slope symbol is placed followed by the change in height and the change in unit distance. This makes the slope dimension a ratio of the height and unit distance.

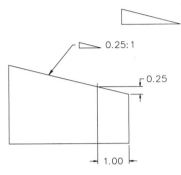

Figure 11–82

APPLYING THE COUNTERSINK DIMENSION SYMBOL

A countersink is a V-shaped conical taper at one end of a hole. The purpose is to accept a flat-head screw and make it flush with the top surface of the part. In Figure 11–83, three holes of 0.50 diameter are first identified. On the next line, the countersink symbol followed by the number of degrees in the countersink is specified. Below this example is yet another method of identifying countersinks although it is considered dated.

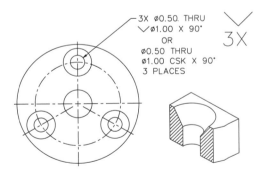

Figure 11–83

APPLYING THE ARC DIMENSION SYMBOL

Illustrated in Figure 11–84 is the difference between dimensioning the chord of an arc and the actual distance of the arc. The 7.00 unit dimension, in addition to defining the width of the part, specifies the distance of the chord of an arc. The 8.48 unit dimension is the distance of the arc and is specified by a small arc symbol above the distance. Accomplish this by drawing a small arc and moving it above the arc distance. Although AutoCAD does not dimension the length of an arc, it does allow you to use the LIST command to obtain the length of an arc. You then enter this value at the dimension distance prompt.

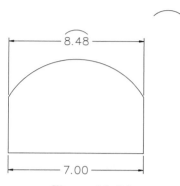

Figure 11–84

APPLYING THE ORIGIN DIMENSION SYMBOL

The origin dimension symbol is used to indicate the origin of the dimension. The small circle is substituted for an arrowhead as the termination of the dimension line, as in Figure 11–85. To construct the open circle and filled arrowhead, go to the Arrowheads group in the Geometry dialog box, click in the edit box next to 1st: and choose the Dot Blanked terminator. Next, click in the edit box next to 2nd: and choose the Closed Filled terminator.

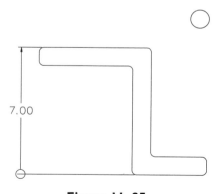

Figure 11–85

APPLYING THE CONICAL TAPER DIMENSION SYMBOL

Similar to the slope dimension symbol, the conical taper symbol is used when the amount of taper per unit of length is desired. The conical taper symbol is placed before the value of the taper. The value of the taper is based on a ratio of 1.00 units in length to a Delta diameter illustrated in Figure 11–86. The Delta diameter is based on the large shaft diameter minus the diameter taken at the area where the 1.00 unit length is located. This value becomes the ratio to 1.00 unit of length.

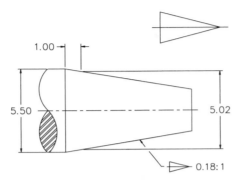

Figure 11–86

LOCATION OF UNIFORMLY SPACED FEATURES

Figure 11–87 shows an example of a plate consisting of eight holes equally spaced at 45° angles about a 7.30-diameter bolthole circle. Eliminating the need to dimension each individual angle, the dimension symbol "X" is used for the number of times a feature is repeated. Notice in Figure 11–87 that "8X" is used to call out the number of times the angle and diameter of the holes repeat.

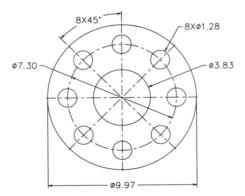

Figure 11–87

Figure 11–88A is similar to the previous illustration. Five 72° angles are used to locate five slots each with a width of 1.50 units.

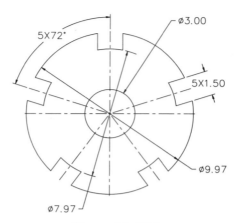

Figure 11–88A

Figure 11–88B shows a different example of the location of features that are uniformly spaced. Six holes of 1.00 diameter are located along a bolt circle of 7.30 diameter. All six holes are spaced at 30° angles. Since the holes are not laid out in a full circle, only five angles are required to locate the six holes. The symbol "5X" is used to identify the number of angles, while the symbol "6X" is used to identify the number of holes.

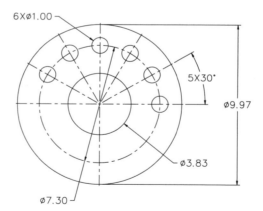

Figure 11–88B

CHARACTER MAPPING FOR DIMENSION SYMBOLS

Illustrated in Figure 11–89 is a typical counterbore operation and the correct dimension layout. The LEADER command was used to begin the diameter dimension. The counterbore and depth symbols were generated with the MTEXT command along with the Unicode Character Mapping dialog box. In any case, the first step in using the dimension symbols is to first create a new text style, for which you can use any name.

First use the QLEADER command to place the leader. All parameters in the Leader Settings dialog box should be properly set, including the Middle of top line radio button for the right attachment location.

Command: **LE** *(For QLEADER)*
Specify first leader point, or [Settings]<Settings>: **Nea**
To *(Select the edge of the circle)*
Specify next point: *(Pick a point to extend the leader)*
Specify next point: *(Press ENTER to continue)*
Specify text width <0.0000>: *(Press ENTER to accept this default value)*
Enter first line of annotation text <Mtext>: *(Press ENTER to activate the Multiline Text Editor dialog box)*

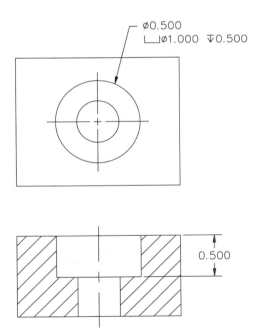

Figure 11–89

Enter the first listing for the hole in the Multiline Text Editor dialog box in Figure 11–90. The characters "%%c" signify the diameter symbol, which is selected from the Symbol menu under Diameter. When finished, press ENTER to drop down to the second line.

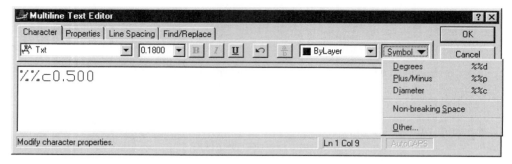

Figure 11–90

The remaining symbols are located in a special area called a Unicode Character Map. Clicking on Other… in Figure 11–91A displays this character mapping dialog box in Figure 11–91B. Notice, in the upper left corner, that the current font is GDT, which holds all geometric tolerancing symbols along with the special dimensioning symbols such as counterbore, deep, and countersink. Be sure your current font is set to GDT. Once you identify a symbol, double-click on it. A box will appear around the symbol. Also, the symbol will appear in the Characters to Copy area in the upper right corner of the dialog box. Click on the Copy button to copy this symbol to the Windows Clipboard.

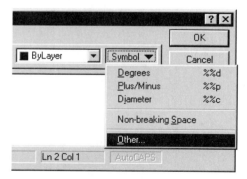

Figure 11–91A

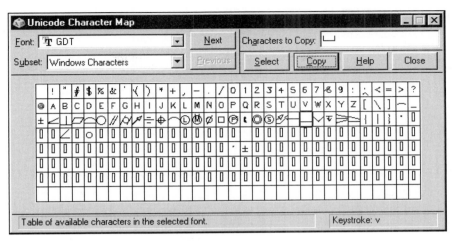

Figure 11–91B

Return to the Mutliline Text Editor dialog box and type CTRL+V, which performs a paste operation. Since the counterbore symbol was copied to the Windows Clipboard, it pastes into the dialog box in Figure 11–91C.

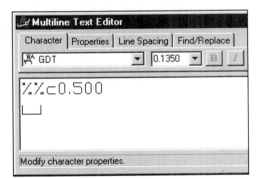

Figure 11–91C

Enter text and repeat the procedure for using the diameter and depth symbols outlined in Figure 11–91D.

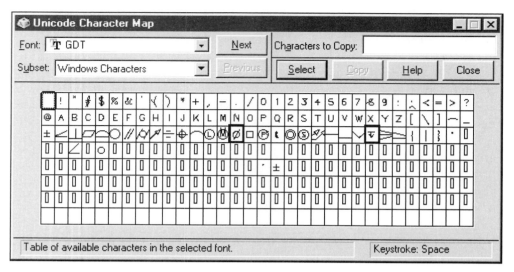

Figure 11–91D

The finished results are displayed in Figure 11–92A. Click the OK button to return back to the drawing. The leader and notes are both displayed as in Figure 11–92B.

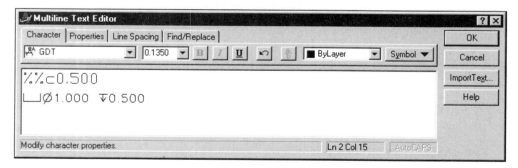

Figure 11–92A

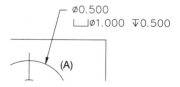

Figure 11–92B

GRIPS AND DIMENSIONS

Grips have a tremendous amount of influence on dimensions. Grips allow dimensions to be moved to better locations; grips also allow the dimension text to be located at a better position along the dimension line. In the example in Figure 11–93, notice the various unacceptable dimension placements. The 2.88 horizontal dimension lies almost on top of another extension line. To relocate this dimension to a better position, click on the dimension. Notice that the grips appear and the entire dimension highlights. Now click on the grip near "A" in the illustration in Figure 11–93. (This grip is located at the left end of the dimension line.) When the grip turns red, the Stretch mode is currently active. Stretch the dimension above the extension line but below the 3.50 dimension. The same results can be accomplished with the 4.50 horizontal dimension as it is stretched closer to the 3.50 dimension.

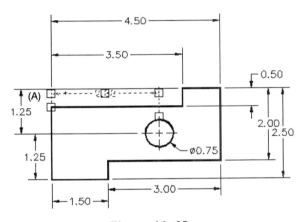

Figure 11–93

Notice that the two vertical dimensions in Figure 11–94 on the left side of the object do not line up with each other; this would be considered poor practice. Pick both dimensions and notice the appearance of the grips, in addition to both dimensions being highlighted. Click on the upper grip at "A" of the 1.25 dimension. When this grip turns red and places you in Stretch mode, select the grip at "B" of the opposite dimension. The result will be that both dimensions now line up with each other. The same can be accomplished with the 1.50 and 3.00 horizontal dimensions.

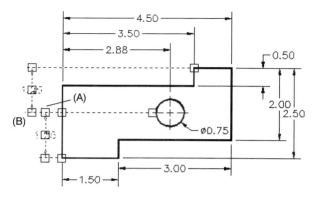

Figure 11–94

In the illustration in Figure 11–95, the 2.50 dimension text is too close to the 2.00 vertical dimension on the right side of the object. Click on the 2.50 dimension; the grips appear and the dimension highlights. Click in the grip representing the text location at "A." When this grip turns red, stretch the dimension text to a better location.

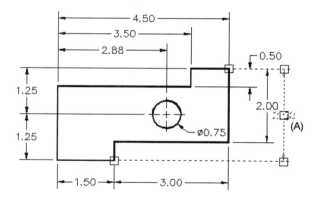

Figure 11–95

It is very easy to use grips to control the placement of diameter and radius text. As shown in Figure 11–96, click on the diameter dimension; the grips appear, in addition to the diameter dimension being highlighted. Click on the grip that locates the dimension text at "A." When this grip turns red, relocate the diameter dimension text to a better location using the Stretch mode.

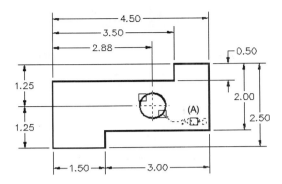

Figure 11–96

The completed object, with dimensions edited through grips, is displayed in Figure 11–97.

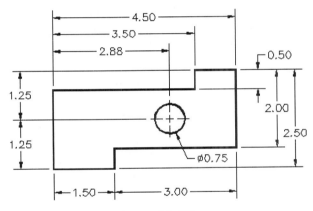

Figure 11–97

OBJECT SNAP AND DIMENSIONS

Proper selection of proper Object snap modes is important in dimensioning. You can create a dimension from the endpoint of an object's corner to what you think is an opposite corner. However incorrect dimensions result from accidentally selecting the endpoints of extension lines of other dimensions. For this reason, it is highly recommended to use the Object Snap Intersection mode for beginning dimensions. This will ensure that valid intersections are only selected.

Other problems involve dimensioning an object that has been crosshatched. Using the Object Snap Endpoint or Intersection modes could select an intersection created by an object line and the crosshatch pattern. It is therefore considered good practice to turn off the layer holding the crosshatch pattern before attempting dimensions.

TUTORIAL EXERCISE: 11_BAS-PLAT.DWG

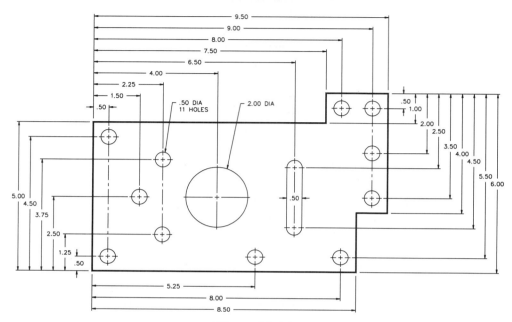

Figure 11–98

Purpose

The purpose of this tutorial is to change the annotation of the drawing of the Bas-plat (Base plate) from the conventional dimensioning style to the ordinate dimensioning style.

System Settings

The drawing in Figure 11–98 is already constructed and dimensioned up to a certain point. Enter AutoCAD and follow the steps in this tutorial for adding dimensions in Ordinate mode. Be sure the following Object Snap modes are currently set: Endpoint, Center, Intersect, and Extension.

Layers

All layers have already been created:

Name	Color	Linetype
Cen	Yellow	Center
Defpoints	White	Continuous
Dim	Yellow	Continuous
Object	White	Continuous

Suggested Commands

Use the DIMORDINATE command for placing ordinate dimensions throughout this tutorial.

Whenever possible, substitute the appropriate command alias in place of the full AutoCAD command in each tutorial step; for example, use "CP" for the COPY command, "L" for the LINE command, and so on. The complete listing of all command aliases is located in Table 1–2.

STEP I

All ordinate dimensions make reference to the current 0,0 location identified by the position of the User Coordinate System (see Figure 11–99). Since this icon is located in the lower left corner of the display screen by default, the coordinate system must be moved to a point on the object where all ordinate dimensions will be referenced from. First use the UCS command to define a new coordinate system with the origin at the lower left corner of the object. Then use the UCSICON command to force the icon to display at the new origin.

Command: **UCS**
Current ucs name: *NO NAME*
Enter an option [New/Move/
　orthoGraphic/Prev/Restore/Save/Del/
　Apply/?/World]
<World>: **O** *(For Origin)*
Specify new origin point <0,0,0>: *(Pick
　the intersection at "A")*
Command: **UCSICON**
Enter an option [ON/OFF/All/Noorigin/
　ORigin] <ON>: **OR** *(For Origin)*

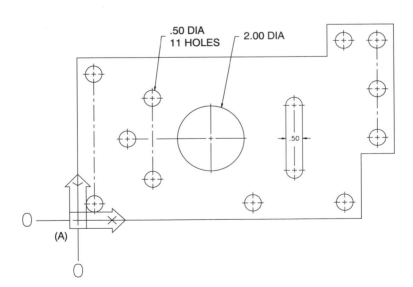

.50 DIA
11 HOLES

2.00 DIA

.50

(A)

Figure 11–99

STEP 2

Use the ZOOM-CENTER command to magnify the screen, similar to the illustration in Figure 11–100.

Command: **Z** *(For ZOOM)*
Specify corner of window, enter a scale factor (nX or nXP), or
[All/Center/Dynamic/Extents/Previous/Scale/Window] <real time>: **C** *(For Center)*
Specify center point: **0,0**
Enter magnification or height <11.25>: **4**

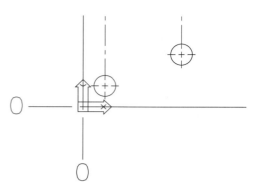

Figure 11–100

STEP 3

Before continuing with this next step, be sure that the grid is turned on. Use F7 to do this. Next, begin to place the first ordinate dimension using the DIMORDINATE command, or choose the DIMORDINATE command button from the Dimension toolbar. Select the endpoint of the center line as a feature. With the Snap and Ortho modes turned on, locate a point two snap distances below the object to identify the leader endpoint. AutoCAD will automatically determine if the dimension is Xdatum or Ydatum. See Figure 11–101.

Command: **DOR** *(For DIMORDINATE)*
Specify feature location: *(Pick the endpoint of the center line near point "A")*
Specify leader endpoint or [Xdatum/Ydatum/Mtext/Text/Angle]: *(Pick a point two snap points below the edge of the object at "B")*
Dimension text = 0.50

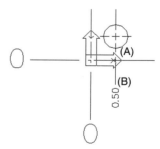

Figure 11–101

STEP 4

Perform a zoom-previous operation to display the overall drawing of the base plate, similar to the illustration in Figure 11–102. Repeat the procedure in Step 3 to place ordinate dimensions at locations "A" through "G." Be sure Ortho mode is on and that the leader location is two snap distances below the bottom edge of the object. The OSNAP-Endpoint mode should be used on the endpoints of each center line to satisfy the prompt "Select feature."

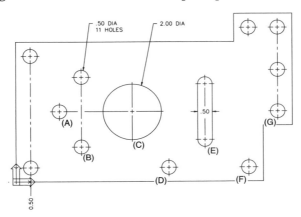

Figure 11–102

STEP 5

Continue placing ordinate dimensions similar to the procedure used in Step 3. Use the OSNAP-Endpoint mode to select features at "A" and "B" in Figure 11–103. Have Ortho mode on and identify the leader endpoint two grid dots below the bottom edge of the object.

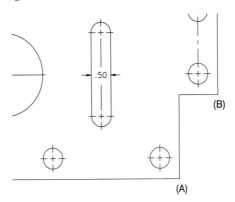

Figure 11–103

STEP 6

Use the UCSICON command to turn off the display of the icon. This will still keep the 0,0 origin at the lower left corner of the object. After you turn off the icon, your display should appear similar to Figure 11–104.

Command: **UCSICON**

Enter an option [ON/OFF/All/Noorigin/
  ORigin] <ON>: **Off**

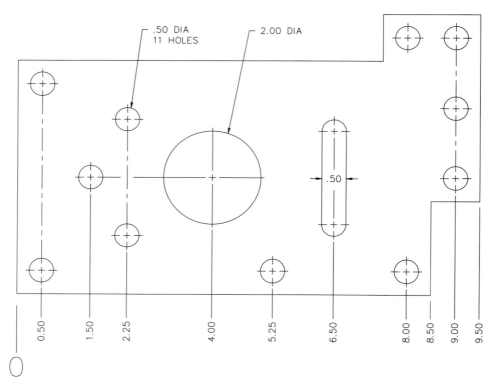

Figure 11–104

STEP 7

With Snap on and Ortho on, begin placing the first vertical ordinate dimension using the DIMORDINATE command. The procedure and prompts are identical to those for placing a horizontal ordinate dimension. See Figure 11–105 and the following prompts to perform this operation. (OSNAP-Endpoint should still be running.)

Command: **DOR** *(For DIMORDINATE)*
Specify feature location: *(Select the endpoint of the center line at "A")*
Specify leader endpoint or [Xdatum/ Ydatum/Mtext/Text/Angle]: *(Pick a point two snap distances to the left of the object at "B")*
Dimension text = 0.50

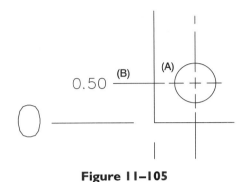

Figure 11–105

STEP 8

Follow the procedure in the previous step to complete the vertical ordinate dimensions along this edge of the object, as in Figure 11–106. Use the OSNAP-Quadrant mode for "A" through "D" and OSNAP-Endpoint for "E." Again, have Ortho on and Snap on. For the leader endpoint, count two grid dots to the left of the object and place the dimensions.

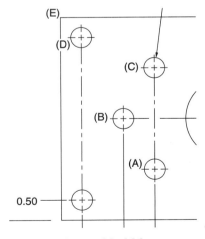

Figure 11–106

STEP 9

Your display should appear similar to the illustration in Figure 11–107.

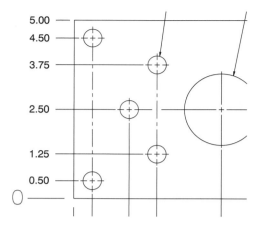

Figure 11–107

STEP 10

Magnify the right portion of the object using the ZOOM-WINDOW command. Use ordinate dimensions and OSNAP-Endpoint to place vertical ordinate dimensions from "A" to "I," as shown in Figure 11–108.

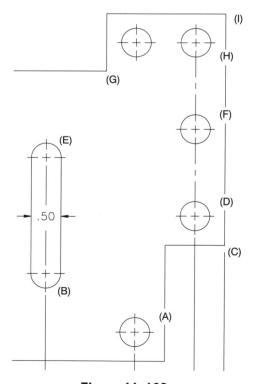

Figure 11–108

STEP 11

Your display should appear similar to the illustration in Figure 11–109.

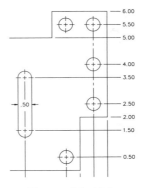

Figure 11–109

STEP 12

Complete the dimensioning by placing two horizontal ordinate dimensions at the locations illustrated in Figure 11–110. As in the earlier steps, have Ortho and Snap modes on, and use OSNAP-Endpoint to select the features. Place the leader end-point two grid dots above the top line of the object. Use the ZOOM-ALL option to return the entire object to your display. See Figure 11–111.

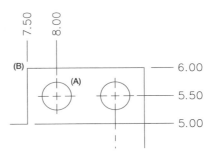

Figure 11–110

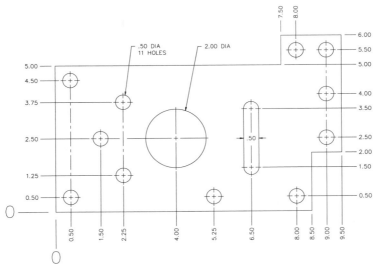

Figure 11–111

PROBLEMS FOR CHAPTER 11

Problems 11–1 through 11–21

1. Open each drawing.
2. Use proper techniques to dimension each drawing.
3. Follow the steps involved in completing Problem 11-1.

PROBLEM 11–1

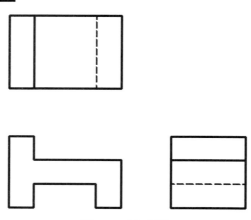

Figure 11–112

STEP 1

Linear dimensions are placed identifying the overall length, width, and depth dimensions. The DIMLINEAR command is used to perform this task. The overall depth in Figure 11–113A is placed above the Right Side view. It could also be placed in the Top view as in Figure 11–113B.

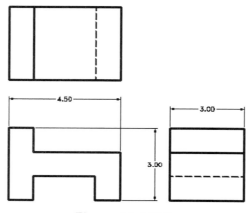

Figure 11–113A

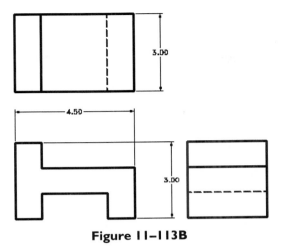

Figure 11–113B

STEP 2

Detail dimensions that identify cuts and slots are placed. Since these cuts are visible in the Front view, the dimensions are placed there. See Figure 11–114.

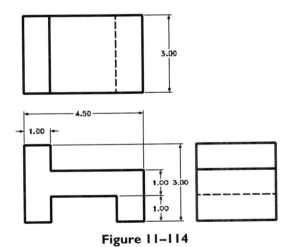

Figure 11–114

STEP 3

Once the spaces between the Front, Top, and Side views are used up by dimensions, the outer areas are used for placing additional dimensions such as the two horizontal dimensions in Figure 11–115.

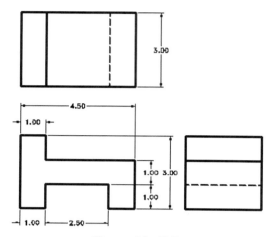

Figure 11–115

STEP 4

All dimensions in Steps 1 through 3 were placed with the DIMLINEAR command. Figure 11–116 shows the effects of using the QDIM command and the Baseline mode for placing the three vertical dimensions located in the Front view.

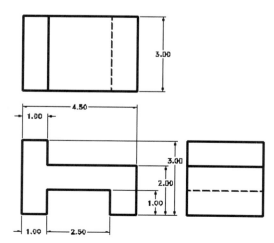

Figure 11–116

PROBLEM 11–2

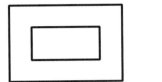

PROBLEM 11–4

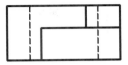

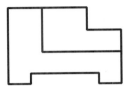

PROBLEM 11–3

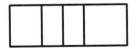

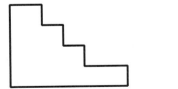

PROBLEM 11–5

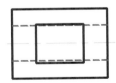

PROBLEM 11-6

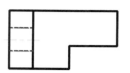

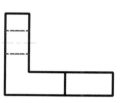

PROBLEM 11-8

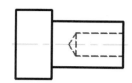

PROBLEM 11-7

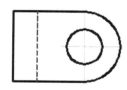

PROBLEM 11-9

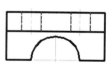

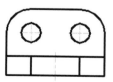

PROBLEM 11–10

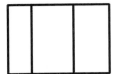

PROBLEM 11–12

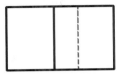

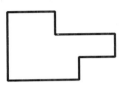

PROBLEM 11–11

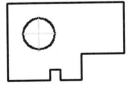

PROBLEM 11–13

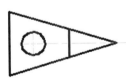

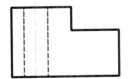

problem EXERCISE

718

PROBLEM 11–14

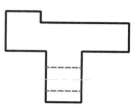
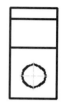

PROBLEM 11–16

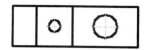

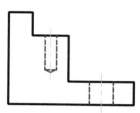
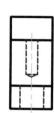

PROBLEM 11–15

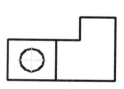

PROBLEM 11–17

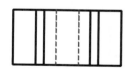

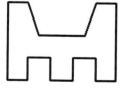

PROBLEM 11-18

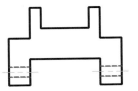

PROBLEM 11-20

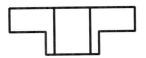

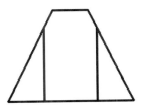

PROBLEM 11-19

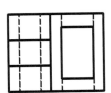

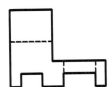

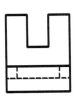

PROBLEM 11-21

Directions for Problem 11–22

Use ordinate dimensioning techniques to dimension this drawing.

PROBLEM 11–22

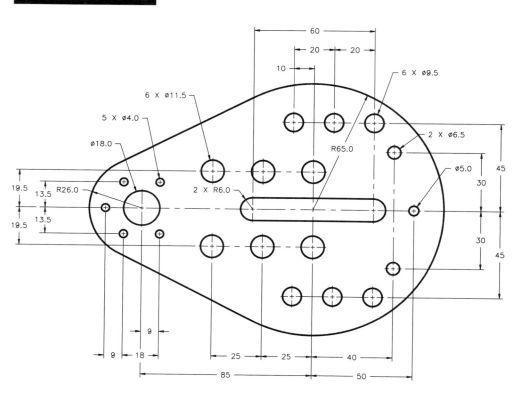

CHAPTER 12

The Dimension Style Manager

THE DIMENSION STYLE MANAGER DIALOG BOX

Dimensions have different settings that affect the group of dimensions. These settings include the control of the dimension text height, the size and type of arrowhead used, and whether the dimension text is centered in the dimension line or placed above the dimension line. These are but a few of the numerous settings available to you. In fact, some settings are used mainly for architectural applications, while other settings are only for mechanical uses. As a means of managing these settings, dimension styles are used to group a series of dimension settings under a unique name to determine the appearance of dimensions.

Begin the process of creating a dimension style by choosing Style… from the Dimension pull-down menu as in Figure 12–1A. This activates the Dimension Style Manager dialog box in Figure 12–1B. Entering the keyboard command DDIM can also activate this dialog box. The current dimension style is listed as "Standard." This style is automatically available when you create any new drawing. In the middle of the dialog box is an image icon that displays how some dimensions will appear based on the current value of the dimension settings. Various buttons are also available to set a dimension style current, create a new dimension style, make modifications to an existing dimension style, display dimension overrides, and compare two dimension styles regarding their differences and similarities.

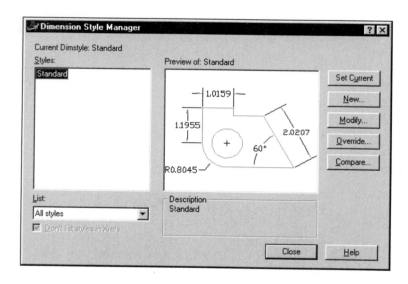

Figure 12–1A   **Figure 12–1B**

To create a new dimension style, click on the New... button. This activates the Create New Dimension Style dialog box in Figure 12–2A. Enter a new name such as "Mechanical" in the New Style Name area. Then click the Continue button. This takes you to the New Dimension Style: Mechanical dialog box in Figure 12–2B, where a number of tabs hold all of the settings needed in dimensioning. The Lines and Arrows tab deals with settings that control dimension lines, extension lines, arrowheads, and center marks used for circles. The Text tab contains the settings that control the appearance, placement, and alignment of dimension text. The Fit tab contains various fit options for placing dimension text and arrows, especially in narrow places. You control the units used in dimensioning through the Primary Units tab. If you need to display primary and secondary units in the same drawing, the Alternate Units tab is used. Finally for mechanical applications, various ways to show tolerances are controlled in the Tolerances tab. Making changes to any of the settings under the tabs will update the preview image, showing these changes. This provides a quick way of previewing how your dimensions will appear in your drawing. All of these areas will be explained in greater detail throughout this chapter.

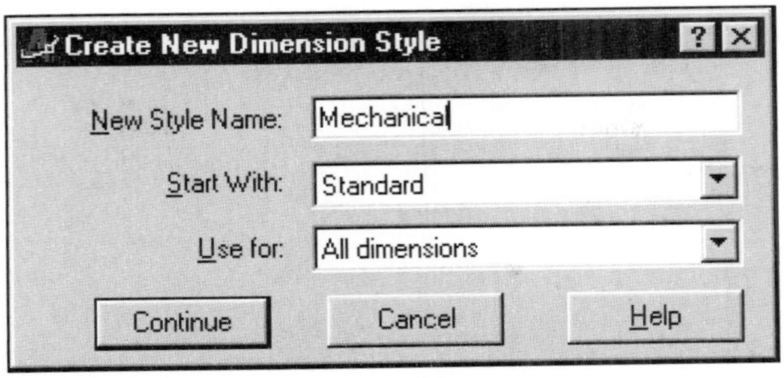

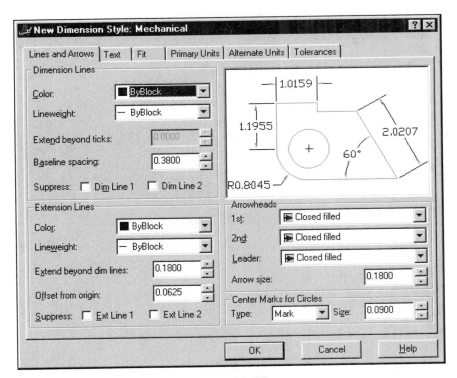

Figure 12–2A

Figure 12–2B

Click the OK button. This returns you to the Dimension Style Manager dialog box in Figure 12–2C. Notice the Mechanical dimension style has been added to the list of styles. Also, any changes made in the tab areas are automatically saved to the dimension style you are creating or modifying. Click on Mechanical and then pick the Set Current button to make this style the current dimension style. Clicking the Close button returns you to your drawing. Let's get a closer look at all of the tabs and their settings. To do this, click on the Modify... button.

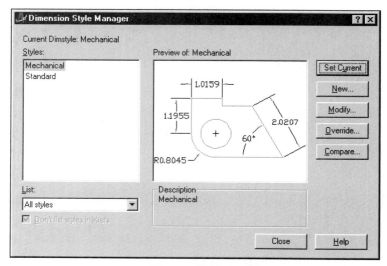

Figure 12–2C

THE LINES AND ARROWS TAB

The Modify Dimension Style: Mechanical dialog box in Figure 12–3 displays the Lines and Arrows tab, which will now be discussed in greater detail. This tab consists of four main areas dealing with dimension lines, extension lines, arrowheads, and center markers.

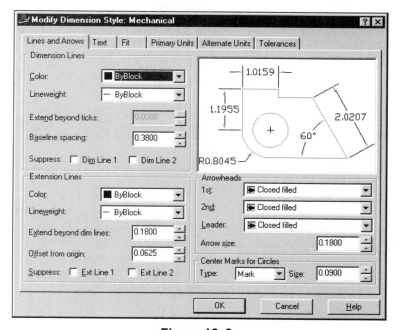

Figure 12–3

DIMENSION LINE SETTINGS

Use the Dimension Lines area, shown in Figure 12–4, to control the color, visibility, and spacing of the dimension line.

Clicking on the Color button of the Dimension Lines area allows you to assign a different color to the dimension line (see Figure 12–5). The color of the current arrowhead terminator will also be affected by the dimension line color. A practical use of this operation is to assign color to the dimension line as a means of controlling line quality by assigning different pen weights for color-dependent plot styles.

The Baseline spacing setting in the Dimension Lines area controls the spacing of baseline dimensions, because they are placed at a distance from each other similar to the illustration in Figure 12–6. This value affects the DIMBASELINE command.

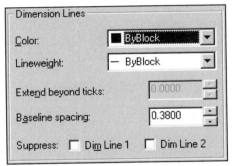

Figure 12–4

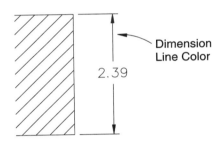

Figure 12–5

By default, dimension line suppression is turned off. To turn on suppression of dimension lines, place a check in the Dim Line 1 or Dim Line 2 box next to Suppress. This operation turns off the display of dimension lines, similar to the illustration in Figure 12–7. This may be beneficial where tight spaces require only the dimension text to be placed.

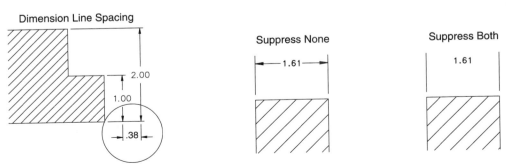

Figure 12–6

Figure 12–7

EXTENSION LINES

The Extension Lines area controls the color, how far the extension extends past the arrowhead, and how far away from the object the extension line begins, and visibility of extension lines. See Figure 12–8.

As with dimension lines, you can set extension lines to a different color to control line quality by clicking on the Color button and assigning a new color that only affects extension lines. See Figure 12–9.

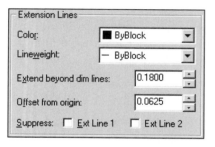

Figure 12–8

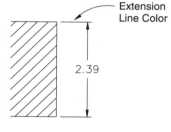

Figure 12–9

The Extend beyond dim lines setting controls how far the extension extends past the arrowhead or dimension line, as shown in Figure 12–10. By default, a value of 0.18 is assigned to this setting.

The Offset from Origin setting controls how far away from the object the extension line will start (see Figure 12–11). By default, a value of 0.06 is assigned to this setting.

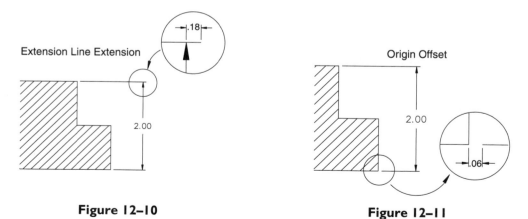

Figure 12–10

Figure 12–11

The Suppress 1st and 2nd check boxes control the visibility of extension lines. They are useful when you dimension to an object line and for avoiding placing the extension line on top of the object line. Placing a check in the 1st box of Suppress turns off the first extension line. The same is true when you place a check in the 2nd box of Suppress—the second extension line is turned off. Study the examples in Figure 12–12 to get a better idea about how suppression of extension lines operates.

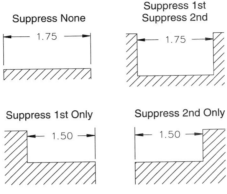

Figure 12–12

ARROWHEAD SETTINGS

Use the Arrowheads area to control the type of arrowhead terminator used for dimension lines and leaders (see Figure 12–13). This dialog box also controls the size of the arrowhead.

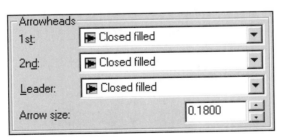

Figure 12–13

Clicking in the 1st: box displays a number of arrowhead terminators. Choose the desired terminator from the partial list in Figure 12–14A. When you choose the desired arrowhead from the 1st: box, the 2nd: box automatically updates to the selection made in the 1st. If you choose an arrowhead from the 2nd: edit box, both first and second arrowheads may be different at opposite ends of the dimension line; this is desired in some applications. Choosing a terminator in the Leader box displays that arrowhead whenever you place a leader.

Illustrated in Figure 12–14B is the complete set of arrowheads, along with their names. The last arrow type is a User Arrow, which allows you to define your own custom arrowhead.

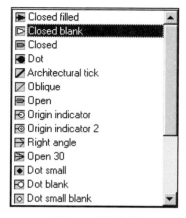

Figure 12–14A

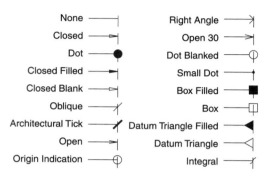

Figure 12–14B

Use the Arrow size setting to control the size of the arrowhead terminator. Arrow types that appear filled in are controlled by the FILL command. With Fill turned on, all arrowheads are filled in as in Figure 12–15. With Fill turned off, only the outline of the arrow is displayed.

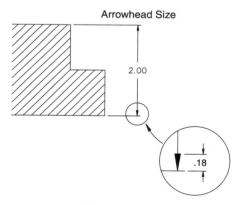

Figure 12–15

CENTER MARK SETTINGS

The Center Marks for Circles area allows you to control the type of center marker used when identifying the centers of circles and arcs. You can make changes to these settings by clicking on the three center mark modes: Mark, None, or Line, as in Figure 12–16. The Size box controls the size of the small plus mark (+).

The three types of center marks are illustrated in Figure 12–17. The Mark option places a plus mark in the center of the circle. The Line option places the plus mark and extends a line past the edge of the circle. The None option will display a circle with no center mark.

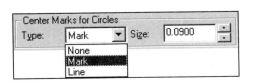

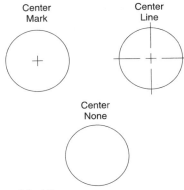

Figure 12–16

Figure 12–17

THE TEXT TAB

Use the Text tab in Figure 12–18 to change the text appearance, such as height, the text placement, such as centered vertically and horizontally, and the text alignment, such as always horizontal or parallel with the dimension line.

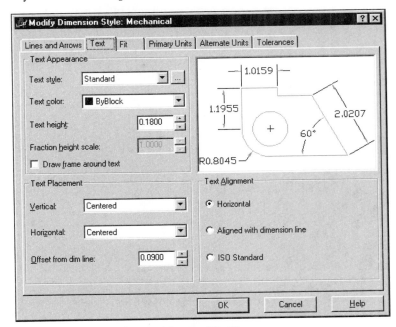

Figure 12–18

TEXT APPEARANCE

Use the Text Appearance area to make changes to settings such as the dimension Text style, dimension Text color, Text height, Fraction height scale, and the ability to draw a frame or box around the text. See Figure 12–19.

The Text color area is used to assign a different color exclusively to the dimension text (see Figure 12–20). This can prove very beneficial especially when you assign a medium lineweight to the dimension text and a thin lineweight to the dimension and extension lines.

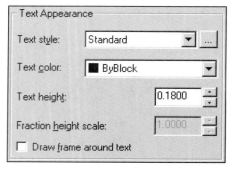

Figure 12–19

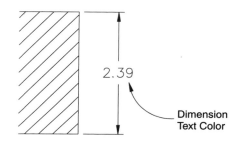

Figure 12–20

The Text height setting controls the size of the dimension text as in Figure 12–21. By default, a value of 0.18 is assigned to this setting.

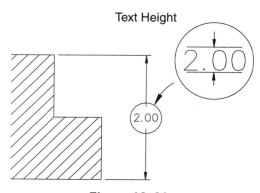

Figure 12–21

If the primary dimension units are set to architectural or fractional, the Fraction height scale activates. Changing this value affects the height of fractions that appear in the dimension. In Figure 12–22, a fractional height scale of 0.5000 will make the fractions as tall as the primary dimension number in the preview window.

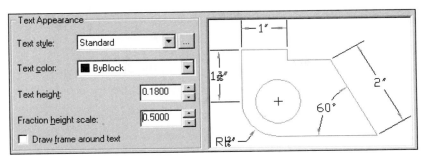

Figure 12–22

Placing a check in the Draw frame around text box will draw a rectangular box around all dimensions, as shown in the preview window in Figure 12–23.

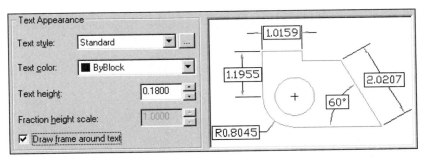

Figure 12–23

TEXT PLACEMENT

The Text Placement area in Figure 12–24 allows you to control the vertical and horizontal placement of dimension text. You can also set an offset distance from the dimension line for the dimension text.

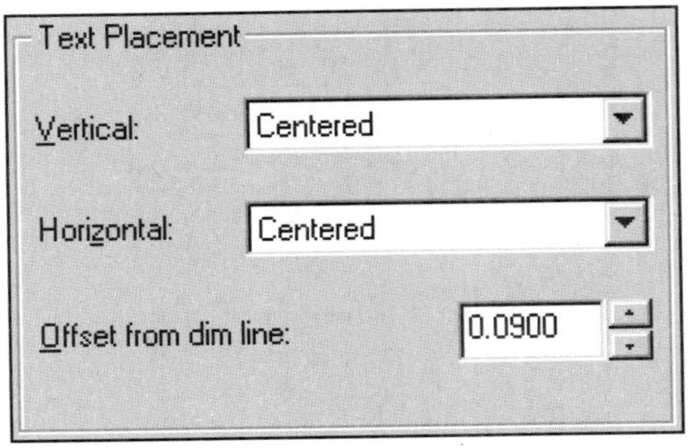

Text Placement	
Vertical:	Centered
Horizontal:	Centered
Offset from dim line:	0.0900

Figure 12–24

VERTICAL TEXT PLACEMENT

The Vertical area of Text Placement controls the vertical justification of dimension text. Clicking in the drop-down edit box allows you to set vertical justification modes. By default, dimension text is centered vertically in the dimension line. Other modes include justifying vertically above the dimension line, justifying vertically outside the dimension line, and using the JIS (Japan International Standard) for placing text vertically. See Figure 12–25A.

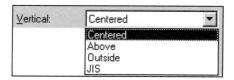

Figure 12–25A

Illustrated in Figure 12–25B is the result of setting the Vertical Justification to Centered. The dimension line will automatically be broken to accept the dimension text.

Illustrated in Figure 12–25C is the result of setting the Vertical Justification to Above. Here the text is placed directly above a continuous dimension line. This mode is very popular for architectural applications.

Figure 12–25B   **Figure 12–25C**

Figure 12–25D illustrates the result of setting the Vertical Justification mode to outside. All text, including those contained in angular and radial dimensions, will be placed outside the dimension lines and leaders.

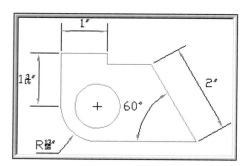

Figure 12–25D

HORIZONTAL TEXT PLACEMENT

At times, dimension text needs to be better located in the horizontal direction; this is the purpose of the Horizontal justification area. Illustrated in Figure 12–26A are the five modes of justifying text horizontally. By default, the horizontal text justification is centered in the dimension.

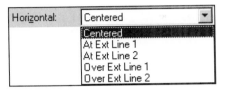

Figure 12–26A

Clicking on the Centered option of the Horizontal Justification area displays the dimension text in Figure 12–26B. This is the default setting because it is the most commonly used text justification mode in dimensioning.

Clicking on the At Ext Line 1 option displays the dimension text in Figure 12–26C, where the dimension text slides over close to the first extension line. Use this option to position the text out of the way of other dimensions. Notice the corresponding option to have the text positioned nearer to the second extension line.

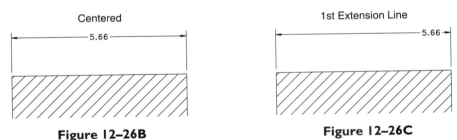

Figure 12–26B   **Figure 12–26C**

Clicking on the Over Ext Line 1 option displays the dimension text parallel to and over the first extension line in Figure 12–26D. Notice the corresponding option to position dimension text over the second extension line.

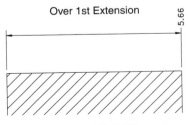

Figure 12–26D

The last item of the Text Placement area deals with setting an offset distance from the dimension line. When you place dimensions, a gap is established between the inside ends of the dimension lines and the dimension text. Entering different values depending on the desired results can control this gap. Study Figures 12–27A and 12–27B, which have different text offset settings.

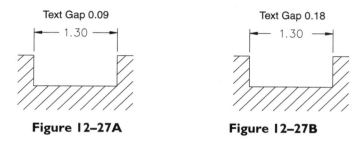

Figure 12–27A   Figure 12–27B

TEXT ALIGNMENT

Use the Text Alignment area in Figure 12–28A to control the alignment of text. Dimension text can either be placed horizontally or parallel (aligned) to the edge of the object being dimensioned. An ISO (International Organization for Standardization) Standard is also available for metric drawings. Click in the appropriate radio button to turn the desired text alignment mode on.

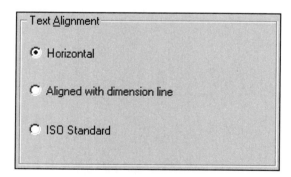

Figure 12–28A

If the Horizontal radio button is clicked, all text will be read horizontally as shown in Figure 12–28B. This includes text located inside and outside the extension lines.

Clicking on the Aligned with dimension line radio button displays the alignment results shown in Figure 12–28C. Here all text is read parallel to the edge being dimensioned. Not only will vertical dimensions align the text vertically, but the 2.06 dimension used to dimension the incline is also parallel to the edge.

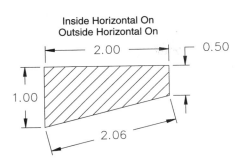

Inside Horizontal On
Outside Horizontal On

Figure 12–28B

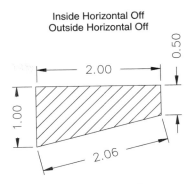

Inside Horizontal Off
Outside Horizontal Off

Figure 12–28C

THE FIT TAB

Use the Fit tab in Figure 12–29 to control how text and/or arrows are displayed if there isn't enough room to place both inside the extension lines. You have extra control of text if is not in the default position. You could place the text over the dimension with or without a leader line. The Scale for Dimension Features area is very important when you dimension in Model Space or when you scale dimensions to Paper Space units. You can even fine tune the placement of text manually and force the dimension line to be drawn between extension lines.

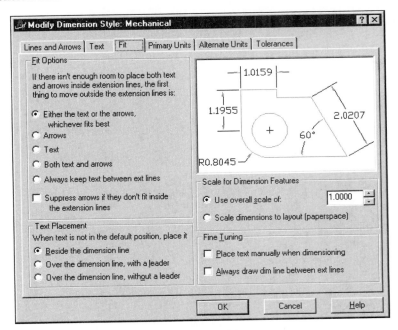

Figure 12–29

FIT OPTIONS

The Fit Options area in Figure 12–30A has the radio button set for Either the text or the arrows, whichever fits best. AutoCAD will decide to move either text or arrows outside extension lines. It will determine the item that fits the best. This setting is illustrated in the preview area in Figure 12–29. It so happens that the preview image is identical when you click on the radio button for Arrows. This tells AutoCAD to move the arrowheads outside the extension lines if there isn't enough room to fit both text and arrows.

Clicking on the Text radio button of the Fit Options area updates the preview image in Figure 12–30B. Here you are moving the dimension text outside the extension lines if the text and arrows do not fit. Since the value 1.0159 is the only dimension that does not fit, it is placed outside the extension lines; but the arrows are drawn inside the extension lines.

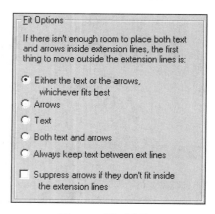

Figure 12–30A

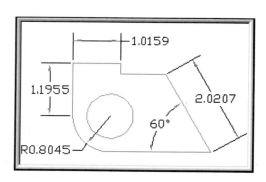

Figure 12–30B

Clicking on the Both text and arrows radio button updates the preview image in Figure 12–30C where the 1.0159 dimension text and arrows are both placed outside of the extension lines.

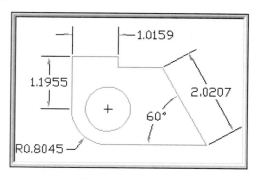

Figure 12–30C

If you click on the radio button for Always keep text between ext lines, the result is illustrated in the preview image in Figure 12–30D. Here all dimension text, including the radius dimension, is placed between the extension lines.

If you click on the radio button for Either the text or the arrows, whichever fits best and you also place a check in the box for Suppress arrows if they don't fit inside the extension lines, the dimension line is turned off only for dimensions that cannot fit the dimension text and arrows. This is illustrated in Figure 12–30E.

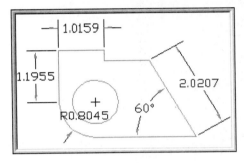

Figure 12–30D   **Figure 12–30E**

TEXT PLACEMENT

You control the placement of the text if it is not in the default position. Your choices, shown in Figure 12–31A, are Beside the dimension line, which is the default, Over the dimension line, with a leader, or Over the dimension line, without a leader.

If you click on the radio button for Text in the Fit Options area and you click on the radio button for Over the dimension line, with a leader, the result is illustrated in Figure 12–31B. For the 1.0159 dimension that does not fit, the text is placed outside the dimension line with the text connected to the dimension line with a leader.

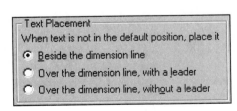

Figure 12–31A   **Figure 12–31B**

If you click on the radio button for Text in the Fit Options area and you click on the radio button for Over the dimension line, without a leader, the result is illustrated in

Figure 12–31C. For the 1.0159 dimension that does not fit, the text is placed outside the dimension. No leader is used.

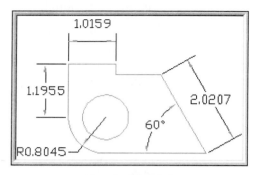

Figure 12–31C

SCALE FOR DIMENSION FEATURES

This setting acts as a multiplier and globally affects all current dimension settings that are specified by sizes or distances. This means that if the current overall scale value is 1.00, other settings that require values will remain unchanged. There is also a radio button allowing you have dimensions automatically scaled to Paper Space units inside a layout. The scale of the viewport will control the scale of the dimensions. (See Figure 12–32A.)

Figure 12–32B shows the effects of an overall scale factor of 1.00 and 2.00. The dimension text, arrows, origin offset, and extension beyond the arrow have all doubled in size.

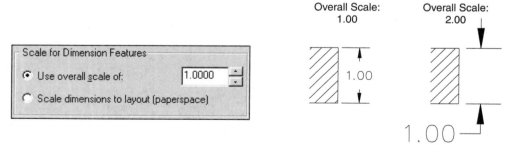

Figure 12–32A

Figure 12–32B

FINE TUNING

You have two options to add further control of the fitting of dimension text in the Fine Tuning area in Figure 12–33A. You can have total control for horizontally justifying dimension text if you place a check in the box for Place text manually when dimensioning. You can also force the dimension line to be drawn between extension lines by placing a check in this box. The results are displayed in Figure 12–33B.

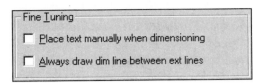

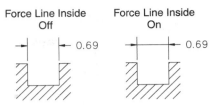

Figure 12–33A   **Figure 12–33B**

THE PRIMARY UNITS TAB

Use the Primary Units tab in Figure 12–34 to control settings affecting the primary units. This includes the type of units the dimensions will be constructed in (decimal, engineering, architectural, and so on), and whether the dimension text requires a prefix or suffix.

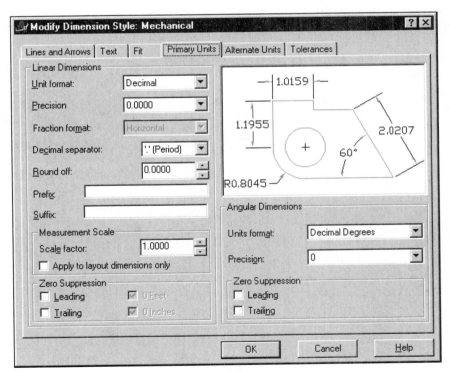

Figure 12–34

The Linear Dimensions area in Figure 12–35A has various settings that deal with primary dimension units. A few of the settings deal with the format when working with fractions. This area activates only if you are working in architectural or fractional units. Even though you may be drawing in architectural units, the dimension units are set by default to decimal. You can also designate the decimal separator as a Period, Comma, or Space.

Clicking in the box for Unit format in Figure 12–35B displays the types of units you can apply to dimensions. You also control the precision of the primary units by clicking in the Precision box.

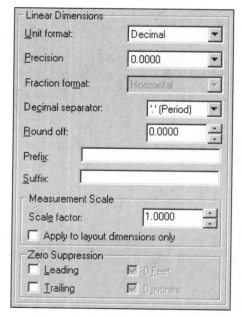

Figure 12–35A

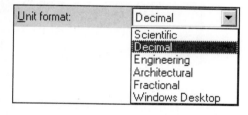

Figure 12–35B

Use a Round Off value to round off all dimension distances to the nearest unit based on the round off value. With a round off value of 0.0000, the dimension text reflects the actual distance being dimensioned, as in Figure 12–35C. With a round off value set to 0.2500, the dimension text reflects the next 0.2500 increment, namely 2.50. See Figure 12–35D.

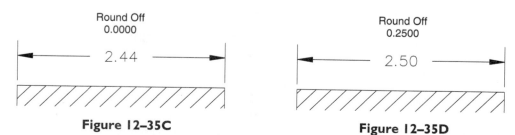

Figure 12–35C

Figure 12–35D

A prefix allows you to specify text that will be placed in front of the dimension text whenever you place a dimension. Use the Suffix box to control the placement of a character string immediately after the dimension value. In the illustration in Figure 12–35E, the letters "IN" have been added to the dimension, signifying "inches."

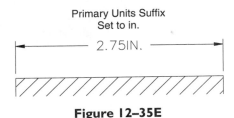

Figure 12–35E

MEASUREMENT SCALE

The Measurement Scale area in Figure 12–36A acts as a multiplier for all linear dimension distances, including radius and diameter dimensions. When a dimension distance is calculated, the current value set in the Scale factor edit box is multiplied by the dimension to arrive at a new dimension value. In the illustration in Figure 12–36B, and with a Linear Scale value of 1.00, the dimension distances are taken at their default values.

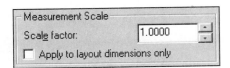

Figure 12–36A

In Figure 12–36C, the Linear Scale value has been changed to 2.00 units. This means 2.00 will multiply every dimension distance; the result is that the previous 3.00 and 2.00 dimensions are changed to 6.00 and 4.00. In a similar fashion, having a Linear Scale value set to 0.50 will cut all dimension values in half.

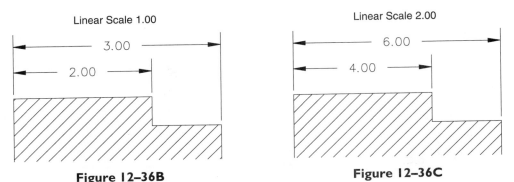

Figure 12–36B   Figure 12–36C

ANGULAR DIMENSIONS

Use the Angular Dimensions area in Figure 12–37 to control unit of measure for angles. Clicking in the Units format box displays four angular measurement modes.

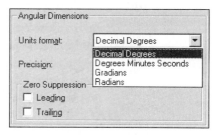

Figure 12–37

THE ALTERNATE UNITS TAB

Use the Alternate Units tab in Figure 12–38 to enable alternate units, set the units and precision of the alternate units, set a multiplier for all units, use a round off distance, and set a prefix and suffix for these units. You also have two placement modes for displaying these units. By default, alternate units are placed beside primary units. The alternate units are enclosed in square brackets. With two sets of units being displayed, your drawing could tend to become very busy. An important application would be to display English and metric units, since some design firms require both.

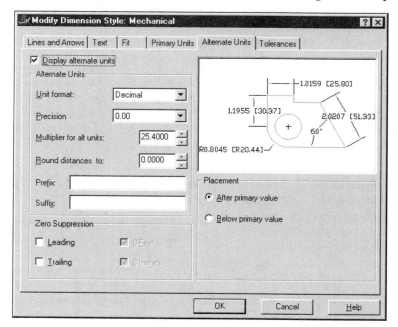

Figure 12–38

ALTERNATE UNITS

Once alternate units are enabled, all items in the Alternate Units area become active, as shown in Figure 12–39. Figure 12–40A shows the effect of turning off Alternate Units. In this example, a single dimension value is calculated and placed between dimension and extension lines. Figure 12–40B shows the effects of Alternate Units being enabled. Next to the calculated dimension value, the alternate value is placed in brackets. This value depends on the current setting in the Multiplier for all units edit box. With this factor set to 25.40, it is used as a multiplier for all calculated alternate dimension values. As with primary units, you can set a round off value, a prefix, and a suffix to be used exclusively for alternate units. For example, if the alternate unit of measure is millimeters, you could enter "mm" for a suffix.

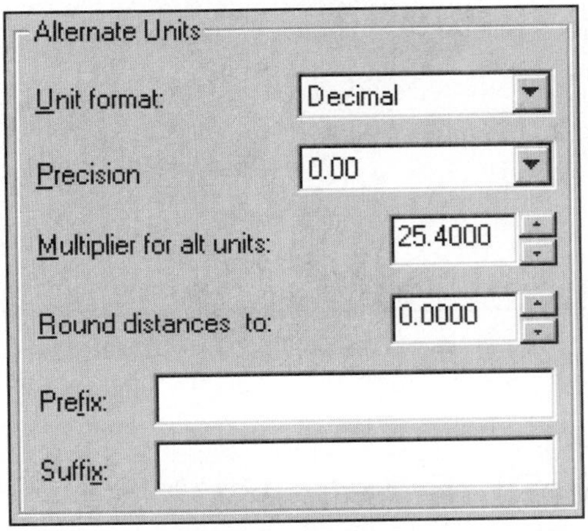

Figure 12–39

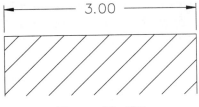

Figure 12–40A

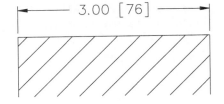

Figure 12–40B

It has already been mentioned that alternate dimensions are placed inside square brackets and to the right of the primary dimension. You could click in the radio button for Below primary value in Figure 12–41. This places the primary dimension above the dimension line and the alternate dimension below it, which may help with the crowded look that alternate dimensions tend to add in a drawing.

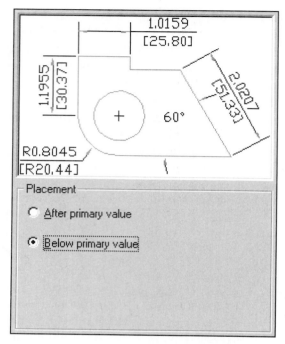

Figure 12–41

THE TOLERANCES TAB

The Tolerances tab in Figure 12–42 consists of various edit boxes used to control the five types of tolerance settings: none, symmetrical, deviation, limits, and basic. Depending on the type of tolerance being constructed, an Upper Value and Lower Value may be set to call out the current tolerance variance. The Vertical position setting allows you to determine where the tolerance will be drawn in relation to the location of the body text. The Scaling for height setting controls the text size of the tolerance. Usually this value is smaller than the body text size.

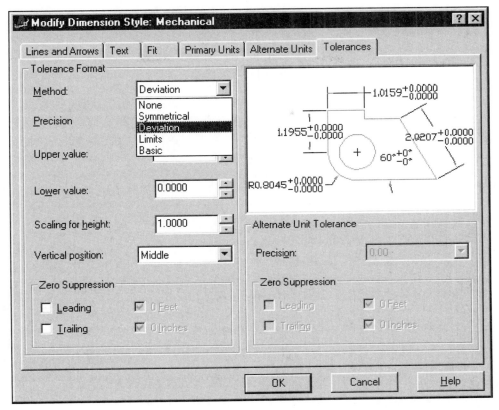

Figure 12–42

TOLERANCE FORMAT

The five tolerance types available in the drop-down list are illustrated in Figure 12–43. A tolerance setting of None uses the calculated dimension value without applying any tolerances. The Symmetrical tolerance uses the same value set in the Upper and Lower Value. The Deviation tolerance setting will have a value set in the Upper Value and an entirely different value set in the Lower Value. The Limits tolerance will use the Upper and Lower values and place the results with the larger limit dimension placed above the smaller limit dimension. The Basic tolerance setting does not add any tolerance value; instead, a box is drawn around the dimension value.

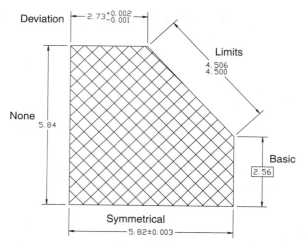

Figure 12–43

The Scaling for height setting affects the height of the tolerance text. In Figure 12–44A, a tolerance height setting of 1.00 has no effect on the height of the tolerance text; in fact the tolerance text is the same height as the main dimension text. Figure 12–44B shows the effect of setting the tolerance height to a value of 0.70 units. Here, the tolerance height is noticeably smaller than the main dimension text height. It is good practice to have the main dimension text value set to a higher value than the tolerance heights, for greater emphasis on the main dimension.

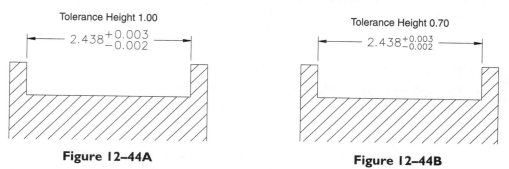

Figure 12–44A   **Figure 12–44B**

DIMASO—THE ASSOCIATIVE DIMENSIONING CONTROL SETTING

Use this dimension setting to turn associative dimensioning on or off; it must be entered from the keyboard. By default, this setting is on where all objects that make up the dimension are considered one object. If this setting is off, all dimension components such as arrowheads, extension lines, dimension lines, and dimension text will be considered single objects (see Figures 12–45A and 12–45B). This has the same effect as using the EXPLODE command on an associative dimension. With DIMASO turned off, commands that normally affect dimensions that are associative will have no effect. In fact, grips have no effect on a dimension considered nonassociative. It is considered poor practice to turn this variable off or to explode dimensions that are associative. The prompt for this dimension setting is as follows:

Command: **DIMASO**
New value for DIMASO<On>: *(Keep the default turned on)*

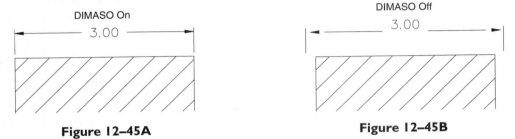

Figure 12–45A

Figure 12–45B

USING DIMENSION TYPES IN DIMENSION STYLES

In addition to creating dimension styles, you can assign dimension types to dimension styles. The purpose of using dimension types is to reduce the number of dimension styles defined in a drawing. For example, the object in Figure 12–46 consists of linear dimensions with three decimal place accuracy, a radius dimension with one decimal place accuracy and an angle dimension with a box surrounding the number. Normally you would have to create three separate dimension styles to create this effect. However, the linear, radius, and angular dimensions consist of what are called dimension types.

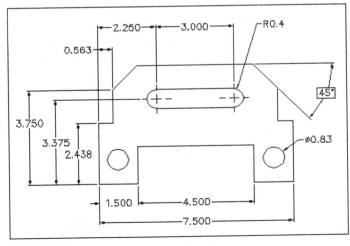

Figure 12–46

Dimension types appear similar to Figure 12–47 when assigned to the Mechanical dimension style. Before creating an angular dimension, you create the dimension type and make changes in various tabs located in the Modify Dimension Style dialog box. These changes will only apply to the dimension type. To expose the dimension types, click on the New button in the main Dimension Style Manager dialog box.

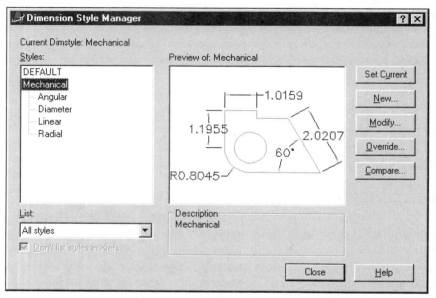

Figure 12–47

When the Create New Dimension Style dialog box appears in Figure 12–48, click in the Use for: box. This usually displays "All dimensions." Notice all dimension types appearing. You can make changes to dimension settings that will apply only for linear, angular, radius, diameter, and ordinate dimensions. Also, a dimension type for Leaders and Tolerances is available. Highlight this option and then click the Continue button.

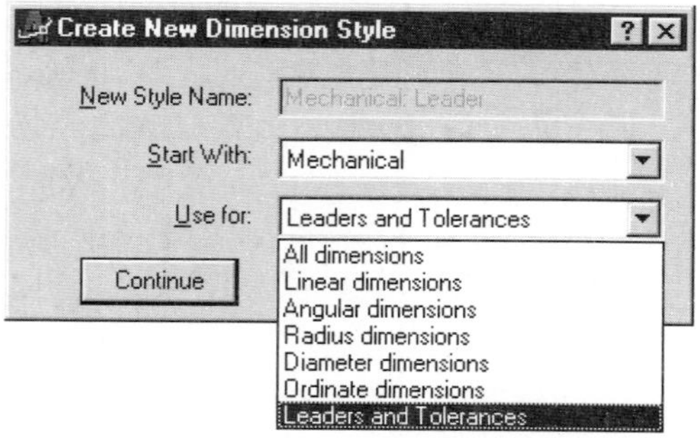

Figure 12–48

This takes you to the Modify Dimension Style: Mechanical: Leader dialog box in Figure 12–49. Any changes you make in the tabs will only apply whenever you place a leader dimension. The use of Dimension Types is an efficient means of keeping the number of dimension styles down to a manageable level.

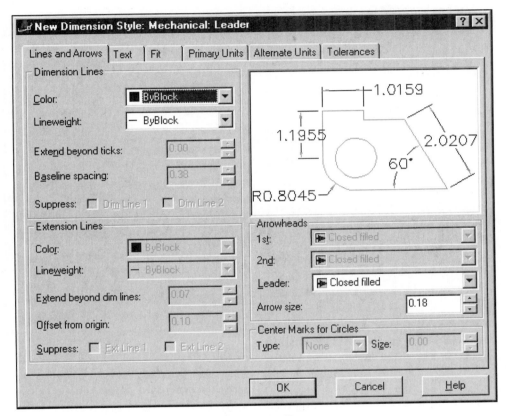

Figure 12–49

OVERRIDING A DIMENSION STYLE

For special cases where you need to change the settings of one dimension, a dimension override would be used. Clicking on the Override button of the main Dimension Style Manager dialog box displays the Override Current Style: Mechanical dialog box in Figure 12–50. The Ext Line 1 and Ext Line 2 boxes in the Extension Lines area have been checked for suppression. One dimension needs to be constructed without displaying extension lines.

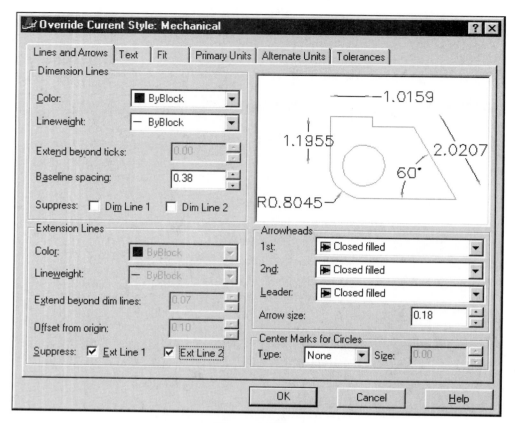

Figure 12–50

Clicking the OK button returns you to the main Dimension Style Manager dialog box in Figure 12–51. Notice that, under the Mechanical style, a new dimension style has been created called <style overrides> (Angular, Diameter, Linear, and Radial were already existing in this example). Also notice that the image in the Preview box shows sample dimensions displayed without extension lines. The <style overrides> dimension style is also the current style. Click the Close button to return to your drawing.

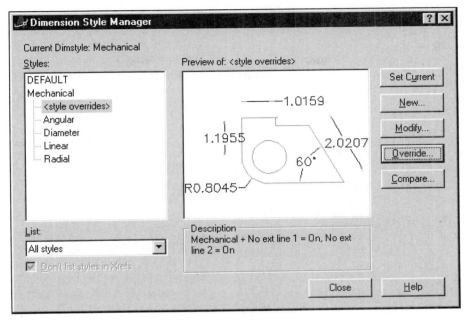

Figure 12–51

In Figure 12–52, a linear dimension is placed identifying the vertical distance of 1.250. To avoid the mistake of placing extension lines on top of existing object lines, the extension lines are not drawn due to the style override.

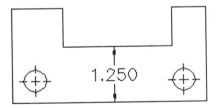

Figure 12–52

Unfortunately, if you place other linear dimensions, these will also lack extension lines. The Dimension Style Manager dialog box is once again activated. Clicking on an existing dimension style such as Mechanical and then clicking on the Set Current button displays the AutoCAD Alert box in Figure 12–53. If you click OK, the style override will disappear from the listing of dimension styles. If you would like to save the overrides under a name, click the Cancel button, right-click on the <style overrides> listing, and rename this style to a new name. This will preserve the settings under this new name.

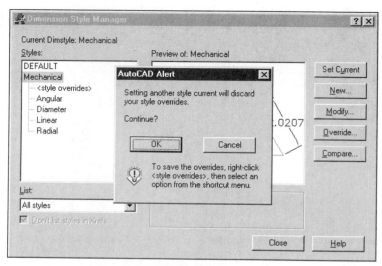

Figure 12–53

COMPARING TWO DIMENSION STYLES

Yet another feature of the Dimension Style Manager dialog box deals with the ability to compare the properties of two dimension styles or view all the properties of one style. In Figure 12–54 when comparing the DEFAULT dimension style with the CONTINUE dimension style, which was previously created, AutoCAD lists the differences between the two styles by providing the description of the dimension setting, the dimension variable name, the setting value of the DEFAULT style, and the setting value of the CONTINUE style. This allows you to quickly preview the differences between two dimension styles.

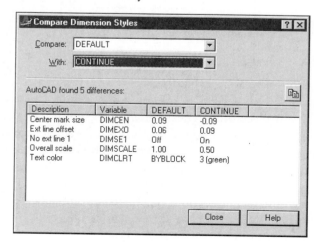

Figure 12–54

You can also obtain a complete listing of all settings and variables for single styles by comparing CONTINUE with CONTINUE in Figure 12–55. Use the scroll bar to view all variables, descriptions, and current values.

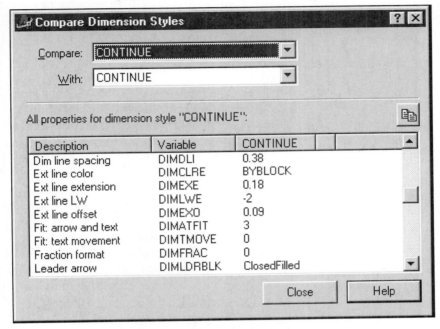

Figure 12–55

MODIFYING THE DIMENSION STYLE OF AN OBJECT

Existing dimensions in a drawing can easily be modified through a number of methods that will be outlined in this section. Figure 12–56 illustrates the first of these methods: the use of the Dim Style Control box, which is part of the Dimension toolbar. First click on a dimension and notice that the dimension highlights and the grips appear. The current dimension style is also displayed in the Dim Style Control box. Click in the control box to display all other styles currently defined in the drawing. Clicking on another style name will change the highlighted dimension to that style. Be careful when docking this toolbar along the side of your screen. In lower screen resolutions, the Dim Style Control box will not display.

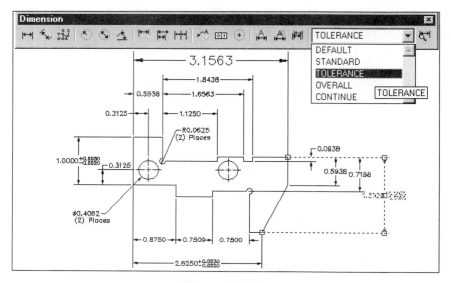

Figure 12–56

The second method of modifying dimensions is illustrated in Figure 12–57A. In this example, another dimension is selected; it highlights and grips appear. Right-clicking the mouse button displays the cursor menu. Choosing Dim Style heading displays the cascading menu of all dimension styles defined in the drawing. Clicking on Other… displays the Apply Dimension Style dialog box in Figure 12–57B. Clicking on the appropriate style and clicking the OK button will change the highlighted dimension to the new style.

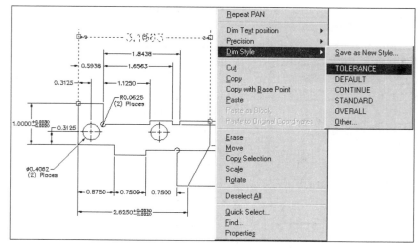

Figure 12–57A

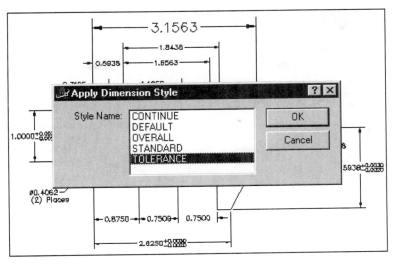

Figure 12–57B

The third method of modifying dimensions begins with selecting the dimension. Again, the dimension highlights and grips appear. Clicking on the Properties button displays the Properties dialog box in Figure 12–58. Click in the Alphabetic tab to display all dimension settings and current values. If you click in the name box next to Dim style, all dimension styles will appear in the edit box. Clicking on the desired style will change the highlighted dimension to the new style.

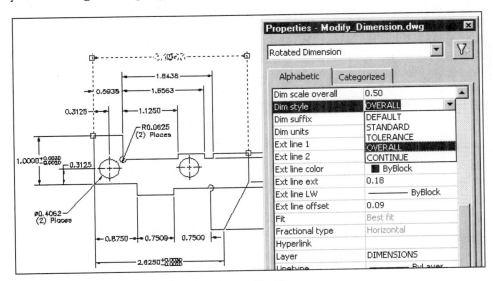

Figure 12–58

TUTORIAL EXERCISE: 12_DIMEX.DWG

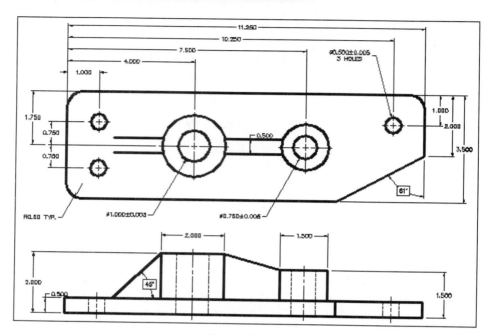

Figure 12–59

Purpose

The purpose of this tutorial is to place dimensions on the drawing of the two-view object illustrated In Figure 12–59.

System Settings

No special system settings need to be made for this drawing file.

Layers

The drawing file Dimex.Dwg has the following layers already created for this tutorial.

Name	Color	Linetype
Object	Magenta	Continuous
Hidden	Red	Hidden
Center	Yellow	Center
Dim	Yellow	Continuous

Suggested Commands

Open the drawing called 12_Dimex.Dwg. The following dimension commands will be used: DIMLINEAR, DIMCONTINUE, DIMBASELINE, DIMCENTER, DIMRADIUS, DIMDIAMETER, DIMANGULAR, and LEADER. All dimension commands may be chosen from the Dimension toolbar or the Dimension pull-down menu, or entered from the keyboard. Use the ZOOM command to get a closer look at details and features that are being dimensioned.

Whenever possible, substitute the appropriate command alias in place of the full AutoCAD command in each tutorial step; for example, use "CO" for the COPY command, "L" for the LINE command, and so on. The complete listing of all command aliases is located in Chapter 1, Table 1–2.

STEP I

To prepare for the dimensioning of the drawing, use the DDIM command to activate the Dimension Style Manager dialog box. Click on the New button, which activates the Create New Dimension Style dialog box in Figure 12–60. In the New Style Name area, enter "MECHANICAL." Click the Continue button to create the style. See Figure 12–60.

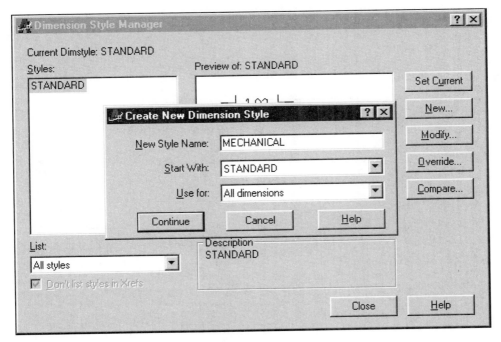

Figure 12–60

STEP 2

The New Dimension Style: ME-CHANICAL dialog box appears in Figure 12–61. Make the following changes in the Lines and Arrows tab: Change the size of the arrowheads from 0.18 to a new value of 0.12; Change the size of the Extension beyond dimension lines from a value of 0.18 to a new value of 0.07. Change the Offset from origin from a value of 0.06 to a new value of 0.12. Finally, be sure the Center Marks for Circles is set to Line. Your display should appear similar to Figure 12–61.

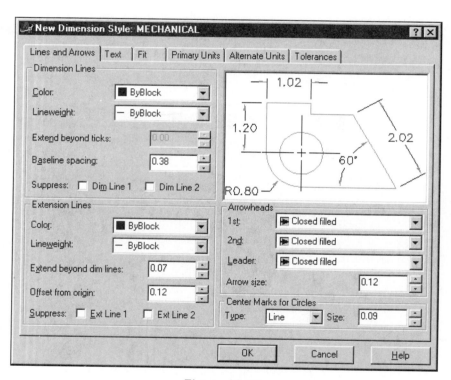

Figure 12–61

STEP 3

Click on the Text tab and change the Text color from Byblock to Green. Also change the Text height from a value of 0.18 to a new value of 0.12. Your display should appear similar to Figure 12–62. Click the OK button to return to the main Dimension Style Manager dialog box.

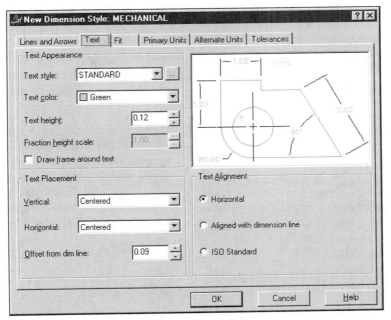

Figure 12–62

STEP 4

You will now create four dimension types and change various settings for each type. The four dimension types are Linear, Angular, Radial, and Diameter. Click the New button to display the Create New Dimen-

sion Style dialog. In the Use for: box, click on Linear dimensions in Figure 12–63 and then click the Continue button. Any changes made to dimension settings will be applied only to linear dimensions.

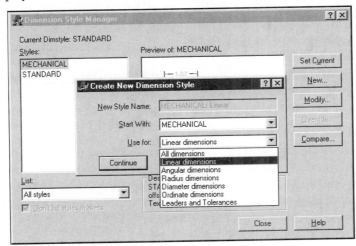

Figure 12–63

STEP 5

Click on the Primary Units tab and change the Precision for linear dimensions to 3 places. Your display should appear similar to Figure 12–64. Notice that the image in the preview box displays linear dimensions only to 3-decimal-place accuracy. Click the OK button to return to the main Dimension Style Manager dialog box.

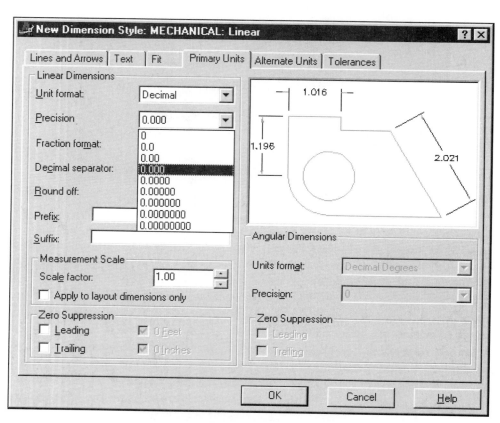

Figure 12–64

STEP 6

Notice in the Styles area of the Dimension Style Manager that a new style type called Linear is listed under MECHANICAL. Click the New button and create another dimension type by selecting Angular dimensions in Figure 12–65. Click the Continue button.

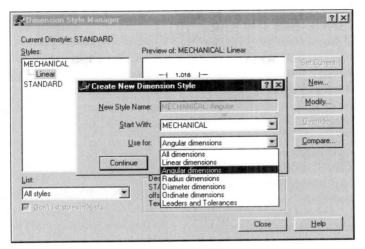

Figure 12–65

STEP 7

A rectangular box will be drawn around all angular dimensions to signify that this value is considered basic. Click on the Text tab and in the Text Appearance area, place a check for Draw frame around text. Notice the image in the Preview box shows only an angular dimension with a box placed around the text. Your display should appear similar to Figure 12–66. Click the OK button to return to the main Dimension Style Manager dialog box.

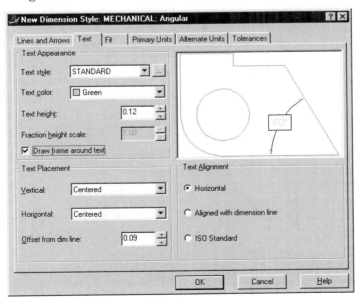

Figure 12–66

STEP 8

Create another new dimension type by clicking on Radius dimensions in Figure 12–67. Click the Continue button.

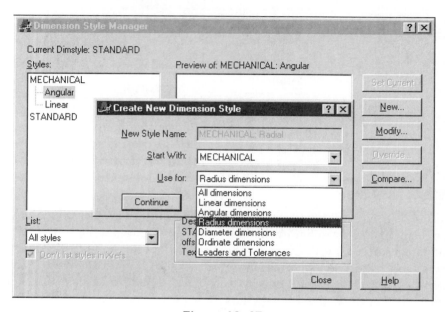

Figure 12–67

STEP 9

Click on the Lines and Arrows tab and change the Center Marks for Circles to None in Figure 12–68A. Then click in the Fit tab and in the Fit options area, click on the radio button for Text in Figure 12–68B to have the text placed outside any radius dimension. Click the OK button and return to the main Dimension Style Manager dialog box.

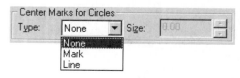

Figure 12–68A

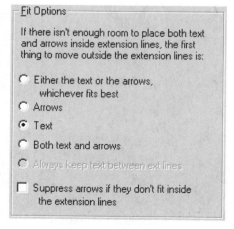

Figure 12–68B

STEP 10

Create the last new dimension type by clicking on Diameter dimensions in Figure 12–69. Click the Continue button.

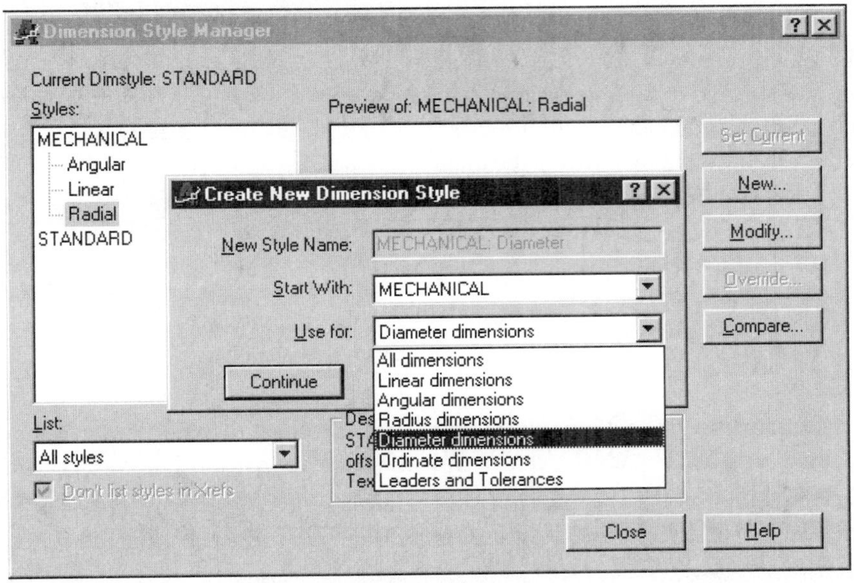

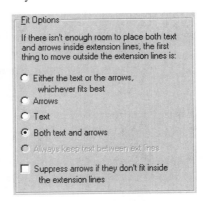

Figure 12–69

STEP 11

As with the Radius dimension type, click on the Lines and Arrows tab and change the Center Marks for Circles to None in Figure 12–70A.

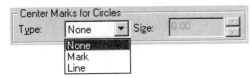

Figure 12–70A

Then click in the Fit tab and in the Fit options area, click on the radio button for Both text and arrows in Figure 12–70B,

to have the text and arrows placed outside any diameter dimension.

Figure 12–70B

Click on the Primary Units tab and change the Precision to 3-decimal-place accuracy in Figure 12–70C.

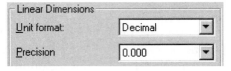

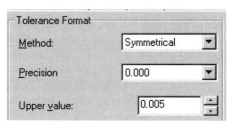

Figure 12–70C

Finally, click on the Tolerances tab and change the Method to Symmetrical and the Upper value to 0.005 in Figure 12–70D. Click the OK button and return to the main Dimension Style Manager dialog box.

Figure 12–70D

STEP 12

In the Dimension Style Manager dialog box, click on MECHANICAL in the Styles area. Then click on the Set Current button to make MECHANICAL the current dimension style. Your display should appear similar to Figure 12–71. Click the Close button to save all changes and return to the drawing.

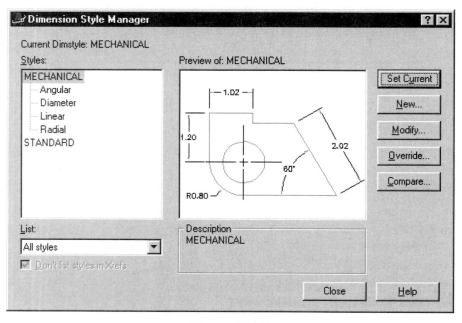

Figure 12–71

STEP 13

Begin placing center markers to identify the centers of all circular features in the Top view. Use the DIMCENTER command to perform this operation on circles "A" through "E" in Figure 12–72.

Command: **DCE** *(For DIMCENTER)*
Select arc or circle: *(Select the edge of circle "A")*
Repeat the above procedure for circles "B," "C," "D," and "E" in Figure 12–72.

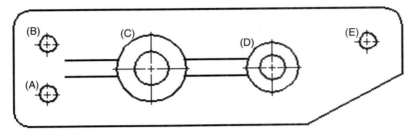

Figure 12–72

STEP 14

Use the QDIM command to place a string of baseline dimensions. First select the individual highlighted lines in Figure 12–73A. When the group of dimensions previews as continued dimensions, change this grouping to Baseline and click a location to place the baseline group of dimensions, as shown in Figure 12–73B.

Command: **QDIM**
Select geometry to dimension: *(Select the lines "A" through "F" in Figure 12–73A)*
Select geometry to dimension: *(Press ENTER to continue)*

Specify dimension line position, or
[Continuous/Staggered/Baseline/
    Ordinate/Radius/Diameter/
    datumPoint/Edit]
<Continuous>: **B** *(For Baseline)*
Specify dimension line position, or
[Continuous/Staggered/Baseline/
    Ordinate/Radius/Diameter/
    datumPoint/Edit]
<Baseline>: *(Pick a point on the screen to locate the baseline dimensions and exit the command)*

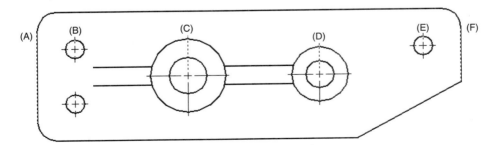

Figure 12–73A

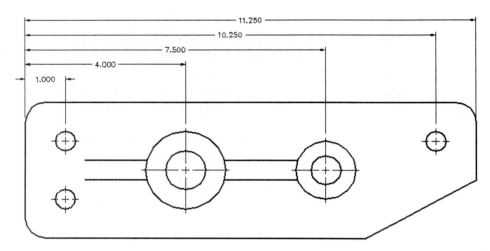

Figure 12–73B

STEP 15

Magnify the left side of the Top view using the ZOOM command. Then use the DIMLINEAR command to place the 0.750 vertical dimension in Figure 12–74.

Command: **DLI** *(For DIMLINEAR)*
Specify first extension line origin or
   <select object>: *(Select the endpoint of
   the centerline at "A")*
Specify second extension line origin:
   *(Select the endpoint of the centerline at
   "B")*
Specify dimension line location or
[Mtext/Text/Angle/Horizontal/Vertical/
   Rotated]: *(Locate the dimension
   approximately at "C")*
Dimension text = 0.750

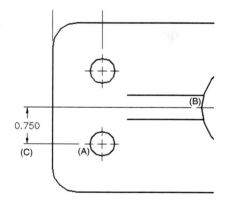

Figure 12–74

STEP 16

Use the DIMCONTINUE command to place the next dimension in line with the previous dimension in Figure 12–75.

Command: **DCO** *(For DIMCONTINUE)*
Specify a second extension line origin or [Undo/Select] <Select>: *(Select the endpoint of the centerline at "A")*
Dimension text = 0.750
Specify a second extension line origin or [Undo/Select] <Select>: *(Press ENTER to exit this mode)*
Select continued dimension: *(Press ENTER to exit this command)*

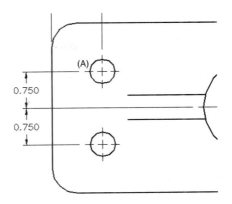

Figure 12–75

STEP 17

Use the DIMLINEAR command to place the 1.750 vertical dimension in Figure 12–76.

Command: **DLI** *(For DIMLINEAR)*
Specify first extension line origin or <select object>: *(Select the endpoint of the extension line at "A")*
Specify second extension line origin: *(Select the endpoint of the line at "B")*
Specify dimension line location or [Mtext/Text/Angle/Horizontal/Vertical/Rotated]: *(Locate the dimension approximately at "C")*
Dimension text = 1.750

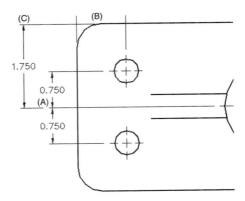

Figure 12–76

STEP 18

Use the PAN command to slide over to the right side of the Top view while keeping the same zoom percentage. Then use the QDIM command and select the highlighted lines in Figure 12–77A. When the group of dimensions previews as continued dimensions, change this grouping to baseline and click a location to place the baseline group of dimensions in Figure 12–77B.

Command: **QDIM**
Select geometry to dimension: *(Select lines "A" through "C" in Figure 12–77A)*
Select geometry to dimension: *(Press ENTER to continue)*
Specify dimension line position, or [Continuous/Staggered/Baseline/ Ordinate/Radius/Diameter/ datumPoint/Edit] <Continuous>: **B** *(For Baseline)*
Specify dimension line position, or [Continuous/Staggered/Baseline/ Ordinate/Radius/Diameter/ datumPoint/Edit] <Baseline>: **P** *(For datumPoint)*

Select new datum point: (Select the endpoint at "A" again to identify a new datum point)
Specify dimension line position, or [Continuous/Staggered/Baseline/ Ordinate/Radius/Diameter/ datumPoint/Edit] <Baseline>: *(Pick a point on the screen to locate the baseline dimensions and exit the command)*

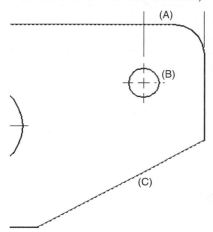

Figure 12–77A

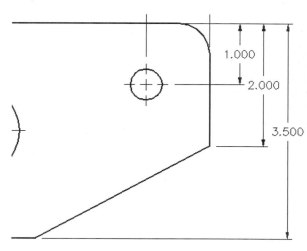

Figure 12–77B

STEP 19

Use the DIMANGULAR command to place the 61° dimension in Figure 12–78.

Command: **DAN** *(For DIMANGULAR)*
Select arc, circle, line, or <specify vertex>: *(Select the line at "A")*
Select second line: *(Select the line at "B")*
Specify dimension arc line location or [Mtext/Text/Angle]: *(Locate the angular dimension at "C")*
Dimension text = 61

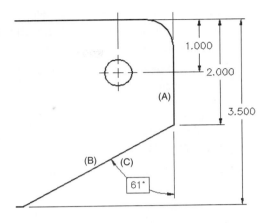

Figure 12–78

STEP 20

Use the ZOOM command and the Extents option to display both the Front and Top views. Then use the DIMDIAMETER command to place two diameter dimensions in Figure 12–79.

Command: **DDI** *(For DIMDIAMETER)*
Select arc or circle: *(Select the circle at "A")*
Dimension text = 1.000

Specify dimension line location or [Mtext/Text/Angle]: *(Locate the diameter dimension at "B")*
Command: **DDI** *(For DIMDIAMETER)*
Select arc or circle: *(Select the circle at "C")*
Dimension text = 0.750
Specify dimension line location or [Mtext/Text/Angle]: *(Locate the diameter dimension at "D")*

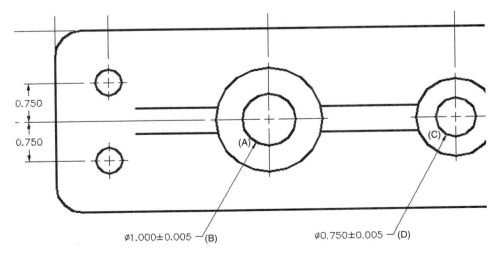

Figure 12–79

STEP 21

Place a radius dimension using the DIMRADIUS command in Figure 12–80A.

Command: **DRA** *(For DIMRADIUS)*
Select arc or circle: *(Select the arc at "A")*
Dimension text = 0.50
Specify dimension line location or [Mtext/
 Text/Angle]: *(Locate the radius
 dimension at "B")*

Because the two other arcs share the same radius value, use the DDEDIT command to edit this dimension value.

Clicking on the radius value activates the Multiline Text Editor dialog box. Add the note "TYP." for TYPICAL as the dimension suffix in Figure 12–80B. Click the OK button to return to your drawing.

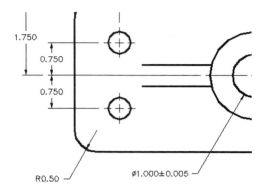

Figure 12–80A

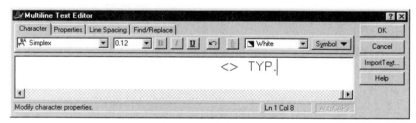

Figure 12–80B

The results are displayed in Figure 12–80C.

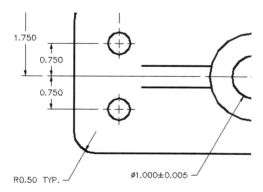

Figure 12–80C

STEP 22

Place a diameter dimension using the DIMDIAMETER command in Figure 12–81A.

Command: **DDI** *(For DIMDIAMETER)*
Select arc or circle: *(Select the circle at "A")*
Dimension text = 0.500
Specify dimension line location or [Mtext/Text/Angle]: *(Locate the diameter dimension at "B")*

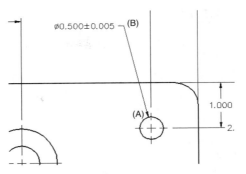

Figure 12–81A

Since the two other smaller holes share the same diameter value, use the DDEDIT command to edit this dimension value. Clicking on the diameter value activates the Multiline Text Editor dialog box. Drop down to the next line and add the note "3 HOLES" in Figure 12–81B. Click the OK button to return to your drawing.

Figure 12–81B

The results are illustrated in Figure 12–81C.

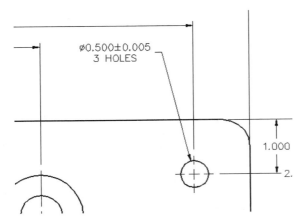

Figure 12–81C

STEP 23

One more dimension needs to be placed in the Top view: the 0.50 width of the rib. Unfortunately, because of the placement of this dimension, extension lines will be drawn on top of the object's lines. This is considered poor practice. To remedy this, a dimension override will be created. To do this, activate the Dimension Style Manager dialog box and click on the Override button. This displays the Override Current Style: MECHANICAL dialog box. In the Lines and Arrows tab, place checks in the boxes to Suppress (turn off) Ext Line 1 and Ext Line 2 in the Extension Lines area (see Figure 12–82A).

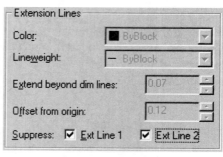

Figure 12–82A

Click the OK button and notice the new <style overrides> listing under MECHANICAL in Figure 12–82B. Notice also the preview image lacks extension lines. Click the Close button to return to the drawing.

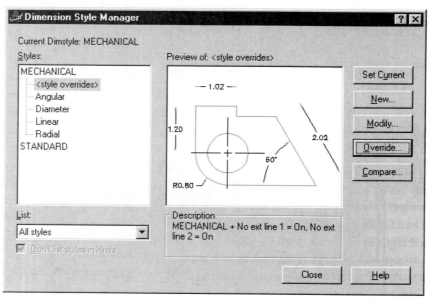

Figure 12–82B

STEP 24

Pan to the middle area of the Top view and place the 0.500 rib width dimension in Figure 12–83 while in the dimension style override. This dimension should be placed without having any extension lines visible.

Command: **DLI** *(For DIMLINEAR)*
Specify first extension line origin or
   <select object>: **Nea**
to *(Select the line at "A')*
Specify second extension line origin: **Per**
to *(Select the line at "B")*
Specify dimension line location or [Mtext/
   Text/Angle/Horizontal/Vertical/
   Rotated]: *(Locate the dimension at "B")*
Dimension text = 0.500

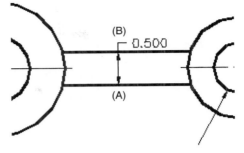

Figure 12–83

STEP 25

The completed Top view, including dimensions, is illustrated in Figure 12–84A. Use this figure to check that all dimensions have been placed and all features such as holes and fillets have been properly identified.

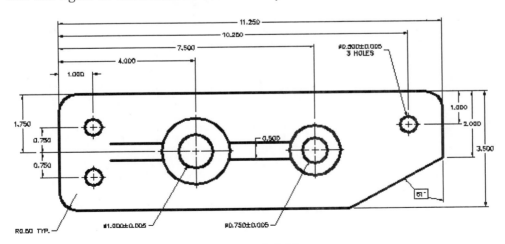

Figure 12–84A

The Front view in Figure 12–84B will now be the focus for the next series of dimensioning steps. Again use the ZOOM and PAN commands whenever you need to magnify or slide to a better drawing view position.

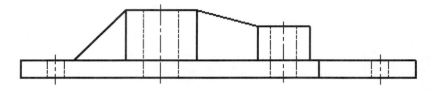

Figure 12–84B

Before continuing, activate the Dimension Style Manager dialog box, click on the MECHANICAL style and then click on the Set Current button. An AutoCAD Alert box displays in Figure 12–84C.

Making MECHANICAL current will discard any style overrides. Click the OK button to discard the changes because you need to return to having extension lines visible in linear dimensions.

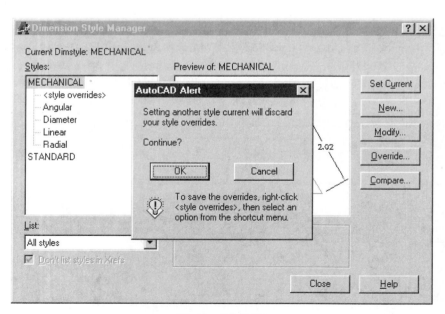

Figure 12–84C

STEP 26

Place a horizontal linear dimension in Figure 12–85 using the DIMLINEAR command.

Command: **DLI** *(For DIMLINEAR)*
Specify first extension line origin or
   <select object>: *(Select the endpoint of the corner at "A")*
Specify second extension line origin:
   (Select the endpoint of the corner at "B")
Specify dimension line location or [Mtext/
   Text/Angle/Horizontal/Vertical/
   Rotated]: *(Locate the dimension at "C")*
Dimension text = 2.000

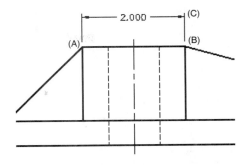

Figure 12–85

STEP 27

Place horizontal and vertical dimensions in Figure 12–86 using the DIMLINEAR command.

Command: **DLI** *(For DIMLINEAR)*
Specify first extension line origin or
   <select object>: *(Select the endpoint of the corner at "A")*
Specify second extension line origin:
   (Select the endpoint of the corner at "B")
Specify dimension line location or [Mtext/
   Text/Angle/Horizontal/Vertical/
   Rotated]: *(Locate the dimension at "C")*

Dimension text = 1.500
Command: **DLI** *(For DIMLINEAR)*
Specify first extension line origin or
   <select object>: *(Select the endpoint of the corner at "A")*
Specify second extension line origin:
   (Select the endpoint of the corner at "D")
Specify dimension line location or [Mtext/
   Text/Angle/Horizontal/Vertical/
   Rotated]: *(Locate the dimension at "E")*
Dimension text = 1.500

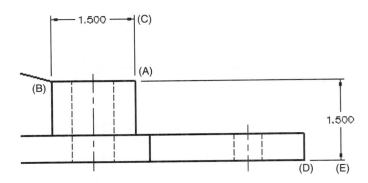

Figure 12–86

STEP 28

Use the QDIM command to place the vertical baseline dimensions in Figure 12–87. Place the 45° angular dimension.

Command: **QDIM**
Select geometry to dimension: *(Select horizontal lines "A", "B", and "C")*
Select geometry to dimension: *(Press ENTER to continue)*
Specify dimension line position, or [Continuous/Staggered/Baseline/Ordinate/Radius/Diameter/datumPoint/Edit] <Baseline>: **P** *(For datumPoint)*
Select new datum point: *(Select the endpoint at "A" again to identify a new datum point)*
Specify dimension line position, or [Continuous/Staggered/Baseline/Ordinate/Radius/Diameter/datumPoint/Edit] <Baseline>: *(Locate the baseline dimensions at "D")*

Command: **DAN** *(For DIMANGULAR)*
Select arc, circle, line, or <specify vertex>: *(Select the line at "E")*
Select second line: *(Select the line at "F")*
Specify dimension arc line location or [Mtext/Text/Angle]: *(Locate the angular dimension)*
Dimension text = 45

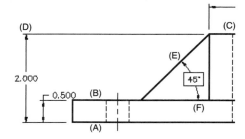

Figure 12–87

Once you have completed all dimensioning steps, your drawing should appear similar to Figure 12–88.

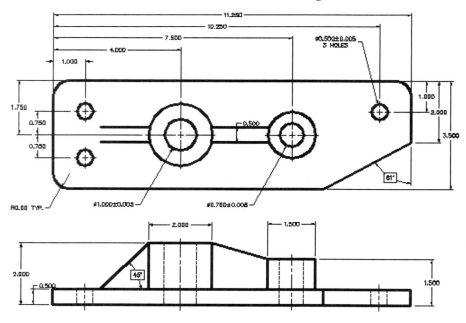

Figure 12–88

PROBLEMS FOR CHAPTER 12

Problems 12–1 through 12–19

1. Use the grid and spacing of 0.50 units to determine all dimensions.
2. Reproduce the views shown and fully dimension the drawings.

PROBLEM 12–1

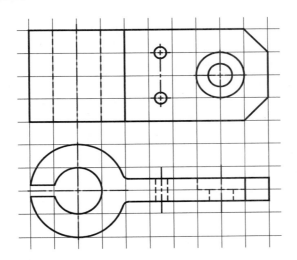

PROBLEM 12–2

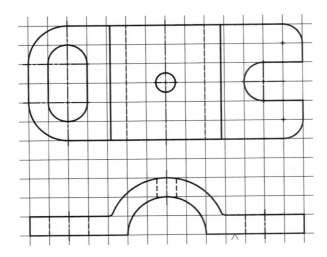

PROBLEM 12–3

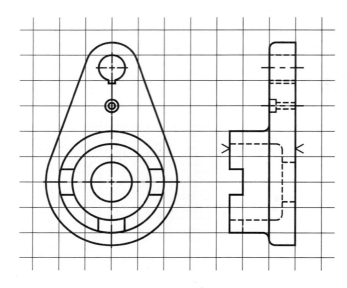

PROBLEM 12–4

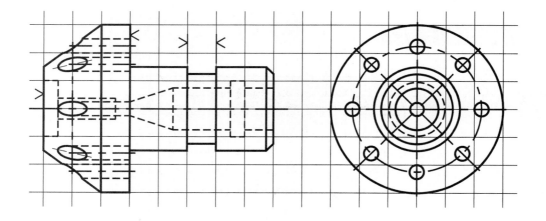

PROBLEM 12–5

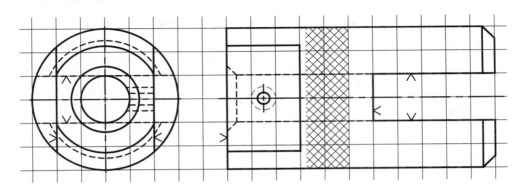

PROBLEM 12–6

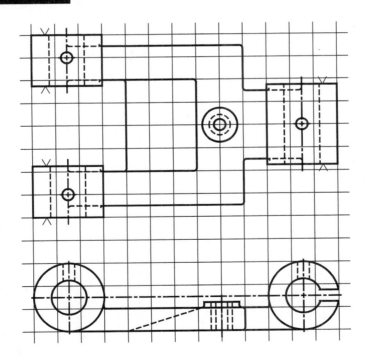

PROBLEM 12–7

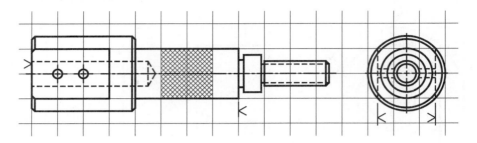

PROBLEM 12–8

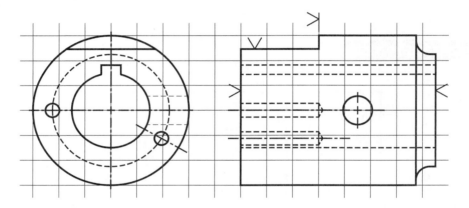

PROBLEM 12–9

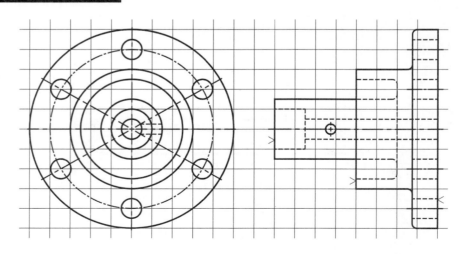

PROBLEM 12–10

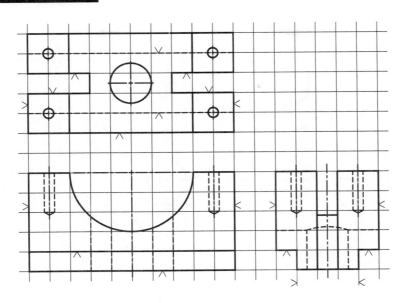

PROBLEM 12–11

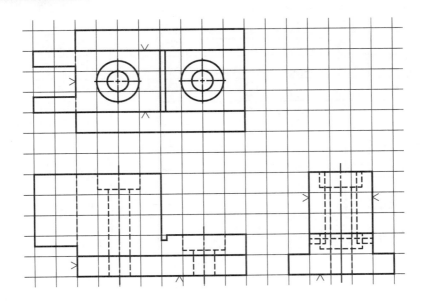

PROBLEM 12–12

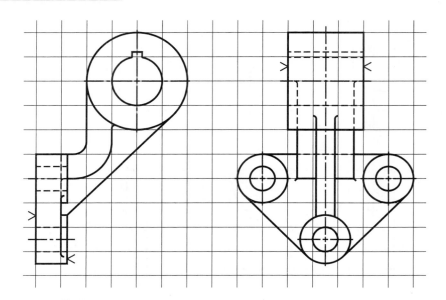

PROBLEM 12–13

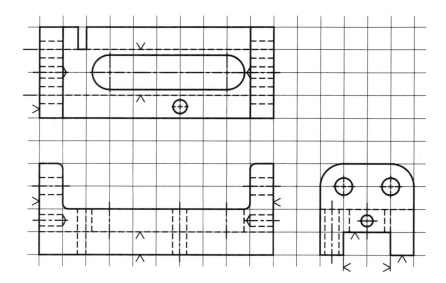

PROBLEM 12–14

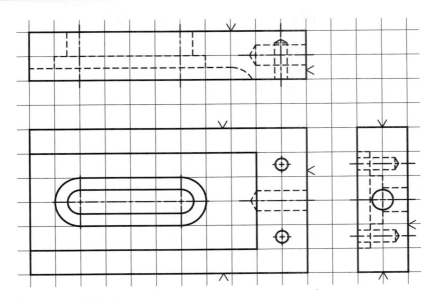

PROBLEM 12–15

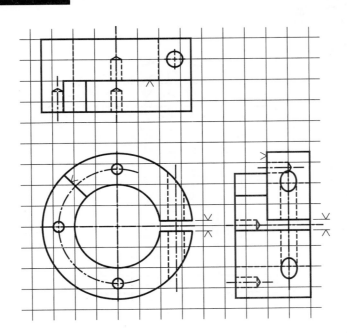

PROBLEM 12–16

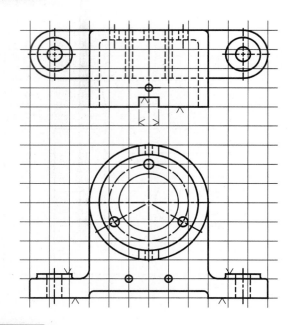

PROBLEM 12–17

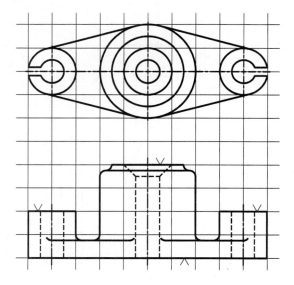

PROBLEM 12–18

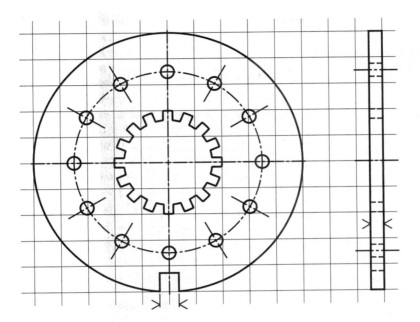

PROBLEM 12–19

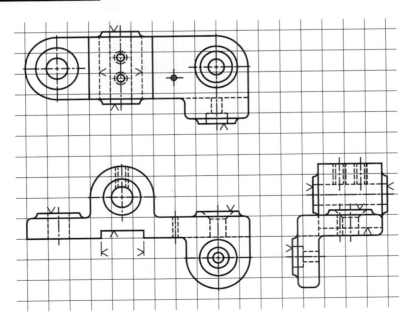

Problems 12–20 through 12–25

1. Convert the isometric drawings provided to orthographic drawings, showing as many views as necessary to communicate the design.

2. Fully dimension your drawings.

PROBLEM 12–20

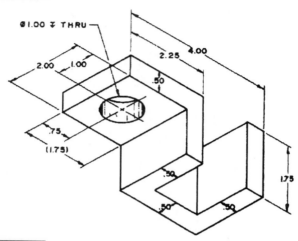

PROBLEM 12–21

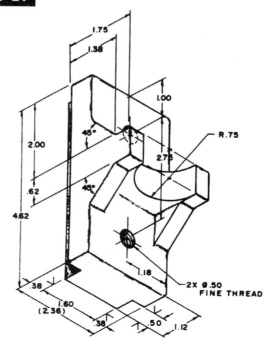

PROBLEM 12–22

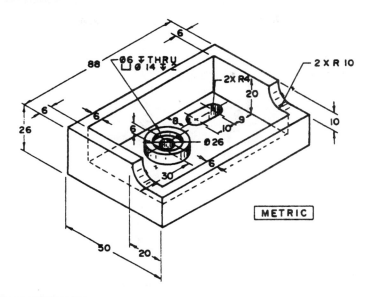

METRIC

PROBLEM 12–23

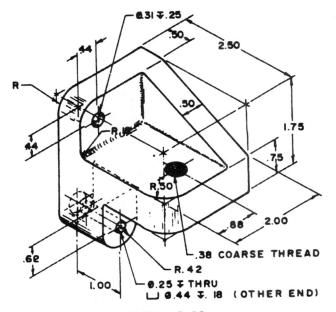

.38 COARSE THREAD

R.42

Ø.25 ⊤ THRU

⊔ Ø.44 ⊤.18 (OTHER END)

ALL UNMARKED RADII = R.09

PROBLEM 12-24

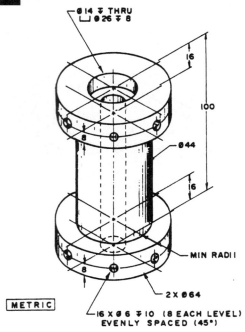

Ø14 ⌲ THRU
⌴ Ø26 ⌲ 8

16

100

Ø44

16

8

MIN RADII

8

2 X Ø64

METRIC

16 X Ø6 ⌲10 (8 EACH LEVEL)
EVENLY SPACED (45°)

PROBLEM 12-25

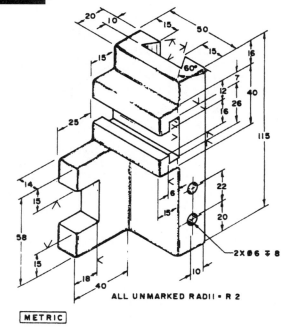

20

10

15

50

15

15

16

60°

7

12

40

25

16

26

115

14

6

15

22

15

58

20

15

2 X Ø6 ⌲ 8

18

40

10

ALL UNMARKED RADII = R 2

METRIC

Analyzing 2D Drawings

USING INQUIRY COMMANDS

This chapter will focus on a series of commands used to calculate distances and angles of selected objects. Surface areas may be calculated on complex geometric shapes. The following pages highlight all of AutoCAD's inquiry commands and show how you can use them to display useful information on an object or group of objects. You can invoke AutoCAD's inquiry commands by using the Inquiry toolbar, by entering the commands at the keyboard, or by choosing them from the Tools menu, as illustrated in Figures 13–1 and 13–2. The following is a list of the inquiry commands with a short description of each:

AREA   calculates the surface area after you specify a series of points or select a polyline or circle. You can add or subtract multiple objects to calculate the area with holes and cutouts.

DIST   calculates the distance between two points. Also provides the delta X,Y,Z coordinate values, the angle in the XY plane, and the angle from the XY plane.

HELP   provides online help for any command. You can type HELP at the keyboard or choose the Help menu.

ID   displays the X,Y,Z absolute coordinate of a selected point.

LIST   displays key information depending on the object selected.

STATUS   displays important information on the current drawing.

TIME   displays the time spent in the drawing editor.

The following pages give a detailed description of how these commands are used in interaction with different AutoCAD objects.

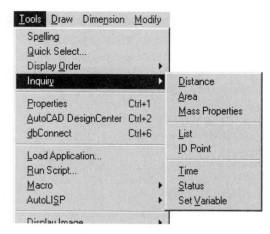

Figure 13–1 **Figure 13–2**

FINDING THE AREA OF AN ENCLOSED SHAPE

The AREA command is used to calculate the area through the selection of a series of points. Select the endpoints of all vertices of the image in Figure 13–3 with the OSNAP-Endpoint option. Once you have selected the first point along with the remaining points in either a clockwise or counterclockwise pattern, respond to the prompt "Next point:" by pressing ENTER to calculate the area of the shape. Along with the area is a calculation for the perimeter. Use Figure 13–3 and the prompt sequence below for finding the area by identifying a series of points.

Command: **AA** (For AREA)
Specify first corner point or [Object/Add/Subtract]: (Select the endpoint at "A")
Specify next corner point or press ENTER for total: (Select the endpoint at "B")
Specify next corner point or press ENTER for total: (Select the endpoint at "C")
Specify next corner point or press ENTER for total: (Select the endpoint at "D")
Specify next corner point or press ENTER for total: (Select the endpoint at "E")
Specify next corner point or press ENTER for total: (Select the endpoint at "A")
Specify next corner point or press ENTER for total: (Press ENTER to calculate the area)
Area = 25.25, Perimeter = 20.35

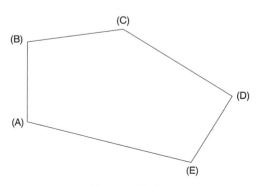

Figure 13–3

FINDING THE AREA OF AN ENCLOSED POLYLINE OR CIRCLE

The previous example showed how to find the area of an enclosed shape using the AREA command and identifying the corners and intersections of the enclosed area by a series of points. For a complex area, this could be a very tedious operation. As a result, the AREA command has a built-in Object option that will calculate the area and perimeter of a polyline and the area and circumference of a circle. Finding the area of a polyline can only be accomplished if one of the following conditions are satisfied:

- The shape must have already been constructed through the PLINE command.

- The shape must have already been converted to a polyline through the PEDIT command if originally constructed from individual objects.

Study Figure 13–4 and the following prompt sequence for finding the area of both shapes.

Command: **AA** *(For AREA)*
Specify first corner point or [Object/Add/Subtract]: **O** *(For Object)*
Select objects: *(Select the polyline at "A")*
Area = 24.88, Perimeter = 19.51
Command: **AA** *(For AREA)*
Specify first corner point or [Object/Add/Subtract]: **O** *(For Object)*
Select objects: *(Select the circle at "B")*
Area = 7.07, Circumference = 9.42

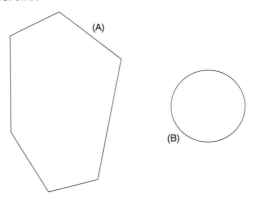

Figure 13–4

FINDING THE AREA OF A SURFACE BY SUBTRACTION

The steps you use to calculate the total surface area: (1) calculate the area of the outline and (2) subtract the objects inside the outline. All individual objects, except circles, must first be converted to polylines through the PEDIT command. Next, find the overall area and add it to the database using the Add mode of the AREA command. Exit the Add mode and remove the inner objects using the Subtract mode of the AREA command. Remember that all objects must be in the form of a circle or polyline. This means that the inner shape at "B" in Figure 13–5 must also be converted to a polyline through the PEDIT command before the area is calculated. Care must be taken when selecting the objects to subtract. If an object is selected twice, it is subtracted twice and may yield an inaccurate area in the final calculation.

For the image in Figure 13–5, the total area with the circle and rectangle removed is 30.4314. See the prompt sequence below to verify this area calculation.

Command: **AA** *(For AREA)*
Specify first corner point or [Object/Add/Subtract]: **A** *(For Add)*
Specify first corner point or [Object/Subtract]: **O** *(For Object)*
(ADD mode) Select objects: *(Select the polyline at "A")*
Area = 47.5000, Perimeter = 32.0000
Total area = 47.5000
(ADD mode) Select objects: *(Press ENTER to exit ADD mode)*
Specify first corner point or [Object/Subtract]: **S** *(For Subtract)*
Specify first corner point or [Object/Add]: **O** *(For Object)*
(SUBTRACT mode) Select objects: *(Select the polyline at "B")*
Area = 10.0000, Perimeter = 13.0000
Total area = 37.5000
(SUBTRACT mode) Select objects: *(Select the circle at "C")*
Area = 7.0686, Circumference = 9.4248
Total area = 30.4314
(SUBTRACT mode) Select objects: *(Press ENTER to exit SUBTRACT mode)*
Specify first corner point or [Object/Add]: *(Press ENTER to exit this command)*

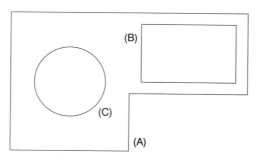

Figure 13–5

THE DIST (DISTANCE) COMMAND

The DIST command calculates the linear distance between two points on an object, whether it be the distance of a line, the distance between two points, or the distance from the quadrant of one circle to the quadrant of another circle. The following information is also supplied when you use the DIST command: the angle in the XY plane, the angle from the XY plane, and the delta X,Y,Z coordinate values. The angle in the XY plane is given in the current angular mode set by the Drawing Units dialog box. The delta X,Y,Z coordinate is a relative coordinate value taken from the first point identified by the DIST command to the second point. Study Figure 13–6 and the prompt sequences below for the DIST command.

Command: **DI** *(For DIST)*
Specify first point: *(Select the endpoint at "A")*
Specify second point: *(Select the endpoint at "B")*
Distance = 6.36, Angle in XY Plane = 45.00, Angle from XY Plane = 0.00
Delta X = 4.50, Delta Y = 4.50,   Delta Z = 0.00
Command: **DI** *(For DIST)*
Specify first point: *(Select the endpoint at "C")*
Specify second point: *(Select the endpoint at "D")*
Distance = 9.14, Angle in XY Plane = 192.75, Angle from XY Plane = 0.00
Delta X = -8.91, Delta Y = -2.02,   Delta Z = 0.00

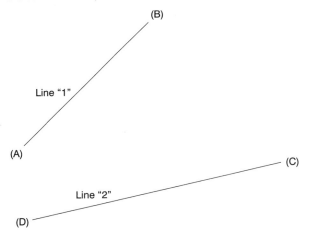

Figure 13–6

INTERPRETATION OF ANGLES USING THE DIST COMMAND

Previously, it was noted that the DIST command yields information regarding distance, delta X,Y coordinate values, and angle information. Of particular interest is the angle in the XY plane formed between two points. In Figure 13–7, picking the endpoint of the line segment at "A" as the first point followed by the endpoint of the line segment at "B" as the second point displays an angle of 42°. This angle is formed from an imaginary horizontal line drawn from the endpoint of the line segment at "A" in the zero direction.

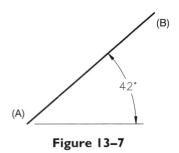

Figure 13–7

Take care when using the DIST command to find an angle on an identical line segment illustrated in Figure 13–8, as with the example in Figure 13–7. However, notice that the two points for identifying the angle are selected differently. With the DIST command, you select the endpoint of the line segment at "B" as the first point, followed by the endpoint of the segment at "A" for the second point. A new angle in the XY plane of 222° is formed. In Figure 13–8, the angle is calculated by the construction of a horizontal line from the endpoint at "B," the new first point of the DIST command. This horizontal line is also drawn in the zero direction. Notice the relationship of the line segment to the horizontal baseline. Be careful when identifying the order of line segment endpoints for extracting angular information.

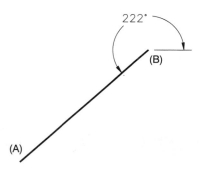

Figure 13–8

THE ID (IDENTIFY) COMMAND

The ID command is probably one of the more straightforward of the inquiry commands. ID stands for "Identify" and allows you to obtain the current absolute coordinate listing of a point along or near an object.

In Figure 13–9, the coordinate value of the center of the circle at "A" was found through the use of ID and the OSNAP-Center mode. The coordinate value of the starting point of text string "B" was found with ID and the OSNAP-Insert mode. The coordinate value of the endpoint of line segment "C" was found with ID and the

OSNAP-Endpoint mode. The coordinate value of the midpoint of line segment at "CD" was found with ID and the OSNAP-Midpoint mode. Finally, the coordinate value of the current position of point "E" was found with ID and the OSNAP-Node mode. Follow the prompt sequences below for calculating the X,Y,Z coordinate point of the objects in Figure 13–9.

Command: **ID**
Specify point: **Cen**
of *(Select the circle "A")*
X = 2.00 Y = 7.00 Z = 0.00

Command: **ID**
Specify point: **Ins**
of *(Select the text at "B")*
X = 5.54 Y = 7.67 Z = 0.00

Command: **ID**
Specify point: **End**
of *(Select the line at "C")*
X = 8.63 Y = 4.83 Z = 0.00

Command: **ID**
Specify point: **Mid**
of *(Select line "CD")*
X = 1.63 Y = 1.33 Z = 0.00

Command: **ID**
Specify point: **Nod**
of *(Select the point at "E")*
X = 9.98 Y = 1.98 Z = 0.00

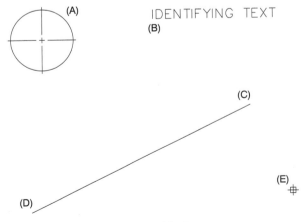

Figure 13–9

THE LIST COMMAND

Use the LIST command to obtain information about an object or group of objects. In Figure 13–10, two rectangles are displayed along with a circle. However, are the rectangles made up of individual line segments or a polyline object? Using the LIST command on each object informs you that the first rectangle at "A" is a polyline, the circle lists as a circle, and the second rectangle is actually a block reference. In addition to the object type, you also can obtain key information such as the layer that the object resides on, area and perimeter information for polylines, and circumference information for circles. Study the prompt sequence below for using the LIST command.

Command: **LI** *(For LIST)*
Select objects: *(Select the objects at "A", "B", and "C" in order)*
Select objects: *(Press ENTER to list the information on each object)*
LWPOLYLINE Layer: "0"
Space: Model space
Handle = 2B
Closed
Constant width   0.0000
area   3.8502
perimeter   8.0116
at point X=   2.3683 Y=   5.6887 Z=   0.0000
at point X=   4.7730 Y=   5.6887 Z=   0.0000
at point X=   4.7730 Y=   7.2898 Z=   0.0000
at point X=   2.3683 Y=   7.2898 Z=   0.0000
CIRCLE   Layer: "0"
Space: Model space
Handle = 36
center point, X=   5.5989 Y=   6.4892 Z=   0.0000
radius   1.5526
circumference   9.7550
area   7.5726
BLOCK REFERENCE Layer: "0"
Space: Model space
Handle = 31
"REC"
at point, X=   6.4248 Y=   5.6887 Z=   0.0000
X scale factor   1.0000
Y scale factor   1.0000
rotation angle   0
Z scale factor   1.0000

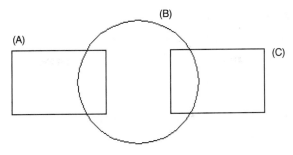

Figure 13–10

USING THE STATUS COMMAND

Once inside a large drawing, sometimes it becomes difficult to keep track of various settings that have been changed from their default values to different values required by the drawing. To obtain a listing of these important settings contained in the current drawing file, use the STATUS command. Once the STATUS command is invoked, the graphics screen changes to a text screen displaying the information shown in the following prompt sequence.

Command: **STATUS**
793 objects in D:\BUILDING.DWG
Model space limits are X: 0'-0" Y: 0'-0" (Off)
                        X: 120'-0" Y: 90'-0"
Model space uses X: 0'-0" Y: 0'-0"
                  X: 120'-0" Y: 90'-0"
Display shows X: -0'-0 3/16" Y: -0'-0 1/8"
               X: 173'-5 3/8" Y: 90'-0 1/8"
Insertion base is X: 0'-0" Y: 0'-0" Z: 0'-0"
Snap resolution is X: 3'-0" Y: 3'-0"
Grid spacing is X: 0'-0" Y: 0'-0"
Current space: Model space
Current layout: Model
Current layer: "0"
Current color: BYLAYER — 7 (white)
Current linetype: BYLAYER — "CONTINUOUS"
Current lineweight: BYLAYER
Current plot style: ByLayer
Current elevation: 0'-0" thickness: 0'-0"
Fill on Grid off Ortho off Qtext off Snap off Tablet off
Object snap modes: Center, Endpoint, Intersection, Quadrant, Extension
Free dwg disk (D:) space: 1119.2 MBytes
Free temp disk (C:) space: 422.0 MBytes
Free physical memory: 6.6 Mbytes (out of 63.4M).
Free swap file space: 76.1 Mbytes (out of 127.7M).

THE TIME COMMAND

The TIME command provides the following information:

- The current date and time.
- The date and time the drawing was created.
- The last time the drawing was updated.
- The total time spent editing the drawing so far.
- The total time spent in AutoCAD (not necessarily in a particular drawing).
- The current automatic save time interval.

CURRENT TIME

Displays the current date and time.

CREATED

This date and time value is set when you use the NEW command to create a new drawing file. This value is also set to the current date and time whenever a drawing is saved under a different name with the SAVE or SAVEAS commands.

LAST UPDATED

This data consists of the date and time the current drawing was last updated. This value updates itself when you use the SAVE command.

TOTAL EDITING TIME

This represents the total time spent editing the drawing. The timer is always updating itself and cannot be reset to a new or different value.

ELAPSED TIMER

This timer runs while AutoCAD is in operation, and you can turn it on or off or reset it.

NEXT AUTOMATIC SAVE IN

This timer displays when the next automatic save will occur. This value is controlled by the system variable SAVETIME. If this system variable is set to zero, the automatic save utility is disabled. If the timer is set to a nonzero value, the timer displays when the next automatic save will take place. The increment for automatic saving is in minutes.

Command: **TIME**
Current time: Wednesday, May 26, 1999 at 10:10:32:122 PM
Times for this drawing:
Created: Friday, November 12, 1993 at 3:03:11:980 PM
Last updated: Sunday, January 24, 1999 at 11:41:01:087 PM
Total editing time:0 days 00:31:32.760
Elapsed timer (on):0 days 00:31:32.760
Next automatic save in: <no modifications yet>

TUTORIAL EXERCISE: EXTRUDE.DWG

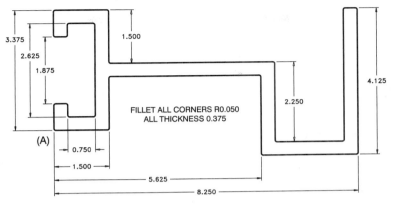

Figure 13–11

Purpose

This tutorial is designed to show you various methods of constructing the extruded pattern in Figure 13–11. The surface area of the extrusion will also be found through the AREA command.

System Settings

Use the Drawing Units dialog box and change the number of decimal places past the zero from four units to three units. Keep the default drawing limits at (0.000,0.000) for the lower left corner and (12.000,9.000) for the upper right corner.

Layers

Create the following layers with the format:

Name	Color	Linetype
Boundary	Magenta	Continuous
Object	Yellow	Center

Suggested Commands

Begin drawing the extrusion with point "A" in Figure 13–11 at absolute coordinate

2.000,3.000. Use either of the following methods to construct the extrusion:

1. Use a series of absolute, relative, and polar coordinates to construct the profile of the extrusion.

2. Construct a few lines; then use the OFFSET command followed by the TRIM command to construct the extrusion profile.

The FILLET command is used to create the 0.050 radius rounds at all corners of the extrusion. Before calculating the area of the extrusion, convert and join all objects into one single polyline. This will allow the AREA command to be used in a more productive way. Do not dimension this drawing.

Whenever possible, substitute the appropriate command alias in place of the full AutoCAD command in each tutorial step; for example, use "CP" for the COPY command, "L" for the LINE command, and so on. The complete listing of all command aliases is located in Chapter 1, Table 1–2.

STEP I

First make the Object layer current. One method of constructing the extrusion in Figure 13–12 is to use the measurements in Figure 13–11 to calculate a series of polar coordinate system distances. The Direct Distance mode could also be used to accomplish this step.

Command: **L** *(For LINE)*
Specify first point: 2.000,3.000 *(Starting at "A")*
Specify next point or [Undo]: **@1.500<0** *(To "B")*
Specify next point or [Undo]: **@1.500<90** *(To "C")*
Specify next point or [Close/Undo]: **@4.125<0** *(To "D")*
Specify next point or [Close/Undo]: **@2.250<270** *(To "E")*
Specify next point or [Close/Undo]: **@2.625<0** *(To "F")*
Specify next point or [Close/Undo]: **@4.125<90** *(To "G")*
Specify next point or [Close/Undo]: **@0.375<180** *(To "H")*
Specify next point or [Close/Undo]: **@3.750<270** *(To "I")*
Specify next point or [Close/Undo]: **@1.875<180** *(To "J")*

Specify next point or [Close/Undo]: **@2.250<90** *(To "K")*
Specify next point or [Close/Undo]: **@4.500<180** *(To "L")*
Specify next point or [Close/Undo]: **@1.500<90** *(To "M")*
Specify next point or [Close/Undo]: **@1.500<180** *(To "N")*
Specify next point or [Close/Undo]: **@0.750<270** *(To "O")*
Specify next point or [Close/Undo]: **@0.375<0** *(To "P")*
Specify next point or [Close/Undo]: **@0.375<90** *(To "Q")*
Specify next point or [Close/Undo]: **@0.750<0** *(To "R")*
Specify next point or [Close/Undo]: **@2.625<270** *(To "S")*
Specify next point or [Close/Undo]: **@0.750<180** *(To "T")*
Specify next point or [Close/Undo]: **@0.375<90** *(To "U")*
Specify next point or [Close/Undo]: **@0.375<180** *(To "V")*
Specify next point or [Close/Undo]: **C** *(To Close)*

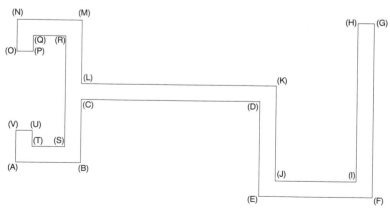

Figure 13–12

STEP 2

Rather than converting individual objects to one polyline using the PEDIT command and Join option, you could use a more efficient means of creating a polyline: the BOUNDARY command. First make the Boundary layer current. Choosing Boundary... from the Draw pull-down menu, as in Figure 13–13A, activates the Boundary Creation dialog box in Figure 13–13B. Click on the Pick Points button in the upper right corner of the dialog box. Then pick an internal point to automatically trace a polyline around a closed shape in the color of the current layer. It must be emphasized that the shape must be completely closed for the BOUNDARY command to function correctly.

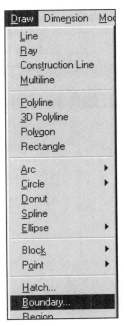

Figure 13–13A

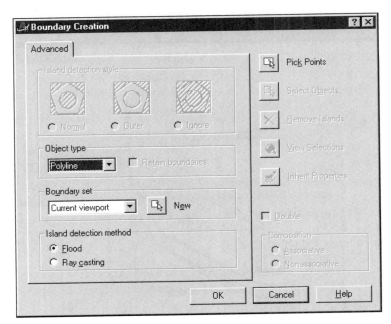

Figure 13–13B

Figure 13–14 shows the extrusion. If you open the Boundary Creation dialog box and click on the Pick Points button, AutoCAD prompts you to pick an internal point. Selecting a point inside the extrusion at "A" traces the polyline in the Boundary layer. Turning off the Object layer containing the individual line segments leaves the polyline for performing various calculations.

Command: **BO** *(For BOUNDARY)*
Select internal point: *(Select a point inside of the extrusion at "A")*
Selecting everything...
Selecting everything visible...
Analyzing the selected data...
Analyzing internal islands...
Select internal point: *(Press ENTER to create the boundary polyline on the Boundary layer)*
BOUNDARY created 1 polyline

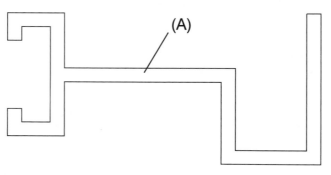

(A)

Figure 13–14

STEP 3

With the entire extrusion converted to a polyline, use the FILLET command, set a radius of 0.050, and use the polyline option of the FILLET command to fillet all corners of the extrusion at once. See Figure 13–15.

Command: **F** *(For FILLET)*
Current settings: Mode = TRIM, Radius = 0.500
Select first object or [Polyline/Radius/ Trim]: **R** *(For Radius)*
Specify fillet radius <0.500>: **0.050**
Command: **F** *(For FILLET)*
Current settings: Mode = TRIM, Radius = 0.050
Select first object or [Polyline/Radius/ Trim]: **P** *(For Polyline)*

Select 2D polyline: *(Select the polyline in Figure 13–15)*
22 lines were filleted

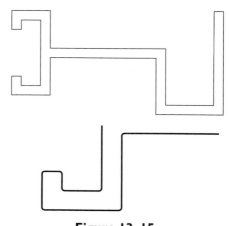

Figure 13–15

CHECKING THE ACCURACY OF EXTRUDE.DWG

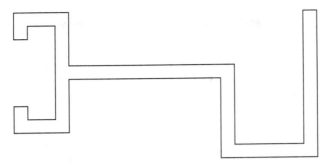

Figure 13–16

Once the extrusion has been constructed, answer the question to determine the accuracy of the drawing in Figure 13–16.

Question 1

The total surface area of the extrusion is closest to:

 (A) 7.020

 (B) 7.070

 (C) 7.120

 (D) 7.170

 (E) 7.220

Use the AREA command to calculate the surface area of the extrusion. This is easily accomplished because the extrusion has already been converted to a polyline.

Command: AA (For AREA)
Specify first corner point or [Object/Add/
 Subtract]: O (For Object)
Select objects: (Select any part of the
 extrusion in Figure 13–16)
Area = 7.170, Perimeter = 38.528

**Total surface area of the extrusion
 is "D," 7.170.**

TUTORIAL EXERCISE: C-LEVER.DWG

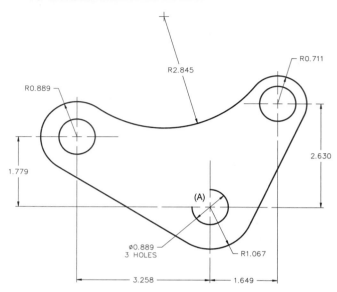

Figure 13–17

Purpose

This tutorial is designed to show you various methods of constructing the C-Lever object in Figure 13–17. Numerous questions will be asked about the object, requiring the use of the AREA, DIST, ID, and LIST commands.

System Settings

Use the Drawing Units dialog box and change the number of decimal places past the zero from four units to three units. Keep the default drawing limits at (0.000,0.000) for the lower left corner and (12.000,9.000) for the upper right corner. Check to see that the following Object Snap modes are already set: Endpoint, Extension, Intersection, Center.

Layers

Create the following layers with the format:

Name	Color	Linetype
Boundary	Magenta	Continuous
Object	Yellow	Center

Suggested Commands

Begin drawing the C-Lever with point "A" illustrated in Figure 13–17 at absolute coordinate 7.000,3.375. Begin laying out all circles. Then draw tangent lines and arcs. Use the TRIM command to clean up unnecessary objects. To prepare to answer the AREA command question, convert the profile of the C-Lever to a polyline using the BOUNDARY command. Other questions pertaining to distances, angles, and point identifications follow. Do not dimension this drawing.

Whenever possible, substitute the appropriate command alias in place of the full AutoCAD command in each tutorial step; for example, use "CP" for the COPY command, "L" for the LINE command, and so on. The complete listing of all command aliases is located in Chapter 1, Table 1–2.

STEP 1

Make the Object layer current. Then construct one circle of 0.889 diameter with the center of the circle at absolute coordinate 7.000,3.375 (see Figure 13–18). Construct the remaining circles of the same diameter by using the COPY command with the Multiple option. Use of the @ symbol for the base point in the COPY command identifies the last known point, which in this case is the center of the first circle drawn at coordinate 7.000,3.375.

Command: **C** *(For CIRCLE)*
Specify center point for circle or [3P/2P/ Ttr (tan tan radius)]: **7.000,3.375**
Specify radius of circle or [Diameter]: **D** *(For Diameter)*

Specify diameter of circle: **0.889**
Command: **CP** *(For COPY)*
Select objects: **L** *(For Last)*
Select objects: *(Press* ENTER *to continue)*
Specify base point or displacement, or [Multiple]: **M** *(For Multiple)*
Specify base point: **@**
Specify second point of displacement or <use first point as displacement>: **@1.649,2.630**
Specify second point of displacement or <use first point as displacement>: **@-3.258,1.779**
Specify second point of displacement or <use first point as displacement>: *(Press* ENTER *to exit this command)*

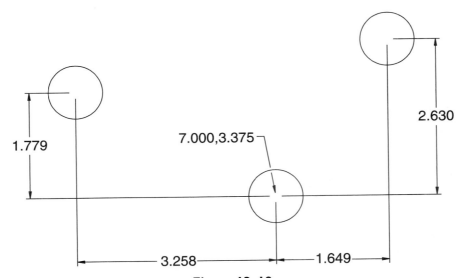

Figure 13–18

STEP 2

Construct three more circles (see Figure 13–19). Even though these objects actually represent arcs, circles will be drawn now and trimmed later to form the arcs.

Command: **C** *(For CIRCLE)*
Specify center point for circle or [3P/2P/
   Ttr (tan tan radius)]: *(Select the edge of
   the circle at "A" to snap to its center)*
Specify radius of circle or [Diameter]
   <0.445>: **1.067**
Command: **C** *(For CIRCLE)*
Specify center point for circle or [3P/2P/
   Ttr (tan tan radius)]: *(Select the edge of
   the circle at "B" to snap to its center)*
Specify radius of circle or [Diameter]
   <1.067>: **0.889**
Command: **C** *(For CIRCLE)*

Specify center point for circle or [3P/2P/
   Ttr (tan tan radius)]: *(Select the edge of
   the circle at "C" to snap to its center)*
Specify radius of circle or [Diameter]
   <0.889>: **0.711**

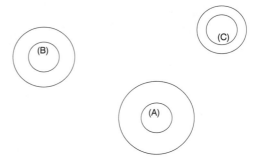

Figure 13–19

STEP 3

Construct lines tangent to the three outer circles as shown in Figure 13–20.

Command: **L** *(For LINE)*
Specify first point: **Tan**
to *(Select the outer circle near "A")*
Specify next point or [Undo]: **Tan**
to *(Select the outer circle near "B")*
Specify next point or [Undo]: *(Press
   ENTER to exit this command)*
Command: **L** *(For LINE)*
Specify first point: **Tan**
to *(Select the outer circle near "C")*
Specify next point or [Undo]: **Tan**
to *(Select the outer circle near "D")*

Specify next point or [Undo]: *(Press
   ENTER to exit this command)*

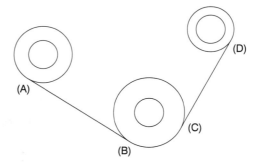

Figure 13–20

STEP 4

Construct a circle tangent to the two circles in Figure 13–21 using the CIRCLE command with the Tangent-Tangent-Radius option (TTR).

Command: **C** *(For CIRCLE)*
Specify center point for circle or [3P/2P/ Ttr (tan tan radius)]: **TTR**
Specify point on object for first tangent of circle: *(Select the outer circle near "A")*
Specify point on object for second tangent of circle: *(Select the outer circle near "B")*
Specify radius of circle <0.711>: **2.845**

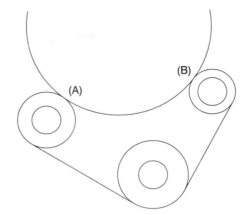

Figure 13–21

STEP 5

Use the TRIM command to clean up and form the finished drawing. Press ENTER at the "Select objects:" prompt will select all objects as cutting edges although they will not highlight (see Figure 13–22). Study the following prompts for selecting the objects to trim.

Command: **TR** *(For TRIM)*
Current settings: Projection=UCS Edge=None
Select cutting edges ...
Select objects: *(Press ENTER which will select all objects cutting edges)*
Select object to trim or [Project/Edge/ Undo]: *(Select the circle at "A")*
Select object to trim or [Project/Edge/ Undo]: *(Select the circle at "B")*
Select object to trim or [Project/Edge/ Undo]: *(Select the circle at "C")*

Select object to trim or [Project/Edge/ Undo]: *(Select the circle at "D")*
Select object to trim or [Project/Edge/ Undo]: *(Press ENTER to exit this command)*

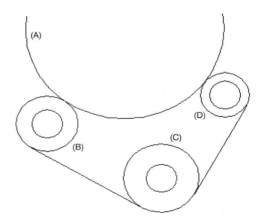

Figure 13–22

CHECKING THE ACCURACY OF C-LEVER.DWG

Once the C-Lever has been constructed, answer the following questions to determine the accuracy of this drawing. Use Figure 13–23 to assist in answering the questions.

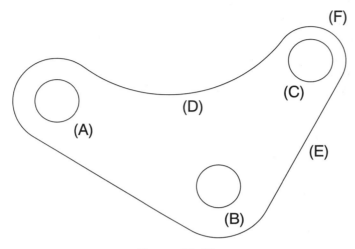

Figure 13–23

1. The total area of the C-Lever with all three holes removed is

   (A) 13.744

   (B) 13.749

   (C) 13.754

   (D) 13.759

   (E) 13.764

2. The total distance from the center of circle "A" to the center of circle "B" is

   (A) 3.692

   (B) 3.697

   (C) 3.702

   (D) 3.707

   (E) 3.712

3. The angle formed in the XY plane from the center of circle "C" to the center of circle "B" is

   (A) 223°

   (B) 228°

   (C) 233°

   (D) 238°

   (E) 243°

4. The delta X,Y distance from the center of circle "C" to the center of circle "A" is

   (A) -4.907,-0.851

   (B) -4.907,-0.856

   (C) -4.907,-0.861

   (D) -4.907,-0.866

   (E) -4.907,-0.871

5. The absolute coordinate value of the center of arc "D" is

   (A) 5.869,8.218

   (B) 5.869,8.223

   (C) 5.869,8.228

   (D) 5.869,8.233

   (E) 5.869,8.238

6. The total length of line "E" is

   (A) 3.074

   (B) 3.079

   (C) 3.084

   (D) 3.089

   (E) 3.094

7. The total length of arc "F" is

   (A) 2.051

   (B) 2.056

   (C) 2.061

   (D) 2.066

   (E) 2.071

A solution for each question follows, complete with the method used to arrive at the answer. Apply these methods to any type of drawing that requires the use of inquiry commands.

SOLUTIONS TO THE QUESTIONS ON C-LEVER

Question 1

The total area of the C-Lever with all three holes removed is

(A) 13.744   (D) 13.759

(B) 13.749   (E) 13.764

(C) 13.754

First make the Boundary layer current. Then use the BOUNDARY command and pick a point inside the object at "A" in Figure 13–24. This will trace a polyline around all closed objects on the Boundary layer.

Command: **BO** *(For BOUNDARY)*
Select internal point: *(Pick a point inside of the object at "Y")*
Selecting everything...
Selecting everything visible...
Analyzing the selected data...
Analyzing internal islands...
Select internal point: *(Press ENTER to create the boundaries)*
BOUNDARY created 4 polylines

Next, turn off the Object layer. All objects on the Boundary layer should be visible. Then use the AREA command to add and subtract objects to arrive at the final area of the object.

Command: **AA** *(For AREA)*
Specify first corner point or [Object/Add/ Subtract]: **A** *(For Add)*
Specify first corner point or [Object/ Subtract]: **O** *(For Object)*
(ADD mode) Select objects: *(Select the edge of the shape near "X")*

Area = 15.611, Perimeter = 17.771
Total area = 15.611
(ADD mode) Select objects: *(Press ENTER to continue)*
Specify first corner point or [Object/ Subtract]: **S** *(For Subtract)*
Specify first corner point or [Object/ Add]: **O** *(For Object)*
(SUBTRACT mode) Select objects: *(Select circle "A")*
Area = 0.621, Perimeter = 2.793
Total area = 14.991
(SUBTRACT mode) Select objects: *(Select circle "B")*
Area = 0.621, Perimeter = 2.793
Total area = 14.370
(SUBTRACT mode) Select objects: *(Select circle "C")*
Area = 0.621, Perimeter = 2.793
Total area = 13.749
(SUBTRACT mode) Select objects: *(Press ENTER to continue)*
Specify first corner point or [Object/ Add]: *(Press ENTER to exit this command)*

The total area of the C-Lever with all three holes removed is (B), 13.749.

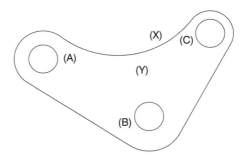

Figure 13–24

Question 2

The total distance from the center of circle "A" to the center of circle "B" is

(A) 3.692   (D) 3.707

(B) 3.697   (E) 3.712

(C) 3.702

Use the DIST (Distance) command to calculate the distance from the center of circle "A" to the center of circle "B" in Figure 13–25. Be sure to use the OSNAP-Center mode for locating the centers of all circles.

Command: **DI** *(For DIST)*
Specify first point: *(Select the edge of the circle at "A")*
Specify second point: *(Select the edge of the circle at "B")*

Distance = 3.712, Angle in XY Plane = 331, Angle from XY Plane = 0
Delta X = 3.258, Delta Y = -1.779, Delta Z = 0.000

The total distance from the center of circle "A" to the center of circle "B" is (E), <u>3.712.</u>

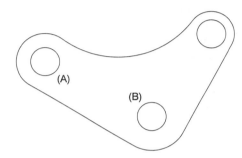

Figure 13–25

Question 3

The angle formed in the XY plane from the center of circle "C" to the center of circle "B" is

(A) 223°   (D) 238°

(B) 228°   (E) 243°

(C) 233°

Use the DIST (Distance) command to calculate the angle from the center of circle "C" to the center of circle "B" in Figure 13–26. Be sure to use the OSNAP-Center mode for locating the centers of all circles.

Command: **DI** *(For DIST)*
Specify first point: *(Select the edge of the circle at "C")*
Specify second point: *(Select the edge of the circle at "B")*

Distance = 3.104, Angle in XY Plane = 238, Angle from XY Plane = 0
Delta X = -1.649, Delta Y = -2.630, Delta Z = 0.000

The angle formed in the XY plane from the center of circle "C" to the center of circle "B" is (D), <u>238°.</u>

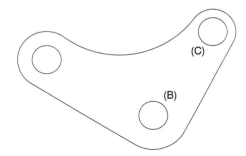

Figure 13–26

Question 4

The delta X,Y distance from the center of circle "C" to the center of circle "A" is

(A) -4.907,-0.851   (D) -4.907,-0.866

(B) -4.907,-0.856   (E) -4.907,-0.871

(C) -4.907,-0.861

Use the DIST (Distance) command to calculate the delta X,Y distance from the center of circle "C" to the center of circle "A" in Figure 13–27. Be sure to use the OSNAP-Center mode. Notice that additional information is given when you use the DIST command. For the purpose of this question, we will only be looking for the delta X,Y distance. The DIST command

will display the relative X,Y,Z distances. Since this is a 2D problem, only the X and Y values will be used.

Command: **DI** *(For DIST)*
Specify first point: *(Select the edge of the circle at "C")*
Specify second point: *(Select the edge of the circle at "A")*
Distance = 4.980, Angle in XY Plane = 190, Angle from XY Plane = 0
Delta X = -4.907, Delta Y = -0.851, Delta Z = 0.000

The delta X,Y distance from the center of circle "C" to the center of circle "A" is (A), -4.907,-0.851.

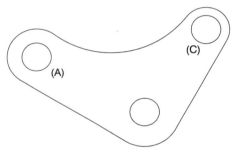

Figure 13–27

Question 5

The absolute coordinate value of the center of arc "D" is

(A) 5.869,8.218   (D) 5.869,8.233

(B) 5.869,8.223   (E) 5.869,8.238

(C) 5.869,8.228

The ID command is used to get the current absolute coordinate information on a desired point (see Figure 13–28). This command will display the X,Y,Z coordi-

nate values. Since this is a 2D problem, only the X and Y values will be used.

Command: **ID**
Specify point: *(Select the edge of the arc at "D"; OSNAP Center mode should be active)*
X = 5.869   Y = 8.223   Z = 0.000

The absolute coordinate value of the center of arc "D" is (B), 5.869,8.223.

? +

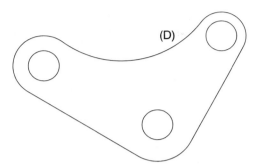

Figure 13–28

Question 6

The total length of line "E" is

    (A) 3.074   (D) 3.089

    (B) 3.079   (E) 3.094

    (C) 3.084

Use the DIST (Distance) command to find the total length of line "E" in Figure 13–29. Be sure to use the OSNAP-Endpoint mode. Notice that additional information is given when you use the DIST command. For the purpose of this question, we will only be looking for the distance.

Command: **DI** *(For DIST)*
Specify first point: (Select the endpoint of the line at "X")
Specify second point: (Select the endpoint of the line at "Y")
Distance = 3.084, Angle in XY Plane = 64, Angle from XY Plane = 0
Delta X = 1.328, Delta Y = 2.783, Delta Z = 0.000

The total length of line "E" is (C), <u>3.084</u>.

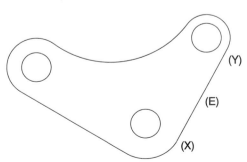

Figure 13–29

Question 7

The total length of arc "F" is

    (A) 2.051   (D) 2.066

    (B) 2.056   (E) 2.071

    (C) 2.061

The LIST command is used to calculate the lengths of arcs. However, a little preparation is needed before you perform this operation. If arc "F" is selected as in Figure 13–30, notice that the entire outline is selected because it is a polyline. Use the EXPLODE command to break the outline into individual objects. Use the LIST command to get a listing of the arc length. See also Figure 13–31.

Command: **X** *(For EXPLODE)*
Select objects: *(Select the edge of the dashed polyline in Figure 13–30)*

Select objects: *(Press* ENTER *to perform the explode operation)*
Command: **LI** *(For LIST)*
Select objects: *(Select the edge of the arc at "F" in Figure 13–31)*
Select objects: *(Press* ENTER *to continue)*
ARC   Layer: "Boundary"
Space: Model space
Handle = 48
center point, X=   8.649 Y=   6.005
   Z=   0.000
radius   0.711
start angle   334
end angle   141
length   2.071

The total length of arc "F" is (E), 2.071.

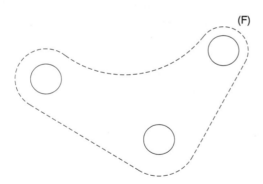

Figure 13–30

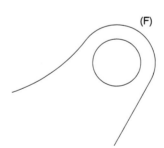

Figure 13–31

PROBLEMS FOR CHAPTER 13

PROBLEM 13–1 ANGLEBLK.DWG

Directions for Angleblk.Dwg

Use the Drawing Units dialog box to set the units to decimal. Set the number of digits to the right of the decimal point from four to two. Be sure the system of angle measure is set to decimal degrees and the number of decimal places for the display of angles is zero. Keep all remaining default unit values. Keep the default settings for the drawing limits.

Begin the drawing in Problem 13–1 by locating the lower left corner of Angleblk identified by "X" at coordinate (2.35,3.17).

Refer to the drawing of the Angleblk in Problem 13–1 to answer the following questions.

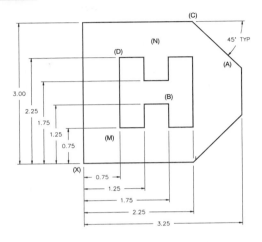

1. The total surface area of the Angleblk with the inner "H" shape removed is
   - (A) 6.66
   - (D) 7.00
   - (B) 6.77
   - (E) 7.11
   - (C) 6.89

2. The total area of the inner "H" shape is
   - (A) 1.75
   - (D) 1.93
   - (B) 1.81
   - (E) 1.99
   - (C) 1.87

3. The total length of line "A" is
   - (A) 1.29
   - (D) 1.47
   - (B) 1.35
   - (E) 1.53
   - (C) 1.41

4. The absolute coordinate value of the endpoint of the line at "B" is
   - (A) 4.04,4.42
   - (D) 4.22,4.42
   - (B) 4.10,4.42
   - (E) 4.28,4.42
   - (C) 4.16,4.42

5. The absolute coordinate value of the endpoint of the line at "C" is
   - (A) 4.60,6.11
   - (D) 4.60,6.29
   - (B) 4.60,6.17
   - (E) 4.60,6.35
   - (C) 4.60,6.23

6. Use the STRETCH command and extend the inner "H" shape a distance of 0.37 units in the 180° direction. Use "N" as the first corner of the crossing window and "M" as the other corner. Use the endpoint of "D" as the base point of the stretching operation. The new surface area of Angleblk with the inner "H" shape removed is closest to
   - (A) 6.63
   - (D) 6.81
   - (B) 6.69
   - (E) 6.87
   - (C) 6.75

7. Use the SCALE command with the endpoint of the line at "D" as the base point. Reduce the size of just the inner "H" using a scale factor of 0.77. The new surface area of Angleblk with the inner "H" removed is
   - (A) 7.48
   - (D) 7.66
   - (B) 7.54
   - (E) 7.72
   - (C) 7.60

PROBLEM 13–2 LEVER1.DWG

Directions for Lever1.Dwg

Use the Drawing Units dialog box to set the units to decimal. Set the number of digits to the right of the decimal point from four to three. Be sure the system of angle measure is set to decimal degrees and the number of decimal places for the display of angles is zero. Keep the remaining default unit values.

Begin the drawing in Problem 13–2 by locating the center of the 2.000-diameter circle at coordinate (4.500,3.250).

Refer to the drawing of Lever1 in Problem 13–2 to answer the following questions.

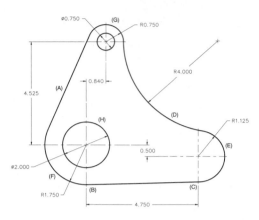

1. The total length of line segment "A" is
   - (A) 4.483
   - (B) 4.492
   - (C) 4.499
   - (D) 4.504
   - (E) 4.515

2. The absolute coordinate value of the center of the 4.00-radius arc "D" is
   - (A) 10.090,7.773
   - (B) 10.090,7.782
   - (C) 10.090,7.791
   - (D) 10.090,7.800
   - (E) 10.090,7.806

3. The length of the 1.125-radius arc segment "E" is
   - (A) 3.299
   - (B) 3.308
   - (C) 3.319
   - (D) 3.329
   - (E) 3.337

4. The distance from the center of the 1.750-radius arc "F" to the center of the 0.750-radius arc "G" is
   - (A) 4.572
   - (B) 4.584
   - (C) 4.596
   - (D) 4.602
   - (E) 4.610

5. The total area of Lever1 with the two holes removed is
   - (A) 26.271
   - (B) 26.282
   - (C) 26.290
   - (D) 26.298
   - (E) 26.307

6. The circumference of the 2.000-diameter circle "H" is
   - (A) 6.269
   - (B) 6.276
   - (C) 6.283
   - (D) 6.289
   - (E) 6.294

7. Use the SCALE command to reduce Lever1 in size by a scale factor of 0.333. Use the center of the 2.000-diameter hole as the base point. The absolute coordinate value of the center of the 0.750 arc "G" is
   - (A) 4.780,4.757
   - (B) 4.780,4.767
   - (C) 4.780,4.777
   - (D) 4.785,4.757
   - (E) 4.793,4.777

PROBLEM 13–3 PLATE1.DWG

Directions for Plate1.Dwg

Use the Drawing Units dialog box and set to decimal units. Set the number of digits to the right of the decimal point from four to three. Be sure the system of angle measure is set to decimal degrees and the number of decimal places for the display of angles is zero. Keep all remaining default unit values. Use the LIMITS command and set the upper right corner of the screen area to a value of (36.000,24.000).

Begin the drawing in Problem 13–3 by placing the center of the 4.000-diameter arc with keyway at coordinate (16.000,13.000).

Refer to the drawing of Plate1 in Problem 13–3 to answer the following questions.

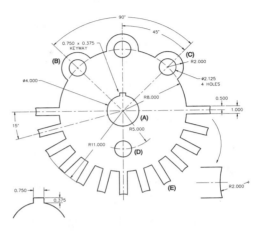

1. The distance from the center of the 2.000-radius arc "B" to the center of the 2.000-radius arc "C" is
   (A) 10.286
   (B) 10.293
   (C) 11.300
   (D) 11.307
   (E) 11.314

2. The absolute coordinate value of the center of arc "C" is
   (A) 21.657,18.657
   (B) 21.657,18.664
   (C) 21.657,18.671
   (D) 21.657,18.678
   (E) 21.657,18.685

3. The angle formed in the XY plane from the center of the 2.000-radius arc "C" to the center of the 2.125-diameter hole "D" is
   (A) 242°
   (B) 244°
   (C) 246°
   (D) 248°
   (E) 250°

4. The total length of arc "E" is
   (A) 0.999
   (B) 1.005
   (C) 1.011
   (D) 1.017
   (E) 1.023

5. The total length of arc "C" is
   (A) 6.766
   (B) 6.772
   (C) 6.778
   (D) 6.784
   (E) 6.790

6. The total area of Plate1 with all holes including keyway removed is
   (A) 232.259
   (B) 232.265
   (C) 232.271
   (D) 232.277
   (E) 232.283

7. The distance from the center of the 2.000-radius arc "B" to the center of the 2.000-radius arc "E" is
   (A) 20.732
   (B) 20.738
   (C) 20.744
   (D) 20.750
   (E) 20.756

PROBLEM 13–4 ANGER.DWG

Directions for Hanger.Dwg

Use the Drawing Units dialog box set to decimal. Set the number of digits to the right of the decimal point from four to three. Be sure the system of angle measure is set to decimal degrees and the number of decimal places for the display of angles is zero. Keep all remaining default unit values. Use the LIMITS command and set the upper right corner of the screen area to a value of (250.000,150.000).

Begin drawing the hanger by locating the center of the 40.000-radius arc at coordinate (55.000,85.000). See Problem 13–4.

Refer to the drawing of the hanger in Problem 13–4 to answer the following questions.

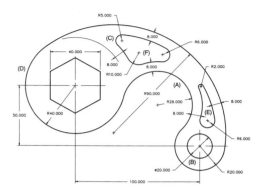

1. The area of the outer profile of the hanger is closest to

   (A) 9970.567   (D) 9975.005

   (B) 9965.567   (E) 9980.347

   (C) 9975.567

2. The area of the hanger with the polygon, circle, and irregular shapes removed is closest to

   (A) 7304.089   (D) 7304.000

   (B) 7305.000   (E) 7306.098

   (C) 7303.890

3. The absolute coordinate value of the center of the 28.000-radius arc "A" is closest to

   (A) 120.000,69.000

   (B) 121.082,68.964

   (C) 121.520,68.237

   (D) 121.082,69.000

   (E) 121.082,66.937

4. The absolute coordinate value of the center of the 20.000-diameter circle "B" is closest to

   (A) 154.000,35.000

   (B) 156.000,36.000

   (C) 156.147,35.256

   (D) 155.000,35.000

   (E) 156.000,37.000

5. The angle formed in the XY plane from the center of the 5.000-radius arc "C" to the center of the 40.000-radius arc "D" is

   (A) 219°   (D) 225°

   (B) 221°   (E) 227°

   (C) 223°

6. The total area of irregular shape "E" is

   (A) 271.613   (D) 271.801

   (B) 271.723   (E) 271.822

   (C) 271.784

7. The total area of irregular shape "F" is

   (A) 698.511   (D) 699.856

   (B) 698.621   (E) 699.891

   (C) 699.817

PROBLEM 13–5 GASKET1.DWG

Directions for Gasket1.Dwg

Use the Drawing Units dialog box to set the units to decimal. Set the number of digits to the right of the decimal point from four to three. Be sure the system of angle measure is set to decimal degrees and the number of decimal places for the display of angles is zero. Keep the remaining default unit values.

Begin the drawing in Problem 13–5 by locating the center of the 1.500-radius circle at coordinate (5.750,4.750).

Refer to the drawing of Gasket1 in Problem 13–5 to answer the following questions.

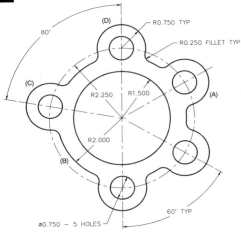

1. The total area of Gasket1 with the all holes removed is closest to
   - (A) 9.918
   - (B) 9.921
   - (C) 9.924
   - (D) 9.927
   - (E) 9.930

2. The absolute coordinate value of the center of the 0.750-radius arc "A" is closest to
   - (A) 7.669,5.875
   - (B) 7.669,5.870
   - (C) 7.666,5.875
   - (D) 7.699,5.875
   - (E) 7.699,5.975

3. The length of arc segment "B" is closest to
   - (A) 1.698
   - (B) 1.704
   - (C) 1.710
   - (D) 1.716
   - (E) 1.722

4. The angle formed in the XY plane from the center of the arc "C" to the center of arc "D" is closest to
   - (A) 30°
   - (B) 35°
   - (C) 40°
   - (D) 45°
   - (E) 50°

5. The total length of the 0.750-radius arc "C" is
   - (A) 2.674
   - (B) 2.680
   - (C) 2.686
   - (D) 2.692
   - (E) 2.698

6. Use the MOVE command to reposition Gasket1 at a distance of 1.832 in the -22245° direction. Use the center of the 1.500-radius circle as the base point of the move. The new absolute coordinate value of the center of the 1.500-radius circle is
   - (A) 7.045,3.437
   - (B) 7.045,3.443
   - (C) 7.045,3.449
   - (D) 7.045,3.455
   - (E) 7.045,3.461

7. The new absolute coordinate value of the center of the 0.750-radius arc "C" is
   - (A) 4.806,3.845
   - (B) 4.812,3.845
   - (C) 4.818,3.845
   - (D) 4.824,3.845
   - (E) 4.830,3.845

PROBLEM 13–6 GASKET2.DWG

Directions for Gasket2.Dwg

Use the Drawing Units dialog box to set the units to decimal. Set the number of digits to the right of the decimal point from four to two. Be sure the system of angle measure is set to decimal degrees and the number of decimal places for the display of angles is zero. Keep the remaining default unit values.

Begin the drawing in Problem 13–6 by locating the center of the 6.00 × 3.00 rectangle at coordinate (6.00,4.75).

Refer to the drawing of Gasket2 in Problem 13–6 to answer the following questions.

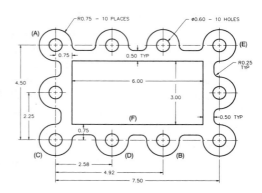

1. The total surface area of Gasket2 with the rectangle and all ten holes removed is closest to
   (A) 21.46   (D) 21.52
   (B) 21.48   (E) 21.54
   (C) 21.50

2. The distance from the center of arc "A" to the center of arc "B" is closest to
   (A) 6.63   (D) 6.75
   (B) 6.67   (E) 6.79
   (C) 6.71

3. The length of arc segment "C" is closest to
   (A) 3.44   (D) 3.53
   (B) 3.47   (E) 3.56
   (C) 3.50

4. The absolute coordinate value of the center of the 0.75-radius arc "D" is closest to
   (A) 4.83,2.50   (D) 4.80,2.50
   (B) 4.83,2.47   (E) 4.83,2.56
   (C) 4.83,2.53

5. The angle formed in the XY plane from the center of the 0.75-radius arc "D" to the center of the 0.75-radius arc "A" is
   (A) 116°   (D) 122°
   (B) 118°   (E) 124°
   (C) 120°

6. The delta X,Y distance from the intersection at "E" to the midpoint of the line at "F" is
   (A) -4.50,3.65   (D) -4.50,3.75
   (B) -4.50,-3.65 (E) -4.50,-3.75
   (C) -4.50,-3.70

7. Use the SCALE command to reduce the size of the inner rectangle. Use the midpoint of the line at "F" as the base point. Use a scale factor of 0.83 units. The new total surface area with the rectangle and all ten holes removed is
   (A) 26.99   (D) 27.14
   (B) 27.04   (E) 27.19
   (C) 27.09

PROBLEM 13–7 LEVER2.DWG

Directions for Lever2.Dwg

Use the Drawing Units dialog box to set the units to decimal. Keep the number of digits to the right of the decimal point at four places. Be sure the system of angle measure is set to decimal degrees and the number of decimal places for the display of angles is zero. Keep the remaining default unit values.

Begin the drawing in Problem 13–7 by locating the center of the 1.0000-diameter circle at coordinate (2.2500,4.0000).

Refer to the drawing of Lever2 in Problem 13–7 to answer the following questions.

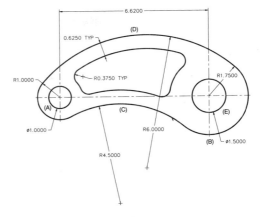

1. The total area of Lever2 with the inner irregular shape and both holes removed is
   - (A) 17.6813
   - (B) 17.6819
   - (C) 17.6825
   - (D) 17.6831
   - (E) 17.6837

2. The absolute coordinate value of the center of the 4.5000-radius arc "C" is
   - (A) 4.8944,-0.0822
   - (B) 4.8944,-0.8226
   - (C) 4.8950,-0.8232
   - (D) 4.8956,-0.8238
   - (E) 4.8962,-0.8244

3. The absolute coordinate value of the center of the 6.0000-radius arc "D" is
   - (A) 6.0828,0.7893
   - (B) 6.0834,0.7899
   - (C) 6.0840,0.7905
   - (D) 6.0846,0.7911
   - (E) 6.0852,0.7917

4. The total length of arc "C" is
   - (A) 5.3583
   - (B) 5.3589
   - (C) 5.3595
   - (D) 5.3601
   - (E) 5.3607

5. The distance from the center of the 1.0000-diameter circle "A" to the intersection of the circle and centerline at "B" is closest to
   - (A) 6.8456
   - (B) 6.8462
   - (C) 6.8474
   - (D) 6.8480
   - (E) 6.8486

6. The angle formed in the XY plane from the upper quadrant of arc "D" to the center of the 1.5000-circle (E) is
   - (A) 313°
   - (B) 315°
   - (C) 317°
   - (D) 319°
   - (E) 321°

7. The delta X,Y distance from the upper quadrant of arc "C" to the center of the 1.0000-hole "A" is
   - (A) -2.6444,0.3220
   - (B) -2.6444,0.3226
   - (C) -2.6444,0.3232
   - (D) -2.6444,0.3238
   - (E) -2.6444,0.3244

problem EXERCISE

PROBLEM 13–8 FLANGE1.DWG

Directions for Flange1.Dwg

Use the Drawing Units dialog box to set the units to decimal. Set the number of digits to the right of the decimal point from four to two. Be sure the system of angle measure is set to decimal degrees and the number of decimal places for the display of angles is zero. Keep the remaining default unit values.

Begin the drawing in Problem 13–8 by locating the center of the 2.00-diameter circle at coordinate (6.00,5.50).

Refer to the drawing of Flange1 in Problem 13–8 to answer the following questions.

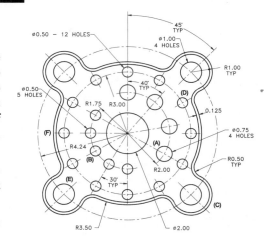

1. The total area of the 0.125 strip around the perimeter of Flange1 is
   - (A) 4.10
   - (D) 4.16
   - (B) 4.12
   - (E) 4.18
   - (C) 4.14

2. The absolute coordinate value of the center of the 0.75-diameter circle "A" is
   - (A) 7.73,4.45
   - (D) 7.75,4.55
   - (B) 7.73,4.50
   - (E) 7.75,4.60
   - (C) 7.73,4.55

3. The absolute coordinate value of the center of the 0.50-diameter circle "B" is closest to
   - (A) 4.45,4.55
   - (D) 4.48,4.69
   - (B) 4.48,4.55
   - (E) 4.50,4.56
   - (C) 4.48,4.62

4. The length of the 1.00-radius arc "C" is
   - (A) 3.82
   - (D) 3.91
   - (B) 3.85
   - (E) 3.97
   - (C) 3.88

5. The total surface area of the inner part of Flange1 with all holes removed is closest to
   - (A) 35.12
   - (D) 35.30
   - (B) 35.18
   - (E) 35.36
   - (C) 35.24

6. The total length of outer arc "F" is
   - (A) 2.91
   - (D) 3.09
   - (B) 2.97
   - (E) 3.15
   - (C) 3.03

7. The angle formed in the XY plane from the center of the 0.50 hole "D" to the center of the 0.50 hole "B" is
   - (A) 204°
   - (D) 210°
   - (B) 206°
   - (E) 212°
   - (C) 208°

PROBLEM 13-9 WEDGE.DWG

Directions for Wedge.Dwg

Use the Drawing Units dialog box to set the units to decimal. Set the number of digits to the right of the decimal point from four to two. Be sure the system of angle measure is set to decimal degrees and the number of decimal places for the display of angles is zero. Keep all remaining default unit values.

Begin constructing the wedge with vertex "A" located at coordinate (30,30) as shown in Problem 13–9.

Refer to the drawing of the wedge in Problem 13–9 to answer the following questions.

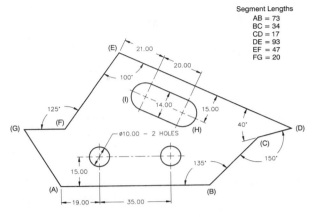

Segment Lengths
AB = 73
BC = 34
CD = 17
DE = 93
EF = 47
FG = 20

1. The total area of the wedge with the two holes and slot removed is
   - (A) 4367.97
   - (B) 4368.54
   - (C) 4370.12
   - (D) 4371.83
   - (E) 4374.91

2. The distance from the intersection of vertex "E" to the intersection of vertex "G" is
   - (A) 60.72
   - (B) 60.74
   - (C) 60.80
   - (D) 60.85
   - (E) 60.87

3. The distance from the intersection of vertex "D" to the intersection of vertex "G" is
   - (A) 131.00
   - (B) 131.12
   - (C) 131.25
   - (D) 131.36
   - (E) 131.48

4. The length of arc "H" is closest to
   - (A) 21.00
   - (B) 21.50
   - (C) 21.99
   - (D) 22.50
   - (E) 22.99

5. The overall height of the wedge from the base of line "AB" to the peak at "E" is
   - (A) 60.72
   - (B) 65.87
   - (C) 67.75
   - (D) 69.08
   - (E) 71.98

6. The distance from the intersection of vertex "A" to the center of arc "I" is
   - (A) 61.09
   - (B) 61.67
   - (C) 61.98
   - (D) 62.93
   - (E) 63.02

7. The length of line "AG" is closest to
   - (A) 31.92
   - (B) 32.47
   - (C) 33.62
   - (D) 34.23
   - (E) 35.33

problem EXERCISE

PROBLEM 13–10 PATTERN1.DWG

Directions for Pattern1.Dwg

Use the Drawing Units dialog box to set the units to decimal. Set the number of digits to the right of the decimal point from four to three. Be sure the system of angle measure is set to decimal degrees and the number of decimal places for the display of angles is zero. Keep all remaining default unit values. Use the LIMITS command and set the upper right corner of the limits to (16.000,12.000).

Begin the drawing in Problem 13–10 by locating the center of the 2.500-radius arc at coordinate (4.250,5.750).

Refer to the drawing of Pattern1 in Problem 13–10 to answer the following questions.

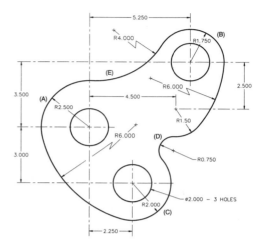

1. The total surface area of Pattern1 with the three holes removed is
   - (A) 47.340
   - (B) 47.346
   - (C) 47.386
   - (D) 47.486
   - (E) 47.586

2. The distance from the center of the 2.500-radius arc "A" to the center of the 1.750-radius arc "B" is
   - (A) 6.310
   - (B) 6.315
   - (C) 6.210
   - (D) 6.321
   - (E) 6.305

3. The absolute coordinate value of the center of the 4.000-radius arc at "E" is
   - (A) 4.580,12.241
   - (B) 4.589,12.249
   - (C) 4.589,12.237
   - (D) 4.480,12.237
   - (E) 4.589,12.241

4. The perimeter of the outline of Pattern1 is
   - (A) 31.741
   - (B) 31.747
   - (C) 31.753
   - (D) 31.759
   - (E) 31.765

5. The total length of arc "A" is
   - (A) 4.633
   - (B) 4.639
   - (C) 4.645
   - (D) 4.651
   - (E) 4.657

6. The angle formed in the XY plane from the center of the 2.500-radius arc "A" to the center of the 2.000-radius arc "C" is
   - (A) 301°
   - (B) 303°
   - (C) 305°
   - (D) 307°
   - (E) 309°

7. Use the MIRROR command to flip but not duplicate Pattern1. Use the center of the 2.000-radius arc "C" as the first point of the mirror line. Use a polar coordinate value of @1.000<90 as the second point. The new absolute coordinate value of the center of the 0.750-radius arc "D" is
   - (A) 4.378,4.504
   - (B) 4.382,4.504
   - (C) 4.386,4.504
   - (D) 4.390,4.504
   - (E) 4.394,4.504

PROBLEM 13–11 BRACKET1.DWG

Directions for Bracket1.Dwg

Use the Drawing Units dialog box to set the units to decimal. Set the number of digits to the right of the decimal point from four to three. Be sure the system of angle measure is set to decimal degrees and the number of decimal places for the display of angles is zero. Keep all remaining default unit values. Keep the default values for the limits.

Begin the drawing in Problem 13–11 by locating the center of the 1.500-radius arc "A" at coordinate (4.000,3.500).

Refer to the drawing of Bracket1 in Problem 13–11 to answer the following questions.

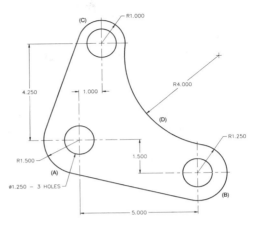

1. The distance from the center of the 1.500-radius arc "A" to the center of the 1.250-radius arc "B" is

   (A) 5.205   (D) 5.220

   (B) 5.210   (E) 5.228

   (C) 5.215

2. The distance from the center of the 1.500-radius arc "A" to the center of the 1.000-radius arc "C" is

   (A) 4.366   (D) 4.378

   (B) 4.370   (E) 4.382

   (C) 4.374

3. The distance from the center of the 1.250-radius arc "B" to the center of the 1.000-radius arc "C" is

   (A) 6.990   (D) 7.000

   (B) 6.995   (E) 7.004

   (C) 6.998

4. The length of arc "B" is

   (A) 3.994   (D) 4.012

   (B) 4.000   (E) 4.018

   (C) 4.006

5. The absolute coordinate value of the center of the 4.000-radius arc "D" is

   (A) 9.965,7.112

   (B) 9.965,7.250

   (C) 9.960,7.161

   (D) 9.965,7.161

   (E) 9.995,1.161

6. The total area of Bracket1 with all three 1.250-diameter holes removed is

   (A) 27.179   (D) 27.198

   (B) 27.187   (E) 28.003

   (C) 27.193

7. The angle formed in the XY plane from the center of the 1.250-radius arc "B" to the center of the 1.000-radius arc "C" is

   (A) 121°   (D) 127°

   (B) 123°   (E) 129°

   (C) 125°

PROBLEM 13–12 LEVER3.DWG

Directions for Lever3.Dwg

Begin the construction of Lever3 in Problem 13–12 by keeping the default units set to decimal but changing the number of decimal places past the zero from four to two. Be sure the system of angle measure is set to decimal degrees and the number of decimal places for the display of angles is zero. Keep the remaining default unit values. Use the LIMITS command to change the drawing limits to (0.00,0.00) for the lower left corner and (15.00,12.00) for the upper right corner.

Begin the drawing in Problem 13–12 by placing the center of the regular hexagon and 2.25-radius arc at coordinate (6.25,6.50).

Refer to the drawing of Lever3 in Problem 13–12 to answer the following questions.

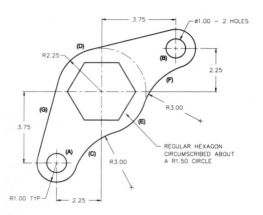

1. The distance from the center of the 1.00-diameter hole "A" to the center of the other 1.00-diameter hole "B" is
   - (A) 8.39
   - (D) 8.49
   - (B) 8.42
   - (E) 8.52
   - (C) 8.46

2. The absolute coordinate value of the center of the 3.00-radius arc "C" is
   - (A) 7.79,1.48
   - (D) 7.76,1.51
   - (B) 7.79,1.51
   - (E) 7.76,1.48
   - (C) 7.79,1.54

3. The total length of line "G" is
   - (A) 4.14
   - (D) 4.29
   - (B) 4.19
   - (E) 4.34
   - (C) 4.24

4. The length of the 2.25-radius arc segment "E" is
   - (A) 2.10
   - (D) 2.19
   - (B) 2.13
   - (E) 2.22
   - (C) 2.16

5. The angle formed in the XY plane from the center of the 1.00-diameter circle "B" to the center of the 2.25-radius arc "D" is closest to
   - (A) 203°
   - (D) 209°
   - (B) 205°
   - (E) 211°
   - (C) 207°

6. The total surface area of Lever3 with the hexagon and both 1.00-diameter holes removed is
   - (A) 20.97
   - (D) 21.06
   - (B) 21.00
   - (E) 21.09
   - (C) 21.03

7. The total length of arc "D" is
   - (A) 2.31
   - (D) 2.46
   - (B) 2.36
   - (E) 2.51
   - (C) 2.41

PROBLEM 13–13 HOUSING1.DWG

Directions for Housing1.Dwg

Begin constructing Housing1 in Problem 13–13 by keeping the default units set to decimal but changing the number of decimal places past the zero from four to three. Be sure the system of angle measure is set to decimal degrees and the number of decimal places for the display of angles is zero. Keep all remaining default unit values. Keep the default drawing limits set to (0.000,0.000) for the lower left corner and (12.000,9.000) for the upper right corner.

Place the center of the 1.500-radius circular centerline at coordinate (6.500,5.250). Before constructing the outer ellipse, set the PELLIPSE command to a value of 1. This will draw the ellipse as a polyline object.

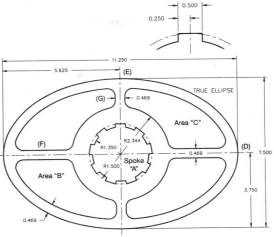

ALL FILLETS AND ROUNDS R0.375

Refer to the drawing of Housing1 in Problem 13–13 to answer the following questions.

1. The perimeter of Spoke "A" is
   - (A) 11.564
   - (B) 11.570
   - (C) 11.576
   - (D) 11.582
   - (E) 11.588

2. The perimeter of Area "B" is
   - (A) 12.513
   - (B) 12.519
   - (C) 12.525
   - (D) 12.531
   - (E) 12.537

3. The total area of Area "C" is closest to
   - (A) 7.901
   - (B) 7.907
   - (C) 7.913
   - (D) 7.919
   - (E) 7.927

4. The absolute coordinate value of the intersection of the ellipse and centerline at "D" is
   - (A) 12.125,5.244
   - (B) 12.125,5.250
   - (C) 12.125,5.256
   - (D) 12.125,5.262
   - (E) 12.125,5.268

5. The total surface area of Housing1 with the spoke and all slots removed is
   - (A) 27.095
   - (B) 28.101
   - (C) 28.107
   - (D) 28.113
   - (E) 28.119

6. The distance from the midpoint of the horizontal line segment at "F" to the midpoint of the vertical line segment at "G" is
   - (A) 4.235
   - (B) 4.241
   - (C) 4.247
   - (D) 4.253
   - (E) 4.259

7. Increase Spoke "A" in size using the SCALE command. Use the center of the 1.500-radius arc as the base point. Use a scale factor of 1.115 units. The new total area of Housing1 with the spoke and all slots removed is
   - (A) 26.535
   - (B) 26.541
   - (C) 26.547
   - (D) 26.553
   - (E) 26.559

problem EXERCISE

PROBLEM 13–14 CAM1.DWG

Directions for Cam1.Dwg

Start a new drawing called Cam1 (see Problem 13–14). Keep the default settings of decimal units but change the number of decimal places past the zero from four to two. Be sure the system of angle measure is set to decimal degrees and the number of decimal places for the display of angles is zero. Keep all remaining default unit values. Keep the default limit settings at (0.00,0.00) by (12.00,9.00).

Begin the drawing in Problem 13–14 by constructing the center of the 2.00-unit-diameter circle at coordinate (3.00,4.00).

Refer to the drawing of Cam1 in Problem 13–14 to answer the following questions.

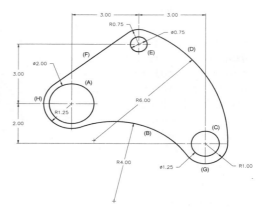

1. The absolute coordinate value of the center of the 4.00-radius arc "B" is
   - (A) 4.92,0.89
   - (B) 4.92,-0.89
   - (C) 4.95,-0.91
   - (D) 4.92,-0.85
   - (E) -4.95,-0.85

2. The angle formed in the XY plane from the center of the 2.00-diameter circle "A" to the center of the 1.25-diameter circle "C" is
   - (A) 334°
   - (B) 336°
   - (C) 338°
   - (D) 340°
   - (E) 342°

3. The length of arc "D" is
   - (A) 7.20
   - (B) 7.23
   - (C) 7.26
   - (D) 7.29
   - (E) 7.32

4. The total area of Cam1 with all three holes removed is
   - (A) 26.78
   - (B) 26.81
   - (C) 26.84
   - (D) 26.87
   - (E) 26.90

5. The total length of line "F" is
   - (A) 4.05
   - (B) 4.09
   - (C) 4.13
   - (D) 4.17
   - (E) 4.21

6. The delta X distance from the quadrant of the 1.00-radius arc at "G" to the quadrant of the 1.25-radius arc at "H" is
   - (A) -7.13
   - (B) -7.16
   - (C) -7.19
   - (D) -7.22
   - (E) -7.25

7. Use the ROTATE command to realign Cam1. Use the center of the 2.00-diameter circle "A" as the base point of the rotation. Rotate Cam1 from this point at a -10-degree angle. The absolute coordinate value of the center of the 0.75-diameter hole "E" is
   - (A) 6.48,6.37
   - (B) 6.48,6.40
   - (C) 6.48,6.43
   - (D) 6.51,6.46
   - (E) 6.54,6.49

PROBLEM 13–15 PATTERN4.DWG

Directions for Pattern4.Dwg

Start a new drawing called Pattern4 (see Problem 13–15). Even though this is a metric drawing, no special limits need be set. Keep the default setting of decimal units but change the number of decimal places past the zero from four to zero. Be sure the system of angle measure is set to decimal degrees and the number of decimal places for the display of angles is zero. Keep all remaining default unit values.

Begin the drawing in Problem 13–15 by constructing Pattern4 with vertex "A" at absolute coordinate (50,30).

Refer to the drawing of Pattern4 in Problem 13–15 to answer the following questions.

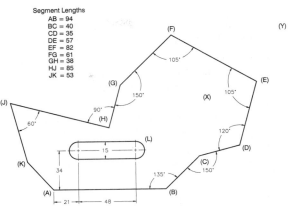

Segment Lengths
AB = 94
BC = 40
CD = 35
DE = 57
EF = 82
FG = 61
GH = 38
HJ = 85
JK = 53

1. The total distance from the intersection of vertex "K" to the intersection of vertex "A" is

    (A) 33   (D) 36
    (B) 34   (E) 37
    (C) 35

2. The total area of Pattern4 with the slot removed is

    (A) 14493   (D) 14539
    (B) 14500   (E) 14620
    (C) 14529

3. The perimeter of the outline of Pattern4 is

    (A) 570   (D) 594
    (B) 578   (E) 602
    (C) 586

4. The distance from the intersection of vertex "A" to the intersection of vertex "E" is

    (A) 186   (D) 198
    (B) 190   (E) 202
    (C) 194

5. The absolute coordinate value of the intersection at vertex "G" is

    (A) 104,117   (D) 107,120
    (B) 105,118   (E) 108,121
    (C) 106,119

6. The total length of arc "L" is

    (A) 20   (D) 23
    (B) 21   (E) 24
    (C) 22

7. Stretch the portion of Pattern4 around the vicinity of angle "E." Use "Y" as the first corner of the crossing box. Use "X" as the other corner. Use the endpoint of "E" as the base point of the stretching operation. For the new point, enter a polar coordinate value of 26 units in the 40° direction. The new degree value of the angle formed at vertex "E" is

    (A) 78°   (D) 81°
    (B) 79°   (E) 82°
    (C) 80°

PROBLEM 13–16 ROTOR.DWG

Directions for Rotor.Dwg

Start a new drawing called Rotor as shown in Problem 13–16. Keep the default setting of decimal units but change the number of decimal places past the zero from four to three. Be sure the system of angle measure is set to decimal degrees and the number of decimal places for the display of angles is zero. Keep all remaining default unit values. Keep the default limit settings at (0.000,0.000) for the lower left corner and (12.000,9.000) for the upper right corner.

Begin the drawing in Problem 13–16 by constructing the center of the 6.250-unit-diameter circle at "A" at coordinate (5.500,5.000).

Refer to the drawing of the rotor in Problem 13–16 to answer the following questions.

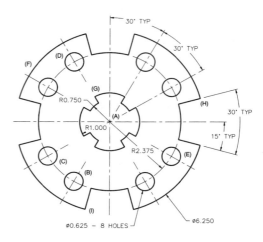

1. The absolute coordinate value of the center of the 0.625-diameter circle "B" is closest to

   (A) 4.294,2.943   (D) 4.312,2.943
   (B) 4.300,2.943   (E) 4.318,2.943
   (C) 4.306,2.943

2. The total area of the rotor with all eight holes and the center slot removed is

   (A) 21.206   (D) 21.218
   (B) 21.210   (E) 21.222
   (C) 21.214

3. The total length of arc "F" is

   (A) 3.260   (D) 3.272
   (B) 3.264   (E) 3.276
   (C) 3.268

4. The distance from the center of the 0.625 circle "C" to the center of the 0.625 circle "D" is

   (A) 3.355   (D) 3.367
   (B) 3.359   (E) 3.371
   (C) 3.363

5. The angle formed in the XY plane from the center of the 0.625 circle "B" to the center of the 0.625 circle "E" is

   (A) 11°   (D) 17°
   (B) 13°   (E) 19°
   (C) 15°

6. The delta X,Y distance from the intersection at "H" to the intersection at "I" is

   (A) -3.827,-3.827   (D) -3.841,3.813
   (B) -3.827,3.827   (E) -3.848,-3.806
   (C) -3.834,3.820

7. Use the SCALE command to increase the size of just the center slot "G." Use the center of arc "F" as the base point. Use a scale factor of 1.500 units. The new surface area of the rotor with the center slot and all eight holes removed is

   (A) 17.861   (D) 17.882
   (B) 17.868   (E) 17.889
   (C) 17.875

PROBLEM 13-17 TEMPLATE.DWG

Directions for Template.Dwg

Use the Drawing Units dialog box to set the units to decimal. Set the number of digits to the right of the decimal point from four to three. Change the system of degrees from Decimal to Degrees/Minutes/Seconds along with the following angular precision: 0d00'00.00". Keep the default values for the limits of the drawing.

Begin the drawing in Problem 13–17 by locating Point "A" at coordinate (4.500,2.750).

Refer to the drawing of the template in Problem 13–17 to answer the following questions.

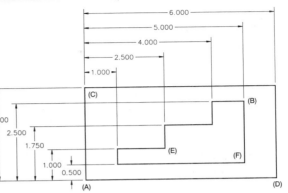

1. The angle formed in the XY plane from the intersection at "A" to the intersection at "B" is closest to
   (A) 26d25'21"   (D) 26d32'37"
   (B) 26d28'3"   (E) 26d33'54"
   (C) 26d30'10"

2. The distance from the intersection at "C" to the intersection at "D" is
   (A) 6.704   (D) 6.716
   (B) 6.708   (E) 6.720
   (C) 6.712

3. The angle formed in the XY plane from the intersection at "E" to the intersection at "F" is closest to
   (A) 347d31'10"   (D) 348d20'20"
   (B) 347d47'32"   (E) 348d41'24"
   (C) 348d1'53"

4. The total area of the template with the step pattern removed is
   (A) 13.367   (D) 13.379
   (B) 13.371   (E) 13.383
   (C) 13.375

5. Using the scale command, select all objects to reduce the size of the template. Use Point "A" as the base point and 0.950 as the scale factor. The perimeter of the inside step pattern is
   (A) 11.384   (D) 11.396
   (B) 11.388   (E) 11.400
   (C) 11.392

6. The new total area of the template with the step pattern removed is
   (A) 12.059   (D) 12.071
   (B) 12.063   (E) 12.075
   (C) 12.067

7. The new absolute coordinate value of the endpoint of the line at "B" is
   (A) 9.200,5.100   (D) 9.260,5.135
   (B) 9.250,5.125   (E) 9.260,5.140
   (C) 9.250,5.130

PROBLEM 13–18 S-CAM.DWG

Directions for S-Cam.Dwg

Begin the construction of S-Cam in Problem 13–18 by keeping the default units set to decimal but changing the number of decimal places past the zero from four to three. Be sure the system of angle measure is set to decimal degrees and the number of decimal places for the display of angles is zero. Keep all remaining default unit values. Keep the default drawing limits set to (0.000,0.000) for the lower left corner and (12.000,9.000) for the upper right corner.

Begin the drawing in Problem 13–18 by placing the center of the 1.750-diameter circle at coordinate (2.500,3.750).

Refer to the drawing of the S-Cam in Problem 13–18 to answer the following questions.

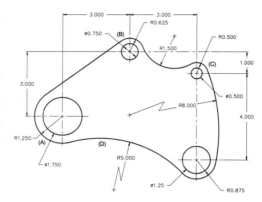

1. The total surface area of the S-Cam with all four holes removed is
   (A) 27.654   (D) 27.672
   (B) 27.660   (E) 28.678
   (C) 27.666

2. The distance from the center of the 1.250-radius arc "A" to the center of the 0.625-radius arc "B" is
   (A) 4.237   (D) 4.255
   (B) 4.243   (E) 4.261
   (C) 4.249

3. The absolute coordinate value of the center of arc "D" is
   (A) 4.208,-2.258   (D) 4.208,-2.270
   (B) 4.208,-2.262   (E) 4.208,-2.274
   (C) 4.208,-2.266

4. The total length of arc "D" is
   (A) 5.480   (D) 5.498
   (B) 5.486   (E) 6.004
   (C) 5.492

5. The angle formed in the XY plane from the center of the 0.625-arc "B" to the center of the 0.500-arc "C" is closest to
   (A) 334°   (D) 340°
   (B) 336°   (E) 342°
   (C) 338°

6. The delta X,Y distance from the center of the 0.500-arc "C" to the quadrant of the 5.000-arc "D" is
   (A) -4.284,-3.012   (D) -4.296-3.012
   (B) -4.288,-3.012   (E) -4.300,-3.012
   (C) -4.292,-3.012

7. Use the SCALE command to reduce S-Cam in size. Use the center of the 1.250-radius arc "A" as the base point. Use a scale factor of 0.822 units. The new surface area of S-Cam with all four holes removed is
   (A) 18.694   (D) 18.706
   (B) 18.698   (E) 18.710
   (C) 18.702

PROBLEM 13–19 HOUSING2.DWG

Directions for Housing2.Dwg

Begin the construction of Housing2 in Problem 13–19 by keeping the default units set to decimal but changing the number of decimal places past the zero from four to three. Be sure the system of angle measure is set to decimal degrees and the number of decimal places for the display of angles is zero. Keep all remaining default unit values. Keep the default drawing limits set to (0.000,0.000) for the lower left corner and (12.000,9.000) for the upper right corner. Before constructing the outer ellipse, set the PELLIPSE command to a value of 1. This will draw the ellipse as a polyline object.

Begin the drawing in Problem 13–19 by placing the center of the ellipse at coordinate (5.500,4.500).

Refer to the drawing of Housing2 in Problem 13–19 to answer the following questions.

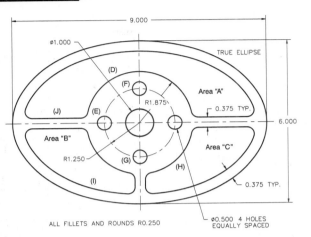

1. The total area of Housing2 with all five holes and Areas "A," "B," and "C" removed is
   (A) 20.082   (D) 20.100
   (B) 20.088   (E) 20.106
   (C) 20.094

2. The perimeter of Area "A" is
   (A) 19.837   (D) 19.855
   (B) 19.843   (E) 19.861
   (C) 19.849

3. The length of arc "D" is
   (A) 5.107   (D) 5.125
   (B) 5.113   (E) 5.131
   (C) 5.119

4. The distance from the center of the 0.500-diameter hole "E" to the center of the 0.500-diameter hole "F" is
   (A) 1.762   (D) 1.780
   (B) 1.768   (E) 1.786
   (C) 1.774

5. The total length of polyarc "I" is
   (A) 4.374   (D) 4.392
   (B) 4.380   (E) 4.398
   (C) 4.386

6. The delta X,Y distance from the center of circle "G" to the midpoint of the horizontal line at "J" is
   (A) -2.943,1.438   (D) -2.931,1.431
   (B) 2.943,1.438   (E) -2.931,-1.425
   (C) 2.937,-1.438

7. Use the MOVE command to relocate Housing2 at a distance of 0.375 units in a 45° direction. The new absolute coordinate value of the center of the 0.500-diameter hole "G" is
   (A) 5.759,3.515   (D) 5.765,3.509
   (B) 5.765,3.503   (E) 5.765,3.515
   (C) 5.759,3.509

PROBLEM 13-20 PATTERN5.DWG

Directions for Pattern5.Dwg

Begin the construction of Pattern5 in Problem 13–20 by keeping the default units set to decimal but changing the number of decimal places past the zero from four to zero. Be sure the system of angle measure is set to decimal degrees and the number of decimal places for the display of angles is zero. Keep all remaining default unit values. Use the LIMITS command to set the drawing limits to (0,0) for the lower left corner and (250,200) for the upper right corner.

Begin the Problem 13–20 drawing by placing Vertex "A" at absolute coordinate (190,30).

Refer to the drawing of Pattern5 in Problem 13–20 to answer the following questions.

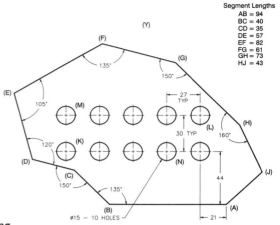

Segment Lengths
AB = 94
BC = 40
CD = 35
DE = 57
EF = 82
FG = 61
GH = 73
HJ = 43

1. The distance from the intersection of vertex "J" to the intersection of vertex "A" is

   (A) 38　　(D) 41
   (B) 39　　(E) 42
   (C) 40

2. The perimeter of Pattern5 is

   (A) 523　　(D) 526
   (B) 524　　(E) 527
   (C) 525

3. The total area of Pattern5 with all 10 holes removed is

   (A) 16369　　(D) 16372
   (B) 16370　　(E) 16373
   (C) 16371

4. The distance from the center of the 15-diameter hole "K" to the center of the 15-diameter hole "L" is

   (A) 109　　(D) 112
   (B) 110　　(E) 113
   (C) 111

5. The angle formed in the XY plane from the center of the 15-diameter hole "M" to the center of the 15-diameter hole "N" is closest to

   (A) 340°　　(D) 346°
   (B) 342°　　(E) 348°
   (C) 344°

6. The absolute coordinate value of the intersection at "F" is

   (A) 90,163　　(D) 96,167
   (B) 92,163　　(E) 98,169
   (C) 94,165

7. Use the STRETCH command to lengthen Pattern5. Use "Y" as the first point of the stretch crossing box. Use "X" as the other corner. Pick the intersection at "F" as the base point and stretch Pattern5 a total of 23 units in the 180 direction. The new total area of Pattern5 with all ten holes removed is

   (A) 17746　　(D) 17767
   (B) 17753　　(E) 17774
   (C) 17760

PROBLEM 13–21 BRACKET5.DWG

Directions for Bracket5.Dwg

Begin the construction of Bracket5 in Problem 13–21 by keeping the default units set to decimal but changing the number of decimal places past the zero from four to two. Be sure the system of angle measure is set to decimal degrees and the number of decimal places for the display of angles is zero. Keep all remaining default unit values. Use the LIMITS command to set the drawing limits to (0.00,0.00) for the lower left corner and (15.00,12.00) for the upper right corner.

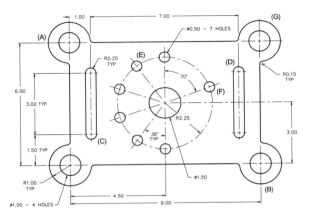

Begin the drawing in Problem 13–21 by placing the center of the 1.50-diameter hole at coordinate (7.50,5.75).

Refer to the drawing of Bracket5 in Problem 13–21 to answer the following questions.

1. The distance from the quadrant of the 1.00-radius arc at "A" to the quadrant of the 1.00-radius arc at "B" is
   (A) 12.09   (D) 12.18
   (B) 12.12   (E) 12.21
   (C) 12.15

2. The distance from the center of the 0.25-radius arc at "C" to the center of the 0.25-radius arc at "D" is
   (A) 7.62   (D) 7.71
   (B) 7.65   (E) 7.74
   (C) 7.68

3. The distance from the center of the 0.50-diameter circle at "E" to the center of the 0.50-diameter circle at "F" is
   (A) 3.50   (D) 3.59
   (B) 3.53   (E) 3.62
   (C) 3.56

4. The length of arc "G" is
   (A) 4.42   (D) 4.51
   (B) 4.45   (E) 4.54
   (C) 4.48

5. The total area of Bracket5 with all holes and slots removed is
   (A) 53.72   (D) 53.81
   (B) 53.75   (E) 53.84
   (C) 53.78

6. The delta X,Y distance from the center of the 0.25-radius arc "C" to the center of the 0.50 circle "E" is
   (A) 2.12,3.38   (D) 2.30,3.20
   (B) 2.18,3.32   (E) 2.36,3.14
   (C) 2.24,3.26

7. The angle formed in the XY plane from the center of the 0.25-radius arc "D" to the center of the 1.00-radius arc "G" is
   (A) 50°   (D) 56°
   (B) 52°   (E) 58°
   (C) 54°

PROBLEM 13–22 PLATE2A.DWG

Directions for Plate2A.Dwg

Begin Plate2A by keeping the default units set to decimal but changing the number of decimal places past the zero from four to two. Keep the system of angle measure set to decimal degrees, but change the fractional places for display of angles to two. Keep the remaining default unit values. Keep the default drawing limits set to (0.00,0.00) for the lower left corner and (12.00,9.00) for the upper right corner.

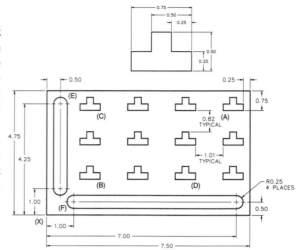

Begin constructing Plate2A by starting the lower left corner "X" at absolute coordinate (2.25,2.25). See Problem 13–22.

Refer to the drawing of Plate2A in Problem 13–22 to answer the following questions.

1. The total perimeter of the horizontal slot is
   - (A) 13.42
   - (B) 13.47
   - (C) 13.52
   - (D) 13.57
   - (E) 13.62

2. The total perimeter of the vertical slot is
   - (A) 7.98
   - (B) 8.03
   - (C) 8.07
   - (D) 8.14
   - (E) 8.20

3. The total area of the horizontal slot is
   - (A) 3.02
   - (B) 3.08
   - (C) 3.14
   - (D) 3.20
   - (E) 3.26

4. The total area of the vertical slot is
   - (A) 1.82
   - (B) 1.88
   - (C) 1.94
   - (D) 2.00
   - (E) 2.06

5. The distance from the endpoint of the line at "C" to the endpoint of the line at "D" is
   - (A) 4.25
   - (B) 4.30
   - (C) 4.35
   - (D) 4.40
   - (E) 4.45

6. The angle formed in the XY plane formed from the center of arc "E" to the endpoint of the line at "B" is
   - (A) 296.90°
   - (B) 296.96°
   - (C) 297.02°
   - (D) 297.08°
   - (E) 297.14°

7. The angle formed in the XY plane from the endpoint of the line at "A" to the center of arc "F" is
   - (A) 212.47°
   - (B) 212.53°
   - (C) 212.59°
   - (D) 212.65°
   - (E) 212.71°

PROBLEM 13–23 PLATE2B.DWG

Directions for Plate2B.Dwg

Change the number of decimal places past the zero from two to three. Change from decimal degrees to degrees/minutes/seconds. Change the number of places for display of angles to four.

Rotate Plate2B at an angle of 32°, 0', 0" angle using absolute coordinate (2.250,2.250) at "X" below as the base point of rotation. Perform a ZOOM-Extents. Also see Problem 13–23.

Refer to the drawing of Plate2B in Problem 13–23 to answer the following questions.

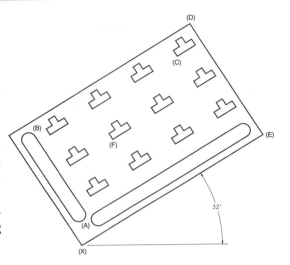

1. The angle in the XY plane from the center of arc "A" to the center of arc "B" is
   - (A) 122d0'0"
   - (B) 122d20'0"
   - (C) 122d40'0"
   - (D) 122d40'30'
   - (E) 123d0'0"

2. The absolute coordinate value of the endpoint of the line at "C" is
   - (A) 5.627,9.071
   - (B) 5.631,9.075
   - (C) 5.635,9.079
   - (D) 5.639,9.083
   - (E) 5.643,9.087

3. The absolute coordinate value of the endpoint of the line at "D" is
   - (A) 6.077,10.237
   - (B) 6.081,10.241
   - (C) 6.085,10.245
   - (D) 6.089,10.249
   - (E) 6.093,10.253

4. The absolute coordinate value of the endpoint of the line at "E" is
   - (A) 8.606,6.220
   - (B) 8.610,6.224
   - (C) 8.614,6.228
   - (D) 8.618,6.232
   - (E) 8.622,6.236

5. The angle formed in the XY plane from the endpoint of the line at "F" to the center of arc "B" is
   - (A) 178d1'2"
   - (B) 178d27'37"
   - (C) 179d15'12"
   - (D) 179d39'49"
   - (E) 180d34'29"

6. The delta X,Y distance from the intersection at "D" to the center of the 0.250-radius arc "A" is
   - (A) -3.937,-6.878
   - (B) -3.941,-6.882
   - (C) -3.945,-6.886
   - (D) -3.949,-6.890
   - (E) -3.953,-6.894

7. Use the SCALE command to reduce Plate2B in size. Use absolute coordinate 2.250,2.250 as the base point. Use 0.777 as the scale factor. The area of Plate2B with both slots and the 12 T-shaped objects removed is
   - (A) 16.659
   - (B) 16.663
   - (C) 16.667
   - (D) 16.671
   - (E) 16.675

PROBLEM 13–24 RATCHET.DWG

Directions for Ratchet.Dwg

Use the Drawing Units dialog box to change the number of decimal places past the zero from four to two. Be sure the system of angle measure is set to decimal degrees and the number of decimal places for the display of angles is zero. Keep the remaining default unit values. Use the LIMITS command to set the upper right corner of the display screen to (13.00,10.00).

Begin by drawing the center of the 1.00-radius arc of the ratchet at absolute coordinate (6.00,4.50), as shown in Problem 13–24.

Refer to the drawing of the ratchet in Problem 13–24 to answer the following questions.

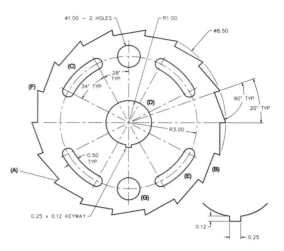

1. The total length of the short line segment "A" is
   (A) 0.22 (D) 0.28
   (B) 0.24 (E) 0.30
   (C) 0.26

2. The total length of line "B" is
   (A) 1.06 (D) 1.45
   (B) 1.19 (E) 1.57
   (C) 1.33

3. The total length of arc "C" is
   (A) 1.81 (D) 2.19
   (B) 1.93 (E) 2.31
   (C) 2.08

4. The perimeter of the 1.00-radius arc "D" with the 0.25 × 0.12 keyway is
   (A) 6.52 (D) 6.89
   (B) 6.63 (E) 7.02
   (C) 6.77

5. The total surface area of the ratchet with all 4 slots, the two 1.00-diameter holes, and the 1.00-radius arc with the keyway removed is
   (A) 42.98 (D) 43.16
   (B) 43.04 (E) 43.22
   (C) 43.10

6. The absolute coordinate value of the endpoint at "F" is
   (A) 2.01,5.10 (D) 2.20,5.95
   (B) 2.01,5.63 (E) 2.37,5.95
   (C) 2.01,5.95

7. The angle formed in the XY plane from the endpoint of the line at "F" to the center of the 1.00-diameter hole "G" is
   (A) 304° (D) 310°
   (B) 306° (E) 312°
   (C) 308°

PROBLEM 13–25 SLIDE.DWG

Directions for Slide.Dwg

Use the current unit settings and limits settings for this drawing (Problem 13–25). The number of decimal places past the zero should already be set to four. Be sure the system of angle measure is set to decimal degrees and the number of decimal places for the display of angles is zero.

Begin by drawing the center of the 0.5000-diameter circle of the slide at absolute coordinate (2.0000,2.2500).

Refer to the drawing of the slide in Problem 13–25 to answer the following questions.

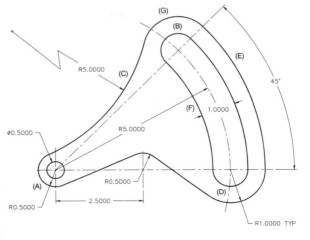

1. The absolute coordinate value of the center of the 5.0000-radius arc "C" is
   - (A) -0.2800,6.5603
   - (B) 0.2887,-5.4393
   - (C) -0.3953,7.2503
   - (D) -0.2819,7.2543
   - (E) -0.3953,6.2543

2. The total length of arc "E" is
   - (A) 4.7264
   - (B) 4.7124
   - (C) 5.4302
   - (D) 4.8710
   - (E) 4.6711

3. The distance from the center of the 0.5000-diameter circle "A" to the center of the 5.0000-radius arc "C" is
   - (A) 5.5000
   - (B) 5.0043
   - (C) 4.8768
   - (D) 4.6691
   - (E) 4.0001

4. The total area of the of the 1.0000-diameter slot is
   - (A) 4.7124
   - (B) 4.7863
   - (C) 6.8370
   - (D) 4.7625
   - (E) 5.9102

5. The angle formed in the XY plane from the upper quadrant of the 1.0000-diameter slot at "B" to the lower quadrant of the slot at "D" is
   - (A) 282°
   - (B) 284°
   - (C) 286°
   - (D) 288°
   - (E) 290°

6. The total area of the slide with the 0.5000-diameter hole and slot "F" removed is
   - (A) 8.9246
   - (B) 13.6370
   - (C) 13.8750
   - (D) 13.8333
   - (E) 14.0297

7. The absolute coordinate value of the center of the 1.0000-radius arc "G" is
   - (A) 5.5120,5.5621
   - (B) 5.5237,5.5551
   - (C) 5.5355,5.5590
   - (D) 5.5355,5.6123
   - (E) 5.5355,5.7855

PROBLEM 13–26 GENEVA.DWG

Directions for Geneva.Dwg

Start a new drawing called Geneva (Problem 13–26). Keep the default settings of decimal units, but change the number of decimal places past the zero from four to two. Be sure the system of angle measure is set to decimal degrees and the number of decimal places for the display of angles is zero. Keep the remaining default unit values.

Begin the drawing by constructing the 1.50-diameter arc at absolute coordinate (7.50,5.50).

Refer to the drawing of the Geneva in Problem 13–26 to answer the following questions.

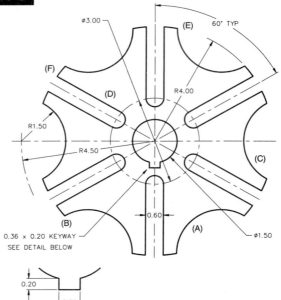

1. The total length of arc "A" is closest to
   (A) 3.00   (D) 3.30
   (B) 3.10   (E) 3.40
   (C) 3.20

2. The angle formed in the XY plane from the intersection at "B" to the center of arc "C" is
   (A) 11°   (D) 17°
   (B) 13°   (E) 19°
   (C) 15°

3. The absolute coordinate value of the midpoint of line "D" is
   (A) 5.27,7.13   (D) 5.31,7.13
   (B) 5.27,7.17   (E) 5.31,7.09
   (C) 5.23,7.13

4. The total area of the Geneva with the 1.50-diameter hole and keyway removed is closest to
   (A) 27.20   (D) 27.50
   (B) 27.30   (E) 27.60
   (C) 27.40

5. The total distance from the midpoint of arc "F" to the center of arc "A" is
   (A) 8.24   (D) 8.39
   (B) 8.29   (E) 8.44
   (C) 8.34

6. The delta X,Y distance from the intersection at "E" to the center of arc "C" is
   (A) 3.93,-3.75   (D) 3.75,3.93
   (B) 3.93,3.75   (E) 3.75,-3.93
   (C) 3.75,3.80

7. Use the SCALE command to reduce the Geneva in size. Use 7.50,5.50 as the base point; use a scale factor of 0.83 units. The absolute coordinate value of the intersection at "E" is
   (A) 8.12,8.71   (D) 8.12,8.86
   (B) 8.12,8.76   (E) 8.12,8.91
   (C) 8.12,8.81

PROBLEM 13–27 ROTOR2.DWG

Directions for Rotor2.Dwg

Start a new drawing called Rotor2 (Problem 13–27). Keep the default settings of decimal units, but change the number of decimal places past the zero from four to three. Be sure the system of angle measure is set to decimal degrees and the number of decimal places for the display of angles is zero. Keep all remaining default unit values.

Begin by constructing the 2.550-diameter circle at absolute coordinate (11.125,9.225).

Refer to the drawing of Rotor2 in Problem 13–27 to answer the following questions.

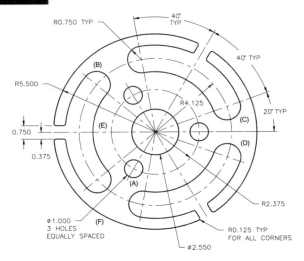

1. The absolute coordinate value of the center of hole "A" is closest to

   (A) 9.937,7.158   (D) 9.942,7.168
   (B) 9.937,7.163   (E) 9.947,7.173
   (C) 9.937,7.168

2. The perimeter of Rotor2 is

   (A) 81.850   (D) 81.865
   (B) 81.855   (E) 81.870
   (C) 81.860

3. The distance from the center of arc "B" to the center of arc "C" is

   (A) 7.125   (D) 7.140
   (B) 7.130   (E) 7.145
   (C) 7.135

4. The absolute coordinate value of the center of arc "D" is closest to

   (A) 15.001,7.814   (D) 15.006,7.829
   (B) 15.001,7.819   (E) 15.011,7.834
   (C) 15.001,7.824

5. The total length of arc "F" is

   (A) 10.474   (D) 10.489
   (B) 10.479   (E) 10.494
   (C) 10.484

6. The total area of Rotor2 with all four holes removed is

   (A) 54.902   (D) 54.917
   (B) 54.907   (E) 54.922
   (C) 54.912

7. Change the diameter of all three 1.000-diameter holes to 0.700 diameter. Change the diameter of the 2.550 hole to a new diameter of 1.625. The new area of Rotor2 with all four holes removed is

   (A) 59.122   (D) 59.137
   (B) 59.127   (E) 59.142
   (C) 59.132

problem EXERCISE

PROBLEM 13–28 ANGLE_GUIDE.DWG

Directions for Angle_Guide.Dwg

Change the number of decimal places past the zero to two places. Change the number of places for display of angles to zero. Line AB is represented as a horizontal line. The centerline of the slot is parallel to vertex DE. Begin this drawing by constructing vertex "A" at absolute coordinate 4.24,2.81 (see Problem 13–28).

Refer to the drawing of the Angle_Guide in Problem 13–28 to answer the following questions.

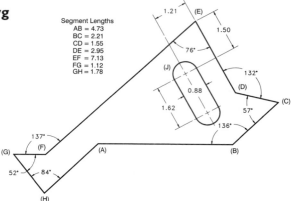

Segment Lengths
AB = 4.73
BC = 2.21
CD = 1.55
DE = 2.95
EF = 7.13
FG = 1.12
GH = 1.78

1. The distance of the line from the intersection of vertex "H" to the intersection vertex "A" is
   (A) 2.49   (D) 2.64
   (B) 2.54   (E) 2.69
   (C) 2.59

2. The angle of Vertex "A" measures closest to
   (A) 128°   (D) 134°
   (B) 130°   (E) 136°
   (C) 132°

3. The angle formed in the X-Y plane from the center of arc "J" to the intersection of vertex "B" is closest to
   (A) 297°   (D) 303°
   (B) 299°   (E) 305°
   (C) 301°

4. The total surface area of the Angle Plate with the slot removed is
   (A) 16.90   (D) 17.05
   (B) 16.95   (E) 17.10
   (C) 17.00

5. The perimeter of the Angle Plate is
   (A) 23.96   (D) 24.11
   (B) 24.01   (E) 24.16
   (C) 24.06

6. The absolute coordinate value of the center of arc "J" is
   (A) 7.24,5.38   (D) 7.34,5.38
   (B) 7.29,5.38   (E) 7.39,5.43
   (C) 7.34,5.33

7. Stretch vertex "E" 1.62 units to the left and 1.47 units straight up. Also, round off all vertex corners of the outline using a radius value of 0.25 units. The new surface area of the Angle Plate with the slot removed is
   (A) 23.33   (D) 23.48
   (B) 23.38   (E) 23.53
   (C) 23.43

problem EXERCISE

PROBLEM 13–29 CONTROL_BEARING.DWG

Directions for Control_Bearing.Dwg

Change the number of decimal places past the zero from four to two. Keep the remaining default values. Begin this drawing by locating the center of the 9.00-diameter circle at (11.50,11.50). See Problem 13–29.

Refer to the drawing of the Control_Bearing in Problem 13–29 to answer the following questions.

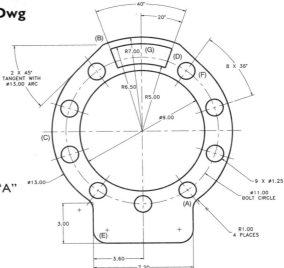

1. The distance from the center of arc "A" to the intersection at "B" is
   (A) 14.26
   (B) 14.30
   (C) 14.34
   (D) 14.38
   (E) 14.42

2. The total length of arc "C" is
   (A) 11.02
   (B) 11.06
   (C) 11.10
   (D) 11.14
   (E) 11.18

3. The perimeter of inner slot "G" is
   (A) 10.87
   (B) 10.91
   (C) 10.95
   (D) 10.99
   (E) 11.03

4. The absolute coordinate value of the center of the 0.44-diameter hole "F" is
   (A) 14.61,15.91
   (B) 14.65,15.91
   (C) 14.69,15.95
   (D) 14.73,15.95
   (E) 14.73,15.99

5. The perimeter of the outermost shape of the Control_Bearing is
   (A) 46.01
   (B) 46.05
   (C) 46.09
   (D) 46.13
   (E) 46.17

6. The angle formed in the XY plane from the intersection at "D" to the center of the arc "E" is
   (A) 250°
   (B) 252°
   (C) 254°
   (D) 256°
   (E) 258°

7. The area of the Control_Bearing with all holes and inner slot removed is
   (A) 71.45   (D) 71.57
   (B) 71.49   (E) 71.61
   (C) 71.53

PROBLEM 13–30

ARCHITECTURAL APPLICATION — APARTMENT.DWG

Directions for Apartment.Dwg

Change from decimal units to architectural units. Keep all remaining default values. The thickness of all walls measures 4". Answer the questions on the next page regarding this drawing (see Problem 13–30).

Refer to the drawing of Apartment in Problem 13–30 to answer the following questions.

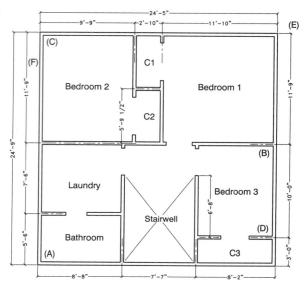

1. The total area in square feet of all bedrooms is closest to

   (A) 296 sq. ft.　(D) 326 sq. ft.

   (B) 306 sq. ft.　(E) 336 sq. ft.

   (C) 316 sq. ft.

2. The distance from the intersection of the corner at "A" to the intersection of the corner at "B" is closest to

   (A) 25'-2"　(D) 26'-5"

   (B) 25'-7"　(E) 26'-10"

   (C) 26'-0"

3. The total area in square feet of all closets (C1, C2, and C3) is closest to

   (A) 46 sq. ft.　(D) 52 sq. ft.

   (B) 48 sq. ft.　(E) 54 sq. ft.

   (C) 50 sq. ft.

4. The total area in square feet of the laundry and bathroom is closest to

   (A) 91 sq. ft.　(D) 97 sq. ft.

   (B) 93 sq. ft.　(E) 99 sq. ft.

   (C) 95 sq. ft.

5. The distance from the intersection of the corner at "C" to the intersection of the corner at "D" is closest to

   (A) 31'-0"　(D) 32'-3"

   (B) 31'-5"　(E) 32'-8"

   (C) 31'-10"

6. The angle in the XY plane from the intersection of the wall corner at "A" to the intersection of the wall corner at "D" is

   (A) 1°　(D) 7°

   (B) 3°　(E) 9°

   (C) 5°

7. Stretch the floor plan up at a distance of 2'-5". Use "E" and "F" as the locations of the crossing box. The new total area of Bedroom #1, Bedroom #2, C1, and C2 is

   (A) 315 sq. ft.　(D) 327 sq. ft.

   (B) 319 sq. ft.　(E) 331 sq. ft.

   (C) 323 sq. ft.

PROBLEM 13–31

MECHANICAL APPLICATION — PULLEYS.DWG

Directions for Pulleys.Dwg

Change the number of decimal places past the zero from four to two. Construct the image of the pulley in Figure A. Use the UCS command and the Origin option to locate a new User Coordinate System at (5.00,5.00). Create a block of this object and call it PULLEY. Insert PULLEY "A" in the current drawing at insertion point (0.00,0.00) and at a scale of 0.50. Use Figure B as a guide for inserting the remainder of the pulleys at the following scales: PULLEY "B" has a scale of 1.00; PULLEY "C" has a scale of 0.75; PUL-

LEY "D" has a scale of 0.50; PULLEY "E" has a scale of 1.00.

The dimensions in Figure B are given as ordinate or datum dimensions, which are all calculated from the center of PULLEY "A" located at 0,0. Construct a continuous belt of thickness of 0.30 around all pulleys in Figure B.

Refer to the drawing of Pulleys in Figure B of Problem 13–31 to answer the questions on the following page.

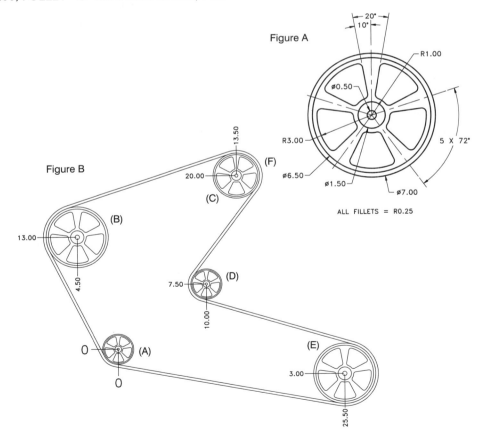

Figure A

Figure B

ALL FILLETS = R0.25

1. The angle in the XY plane from the center of Pulley "B" to the center of Pulley "E" is
   - (A) 320°
   - (B) 323°
   - (C) 326°
   - (D) 329°
   - (E) 332°

2. The distance from the quadrant of the belt at "F" to the center of Pulley "D" is
   - (A) 14.00
   - (B) 14.05
   - (C) 14.10
   - (D) 14.15
   - (E) 14.20

3. The total area of the belt is
   - (A) 35.38
   - (B) 35.43
   - (C) 35.48
   - (D) 35.53
   - (E) 35.58

4. The total area of the shape inside of Pulley "B" at "G" is
   - (A) 3.38
   - (B) 3.43
   - (C) 3.48
   - (D) 3.53
   - (E) 3.58

5. The angle in the XY plane from the center of Pulley "E" to the center of Pulley "A" is
   - (A) 167°
   - (B) 170°
   - (C) 173°
   - (D) 176°
   - (E) 179°

6. The distance from the center of Pulley "A" to the center of Pulley "D" is
   - (A) 12.45
   - (B) 12.50
   - (C) 12.55
   - (D) 12.60
   - (E) 12.65

7. Move Pulley "E" 5.25 units to the right and reconstruct the belt. The new area of the belt is closest to
   - (A) 38.56
   - (B) 38.61
   - (C) 38.66
   - (D) 38.71
   - (E) 38.76

PROBLEM 13–32

FACILITIES APPLICATION BUILDING.DWG

Directions for Building.Dwg

Change from decimal units to architectural units. Keep all remaining default values. All block wall thicknesses identified by crosshatching measure 12". All other interior wall thicknesses measure 6". Do not add any dimensions to this drawing (see Problem 13–32).

Refer to the drawing of the building in Problem 13–32 to answer the questions on the following page.

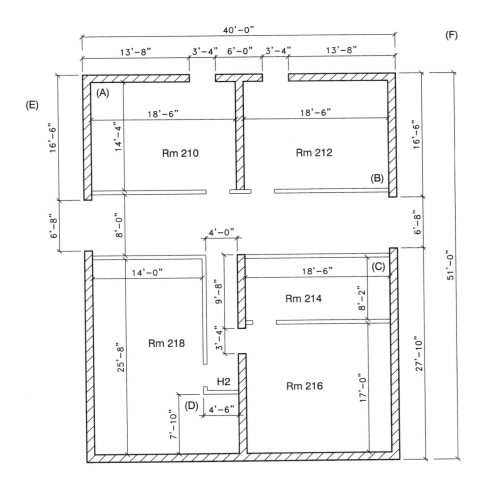

1. The total area of all concrete block walls identified by the crosshatching pattern is closest to
   (A) 183 sq. ft.
   (B) 186 sq. ft.
   (C) 189 sq. ft.
   (D) 192 sq. ft.
   (E) 195 sq. ft.

2. The total area of rooms 210, 216, and 218 is closest to
   (A) 971 sq. ft.
   (B) 974 sq. ft.
   (C) 977 sq. ft.
   (D) 980 sq. ft.
   (E) 983 sq. ft.

3. The angle formed in the XY plane from the inside corner intersection at "A" to the inside corner intersection at "B" is closest to
   (A) 333°
   (B) 335°
   (C) 337°
   (D) 339°
   (E) 341°

4. The delta X,Y distance from the inside corner intersection at "C" to the outside intersection of the hallway corner at "D" is closest to
   (A) -24'-0",-17'-10"
   (B) -24'-0",-18'-1"
   (C) -24'-0",-18'-4"
   (D) -24'-0",-18'-7"
   (E) -24'-0",-18'-10"

5. The total area of rooms 212, 214, and 218 is closest to
   (A) 802 sq. ft.
   (B) 805 sq. ft.
   (C) 808 sq. ft.
   (D) 811 sq. ft.
   (E) 814 sq. ft.

6. The total distance from the intersection of the wall corner at "A" to the intersection of the interior wall corner at "D" is closest to
   (A) 43'-2"
   (B) 43'-5"
   (C) 43'-8"
   (D) 43'-11"
   (E) 44'-2"

7. Stretch the building straight up using a crossing box from "E" to "F" and at a distance of 5'-4". The total area of rooms 212, 214, and 218 is closest to
   (A) 900 sq. ft.
   (B) 903 sq. ft.
   (C) 906 sq. ft.
   (D) 909 sq. ft.
   (E) 912 sq. ft.

PROBLEM 13–33

STRUCTURAL APPLICATION GUSSET.DWG

Directions for Gusset.Dwg

Begin the construction of the gusset illustrated in Problem 13–33 by keeping the default units set to decimal but changing the number of decimal places past the zero from four to two. No special limits need be set for this drawing (see Problem 13–33).

Refer to the drawing of the gusset in Problem 13–33 to answer the following questions.

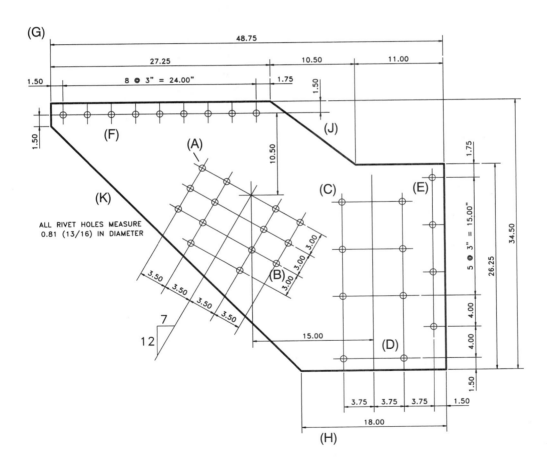

1. The total area of the gusset plate with all 35-rivet holes removed is closest to
   - (A) 1045.46
   - (B) 1050.67
   - (C) 1055.21
   - (D) 1060.40
   - (E) 1065.88

2. The delta X,Y distance from the center of hole "A" to the center of hole "B" is
   - (A) 9.07,-12.19
   - (B) 9.07,-12.24
   - (C) 9.07,-12.29
   - (D) 9.07,-12.34
   - (E) 9.07,-12.39

3. The angle formed in the XY plane from the center of hole "C" to the center of hole "D" is
   - (A) 285°
   - (B) 288°
   - (C) 291°
   - (D) 294°
   - (E) 297°

4. The angle formed in the XY plane from the center of hole "E" to the center of hole "F" is
   - (A) 165°
   - (B) 168°
   - (C) 171°
   - (D) 174°
   - (E) 177°

5. The total length of line "J" is
   - (A) 13.15
   - (B) 13.20
   - (C) 13.25
   - (D) 13.30
   - (E) 13.35

6. The total length of line "K" is
   - (A) 44.02
   - (B) 44.07
   - (C) 44.12
   - (D) 44.17
   - (E) 44.22

7. Stretch the gusset plate directly to the left using a crossing box from "H" to "G" and at a distance of 3.00 units. The new area of the gusset plate with all 35-rivet holes removed is
   - (A) 1121.98
   - (B) 1126.72
   - (C) 1131.11
   - (D) 1136.59
   - (E) 1141.34

PROBLEM 13–34

CIVIL APPLICATION PLATPLAN.DWG

Directions for Platplan.Dwg

Use the UNITS Control dialog box to change the system of units from decimal to engineering; change the number of decimal places past the zero from four to two. Change the system of angle measurement to Surveyors units. Keep the remaining UNITS Control dialog box default values. No special limits need be set for this drawing. Do not add any dimensions to this drawing.

Use the detail measurements at the left to construct the house and all decks outlined by the crosshatching. The house is aligned parallel with line AB of the plat. Answer the questions on the next page regarding this drawing (see Problem 13–34).

Refer to the drawing of Platplan in Problem 13–34 to answer the questions on the following page.

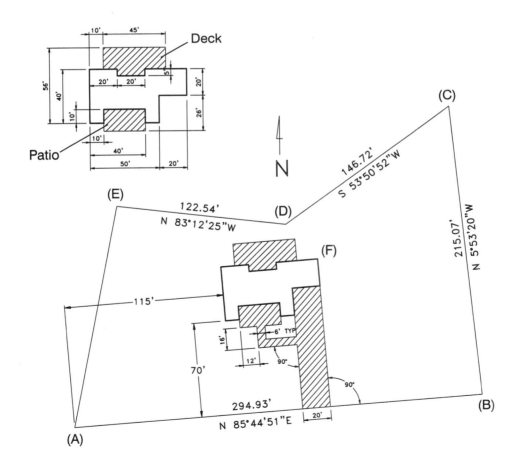

852

1. The correct direction of the line segment from the endpoint at "E" to the endpoint at "A" is
    (A) S 10d30'20" W
    (B) S 11d1'30" W
    (C) S 12d52'10" W
    (D) S 13d23'17" W
    (E) S 14d9'24" W

2. The total area of the plat with only the house removed is closest to
    (A) 42082.00 sq. ft.
    (B) 42086.00 sq. ft.
    (C) 42090.00 sq. ft
    (D) 42094.00 sq. ft.
    (E) 42098.00 sq. ft.

3. The total area of the deck, patio, sidewalks, and driveway is closest to
    (A) 3328.00 sq. ft.
    (B) 3332.00 sq. ft.
    (C) 3336.00 sq. ft.
    (D) 3340.00 sq. ft.
    (E) 3344.00 sq. ft.

4. The distance from the intersection of the corner of the house at "F" to the intersection of vertex "B" is closest to
    (A) 154'-6.00"
    (B) 154'-9.00"
    (C) 155'-0.00"
    (D) 155'-3.00"
    (E) 155'-6.00"

5. The distance from the intersection of vertex "A" to the intersection of vertex "C" is closest to
    (A) 352'-0.00
    (B) 356'-0.00
    (C) 360'-0.00
    (D) 364'-0.00
    (E) 368'-0.00

6. The angle formed in the XY plane from the intersection of vertex "E" to the intersection of vertex "B" is closest to
    (A) S 53d55'9" E
    (B) S 55d49'7" E
    (C) S 57d27'5" E
    (D) S 59d10'4" E
    (E) S 61d35'2" E

7. An error has been discovered in the original plat. Vertex "C" needs to be stretched straight down a distance of 35.00'. The total area of the plat with the house, deck, patio, sidewalks, and driveway removed is closest to
    (A) 36302.00 sq. ft.
    (B) 36306.00 sq. ft.
    (C) 36310.00 sq. ft.
    (D) 36314.00 sq. ft.
    (E) 36318.00 sq. ft.

PROBLEM 13–35

ARCHITECTURAL APPLICATION FOUNDATION.DWG

Directions for Foundation.Dwg

Start a new drawing called Foundation. Change the default units from decimal to architectural. Do not add dimensions to this drawing. Unless otherwise noted, the thickness of all concrete block walls is 8" and the thickness of all footings is 16" (see Problem 13–35).

Refer to the drawing of Foundation in Problem 13–35 to answer the questions on the following page.

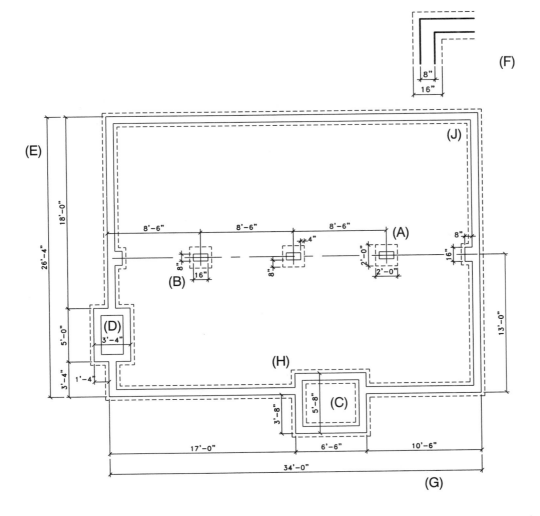

1. The total area of all concrete block regions including the three piers located in the center of the foundation plan is closest to
   (A) 91 sq. ft.
   (B) 93 sq. ft.
   (C) 95 sq. ft.
   (D) 97 sq. ft.
   (E) 99 sq. ft.

2. The total distance from the intersection of the footing at "A" to the intersection of the footing at "B" is closest to
   (A) 13 ft.
   (B) 15 ft.
   (C) 17 ft.
   (D) 19 ft.
   (E) 21 ft.

3. The total area of all footing regions including the three footings that support the piers located in the center of the foundation plan is closest to
   (A) 206 sq. ft.
   (B) 208 sq. ft.
   (C) 210 sq. ft.
   (D) 212 sq. ft.
   (E) 214 sq. ft.

4. The total distance from the middle of the concrete slab area at "C" to the middle of the fireplace area at "D" is closest to
   (A) 15 ft.
   (B) 17 ft.
   (C) 19 ft.
   (D) 21 ft.
   (E) 23 ft.

5. The angle formed in the XY plane from the intersection of the footing at "A" to the intersection of the footing at "H" is closest to
   (A) 228°
   (B) 230°
   (C) 232°
   (D) 234°
   (E) 236°

6. The delta Y distance from the intersection of the footing at "B" to the intersection of the footing at "J" is
   (A) 11'-4"
   (B) 11'-10"
   (C) 12'-4"
   (D) 12'-10"
   (E) 13'-4"

7. Increase the size of the foundation plan using the STRETCH command according to the following specifications:

   (a) Use a crossing box from "F" to "E" to stretch the foundation plan straight up at a distance of 2'-6".

   (b) Use a crossing box from "F" to "G" to stretch the foundation plan directly to the left at a distance of 18".

   The new area of the concrete block region **not including the three piers** located at the middle of the foundation plan is closest to
   (A) 94 sq. ft.
   (B) 96 sq. ft.
   (C) 98 sq. ft.
   (D) 100 sq. ft.
   (E) 102 sq. ft.

CHAPTER 14

Section Views

SECTION VIEW BASICS

Principles of orthographic projections remain the key method for the production of engineering drawings. As these drawings get more complicated in nature, the job of the designer becomes more challenging in the interpretation of views, especially where hidden features are involved. The concept of slicing a view to expose these interior features is the purpose of performing a section. Figure 14–1 is a pictorial representation of a typical flange consisting of eight bolt holes and counterbore hole in the center.

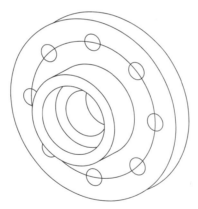

Figure 14–1

The drawings in Figure 14–2 show a typical solution to a multiview problem complete with Front and Side views. The Front view displaying the eight bolt holes is obvious to interpret; however, the numerous hidden lines in the Side view make the drawing difficult to understand, and this is considered a relatively simple drawing. To relieve the confusion associated with a drawing too difficult to understand because of numerous hidden lines, a section is made of the part. Orthographic methods are followed up to the creation of the Side view.

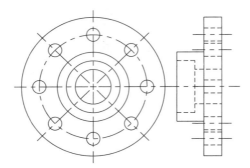

Figure 14–2

To understand section views better, see Figure 14–3. Creating a section view slices an object in such a way as to expose what used to be hidden features and convert them to visible features. This slicing or cutting operation can be compared to that of using a glass plate or cutting plane to perform the section. In the object in Figure 14–3, the glass plate cuts the object in half. It is the responsibility of the designer or CAD operator to convert one half of the object to a section and to discard the other half. Surfaces that come in contact with the glass plane are crosshatched to show where the actual cutting took place.

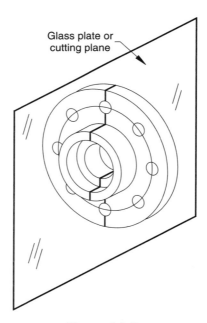

Figure 14–3

A completed section view drawing is shown in Figure 14–4. Two new types of lines are also shown; namely a cutting plane line and section lines. The cutting plane line performs the cutting operation on the Front view. In the Side view, section lines show the surfaces that were cut. Notice that holes are not section lined because the cutting plane passes across the center of the hole. Notice also that hidden lines are not displayed in the Side view. It is poor practice to merge hidden lines into a section view, although there are always exceptions. The arrows of the cutting plane line tell the designer to view the section in the direction of the arrows and discard the other half.

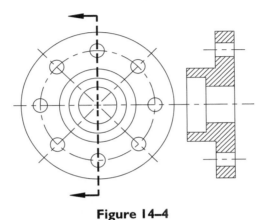

Figure 14–4

The cutting plane line consists of a very thick line with a series of dashes approximately 0.25" in length (see Figure 14–5A). A polyline is used to create this line of 0.05 thickness. The arrows point in the direction of sight used to create the section, with the other half generally discarded. Assign this line one of the dashed linetypes; the hidden linetype is reserved for detailing invisible features in views. The section line, in contrast with the cutting plane line, is a very thin line (see Figure 14–5B). This line identifies the surfaces being cut by the cutting plane line. The section line is usually drawn at an angle and at a specified spacing.

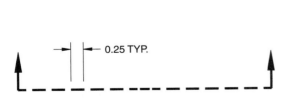

Figure 14–5A

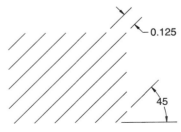

Figure 14–5B

A wide variety of hatch patterns are already supplied with the software. One of these patterns, ANSI31, is displayed in Figure 14–6. This is one of the more popular patterns, with lines spaced 0.125 units apart from each other and at a 45° angle.

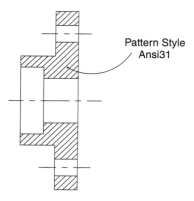

Figure 14–6

The object in Figure 14–7 illustrates proper section lining techniques. Much of the pain of spacing the section lines apart from each other and at angles has been eased considerably with the use of the computer as a tool. However, the designer must still practice proper section lining techniques at all times for clarity of the section.

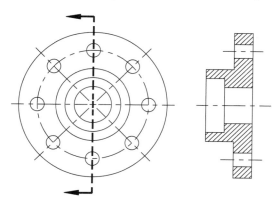

Figure 14–7

The next four examples illustrate common errors involved with section lines. In Figure 14–8A, section lines run in the correct direction and at the same angle; however, the hidden lines have not been converted to object lines. This will confuse the more experienced designer because the presence of hidden lines in the section means more complicated invisible features.

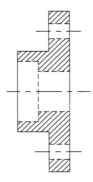

Figure 14–8A

Figure 14–8B shows yet another error encountered when sections are created. Again, the section lines are properly placed; however, all surfaces representing holes have been removed, which displays the object as a series of sectioned blocks unconnected, implying four separate parts.

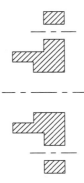

Figure 14–8B

Figure 14–8C appears to be a properly sectioned object; however, upon closer inspection, we see that the angle of the crosshatch lines in the upper half differs from the angle of the same lines in the lower left half. This suggests two different parts when, in actuality, it is the same part.

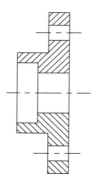

Figure 14–8C

In Figure 14–8D, all section lines run in the correct direction. The problem is that the lines run through areas that were not sliced by the cutting plane line. These areas in Figure 14–8D represent drill and counterbore holes and are left unsectioned.

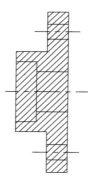

Figure 14–8D

These have been identified as the most commonly made errors when an object is crosshatched. Remember just a few rules to follow:

- Section lines are present only on surfaces that are cut by the cutting plane line.

- Section lines are drawn in one direction when the same part is crosshatched.

- Hidden lines are usually omitted when a section view is created.

- Areas such as holes are not sectioned because the cutting line only passes across this feature.

FULL SECTIONS

When the cutting plane line passes through the entire object, a full section is formed. In Figure 14–9, a full section would be the same as taking an object and cutting it completely in half. Depending on the needs of the designer, one half is kept while the other half is discarded. The half that is kept is section lined.

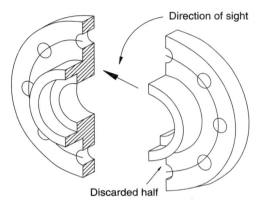

Figure 14–9

The multiview solution to the problem in Figure 14–9 is shown in Figure 14–10. The Front view is drawn with lines projected across to form the Side view. To show that the Side view is in section, a cutting plane line is added to the Front view. This line performs the physical cut. You have the option of keeping either half of the object. This is the purpose of adding arrowheads to the cutting plane line. The arrowheads define the direction of sight in which you view the object to form the section. You must then interpret what surfaces are being cut by the cutting plane line in order to properly add crosshatching lines to the section that is located in the Right Side view. Hidden lines are not necessary once a section has been made.

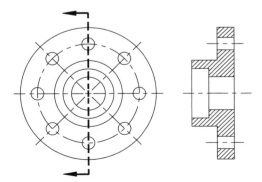

Figure 14–10

Numerous examples illustrate a cutting plane line with the direction of sight off to the left. This does not mean that a cutting plane line cannot have a direction of sight going to the right as in Figure 14–11. In this example, the section is formed from the Left Side view if the circular features are located in the Front view.

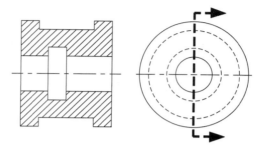

Figure 14–11

HALF SECTIONS

When symmetrical-shaped objects are involved, sometimes it is unnecessary to form a full section by cutting straight through the object. Instead, the cutting plane line passes only halfway through the object, which makes the drawing in Figure 14–12 a half section. The rules for half sections are the same as for full sections; namely, a direction of sight is established, part of the object is kept, and part is discarded.

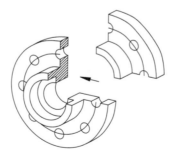

Figure 14–12

The views are laid out in Figure 14–13 in the usual multiview format. To prepare the object as a half section, the cutting plane line passes halfway through the Front view before being drawn off to the right. The Right Side view is converted to a half section by the crosshatching of the upper half of the Side view while hidden lines remain in the lower half.

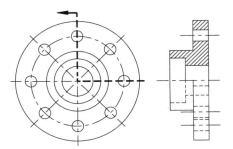

Figure 14–13

Depending on office practices, some designers prefer to omit hidden lines entirely from the Side view similar to the drawing in Figure 14–14. In this way, the lower half is drawn showing only what is visible.

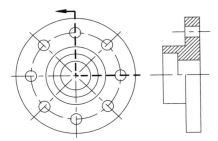

Figure 14–14

Figure 14–15 shows another way of drawing the cutting plane line to conform to the Right Side view drawn in section. Hidden lines have been removed from the lower half; only those lines visible are displayed.

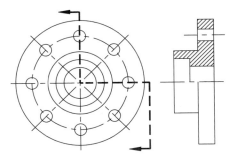

Figure 14–15

ASSEMBLY SECTIONS

It would be unfair to give designers the impression that section views are only used for displaying internal features of individual parts. Another advantage of using section views is that it permits the designer to create numerous objects, assemble them, and then slice the assembly to expose internal details of all parts. This type of section is an assembly section similar to Figure 14–16. For all individual parts, notice the section lines running the same directions. This follows one of the basic rules of section views: keep section lines at the same angle for each individual part.

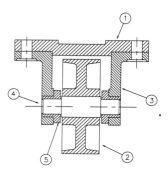

Figure 14–16

Figure 14–17 shows the difference between assembly sections and individual parts that have been sectioned. For parts in an assembly that contact each other, it is good practice to alternate the directions of the section lines to make the assembly much clearer and to distinguish the parts from each other. You can accomplish this by changing the angle of the hatch pattern or even the scale of the pattern.

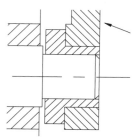

Figure 14–17

To identify parts in an assembly, an identifying part number along with a circle and arrowhead line are used, as shown in Figure 14–18. The line is very similar to a leader line used to call out notes for specific parts on a drawing. The addition of the circle highlights the part number. Sometimes this type of callout is referred to as a "bubble."

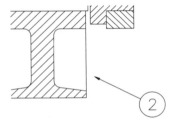

Figure 14–18

In the enlarged assembly shown in Figure 14–19, the large area in the middle is actually a shaft used to support a pulley system. With the cutting plane passing through the assembly including the shaft, refrain from crosshatching features such as shafts, screws, pins, or other types of fasteners. The overall appearance of the assembly is actually enhanced when these items are not crosshatched.

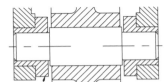

Figure 14–19

ALIGNED SECTIONS

Aligned sections take into consideration the angular position of details or features of a drawing. Instead of the cutting plane line being drawn vertically through the object as in Figure 14–20, the cutting plane is angled or aligned with the same angle as the elements. Aligned sections are also made to produce better clarity of a drawing. In Figure 14–20, with the cutting plane forming a full section of the object, it is difficult to obtain the true size of the angled elements. In the Side view, they appear foreshortened or not to scale. Hidden lines were added as an attempt to better clarify the view.

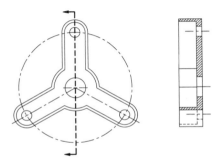

Figure 14–20

Instead of the cutting plane line being drawn all the way through the object, the line is bent at the center of the object before being drawn through one of the angled legs (see Figure 14–21). The direction of sight arrows on the cutting plane line not only determines the direction in which the view will be sectioned, but it also shows another direction for rotating the angled elements so they line up with the upper elements. This rotation is usually not more than 90°. As lines are projected across to form the Side view, the section appears as if it were a full section. This is only because the features were rotated and projected in section for greater clarity of the drawing.

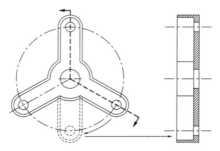

Figure 14–21

OFFSET SECTIONS

Offset sections take their name from the offsetting of the cutting plane line to pass through certain details in a view (see Figure 14–22). If the cutting plane line passes straight through any part of the object, some details would be exposed while others would remain hidden. By offsetting the cutting plane line, the designer controls its direction and which features of a part it passes through. The view to section follows the basic section rules.

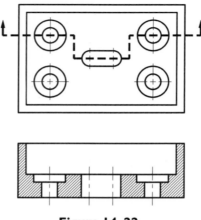

Figure 14–22

SECTIONING RIBS

Contrary to section view principles, parts made out of cast iron with webs or ribs used for reinforcement do not follow basic rules of sections. In Figure 14–23, the Front view has the cutting plane line passing through the entire view; the Side view at "A" is crosshatched according to section view basics. However, it is difficult to read the thickness of the base because the crosshatching includes the base along with the web. A more efficient method is to leave off crosshatching webs as in "B." Therefore, not crosshatching the web exposes other important details such as thickness of bases and walls.

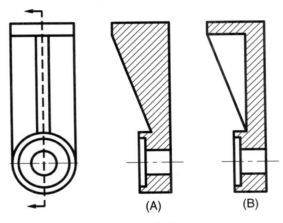

(A) (B)

Figure 14–23

The object in Figure 14–24 is another example of performing a full section on an area consisting of webbed or ribbed features. If you do not crosshatch the webbed areas, more information is available, such as the thickness of the base and wall areas around the cylindrical hole. This may not be considered true projection; however, it is considered good practice.

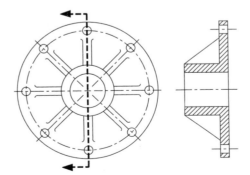

Figure 14–24

BROKEN SECTIONS

At times, only a partial section of an area needs to be created. For this reason, a broken section might be used. The object in Figure 14–25A shows a combination of sectioned areas and conventional areas outlined by the hidden lines. When converting an area to a broken section, you create a break line, crosshatch one area, and leave the other area in a conventional drawing. Break lines may take the form of short freehanded line segments as shown in Figure 14–25B or a series of long lines separated by break symbols as in Figure 14–25C. You can use the LINE command with Ortho off to produce the desired effect as in Figures 14–25A and 14–25B.

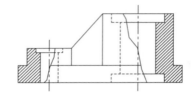

Figure 14–25A

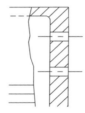

Figure 14–25B

Figure 14–25C

REVOLVED SECTIONS

Section views may be constructed as part of a view with revolved sections. In Figure 14–26, the elliptical shape is constructed while it is revolved into position and then crosshatched.

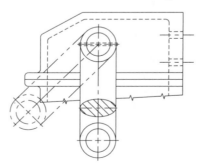

Figure 14–26

Figure 14–27 is another example of a revolved section where a cross section of the C-clamp was cut away and revolved to display its shape.

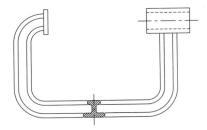

Figure 14–27

REMOVED SECTIONS

Removed sections are very similar to revolved sections except that instead of the section being drawn somewhere inside the object, as is the case with a revolved section, the section is placed elsewhere or removed to a new location in the drawing. The cutting plane line is present, with the arrows showing the direction of sight. Identifying letters are placed on the cutting plane and underneath the section to keep track of the removed sections, especially when there are a number of them on the same drawing sheet. See Figure 14–28.

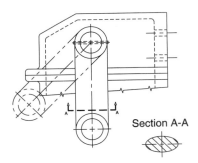

Section A-A

Figure 14–28

Another way of displaying removed sections is to use centerlines as a substitute for the cutting plane line. In Figure 14–29, the centerlines determine the three shapes of the chisel and display the basic shapes from circle to octagon to rectangle. Identification numbers are not required in this particular example.

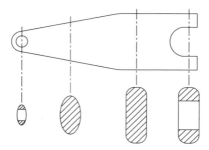

Figure 14–29

ISOMETRIC SECTIONS

Section views may be incorporated into pictorial drawings that appear in Figure 14–30A and 14–30B. Figure 14–30A is an example of a full isometric section with the cutting plane passing through the entire object. In keeping with basic section rules, only those surfaces sliced by the cutting plane line are crosshatched. Isometric sections make it easy to view cut edges compared to holes or slots. Figure 14–30B is an example of an isometric drawing converted to a half section.

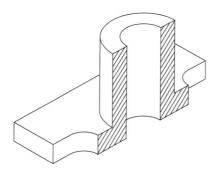

Figure 14–30A

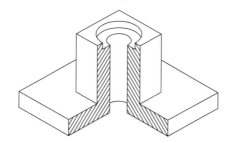

Figure 14–30B

ARCHITECTURAL SECTIONS

Mechanical representations of machine parts are not the only type of drawings where section views are used. Architectural drawings rely on sections to show the type of building materials that go into the construction of foundation plans, roof details, or wall sections, as shown in Figure 14–31. Here numerous types of crosshatching symbols are used to call out the different types of building materials, such as brick veneer at "A," insulation at "B," finished flooring at "C," floor joists at "D," concrete block at "E," earth at "F," and poured concrete at "G." Section symbols that are provided in AutoCAD were used to crosshatch most of the building components.

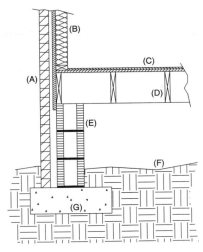

Figure 14–31

THE BHATCH COMMAND

Choosing Hatch… from the Draw pull-down menu in Figure 14–32A activates the main Boundary Hatch dialog box in Figure 14–32B. Use the Quick tab to pick points identifying the area or areas to be crosshatched, select objects using one or more of the popular selection set modes, select hatch patterns supported in the software, and change the scale and angle of the hatch pattern. All areas grayed out in Figure 14–32B, are currently inactive. Only when additional parameters have been satisfied will these buttons activate for use.

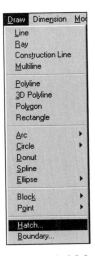

Figure 14–32A

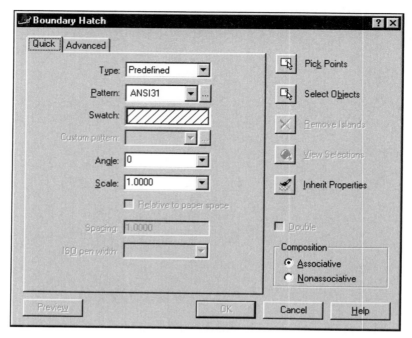

Figure 14–32B

The object shown in Figure 14–33 will be used to demonstrate the boundary cross-hatching method. The object needs to have areas "A," "B," and "C" crosshatched.

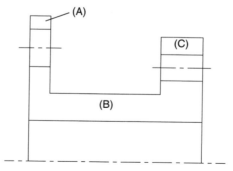

Figure 14–33

In the past, the BREAK command was one technique used to isolate the three areas shown in Figure 14–34. Another technique to use in crosshatching is to trace a closed polyline on top of the three areas, perform the hatch operation, and then erase each polyline boundary. The BHATCH command automatically highlights the areas to be crosshatched.

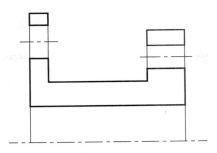

Figure 14–34

The BHATCH command automatically outlines the boundaries to be crosshatched. Figure 14–35 illustrates three areas with internal points identified by the "X's." The dashed areas identify the highlighted areas to be crosshatched.

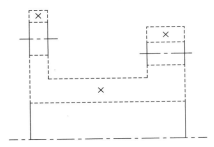

Figure 14–35

In the final step of the BHATCH command, as the crosshatch pattern is applied to the areas, the highlighted outlines deselect, leaving the crosshatch patterns. See Figure 14–36.

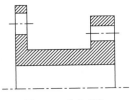

Figure 14–36

Choosing the Advanced tab in the main Boundary Hatch dialog box activates the dialog box shown in Figure 14–37. This dialog box holds a number of crosshatching options, such as Make a New Boundary Set, Boundary Style, Boundary Options, and an Island Detection check box.

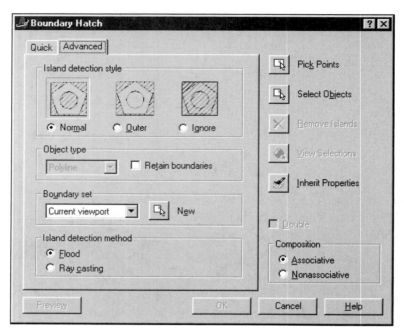

Figure 14–37

Figure 14–38A shows three boundary styles and how they affect levels of crosshatching. The Normal boundary style is the default hatching method in which, when you select all objects within a window, the hatching begins with the outermost boundary, skips the next inside boundary, hatches the next innermost boundary, and so on. Notice the hatching pattern still exposing the text objects for easy reading. The outermost boundary style hatches only the outermost boundary of the object. The Ignore style ignores the default hatching methods of alternating crosshatching and hatches the entire object.

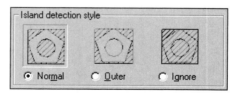

Figure 14–38A

You have the option of retaining the hatch boundary by clicking in the check box for Retain Boundaries. By default, Flood mode in the Island Detection area is turned on. In this way, selecting an internal point identifies all boundaries including islands (see Figure 14–38B).

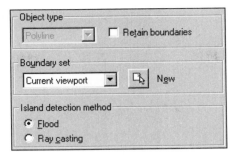

Figure 14–38B

The Quick tab holds numerous hatch patterns in a drop-down edit box in Figure 14–39. Clicking on a pattern in this listing makes the pattern current.

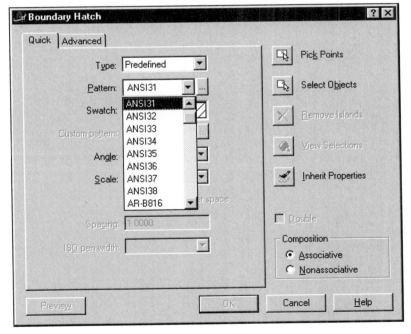

Figure 14–39

Selecting the pattern (…) button next to the pattern name in Figure 14–39 activates a series of tabs that display a number of crosshatching patterns already created and ready for use in Figure 14–40. Select a particular pattern by clicking on the pattern itself. If the wrong pattern was selected, simply choose the correct one or move to another tab to view other patterns for use.

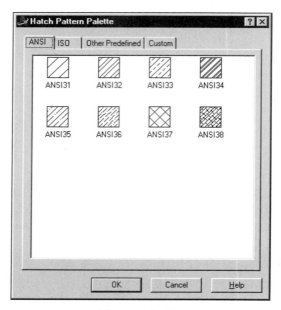

Figure 14–40

Once a hatch pattern is selected, the name is displayed in the Boundary Hatch dialog box in Figure 14–41. If the scaling and angle settings look favorable, click on the Pick Points button. The command prompt requires you to select an internal point or points.

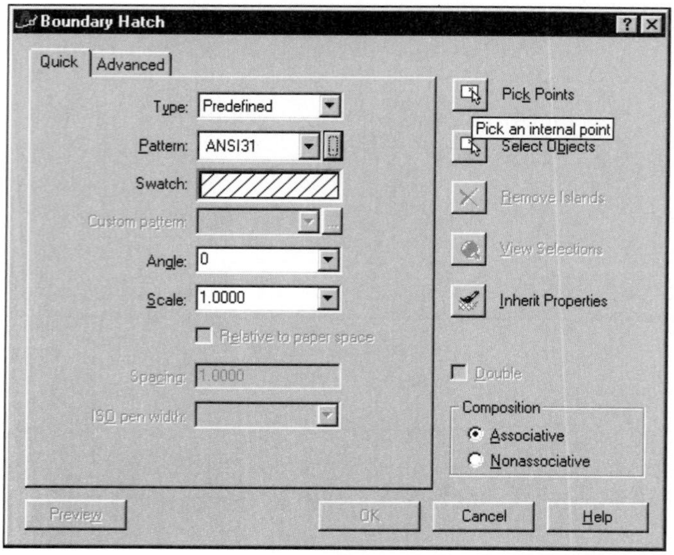

Figure 14–41

In Figure 14–42, click in the three areas marked by the "X." Notice that these areas highlight. If these areas are correct, press ENTER to return to the Boundary Hatch dialog box.

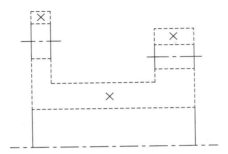

Figure 14–42

When you're back in the Boundary Hatch dialog box, you now have the option of first previewing the hatch to see if all settings and the appearance of the pattern are desirable. Click on the Preview button in the lower left corner of the dialog box to accomplish this. The results appear in Figure 14–43. At this point in the process, the pattern is still not permanently placed on the object. Pressing the ENTER key to exit preview mode returns you to the main Boundary Hatch dialog box. If the hatch pattern is correct in appearance, click the OK button to place the pattern with the drawing.

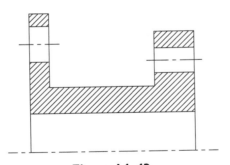

Figure 14–43

SOLID FILL HATCH PATTERNS

Activating the Boundary Hatch dialog box and clicking on the pattern (...) button followed by the Other Predefined tab displays the dialog box in Figure 14–44A. Additional patterns are available and include brick, roof, and other construction related items. One powerful pattern is called SOLID.

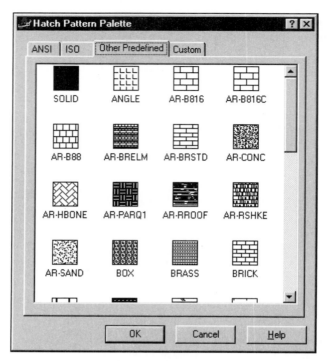

Figure 14–44A

The object in Figure 14–44B illustrates a thin wall that needs to be filled in. The BHATCH command used with the SOLID hatch pattern can perform this task very efficiently. When you pick an internal point, the entire closed area will be filled with a solid pattern and placed on the current layer. The color of the pattern will take on the same color as the one assigned to the layer.

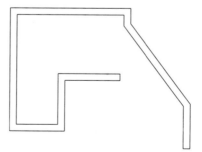

Figure 14–44B

The solid hatch pattern in Figure 14–44C was placed with one internal point pick.

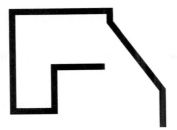

Figure 14–44C

Applying the SOLID hatch pattern to the object in Figure 14–45A can easily be accomplished with the BHATCH command. This object also is a thin wall example with the addition of various curved edges and fillets. In the past, you had to simulate a solid fill using a hatch pattern such as ANSI31, where the scale of the pattern was reduced to very small. The end result was a solid fill; however, upon closer inspection, the crosshatch lines were placed at very small increments from each other. This tended to increase the size of the drawing file for even the simplest of fill operations.

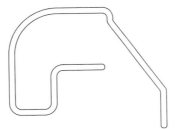

Figure 14–45A

The solid hatch pattern also works for curved outlines as well as outlines involving known vertex corners. Figure 14–45B displays the solid pattern, completely filling in the outline where all corners consist of some type of curve generated by the FILLET command.

Figure 14–45B

USING BHATCH TO HATCH ISLANDS

When charged with the task of crosshatching islands, first activate the Boundary Hatch dialog box in Figure 14–46 and select a pattern. Next pick a point at "A" in Figure 14–47, which will define not only the outer perimeter of the object but the inner shapes as well. This result is due to Flood mode in the Island Detection area being activated in the Advanced tab of the Boundary Hatch dialog box.

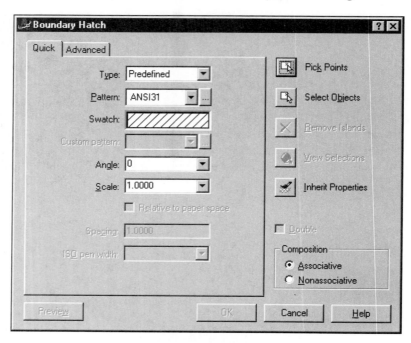

Figure 14–46

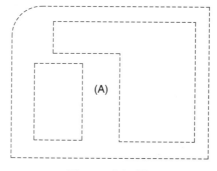

Figure 14–47

Selecting the Preview button will display the hatch pattern shown in Figure 14–48. If changes need to be made, such as a change in the hatch scale or angle, the preview allows these changes to be made. After making changes, be sure to preview the pattern once again to check if the results are desirable. Choosing Apply places the pattern and exits to the command prompt.

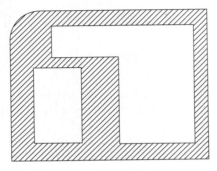

Figure 14–48

HATCH PATTERN SCALING

Hatching patterns are already predefined in size and angle. When you use the BHATCH command, the pattern used is assigned a scale value of 1.00, which will draw the pattern exactly the way it was originally created. Figure 14–49 shows the effects of the BHATCH command when accepting the default value for the pattern scale.

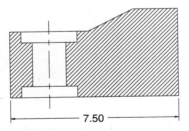

Figure 14–49

Entering a different scale value for the pattern in the Boundary Hatch dialog box will either increase or decrease the spacing between crosshatch lines. Figure 14–50 is an example of the ANSI31 pattern with a new scale value of 0.50.

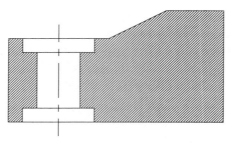

Figure 14–50

As you can decrease the scale of a pattern to hatch small areas, so also can you scale the pattern up for large areas. Figure 14–51 has a hatch scale of 2.00, which doubles all distances between hatch lines.

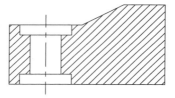

Figure 14–51

Use care when hatching large areas. In Figure 14–52, the distance measures 190.50 millimeters. If the hatch scale of 1.00 were used, the pattern would take on a filled appearance. This results in numerous lines being generated, which will increase the size of the drawing file. A value of 25.4 is used to scale hatch lines for metric drawings.

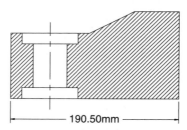

Figure 14–52

HATCH PATTERN ANGLE MANIPULATION

As with the scale of the hatch pattern, depending on the effect, you can control the angle for the hatch pattern within the area being hatched. By default, the BHATCH command displays a 0° angle for all patterns. In Figure 14–53, the angle for "ANSI31" is drawn at 45°—the angle in which the pattern was originally created.

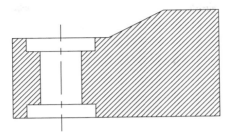

Figure 14–53

Entering any angle different from the default value of "0" will rotate the hatch pattern by that value. This means that if a pattern were originally designed at a 45° angle like "ANSI31," entering a new angle for the pattern would begin rotating the pattern starting at the 45° position. In Figure 14–54, a new angle of 45° is entered in the Boundary Hatch dialog box. Since the original angle was already 45°, this new angle value is added to the original to obtain a vertical crosshatch pattern; positive angles rotate the hatch pattern in the counterclockwise direction.

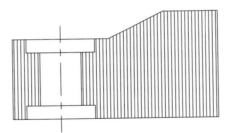

Figure 14–54

Again, entering an angle other than the default rotates the pattern from the original angle to a new angle. In Figure 14–55, an angle of 90° has been applied to the "ANSI31" pattern in the Boundary Hatch dialog box. Providing different angles for patterns is useful when you create section assemblies where different parts are in contact with each other and patterns are placed at different angles, because it makes the parts easy to see.

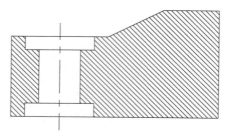

Figure 14–55

INHERIT HATCHING PROPERTIES

The image in Figure 14–56 consists of a simple assembly drawing. At least three different hatch patterns are displayed to define the different parts of the assembly. Unfortunately, a segment of one of the parts was not crosshatched and it is unclear what pattern, scale, and angle were used to place the pattern.

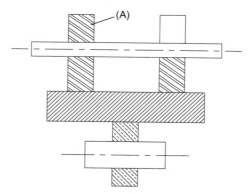

Figure 14–56

Whenever you are faced with this problem, click on the Inherit Properties button in the main Boundary Hatch dialog box in Figure 14–57. Clicking on the pattern at "A" in Figure 14–56 returns you to the Boundary Hatch dialog box where the pattern, scale, and angle are automatically set from the selected pattern.

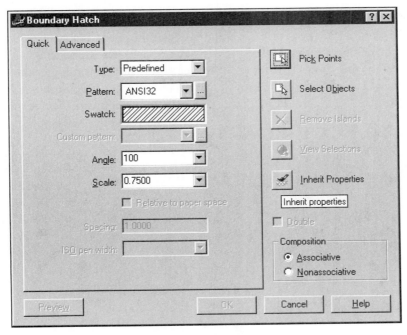

Figure 14–57

To complete the assembly, click on the Pick Points button of the Boundary Hatch dialog box and click an internal point in the empty area. The hatch pattern is placed in this area and it matches that of the other patterns to identify the common parts, as shown in Figure 14–58.

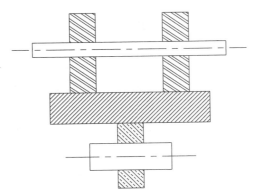

Figure 14–58

PROPERTIES OF ASSOCIATIVE CROSSHATCHING

The following is a review of associative crosshatching. In Figure 14–59, the plate needs to be hatched in the ANSI31 pattern at a scale of one unit and an angle of zero; the two slots and three holes are to be considered islands. Enter the BHATCH command, make the necessary changes in the Boundary Hatch dialog box, click on the OK button, and mark an internal point somewhere inside the object (such as at "A").

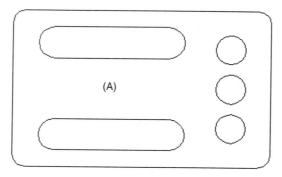

Figure 14–59

When all objects inside and including the outline highlight, press ENTER to return to the Boundary Hatch dialog box. It is very important to realize that all objects highlighted are tied to the associative crosshatch object. Each shape works directly with the hatch pattern to ensure that the outline of the object is being read by the hatch pattern and that the hatching is performed outside the outline. You have the option of first previewing the hatch pattern or applying the hatch pattern. In either case, the results appear similar to the object in Figure 14–60.

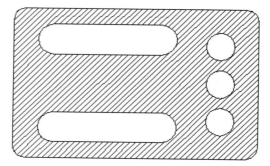

Figure 14–60

Associative crosshatch objects may be edited, and the results of this editing will have an immediate impact on the future appearance of the hatch pattern. For example, the two outer holes need to be increased in size by a factor of 1.5; also, the middle hole needs to be repositioned to the other side of the object (see Figure 14–61). Using the MOVE command not only allows you to reposition the hole, but when the move is completed, the crosshatch pattern mends itself automatically to the moved circle. In the same manner, using the SCALE command to increase the size of the two outer circles by a value of 1.5 units makes the circles large and automatically mends the hatch pattern.

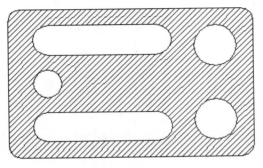

Figure 14–61

In Figure 14–62, the length of the slots needs to be decreased. Also, hole "A" needs to be deleted. Use the STRETCH command to shorten the slots. Use the crossing box at "B" to select the slots and stretch them one unit to the right. Use the ERASE command to delete hole "A."

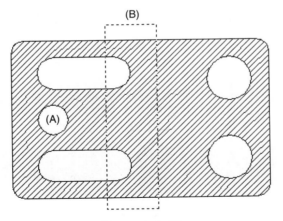

Figure 14–62

The result of the editing operations is shown in Figure 14–63. In past uses of associative crosshatching, if a member of the group of objects that helped define the boundary of the hatch pattern was erased, the hatch pattern would remain in the drawing;

however, it would lose its associativity. Also, the crosshatch pattern would not mend to the erased hole, resulting in a blank spot on the drawing where the hole once existed. In Figure 14–63, associative crosshatching has been enhanced for members to be erased while the hatch pattern still maintains its associativity. The inner hole was deleted with the ERASE command. After the ERASE command was executed, the hatch pattern updated itself and hatched through the area once occupied by the hole.

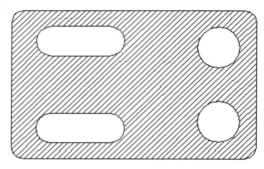

Figure 14–63

Once a hatch pattern is placed in the drawing, it will not update itself to any new additions in the form of closed shapes in the drawing. In Figure 14–64, a rectangle was added at the center of the object. Notice, however, that the crosshatch pattern cuts directly through the rectangle. Since the crosshatch pattern does not have the intelligence to recognize the new boundary, the entire hatch pattern must be deleted and crosshatched again. In this way, the rectangular boundary will be recognized.

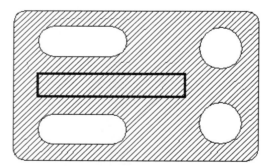

Figure 14–64

THE HATCHEDIT COMMAND

In Figure 14–65, the pattern needs to be increased to a new scale factor of 3 units and the angle of the pattern needs to be rotated 90°. The current scale value of the pattern is 1 unit and the angle is 0°. Issuing the HATCHEDIT command prompts you to select the hatch pattern to edit. Clicking on the hatch pattern anywhere inside the object in Figure 14–65 displays the Hatch Edit dialog box, shown in Figure 14–66.

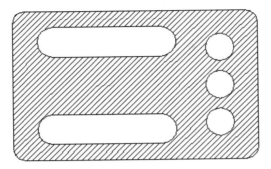

Figure 14–65

With the dialog box displayed as in Figure 14–66, click in the Scale edit box to change the scale from the current value of 1 unit to the new value of 3 units. Next, click in the Angle edit box and change the angle of the hatch pattern from the current value of 0° to the new value of 90°. This will increase the spacing between hatch lines and rotate the pattern 90° in the counterclockwise direction.

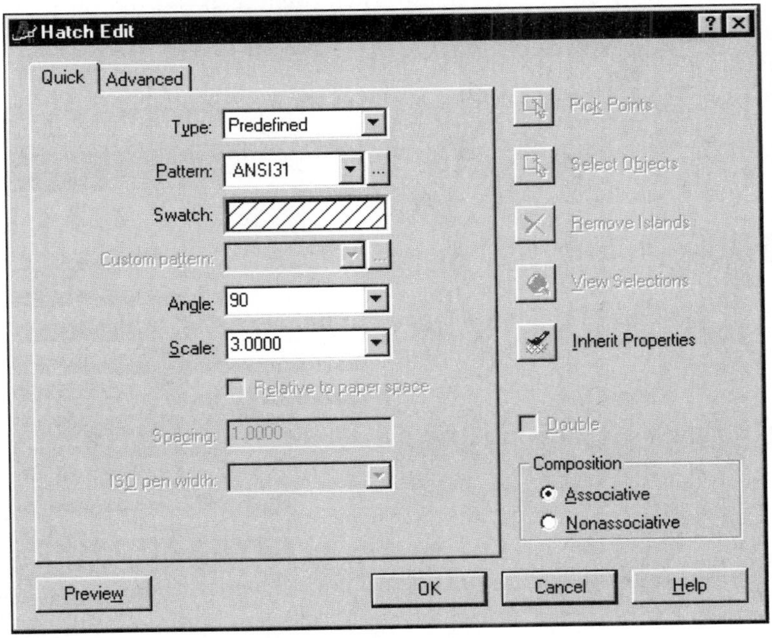

Figure 14–66

Clicking the OK button in the Hatch Edit dialog box returns you to the drawing editor and updates the hatch pattern to these changes (see Figure 14–67). Be aware of the following rules for the HATCHEDIT command to function: it only works on an associative hatch pattern. If the pattern loses associativity through the use of the MIRROR command or if the hatch pattern is exploded, the HATCHEDIT command will cease to function.

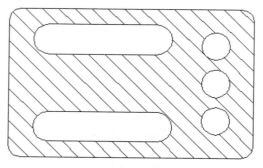

Figure 14–67

EXPRESS TOOLS APPLICATION FOR CROSSHATCHING

THE SUPERHATCH COMMAND

Choosing Draw from the Express pull-down menu exposes the Super Hatch command in Figure 14–68A. Clicking on Super Hatch activates the SuperHatch dialog box in Figure 14–68B. Notice in this dialog box that you can create an area fill using an image file, block, attached external reference, or even a wipeout pattern.

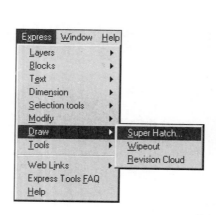

Figure 14–68A

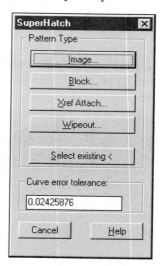

Figure 14–68B

Clicking on the Image button takes you to the main AutoCAD 2000 folder system. A folder called Textures is available that holds numerous TGA files. Double-clicking on this folder activates the Select Image File dialog box. The image "checker.tga" will be used as the image to fill in a rectangle. Selecting "checker.tga" displays a preview of this image in Figure 14–69. Click the OK button when finished.

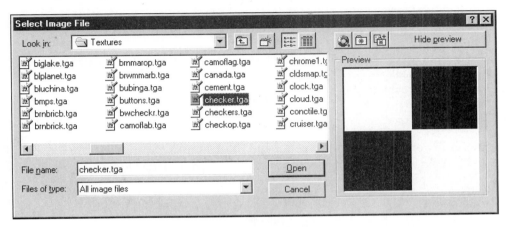

Figure 14–69

The Image dialog box illustrated in Figure 14–70 allows you to control the insertion point, scale, or rotation angle of the image soon to be used as the fill pattern. Place checks in the appropriate boxes to specify these parameters on the screen. When you click the OK button in the Insert dialog box, you are given the following prompts before the image is placed:

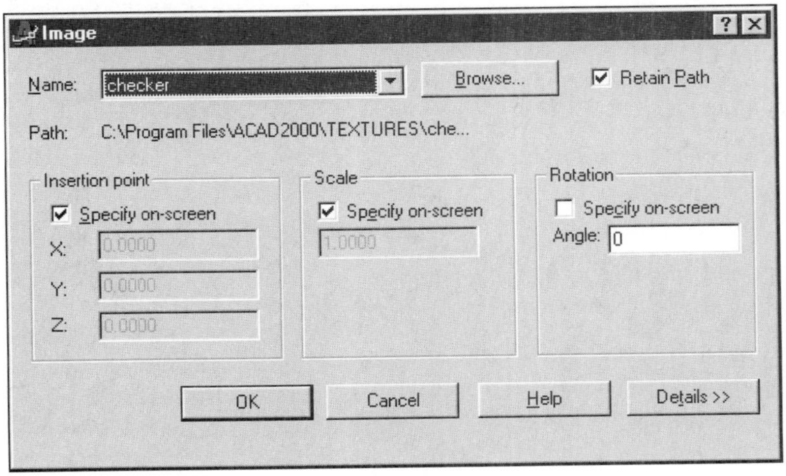

Figure 14–70

Command: **SUPERHATCH**
(Select a valid image file)
Insertion point <0,0>: *(Select the endpoint of the rectangle at "A" in Figure 14–71)*
Base image size: Width: 1.000000, Height: 0.950311, Inches
Specify scale factor <1>: *(Press ENTER for the default)*
Is the placement of this IMAGE acceptable? [Yes/No] <Yes>: *(Press ENTER if ok)*
Selecting visible objects for boundary detection...Done.
Specify an option [Advanced options] <Internal point>: *(Pick a point inside of the area to be hatched)*
Specify an option [Advanced options] <Internal point>: *(Press ENTER to perform the hatch operation)*
Preparing hatch objects for display...
Done.
Use TFRAMES to toggle object frames on and off.

The results of the SUPERHATCH command on the rectangle are illustrated in Figure 14–71. A circle is also displayed with a stone image pattern used to fill in its boundaries.

(A)

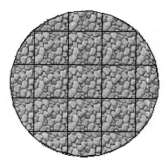

Figure 14–71

TUTORIAL EXERCISE: COUPLER.DWG

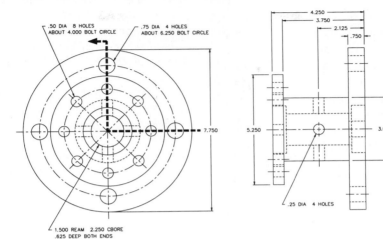

Figure 14–72

Purpose

This tutorial is designed to use the MIRROR and BHATCH commands to convert the coupler to a half section.

System Settings

Since this drawing is provided on CD, edit an existing drawing called "Coupler." Follow the steps in this tutorial for converting the upper half of the object to a half section. All Units, Limits, Grid, and Snap values have been previously set.

Layers

The following layers are already created:

Name	Color	Linetype
Object	White	Continuous
Center	Yellow	Center
Hidden	Red	Hidden
Hatch2	Magenta	Continuous
Cutting Plane Line	Yellow	Dashed
Dimension	Yellow	Continuous

Suggested Commands

Convert one-half of the object to a section by erasing unnecessary hidden lines. Use the Layer Control box and change the remaining hidden lines to the Object layer. Issue the BHATCH command and use the ANSI31 hatch pattern to hatch the upper half of the coupler on the Hatch layer.

Whenever possible, substitute the appropriate command alias in place of the full AutoCAD command in each tutorial step. For example, use "CP" for the COPY command, "L" for the LINE command, and so on. The complete listing of all command aliases is located in Chapter 1, Table 1–2.

STEP I

Use the MIRROR command to copy and flip the upper half of the Side view and form the lower half. When in object selection mode, use the Remove option to deselect the main centerline and hole. If these objects are included in the mirror operation, a duplicate copy of these objects will be created. See Figure 14–73.

Command: **MI** *(For MIRROR)*
Select objects: *(Pick a point at "A")*
Specify opposite corner: (Pick a point at "B")
Select objects: *(Hold down the SHIFT key and pick centerlines "C" and "D" to remove them from the selection set)*

Select objects: *(With the SHIFT key held down, pick a point at "E")*
Specify opposite corner: *(With the SHIFT key held down, pick a point at "F")*
Select objects: *(Press ENTER to continue)*
Specify first point of mirror line: *(Select the endpoint of the centerline near "C")*
Specify second point of mirror line: *(Select the endpoint of the centerline near "D")*
Delete source objects? [Yes/No] <N>: *(Press ENTER to perform the mirror operation)*

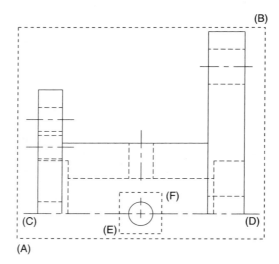

Figure 14–73

STEP 2

Begin converting the upper half of the Side view to a half section by using the ERASE command to remove any unnecessary hidden lines and centerlines from the view. See Figure 14–74.

Command: **E** *(For ERASE)*
Select objects: *(Carefully select the hidden lines labeled "A," "B," "C," and "D")*
Select objects: *(Select the centerline labeled "E")*
Select objects: *(Press ENTER to execute the ERASE command)*

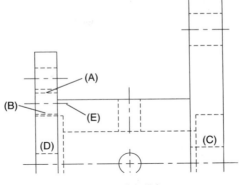

Figure 14–74

STEP 3

Since the remaining hidden lines actually represent object lines when shown in a full section, use the Layer Control box in Figure 14–75 to convert all highlighted hidden lines from the Hidden layer to the Object layer.

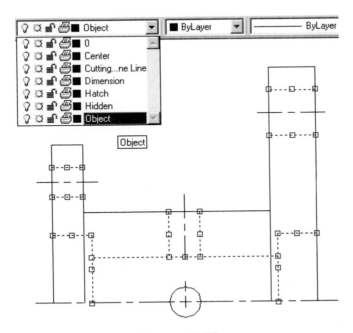

Figure 14–75

STEP 4

Remove unnecessary line segments from the upper half of the converted section using the TRIM command. Use the horizontal line at "A" as the cutting edge, and select the two vertical segments at "B" and "C" as the objects to trim. See Figure 14–76.

Command: **TR** *(For TRIM)*
Current settings: Projection=UCS
 Edge=None
Select cutting edges ...
Select objects: *(Select the horizontal line at "A")*
Select objects: *(Press ENTER to continue)*
Select object to trim or [Project/Edge/ Undo]: *(Select the vertical line at "B")*
Select object to trim or [Project/Edge/ Undo]: *(Select the vertical line at "C")*

Select object to trim or [Project/Edge/ Undo]: *(Press ENTER to exit this command)*

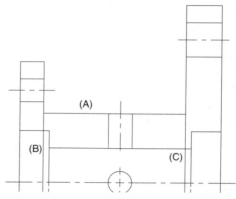

Figure 14–76

STEP 5

Make the Hatch layer the current layer. Then, use the BHATCH command to display the Boundary Hatch dialog box shown in Figure 14–77. Use the pattern "ANSI31" and keep all default settings. Click on the Pick Points button. When the drawing reappears, click inside areas "A," "B," "C," and "D" in Figure 14–78. When finished selecting these internal points, press ENTER to return to the Boundary Hatch dialog box.

Command: **BH** *(For BHATCH)*
(The Boundary Hatch dialog box appears. Make changes to match Figure 14–77. When finished, click on the Pick Points < button.)

Select internal point: *(Pick a point at "A")*
Selecting everything...
Selecting everything visible...
Analyzing the selected data...
Analyzing internal islands...

Select internal point: *(Pick a point at "B")*
Analyzing internal islands...
Select internal point: *(Pick a point at "C")*
Analyzing internal islands...
Select internal point: *(Pick a point at "D")*
Analyzing internal islands...
Select internal point: *(Press ENTER to exit this area and return to the Boundary Hatch dialog box)*

While in the Boundary Hatch dialog box, click on the Preview button to view the results. If the hatch results are accept- able, click the OK button to place the hatch pattern.

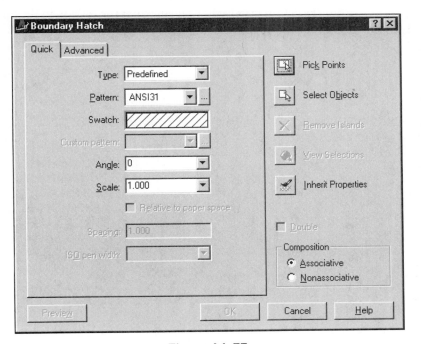

Figure 14–77

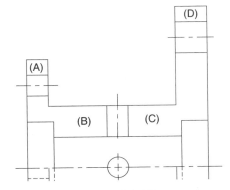

Figure 14–78

STEP 6

The complete crosshatched view is shown in Figure 14–79. The full drawing is also shown in Figure 14–80, along with dimensions. As with other half section examples, it is the option of the operator or designer to show all hidden lines in the lower half or delete all hidden and centerlines from the lower half and simply interpret the section in the upper half.

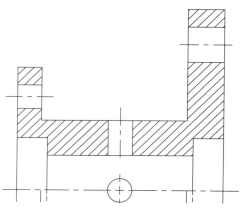

Figure 14–79

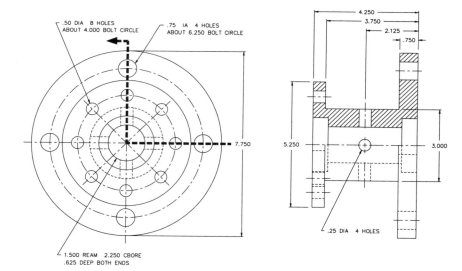

Figure 14–80

TUTORIAL EXERCISE: ASSEMBLY.DWG

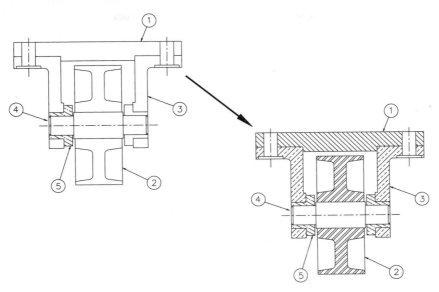

Figure 14–81

Purpose

This tutorial is designed to use the BHATCH command to crosshatch an assembly drawing.

System Settings

Since this drawing is provided on CD, edit an existing drawing called "Assembly." Follow the steps in this tutorial for converting the object to an assembly section. All Units, Limits, Grid, and Snap values have been previously set.

Layers

Layers have already been created for this tutorial exercise.

Name	Color	Linetype
CENTER	Yellow	Center
LEADER	Cyan	Continuous
OBJECT	White	Continuous
SECTION	Magenta	Continuous

Suggested Commands

The BHATCH command will be used exclusively during this tutorial exercise.

Whenever possible, substitute the appropriate command alias in place of the full AutoCAD command in each tutorial step. For example, use "CP" for the COPY command, "L" for the LINE command, and so on. The complete listing of all command aliases is located in Chapter 1, Table 1–2.

tutorial EXERCISE

STEP 1

Before beginning the crosshatching operations, use the Layer Control box in Figure 14–82 to ensure that the SECTION layer is the current layer, and then turn off the layers called LEADER and CENTER. The LEADER layer holds all bubbles identifying numbers and leaders pointing to the specific parts. The CENTER layer will turn off all centerlines in the drawing.

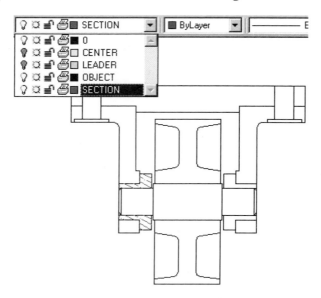

Figure 14–82

STEP 2

Issue the BHATCH command and begin crosshatching the assembly; first crosshatch Part 1, the Plate. Use the default hatch pattern of ANSI31 in addition to a scale factor of 1.0000. Change the angle to 90° in Figure 14–83A. Pick three internal points "A," "B," and "C" to identify the areas to crosshatch (see Figure 14–83B).

Command: **BH** (For BHATCH)
(The Boundary Hatch dialog box appears. Make changes to match Figure 14–83A. When finished, click on the Pick Points button.)

Select internal point: (Pick a point at "A")
Selecting everything...
Selecting everything visible...
Analyzing the selected data...
Analyzing internal islands...

Select internal point: (Pick a point at "B")
Analyzing internal islands...
Select internal point: (Pick a point at "C")
Analyzing internal islands...

Select internal point: (Press ENTER to exit this area and return to the Boundary Hatch dialog box)

When the Boundary Hatch dialog box reappears, click on the Preview button to view the results. If the preview appears correct, press ENTER to return to the Boundary Hatch dialog box. If the hatch results are acceptable, click on the Apply button to place the hatch pattern.

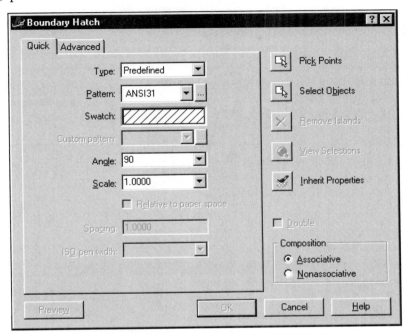

Figure 14–83A

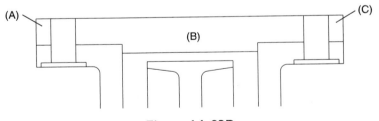

Figure 14–83B

STEP 3

Next, crosshatch the pulley in the assembly drawing. Execute the BHATCH command; when the Boundary Hatch dialog box appears, choose the pattern "ANSI32" and change the angle to 0° in Figure 14–84A. Click on two internal points inside the pulley at "A" and "B" in Figure 14–84B. If the proper boundaries highlight, press ENTER to return to the Boundary Hatch dialog box.

Command: **H** *(For BHATCH)*
(The Boundary Hatch dialog box appears. Make changes to match Figure 14–84A. When finished, click on the Pick Points button.)

Select internal point: *(Pick a point at "A")*
Selecting everything...
Selecting everything visible...
Analyzing the selected data...
Analyzing internal islands...

Select internal point: *(Pick a point at "B")*
Analyzing internal islands...

Select internal point: *(Press ENTER to exit this area and return to the Boundary Hatch dialog box)*

Click on the Preview button to preview the hatch pattern; if the results are desirable, press ENTER to return to the Boundary Hatch dialog box and click the OK button to place the hatch pattern.

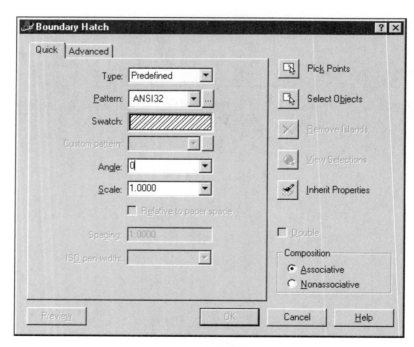

Figure 14–84A

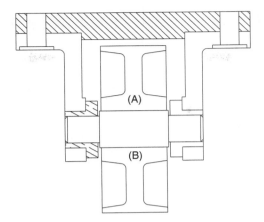

Figure 14–84B

STEP 4

Next, crosshatch the two brackets that support the pulley axle and are connected to the plate. Execute the BHATCH command and select the "ANSI33" hatch pattern; keep all other defaults in Figure 14–85A. When prompted to select internal points, click inside the areas identified by "A" through "F" in Figure 14–85B.

Command: **BH** *(For BHATCH)*
(The Boundary Hatch dialog box appears. Make changes to match Fig. 14–85A. When finished, click on the Pick Points button.)
Select internal point: *(Pick a point at "A")*
Selecting everything...
Selecting everything visible...
Analyzing the selected data...
Analyzing internal islands...

Select internal point: *(Pick a point at "B")*
Analyzing internal islands...

Select internal point: *(Pick a point at "C")*
Analyzing internal islands...

Select internal point: *(Pick a point at "D")*
Analyzing internal islands...

Select internal point: *(Pick a point at "E")*
Analyzing internal islands...

Select internal point: *(Pick a point at "F")*
Analyzing internal islands...

Select internal point: *(Press ENTER to exit this area and return to the Boundary Hatch dialog box)*

Click on the Preview button to preview the hatch pattern; if the results are desirable, press ENTER to return to the Boundary Hatch dialog box and click the OK button to place the hatch pattern.

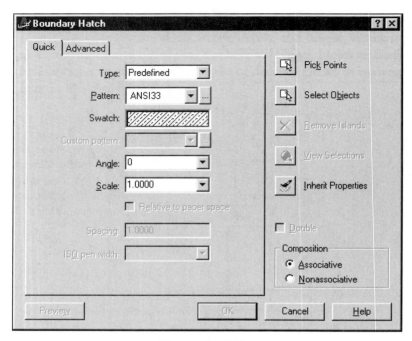

Figure 14–85A

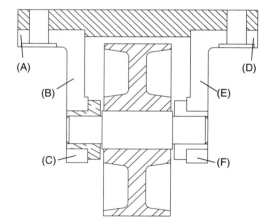

Figure 14–85B

STEP 5

Notice that, although one of the bushings has already been crosshatched, it is difficult to determine the hatch pattern used along with scale and angle. Issue the BHATCH command, click on the Inherit Properties button in Figure 14–86A, and select the existing hatch pattern at "A" using the Property paintbrush. This will transfer the hatch properties of the bushing to the Boundary Hatch dialog box. Pick internal points at "A" and "B" in Figure 14–86B in the drawing, and the object will be crosshatched to these new current parameters.

Command: **BH** (For BHATCH)
(When the Boundary Hatch dialog box appears, click on the Inherit Properties button.)
Select associative hatch object: (Select the hatch pattern already visible in the bushing at "A" in Figure 14–86B. Press

ENTER *when finished. When the Boundary Hatch dialog box reappears, click on the Pick Points < button.)*
Inherited Properties: Name <ANSI35>, Scale <1.0000>, Angle <105>
Select internal point: (Pick a point at "B")
Select internal point: Selecting everything...
Selecting everything visible...
Analyzing the selected data...
Analyzing internal islands...
Select internal point: (Pick a point at "C")
Analyzing internal islands...
Select internal point: (Press ENTER to return to the Boundary Hatch dialog box)

Click on the Preview button to preview the hatch pattern; if the results are desirable, press ENTER to return to the Boundary Hatch dialog box and click the OK button to place the hatch pattern.

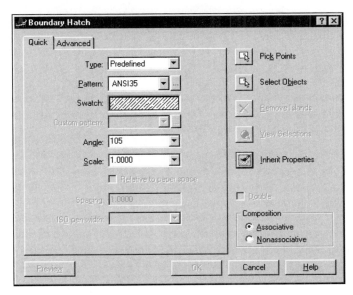

Figure 14–86A

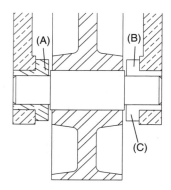

Figure 14–86B

STEP 6

The hatch pattern currently displayed in the pulley is too large for the area it occupies; the spacing needs to be scaled to half. Issue the HATCHEDIT command and select the hatch pattern anywhere in the pulley. When the Hatch Edit dialog box appears, change the scale from a value of 1.0000 to 0.5000 units in Figure 14–87A. Click the OK button to update the hatch pattern to the new changes. The results are illustrated in Figure 14–87B.

Command: **HE** (For HATCHEDIT)
Select hatch object: *(Select the hatch pattern inside of the pulley)*

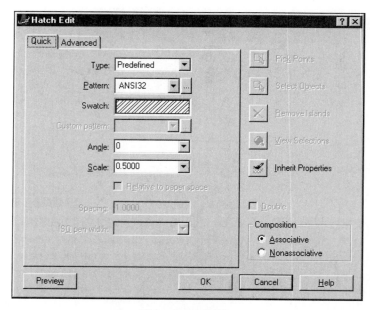

Figure 14–87A

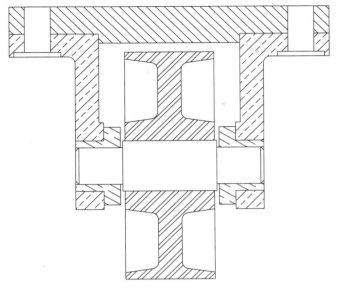

Figure 14–87B

STEP 7

Stretch the base to make it 0.25 units longer on both sides. With Ortho mode turned on, use the Direct Distance mode to accomplish this task. When selecting the objects to stretch, take care not to accidentally select the edge of the counterbore hole. Use the ZOOM command to magnify this part of the screen. See Figure 14–88.

Command: **S** *(For STRETCH)*
Select objects to stretch by crossing-window or crossing-polygon...
Select objects: *(Pick a point at "A")*
Specify opposite corner: *(Pick a point at "B")*
Select objects: *(Press* ENTER *to continue)*

Specify base point or displacement: *(Pick a blank part of the display screen at "C")*
Specify second point of displacement: *(With Ortho on, move your cursor directly to the left and enter a value of 0.25)*
Command: **S** *(For STRETCH)*
Select objects to stretch by crossing-window or crossing-polygon...
Select objects: *(Pick a point at "D")*
Specify opposite corner: *(Pick a point at "E")*
Select objects: *(Press* ENTER *to continue)*
Specify base point or displacement: *(Pick a blank part of the display screen at "F")*
Specify second point of displacement: *(With Ortho on, move your cursor directly to the right and enter a value of 0.25)*

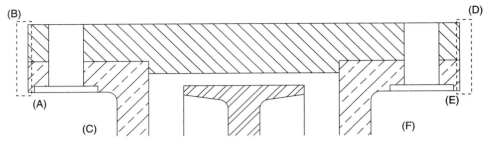

Figure 14–88

STEP 8

Turn all layers back on to complete this
crosshatching exercise (see Figure 14–89).

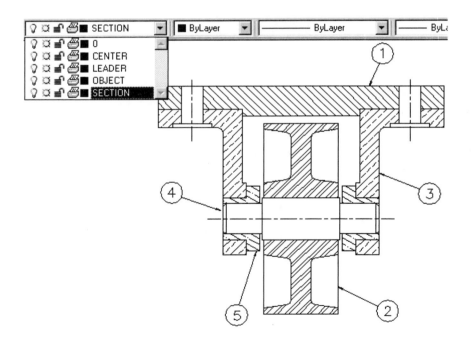

Figure 14–89

PROBLEMS FOR CHAPTER 8

PROBLEM 14–1

Center a three-view drawing and make the Front view a full section.

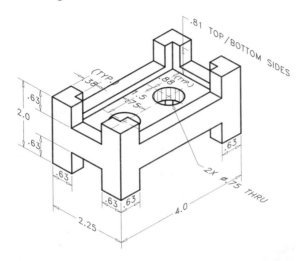

PROBLEM 14–2

Center two views within the work area, and make one view a full section. Use correct drafting practices for the ribs.

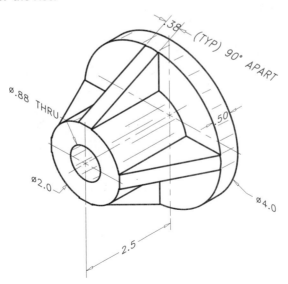

PROBLEM 14–3

Center two views within the work area, and make one view a full section.

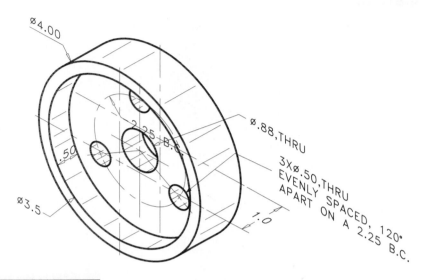

PROBLEM 14–4

Center the Front view and Ttop view within the work area. Make one view a full section.

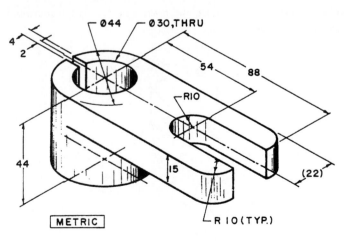

PROBLEM 14–5

Center the required views within the work area, and add removed section A-A.

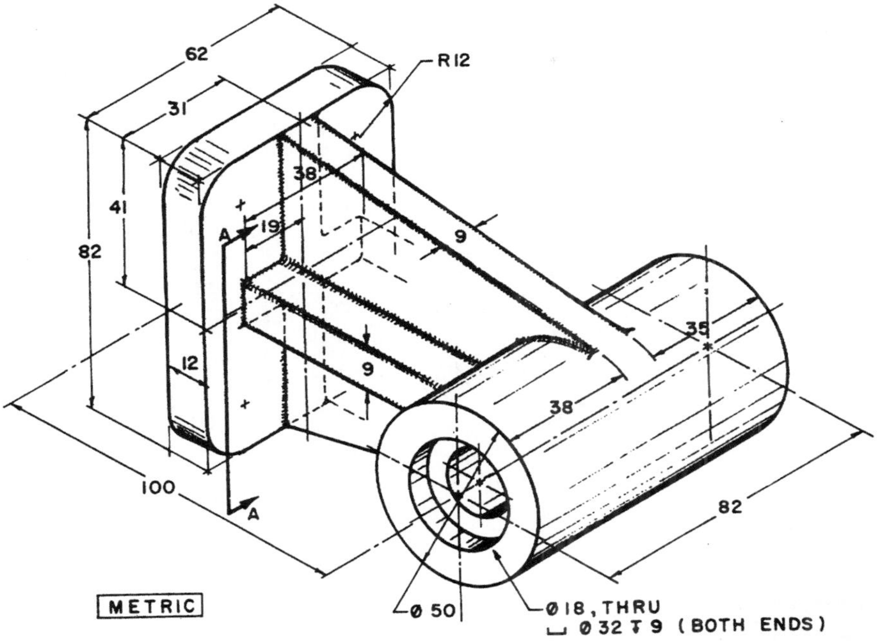
62
31
R12
38
41
82
A
19
9
35
12
9
38
100
A
82
METRIC
Ø 50
Ø 18, THRU
⌴ Ø 32 ↧ 9 (BOTH ENDS)

PROBLEM 14–6

Center three views within the work area, and make one view an offset section.

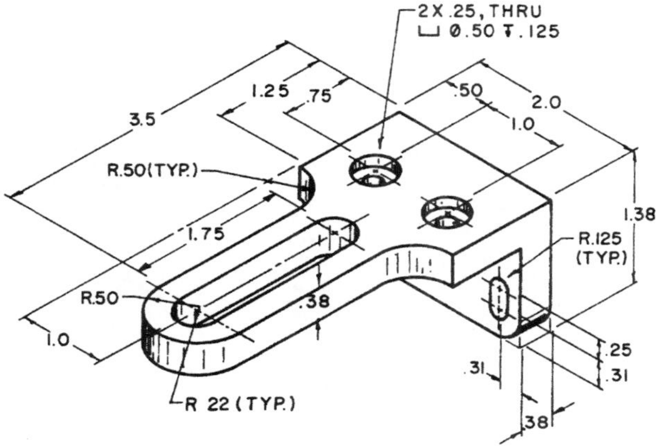
2X .25, THRU
⌴ Ø.50 ↧.125
1.25
.75
.50
2.0
1.0
3.5
R.50(TYP.)
1.38
1.75
R.125
(TYP.)
R.50
.38
1.0
.25
.31
.31
R 22 (TYP.)
.38

PROBLEM 14–7

Center three views within the work area, and make one view an offset section.

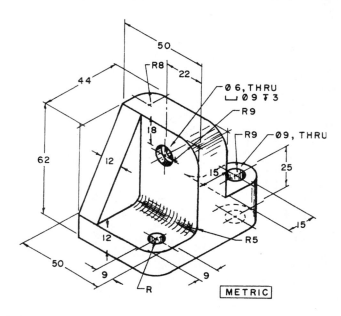

METRIC

PROBLEM 14–8

Center three views within the work area, and make one view an offset section.

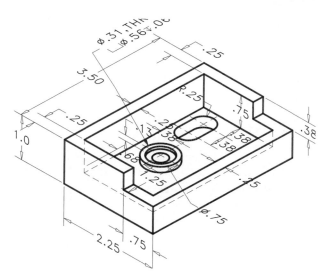

PROBLEM 14–9

Center three views within the work area, and make one view an offset section.

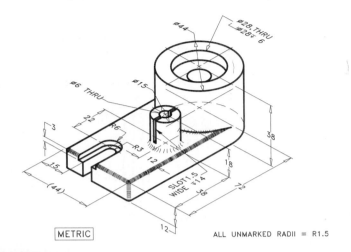

ø44
ø28 THRU
⌴ø28▽ 6
ø6 THRU
ø15
22
R6
R3
3
1.5
(44)
12
38
18
72
38
SLOT 1.5
WIDE ▽ 14
12

METRIC

ALL UNMARKED RADII = R1.5

PROBLEM 14–10

Center two views within the work area, and make one view an offset section.

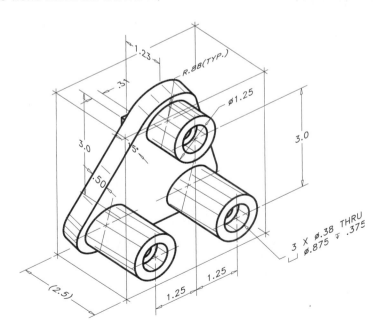

1.23
.31
R.88(TYP.)
ø1.25
3.0
3.0
.50
5°
3 X ø.38 THRU
⌴ø.875 ▽ .375
(2.5)
1.25
1.25
1.25

PROBLEM 14–11

Center the Front view and Top view within the work area. Make one view a half section.

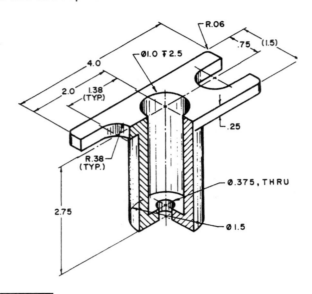

PROBLEM 14–12

Center two views within the work area, and make one view a half section.

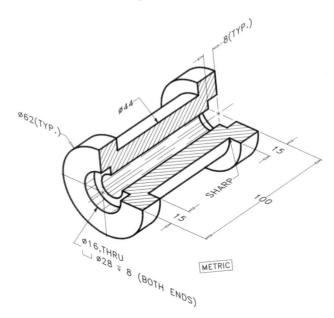

PROBLEM 14–13

Center the two views within the work area, and make one view a half section.

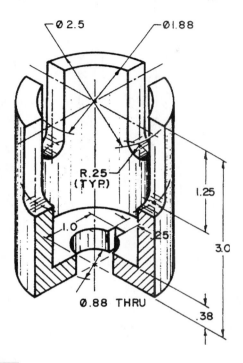

PROBLEM 14–14

Center two views within the work area, and make one view a half section.

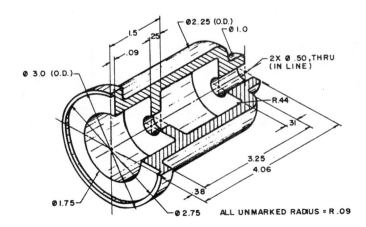

PROBLEM 14–15

Center two views within the work area, and make one view a half section.

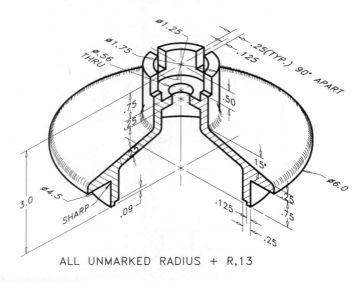

ALL UNMARKED RADIUS + R,13

PROBLEM 14–16

Center the required views within the work area, and make one view a broken-out section to illustrate the complicated interior area.

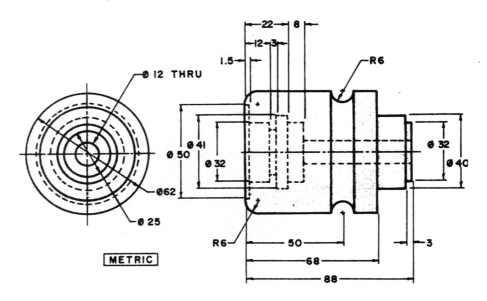

METRIC

PROBLEM 14–17

Center the required views within the work area, and add removed section A-A.

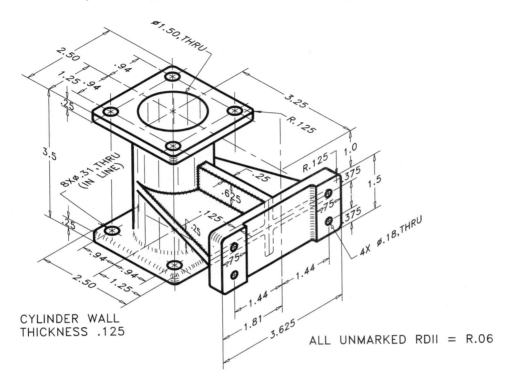

CYLINDER WALL
THICKNESS .125

ALL UNMARKED RDII = R.06

PROBLEM 14–18

Center the required views within the work area, and add removed section A-A.

problem EXERCISE

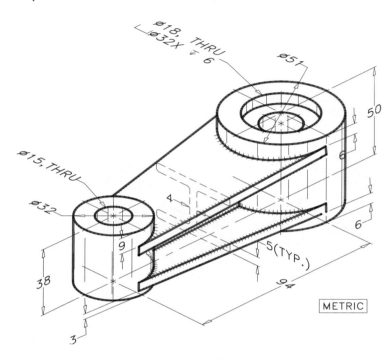

Directions for Problems 14–19 through 14–21:

Center required views within the work area. Leave a 1-inch or 25 mm space between views. Make one view a section view to fully illustrate the object. Use a full half, offset, broken-out, revolved, or removed section Consult your instructor if you need to add dimensions.

PROBLEM 14–19

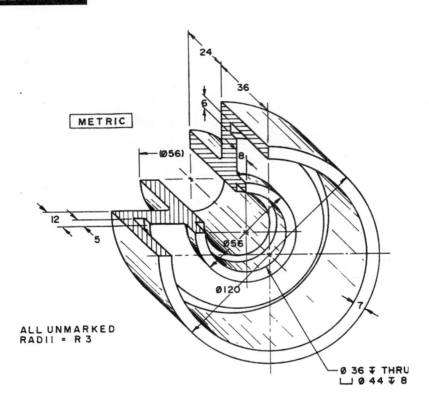

METRIC

(Ø56)

24

36

6

8

12

5

Ø56

Ø120

7

ALL UNMARKED
RADII = R 3

Ø 36 ⛛ THRU
⊔ Ø 44 ⛛ 8

PROBLEM 14–20

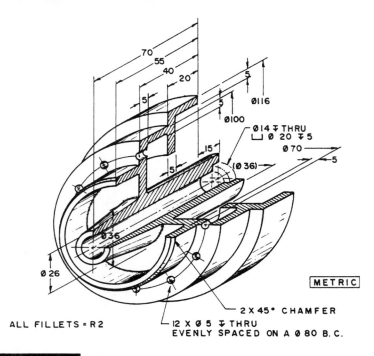

70
55
40
20

5

5

5

Ø116

Ø100

Ø14 ⍕ THRU
⊔ Ø 20 ⍕ 5

Ø 70

5

15

(Ø 36)

Ø 36

Ø 26

METRIC

2 X 45° CHAMFER

ALL FILLETS = R2

12 X Ø 5 ⍕ THRU
EVENLY SPACED ON A Ø 80 B.C.

PROBLEM 14–21

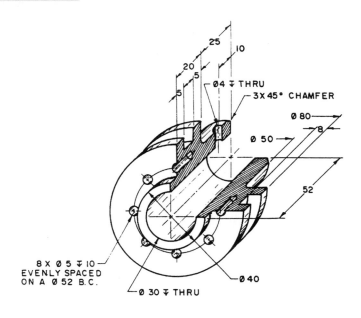

25
10
20
5
5

Ø4 ⍕ THRU

3 X 45° CHAMFER

Ø 80
8

Ø 50

52

8 X Ø 5 ⍕ 10
EVENLY SPACED
ON A Ø 52 B.C.

Ø 40

Ø 30 ⍕ THRU

CHAPTER 15

Auxiliary Views

AUXILIARY VIEW BASICS

During the discussion of multiview drawings, we discovered that you need to draw enough views of an object to accurately describe it. In most cases, this requires a Front, Top, and Right Side view. Sometimes additional views are required, such as Left Side, Bottom, and Back views, to show features not visible in the three primary views. Other special views, like sections, are taken to expose interior details for better clarity. Sometimes all of these views are still not enough to describe the object, especially when features are located on an inclined surface. To produce a view perpendicular to this inclined surface, an auxiliary view is drawn. This chapter will describe where auxiliary views are used and how they are projected from one view to another. A tutorial exercise is presented to show the steps in the construction of an auxiliary view. Additional problems are provided at the end of this chapter for further study of auxiliary views.

Figure 15–1 presents interesting results if constructed as a multiview drawing or orthographic projection.

Figure 15–1

Figure 15–2 should be quite familiar; it represents the standard glass box with object located in the center. Again, the purpose of this box is to prove how orthographic views are organized and laid out. Figure 15–2 is no different. First the primary views, Front, Top, and Right Side views are projected from the object to intersect perpendicular with the glass plane. Under normal circumstances, this procedure would satisfy most multiview drawing cases. Remember, only those views necessary to describe the object are drawn. However, upon closer inspection, we notice that the object in the Front view consists of an angle forming an inclined surface.

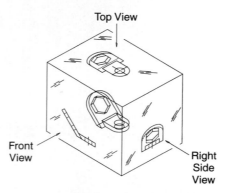

Figure 15–2

When you lay out the Front, Top, and Right Side views, a problem occurs. The Front view shows the basic shape of the object, the angle of the inclined surface (see Figure 15–3). The Top view shows the true size and shape of the surface formed by the circle and arc. The Right Side view shows the true thickness of the hole from information found in the Top view. However, there does not exist a true size and shape of the features found in the inclined surface at "A." We see the hexagonal hole going through the object in the Top and Right Side views. These views, however, show the detail not to scale, or foreshortened. For that matter, the entire inclined surface is foreshortened in all views. This is one case where the Front, Top, and Right Side views are not enough to describe the object. An additional view, or auxiliary view, is used to display the true shape of surfaces along an incline.

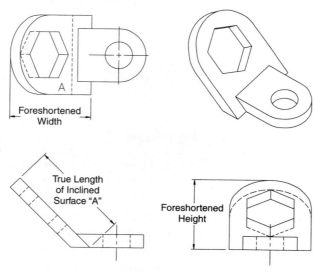

Figure 15–3

To prove the formation of an auxiliary view, let's create another glass box; this time an inclined plane is formed. This plane is always parallel to the inclined surface of the object. Instead of just the Front, Top, and Right Side views being projected, the geometry describing the features along the inclined surface is projected to the auxiliary plane similar to Figure 15–4A.

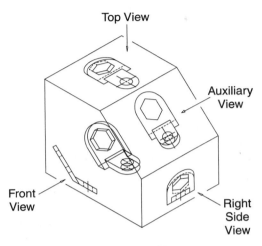

Figure 15–4A

As in multiview projection, the edges of the glass box are unfolded with the edges of the Front view plane as the pivot. See Figure 15–4B.

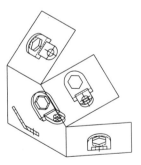

Figure 15–4B

All planes are extended perpendicular to the Front view where the rotation stops. The result is the organization of the multiview drawing complete with an auxiliary viewing plane, as shown in Figure 15–4C.

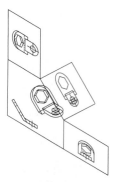

Figure 15–4C

Figure 15–5 represents the final layout complete with auxiliary view. This figure shows the auxiliary being formed as a result of the inclined surface present in the Front view. An auxiliary view may be made in relation to any inclined surface located in any view. Also, the figure displays circles and arcs in the Top view that appear as ellipses in the auxiliary view.

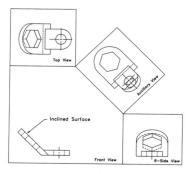

Figure 15–5

It is usually not required that elliptical shapes be drawn in one view where the feature is shown true size and shape in another. The resulting view minus these elliptical features is called a partial view, used extensively in auxiliary views. An example of the Top view converted to a partial view is shown in Figure 15–6.

Figure 15–6

A few rules to follow when constructing auxiliary views are displayed pictorially in Figure 15–7. First, the auxiliary plane is always constructed parallel to the inclined surface. Once this is established, visible as well as hidden features are projected from the incline to the auxiliary view. These projection lines are always drawn perpendicular to the inclined surface and the auxiliary view.

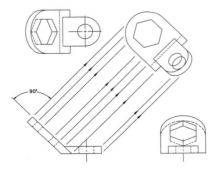

Figure 15–7

CONSTRUCTING AN AUXILIARY VIEW

Figure 15–8 is a multiview drawing consisting of Front, Top, and Right Side views. The inclined surface in the Front view is displayed in the Top and Right Side views; however, the surface appears foreshortened in both adjacent views. An auxiliary view of the incline needs to be made to show its true size and shape. Currently the display screen has the grid on in addition to the position of the typical AutoCAD cursor. Follow Figures 15–9 through 15–18 for one suggested method for projecting to find auxiliary views.

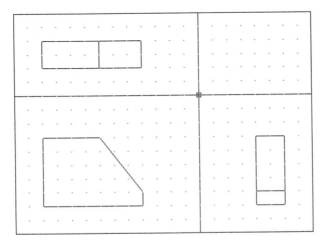

Figure 15–8

To assist with the projection process, it would help if the current grid display could be rotated parallel and perpendicular to the inclined surface (see Figure 15–9). This is accomplished through the SNAP command and the Rotate option.

Command: **SN** *(For SNAP)*
Specify snap spacing or [ON/OFF/Aspect/Rotate/Style/Type] <0.5000>: **R** *(For Rotate)*
Specify base point <0.0000,0.0000>: *(Select the endpoint of the line at "A")*
Specify rotation angle <0>: *(Select the endpoint of the line at "B")*

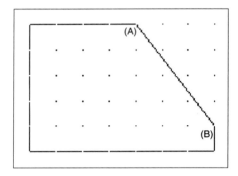

Figure 15–9

Figure 15–10 shows the result of rotating the grid through the use of the SNAP command. This operation has no effect on the already existing views; however, the grid is now placed rotated in relation to the incline located in the Front view. Notice that the appearance of the standard AutoCAD cursor has also changed to conform to the new grid orientation. In addition to snapping to these new grid dots, lines are easily drawn perpendicular to the incline with Ortho on, which will draw lines in relation to the current cursor.

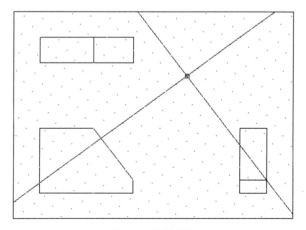

Figure 15–10

Use the OFFSET command to construct a reference line at a specified distance from the incline in the Front view (see Figure 15–11). This reference line becomes the start for the auxiliary view.

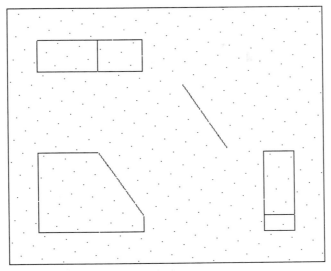

Figure 15–11

Use the LENGTHEN command to extend the two endpoints of the previous line (see Figure 15–12). The exact distances are not critical; however, the line should be long enough to accept projector lines from the Front view.

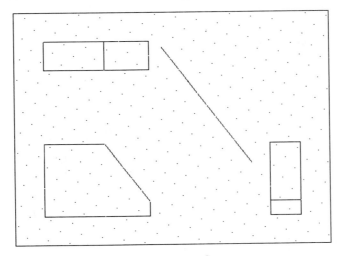

Figure 15–12

Use the OFFSET command to copy the auxiliary reference line the thickness of the object (see Figure 15–13). This distance may be retrieved from the depth of the Top or Right Side views because they both contain the depth measurement of the object.

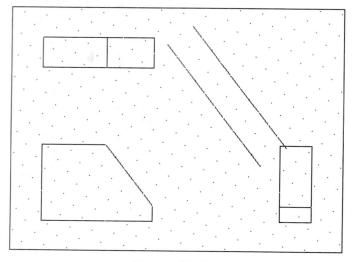

Figure 15–13

Use the LINE command and connect each intersection on the Front view perpendicular to the outer line on the auxiliary view (see Figure 15–14). Draw the lines starting with the OSNAP-Intersect option and ending with the OSNAP-Perpend option.

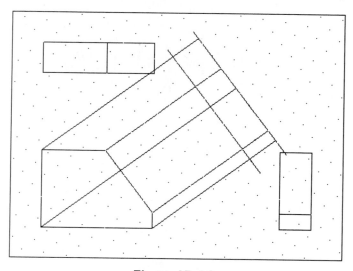

Figure 15–14

Before editing the auxiliary view, analyze the drawing to see if any corners in the Front view are to be represented as hidden lines in the auxiliary view (see Figure 15–15). It turns out that the lower left corner of the Front view is hidden in the auxiliary view. Use the Layer Control box to convert this projection line from a linetype of continuous to hidden.

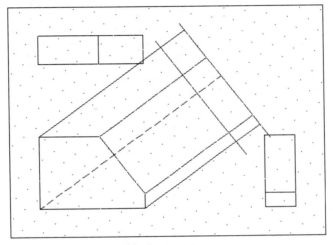

Figure 15–15

Use the TRIM command to partially delete all projection lines and corners of the auxiliary view. Another method would be to use the FILLET command set to a radius of 0. Selecting two lines in a corner automatically trims the excess lines away (see Figure 15–16).

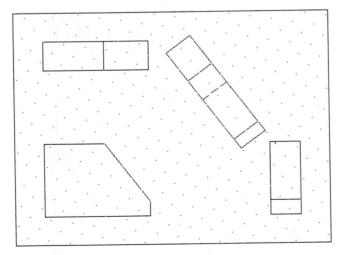

Figure 15–16

The result is a multiview drawing complete with auxiliary view displaying the true size and shape of the inclined surface. See Figure 15–17.

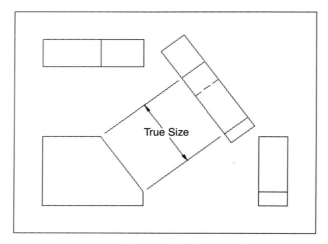

Figure 15–17

For dimensioning purposes, use the SNAP command and set the grid and cursor appearance back to normal. See Figure 15–18.

Command: **SN** *(For SNAP)*
Specify snap spacing or [ON/OFF/Aspect/Rotate/Style/Type] <0.5000>: **R** *(For Rotate)*
Specify base point <Current default>: **0,0**
Specify rotation angle <Current default>: **0**

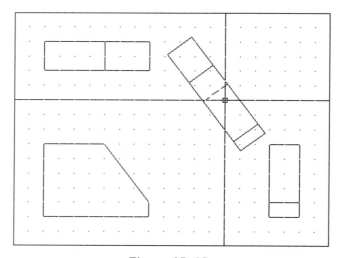

Figure 15–18

TUTORIAL EXERCISE: BRACKET.DWG

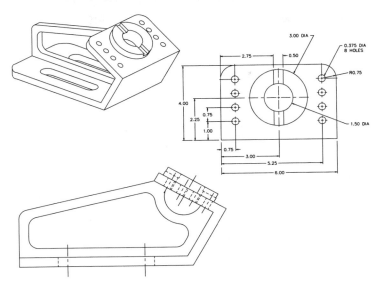

Figure 15–19

Purpose

This tutorial is designed to allow you to construct an auxiliary view of the inclined surface in the bracket shown in Figure 15–19.

System Settings

Since this drawing is provided on CD, edit an existing drawing called "Bracket." Follow the steps in this tutorial for the creation of an auxiliary view. All Units, Limits, Grid, and Snap values have been previously set.

Layers

The following layers have already been created with the following format:

Name	Color	Linetype
CEN	Yellow	Center
DIM	Yellow	Continuous
HID	Red	Hidden
OBJ	Cyan	Continuous

Suggested Commands

Begin this tutorial by using the OFFSET command to construct a series of lines parallel to the inclined surface containing the auxiliary view. Next construct lines perpendicular to the inclined surface. Use the CIRCLE command to begin laying out features that lie in the auxiliary view. Use ARRAY to copy the circle in a rectangular pattern. Add centerlines using the DIMCENTER command. Insert a predefined view called "Top." A three-view drawing consisting of Front, Top, and auxiliary views is completed.

Whenever possible, substitute the appropriate command alias in place of the full AutoCAD command in each tutorial step. For example, use "CP" for the COPY command, "L" for the LINE command, and so on. The complete listing of all command aliases is located in Table 1–2.

STEP 1

Before you begin, understand that an auxiliary view will be taken from a point of view illustrated in Figure 15–20. This direction of sight is always perpendicular to the inclined surface. This perpendicular direction ensures that the auxiliary view of the inclined surface will be of true size and shape. Begin this tutorial by checking to see that Ortho mode is turned off for the next few steps using the command from the keyboard or by clicking on the ORTHO button located in the status bar of the display screen. Also, make the OBJ layer current. Restore a previously saved view called "Front" using the -VIEW command or View dialog box.

Command: **ORTHO**
Enter mode [ON/OFF] <OFF>: *(Press* ENTER *to accept this default)*
Command: **-V** *(For -VIEW)*
Enter an option [?/Orthographic/Delete/ Restore/Save/Ucs/Window]: **R** *(For Restore)*
Enter view name to restore: **Front**

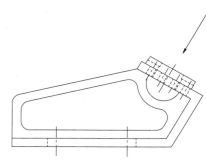

Figure 15–20

STEP 2

Use the SNAP command to rotate the grid perpendicular to the inclined surface. For the base point, identify the endpoint of the line at "A" in Figure 15–21. For the rotation angle, use the rubber-band cursor and mark a point at the endpoint of the line at "B." The grid should change along with the appearance of the standard cursor. Restore the view called "Overall" using the -VIEW command or View dialog box.

Command: **SN** *(For SNAP)*
Specify snap spacing or [ON/OFF/Aspect/ Rotate/Style/Type] <0.50>: **R** *(For Rotate)*
Specify base point <0.00,0.00>: *(Select the endpoint at "A")*

Specify rotation angle <0>: *(Select the endpoint at "B")*
Angle adjusted to 330
Command: **-V** *(For -VIEW)*
Enter an option [?/Orthographic/Delete/ Restore/Save/Ucs/Window]: **R** *(For Restore)*
Enter view name to restore: **Overall**

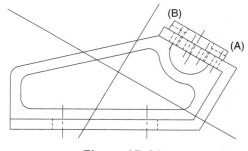

Figure 15–21

STEP 3

Turn snap off by pressing F9. Begin the construction of the auxiliary view by using the OFFSET command to copy a line parallel to the inclined line (see Figure 15–22). Use an offset distance of 8.50 as the distance between the Front and auxiliary views.

Command: **O** *(For OFFSET)*
Specify offset distance or [Through]
   <Through>: **8.50**
Select object to offset or <exit>: *(Select the inclined line at "A")*
Specify point on side to offset: *(Pick a point anywhere near "B")*
Select object to offset or <exit>: *(Press ENTER to exit this command)*

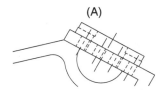

(A)

(B)

Figure 15–22

STEP 4

Refer to the working drawing in Figure 15–19 for the necessary dimensions required to construct the auxiliary view. Use the OFFSET command again to begin constructing the depth of the auxiliary view in Figure 15–23. Remember that the depth of the auxiliary view is the same dimension as found in the Top and Right Side views. Set the offset distance to 6.00. Then set the current layer to OBJ using the Layer Control box.

Command: **O** *(For OFFSET)*
Specify offset distance or [Through]
   <8.50>: *6.00*
Select object to offset or <exit>: *(Select the inclined line at "A")*
Specify point on side to offset: *(Pick a point anywhere near "B")*
Select object to offset or <exit>: *(Press ENTER to exit this command)*

(A)

(B)

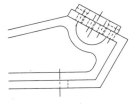

Figure 15–23

STEP 5

Project two lines from the endpoints of the Front view at "A" and "C" in Figure 15–24. These lines should extend past the outer line of the auxiliary view. Turn the Snap off and Ortho on. This should aid in this operation.

Command: **ORTHO**
Enter mode [ON/OFF] <OFF>: **ON**
Command: **L** *(For LINE)*
Specify first point: *(Select the endpoint of the line at "A")*
Specify next point or [Undo]: *(Pick a point anywhere near "B")*
Specify next point or [Undo]: *(Press ENTER to exit this command)*
Command: **L** *(For LINE)*

Specify first point: *(Select the endpoint of the line at "C")*
Specify next point or [Undo]: *(Pick a point anywhere near "D")*
Specify next point or [Undo]: *(Press ENTER to exit this command)*

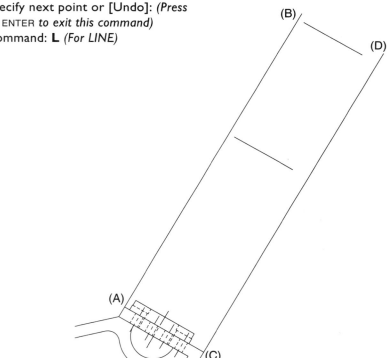

Figure 15–24

STEP 6

Use the ZOOM-Window option to magnify the display of the auxiliary view similar to Figure 15–25. Then use the MULTIPLE command and enter FILLET to create four corners of the view in Figure 15–25. When finished with this operation, save the display to a new name of "Aux" using the -VIEW command or View dialog box.

Command: **MULTIPLE**
Enter command name to repeat: **F** *(For FILLET)*
Current settings: Mode = TRIM, Radius = 0.00
Select first object or [Polyline/Radius/Trim]: *(Select line "A")*
Select second object: *(Select line "B")*
FILLET
Current settings: Mode = TRIM, Radius = 0.00
Select first object or [Polyline/Radius/Trim]: *(Select line "B")*
Select second object: *(Select line "C")*
FILLET

Current settings: Mode = TRIM, Radius = 0.00
Select first object or [Polyline/Radius/Trim]: *(Select line "C")*
Select second object: *(Select line "D")*
FILLET
Current settings: Mode = TRIM, Radius = 0.00
Select first object or [Polyline/Radius/Trim]: *(Select line "D")*
Select second object: *(Select line "A")*
FILLET
Current settings: Mode = TRIM, Radius = 0.00
Select first object or [Polyline/Radius/Trim]: *(Press ESC to exit Multiple mode)*
Command: **-V** *(For -VIEW)*
Enter an option [?/Orthographic/Delete/Restore/Save/Ucs/Window]: **S** *(For Save)*
Enter view name to save: **Aux**
UCSVIEW = I UCS will be saved with view

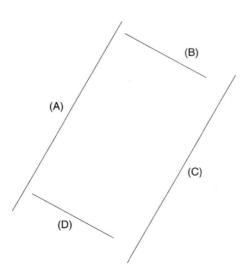

Figure 15–25

STEP 7

Use the ZOOM-Previous option to demagnify the screen back to the original display. Then draw a line from the endpoint of the centerline in the Front view at "A" to a point past the auxiliary view at "B" in Figure 15–26. Check to see that Ortho mode is on. This line will assist in constructing this line in the auxiliary view.

Command: **Z** *(For ZOOM)*
Specify corner of window, enter a scale factor (nX or nXP), or
[All/Center/Dynamic/Extents/Previous/ Scale/Window] <real time>: **P** *(For Previous)*
Command: **L** (For LINE)
Specify first point: *(Select the endpoint of the centerline at "A")*
Specify next point or [Undo]: *(Pick a point anywhere near "B")*
Specify next point or [Undo]: *(Press* ENTER *to exit this command)*

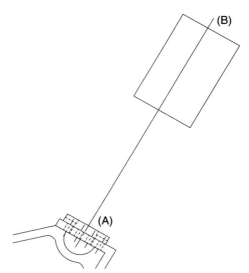

Figure 15–26

STEP 8

Use the OFFSET command and offset the line at "A" a distance of 3.00 units in Figure 15–27. The intersection of this line and the previous line form the center for placing two circles.

Command: **O** *(For OFFSET)*
Specify offset distance or [Through] <6.00>: **3.00**
Select object to offset or <exit>: *(Select the line at "A")*
Specify point on side to offset: *(Pick a point anywhere near "B")*
Select object to offset or <exit>: *(Press* ENTER *to exit this command)*

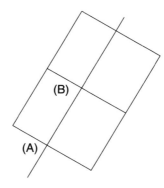

Figure 15–27

STEP 9

Use the -VIEW command or View dialog box and restore the view "Aux." Then, draw two circles of diameters 3.00 and 1.50 from the center at "A" in Figure 15–28 using the CIRCLE command. For the center of the second circle, you can use the @ option to pick up the previous point that was the center of the 3.00-diameter circle.

Command: **-V** *(For -VIEW)*
Enter an option [?/Orthographic/Delete/
   Restore/Save/Ucs/Window]: **R** *(For
   Restore)*
Enter view name to restore: **Aux**
Command: **C** *(For CIRCLE)*
Specify center point for circle or [3P/2P/
   Ttr (tan tan radius)]: *(Select the
   intersection at "A")*
Specify radius of circle or [Diameter]: **D**
   (For Diameter)
Specify diameter of circle: **3.00**
Command: **C** *(For CIRCLE)*
Specify center point for circle or [3P/2P/
   Ttr (tan tan radius)]: **@** *(For the last
   point)*
Specify radius of circle or [Diameter]
   <1.50>: **D** *(For Diameter)*
Specify diameter of circle <3.00>: **1.50**

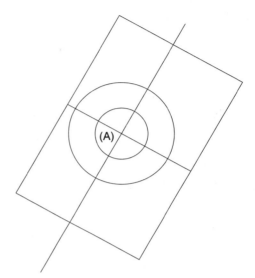

Figure 15–28

STEP 10

Use the OFFSET command to offset the centerline the distance of 0.25 units in Figure 15–29. Perform this operation on both sides of the centerline. Both offset lines form the width of the 0.50 slot.

Command: **O** *(For OFFSET)*
Specify offset distance or [Through]
   <3.00>: **0.25**
Select object to offset or <exit>: *(Select
   the middle line at "A")*
Specify point on side to offset: *(Pick a
   point anywhere near "B")*

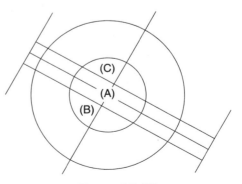

Figure 15–29

Select object to offset or <exit>: *(Select the middle line at "A" again)*
Specify point on side to offset: *(Pick a point anywhere near "C")*
Select object to offset or <exit>: *(Press ENTER to exit this command)*

STEP 11

Use the TRIM command to trim away portions of the lines using the circles as cutting edges in Figure 15–30.

Command: **TR** *(For TRIM)*
Current settings: Projection=UCS
   Edge=None
Select cutting edges ...
Select objects: *(Select both circles as cutting edges)*
Select objects: *(Press ENTER to continue)*
Select object to trim or [Project/Edge/Undo]: *(Select the line at "A")*
Select object to trim or [Project/Edge/Undo]: *(Select the line at "B")*
Select object to trim or [Project/Edge/Undo]: *(Select the line at "C")*
Select object to trim or [Project/Edge/Undo]: *(Select the line at "D")*
Select object to trim or [Project/Edge/Undo]: *(Select the line at "E")*
Select object to trim or [Project/Edge/Undo]: *(Select the line at "F")*
Select object to trim or [Project/Edge/Undo]: *(Press ENTER to exit this command)*

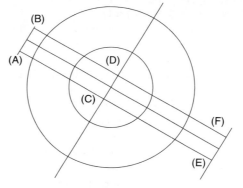

Figure 15–30

STEP 12

Use the ERASE command to delete the two lines at "A" and "B" in Figure 15–31. Standard centerlines will be placed here later, marking the center of both circles.

Command: **E** *(For ERASE)*
Select objects: *(Select the lines at "A" and "B")*
Select objects: *(Press ENTER to execute this command)*

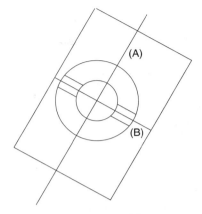

Figure 15–31

STEP 13

To identify the center of the small 0.375-diameter circle, use the OFFSET command to copy parallel the line at "A" a distance of 0.75 and the line at "C" the distance of 1.00 in Figure 15–32.

Command: **O** *(For OFFSET)*
Specify offset distance or [Through] <0.25>: **0.75**
Select object to offset or <exit>: *(Select the line at "A")*
Specify point on side to offset: *(Pick a point anywhere near "B")*
Select object to offset or <exit>: *(Press ENTER to exit this command)*
Command: **O** *(For OFFSET)*
Specify offset distance or [Through] <0.75>: **1.00**
Select object to offset or <exit>: *(Select the line at "C")*
Specify point on side to offset: *(Pick a point anywhere near "D")*
Select object to offset or <exit>: *(Press ENTER to exit this command)*

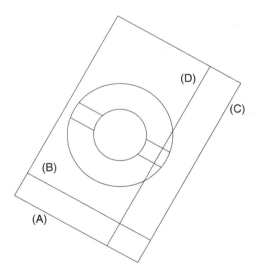

Figure 15–32

STEP 14

Draw a circle of 0.375 diameter from the intersection of the two lines created in the last OFFSET command in Figure 15–33. Use the ERASE command to delete the two lines labeled "B" and "C." A standard center marker will be placed at the center of this circle.

Command: **C** *(For CIRCLE)*
Specify center point for circle or [3P/2P/ Ttr (tan tan radius)]: *(Select the intersection at "A")*
Specify radius of circle or [Diameter] <0.75>: **D** *(For Diameter)*
Specify diameter of circle <1.50>: **0.375**
Command: **E** *(For ERASE)*
Select objects: *(Select the two lines at "B" and "C")*
Select objects: *(Press ENTER to execute this command)*

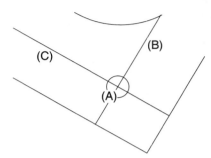

Figure 15–33

STEP 15

Set the current layer to CEN using the Layer Control box. Prepare the following parameters before placing a center marker at the center of the 0.375-diameter circle in Figure 15–34. Set the dimension variable DIMCEN to a value of -0.07 units. The negative value will construct lines that are drawn outside the circle. Use the DIMCENTER command to place the center marker.

Command: **DIMCEN**
Enter new value for DIMCEN <0.09>: **-0.07**
Command: **DCE** *(For DIMCENTER)*
Select arc or circle: *(Select the small circle at "A")*

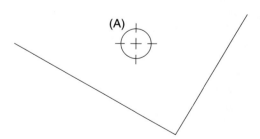

Figure 15–34

STEP 16

Use the ROTATE command to rotate the center marker parallel to the edges of the auxiliary view in Figure 15–35. Select the center marker and circle as the objects to rotate. Check to see that Ortho is on.

Command: **RO** *(For ROTATE)*
Current positive angle in UCS:
ANGDIR=counterclockwise
ANGBASE=0
Select objects: *(Select the small circle and all objects that make up the center marker; the Window option is recommended here)*
Select objects: *(Press ENTER to continue)*
Specify base point: *(Select the center of the small circle)*
Specify rotation angle or [Reference]: *(Pick a point anywhere near "B")*

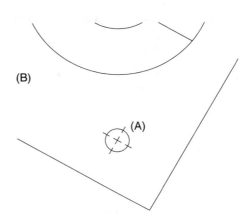

(B)

(A)

Figure 15–35

STEP 17

Since the remaining seven holes form a rectangular pattern, use the ARRAY command and perform a rectangular array in Figure 15–36. The number of rows is two and number of columns four. Distance between rows is 4.50 units and between columns is -0.75 units; the negative distance this will force the circles to be patterned to the left.

Command: **AR** *(For ARRAY)*
Select objects: *(Select the small circle and center marker)*
Select objects: *(Press ENTER to continue)*
Enter the type of array [Rectangular/ Polar] <R>: *(Accept the default as rectangular)*
Enter the number of rows (—) <1>: **2**
Enter the number of columns (|||) <1> **4**
Enter the distance between rows or specify unit cell (—): **4.50**
Specify the distance between columns (|||): **-0.75**

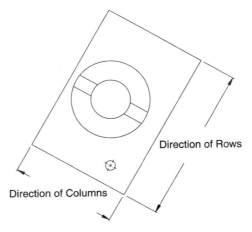

Direction of Rows

Direction of Columns

Figure 15–36

STEP 18

Use the FILLET command set to a radius of 0.75 to place a radius along the two corners of the auxiliary, following the prompts and Figure 15–37.

Command: **F** *(For FILLET)*
Current settings: Mode = TRIM, Radius = 0.00
Select first object or [Polyline/Radius/ Trim]: **R** (For Radius)
Specify fillet radius <0.00>: **0.75**
Command: **F** *(For FILLET)*
Current settings: Mode = TRIM, Radius = 0.75
Select first object or [Polyline/Radius/ Trim]: *(Select line "A")*
Select second object: *(Select line "B")*
Command: **F** *(For FILLET)*
Current settings: Mode = TRIM, Radius = 0.75
Select first object or [Polyline/Radius/ Trim]: *(Select line "B")*
Select second object: *(Select line "C")*

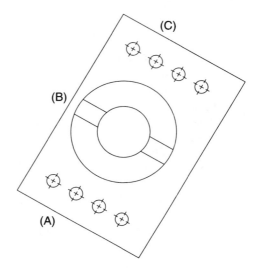

Figure 15–37

STEP 19

Place a center marker in the center of the two large circles using the existing value of the dimension variable DIMCEN (see Figure 15–38). Since the center marker is placed in relation to the World coordinate system, use the ROTATE command to rotate it parallel to the auxiliary view. Ortho should be on.

Command: **DCE** *(For DIMCENTER)*
Select arc or circle: *(Select the large circle at "A")*
Command: **RO** *(For ROTATE)*
Current positive angle in UCS: ANGDIR=counterclockwise ANGBASE=0

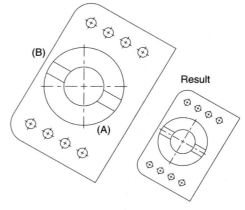

Figure 15–38

Select objects: *(Select all lines that make
up the large center marker)*
Select objects: *(Press* ENTER *to continue)*
Specify base point: **Cen**
of *(Select the edge of the large circle at "A")*
Specify rotation angle or [Reference]:
(Pick a point anywhere near "B")

STEP 20

Restore the view named "Overall" using
the -VIEW command or View dialog box.
Complete the multiview drawing of the
bracket by inserting an existing block
called "Top" in this drawing in Figure
15–39. This block represents the complete Top view of the drawing. Use an
insertion point of 0,0 for placing this view
in the drawing.

Command: **-V** *(For -VIEW)*
Enter an option [?/Orthographic/Delete/
Restore/Save/Ucs/Window]: **R** *(For
Restore)*
Enter view name to restore: **Overall**
Command: **-I** *(For -INSERT)*
Enter block name or [?]: **TOP**
Specify insertion point or [Scale/X/Y/Z/
Rotate/PScale/PX/PY/PZ/PRotate]: **0,0**
Enter X scale factor, specify opposite
corner, or [Corner/XYZ] <1>: *(Press
ENTER)*
Enter Y scale factor <use X scale factor>:
(Press ENTER*)*
Specify rotation angle <0>: *(Press* ENTER*)*

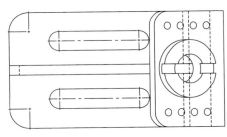

Figure 15–39

STEP 21

Return the grid to its original ortho-graphic form using the Rotate option of the SNAP command. Use a base point of 0,0 and a rotation angle of 0 degrees. The completed drawing should appear similar to Figure 15–40.

Command: **SN** *(For SNAP)*
Specify snap spacing or [ON/OFF/Aspect/
  Rotate/Style/Type] <0.50>: **R** *(For
  Rotate)*
Specify base point <17.33,5.85>: **0,0**
Specify rotation angle <330>: **0**

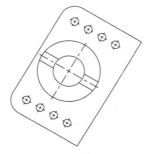

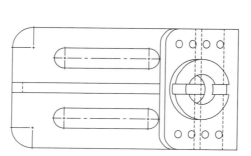

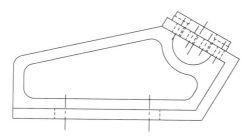

Figure 15–40

PROBLEMS FOR CHAPTER 15

Directions for Problems 15–1 through 15–9

Draw the required views to fully illustrate each object. Be sure to include an auxiliary view.

PROBLEM 15–1

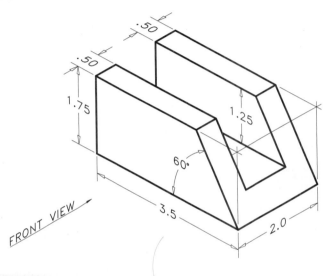

FRONT VIEW

PROBLEM 15–2

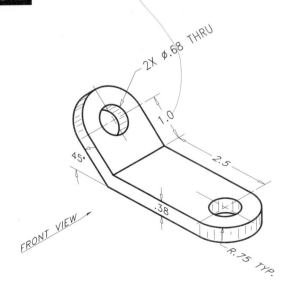

FRONT VIEW

PROBLEM 15–3

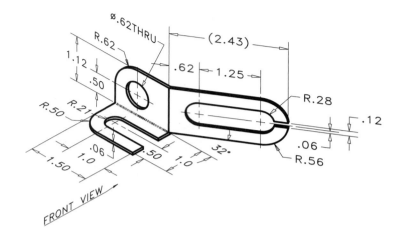

Ø.62 THRU

R.62

(2.43)

1.12

.50

.62

1.25

R.28

R.21

R.50

.06

1.0

.50

1.0

1.50

32°

.12

.06

R.56

FRONT VIEW

PROBLEM 15–4

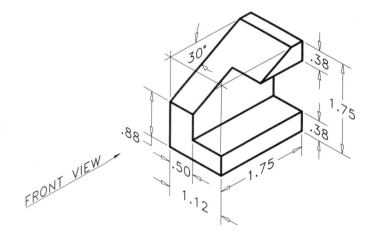

30°

.38

.88

1.75

.38

.50

1.75

1.12

FRONT VIEW

PROBLEM 15–5

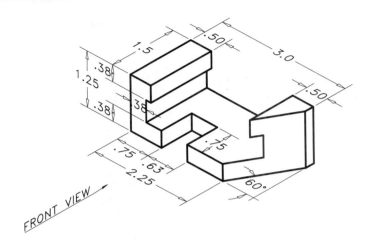

FRONT VIEW

PROBLEM 15–6

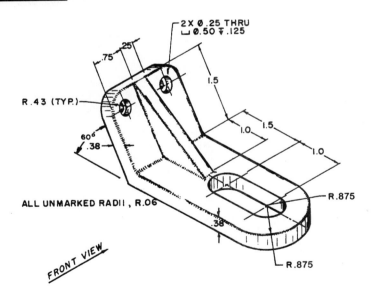

2X ⌀.25 THRU
⌴ ⌀.50 ⍗.125

R.43 (TYP.)

60°

.38

ALL UNMARKED RADII, R.06

R.875

R.875

FRONT VIEW

PROBLEM 15–7

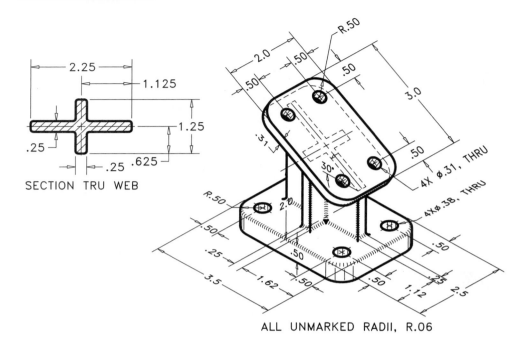

2.25

1.125

1.25

.25

.25   .625

SECTION TRU WEB

R.50

2.0   .50   .50

.50

3.0

.31

30°

.50

4X Ø.31, THRU

4X Ø.38, THRU

R.50   2.0

.50

.50

.25

.50

.50

.50

3.5

1.62   .50   .50

.25

1.12

2.5

ALL UNMARKED RADII, R.06

PROBLEM 15-8

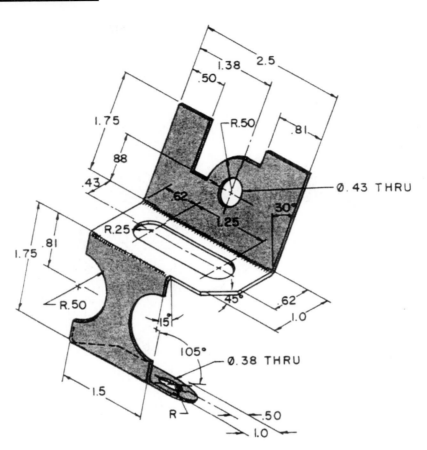

Thickness - 0.0625

PROBLEM 15–9

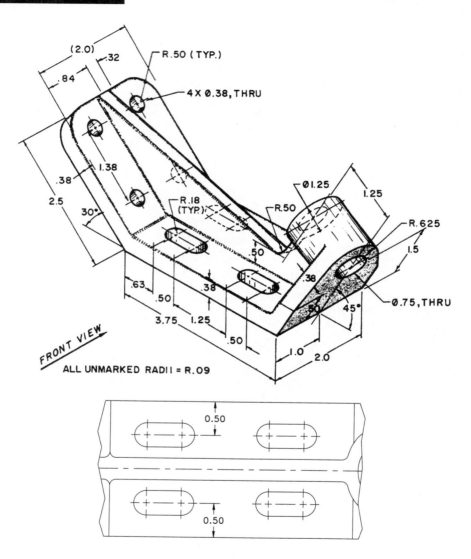

(2.0)
.32
.84
R.50 (TYP.)
4 X Ø.38, THRU
1.38
.38
2.5
30°
R.18 (TYP.)
R.50
Ø1.25
1.25
R.625
1.5
.50
.38
Ø.75, THRU
30°
45°
.63
.50
.38
3.75 1.25
.50
1.0
2.0

FRONT VIEW

ALL UNMARKED RADII = R.09

0.50

0.50

CHAPTER 16

Isometric Drawings

ISOMETRIC BASICS

Isometric drawings are a means of drawing an object in picture form for better clarifying the object's appearance. These types of drawings resemble a picture of an object that is drawn in two dimensions. As a result, existing AutoCAD commands such as LINE and COPY are used for producing isometric drawings. This chapter will explain isometric basics including how regular, angular, and circular objects are drawn in isometric. Numerous isometric aids such as snap and isometric axes will be explained to assist in the construction of isometric drawings.

Isometric drawings consist of two-dimensional drawings that are tilted at some angle to expose other views and give the viewer the illusion that what he or she is viewing is a three-dimensional drawing. The tilting occurs with two 30° angles that are struck from the intersection of a horizontal baseline and a vertical line (see Figure 16–1). The directions formed by the 30° angles represent actual dimensions of the object; this may be either the width or depth. The vertical line in most cases represents the height dimension.

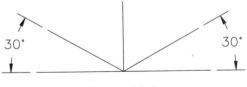

Figure 16–1

Figure 16–2 is a very simple example of how an object is aligned to the isometric axis. Once the horizontal baseline and vertical line are drawn, the 30° angles are projected from this common point, which becomes the reference point of the isometric view. In this example, once the 30° lines are drawn, the baseline is no longer needed and it is usually discarded through erasing. Depending on how the object is to be viewed, width and depth measurements are made along the 30° lines. Height is measured along the vertical line. Figure 16–2 has the width dimension measured off to the left

30° line while the depth dimension measures to the right along the right 30° line. Once the object is blocked with overall width, depth, and height, details are added, and lines are erased and trimmed, leaving the finished object. Holes no longer appear as full circles but rather as ellipses. Techniques of drawing circles in isometric will be discussed later in this chapter.

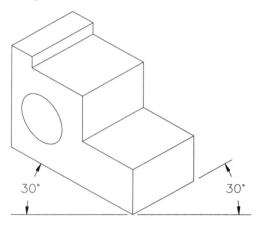

Figure 16–2

Notice that the objects in Figure 16–3 both resemble the object in Figure 16–2 except they appear from a different vantage point. The problem with isometric drawings is that if an isometric of an object is drawn from one viewing point and you want an isometric from another viewing point, an entirely different isometric drawing must be generated from scratch. Complex isometric drawings from different views can be very tedious to draw. Another interesting observation concerning the objects in Figure 16–3 is that one has hidden lines while the other does not. Usually only the visible surfaces of an object are drawn in isometric, with hidden lines left out. Although this is considered the preferred practice, there are always times when hidden lines are needed on very complex isometric drawings.

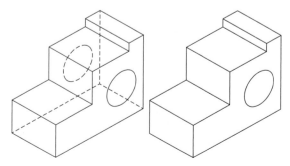

Figure 16–3

CREATING AN ISOMETRIC GRID

Figure 16–4 shows the current AutoCAD screen complete with cursor and grid on. In manual drawing and sketching days, an isometric grid was used to lay out all lines before the lines were transferred to paper or Mylar for pen and ink drawings. An isometric grid may be defined in an AutoCAD drawing through the SNAP command. This would be the same grid found on isometric grid paper.

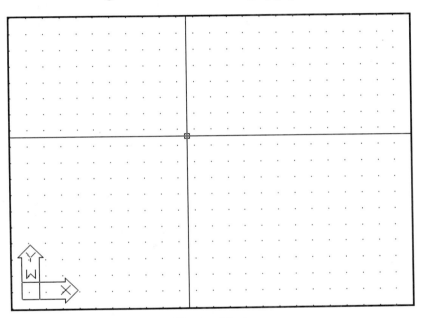

Figure 16–4

The display screen in Figure 16–5 reflects the use of the SNAP command and how this command affects the current grid display:

Command: **SN** *(For SNAP)*
Snap spacing or ON/OFF/Aspect/Rotate/Style <0.2500>: **S** *(for Style)*
Standard/Isometric <S>: **I** *(for Isometric)*
Vertical spacing <0.5000>: *(Press* ENTER *to accept default value)*

Choosing an isometric style of snap changes the grid display from orthographic to isometric, shown in Figure 16–5. The grid distance conforms to a vertical spacing height that you specify. As the grid changes, notice the display of the typical AutoCAD cursor; it conforms to an isometric axis plane and is used as an aid in constructing isometric drawings.

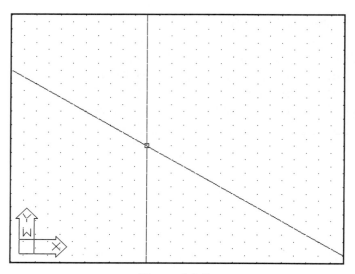

Figure 16–5

To see how this vertical spacing distance affects the grid, changing it to isometric, see Figure 16–6. The grid dot at "A" becomes the reference point where the horizontal baseline is placed followed by the vertical line represented by the dot at "B." At dots "A" and "B," 30° lines are drawn; points "C" and "D" are formed where they intersect. This is how an isometric screen display is formed.

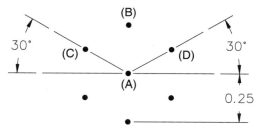

Figure 16–6

ISOPLANE MODES

The AutoCAD cursor has always been the vehicle for drawing objects or constructing windows for object selection mode. Once in isometric snap mode, AutoCAD supports three axes to assist in the construction of isometric drawings. The first axis is the Left axis and may control that part of an object falling into the left projection plane. The Left axis cursor displays a vertical line intersected by a 30°-angle line, which is drawn to the left. This axis is displayed in the illustration in Figure 16–7 and in the following drawing:

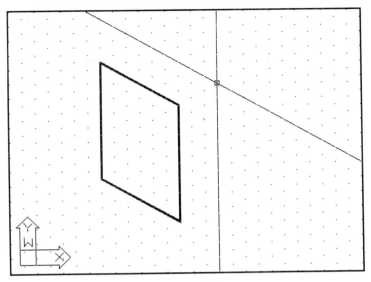

Figure 16–7

The next isometric axis is the Top mode. Objects falling into the top projection plane may be drawn using this isometric axis. This cursor consists of two 30°-angle lines intersecting each other forming the center of the cursor. This mode is displayed in Figure 16–8 and in the following drawing:

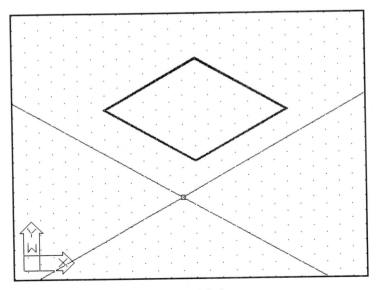

Figure 16–8

The final isometric axis is called the Right mode and is formed by the intersection of a vertical line and a 30° angle drawn off to the right. As with the previous two modes, objects that fall along the right projection plane of an isometric drawing may be drawn with this cursor. It is displayed in Figure 16–9 and in the following drawing. The current Ortho mode affects all three modes. If Ortho is on, and the current isometric axis is Right, lines and other operations requiring direction will be forced to be drawn vertical or at a 30° angle to the right, as shown in the following drawing:

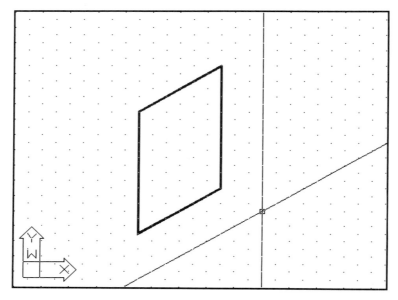

Figure 16–9

RESETTING GRID AND SNAP TO THEIR DEFAULT VALUES

Once an isometric drawing is completed, it may be necessary to change the grid, snap, and cursor back to normal. This might result from the need to place text on the drawing, and the isometric axis now confuses instead of assists in the drawing process. Use the following prompts to reset the snap back to the Standard spacing. See Figure 16–10.

Command: **SN** *(For SNAP)*
Snap spacing or ON/OFF/Rotate/Style <0.2500>: **S** *(for Style)*
Standard/Isometric <I>: **S** *(for Standard)*
Spacing/Aspect <0.5000>: *(Press* ENTER *to accept the default value)*

Notice that when you change the snap style to Standard, the AutoCAD cursor reverts to its original display.

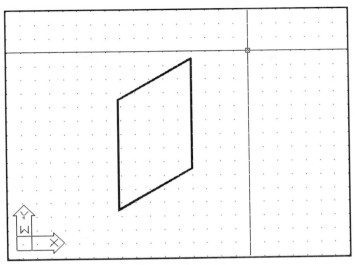

Figure 16–10

ISOMETRIC AIDS IN THE DRAFTING SETTINGS DIALOG BOX

Choosing Drafting Settings… from the Tools pull-down menu in Figure 16–11A displays the Drafting Settings dialog box in Figure 16–11B. Here, in addition to Snap and Grid, an Isometric area exists to automatically set up an isometric grid by clicking in the radio button next to Isometric Snap. This has the same effect as using the SNAP-Style-Isometric option.

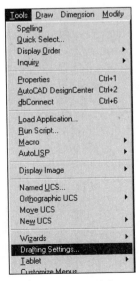

Figure 16–11A

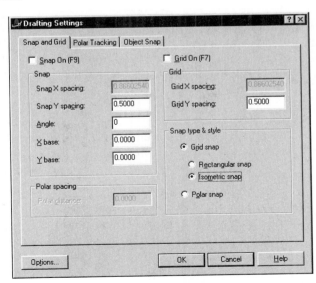

Figure 16–11B

Various methods exist to switch from one isometric axis mode to another. By default, after you set up an isometric grid, the Left isometric axis mode is active (see Figure 16–12). When you press CTRL+E, the Left axis mode changes to the Top axis mode. Pressing CTRL+E again changes from the Top axis mode to the Right axis mode. Pressing CTRL+E a third time changes from the Right axis mode back to the Left axis mode, and the pattern repeats from here. Using this keyboard entry, it is possible to switch or toggle from one mode to another.

The F5 function key also allows you to scroll through the different Isoplane modes.

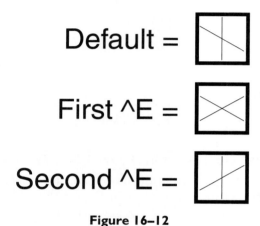

Default =

First ^E =

Second ^E =

Figure 16–12

CREATING ISOMETRIC CIRCLES

Circles appear as ellipses when drawn in any of the three isometric axes. The ELLIPSE command has a special Isocircle option to assist in drawing isometric circles; the Isocircle option will appear in the ELLIPSE command only if the current SNAP-Style is Isometric (see Figure 16–13). The prompt sequence for this command is:

Command: **EL** *(For ELLIPSE)*
Arc/Center/Isocircle/<Axis endpoint 1>: **I** *(for Isocircle)*
Center of circle: *(Select a center point)*
  <Circle radius>/Diameter: *(Enter a value for the radius or type "D" for diameter and enter a value)*

When you draw isometric circles using the ELLIPSE command, it is important to match the isometric axis with the isometric plane the circle is to be drawn in. Figure 16–14 shows a cube displaying all three axes.

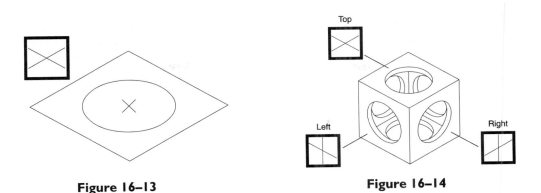

Figure 16–13

Figure 16–14

Figure 16–15 shows the result of drawing an isometric circle using the wrong isometric axis. The isometric box is drawn in the Top isometric plane while the current isometric axis mode is Left. An isometric circle can be drawn to the correct size, but notice that it does not match the box it was designed for. If you notice halfway through the ELLIPSE command that you are in the wrong isometric axis, press CTRL+E until the correct axis appears.

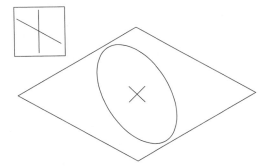

Figure 16–15

BASIC ISOMETRIC CONSTRUCTION

Any isometric drawing, no matter how simple or complex, has an overall width, height, and depth dimension. Start laying out the drawing with these three dimensions to create an isometric box illustrated in the example in Figure 16–16A. Some techniques rely on piecing the isometric drawing together by views; unfortunately, it is very easy to get lost in all of the lines using this method. Once a box is created from overall dimensions, somewhere inside the box is the object.

With the box as a guide, begin laying out all visible features in the primary planes. Use the Left, Top, or Right isometric axis modes to assist you in this construction process. See Figure 16–16B.

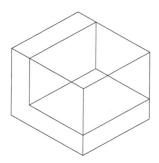 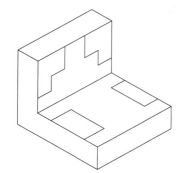

Figure 16–16A Figure 16–16B

You may use existing AutoCAD editing commands, especially COPY, to duplicate geometry to show depth of features, as shown in Figure 16–16C. The OFFSET command should not be attempted for performing isometric drawings. Next, use the TRIM command to partially delete geometry where objects are not visible. Remember, most isometric objects do not require hidden lines.

Use the LINE command to connect intersections of surface corners. The resulting isometric drawing is illustrated in Figure 16–16D.

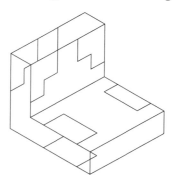 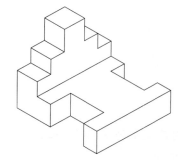

Figure 16–16C   **Figure 16–16D**

CREATING ANGLES IN ISOMETRIC—METHOD #1

Drawing angles in isometric is a little tricky but not impossible. The two-view drawing in Figure 16–17 has an angle of unknown size; however, one endpoint of the angle measures 2.06 units from the top horizontal line of the Front view at "B," and the other endpoint measures 1.00 unit from the vertical line of the Front view at "A." This is more than enough information to lay out the endpoints of the angle in isometric. You can use the MEASURE command to lay out these distances easily. The LINE command is then used to connect the points to form the angle.

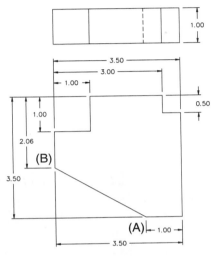

Figure 16–17

Before using the MEASURE command, set the PDMODE variable to a new value of 3. Points will appear as an X instead of a dot, as shown in Figure 16–18A. Points may also be set through the Point Style dialog box found in the Format menu. Now use the MEASURE command to set off the two distances.

Command: **PDMODE**
New value for PDMODE <0>: **3**

Command: **ME** *(For MEASURE)*
Select object to measure: *(Select the inclined line at "A")*
<Segment length>/Block: **1.00**

Command: **ME** *(For MEASURE)*
Select object to measure: *(Select the vertical line at "B")*
 <Segment length>/Block: **2.06**

The interesting point about the MEASURE command is that measuring will occur at the nearest endpoint of the line where the line was selected. Therefore, it is important which endpoint of the line is selected. Once the points have been placed, the LINE command is used to draw a line from one point to the other using the OSNAP-Node option (see Figure 16–18B).

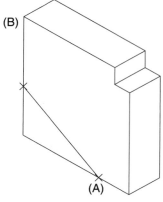

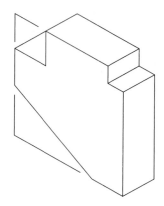

Figure 16–18A

Figure 16–18B

CREATING ANGLES IN ISOMETRIC—METHOD #2

The same two-view drawing is illustrated in Figure 16–19. This time, one distance is specified along with an angle of 30°. Even with the angle given, the position of the isometric axes makes any angle construction by degrees inaccurate. The distance XY is still needed for constructing the angle in isometric. Use the MEASURE command to find distance XY, place a point, and connect the first distance with the second to form the 30° angle in isometric. It is always best to set the PDMODE system variable to a new value in order to visibly view the point. Points may also be set through the Point Style dialog box found in the Format menu. A new value of 3 will assign the point as an X.

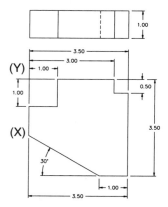

Figure 16–19

See Figures 16–20A, 16–20B, and the following prompt sequences to change the point mode and measure the appropriate distances in the previous step.

Command: **PDMODE**
New value for PDMODE <0>: **3**

Command: **ME** *(For MEASURE)*
Select object to measure: *(Select the inclined line at "A" in Figure 16-20A)*
 <Segment length>/Block: **1.00**

Command: **ME** *(For MEASURE)*
Select object to measure: *(Select the vertical line at "B" in Figure 16-20A)*
 <Segment length>/Block: **End**
 of *(Select the endpoint of the line at X in the 2-view drawing in Figure 16–19)*
Second point: **Int**
 of *(Select the intersection at Y in the 2-view drawing in Figure 16–19)*

Line "B" is selected as the object to measure (see Figure 16–20A). Since this distance is unknown, the MEASURE command may be used to set off the distance XY by identifying an endpoint and intersection from the Front view in Figure 16–19. This means the view must be constructed only enough to lay out the angle and project the results to the isometric using the MEASURE command and the preceding prompts. The result is illustrated in Figure 16–20B.

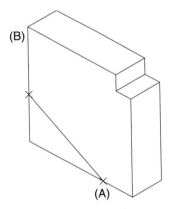

Figure 16–20A

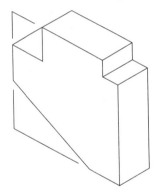

Figure 16–20B

ISOMETRIC CONSTRUCTION USING ELLIPSES

Constructing circles as part of isometric drawing is possible using one of the three isometric axis positions. It is up to you to decide which axis to use. Before this, however, an isometric box consisting of overall distances is first constructed, as shown in Figure 16–21A. Use the ELLIPSE command to place the isometric circle at the base. To select the correct axis, press CTRL+E or F5 until the proper axis appears in the form of the cursor. Place the ellipses. Lay out any other distances.

For ellipses at different positions, press CTRL+E or F5 to select another isometric axis (see Figure 16–21B). Remember that these axis positions may also be selected from the Drawing Aids dialog box, which you access from the Tools menu.

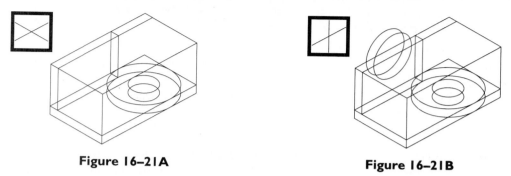

Figure 16–21A   **Figure 16–21B**

Use the TRIM command to trim away any excess objects that are considered unnecessary, as shown in Figure 16–21C.

Use the LINE command to connect endpoints of edges that form surfaces, as shown in Figure 16–21D.

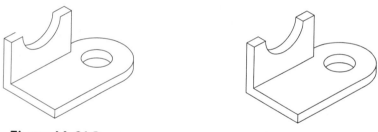

Figure 16–21C   **Figure 16–21D**

CREATING ISOMETRIC SECTIONS

In some cases, it is necessary to cut an isometric drawing to expose internal features. The result is an isometric section similar to a section view formed from an orthographic drawing. The isometric section differs, however, because the cutting plane is usually along one of the three isometric axes. Figure 16–22A displays an orthographic section in addition to the isometric drawing.

The isometric in Figure 16–22B has additional lines representing surfaces cut by the cutting plane line. The lines to define these surfaces are formed through the LINE command in addition to a combination of Top and Right isometric axis modes. During this process, Ortho mode is toggled on and off numerous times depending on the axis direction. Use the TRIM command to remove objects from the half that will eventually be discarded.

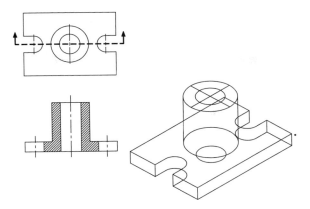

Figure 16–22A

Once ellipses are trimmed, the remaining objects representing the front half of the isometric are removed, exposing the back half. The front half is then discarded. This has the same effect as conventional section views where the direction of sight dictates which half to keep. For a full section, the BHATCH command is used to crosshatch the surfaces being cut by the cutting plane line, as in Figure 16–22C. Surfaces designated holes or slots not cut are not crosshatched. This same procedure is followed for converting an object to a half section.

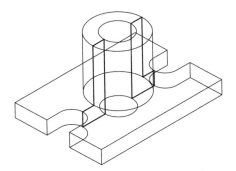

Figure 16–22B

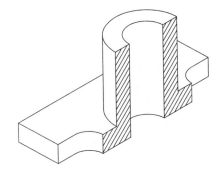

Figure 16–22C

EXPLODED ISOMETRIC VIEWS

Isometric drawings are sometimes grouped together to form an exploded drawing showing how a potential or existing product is assembled (see Figure 16–23). This involves aligning parts that fit with line segments, usually in the form of centerlines. Bubbles identifying the part number are attached to the drawing. Exploded isometric drawings come in handy for creating bill of material information and, for this purpose, they have an important application to manufacturing. Once the part information is identified in the drawing and title block area, this information is extracted and brought into a third-party business package where important data collection information is used to actually track the status of parts in production in addition to the shipping date for all finished products.

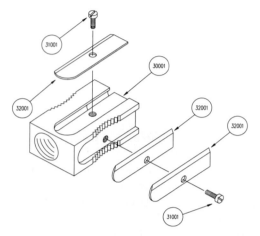

Figure 16–23

ISOMETRIC ASSEMBLIES

Assembly drawings show the completed part as if it were to be assembled (see Figure 16–24). Sometimes this drawing has an identifying number placed with a bubble for bill of material needs. Assembly drawings are commonly placed on the same display screen as the working drawing. With the assembly alongside the working drawing, you have a pictorial representation of what the final product will look like, and this can help you to understand the orthographic views.

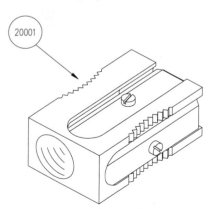

Figure 16–24

TUTORIAL EXERCISE: PLATE.DWG

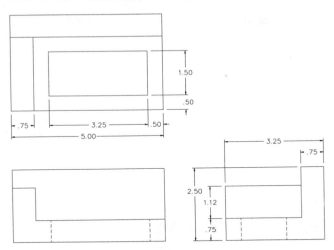

Figure 16–25

Purpose

The purpose of this tutorial exercise is to use a series of coordinates along with AutoCAD editing commands to construct an isometric drawing of the plate in Figure 16–25.

System Settings

Begin a new drawing called "Plate." Use the Drawing Units dialog box to change the number of decimal places past the zero from four to two. Keep the remaining default unit values. Using the LIMITS command, keep (0,0) for the lower left corner and change the upper right corner from (12.00,9.00) to (10.50,8.00). Use the GRID command and change the grid spacing from 1.00 to 0.25 units. Do not turn the Snap or Ortho modes on. Check to see that the following Object Snap modes are already set: Endpoint, Extension, Intersection, and Center.

Layers

Create a layer called "Object" and assign this layer the color Green.

Suggested Commands

Begin this exercise by changing the grid from the standard display to an isometric display using the SNAP-Style option. Remember both the Grid and Snap can be manipulated by the SNAP command only if the current Grid value is 0. Use absolute and polar coordinates to lay out the base of the plate. Then begin using the COPY command followed by TRIM to duplicate objects and clean up or trim unnecessary geometry.

Whenever possible, substitute the appropriate command alias in place of the full AutoCAD command in each tutorial step. For example, use "CP" for the COPY command, "L" for the LINE command, and so on. The complete listing of all command aliases is located in Chapter 1, Table 1–2.

STEP 1

Set a new polar tracking angle of 30° through the Polar Tracking tab of the Drafting Settings dialog box in Figure 16–26A. Make the Object layer current. Then use the LINE command to draw the object in Figure 16–26B. Notice, in the prompt sequence, that the Direct Distance mode is used because of the polar tracking angle setting of 30°. Also, be sure to turn Polar mode on by clicking on the POLAR button in the status bar.

Command: **L** *(For LINE)*
Specify first point: **5.50,0.75**

Specify next point or [Undo]: *(Move your cursor up and to the right until the polar angle tooltip reads 30 degrees and enter* **3.25***)*
Specify next point or [Undo]: *(Move your cursor up and to the left until the polar angle tooltip reads 150 degrees and enter* **5.00***)*
Specify next point or [Close/Undo]: *(Move your cursor down and to the left until the polar angle tooltip reads 210 degrees and enter* **3.25***)*
Specify next point or [Close/Undo]: **C** *(To close the shape and exit the command)*

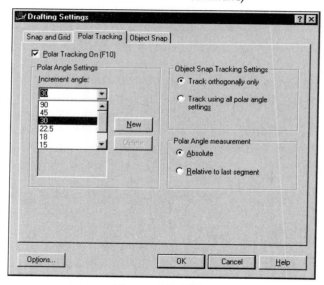

Figure 16–26A

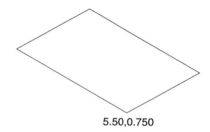

5.50,0.750

Figure 16–26B

STEP 2

Copy the four lines drawn in Step 1 up a distance of 2.50 units in the 90° direction in Figure 16–27.

Command: **CP** *(For COPY)*
Select objects: *(Select lines "A", "B","C", and "D")*
Select objects: *(Press* ENTER *to continue)*
Specify base point or displacement, or [Multiple]: *(Pick the endpoint of the line at "A")*
Specify second point of displacement or <use first point as displacement>: *(Move your cursor up until the polar angle tooltip reads 90° and enter* **2.50***)*

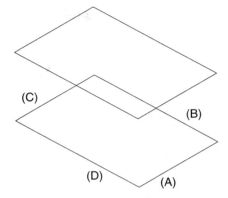

Figure 16–27

STEP 3

Connect the top and bottom isometric boxes with line segments, as shown in Figure 16–28. Draw one segment using the LINE command. Repeat this process using the LINE command at points "C" and "D," or use the COPY command and Multiple option to duplicate and form the remaining segments at "C" and "D." Erase the two dashed lines because they are not visible in an isometric drawing.

Command: **L** *(For LINE)*
Specify first point: *(Select the endpoint of the line at "A")*
Specify next point or [Undo]: *(Select the endpoint of the line at "B")*
Specify next point or [Undo]: *(Press* ENTER *to exit this command)*
Command: **E** *(For ERASE)*
Select objects: *(Select the two dashed lines in Figure 16–28)*
Select objects: *(Press* ENTER *to execute this command)*

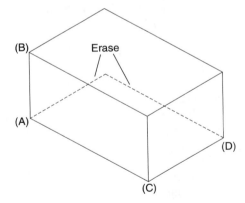

Figure 16–28

STEP 4

Copy the two dashed lines in Figure 16–29 a distance of 0.75 in the 210° direction.

Command: **CP** *(For COPY)*
Select objects: *(Select the two dashed lines in Figure 16–29)*
Select objects: *(Press ENTER to continue)*
Specify base point or displacement, or [Multiple]: *(Select the endpoint at "A")*
Specify second point of displacement or <use first point as displacement>: *(Move your cursor down and to the left until the polar angle tooltip reads 210° and enter 0.75)*

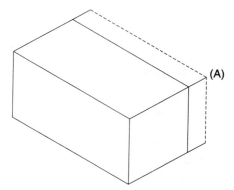

Figure 16–29

STEP 5

Copy the two dashed lines in Figure 16–30 a distance of 0.75 in the 90° direction. This forms the base of the plate.

Command: **CP** *(For COPY)*
Select objects: *(Select the two dashed lines in Figure 16–30)*
Select objects: *(Press ENTER to continue)*
Specify base point or displacement, or [Multiple]: *(Select the endpoint at "A")*
Specify second point of displacement or <use first point as displacement>: *(Move your cursor up until the polar angle tooltip reads 90° and enter 0.75)*

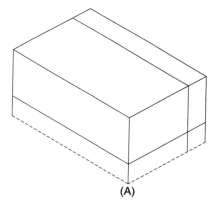

Figure 16–30

STEP 6

Copy the dashed line in Figure 16–31 a distance of 0.75 in the 330° direction.

Command: **CP** *(For COPY)*
Select objects: *(Select the dashed line in Figure 16–31)*
Select objects: *(Press* ENTER *to continue)*
Specify base point or displacement, or [Multiple]: *(Select the endpoint at "A")*
Specify second point of displacement or <use first point as displacement>: *(Move your cursor down and to the right until the polar angle tooltip reads 330° and enter* **0.75***)*

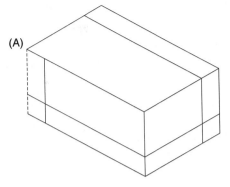

Figure 16–31

STEP 7

Use the FILLET command to place a corner between the two dashed lines at "A" and "B" and at "C" and "D" in Figure 16–32. Set the fillet radius to a value of 0 to accomplish this.

Command: **F** *(For FILLET)*
Current settings: Mode = TRIM, Radius = 0.50
Select first object or [Polyline/Radius/ Trim]: **R** *(For Radius)*
Specify fillet radius <0.50>: **0**
Command: **F** *(For FILLET)*
Current settings: Mode = TRIM, Radius = 0.00
Select first object or [Polyline/Radius/ Trim]: *(Select line "A")*
Select second object: *(Select line "B")*
Command: **F** *(For FILLET)*
Current settings: Mode = TRIM, Radius = 0.00
Select first object or [Polyline/Radius/ Trim]: *(Select line "C")*
Select second object: *(Select line "D")*

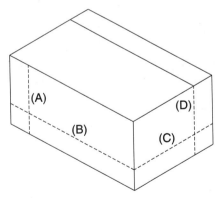

Figure 16–32

STEP 8

Copy the dashed line in Figure 16–33 to begin forming the top of the base.

Command: **CP** *(For COPY)*
Select objects: *(Select the dashed line in Figure 16–33)*
Select objects: *(Press* ENTER *to continue)*
Specify base point or displacement, or [Multiple]: *(Select the endpoint at "A")*
Specify second point of displacement or <use first point as displacement>: *(Select the endpoint at "B")*

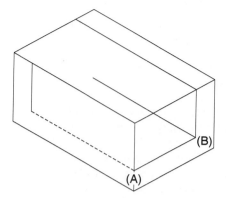

Figure 16–33

STEP 9

Copy the dashed line in Figure 16–34. This forms the base of the plate.

Command: **CP** *(For COPY)*
Select objects: *(Select the dashed line in Figure 16–34)*
Select objects: *(Press* ENTER *to continue)*
Specify base point or displacement, or [Multiple]: *(Select the endpoint at "A")*
Specify second point of displacement or <use first point as displacement>: *(Select the endpoint at "B")*

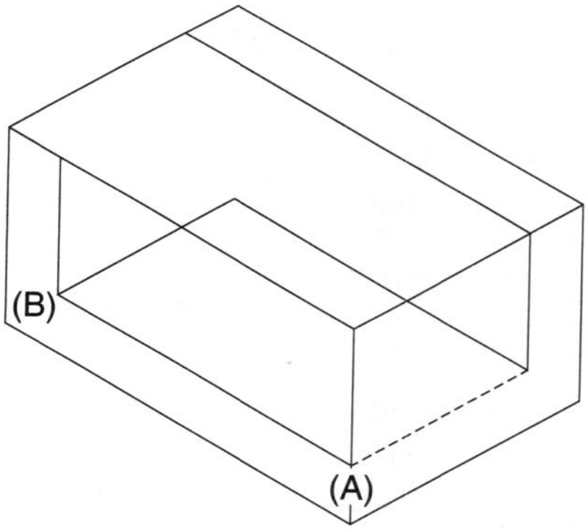

Figure 16–34

STEP 10

Use the TRIM command to clean up the excess lines in Figure 16–35.

Command: **TR** *(For TRIM)*
Current settings: Projection=UCS
  Edge=None
Select cutting edges ...
Select objects: *(Select the three dashed lines shown in Figure 16–35)*
Select objects: *(Press ENTER to continue)*
Select object to trim or [Project/Edge/ Undo]: *(Select the inclined line at "A")*
Select object to trim or [Project/Edge/ Undo]: *(Select the inclined line at "B")*
Select object to trim or [Project/Edge/ Undo]: *(Select the inclined line at "C")*
Select object to trim or [Project/Edge/ Undo]: *(Press ENTER to exit this command)*

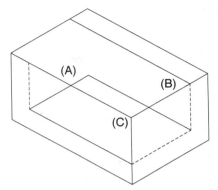

Figure 16–35

STEP 11

Your display should be similar to Figure 16–36.

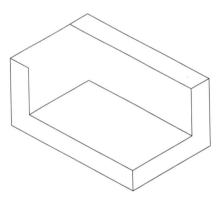

Figure 16–36

STEP 12

Copy the dashed line in Figure 16–37 a distance of 1.12 units in the 90° direction.

Command: **CP** *(For COPY)*
Select objects: *(Select the dashed line in Figure 16–37)*
Select objects: *(Press ENTER to continue)*
Specify base point or displacement, or [Multiple]: *(Select the endpoint at "A")*
Specify second point of displacement or <use first point as displacement>: *(Move your cursor up until the polar angle tooltip reads 90°and enter 1.12)*

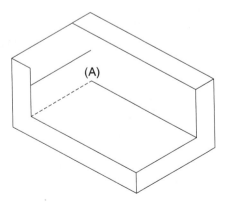

Figure 16–37

STEP 13

As shown in Figure 16–38, copy the dashed line at "A" to new positions at "B" and "C." Then delete the line at "A" using the ERASE command.

Command: **CP** (For COPY)
Select objects: (Select the dashed line in Figure 16–38)
Select objects: (Press ENTER to continue)
Specify base point or displacement, or [Multiple]: **M** (For Multiple)
Specify base point: (Select the endpoint at "A")
Specify second point of displacement or <use first point as displacement>: (Select the endpoint at "B")
Specify second point of displacement or <use first point as displacement>: (Select the endpoint at "C")

Specify second point of displacement or <use first point as displacement>: (Press ENTER to exit this command)
Command: **E** (For ERASE)
Select objects: (Select the dashed line at "A")
Select objects: (Press ENTER to execute this command)

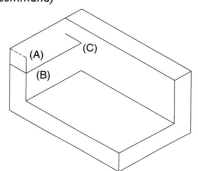

Figure 16–38

STEP 14

Use the TRIM command to clean up the excess lines in Figure 16–39.

Command: **TR** (For TRIM)
Current settings: Projection=UCS Edge=None
Select cutting edges ...
Select objects: (Select the two dashed lines in Figure 16–39)
Select objects: (Press ENTER to continue)
Select object to trim or [Project/Edge/Undo]: (Select the vertical line at "A")
Select object to trim or [Project/Edge/Undo]: (Select the vertical line at "B")
Select object to trim or [Project/Edge/Undo]: (Select the inclined line at "C")
Select object to trim or [Project/Edge/Undo]: (Press ENTER to exit this command)

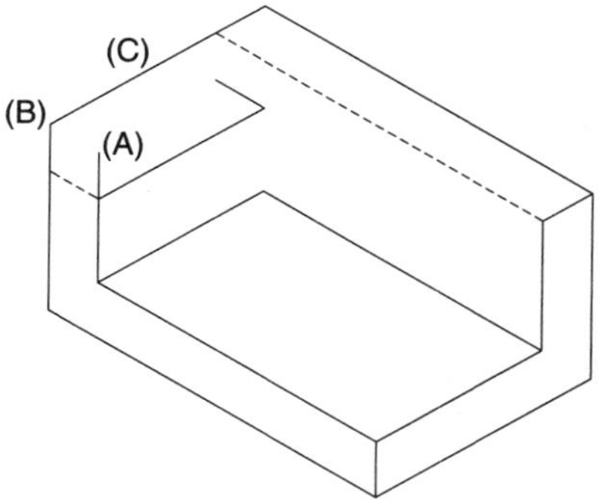

Figure 16–39

STEP 15

Use the COPY command to duplicate the dashed line in Figure 16–40 from the endpoint of "A" to the endpoint at "B."

Command: **CP** *(For COPY)*
Select objects: *(Select the dashed line in Figure 16–40)*
Select objects: *(Press* ENTER *to continue)*
Specify base point or displacement, or [Multiple]: *(Select the endpoint at "A")*
Specify second point of displacement or <use first point as displacement>: *(Select the endpoint at "B")*

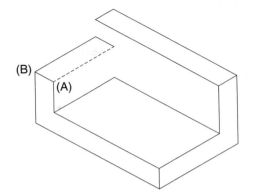

Figure 16–40

STEP 16

Use the COPY command to duplicate the dashed line in Figure 16–41 from the endpoint of "A" to the endpoint at "B."

Command: **CP** *(For COPY)*
Select objects: *(Select the dashed line in Figure 16–41)*
Select objects: *(Press* ENTER *to continue)*
Specify base point or displacement, or [Multiple]: *(Select the endpoint at "A")*
Specify second point of displacement or <use first point as displacement>: *(Select the endpoint at "B")*

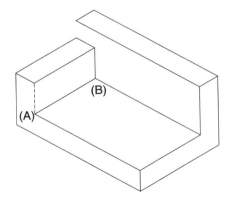

Figure 16–41

STEP 17

Use the LINE command to connect the endpoints of the segments at "A" and "B" in Figure 16–42.

Command: **L** *(For LINE)*
Specify first point: *(Select the endpoint at "A")*
Specify next point or [Undo]: *(Select the endpoint at "B")*
Specify next point or [Undo]: *(Press ENTER to exit this command)*

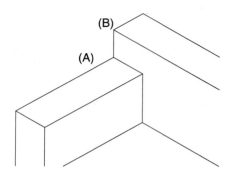

Figure 16–42

STEP 18

Use the COPY command to duplicate the dashed line in Figure 16–43 from the endpoint of "A" a distance of 0.50 units in the 210° direction. This value begins the outline of the rectangular hole through the object.

Command: **CP** *(For COPY)*
Select objects: *(Select the dashed line in Figure 16–43)*
Select objects: *(Press ENTER to continue)*
Specify base point or displacement, or [Multiple]: *(Select the endpoint at "A")*
Specify second point of displacement or <use first point as displacement>: *(Move your cursor down and to the left until the polar angle tooltip reads 210° and enter 0.50)*

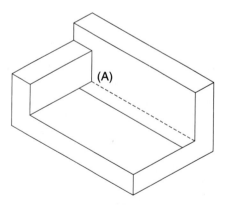

Figure 16–43

STEP 19

Use the COPY command to duplicate the dashed line in Figure 16–44 from the endpoint of "A" a distance of 0.50 in the 330° direction.

Command: **CP** *(For COPY)*
Select objects: *(Select the dashed line in Figure 16–44)*
Select objects: *(Press ENTER to continue)*
Specify base point or displacement, or [Multiple]: *(Select the endpoint at "A")*
Specify second point of displacement or <use first point as displacement>: *(Move your cursor down and to the right until the polar angle tooltip reads 330° and enter 0.50)*

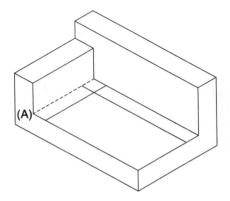

Figure 16–44

STEP 20

Use the COPY command to duplicate the dashed line in Figure 16–45 from the endpoint of "A" a distance of 0.50 in the 30° direction.

Command: **CP** *(For COPY)*
Select objects: *(Select the dashed line in Figure 16–45)*
Select objects: *(Press ENTER to continue)*
Specify base point or displacement, or [Multiple]: *(Select the endpoint at "A")*
Specify second point of displacement or <use first point as displacement>: *(Move your cursor up and to the right until the polar angle tooltip reads 30° and enter 0.50)*

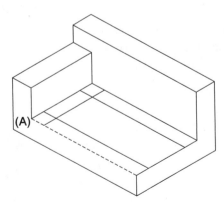

Figure 16–45

STEP 21

Use the COPY command to duplicate the dashed line in Figure 16–46 from the endpoint of "A" a distance of 0.50 in the 150° direction.

Command: **CP** *(For COPY)*
Select objects: *(Select the dashed line in Figure 16–46)*
Select objects: *(Press ENTER to continue)*
Specify base point or displacement, or [Multiple]: *(Select the endpoint at "A")*
Specify second point of displacement or <use first point as displacement>: *(Move your cursor up and to the right until the polar angle tooltip reads 150° and enter **0.50**)*

Figure 16–46

STEP 22

Use the FILLET command with a radius of 0 to corner the four dashed lines in Figure 16–47. Use the MULTIPLE command to remain in the FILLET command. To exit the FILLET command prompts, press ESC to cancel the command when finished.

Command: **MULTIPLE**
Enter command name to repeat: **F** *(For FILLET)*
Current settings: Mode = TRIM, Radius = 0.0000
Select first object or [Polyline/Radius/Trim]: *(Select the line at "A")*
Select second object: *(Select the line at "B")*
FILLET
Current settings: Mode = TRIM, Radius = 0.0000
Select first object or [Polyline/Radius/Trim]: *(Select the line at "B")*
Select second object: *(Select the line at "C")*
FILLET
Current settings: Mode = TRIM, Radius = 0.0000
Select first object or [Polyline/Radius/Trim]: *(Select the line at "C")*

Select second object: *(Select the line at "D")*
FILLET
Current settings: Mode = TRIM, Radius = 0.0000
Select first object or [Polyline/Radius/Trim]: *(Select the line at "D")*
Select second object: *(Select the line at "A")*
FILLET
Current settings: Mode = TRIM, Radius = 0.0000
Select first object or [Polyline/Radius/Trim]: *(Press ESC to cancel this command)*

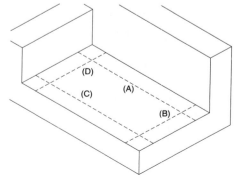

Figure 16–47

STEP 23

Use the COPY command to duplicate the dashed line in Figure 16–48. This will begin forming the thickness of the base inside the rectangular hole.

Command: **CP** *(For COPY)*
Select objects: *(Select the dashed line in Figure 16–48)*
Select objects: *(Press ENTER to continue)*
Specify base point or displacement, or [Multiple]: *(Select the endpoint at "A")*
Specify second point of displacement or <use first point as displacement>: *(Select the endpoint at "B")*

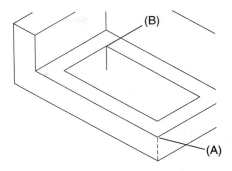

Figure 16–48

STEP 24

Use the COPY command to duplicate the dashed lines in Figure 16–49. These lines form the inside surfaces to the rectangular hole.

Command: **CP** *(For COPY)*
Select objects: *(Select the dashed lines in Figure 16–49)*
Select objects: *(Press ENTER to continue)*
Specify base point or displacement, or [Multiple]: *(Select the endpoint at "A")*
Specify second point of displacement or <use first point as displacement>: *(Select the endpoint at "B")*

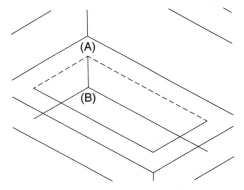

Figure 16–49

STEP 25

Use the TRIM command to clean up excess lines in Figure 16–50.

Command: **TR** *(For TRIM)*
Current settings: Projection=UCS Edge=None
Select cutting edges ...
Select objects: *(Select the two dashed lines in Figure 16–50)*
Select objects: *(Press ENTER to continue)*
Select object to trim or [Project/Edge/ Undo]: *(Select the line at "A")*
Select object to trim or [Project/Edge/ Undo]: *(Select the line at "B")*
Select object to trim or [Project/Edge/ Undo]: *(Press ENTER to exit this command)*

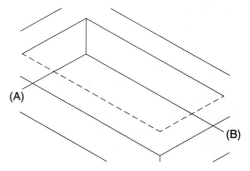

Figure 16–50

STEP 26

The completed isometric is illustrated in Figure 16–51. This drawing may be dimensioned with the Oblique option of the DIMEDIT command. Consult your instructor if this next step is necessary.

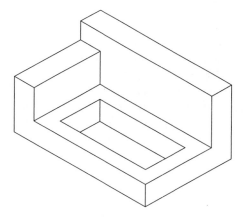

Figure 16–51

TUTORIAL EXERCISE: HANGER.DWG

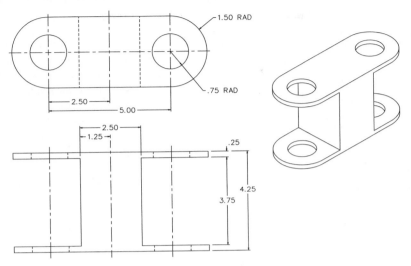

Figure 16–52

Purpose

The purpose of this tutorial exercise is to use a series of coordinates along with AutoCAD editing commands to construct an isometric drawing of the hanger in Figure 16–52.

System Settings

Begin a new drawing called "Hanger." Use the Drawing Units dialog box to change the number of decimal places past the zero from four to two. Keep the remaining default unit values. Using the LIMITS command, keep (0,0) for the lower left corner and change the upper right corner from (12,9) to (15.50,9.50). Use the GRID command and change the grid spacing from 1.00 to 0.25 units. Turn Snap mode off; turn Ortho mode on. Check to see that the following Object Snap modes are already set: Endpoint, Extension, Intersection, Center.

Layers

Create a layer called "Object" and assign this layer the color Green.

Suggested Commands

Begin this exercise by changing the grid from the standard display to an isometric display using the SNAP-Style option. Remember, both the grid and snap can be manipulated by the SNAP command only if the current grid value is 0. Use absolute and polar coordinates to lay out the base of the hanger. Then begin using the COPY command followed by TRIM to duplicate objects and clean up or trim unnecessary geometry.

Whenever possible, substitute the appropriate command alias in place of the full AutoCAD command in each tutorial step. For example, use "CP" for the COPY command, "L" for the LINE command, and so on. The complete listing of all command aliases is located in Chapter 1, Table 1–2.

STEP 1

Make the Object layer current. Then set the SNAP-Style option to Isometric with a vertical spacing of 0.25 units. Press CTRL+E or F5 to switch to the Top Isoplane mode. Use the LINE command to draw the rectangular isometric box representing the total depth of the object along with the center-to-center distance of the holes and arcs that will be placed in the next step. See Figure 16–53.

Command: **SN** (For SNAP)
Specify snap spacing or [ON/OFF/Aspect/ Rotate/Style/Type] <0.50>: **S** (For Style)
Enter snap grid style [Standard/Isometric] <S>: **I** (For Isometric)
Specify vertical spacing <0.50>: **0.25**
Command: (Press CTRL+E to switch to the Top Isoplane mode)
Command: **L** (For LINE)
Specify first point: **6.00,0.50**

Specify next point or [Undo]: (Move your cursor up and to the right and enter 5.00)
Specify next point or [Undo]: (Move your cursor up and to the left and enter 3.00)
Specify next point or [Close/Undo]: (Move your cursor down and to the left and enter 5.00)
Specify next point or [Close/Undo]: **C** (To close the shape and exit the command)

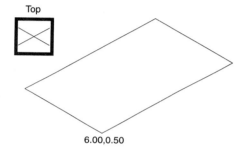

6.00,0.50

Figure 16–53

STEP 2

While in the Top Isoplane mode, use the ELLIPSE command to draw two isometric ellipses of 0.75 and 1.50 radii each in Figure 16–54. Identify the midpoint of the inclined line at "A" as the center of the first ellipse. To identify the center of the second ellipse, use the @ option, which stands for "last point" and will identify the center of the small circle as the same center as the large circle.

Command: **EL** (For ELLIPSE)
Specify axis endpoint of ellipse or [Arc/ Center/Isocircle]: **I** (For Isocircle)
Specify center of isocircle: **Mid**
of (Pick the midpoint of the line at "A")
Specify radius of isocircle or [Diameter]: **0.75**
Command: **EL** (For ELLIPSE)
Specify axis endpoint of ellipse or [Arc/ Center/Isocircle]: **I** (For Isocircle)

Specify center of isocircle: **@** *(For the last point)*
Specify radius of isocircle or [Diameter]: **1.50**

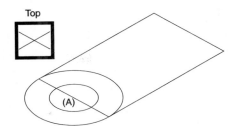

Figure 16–54

STEP 3

Copy both ellipses from the endpoint of the inclined line at "A" to the endpoint of the inclined line at "B" in Figure 16–55. Be sure Snap is turned off.

Command: **CP** *(For COPY)*
Select objects: *(Select both ellipses in Figure 16–55)*
Select objects: *(Press ENTER to continue)*
Specify base point or displacement, or [Multiple]: *(Select the endpoint at "A")*
Specify second point of displacement or <use first point as displacement>: *(Select the endpoint at "B")*

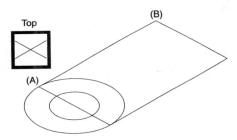

Figure 16–55

STEP 4

Use the TRIM command to delete segments of the ellipses in Figure 16–56.

Command: **TR** *(For TRIM)*
Current settings: Projection=UCS Edge=None
Select cutting edges ...
Select objects: *(Select dashed lines "A" and "B")*
Select objects: *(Press ENTER to continue)*
Select object to trim or [Project/Edge/Undo]: *(Select the ellipse at "C")*
Select object to trim or [Project/Edge/Undo]: *(Select the ellipse at "D")*
Select object to trim or [Project/Edge/Undo]: *(Press ENTER to exit this command)*

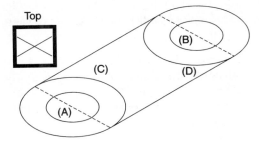

Figure 16–56

STEP 5

Copy all objects in Figure 16–57 up the distance of 0.25 units to form the bottom base of the hanger. Pressing CTRL+E or F5 can activate the Right Isoplane mode.

Command: *(Press CTRL+E To switch to the Right Isoplane mode)*
Right Isoplane
Command: **CP** *(For COPY)*
Select objects: *(Select all objects in Figure 16–57)*
Select objects: *(Press ENTER to continue)*
Specify base point or displacement, or [Multiple]: *(Select the endpoint at "A")*
Specify second point of displacement or <use first point as displacement>: *(Move your cursor up and enter a value of 0.25)*

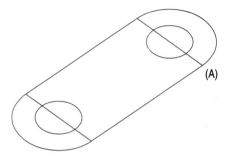

Figure 16–57

STEP 6

Use the ERASE command to delete the three dashed lines in Figure 16–58. These lines are not visible from this point of view and should be erased.

Command: **E** *(For ERASE)*
Select objects: *(Select the three dashed lines in Figure 16–58)*
Select objects: *(Press ENTER to execute this command)*

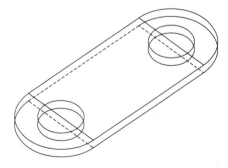

Figure 16–58

STEP 7

Your display should appear similar to the illustration in Figure 16–59. Begin partially deleting other objects to show only visible features of the isometric drawing. Check that the snap has been turned off to better assist in the next series of operations. The next few steps refer to the area outlined in Figure 16–59. Use the ZOOM-Window option to magnify this area.

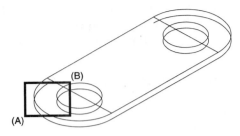

Figure 16–59

Command: **Z** *(For ZOOM)*
Specify corner of window, enter a scale
    factor (nX or nXP), or
[All/Center/Dynamic/Extents/Previous/
    Scale/Window] <real time>: *(Pick a
    point at "A")*
Specify opposite corner: *(Pick a point at
    "B")*

STEP 8

Carefully draw a line tangent to both ellipses. Use the OSNAP-Quadrant option to assist in constructing the line. See Figure 16–60.

Command: **L** *(For LINE)*
Specify first point: **Qua**
of *(Select the quadrant at "A")*
Specify next point or [Undo]: *(Pick the
    polar intersection at "B")*
Specify next point or [Undo]: *(Press
    ENTER to exit this command)*

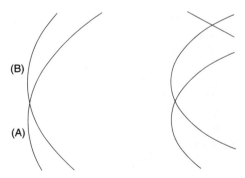

Figure 16–60

STEP 9

Use the TRIM command, select the short dashed line as the cutting edge, and select the arc segment in Figure 16–61 to trim.

Command: **TR** *(For TRIM)*
Current settings: Projection=UCS
    Edge=None
Select cutting edges ...
Select objects: *(Select the dashed line at "A")*
Select objects: *(Press ENTER to continue)*
Select object to trim or [Project/Edge/ Undo]: *(Select the ellipse at "B")*
Select object to trim or [Project/Edge/ Undo]: *(Press ENTER to exit this command)*

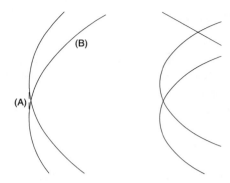

Figure 16–61

STEP 10

The completed operation is in Figure 16–62. Use ZOOM command and the Previous option to return to the previous display.

Command: **Z** *(For ZOOM)*
Specify corner of window, enter a scale factor (nX or nXP), or
[All/Center/Dynamic/Extents/Previous/ Scale/Window] <real time>: **P** *(For Previous)*

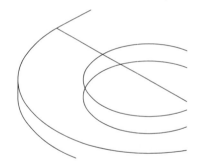

Figure 16–62

STEP 11

Use ZOOM-Window to magnify the right half of the base in Figure 16–63. Prepare to construct the tangent edge to the object using the procedure in Step 8.

Command: **Z** *(For ZOOM)*
Specify corner of window, enter a scale factor (nX or nXP), or
[All/Center/Dynamic/Extents/Previous/ Scale/Window] <real time>: *(Pick a point at "A")*
Specify opposite corner: *(Pick a point at "B")*

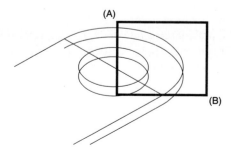

Figure 16–63

tutorial EXERCISE

STEP 12

Follow the same procedure as in Step 8 to construct a line from the quadrant point on the top ellipse to the polar intersection on the bottom ellipse. Then use the TRIM command in Step 9 to clean up any excess objects. Use the ERASE command to delete any elliptical arc segments that may have been left untrimmed. (See Figure 16–64.)

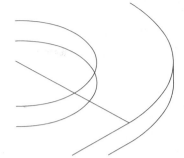

Figure 16–64

STEP 13

Use the TRIM command to partially delete the ellipses in Figure 16–65 to expose the thickness of the base.

Command: **TR** *(For TRIM)*
Current settings: Projection=UCS
    Edge=None
Select cutting edges ...
Select objects: *(Select dashed ellipses "A" and "B")*
Select objects: *(Press ENTER to continue)*
Select object to trim or [Project/Edge/
    Undo]: *(Select the lower ellipse at "C")*
Select object to trim or [Project/Edge/
    Undo]: *(Select the lower ellipse at "D")*
Select object to trim or [Project/Edge/
    Undo]: *(Press ENTER to exit this command)*

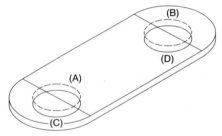

Figure 16–65

STEP 14

Use the COPY command to duplicate the bottom base and form the upper plate of the hanger as shown in Figure 16–66. Copy the base a distance of 4 units straight up.

Command: **CP** *(For COPY)*
Select objects: *(Select all dashed objects in Figure 16–66)*
Select objects: *(Press ENTER to continue)*
Specify base point or displacement, or [Multiple]: *(Select the endpoint of the ellipse at "A")*
Specify second point of displacement or <use first point as displacement>: *(Move your cursor up and enter a value of 4.00)*

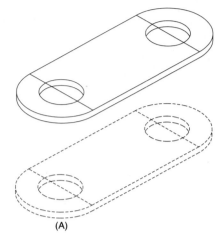

Figure 16–66

STEP 15

Use the COPY command to duplicate the inclined line at "A" the distance of 2.50 units to form the line represented by a series of dashes in Figure 16–67. This line happens to be located at the center of the object.

Command: **CP** *(For COPY)*
Select objects: *(Select the line at "A")*
Select objects: *(Press ENTER to continue)*
Specify base point or displacement, or [Multiple]: *(Select the endpoint at "A")*
Specify second point of displacement or <use first point as displacement>: *(Move your cursor up and to the right and enter 2.50)*

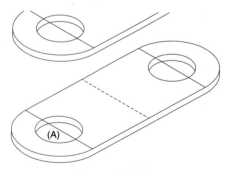

Figure 16–67

STEP 16

Duplicate the line represented by dashes in Figure 16–68 to form the two inclined lines at "B" and "C." These lines will begin the construction of the sides of the hanger. Use the COPY command along with the Multiple option to accomplish this.

Command: **CP** *(For COPY)*
Select objects: *(Select the dashed line in Figure 16–68)*
Select objects: *(Press ENTER to continue)*
Specify base point or displacement, or [Multiple]: **M** *(For Multiple)*
Specify base point: *(Select the endpoint at "A")*
Specify second point of displacement or <use first point as displacement>: *(Move your cursor up and to the right and enter 1.25)*
Specify second point of displacement or <use first point as displacement>: *(Move your cursor down and to the left and enter 1.25)*
Specify second point of displacement or <use first point as displacement>: *(Press ENTER to exit this command)*

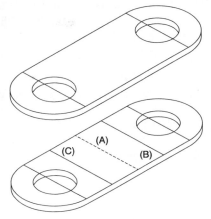

Figure 16–68

STEP 17

Use the COPY command to duplicate the two dashed lines in Figure 16–69 straight up at a distance of 3.75 units. Use the Polar coordinate mode to accomplish this.

Command: **CP** *(For COPY)*
Select objects: *(Select both dashed lines in Figure 16–69)*
Select objects: *(Press ENTER to continue)*
Specify base point or displacement, or [Multiple]: *(Select the endpoint at "A")*
Specify second point of displacement or <use first point as displacement>: *(Pick the Polar Intersection at "B")*

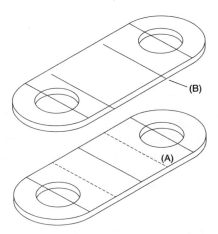

Figure 16–69

990

STEP 18

Use the LINE command along with the
OSNAP-Endpoint option to draw a line
from endpoint "A" to endpoint "B" in
Figure 16–70.

Command: **L** *(For LINE)*
Specify first point: *(Select the endpoint at
"A")*
Specify next point or [Undo]: *(Select the
endpoint at "B")*
Specify next point or [Undo]: (Press
ENTER to exit this command)

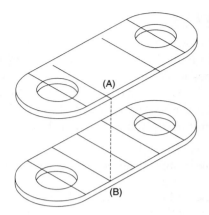

Figure 16–70

STEP 19

Use the COPY command and the Multiple
option to copy the dashed line at "A" to
"B," "C," and "D" in Figure 16–71.

Command: **CP** *(For COPY)*
Select objects: *(Select the dashed line in
Figure 16–71)*
Select objects: *(Press ENTER to continue)*
Specify base point or displacement, or
[Multiple]: **M** *(For Multiple)*
Specify base point: *(Select the endpoint of
the vertical line at "A")*
Specify second point of displacement or
<use first point as displacement>:
(Select the endpoint of the line at "B")
Specify second point of displacement or
<use first point as displacement>:
(Select the endpoint of the line at "C")
Specify second point of displacement or
<use first point as displacement>:
(Select the endpoint of the line at "D")
Specify second point of displacement or
<use first point as displacement>:
(Press ENTER to exit this command)

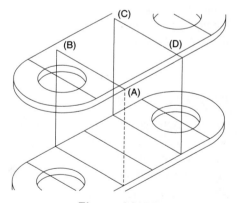

Figure 16–71

STEP 20

Use the ERASE command to delete all lines represented as dashed lines in Figure 16–72.

Command: **E** *(For ERASE)*
Select objects: *(Select all dashed objects in Figure 16–72)*
Select objects: *(Press ENTER to execute this command)*

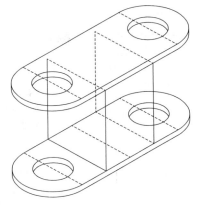

Figure 16–72

STEP 21

Use the TRIM command to partially delete the vertical line in Figure 16–73. The segment to be deleted is hidden and not shown in an isometric drawing.

Command: **TR** *(For TRIM)*
Current settings: Projection=UCS Edge=None
Select cutting edges ...
Select objects: *(Select the dashed objects at "A" and "B")*
Select objects: *(Press ENTER to continue)*
Select object to trim or [Project/Edge/Undo]: *(Select the vertical line at "C")*
Select object to trim or [Project/Edge/Undo]: *(Press ENTER to exit this command)*

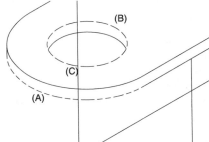

Figure 16–73

STEP 22

Use the TRIM command to partially delete the vertical line in Figure 16–74. The segment to be deleted is hidden and not shown in an isometric drawing.

Command: **TR** *(For TRIM)*
Current settings: Projection=UCS
   Edge=None
Select cutting edges ...
Select objects: *(Select the dashed elliptical arc at "A")*
Select objects: *(Press ENTER to continue)*
Select object to trim or [Project/Edge/
   Undo]: *(Select the vertical line at "B")*
Select object to trim or [Project/Edge/
   Undo]: *(Press ENTER to exit this command)*

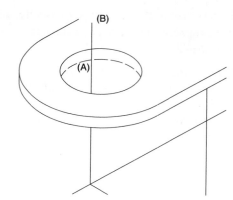

Figure 16–74

STEP 23

Use the TRIM command to partially delete the inclined line in Figure 16–75.

Command: **TR** *(For TRIM)*
Current settings: Projection=UCS
   Edge=None
Select cutting edges ...
Select objects: *(Select the dashed lines "A" and "B")*
Select objects: *(Press ENTER to continue)*
Select object to trim or [Project/Edge/
   Undo]: *(Select the line at "C")*
Select object to trim or [Project/Edge/
   Undo]: *(Select the line at "D")*
Select object to trim or [Project/Edge/
   Undo]: *(Press ENTER to exit this command)*

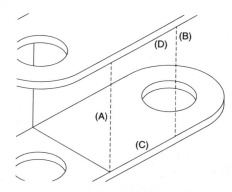

Figure 16–75

STEP 24

Use the TRIM command to partially delete the objects in Figure 16–76. The segments to be deleted are hidden and not shown in an isometric drawing. Use ERASE to delete any leftover elliptical arc segments.

Command: **TR** *(For TRIM)*
Current settings: Projection=UCS Edge=None
Select cutting edges ...
Select objects: *(Select the dashed lines "A" and "B")*
Select objects: *(Press ENTER to continue)*
Select object to trim or [Project/Edge/ Undo]: *(Select the line at "C")*
Select object to trim or [Project/Edge/ Undo]: *(Select the elliptical arc at "D")*
Select object to trim or [Project/Edge/ Undo]: *(Select the ellipse at "E")*
Select object to trim or [Project/Edge/ Undo]: *(Select the elliptical arc at "F")*
Select object to trim or [Project/Edge/ Undo]: *(Press ENTER to exit this command)*

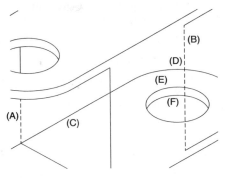

Figure 16–76

STEP 25

The completed isometric appears in Figure 16–77. This drawing may be dimensioned with the Oblique option of the DIMEDIT command. Consult your instructor if this next step is necessary.

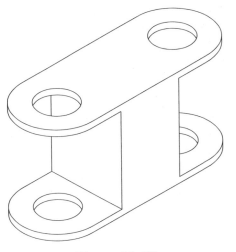

Figure 16–77

STEP 26

As an extra step, convert the completed hanger to an object with a rectangular hole through its center in Figure 16–78. Begin by copying lines "A," "B," "C," and "D" a distance of 0.25 units to form the inside rectangle using polar coordinates. Since the lines will overlap at the corners, use the FILLET command set to a radius of 0 to create corners of the hole. Use the LINE command to draw the inclined line "E" at any distance with a 150° angle. Use either TRIM or EXTEND to complete the new version of the hanger.

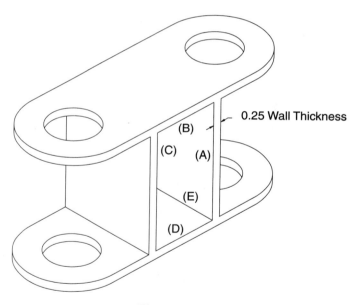

Figure 16–78

PROBLEMS FOR CHAPTER 16

Directions for Problem 16–1:

Construct an isometric drawing of the object.

PROBLEM 16–1

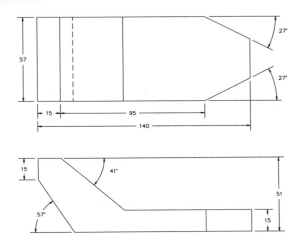

Directions for Problem 16–2:

Construct an isometric drawing of the object. Begin the corner of the isometric at "A."

PROBLEM 16–2

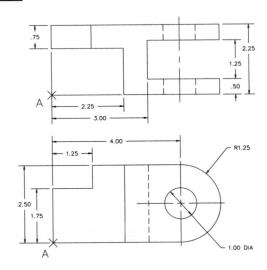

Directions for Problems 16–3 through 16–17:

Construct an isometric drawing of the object.

PROBLEM 16–3

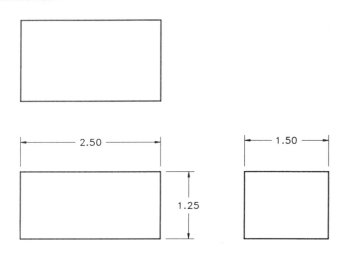

PROBLEM 16–4

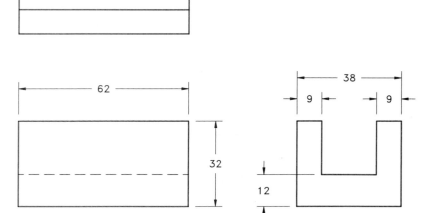

PROBLEM 16-5

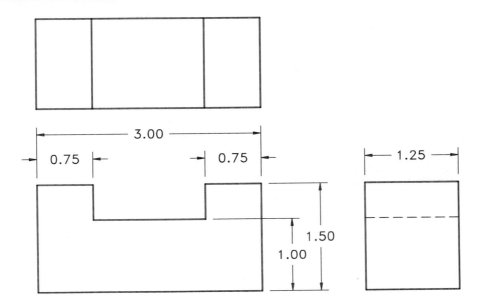

PROBLEM 16-6

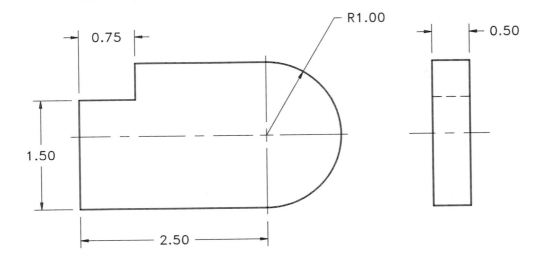

PROBLEM 16–7

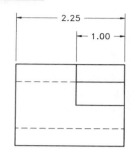

PROBLEM 16–8

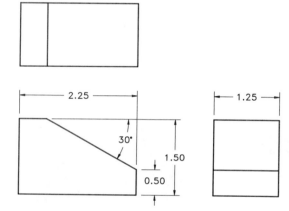

PROBLEM 16–9

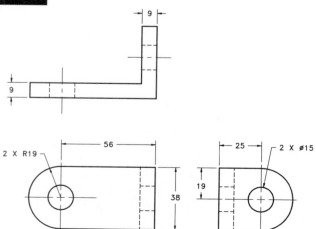

PROBLEM 16–10

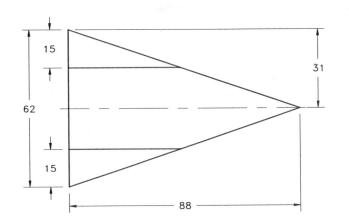

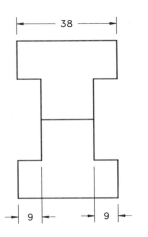

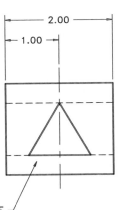

PROBLEM 16–11

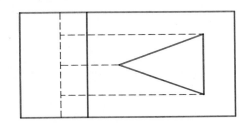

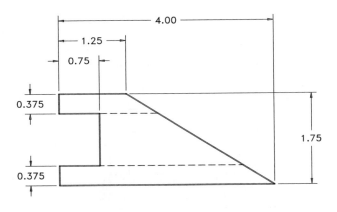

EQUILATERAL TRIANGLE

PROBLEM 16-12

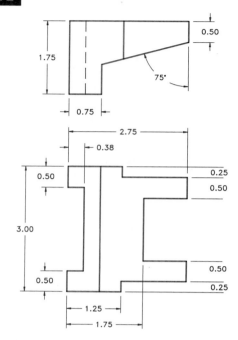

PROBLEM 16-13

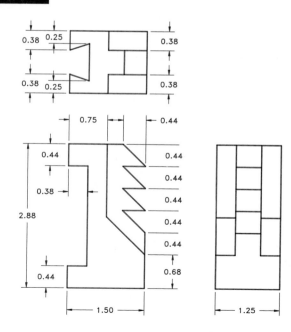

PROBLEM 16-14

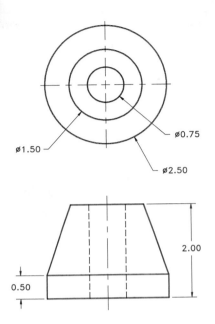

Ø1.50

Ø0.75

Ø2.50

2.00

0.50

PROBLEM 16-15

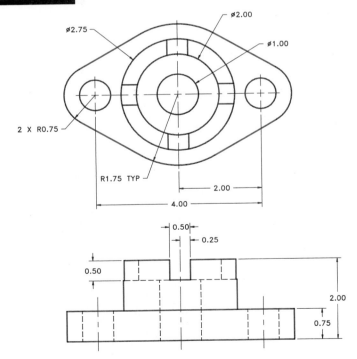

Ø2.00

Ø2.75

Ø1.00

2 X R0.75

R1.75 TYP

2.00

4.00

0.50

0.25

0.50

2.00

0.75

PROBLEM 16–16

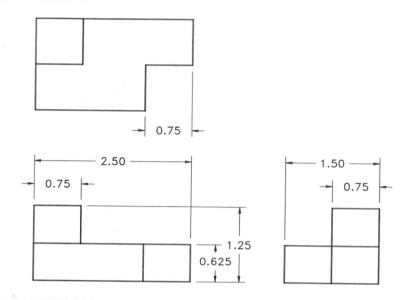

PROBLEM 16–17

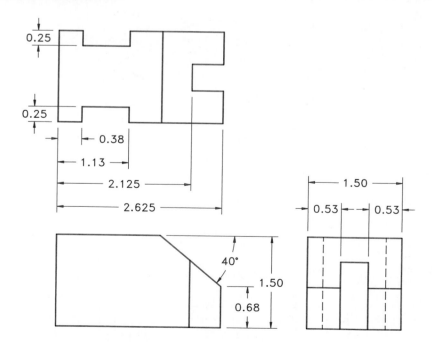

Block Creation, AutoCAD DesignCenter, and MDE

This chapter begins the study of the how blocks are created and merged into drawing files. This is a major productivity enhancement and is considered an electronic template similar to those methods used with block templates in manual drafting practices. The first segment of this chapter will discuss what blocks are and how they are created. Blocks can be inserted either in the current drawing or in any drawing, if created through the proper command. Next, this chapter continues with a discussion about using the Insert dialog box and the AutoCAD DesignCenter to bring blocks into drawings. The DesignCenter is a special feature of AutoCAD 2000 that allows blocks to be inserted in drawings with drag and drop techniques. In addition, blocks found in other drawings can easily be inserted in the current drawing from the DesignCenter. Finally, the chapter will discuss the use of MDE (Multiple Design Environment). This feature allows the opening of multiple drawings within a single AutoCAD session and provides a convenient method of exchanging data, such as blocks, between one drawing and another. As an added bonus to AutoCAD users, a series of block libraries is supplied with the package. These block libraries include such application areas as mechanical, architectural, electrical, piping, and welding just to name a few.

WHAT ARE BLOCKS?

Blocks usually consist of smaller components of a larger drawing. Typical examples include doors and windows for floor plans, nuts and bolts for mechanical assemblies, and resistors and transistors for electrical schematics. In Figure 17–1 showing an electrical schematic, all resistors, capacitors, tetrodes, and diodes are considered blocks that make up the total drawing of the electrical schematic. The capacitor is highlighted as one of these components.

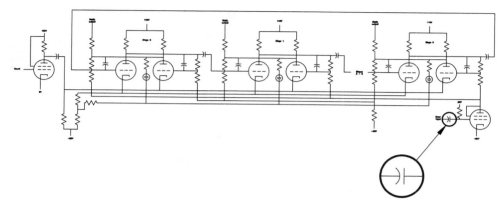

Figure 17–1

Blocks are created and then inserted in a drawing. When creating the block, you must first provide a name for the block. The capacitor in Figure 17–2 was assigned the name "CAPACITOR." Also, when you create a block, an insertion point is required. This acts as a reference point where the block will be inserted. In Figure 17–2, the insertion point of the block is the left end of the line.

BLOCK NAME: CAPACITOR

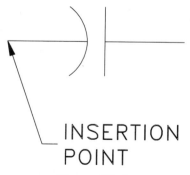

INSERTION POINT

Figure 17–2

At times, blocks have to be rotated into position. In Figure 17–3, notice that one capacitor is rotated at a 45° angle while the other capacitor is rotated 270°. In this way the same block can be used numerous times even though it is positioned differently in the drawing.

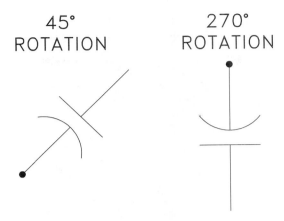

Figure 17–3

METHODS OF SELECTING BLOCK COMMANDS

Commands that deal with blocks can be selected from one of two pull-down menu areas. The first one is illustrated in Figure 17–4; selecting Block from the Draw pull-down menu exposes Make (the Block Definition dialog box) and Base, which defines a new base or insertion point for a block.

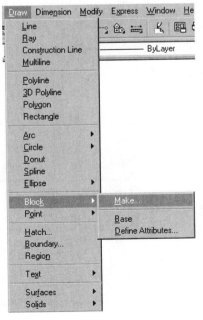

Figure 17–4

The second area that contains block-related commands is found on the Insert pull-down menu in Figure 17–5. This exposes Block (the Insert dialog box).

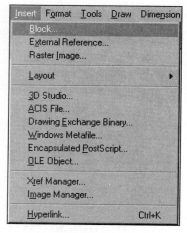

Figure 17–5

Another convenient way of selecting Block-related commands is through the Draw toolbox shown in Figure 17–6. You can also enter block-related commands from the keyboard, either using their entire name or through command aliases, as in the following examples:

Enter B to activate the Block Definition dialog box (BLOCK).

Enter I to activate the Insert dialog box (INSERT).

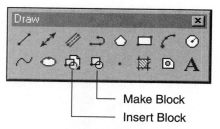

Figure 17–6

For a complete listing of other command aliases, refer to Chapter 1, Table 1–2.

The following commands will be discussed in greater detail in the following pages:

BLOCK	(Block Definition dialog box)
WBLOCK	(Write Block dialog box)
INSERT	(Insert dialog box)
ADCENTER	(AutoCAD DesignCenter)

CREATING A LOCAL BLOCK USING THE BLOCK COMMAND

Figure 17–7 is a drawing of a hex head bolt. This drawing consists of one polygon representing the hexagon, a circle indicating that the hexagon is circumscribed about the circle, and two centerlines. The centerlines were constructed with the dimension variable DIMCEN set to a -0.09 value and the DIMCENTER command. Rather than copy these individual objects numerous times throughout the drawing, you can create a block using the BLOCK command. Selecting Block from the Draw pull-down menu and then selecting Make, as in Figure 17–8A, activates the Block Definition dialog box illustrated in Figure 17–8B. Entering the letter B from the keyboard also activates the Block Definition dialog box.

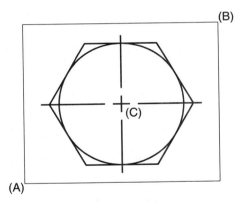

Figure 17–7

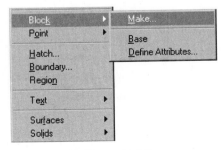

Figure 17–8A

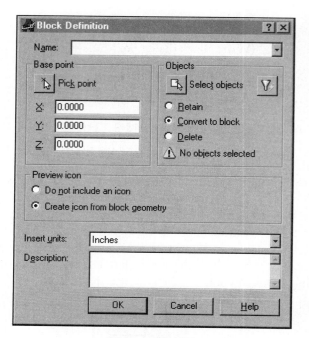

Figure 17–8B

When you create a block, numerous objects that make up the block are all considered a single object. This dialog box allows for the newly created block to be merged into the current drawing file. This means that a block consisting of a bolt can only be used in the assembly drawing it was created in; it cannot be shared directly with other drawings. (The WBLOCK command is used to create global blocks that can be inserted in any drawing file. This command will be discussed later on in this chapter.)

To create a block through the Block Definition dialog box, enter the name of the block in the Name edit box, such as "Hexbolt" (see Figure 17–9). You can use up to 255 alphanumeric characters (only 31 if the system variable EXTNAMES is set to 0) when naming a block.

Figure 17–9

Next, click on the Select objects button; this will return you to the drawing editor (the Quick Select button provides a means of using filters to select objects). Create a window from "A" to "B" around the entire hex bolt to select all objects (refer to Figure 17–7). When finished, press ENTER. This will return you to the Block Definition dialog box, where a previewed image of the block you are about to create is present at the right of the dialog box, illustrated in Figure 17–10. The image will appear only if the Create icon from block geometry radio button is selected. Whenever you create a block, you can elect to allow the original objects that made up the hex bolt to remain on the screen (select Retain button), to be replaced with an instance of the block (select Convert to Block button) or to be removed after the block is created (select Delete button). Select the Delete radio button if you want the original objects to be erased after the block is created. If the original objects that made up the hex bolt are unintentionally removed from the screen during this creation process, the OOPS command can be used to retrieve all original objects to the screen.

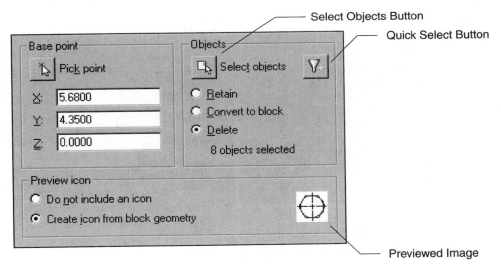

Figure 17–10

The next step is to create a base point or insertion point for the block to be brought into the drawing. This is considered a point of reference and should be identified at a key location along the block. By default, the Base point in the Block Definition dialog box is located at 0,0,0. The block will still be merged into the current drawing; however, 0,0,0 may not represent the correct base point location. Click on the Pick point button; this will return to the drawing editor and allow you to pick the center of the hex bolt at "C" in Figure 17–7 using the OSNAP-Intersect mode. Once selected, this returns you to the Block Definition dialog box; notice how the Base Point information now reflects the key location along the block (see Figure 17–11).

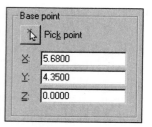

Figure 17–11

Yet another feature of the Block Definition dialog box allows you to add a description for the block. Many times, the name of the block hides the real meaning of the block, especially if the block name consists of only a few letters. Click in the Description edit box and add the following statement: "This is a hexagonal head bolt." This will allow you to refer back to the description in case the block name does not indicate the true intended purpose for the block (see Figure 17–12). Specifying the Insert units determines the type of units utilized for scaling the block when it is inserted from the AutoCAD DesignCenter.

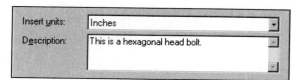

Figure 17–12

The complete dialog box is illustrated in Figure 17–13. If you are satisfied with the block name, base point, and description, click on the OK button at the bottom of the dialog box to add the block to the database of the drawing.

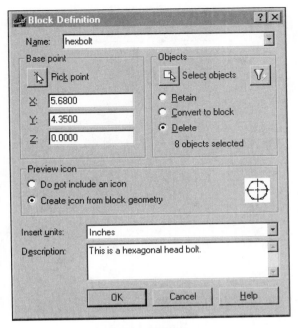

Figure 17–13

As the block is written to the database of the current drawing, the individual objects used to create the block automatically disappear from the screen (due to our earlier selection or the "Delete from drawing" radio button). When the block is later inserted in a drawing, it will be placed in relation to the insertion point.

CREATING A GLOBAL BLOCK USING THE WBLOCK COMMAND

In the previous example of using the BLOCK command, several objects were grouped into a single object for insertion in a drawing. As this type of block is created, its intended purpose is for insertion in the drawing it was created in. Using the WBLOCK (Write block) command creates a block similar to the process used in the BLOCK command. However, this will write the objects directly to a location on the hard disk, which allows the block to be inserted in any type of drawing file. Entering WBLOCK at the command prompt activates the Write Block dialog box illustrated in Figure 17–14. Entering W from the keyboard also activates the dialog box.

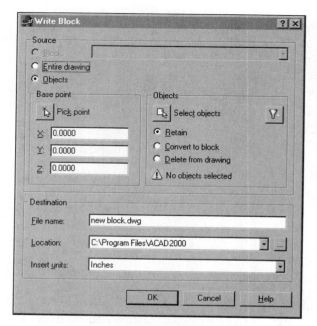

Figure 17–14

With the Objects radio button selected, follow the same steps demonstrated earlier for creating blocks by selecting the objects that will make up the global block. The Block radio button (grayed out if no blocks exist in the current drawing, as in Figure 17–15) allows you to select an existing block to create the global block. The Entire drawing radio button allows you to select the complete current drawing for creation of the block file. Next, click on the Base point area's Pick point button to select an appropriate insertion point for the global block.

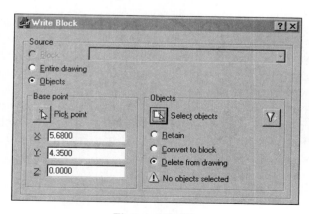

Figure 17–15

It may be unclear where the block is actually being written to on the hard disk. Be sure to carefully provide the destination information (see Figure 17–16) for the Write Block dialog box. Provide an appropriate file name in the edit box. To give a location for the block, click on the Browse button and choose a suitable folder in which to store the file (see Figure 17–17). Clicking on the OK button in the Write Block dialog box writes the objects to the drive location with the file name specified by the user.

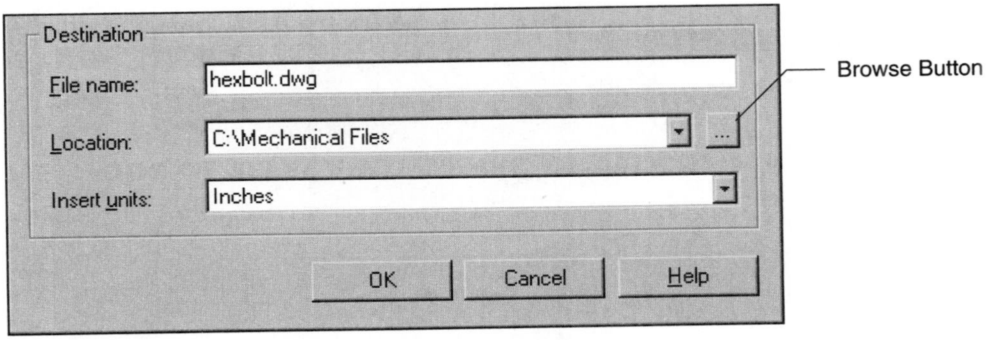

Figure 17–16

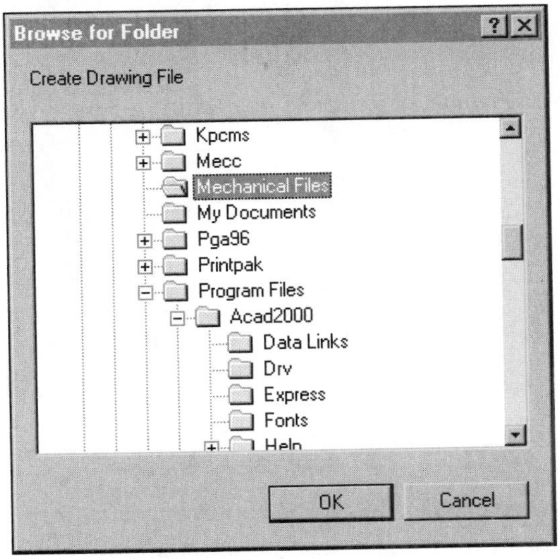

Figure 17–17

THE INSERT DIALOG BOX

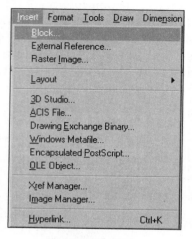

Figure 17–18A

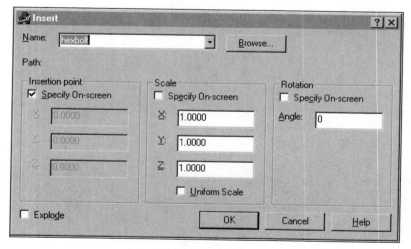

Figure 17–18B

Once blocks are created through the BLOCK and WBLOCK commands, they are merged or inserted in the drawing through the INSERT command. Enter INSERT or I from the keyboard or choose Block… from the Insert pull-down menu, as shown in Figure 17–18A, to activate the Insert dialog box in Figure 17–18B. This dialog box is used to dynamically insert blocks. First, by clicking on the Name drop-down list box, select a block from the current drawing (clicking on the Browse button locates global blocks or drawing files). After you identify the name of the block to insert, the point where the block will be inserted must be specified, along with its size and rotation angle. By default, the Insertion point area's Specify on Screen box is checked. This means you will be prompted for the insertion point at the command prompt area of the drawing editor. The default values for the scale and rotation insert the block at their original size and orientation. It was already mentioned that a block consists of several objects combined into a single object. The Explode box determines if the block is inserted as one object or if the block is inserted and then exploded back to its individual objects. Once the name of the block is selected, such as Hexbolt, click on the OK button; the following prompts complete the block insertion operation:

Specify insertion point or [Scale/X/Y/Z/Rotate/PScale/PX/PY/PZ/PRotate]: *(Mark a point at "A" in Figure 17–19 to insert the block)*

If the Specify on Screen boxes are also checked for scale and rotation, the following prompts will complete the block insertion operation.

Specify insertion point or [Scale/X/Y/Z/Rotate/PScale/PX/PY/PZ/PRotate]: *(Mark a point at "A" in Figure 17–19 to insert the block)*
Enter X scale factor, specify opposite corner, or [Corner/XYZ] <1>: *(Press ENTER to accept default X scale factor)*
Enter Y scale factor <use X scale factor>:): *(Press ENTER to accept default)*
Specify rotation angle <0>: *(Press ENTER to accept the default rotation angle and insert the block in Figure 17–19)*

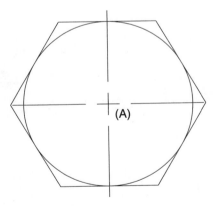

Figure 17–19

If blocks are defined as part of the database of the current drawing, they may be selected from the name drop-down list box (see Figure 17–20).

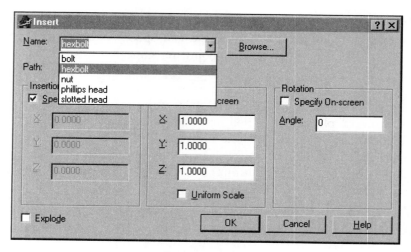

Figure 17–20

For inserting global blocks in a drawing, select the Insert dialog box Browse... button shown in Figure 17–18B, which displays the Select Drawing File dialog box illustrated in Figure 17–21. This is the same dialog box associated with opening drawing files. Because the WBLOCK command writes the group of objects to disk, this makes it possible to insert any valid AutoCAD drawing file in the current drawing. Select the desired subdirectory where the global block or drawing file is located followed by the name of the drawing. This will return you to the main Insert dialog box shown in Figure 17–18B.

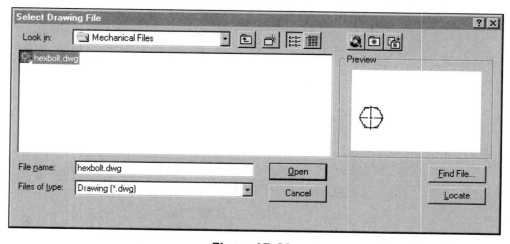

Figure 17–21

APPLICATIONS OF BLOCK INSERTIONS

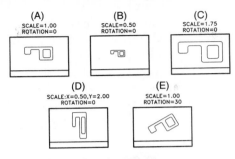

Figure 17–22

Figure 17–22 shows the results of entering different scale factors and rotation angles when blocks are inserted in a drawing file. The image at "A" shows the block inserted in a drawing with its default scale and rotation angle values. The image at "B" shows the result of inserting the block with a scale factor of 0.50 and a rotation angle of 0°. The image appears half its normal size. The image at "C" shows the result of inserting the block with a scale factor of 1.75 and a rotation angle of 0°. In this image, the scale factor increases the block in size along with keeping the same proportions. The image at "D" shows the result of inserting the block with different scale factors. In this image, the X scale factor is 0.50 while the Y scale factor is 2.00 units. Notice how out of proportion the block appears. There are certain applications where different scale factors are required to produce the desired effect. The image at "E" shows the result of inserting the block with the default scale factor and a rotation angle of 30°. As with all rotations, a positive angle rotates the block in the counterclockwise direction; negative angles rotate in the clockwise direction.

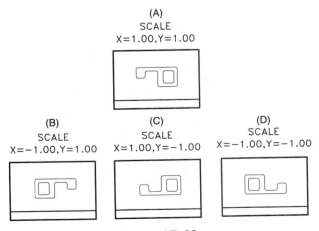

Figure 17–23

Figure 17–23 shows the results of entering different negative scale factors when inserting blocks in a drawing file. The image at "A" shows the block inserted in a drawing with its default scale and rotation angle values. The image at "B" shows the result of inserting the block with an X scale factor of –1.00 units; notice how the image flips in the X direction. This result is similar to using the MIRROR command with the Y-axis as the axis to mirror along. The image at "C" shows the result of inserting the block with a Y scale factor of –1.00 units; this time, notice how the image flips in the Y direction. The image at "D" shows the result of inserting the block with both X and Y scale factors of –1.00 units. Here, the object is flipped in both the X and Y directions.

EXPRESS TOOLS APPLICATIONS FOR BLOCKS

Express tool commands that deal with blocks can be entered at the keyboard, selected from the Express pull-down menu (illustrated in Figure 17–24) or selected from an express toolbox (shown in Figure 17–25).

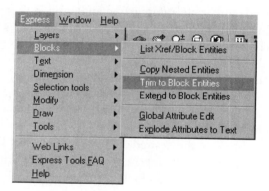

Figure 17–24

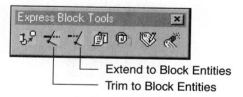

Figure 17–25

The BTRIM and BEXTEND commands will be discussed in greater detail in the following pages:

TRIMMING TO BLOCK OBJECTS

The ability to trim to a block object is now possible through the BTRIM command. As you select cutting edges to trim to, temporary objects are constructed on top of the block. This allows objects to be trimmed to these cutting edges. In Figure 17–26A, the outside edges of the bolt act as cutting edges.

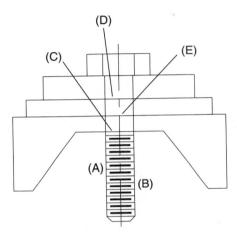

Figure 17–26A

Command: **BTRIM**
Select cutting edges: *(Pick the edge of the bolt at "A")*
Select cutting edges: *(Pick the edge of the bolt at "B")*
Select cutting edges: *(Press* ENTER *to continue with this command)*
Select objects:
Select object to trim or [Project/Edge/Undo]: *(Pick the line at "C")*
Select object to trim or [Project/Edge/Undo]: *(Pick the line at "D")*
Select object to trim or [Project/Edge/Undo]: *(Pick the line at "E")*
Select object to trim or [Project/Edge/Undo]: *(Press* ENTER *to exit this command)*

The result of using the BTRIM command is illustrated in Figure 17–26B. Once this command has been completed, the temporary cutting edge objects disappear and the block returns to its original state.

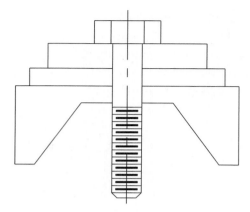

Figure 17–26B

EXTENDING TO BLOCK OBJECTS

The ability to extend to a block object is also possible through the BEXTEND command. As you select boundary edges to extend to, temporary objects are constructed on top of the block. This allows objects to be trimmed to these boundary edges. In Figure 17–27A, the outside edges of the bolt act as cutting edges.

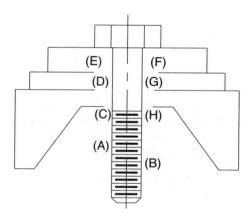

Figure 17–27A

Command: **BEXTEND**
Select edges for extend: *(Pick the edge of the bolt at "A")*
Select edges for extend: *(Pick the edge of the bolt at "B")*
Select edges for extend: *(Press* ENTER *to continue with this command)*
Select objects:

Select object to extend or [Project/Edge/Undo]: *(Pick the line at "C")*
Select object to extend or [Project/Edge/Undo]: *(Pick the line at "D")*
Select object to extend or [Project/Edge/Undo]: *(Pick the line at "E")*
Select object to extend or [Project/Edge/Undo]: *(Pick the line at "F")*
Select object to extend or [Project/Edge/Undo]: *(Pick the line at "G")*
Select object to extend or [Project/Edge/Undo]: *(Pick the line at "H")*
Select object to extend or [Project/Edge/Undo]: *(Press* ENTER *to exit this command)*

The result of using the BEXTEND command is illustrated in Figure 17–27B. Once this command has been completed, the temporary boundary edge objects disappear and the block returns to its original state.

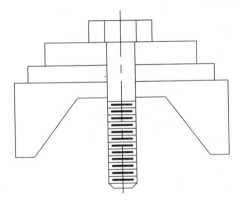

Figure 17–27B

EXPLODING BLOCKS

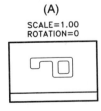 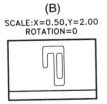 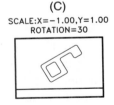

Figure 17–28

It has already been mentioned that as blocks are inserted in a drawing file, they are considered one object even though they consist of numerous individual objects. At times it is necessary to break a block up into its individual parts. The EXPLODE command is used for this. Figure 17–28 shows three blocks that have been inserted with different scale factors. The block at "A" was inserted with the default scale and rotation angle values. The block at "B" was inserted with an X scale factor of 0.50 and a Y scale factor of 2.00 units. Finally, the block at "C" was inserted with an X scale factor of −1.00 and a rotation angle of 30°. In all three cases, the blocks may be broken up into individual objects through the EXPLODE command.

Many times when blocks or drawing files are inserted in other drawing files and exploded, the original definition of the block remains with the drawing. This means you can edit the exploded block, make changes, and even erase certain areas; however, the definition of the block still remains in the drawing database. When this happens, the file size usually increases dramatically. To compress the drawing file by getting rid of unused block definitions, use the PURGE command. When it is entered from the keyboard, the prompt sequence is similar to the following:

Command: **PU** *(For PURGE)*

Enter type of unused objects to purge

[Blocks/Dimstyles/LAyers/LTypes/Plotstyles/SHapes/textSTyles/Mlinestyles/All]: **B** *(For Block*

Enter name(s) to purge <*>: *(Press* ENTER *to accept this default)*
Verify each name to be purged? [Yes/No] <Y>: *(Press* ENTER *to accept this default)*
Purge block "SLIDE"? <N> **Y** *(For Yes)*

If all insertions of a block have been exploded, the block definition that remains with the drawing is considered unused. The PURGE command will delete only the block definitions of unused blocks. In addition, if dimension styles, layers, linetypes, and text styles have been defined in a drawing but not used, the PURGE command will delete them from the database as well. The key to using the PURGE command is in understanding that it will only get rid of block definitions if these definitions are unused.

GRIPS AND BLOCKS

Grips have interesting results when blocks are selected. In Figure 17–29A, when the block is selected, only one grip appears. The location of this grip happens to be at the insertion point of the block. The system variable GRIPBLOCK controls this and can be set in the Options dialog box, Selection tab (see Figure 17–29B). Notice that, in the Grips area, the check box for Enable grips within blocks is not checked. A grip will only appear at the insertion point of the block as a result of this setting in the dialog box.

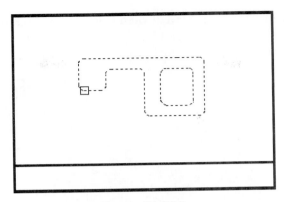

Figure 17–29A

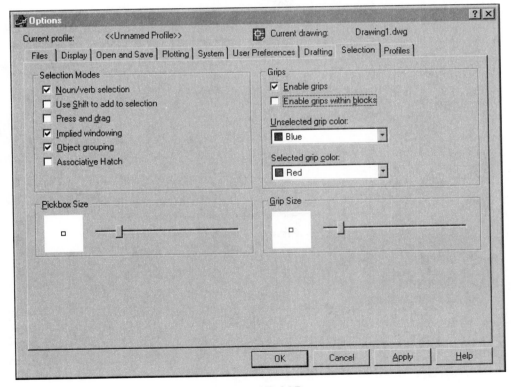

Figure 17–29B

An entirely different result is achieved when you place a check in the Enable Grips within blocks check box, as in Figure 17–30A. Now, whenever a block is selected, the grips will appear on all objects that make up the block (see Figure 17–30B).

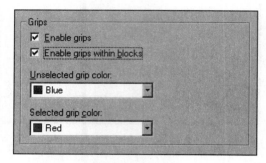

Figure 17–30A

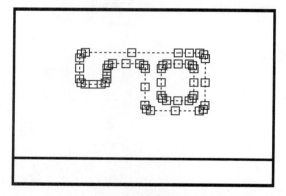

Figure 17–30B

REDEFINING BLOCKS

At times, a block needs to be edited. Rather than erase all occurrences of the block in a drawing, you can redefine it. This is considered a major productivity technique because all blocks that share the same name as the block being redefined will automatically update to the latest changes. Figure 17–31 shows various blocks inserted in a drawing; one is inserted at normal size, another at half size, another with different X and Y scale factors, and two others rotated at different angles. The block name is "Brick" and the insertion point of the brick is at the center of the middle circle.

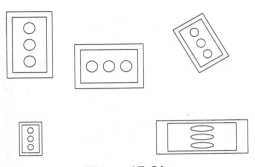

Figure 17–31

In order for a block in the current drawing to be redefined, another block is inserted and exploded to break it down into individual objects. In the example of the Brick, the inside rectangle needs to be deleted. The name of the block will still be "Brick" and its insertion point will remain at the center of the middle circle. Perform the exploding and erasing operations. The new brick is illustrated in Figure 17–32.

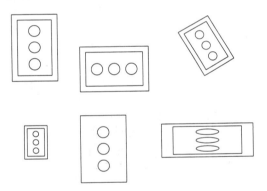

Figure 17–32

Activate the Block Definition dialog box through the BLOCK command; name the block "Brick," select the proper objects needed to define the new brick, identify the insertion point at the center of the middle circle, and give a description of the brick as illustrated in Figure 17–33A.

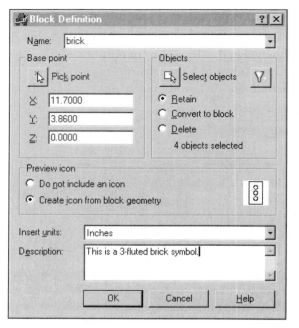

Figure 17–33A

After you click on the OK button, another dialog box appears similar to Figure 17–33B. A block called "Brick" is already defined in the drawing. This dialog box needs confirmation to redefine all other blocks in the drawing.

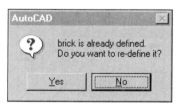

Figure 17–33B

Clicking the Yes button in the AutoCAD Alert box redefines all blocks in the drawing to reflect the changes to the new brick design. These changes are illustrated in Figure 17–34. However, notice one of the bricks did not update to the latest changes. This brick was exploded sometime during the drawing process. Since it already consists of individual objects and is no longer a block definition, it will not update to any new changes made to the brick. While it is very popular to merge complete drawings with other drawings and then explode them, care should be taken when exploding small component blocks, especially when global changes are desired.

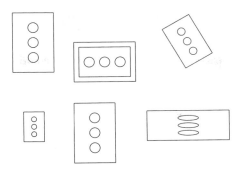

Figure 17–34

THE BASE COMMAND

Entire drawings can easily be inserted in other drawings through the Insert dialog box. This is very popular when you want to use only a portion of the larger drawing. Since the drawing file is considered one object when inserted, it must first be exploded before segments can be isolated. Also, by default, a drawing file is inserted in relation to the 0,0 screen location (usually the lower left corner of the display screen). In Figure 17–35, the title block was constructed from point 2,2; however, when it is inserted in another drawing, the title will be inserted in relation to point 0,0. To provide better control over the insertion of drawing files, the BASE command could be used. This command identifies a new insertion point. In Figure 17–36, a new base point is identified at the lower left corner of the title block. Now when the title block is inserted, it will be brought into the drawing in relation to its lower left corner rather than point 0,0. Study the figures and the following command sequence for using the BASE command:

Command: **BASE**
Enter base point <0.0000,0.0000,0.0000>: **End**
of *(Pick the endpoint of the title block at "A" in Figure 17-36)*

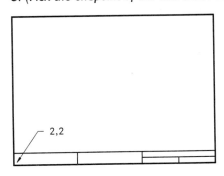

Figure 17–35

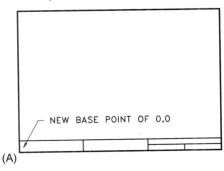

NEW BASE POINT OF 0,0

(A)

Figure 17–36

BLOCKS AND THE DIVIDE COMMAND

The DIVIDE command was already covered in Chapter 5. It allows you to select an object, give the number of segments, and allow the command to place point objects at equally spaced distances depending on the number of segments. In Figure 17–37, a counter-bore hole and centerline needs to be copied 12 times around the elliptical centerline so that each hole is equally spaced from others. (The illustration of the block "Cbore" is displayed at twice its normal size.) Because the ARRAY command is used to copy objects in a rectangular or circular pattern, that command cannot be used for an ellipse. The DIVIDE command has a Block option that allows you to specify the name of the block and the number of segments. In Figure 17–37, the elliptical centerline is identified as the object to divide. Follow the command sequence to place the block "Cbore" in the elliptical pattern:

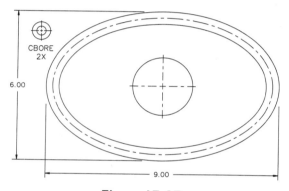

Figure 17–37

Command: **DIV** *(For DIVIDE)*
Select object to divide: *(Select the elliptical centerline)*
Enter the number of segments or [Block]: **B** *(For Block)*
Enter name or block to insert: **CBORE**
Align block with object? [Yes/No] <Y>:*(Press ENTER to accept)*
Enter the number of segments: **12**

The results are illustrated in Figure 17–38. Notice how the elliptical centerline is divided into 12 equal segments by 12 blocks called "Cbore." In this way, any object may be divided through the use of blocks.

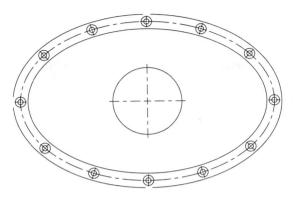

Figure 17–38

BLOCKS AND THE MEASURE COMMAND

As with the DIVIDE command, the MEASURE command offers increased productivity when you measure an object and insert blocks at the same time. In Figure 17–39, a polyline path will be divided into 0.50 length segments using the block "Chain2." The perimeter of the polyline was calculated by the LIST command to be 22.00 units. Follow the command prompt sequence below for placing a series of blocks called "Chain2" around the polyline path to create a linked chain:

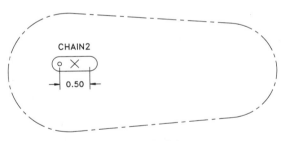

Figure 17–39

Command: **ME** *(For MEASURE)*
Select object to measure: *(Pick the polyline path in Figure 3-35)*
Specify length of segment or [Block]: **B** *(For Block)*
Enter name of block to insert: **CHAIN2**
Align block with object? [Yes/No] <Y> *(Press ENTER to accept)*
Specify length of segment: **0.50**

The result is illustrated in Figure 17–40, with all chain links being measured along the polyline path at increments of 0.50 units.

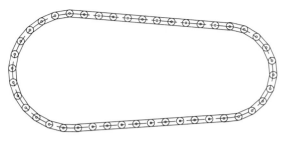

Figure 17–40

Answering "No" to the prompt "Align block with object? <Y>" displays the results in Figure 17–41. Here all blocks are inserted horizontally and travel in 0.50 increments. While the polyline path has been successfully measured, the results are not acceptable for creating the chain in Figure 17–40.

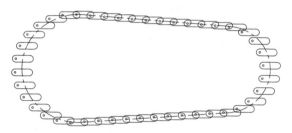

Figure 17–41

RENAMING BLOCKS

Blocks can be renamed to make their meanings more clear through the RENAME command. If it is selected from the Format pull-down menu in Figure 17–42A, a dialog box similar to the one illustrated in Figure 17–42B will appear. Clicking on Blocks in the dialog box lists all blocks defined in the current drawing. One block with the name "REFRIG" was abbreviated and we wish to give it a full name. Clicking on the name "REFRIG" will paste it in the Old Name edit box. Type the desired full name REFRIGERATOR and click on the Rename To button to rename the block.

Figure 17–42A

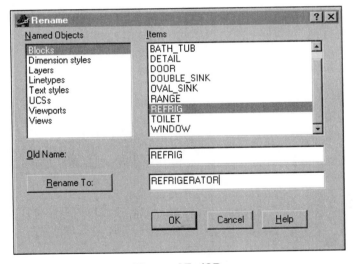

Figure 17–42B

ADDITIONAL TIPS FOR WORKING WITH BLOCKS

CREATE BLOCKS ON LAYER 0

Blocks are best controlled, when dealing with layers, colors, and linetypes, by being drawn on Layer 0, because it is considered a neutral layer. By default, Layer 0 is assigned the color White and the Continuous linetype. Objects drawn on Layer 0 and then converted to blocks will take on the current layer when inserted in the drawing. The current layer will control color and linetype.

CREATE BLOCKS FULL SIZE IF APPLICABLE

Figure 17–43 illustrates a drawing of a refrigerator complete with dimensions. In keeping with the concept of drawing in real world units in CAD or at full size, individual blocks must also be drawn at full size in order for them to be inserted in the drawing at the correct proportions. For this block, construct a rectangle 28 units in the X direction and 24 units in the Y direction. Create a block called Refrigerator by picking the rectangle. When testing out this block, be sure to be in a drawing that is set to the proper units based on a scale such as ½"=1'-0" or ¼"=1'-0". These are typical architectural scales and the block of the refrigerator will be in the correct proportions on these types of sheet sizes.

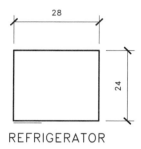

REFRIGERATOR

Figure 17–43

USE GRID WHEN PROPORTIONALITY, BUT NOT SCALE, IS IMPORTANT

Sometimes blocks represent drawings where the scale of each block is not important. In the previous example of the Refrigerator, scale was very important in order for the refrigerator to be drawn according to its full size dimensions. This is not the case in Figure 17–44A of the drawing of the resistor block. Electrical schematic blocks are generally not drawn to any specific scale: however, it is important that all blocks are proportional to each other. Setting up a grid is considered good practice in keeping all blocks of the same proportions. Whatever the size of the grid, all blocks are designed around the same grid size. The result is in Figure 17–44B; with four blocks being drawn with the same grid, their proportions look acceptable.

RESISTOR

Figure 17–44A

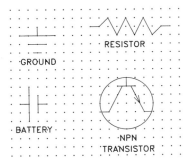

Figure 17–44B

CREATE BLOCK OUT OF 1-UNIT PROPORTIONS

Figure 17–45 shows a door block with its insertion point at the bottom of the line that represents the door. Rather than create each door block separately to account for different door sizes, create the door in Figure 17–45 to fit into a 1-unit by 1-unit square. The purpose of drawing the door block inside a 1-unit square is to create only one block of the door and insert it at a scale factor matching the required door size. For example, for a 2'-8" door, enter 2'8 or 32 when prompted for the X and Y scale factors. For a 3'-0" door, enter 3' or 36 when prompted for the scale factors. Numerous doors of different types can be inserted in a drawing using only one block of the door. Not all blocks can be treated this way; but it does apply in this case to doors.

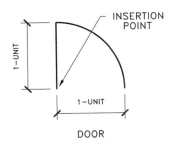

Figure 17–45

LISTING BLOCK INFORMATION

Using the LIST command on a block displays the following information:

BLOCK REFERENCE Layer: "0"
Space: Model space
Handle = 760
"RESISTOR"
at point, X=6.86 Y=2.33 Z=0.00

X scale factor 1.00
Y scale factor 1.00
rotation angle   0
Z scale factor 1.00

Listing a block identifies the object as a Block Reference. Notice that the name of the block is listed, along with insertion point, scale factors, and rotation angle.

THE AUTOCAD DESIGNCENTER

A revolutionary tool called the AutoCAD DesignCenter has been developed as a simple, efficient device for sharing drawing data. It provides a means of inserting blocks and drawing even more efficiently than through the Insert dialog box. This feature not only has the capability of inserting drawing files in another drawing, but it has the distinct advantage of inserting blocks internal to one drawing in another. If the DesignCenter is not present, you can load it by choosing AutoCAD DesignCenter from the Tools pull-down menu, as in Figure 17–46A, selecting the AutoCAD DesignCenter button from the Standard toolbar in Figure 17–46B or typing ADCENTER at the keyboard. In the original configuration, the DesignCenter loads docked to the left side of the AutoCAD drawing screen (see Figure 17–46C). The DesignCenter can be undocked and resized as the user requires and if additional drawing screen area is required, you can unload it by clicking the X in the upper right corner of the DesignCenter box, selecting DesignCenter from the Tools pull-down menu, clicking on the Design Center button on the Standard toolbar, or entering the ADCCLOSE command at the keyboard.

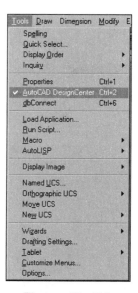

Figure 17–46A

AutoCAD DesignCenter

Figure 17–46B

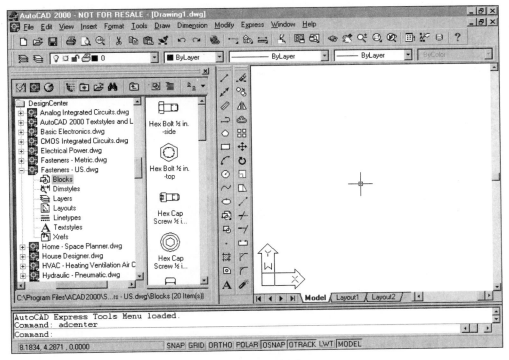

Figure 17–46C

Block libraries may be prepared in different formats in order for them to be used through the AutoCAD DesignCenter. One method is to place all blocks in one subdirectory. The AutoCAD DesignCenter identifies this subdirectory and graphically lists all drawing files to be inserted. Another method of organizing blocks is to create one drawing containing all blocks. When this drawing is identified through the AutoCAD DesignCenter, all blocks internal to this drawing display in the AutoCAD DesignCenter palette area.

AUTOCAD DESIGNCENTER COMPONENTS

The AutoCAD DesignCenter is isolated in Figure 17–47A. The following components of the dialog box are identified below and in the figure: Control buttons, Tree View or Navigation Pane, Palette or Content Pane, and Current Path.

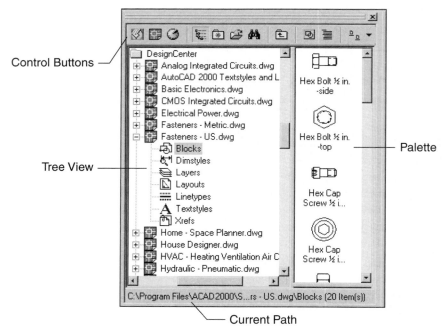

Figure 17–47A

A more detailed illustration of the DesignCenter Control buttons is found in Figure 17–47B. The following buttons are identified: Desktop, Open Drawings, History, Tree View Toggle, Favorites, Load, Find, Up, Preview, Descriptions, and Views. It may be necessary to resize the DesignCenter to see all the buttons.

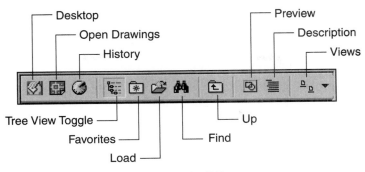

Figure 17–47B

In Figure 17–47A, the Desktop button has been selected to list the local and network drives in the tree view. If you select the Open Drawings button (see Figure 17–48), only the drawings that are currently open in AutoCAD will be displayed. The History button lists the last 20 locations accessed through the DesignCenter.

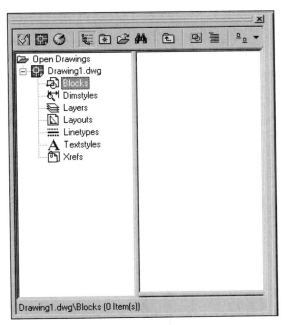

Figure 17–48

In Figure 17–47A, the Tree View Toggle button is selected to show all the folders in a typical Windows Explorer style. If the toggle is off, the Tree View is no longer displayed (see Figure 17–49). The Favorites, Load, Find, and Up buttons provide additional methods to assist with locating and selecting drawings.

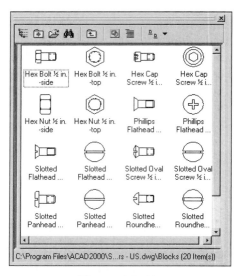

Figure 17–49

Clicking on the Preview button displays an enlarged version of the highlighted block located in the Palette area (see Figure 17–50). This gives you a better idea of what the block will look like once it is inserted in the drawing. The Description button displays a description for the selected item in the Palette area. The View button changes the display in the palette area (large icons, small icons, list or details). Large icons are displayed in Figure 17–50.

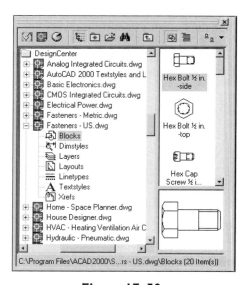

Figure 17–50

USING THE TREE VIEW

As demonstrated earlier, clicking the Tree View button expands the AutoCAD DesignCenter to look similar to the illustration in Figure 17–51. The AutoCAD DesignCenter divides into two major areas; on the right is the familiar Palette area; on the left is the subdirectory layout (tree view) of your hard drive. Clicking on a valid drive or subdirectory displays the contents in the right side of the AutoCAD DesignCenter (as shown in Figure 17–51). Clicking on a drawing in the Tree View displays the drawing objects, as shown in Figure 17–52, that can be shared through the DesignCenter (Blocks, Dimstyles, Layers, Layouts, Linetypes, Textstyles, and Xrefs). Double-click the Blocks icon (or click the "+" symbol and choose Blocks in the Tree View) to display the blocks available in the selected drawing (see Figure 17–53).

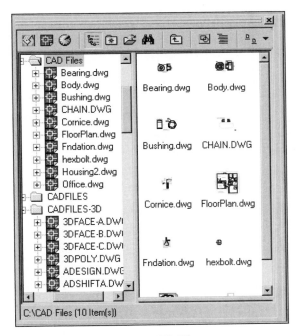

Figure 17–51

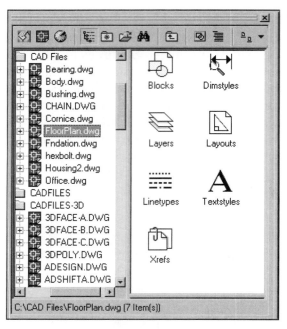

Figure 17–52

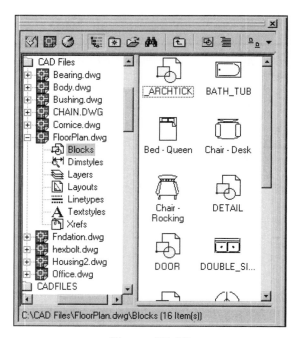

Figure 17–53

INSERTING BLOCKS THROUGH THE AUTOCAD DESIGNCENTER

Figure 17–54 displays a typical floor plan drawing along with the AutoCAD DesignCenter showing the blocks identified in the current drawing. If no blocks are found internal to the drawing, the Palette area will be empty.

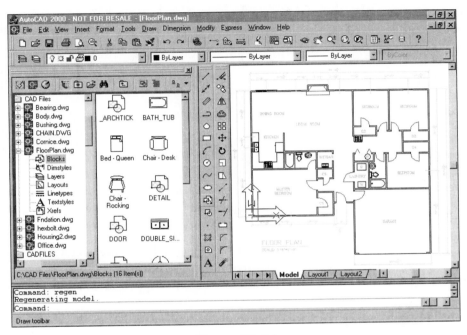

Figure 17–54

The AutoCAD DesignCenter operates on the "drag and drop" principle. Select the desired block located in the palette area of the DesignCenter, drag out and drop it into the desired location of the drawing. Figure 17–55 shows a queen-size bed dragged and dropped into the bedroom area of the floor plan.

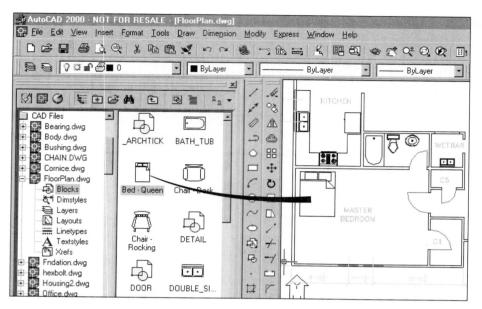

Figure 17–55

When performing a basic drag and drop operation with the left mouse key, all you have to do is identify where the block is located and drop it into that location (the use of running object snaps can ensure that the blocks are dropped in a specific location). What if the block needs to be scaled or rotated? If the drag and drop method is performed with the right mouse key, a shortcut menu is provided (see Figure 17–56A). Selecting Insert Block from the shortcut menu provides the Insert dialog box in Figure 17–56B, which allows you to specify different scale and rotation values. Figure 17–57 shows a rocking chair inserted and rotated into position in the corner of the wall. Instead of dragging the block with the right mouse key, double-click on the block in the palette area and the Insert dialog box will be provided immediately.

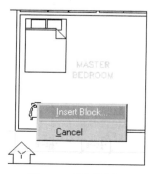

Figure 17–56A

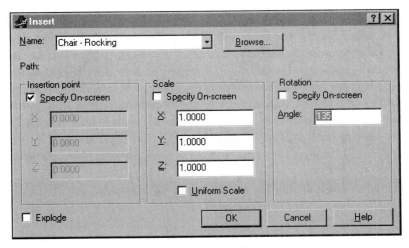

Figure 17–56B

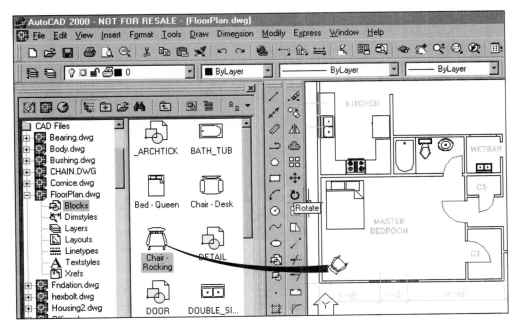

Figure 17–57

INSERTING GLOBAL BLOCKS THROUGH THE AUTOCAD DESIGNCENTER

Once a block or drawing file is available in the palette area of the AutoCAD DesignCenter, it can easily be inserted in the current AutoCAD drawing. In the Tree View area of the DesignCenter, select the appropriate drive or folder where your file is located. The drawing files in this folder will display in the palette area. Now drag the desired file from the palette area into the current AutoCAD drawing and drop it (see Figure 17–58). Drag and drop with the right mouse button, just as with blocks, to make the Insert dialog box available for changing the scale or rotation values.

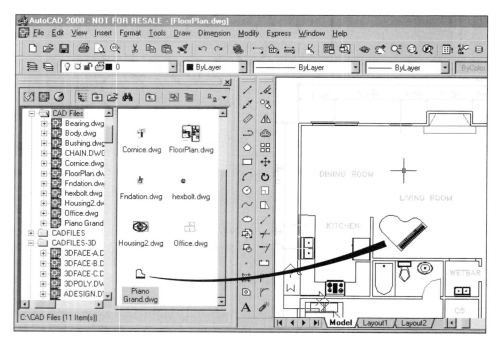

Figure 17–58

INSERTING BLOCKS INTERNAL TO OTHER DRAWINGS

Numerous times throughout the design process, you need to insert a block that belongs to another drawing file. This has not been a problem in past versions of AutoCAD as long as you realized you had to insert the whole drawing in order to get to the internal blocks. Then, you had to remember to use the PURGE command or risk the threat of increasing the size of the drawing database tremendously.

Inserting blocks internal to other drawings with the AutoCAD DesignCenter is again a simple drag and drop operation. The drawing selected in the DesignCenter Tree

View does not have to be the same drawing opened in AutoCAD. That means that any drawings blocks can be made available in the palette area and can be dragged and dropped into another drawing. In Figure 17–59, FloorPlan is the current drawing open in AutoCAD. Blocks from a Landscape drawing are inserted in the FloorPlan drawing, with the blocks dragged from the palette area and dropped into the FloorPlan drawing. The desired internal block can now be inserted in other drawings much more efficiently that in past versions of AutoCAD.

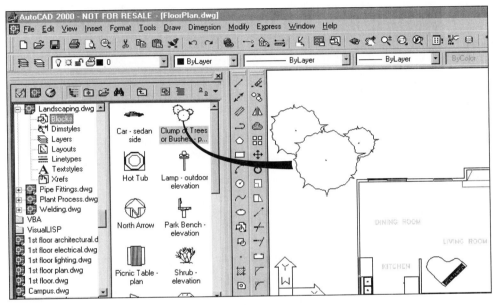

Figure 17–59

PERFORMING SEARCHES FOR INTERNAL BLOCKS

Sometimes, you know a block exists in a particular drawing file somewhere on the hard drive. You do not know the name of the drawing file; however, you think the name of the block is "COL" or "COLUMN."

AutoCAD has a remarkable search and find mechanism that can search out blocks that are internal to a drawing. Click on the Find button (Binoculars icon) in Figure 17–47B to display the Find: dialog box illustrated in Figure 17–60A. Select the type of object you are looking for. You may select "Block" if you know the exact block name. Select "Drawing" to have access to the Advanced tab. You can enter the name of the drawing file to search for if you know it. Enter an asterisk (wild card) to search all drawing file names. You also specify the drive and folder(s) to look in.

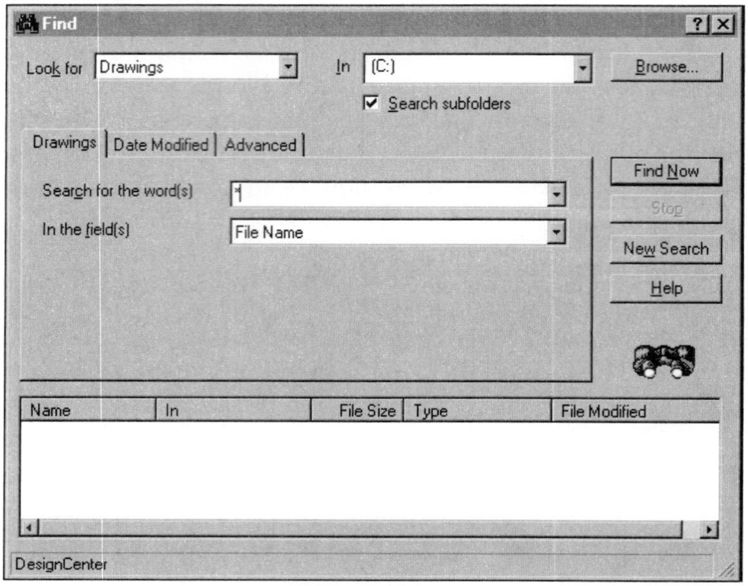

Figure 17–60A

If you do not know the name of the drawing file but have some idea of what the block name is, you can click on the Advanced tab in Figure 17–60A. This displays the Advanced tab dialog box in Figure 17–60B. Click in the Containing edit box and change the category to block name. Click in the Containing text edit box and add the letters "COL." Click the Find Now button, and AutoCAD will search all drawing files having internal block names beginning with "COL."

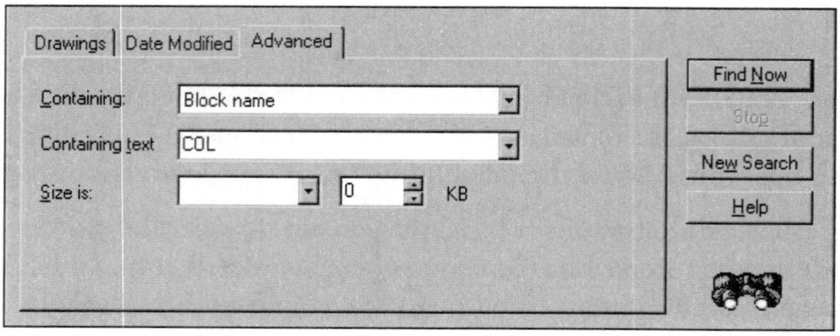

Figure 17–60B

Figure 17–60C displays the results of the search on a block name beginning with "COL." Seven drawings have been identified in the edit box at the bottom of the AutoCAD Find: dialog box.

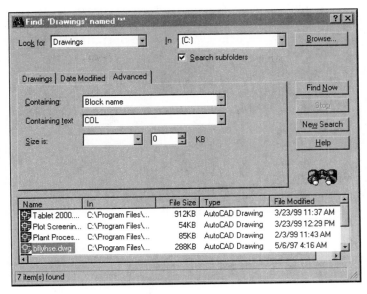

Figure 17–60C

Double-clicking on the drawing name at the bottom of the AutoCAD Find: dialog box in Figure 17–60C expands the AutoCAD DesignCenter. Select the Blocks object type (see Figure 17–61) to display all internal blocks in the drawing. The block "COL" can now be dragged and dropped into the current drawing file.

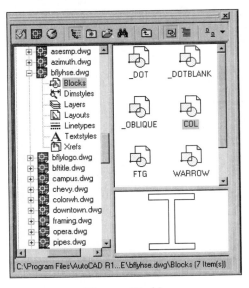

Figure 17–61

MDE – MULTIPLE DESIGN ENVIRONMENT

The Multiple Design Environment allow users to open multiple drawings within a single session of AutoCAD (see Figure 17–62). This feature, like AutoCAD DesignCenter, allows sharing of data between drawings. You can easily copy and move objects, such as blocks, from one drawing to another.

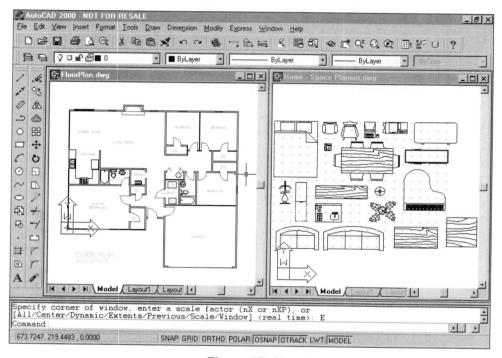

Figure 17–62

OPENING MULTIPLE DRAWINGS

Repeat the OPEN command as many times as necessary to open all drawings you will need. In fact, you can select multiple drawings in the Select File dialog box by holding down CTRL or SHIFT as you select the files (see Figure 17–63).

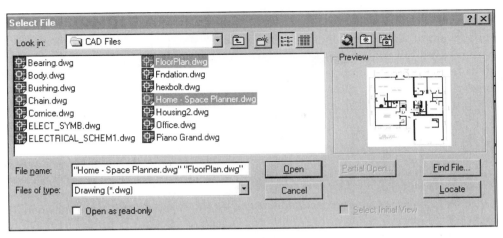

Figure 17–63

Once the drawings are open, use CTRL+F6 or CTRL+TAB to switch back and forth between the drawings. To efficiently work between drawings, you may wish to tile (see Figure 17–62) or cascade (see Figure 17–64A) the drawing windows utilizing the Windows pull-down menu (see Figure 17–64B). Remember to use the CLOSE command to close any drawings that are not being used.

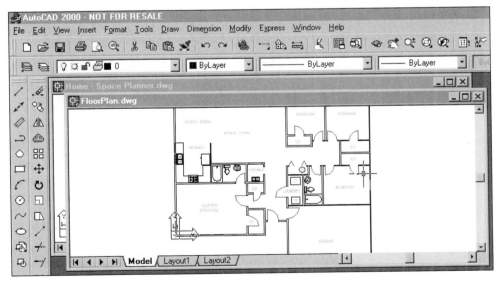

Figure 17–64A

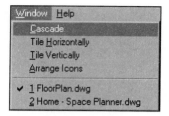

Figure 17–64B

WORKING BETWEEN DRAWINGS

Once your drawings are opened and arranged on the screen, you are now ready to cut and paste, copy and paste, or drag and drop objects between drawings. The first step in any cut and paste or copy and paste operation is to cut or copy objects from a drawing. The object information is stored on the Windows clipboard until you are ready for the second step, which is to paste the objects into that same drawing or any other open drawing. These operations are not limited to AutoCAD. In fact, you can cut, copy and paste between different Windows applications.

Use one of the following commands to cut and copy your objects:

CUTCLIP	(to remove selected objects from a drawing and store them on the clipboard)
COPYCLIP	(to copy selected objects from a drawing and store them on the clipboard)
COPYBASE	(similar to the COPYCLIP command but allows the selection of a base point for locating your objects when they are pasted)

Use one of the following commands to paste your objects:

PASTECLIP	(pastes the objects at the location selected)
PASTEBLOCK	(similar to PASTECLIP command but objects are inserted as a block and an arbitrary block name is assigned)

The commands listed can be typed at the keyboard, selected from the Edit pull down menu (see Figure 17–65A), selected from the Standard toolbar (see Figure 17–65B), or selected from a shortcut menu (activated by a right-click in the drawing screen area, see Figure 17–65C).

Figure 17–65A

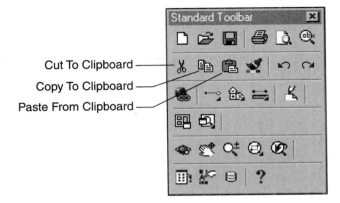

Figure 17–65B

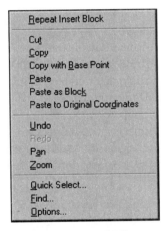

Figure 17–65C

Objects may also be copied between drawings with drag and drop operations. After selecting the objects, place the cursor over the objects (without selecting a grip) and then drag and drop the objects in the new location. Dragging with the right mouse key depressed will provide a shortcut menu allowing additional control over pasting operations (see Figure 17–66).

Figure 17–66

TUTORIAL EXERCISE: ELECTRICAL SCHEMATIC

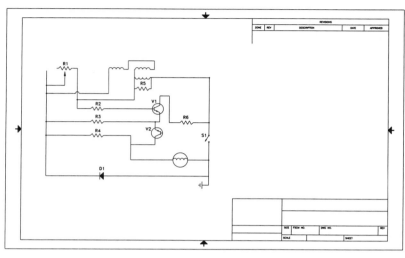

Figure 17–67

Purpose

This tutorial is designed to lay out electrical blocks such as resistors, transistors, diodes, and switches to form an electrical schematic.

System Settings

Create a new drawing called ELECT_SYMB to hold the electrical blocks. Keep all default units and limits settings. Set the GRID and SNAP commands to 0.0750 units. This will be used to assist in the layout of the blocks. Once this drawing is finished, create another new drawing called ELECTRICAL_SCHEM1. This drawing will show the layout of an electrical schematic. Keep the default units but use the LIMITS command to set the upper right corner of the display screen to (17.0000,11.0000). Grid and snap do not have to be set for this drawing.

Layers

Create the following layer for ELECT_SYMB with the format:

Name	Color	Linetype
0	White	Continuous

Create the following layers for ELECTRICAL_SCHEM1 with the format:

Name	Color	Linetype
Blocks	White	Continuous
Wires	White	Continuous

Suggested Commands

Begin this tutorial by creating a new drawing called ELECT_SYMB, which will hold all electrical blocks. Use Figure 17–68 as a guide in drawing all electrical blocks. A grid with spacing 0.0750 would provide further assistance with the drawing of the blocks. Once all blocks are drawn, the BLOCK command is used to create blocks out of the individual blocks. Save this drawing and create a new drawing file called ELECTRICAL_SCHEM1. Use the AutoCAD DesignCenter to drag and drop the internal blocks from the drawing ELECT_SYMB into the new drawing ELECTRICAL_SCHEM1. Connect all blocks with lines that represent wires and electrical connections. Add block identifiers, insert a B-size title block, and save the drawing.

Whenever possible, substitute the appropriate command alias in place of the full AutoCAD command in each tutorial step. For example, use "CP" for the COPY command, "L" for the LINE command, and so on. The complete listing of all command aliases is located in Chapter 1, Table 1–2.

STEP 1

Begin a new drawing and call it ELECT_SYMB. Be sure to save this drawing to a convenient location. It will be used along with the AutoCAD DesignCenter to bring the internal blocks into another drawing file.

STEP 2

Using a grid of 0.075 units, construct each block using Figure 17–68 as a guide. The grid will help keep all blocks proportional to each other. Also, the "X" located on each block signifies its insertion point. Another hint: create all blocks on neutral layer 0.

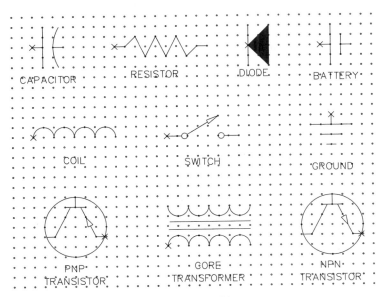

Figure 17–68

STEP 3

For a more detailed approach to creating the blocks, the resistor block will be created through the use of the BLOCK command. Using Figure 17–69A as a guide, first construct the resistor using the grid to connect points until the resistor is created. Be sure Snap mode is active for this procedure. Then issue the BLOCK command, which will bring up the dialog box illustrated in Figure 17–69B. This will be an internal block to the drawing ELECT_SYMB; name the block RESISTOR. In the Select objects area of the dialog box, select all lines that represent the resistor. Returning to the dialog box will show the objects selected in a small image icon on the right of the dialog box. Next, identify the insertion point of the resistor at the "X" located in Figure 17–69A. It is very important to use an Object Snap mode when picking the insertion point. Finally, add a description for the block. When finished with this dialog box, click on the OK button to create the internal block. Repeat this procedure for the remainder of the blocks in Figure 17–68.

Figure 17–69A

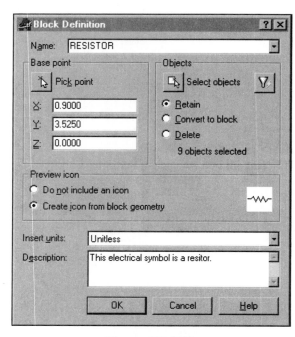

Figure 17–69B

STEP 4

Begin another new drawing called ELECTRICAL_SCHEM1. With the blocks identified inside the drawing named ELECT_SYMB, activate the AutoCAD DesignCenter. Click on the Tree View icon and expand the AutoCAD DesignCenter to include the subdirectory system of your hard drive.

Locate the correct subdirectory the drawing was saved to and click on it. Double-click on ELECT_SYMB and select Blocks to display the internal blocks illustrated in Figure 17–70. These internal blocks can now be inserted in any type of AutoCAD drawing file.

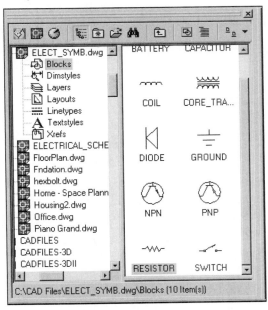

Figure 17–70

STEP 5

If you cannot find the drawing file, an alternate step would be to click on the Find button that has the image of the Binoculars in the AutoCAD DesignCenter. This will activate the AutoCAD Find: dialog box. Enter ELECT_SYMB.DWG for the Drawing name and click on the Find Now button. The result, illustrated in Figure 17–71A, shows the subdirectory

location of the drawing file. Double-clicking on this drawing file expands the AutoCAD DesignCenter to display the subdirectory location of the drawing file. The DesignCenter will display ELECT_SYMB.DWG in the Tree View. Double-clicking on this drawing file and selecting Blocks will display all blocks internal to it (see Figure 17–71B.)

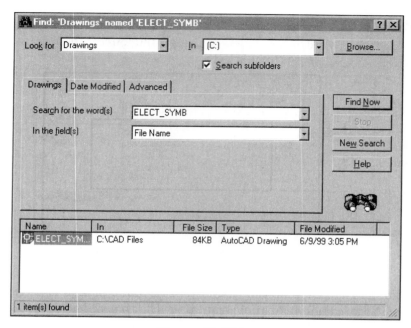

Figure 17–71A

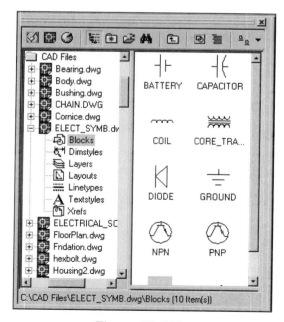

Figure 17–71B

STEP 6

Begin dragging and dropping the blocks into the drawing in Figure 17–72. In the case of the electrical schematic, it is not critical to exactly place the block, because they are moved to better locations when connected to lines that represent wires.

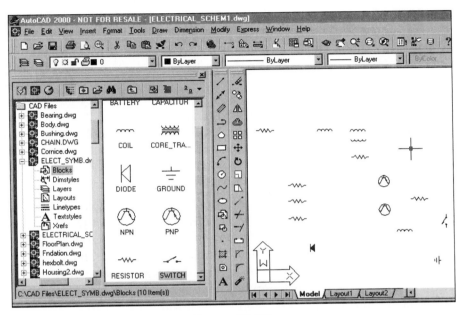

Figure 17–72

STEP 7

Use Figure 17–73 as a guide for connecting all blocks to form the schematic. Some type of Object Snap mode such as Endpoint must be used for the wire lines to connect exactly with the endpoints of the blocks. If spaces get tight or if the blocks look too crowded, move the block to a better location and continue drawing lines to form the schematic.

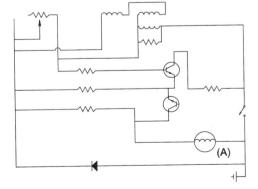

Figure 17–73

STEP 8

Add text identifying all resistors, transistors, diodes, and switches by number. The text height used was 0.12; however, this could vary depending on the overall size of your schematic. Also, certain areas of the schematic show connections with the presence of a dot at the intersection of the connection. Use the DONUT command with an inside diameter of 0.00 and an outside diameter of 0.025. Place donuts in all locations shown in Figure 17–74.

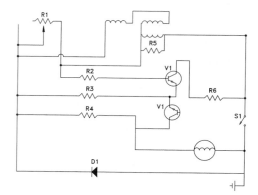

Figure 17–74

STEP 9

Complete the drawing of the electrical schematic by inserting a B-size title block. Use the Insert dialog box in Figure 17–75A to accomplish this. Click on the Browse button to expose the subdirectory system. AutoCAD is supplied with numerous title blocks in the "template" subdirectory in Figure 17–75B. Choose the file "ANSI B title block" as the title block to insert. After returning to the main Insert dialog box, click inside the box next to "Specify on Screen" to remove the check. Because this is a title block, the default insertion point, scale, and rotation values will be used, which will place the title block at an insertion point of 0,0.

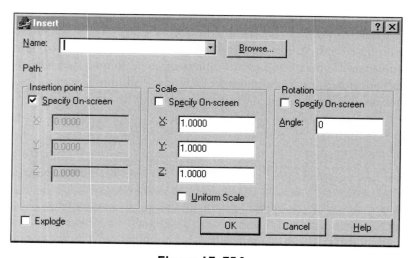

Figure 17–75A

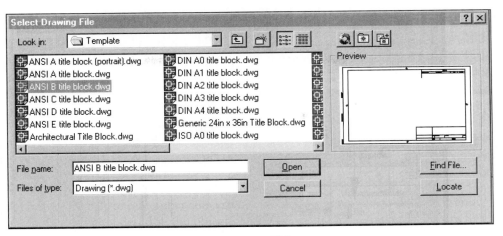

Figure 17–75B

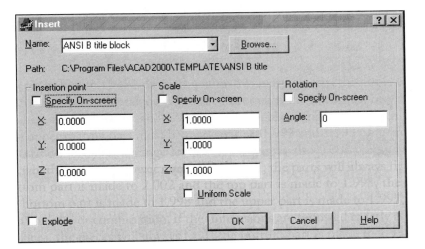

Figure 17–75C

STEP 10

If, after the title block is inserted, the electrical schematic is not centered, move the schematic to a better location. Save the drawing under the name ELECTRICAL_SCHEM1. Your image should appear similar to Figure 17–76.

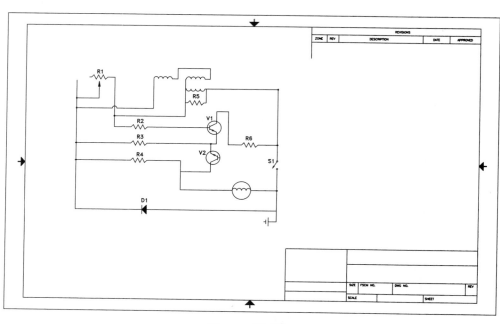

Figure 17–76

PROBLEMS FOR CHAPTER 17

Directions for Problem 17–1

Construct the floor plan of the house using the dimensions provided. Create the appropriate layers to separate the walls from the dimensions, etc. Once the floor plan is completed, create an interior plan consisting of furniture and appliances. Place the blocks in their appropriate spaces using the designated room titles. Use the following block libraries supplied with AutoCAD 2000: House Designer.Dwg, Home – Space Planner.Dwg, and Kitchens.Dwg.

Follow the following notes for the creation of doors and windows:

All windows measure 2'-8" wide

All bedroom doors measure 2'-6" wide

The bathroom door measures 2'-0" wide

The main entrance door measures 3'-0" wide

The kitchen door measures 2'-8" wide

The laundry door measures 2'-8" wide

All closet openings measure 4'-0" wide

All interior walls measure 4"

The exterior brick veneer measures 5"

problem EXERCISE

PROBLEM 17–1

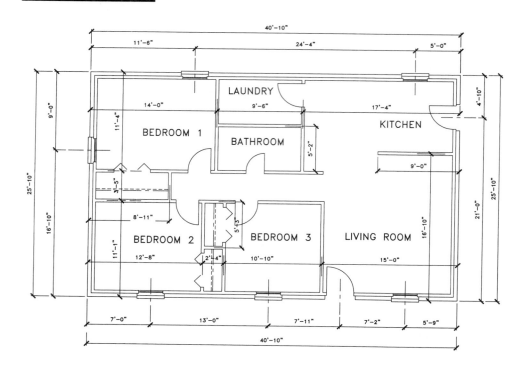

FLOOR PLAN
1/4" = 1'-0"

Directions for Problem 17–2

Construct the first battery pack using the following block library supplied with AutoCAD 2000: Basic Electronics.Dwg. This drawing is not to scale.

PROBLEM 17–2

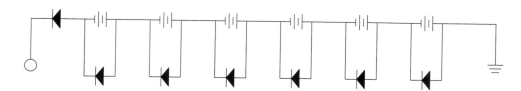

Directions for Problem 17–3

Construct the second battery pack using the following block library supplied with AutoCAD 2000: Basic Electronics.Dwg. This drawing is not to scale.

PROBLEM 17–3

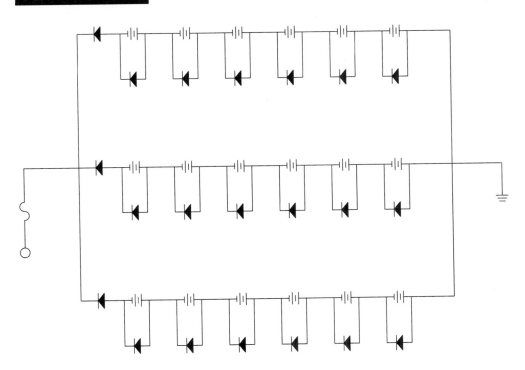

Directions for Problem 17–4

Construct the instrumentation pipe diagram using the following block library supplied with AutoCAD 2000: Pipe Fittings.Dwg. This drawing is not to scale.

PROBLEM 17–4

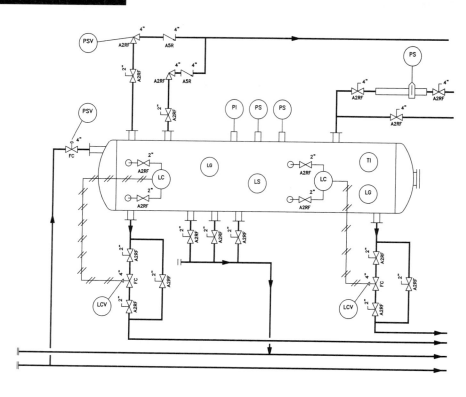

Directions for Problem 17–5

Construct the electrical schematic using the following block library supplied with AutoCAD 2000: Basic Electronics.Dwg. This drawing is not to scale.

PROBLEM 17–5

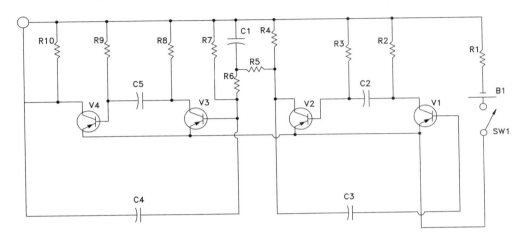

Directions for Problem 17–6

Construct the electrical schematic using the following block library supplied with AutoCAD 2000: Basic Electronics.Dwg. This drawing is not to scale.

PROBLEM 17–6

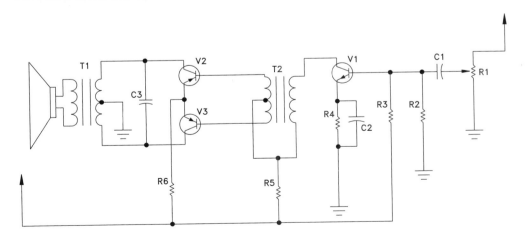

Directions for Problem 17–7

Construct the logic gate schematic using the following block libraries supplied with AutoCAD 2000: CMOS Integrated Circuits.DWG and Basic Electronics.Dwg. This drawing is not to scale.

PROBLEM 17–7

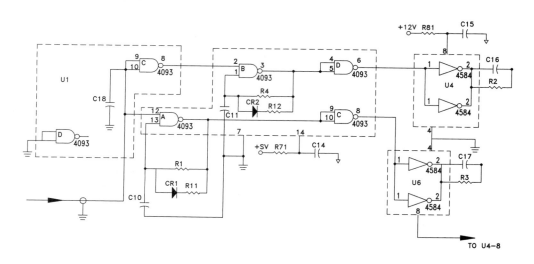

External References

This chapter begins the study of External References and how they differ from blocks. When changes are made to a drawing that has been externally referenced into another drawing, these changes update automatically. There is no need to redefine the external reference in the same way blocks are redefined. Finally, using the AutoCAD DesignCenter to attach external references will be discussed. As an added bonus to AutoCAD users, a series of sample symbol libraries is supplied with the package. These symbol libraries include such application areas as mechanical, architectural, electrical, piping, and welding, just to name a few.

EXTERNAL REFERENCES AND BLOCKS

It has been seen so far in Chapter 17 that it is very popular to insert blocks and drawing files in other drawing files. The advantages in performing these insertions include smaller file size, because all objects connected with the block are grouped into a single object. It is only when blocks are exploded that the drawing size once again increases because the objects that made up the block have been separated into their individual objects. Another advantage of blocks is their ability to be redefined in a drawing.

External references are similar to blocks in that they appear to be considered one object. However, external references are attached to the drawing file, whereas blocks are inserted in the drawing. This attachment actually sets up a relationship between the current drawing file and the external referenced drawing. For example, take a floor plan and externally reference it into a current drawing file. All objects associated with the floor plan are brought in; objects also carry their current layer and linetype qualities. With the floor plan acting as a guide, such items as electrical symbols, furniture, and dimensions can be added to the floor plan and saved to the current drawing file. Now a design change needs to be made to the floor plan. You open the floor plan, stretch a few doors and walls to new locations, and save the file. The next time the drawing holding the electrical symbols and furniture arrangement is opened, the changes to the floor plan are made automatically to the current drawing file. This is one of the primary advantages of external references over blocks. Also, blocks are known to reduce the file size; however using external references reduces the file size even more than using blocks. Items such as layers and other blocks that belong to the external reference can be viewed but they have limited possibilities for manipulation.

CHOOSING EXTERNAL REFERENCE COMMANDS

One way to select external reference commands is from the toolbars. Select External Reference from the Insert or Reference toolbar, illustrated in Figures 18–1A and 18–1B. Yet another way is through the keyboard by entering XREF or XR. A third way is to choose Xref Manager from the Insert pull-down menu, as shown in Figure 18–2A. Whichever method you use, the Xref Manager dialog box displays, as illustrated in Figure 18–2B.

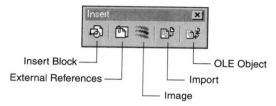

Figure 18–1A

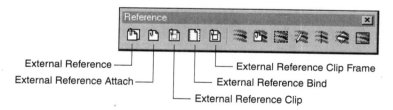

Figure 18–1B

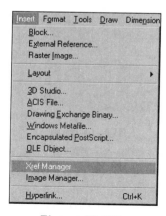

Figure 18–2A

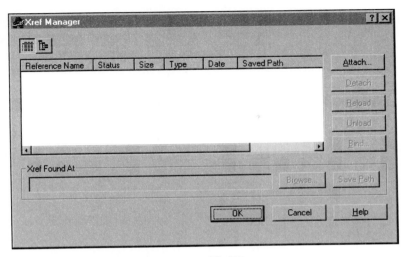

Figure 18–2B

ATTACHING AN EXTERNAL REFERENCE

The primary mode in the Xref Manager dialog box in Figure 18–2B is the Attach option. This can be compared with the INSERT command for merging blocks into drawing files. The Attach option sets up a path, which looks for the external reference every time the drawing containing it is loaded. Clicking on the Attach button, shown in Figure 18–2B, activates the Select Reference File dialog box illustrated in Figure 18–3. This dialog box is very similar to the one for selecting drawing files to initially load into AutoCAD.

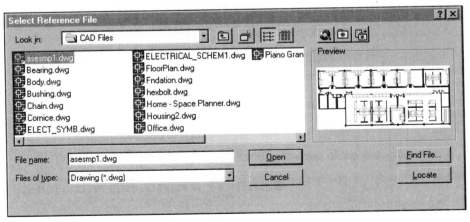

Figure 18–3

Once the desired file is found, the External Reference dialog box appears, as in Figure 18–4. Some of the information contained in this dialog box is similar to the information in the Insert dialog box. The insertion point, scale and rotation angle of the external reference can be specified in this dialog box or on the screen. Of importance in the upper part of the dialog box is the path information associated with the external reference. If, during file management, an externally referenced file is moved to a new location, the new path of the external reference must be re-established. Otherwise it does not load into the drawing it was attached to. Clicking the OK button returns you to the drawing editor. Identifying an insertion point for the external reference attaches the file to the current drawing (in Figure 18–4, the Insertion point area's Specify on-screen box is checked).

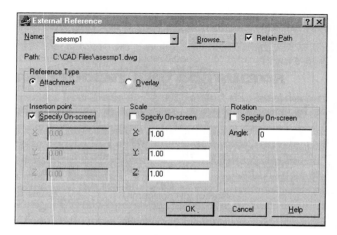

Figure 18–4

The resulting operation is illustrated in Figure 18–5.

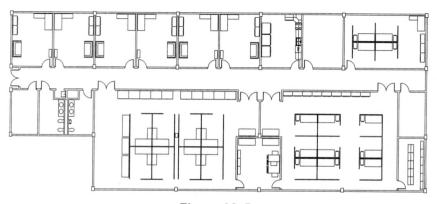

Figure 18–5

With the external reference attached to the drawing file, layers and blocks that belong to the external reference take on a new meaning. Illustrated in Figure 18–6 are the current layers that are part of the drawing. However, notice how a number of layers begin with the same name (ASESMP1); also, what appears to be a vertical bar separates ASESMP1 from the names of the layers. Actually, the vertical bar represents the "Pipe" symbol on the keyboard, and it designates that the layers belong to the external reference, namely ASESMP1. These layers can be turned off, frozen, or even have the color changed. However, you cannot make this layer current for drawing on because it belongs to the external reference. Also, any changes to the layers are only good while the drawing is displayed. Once the drawing is reopened and the external automatically loaded, the layers will revert to their original state, which is governed by the external reference.

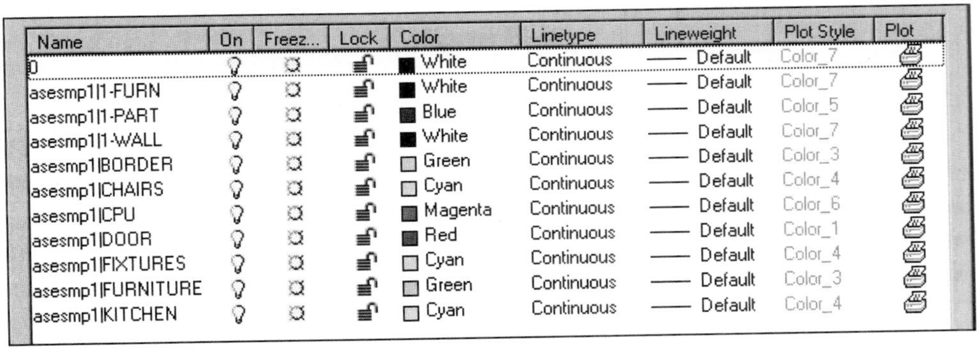

Name	On	Freez...	Lock	Color	Linetype	Lineweight	Plot Style	Plot
0	♀	♻	📭	■ White	Continuous	—— Default	Color_7	🖨
asesmp1|1-FURN	♀	♻	📭	■ White	Continuous	—— Default	Color_7	🖨
asesmp1|1-PART	♀	♻	📭	■ Blue	Continuous	—— Default	Color_5	🖨
asesmp1|1-WALL	♀	♻	📭	■ White	Continuous	—— Default	Color_7	🖨
asesmp1|BORDER	♀	♻	📭	☐ Green	Continuous	—— Default	Color_3	🖨
asesmp1|CHAIRS	♀	♻	📭	☐ Cyan	Continuous	—— Default	Color_4	🖨
asesmp1|CPU	♀	♻	📭	■ Magenta	Continuous	—— Default	Color_6	🖨
asesmp1|DOOR	♀	♻	📭	■ Red	Continuous	—— Default	Color_1	🖨
asesmp1|FIXTURES	♀	♻	📭	☐ Cyan	Continuous	—— Default	Color_4	🖨
asesmp1|FURNITURE	♀	♻	📭	☐ Green	Continuous	—— Default	Color_3	🖨
asesmp1|KITCHEN	♀	♻	📭	☐ Cyan	Continuous	—— Default	Color_4	🖨

Figure 18–6

Just as layers are affected by the presence of an external reference, blocks display the same behavior. When blocks are listed, as in Figure 18–7, the name of the external reference is first given, with the "Pipe" symbol following, and finally the actual name of the block, as in ASESMP1|DESK2. As with layers, these blocks cannot be used in the drawing because of the presence of the "Pipe" symbol, which is not supported in the name of the block. This does, however, provide a quick way of identifying the valid blocks and those belonging to an external reference. Study the following prompt sequence and Figure 18–7 for identifying blocks that belong to external references.

Command: **-BLOCK** *(The "-" preceding the command provides a command line version of the block command)*
Enter block name or [?]: **?**
Enter block(s) to list <*>: *(Press ENTER to accept the default and display the list in Figure 18–7)*

```
Command: -block
Enter block name or [?]: ?
Enter block(s) to list <*>: (Press Enter to accept default)

Defined blocks.
  "asesmp1"                       Xref: resolved
  "asesmp1|ADCADD_ZZ"             Xdep: "asesmp1"
  "asesmp1|CB30"                  Xdep: "asesmp1"
  "asesmp1|CB36"                  Xdep: "asesmp1"
  "asesmp1|CC30"                  Xdep: "asesmp1"
  "asesmp1|CCBASE"                Xdep: "asesmp1"
  "asesmp1|CHAIR7"                Xdep: "asesmp1"
  "asesmp1|DESK2"                 Xdep: "asesmp1"
  "asesmp1|DESK3"                 Xdep: "asesmp1"
  "asesmp1|DESK4"                 Xdep: "asesmp1"
  "asesmp1|FC15X27A"              Xdep: "asesmp1"
  "asesmp1|FC42X18D"              Xdep: "asesmp1"
  "asesmp1|FNCASE3"               Xdep: "asesmp1"
  "asesmp1|PNL24X60"              Xdep: "asesmp1"
  "asesmp1|PNL36X60"              Xdep: "asesmp1"
  "asesmp1|PNL48X60"              Xdep: "asesmp1"
  "asesmp1|RANGE"                 Xdep: "asesmp1"
  "asesmp1|RECTANG"               Xdep: "asesmp1"
  "asesmp1|VENDING"               Xdep: "asesmp1"

User        External      Dependent     Unnamed
Blocks      References    Blocks        Blocks
  0             1           18             0
```

Figure 18–7

One of the real advantages of using external references is the way they affect drawing changes. In Figure 18–8, the original drawing file ASESMP1 is brought up again and changes made to it. Notice how room tags and chairs have been added to the floor plan. These changes are saved and the drawing containing the external reference is opened.

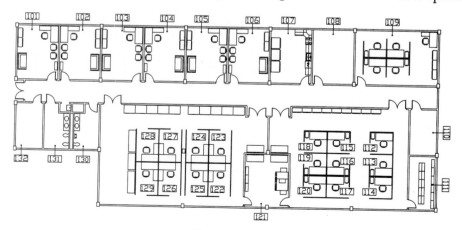

Figure 18–8

Once the drawing holding the external reference is loaded, notice how all chairs and room tags, which belong to the external reference, are automatically displayed.

THE XBIND COMMAND

Earlier, it was mentioned that blocks and layers belonging to external references cannot be used in the drawing they were externally referenced into. However, there is a way to convert a block or layer to a useful object through the XBIND command. Select this command from the Reference toolbar (in Figure 18-1B) or choose Object from the Modify pull-down menu and then External Reference and finally Bind, as in Figure 18–9A. This activates the Xbind dialog box, which lists the external reference ASESMP1. Clicking on the "+" sign expands the listing of all named objects such as blocks and layers associated with the external reference. Clicking on the "+" sign in Block lists all individual blocks associated with the external reference (see Figure 18–9B).

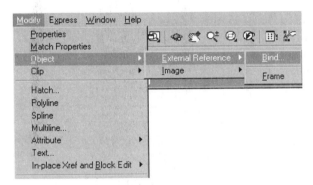

Figure 18–9A

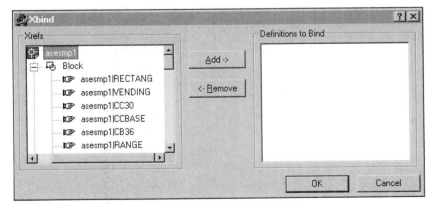

Figure 18–9B

Click on the name of the block in the listing on the left; then click on the Add-> button. This moves the block name over to the right under the listing of Definitions to Bind. Two block definitions have already been added and ASESMP1|DESK3 will be the third item added to the list of items to bind. When finished selecting these items, click on the OK button to bind the blocks to the current drawing file (see Figure 18–10).

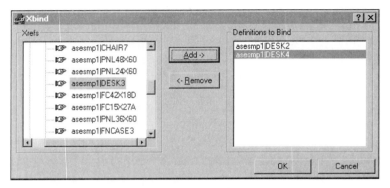

Figure 18–10

Test to see that new blocks have in fact been bound to the current drawing file. Activate the Insert dialog box, click on the Name drop-down list box, and notice the display of the blocks, as in Figure 18–11. The three symbols just bound from the external reference still have the name of the external reference; namely ASESMP1. However instead of the "Pipe" symbol separating the name of the external reference and block names, the characters 0 have been added in place of the "Pipe" symbol. This is what makes the blocks valid in the drawing. Now these three blocks can be inserted in the drawing file even though they used to belong to the external reference ASESMP1.

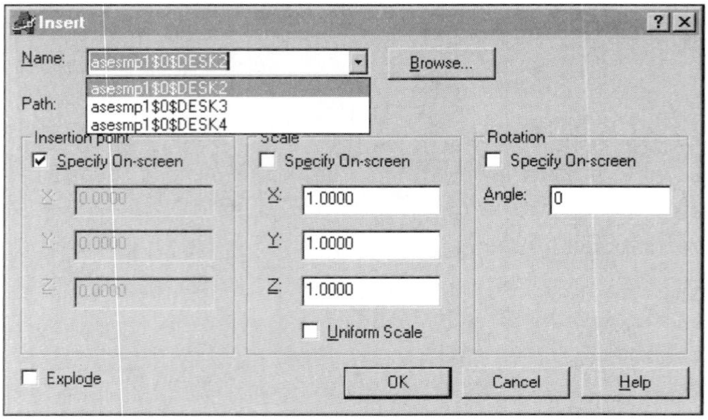

Figure 18–11

IN-PLACE REFERENCE EDITING

In previous releases of AutoCAD, you were required to open the referenced drawing (ASESMP1.DWG in our example) to make any changes. With "in-place reference editing" you can edit a reference drawing from the current drawing externally referencing it. You can access the REFEDIT command, which makes this possible, from the Refedit toolbar (see Figure 18–12) or by choosing In-place Xref and Block Edit from the Modify pull down menu, followed by Edit Reference (see Figure 18–13A).

Edit Block or Xref

Figure 18–12

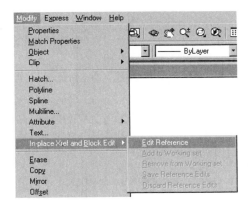

Figure 18–13A

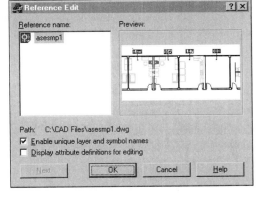

Figure 18–13B

Once in the REFEDIT command, you are prompted to select the reference you wish to modify. Next, the Reference Edit dialog box is displayed, as in Figure 18–13B. The reference to be edited, which is ASESMP1 in our case, should be selected. Nested references may also be displayed; one of these could be selected for editing instead, if desired. Clicking on the OK button returns you to the screen and prompts you to select nested objects. The kitchen cabinets and appliances in Figure 18–14 are selected. The selection set chosen makes up a "working set" of objects, which can be edited. Remaining objects are grayed out and cannot be modified. The Refedit toolbar is automatically displayed (see Figure 18–15) and you are returned to the command prompt. You can now complete modifications to the kitchen objects.

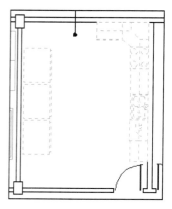

Figure 18–14

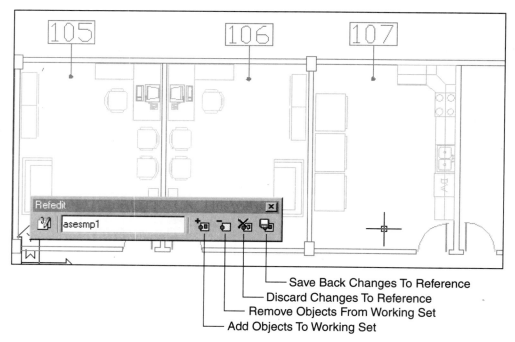

Figure 18–15

When satisfied with the changes, select the Save Back Changes to Reference button on the toobar. An AutoCAD alert box (see Figure 18–16) will ask you to confirm the saving of reference changes. After you click OK, the results can be seen in Figure 18–17.

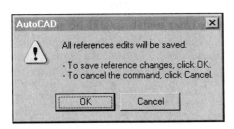

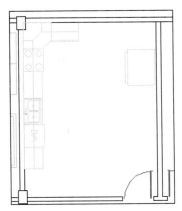

Figure 18–16

Figure 18–17

OTHER OPTIONS OF THE XREF COMMAND

The following additional options of the External Reference dialog box will now be discussed, using Figure 18–18 as a guide:

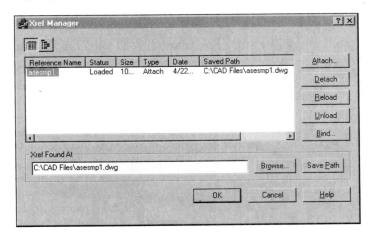

Figure 18–18

DETACH

Use this option to permanently detach or remove an external reference from the database of a drawing.

RELOAD

This option loads the most current version of an external reference. It is used when changes to the external reference are made while the external reference is currently being used in another drawing file. This option works well in a networked environment where all files reside on a file server.

UNLOAD

Unload is similar to the Detach option with the exception that the external reference is not permanently removed from the database of the drawing file. Since this option suppresses the external reference from any drawing regenerations, it is used as a productivity technique.

BIND

This option activates the Bind Xrefs dialog box illustrated in Figure 18–19. Two options are available inside this dialog box: Bind and Insert.

The Bind option binds to the current drawing file all blocks, layers, dimension styles, etc. that belonged to an external reference. After you perform this operation, layers can be drawn on and blocks inserted in the drawing. For example, a typical block definition belonging to an external reference is listed in the symbol table as XREFname|BLOCKname. Once the external reference is bound, all block definitions are converted to XREFname\$0\$BLOCKname. The same naming convention is true for layers, dimension styles, and other named items. A bound external reference is similar to a drawing that was inserted in another drawing.

The Insert option of the Bind Xref dialog box in Figure 18–19 is similar to the Bind option. However instead of named items such as blocks and layers being converted to the format XREFname\$0\$BLOCKname, the name of the external reference is stripped, leaving just the name of the block, layer, or other named item (BLOCKname, LAYERname, etc.)

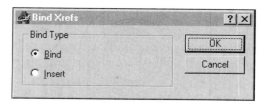

Figure 18–19

The Xref Manager dialog box makes it easy to keep track of external references in a drawing. By default, external references are listed as shown in Figure 18–20A. Clicking on the Tree View button lists external references in hierarchical form similar to the illustration in Figure 18–20B.

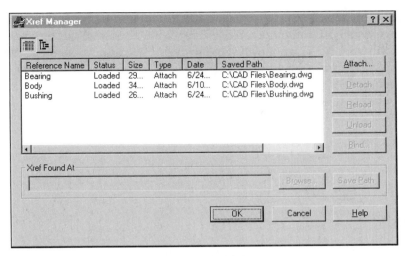

Figure 18–20A

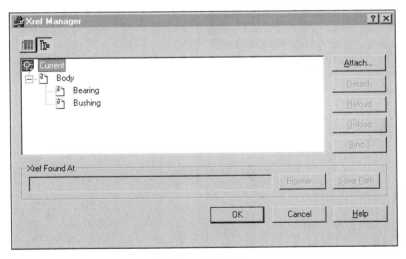

Figure 18–20B

ATTACHING EXTERNAL REFERENCES WITH AUTOCAD DESIGNCENTER

In Chapter 17, the AutoCAD DesignCenter was used to insert blocks and drawing files in the current drawing file. Just as blocks can be inserted, so also can External References be attached to a drawing file. The same rules apply for external references as for blocks. In the AutoCAD DesignCenter, select the desired subdirectory in the tree view (the left column) to load the palette (the right column) with the file you want to externally reference (see Figure 18–21). Using the right mouse key, drag and drop the file symbol into the drawing. The shortcut menu shown in Figure 18–22A lets you insert the file as a block or attach it as an external reference. Clicking on Attach as Xref will display the External Reference dialog box in Figure 18-22B. Click OK and select an insertion point to complete attaching the external reference to the drawing.

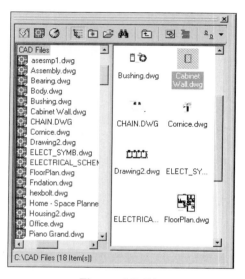

Figure 18–21

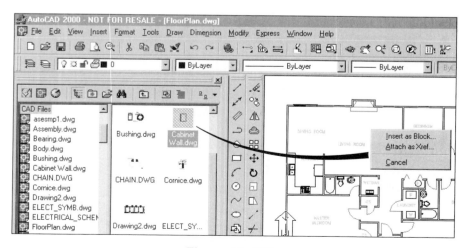

Figure 18–22A

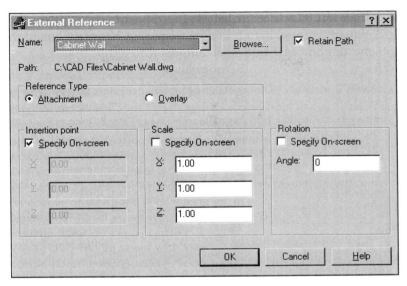

Figure 18–22B

TUTORIAL EXERCISE: EXTERNAL REFERENCES

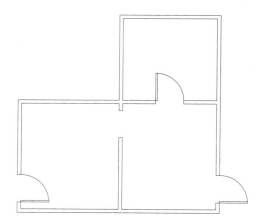

Figure 18–23A

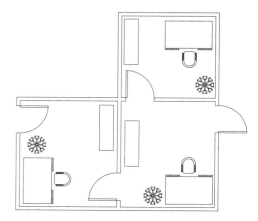

Figure 18–23B

Purpose

This tutorial is designed to use the office floor plan in Figure 18–23A to create an interior plan in Figure 18–23B consisting of various interior symbols, such as desks, chairs, shelves, and plants. The office floor plan will be referenced into another drawing file through the External Reference dialog box.

System Settings

Since these drawings are provided on the CD, all system settings have been made.

Layers

The creation of layers is not necessary, because layers already exist for both drawing files you will be working on.

Suggested Commands

Begin this tutorial by opening the drawing Office.Dwg, which should be located on the CD, and view its layers and internal block definitions. Then open the drawing Interiors.Dwg, also located on the CD, and

view its layers and internal blocks. The file Office.Dwg will now be attached to Interiors.Dwg. Once this is accomplished, chairs, desks, shelves, and plants will be inserted in the office floor plan for laying out the office furniture. Once Interiors.Dwg is saved, a design change needs to be made to the original office plan; open Office.Dwg and stretch a few doors to new locations. Save this file and open Interiors.Dwg; notice how the changes to the doors are automatically made. The Xbind dialog box will also be shown as a means for making a symbol that had previously belonged to an external reference usable in the file Interiors.Dwg.

Whenever possible, substitute the appropriate command alias in place of the full AutoCAD command in each tutorial step. For example, use "CP" for the COPY command, "L" for the LINE command, and so on. The complete listing of all command aliases is located in Chapter 1, Table 1–2.

STEP 1

Open Office.Dwg, which can be found on the CD, and observe a simple floor plan consisting of three rooms. Furniture will be laid out using the floor plan as a template (see Figure 18–24).

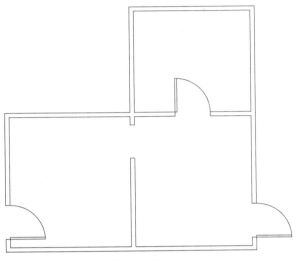

Figure 18–24

STEP 2

While in Office.Dwg, use the Layer Properties Manager dialog box and observe the layers that exist in the drawing for such items as doors, walls, and floor (see Figure 18–25). These layers will change once the office plan is attached to another drawing through the XREF command.

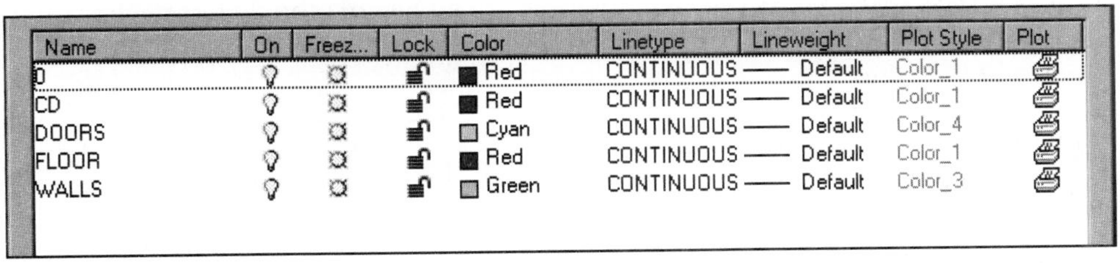

Name	On	Freez..	Lock	Color	Linetype	Lineweight	Plot Style	Plot
0	♀	♡	🔓	■ Red	CONTINUOUS	—— Default	Color_1	🖨
CD	♀	♡	🔓	■ Red	CONTINUOUS	—— Default	Color_1	🖨
DOORS	♀	♡	🔓	□ Cyan	CONTINUOUS	—— Default	Color_4	🖨
FLOOR	♀	♡	🔓	■ Red	CONTINUOUS	—— Default	Color_1	🖨
WALLS	♀	♡	🔓	□ Green	CONTINUOUS	—— Default	Color_3	🖨

Figure 18–25

STEP 3

While in the office plan, activate the Insert dialog box through the INSERT command. At times, this dialog box is useful for displaying all valid blocks in a drawing. Clicking on the Name drop-down list box displays the results in Figure 18–26.

Two blocks are currently defined in this drawing; as with the layers, once the office plan is merged into another drawing through the XREF command, these block names will change. When finished viewing the defined blocks, close Office.Dwg.

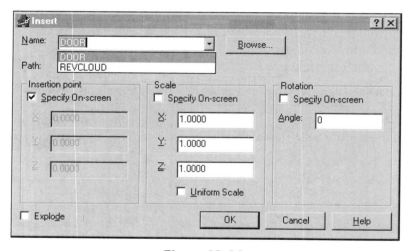

Figure 18–26

STEP 4

This next step involves opening Interiors.Dwg, also found on the CD, and looking at the current layers found in this drawing. Once this drawing is open, use the Layer Properties Manager dialog box to observe that layers exist in this draw-

ing for such items as floor and furniture (see Figure 18–27). Symbols such as desks, chairs, shelves, and plants will be inserted on the Furniture layer. The Office.Dwg file should be attached on the Floor layer.

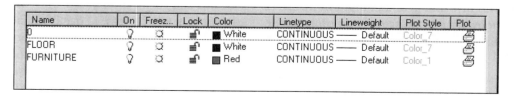

Name	On	Freez...	Lock	Color	Linetype	Lineweight	Plot Style	Plot
0	♀	♧	⬓	■ White	CONTINUOUS	—— Default	Color_7	🖨
FLOOR	♀	♧	⬓	■ White	CONTINUOUS	—— Default	Color_7	🖨
FURNITURE	♀	♧	⬓	■ Red	CONTINUOUS	—— Default	Color_1	🖨

Figure 18–27

STEP 5

As with the office plan, activate the Insert dialog box through the INSERT command to view the blocks internal to the drawing (see Figure 18–28). The four blocks listed consist of various furniture items and will be used to lay out the interior plan.

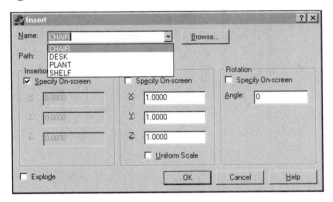

Figure 18–28

STEP 6

The office floor plan will now be merged into the interior plan in order for the furniture to be properly laid out. Rather than insert the office plan as a block, use the Xref Manager dialog box to accomplish this task. This dialog box will activate when you enter the XREF command from the keyboard or choose Xref Manager from the Insert pull-down menu, as in Figure 18–29A. After the dialog box displays, click on the Attach button, shown in Figure 18–29B.

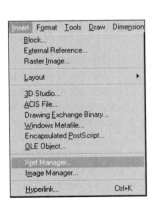

Figure 18–29A

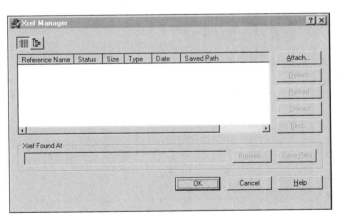

Figure 18–29B

STEP 7

Clicking on the Attach button shown in Figure 18–29B displays the Select Reference File dialog box in Figure 18–30. Find the appropriate subdirectory and click on the drawing file Office.Dwg, which will attach this drawing to the current drawing, Interiors.Dwg.

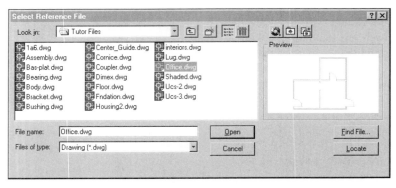

Figure 18–30

STEP 8

Selecting Office.Dwg back in Figure 18–30 displays the External Reference dialog box in Figure 18–31. Notice that OFFICE is the name of the external reference file chosen for attachment in the current drawing. Keeping all default values in the dialog box will allow you to attach the external reference at any location in the drawing. Also, the external reference will be brought in at the default scale factors. In the External Reference dialog box, click on the OK button shown in Figure 18–31 to attach Office.Dwg to Interiors.Dwg.

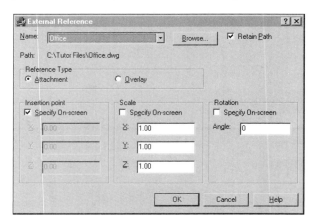

Figure 18–31

STEP 9

Give an insertion point of 0,0 as the location for placing the external reference file of Office.Dwg (see Figure 18–32A). This file looks identical to the original Office.Dwg file; one way to determine the status of the external reference is to use the LIST command and select the edge of one of the walls that make up the office plan. The result is illustrated in Figure 18–32B. Upon closer inspection, the image of OFFICE is actually an External Reference, as verified by the LIST command.

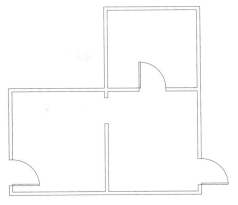

Figure 18–32A

```
BLOCK REFERENCE   Layer: "FLOOR"
   Space: Model space
   Handle = 221
   "Office"
   External reference
   at point, X=    0.0000  Y=    0.0000  Z=    0.0000
   X scale factor     1.0000
   Y scale factor     1.0000
   rotation angle       0
   Z scale factor     1.0000
```

Figure 18–32B

STEP 10

Once again, activate the Layer Properties Manager dialog box, paying close attention to the display of the layers. Using Figure 18–33 as a guide, you can see the familiar layers of Floor and Furniture. However, notice the group of layers beginning with Office; the layers actually belonging to the external reference file have the designation of XREF|LAYER. For example, the layer Office|doors represents a layer located in the file Office.Dwg that holds all door symbols. The "|" or pipe symbol is used to separate the name of the external reference from the layer. The layers belonging to the external reference file may be turned on or off, locked, or even made frozen. However, these layers cannot be made current for drawing on.

Name	On	Freez...	Lock	Color	Linetype	Lineweight	Plot Style	Plot
0	♀	☼	⊞	■ White	CONTINUOUS	—— Default	Color_7	
FLOOR	♀	☼	⊞	■ White	CONTINUOUS	—— Default	Color_7	⊜
FURNITURE	♀	☼	⊞	■ Red	CONTINUOUS	—— Default	Color_1	⊜
Office|CD	♀	☼	⊞	■ Red	CONTINUOUS	—— Default	Color_1	⊜
Office|DOORS	♀	☼	⊞	□ Cyan	CONTINUOUS	—— Default	Color_4	⊜
Office|FLOOR	♀	☼	⊞	■ Red	CONTINUOUS	—— Default	Color_1	⊜
Office|WALLS	♀	☼	⊞	□ Green	CONTINUOUS	—— Default	Color_3	⊜

Figure 18–33

STEP 11

Look at the blocks that are currently defined in the drawing and notice how they appear. The Insert dialog box will show only those blocks that can be inserted in the drawing; instead, use the -BLOCK command, which will identify all blocks in the drawing; those you can and cannot insert (see Figure 18–34.) Notice how the DOOR and REVCLOUD appear in the list of blocks. As with the layers in the

last step, the presence of the "|" or "pipe" symbol means the block is not valid and therefore cannot be inserted in the drawing file.

Command: **-BLOCK** *(The "-" preceding the command provides a command line version of the block command)*
Enter block name or [?]: **?**
Enter block(s) to list <*>: *(Press ENTER to list all blocks)*

```
Command: -BLOCK
Enter block name or [?]: ?
Enter block(s) to list <*>: (Press Enter for all)

Defined blocks.
  "CHAIR"
  "DESK"
  "Office"                        Xref: resolved
  "Office|DOOR"                   Xdep: "Office"
  "Office|REVCLOUD"               Xdep: "Office"
  "PLANT"
  "SHELF"

User        External        Dependent       Unnamed
Blocks      References      Blocks          Blocks
  4             1              2               0
```

Figure 18–34

STEP 12

Begin inserting the desk, chair, shelf, and plant symbols in the drawing using the Insert dialog box or through the AutoCAD DesignCenter. The external reference file OFFICE is to be used as a guide throughout this layout. It is not

important that your drawing match exactly the image in Figure 18–35. After positioning all symbols in the floor plan, close and save your drawing under its original name of Interiors.Dwg.

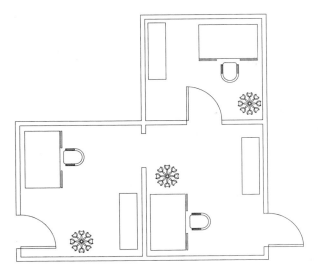

Figure 18–35

STEP 13

Now open the original drawing file Office.Dwg and make the following modifications: Move all three doors as indicated to new locations and mirror one of the doors so it is positioned closer to the wall; Stretch the wall opening over to the other end of the room. Finally, close and save these changes under the original name of Office.Dwg.

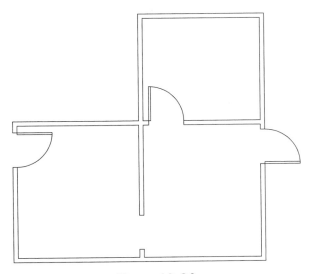

Figure 18–36

STEP 14

Now open the drawing file Interiors.Dwg and notice the results in Figure 18–37. Because Office.Dwg was attached through the External Reference dialog box, any changes to the original drawing file are automatically reflected. In this case, observe how some of the furniture is now in the way of the doorways. If the office plan were inserted in the interiors drawing as a block, these changes would not occur automatically.

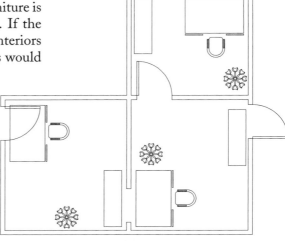

Figure 18–37

STEP 15

Because of the changes in the door openings, edit the drawing by moving the office furniture to better locations. Figure 18–38 can be used as a guide although your drawing may appear different.

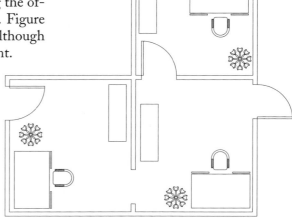

Figure 18–38

STEP 16

A door needs to be added to the gap between the walls of the office plan. However, the door symbol belongs to the externally referenced file Office.Dwg. The door block is defined as OFFICE|DOOR; the "|" character is not valid in the naming of the block and therefore cannot be used in the current drawing. The block must first be bound to the current drawing before it can be used. To use a block that belongs to an external reference, choose Object from the Modify pull-down menu followed by External Reference, and then Bind, as in Figure 18–39A. This displays the Xbind dialog box in Figure 18–39B. While in this dialog box, click on the + symbol next to the file OFFICE and then click on the + symbol next to Block. This will display all blocks that belong to the external reference (see Figure 18–39B).

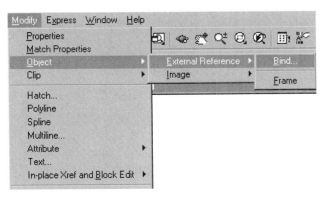

Figure 18–39A

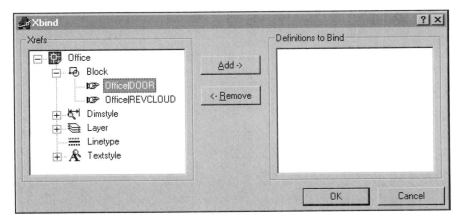

Figure 18–39B

STEP 17

Clicking on OFFICE|DOOR followed by the Add-> button in Figure 18–39B moves the block of the door to the Definitions to Bind area in Figure 18–40.

Click OK to dismiss the dialog box, and the door symbol is now a valid block that can be inserted in the drawing.

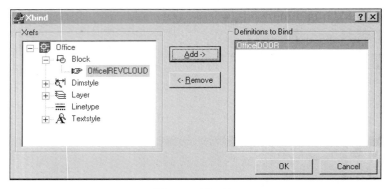

Figure 18–40

STEP 18

Activate the Insert dialog box through the INSERT command, click on the Name drop-down list box, and notice the block of the door in Figure 18–41. It is now listed as OFFICE0DOOR; the "|" character was replaced by the "0," making the block valid in the current draw-

ing. This is AutoCAD's standard way of converting blocks that belong to external references to blocks that can be used in the current drawing file. This same procedure works on layers belonging to external references as well.

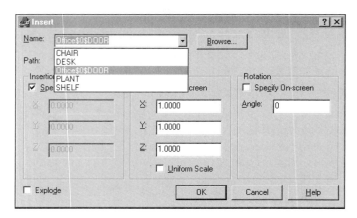

Figure 18–41

STEP 19

Insert the door symbol in the open gap as illustrated in Figure 18–42 to complete the drawing.

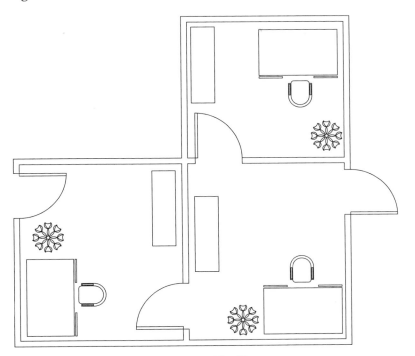

Figure 18–42

CHAPTER 19

Multiple Viewport Drawing Layouts

ARRANGING DIFFERENT VIEWS OF THE SAME DRAWING

Begin this chapter on multiple viewport layouts by opening the drawing 1a6.Dwg. This drawing in Figure 19–1 represents a roof plan of a structure and is drawn in Model Space at full size. The roof plan needs to be plotted out at a scale of 1/8" = 1'-0". Figure 19–2 illustrates the Fit tab of the Modify Dimension Style: dialog box that appears when you click on the Modify button of the Dimension Style Manager dialog box. The important point to focus on in this illustration is the Overall Scale value in the dialog box. Notice that it is set to a value of 96, which is arrived at when 1' or 12 is divided by 1/8" or 0.125. This value will also be used shortly to scale the roof plan once it is brought into Paper Space. One thing to add to this layout is a second viewport, and we need to arrange Area "A" in Figure 19–1 at a scale of 1/2" = 1'-0".

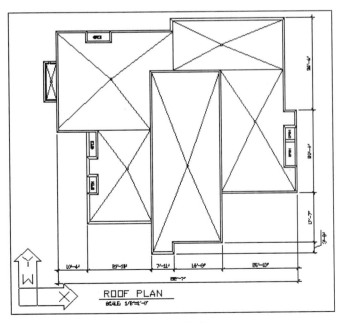

Figure 19–1

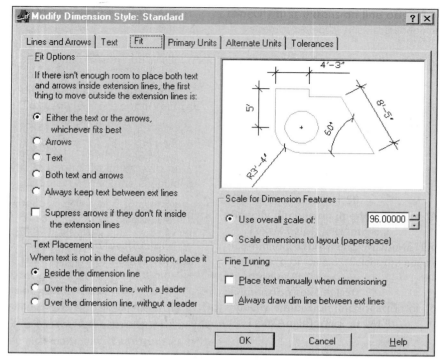

Figure 19–2

To begin the layout, select the Layout1 tab at the bottom of the AutoCAD screen. The Page Setup dialog box in Figure 19–3 will be provided. Change the layout name in the edit box if desired. In this example, Layout 1 was changed to Roof Plan. With the Layout Setting tab selected, choose a "C" size sheet of paper from the drop-down list box. If "C" size is not available in the list box, select the Plot Device tab and choose a printer that supports the plotting of "C" size drawings. Verify that the plotting scale is 1:1 and click OK to accept the settings. You will be returned to the screen and provided a view of your new drawing sheet layout (see Figure 19–4). The "C" size sheet is shown with a dashed line representing the printable area. You have also been provided with a floating viewport, which allows you to see the roof plan.

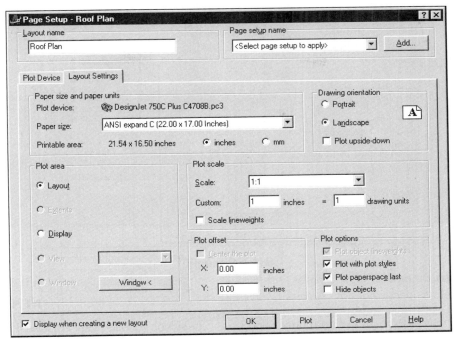

Figure 19–3

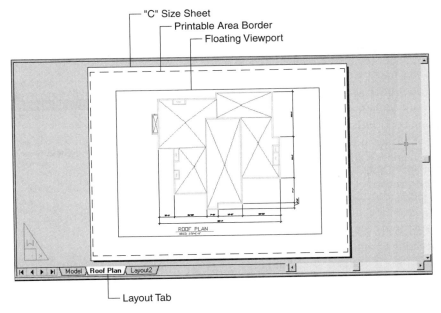

Figure 19–4

Verify from the status bar or the UCS icon, that you are now in Paper Space, and then use the INSERT command or AutoCAD DesignCenter to insert a "C" size title block (see Figure 19–5). This title block will be used to layout both views of the roof plan at different scales.

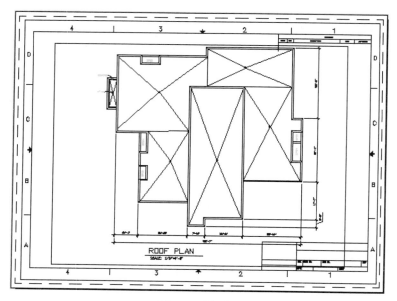

Figure 19–5

The total roof plan will fit inside the original large viewport; the detail area will fit in a new smaller viewport. Use grips to change the shape of the original large viewport. Locate the lower left corner near point "A" and the upper right corner near point "B" as shown in Figure 19–6. Next use the MVIEW command to create the small viewport. Do not worry about the exact size of the viewports. The images may not fit inside the viewports when they are scaled later and grips can stretch the viewports to the desired size at any time.

Command: **MV** *(For MVIEW)*
Specify corner of viewport or
[ON/OFF/Fit/Hideplot/Lock/Object/Polygonal/Restore/2/3/4] <Fit>: *(Pick at "C" in Figure 19–6)*
Specify opposite corner: *(Pick at "D" in Figure 19–6)*
Regenerating model.

Notice that the image of the roof plan is visible in both viewports. This is typical of the Paper Space environment. Demonstrated later on in this chapter is a way to freeze certain layers in one viewport while keeping layers in other viewports visible. This control is not needed for this segment.

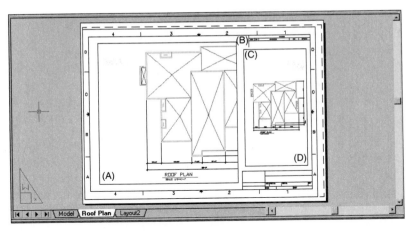

Figure 19–6

Switch to Model Space by clicking on the PAPER button on the status bar (to change it to MODEL) or use the MSPACE command at the keyboard.

Command: **MS** *(For MSPACE)*
In floating Model Space, double click inside of the large viewport to make it current. Because the scale of this image is 1/8" = 1'-0", use the ZOOM command and a factor of 1/96XP to scale the image to Paper Space units (see Figure 19–7). You could also use the Viewports toolbar and click on the 1/8" = 1' scale.

Command: **Z** *(For ZOOM)*
Specify corner of window, enter a scale factor (nX or nXP), or [All/Center/Dynamic/Extents/Previous/Scale/Window] <real time>: **1/96XP**

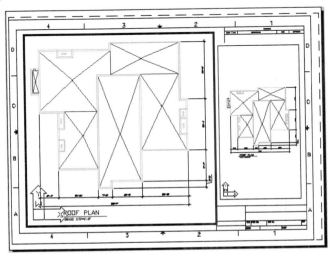

Figure 19–7

The image may get larger or smaller depending on the scale factor and the size of the viewport. Be prepared to use grips to size the viewport relative to the image of the roof plan. Use the PAN command to center the drawing in the viewport. Next, while still in floating Model Space, activate the smaller viewport. It is considered good practice to pan to the center of the area that is to be enlarged (see Figure 19–8).

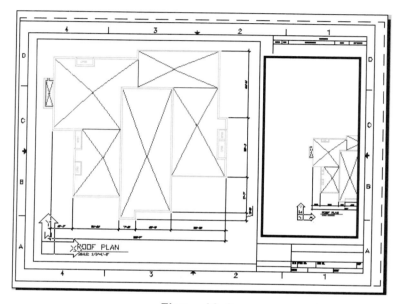

Figure 19–8

This area will be enlarged; to do this, use a larger scale, such as 1/2" = 1'-0". Instead of using the ZOOM command and a scale factor of 1/24XP (24 is the value found when 1' or 12 is divided by 1/2" or 0.50), use the Viewports toolbar to scale the detail. Once the Viewports toolbar is displayed (see Figure 19–9), note that the scale of the active viewport is displayed in the drop-down list box. With the small viewport active, click on the drop-down list box and select 1/2" = 1'-0".

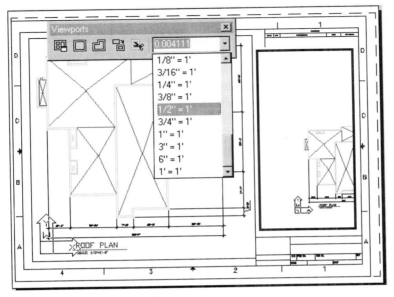

Figure 19–9

The image is panned over further to the right, as in Figure 19–10, to accommodate dimensions. This has become a quick method of arranging multiple details of the same drawing. Save the drawing 1a6.Dwg at this point.

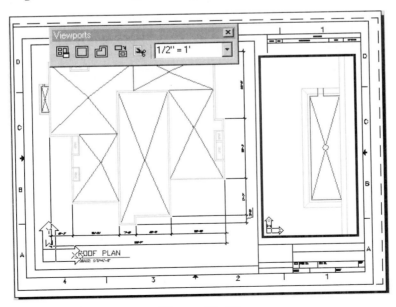

Figure 19–10

SCALING DIMENSIONS TO PAPER SPACE VIEWPORTS

Continue on with this next segment by using the drawing file 1a6.Dwg. Back in Figure 19–10, two images of the same drawing were laid out in Paper Space at two different scale factors. Now, the task is to have the dimension text height match throughout the entire drawing. The roof plan in the larger viewport already has dimensions. The dimension scale of 96 is used for the drawing scale of 1/8" = 1'-0". However, the image in the smaller viewport is scaled differently. The dimension scale must be adjusted to reflect the new drawing scale.

First a new layer is created and made current in the Layer Properties Manager dialog box; this new layer will be called Detail Dim, and it will hold all dimensions that pertain to the detail of the roof plan (see Figure 19–11).

Name	On	Freez...	Lock	Color	Linetype	Lineweight	Plot Style	Plot
0				■ White	Continuous	—— Default	Color_7	
A-SYM-2				■ Blue	Continuous	—— Default	Color_5	
A-SYM-5				□ 9	Continuous	—— Default	Color_9	
A-TX-1				■ Red	Continuous	—— Default	Color_1	
A-TX-5				□ 9	Continuous	—— Default	Color_9	
C-ROOF-4				□ Green	Continuous	—— Default	Color_3	
Roof Dim				■ White	Continuous	—— Default	Color_7	
Title Block				■ White	Continuous	—— Default	Color_7	
Detail Dim				■ White	Continuous	—— Default		

Figure 19–11

Next, a new dimension style called DETAIL is created as in Figure 19–12. Clicking on the New... button in the Dimension Style Manager dialog box displays the Create New Dimension Style dialog box shown in Figure 19–13. Enter " Detail" in the New Style Name edit box and click the Continue button.

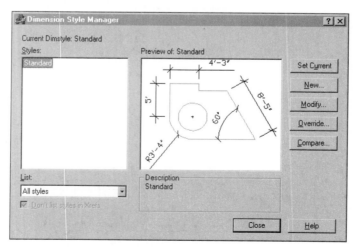

Figure 19–12

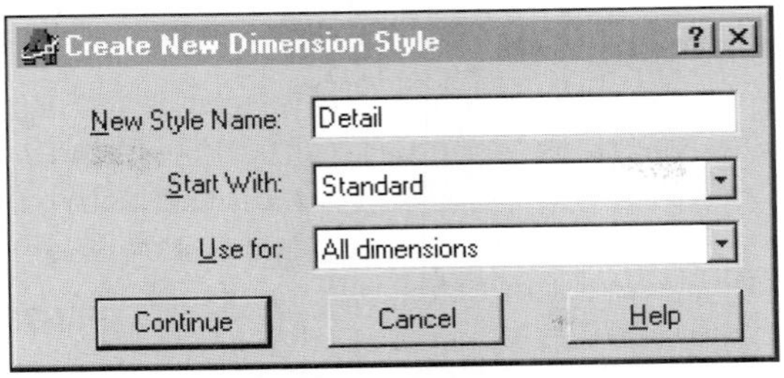

Figure 19–13

Click on the Fit tab in the New Dimension Style: dialog box to display the Scale for Dimension Features area shown in Figure 19–14. Notice in this illustration, that the original Overall scale factor is still set to 96 to match the scale of 1/8" = 1'-0". The scale of the roof plan detail is 1/2" = 1'-0". One technique would be to change the Overall scale value from 96 to a new value of 24. A better method (see Figure 19–15) is to select the radio button for Scale dimensions to layout (paperspace). This will automatically set the Overall dimension scale based on the scale inside the current floating Model Space viewport. After selecting the radio button, click on the OK button to return to the main Dimension Style Manager dialog box, shown in Figure 19–16. With the Detail style selected, click the Set Current button and then click on the Close button. The new dimension style has been saved and is current.

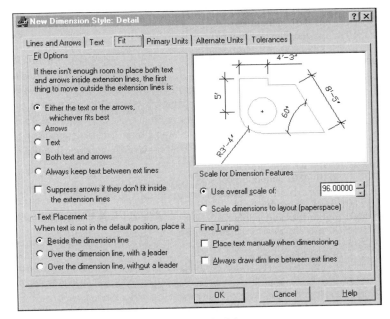

Figure 19–14

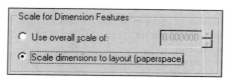

Figure 19–15

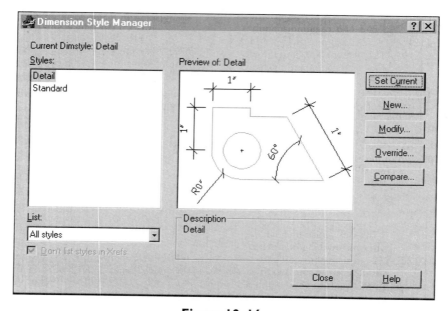

Figure 19–16

Set the Detail Dim Layer as the current layer and add vertical and horizontal dimensions using the DIMLINEAR command in Figure 19–17.

Command: **DLI** *(For DIMLINEAR)*
Specify first extension line origin or <select object>: **Int**
of *(Pick the intersection at "A")*
Specify second extension line origin: **Int**
of *(Pick the intersection at "B")*
Specify dimension line location or

[Mtext/Text/Angle/Horizontal/Vertical/Rotated]: *(Locate the dimension)*
Dimension text = 14'-6"
Command: **DLI** *(For DIMLINEAR)*
Specify first extension line origin or <select object>: **Int**
of *(Pick the intersection at "A")*
Specify second extension line origin: **Int**
of *(Pick the intersection at "C")*

Specify dimension line location or

[Mtext/Text/Angle/Horizontal/Vertical/Rotated]: *(Locate the dimension)*
Dimension text = 4'-5"

Notice in Figure 19–17 that all dimensions in both viewports have the same height even though the scales of both images are different. The ability to scale dimensions to Paper Space viewports through the Dimension Styles dialog box is a very important and powerful feature of AutoCAD. Save your drawing at this point.

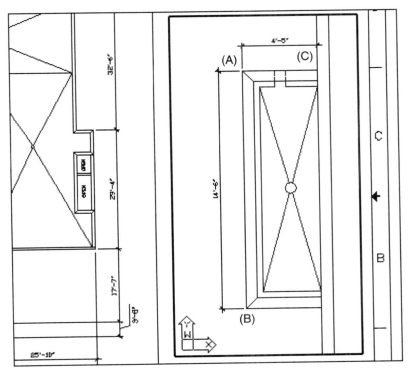

Figure 19–17

CONTROLLING THE VISIBILITY OF LAYERS IN PAPER SPACE VIEWPORTS

Continue on with this segment using 1a6.Dwg. It appears that the drawing in Figure 19–18 is ready to be plotted. However, notice the upper left corner of the large viewport; as dimensions were added to the detail of the roof plan in the smaller viewport, these same dimensions also appear in the overall roof plan in the large viewport. Again, this is a typical occurrence in the Paper Space environment.

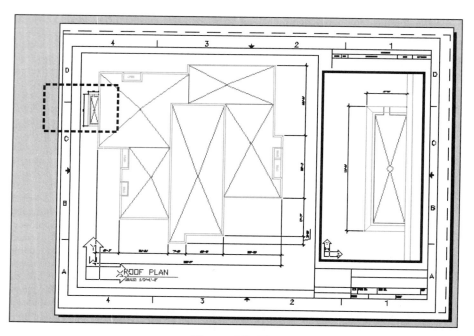

Figure 19–18

Another powerful tool while you work inside Paper Space is the ability to freeze layers in certain viewports. In Figure 19–18, the two dimensions placed in the detail need to be frozen in the viewport of the overall roof plan. Before continuing, be sure the large viewport holding the overall roof plan is the current viewport. Activate the Layer Properties Manager dialog box in Figure 19–19 and notice the various layer states such as Name, On, and Lock. Also notice three states that deal with freezing viewports. (You may have to resize the dialog box to see all the states available.)

Name	On	Freez...	Lock	Color	Linetype	Lineweight	Plot Style	Plot	Active ...	New ...
0				White	Continuous	—— Default	Color_7			
A-SYM-2				Blue	Continuous	—— Default	Color_5			
A-SYM-5				9	Continuous	—— Default	Color_9			
A-TX-1				Red	Continuous	—— Default	Color_1			
A-TX-5				9	Continuous	—— Default	Color_9			
C-ROOF-4				Green	Continuous	—— Default	Color_3			
DefPoints				White	Continuous	—— Default	Color_7			
Detail Dim				White	Continuous	—— Default	Color_7			
Roof Dim				White	Continuous	—— Default	Color_7			
Title Block				White	Continuous	—— Default	Color_7			

Figure 19–19

Expanding the heading in the third column in Figure 19–20 displays the title and purpose of the function of this state, namely to freeze layers in all viewports. This is the Layer state most commonly used in Model Space for freezing layers and then thawing them. It would not be a good idea to use this for the layouts of the roof plan and detail since the freezing the dimension layer would freeze the dimensions in both viewports. We want the dimensions to be visible only in the detail viewport.

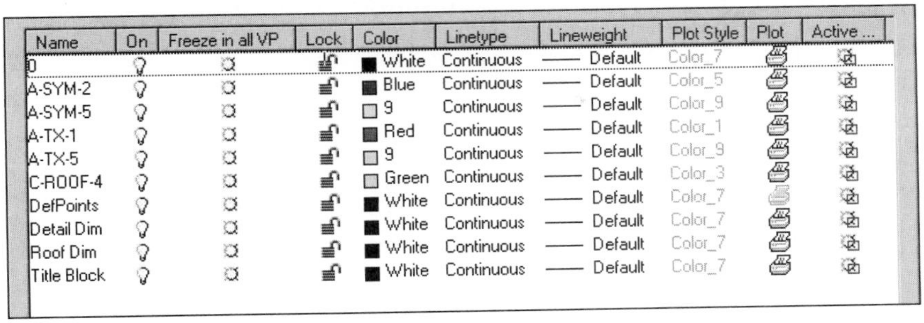

Name	On	Freeze in all VP	Lock	Color	Linetype	Lineweight	Plot Style	Plot	Active ...
0	♀	☼	➡	■ White	Continuous	—— Default	Color_7	🖨	🖰
A-SYM-2	♀	☼	➡	■ Blue	Continuous	—— Default	Color_5	🖨	🖰
A-SYM-5	♀	☼	➡	□ 9	Continuous	—— Default	Color_9	🖨	🖰
A-TX-1	♀	☼	➡	■ Red	Continuous	—— Default	Color_1	🖨	🖰
A-TX-5	♀	☼	➡	□ 9	Continuous	—— Default	Color_9	🖨	🖰
C-ROOF-4	♀	☼	➡	□ Green	Continuous	—— Default	Color_3	🖨	🖰
DefPoints	♀	☼	➡	■ White	Continuous	—— Default	Color_7	🖨	🖰
Detail Dim	♀	☼	➡	■ White	Continuous	—— Default	Color_7	🖨	🖰
Roof Dim	♀	☼	➡	■ White	Continuous	—— Default	Color_7	🖨	🖰
Title Block	♀	☼	➡	■ White	Continuous	—— Default	Color_7	🖨	🖰

Figure 19–20

Expanding the heading in the tenth column in Figure 19–21 displays the title and purpose of the next function, namely to freeze layers in the active viewport. This is the Layer state that will be used to freeze the Detail Dim layer only in the active viewport, which should be the large viewport. As shown in Figure 19–22, clicking on the icon to freeze the Detail Dim layer in the current viewport displays a snowflake, signifying that the layer is frozen only in the active viewport.

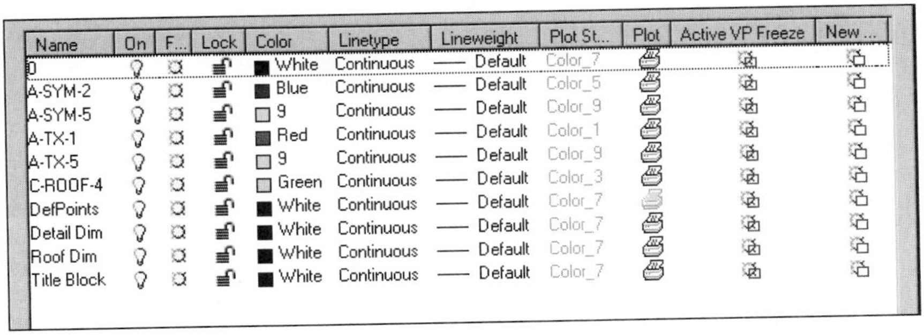

Name	On	F...	Lock	Color	Linetype	Lineweight	Plot St...	Plot	Active VP Freeze	New...
0	♀	☼	➡	■ White	Continuous	—— Default	Color_7	🖨	🖰	🖰
A-SYM-2	♀	☼	➡	■ Blue	Continuous	—— Default	Color_5	🖨	🖰	🖰
A-SYM-5	♀	☼	➡	□ 9	Continuous	—— Default	Color_9	🖨	🖰	🖰
A-TX-1	♀	☼	➡	■ Red	Continuous	—— Default	Color_1	🖨	🖰	🖰
A-TX-5	♀	☼	➡	□ 9	Continuous	—— Default	Color_9	🖨	🖰	🖰
C-ROOF-4	♀	☼	➡	□ Green	Continuous	—— Default	Color_3	🖨	🖰	🖰
DefPoints	♀	☼	➡	■ White	Continuous	—— Default	Color_7	🖨	🖰	🖰
Detail Dim	♀	☼	➡	■ White	Continuous	—— Default	Color_7	🖨	🖰	🖰
Roof Dim	♀	☼	➡	■ White	Continuous	—— Default	Color_7	🖨	🖰	🖰
Title Block	♀	☼	➡	■ White	Continuous	—— Default	Color_7	🖨	🖰	🖰

Figure 19–21

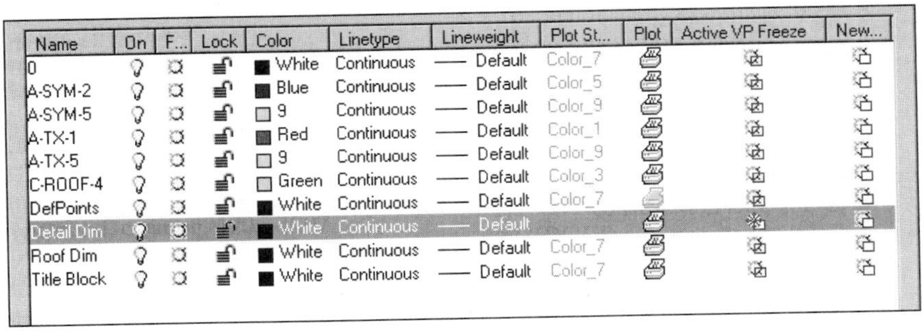

Name	On	F...	Lock	Color	Linetype	Lineweight	Plot St...	Plot	Active VP Freeze	New...
0	♀	☼	➡	■ White	Continuous	—— Default	Color_7	🖨	🖰	🖰
A-SYM-2	♀	☼	➡	■ Blue	Continuous	—— Default	Color_5	🖨	🖰	🖰
A-SYM-5	♀	☼	➡	□ 9	Continuous	—— Default	Color_9	🖨	🖰	🖰
A-TX-1	♀	☼	➡	■ Red	Continuous	—— Default	Color_1	🖨	🖰	🖰
A-TX-5	♀	☼	➡	□ 9	Continuous	—— Default	Color_9	🖨	🖰	🖰
C-ROOF-4	♀	☼	➡	□ Green	Continuous	—— Default	Color_3	🖨	🖰	🖰
DefPoints	♀	☼	➡	■ White	Continuous	—— Default	Color_7	🖨	🖰	🖰
Detail Dim	♀	◉	➡	■ White	Continuous	—— Default		🖨	❄	🖰
Roof Dim	♀	☼	➡	■ White	Continuous	—— Default	Color_7	🖨	🖰	🖰
Title Block	♀	☼	➡	■ White	Continuous	—— Default	Color_7	🖨	🖰	🖰

Figure 19–22

Clicking the OK button at the bottom of the Layer Properties Manager dialog box exits the dialog box and returns you to the drawing. Notice, in Figure 19–23, that the layer holding the detail dimensions has disappeared from the large viewport; however, the Detail Dim layer is still present in the detail viewport.

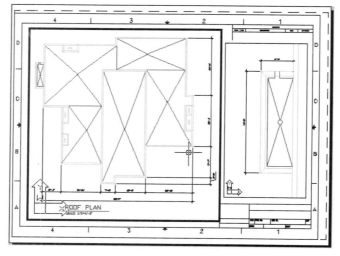

Figure 19–23

Expanding the heading in the eleventh column in Figure 19–24 displays the title and purpose of the next function, namely to freeze layers in any new viewports. This Layer function will not be used for this drawing. However it too is very useful if more viewports need to be created. It has already been demonstrated that when you set up multiple viewports, the same image appears in all viewports. What this function does is freeze selected layers in any newly created viewports. For example, if you don't want any dimensions showing up when a new viewport is created, you could click on the New Viewport Freeze icon associated with the Roof Dim and Detail Dim layers. Then, when any new viewports are created, the image of the roof plan will appear without the dimensions.

Name	On	F...	Lock	Color	Linetype	Lineweight	Plot St...	Plot	Active...	New VP Freeze
0	♀	♢	⬛	■ White	Continuous	—— Default	Color_7			
A-SYM-2	♀	♢	⬛	■ Blue	Continuous	—— Default	Color_5			
A-SYM-5	♀	♢	⬛	☐ 9	Continuous	—— Default	Color_9			
A-TX-1	♀	♢	⬛	■ Red	Continuous	—— Default	Color_1			
A-TX-5	♀	♢	⬛	☐ 9	Continuous	—— Default	Color_9			
C-ROOF-4	♀	♢	⬛	☐ Green	Continuous	—— Default	Color_3			
DefPoints	♀	♢	⬛	■ White	Continuous	—— Default	Color_7			
Detail Dim	♀	♢	⬛	■ White	Continuous	—— Default	Color_7			
Roof Dim	♀	♢	⬛	■ White	Continuous	—— Default	Color_7			
Title Block	♀	♢	⬛	■ White	Continuous	—— Default	Color_7			

Figure 19–24

With the Detail Dim layer frozen in the overall roof plan viewport, you could switch back to Paper Space, turn off the layer holding all viewports (see Figure 19–25), and plot the drawing out at a scale of 1=1.

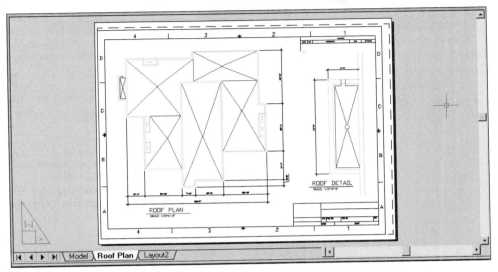

Figure 19–25

Layers have always had the capability of being frozen or thawed. However, this was accomplished globally while in Model Space. Freezing layers in active and new viewports allows freeze and thaw operations to be viewport-specific.

CREATING A DETAIL PAGE IN PAPER SPACE WITH THE AID OF EXTERNAL REFERENCES

This next discussion focuses on laying out on the same sheet a series of details composed of different drawings, which can be at different scales. As the viewports are laid out in Paper Space and images of the drawings are inserted in the Paper Space viewports, all images will appear in all viewports. This is not a major problem, because Paper Space allows for layers to be frozen in certain viewports and not in others. What if you are not sure of the exact layer names to freeze? This becomes a big problem. If images of drawings are not inserted in viewports; but instead attached as External References, this affords greater control over layers, as shown in the next Paper Space example.

Figure 19–26 shows three objects: a body, a bushing, and a bearing. The body and bearing will be laid out at a scale of 1=1. The bushing will be laid out at a scale of 2=1 (enlarged to twice its normal size). The dimension scales have been set for all objects in order for all dimension text to be displayed at the same height. Also, layers have been created for each object. Each object exists as a single file on the hard drive. The three files you will use to complete this segment are Body.Dwg, Bearing.Dwg, and Bushing.Dwg.

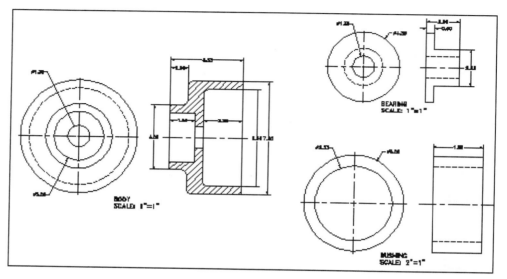

Figure 19–26

When creating a sheet of details consisting of different drawing files, you first start out with a new drawing file. It is highly recommended that you create two new layers. The first layer will hold all viewport information and is called Vports. It is also considered good practice to assign a color to the Vports layer such as Green. The second layer will hold all external reference information and is called Xref. When the drawing is completed, the Vports layer can be frozen, or turned off, leaving the images to be displayed.

Once the new drawing has been created and layers created, switch to Paper Space using the layout methods previously described. Choose the ANSI D Expanded layout sheet. You could also enter the Display tab of the Options dialog box and remove the check in the box adjacent to "Create viewport in new layouts". Also, insert a "D" size title block for this drawing file. Make the Vports layer current and use the MVIEW command to construct viewports that will hold the three separate drawing files (see Figure 19–27). Again, it is not critical that the viewports be constructed to accommodate the entire image because grips can be used later to size the viewports to the image. If you notice the appearance of the Layout2 tab, this can be deleted at this point.

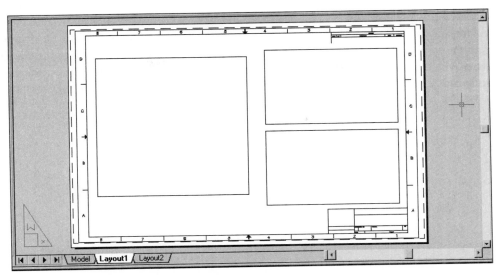

Figure 19–27

In Figure 19–27, notice how all viewports are blank or empty, because the drawing database contains no objects. This will change shortly—the drawing files will now be attached, one to each viewport. Before beginning this process, first switch to floating Model Space, make the large viewport at the left current, and make the Xref layer the new current layer. The display should be similar to Figure 19–28.

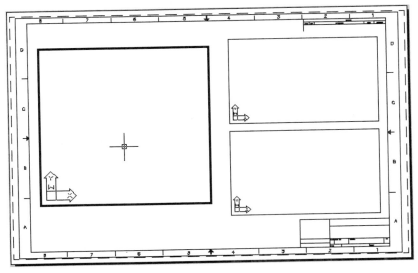

Figure 19–28

Activate the Xref Manager dialog box through the XREF command. Click on the Attach button shown in Figure 19–29 and search for the desired file to attach inside the current viewport; the drawing file Body.Dwg will be used. After you locate this drawing file, it appears in the External Reference dialog box in Figure 19–30. Click the OK button and click inside the viewport at a convenient location to attach the drawing file.

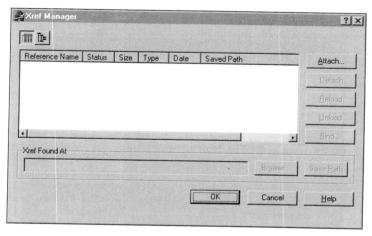

Figure 19–29

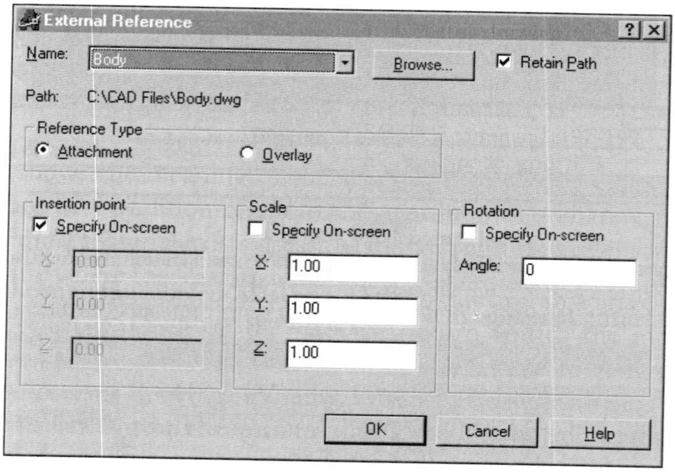

Figure 19–30

Notice how the attached drawing of Body.Dwg appears in all three viewports in Figure 19–31. The Layer Properties Manager dialog box could be used to freeze the layers; however, each viewport must be made active before the layers can be frozen, which would take two steps. Another method, the VPLAYER command, will now be introduced. This command stands for Viewport Layer and is another way to freeze layers in one viewport and not in others. However, for this command to function properly, you must know all layers to freeze. This was the purpose of using External References to attach to the viewports; the External Reference file controls all layers. To freeze layers using the VPLAYER command, all you have to know is the name of the External Reference file. Since the name of this file controls all layers, you can freeze all layers that begin with the name of the External Reference at the same time.

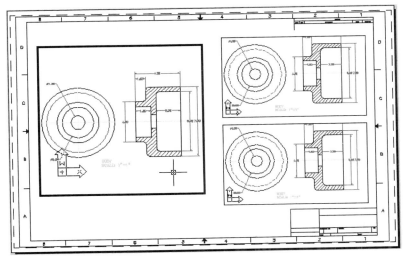

Figure 19–31

Before using the VPLAYER command, the first activate Layer Properties Manager dialog box just to view the layers created through the XREF command and the Attach option (see Figure 19–32). Notice the grouping of layers that begins with Body and the pipe (|) symbol. The Layer Properties Manager dialog box can be closed at this point. These layers need to be frozen in the other two viewports through the VPLAYER command and the following sequence:

Command: **VPLAYER**
Enter an option [?/Freeze/Thaw/Reset/Newfrz/Vpvisdflt]: **F** *(For Freeze)*
Enter layer name(s) to freeze: **Body*** *(For all layers that begin with Body)*
Enter an option [All/Select/Current] <Current>: **S** *(For Select)*
Switching to Paper space.
Select objects: *(Pick the upper right viewport in Figure 19–31)*
Select objects: *(Pick the lower right viewport in Figure 19–31)*
Select objects: *(Press ENTER to continue)*
Switching to Model space.
Enter an option [?/Freeze/Thaw/Reset/Newfrz/Vpvisdflt]: *(Press ENTER to exit this command)*

Name	On	F...	Lock	Color	Linetype	Lineweight	Plot ...	Plot	Active ...	New...
0				■ White	Continuous	—— Default	Color_7			
BodyICENTER				□ Yellow	BodyICENTER	—— Default	Color_2			
BodyICPL				■ White	Continuous	—— Default	Color_7			
BodyIDIM				■ Magenta	Continuous	—— Default	Color_7			
BodyIHIDDEN				■ Red	BodyIHIDDEN	—— Default	Color_1			
BodyINOTES				□ Green	Continuous	—— Default	Color_3			
BodyIOBJECT				■ White	Continuous	—— Default	Color_7			
BodyISECTION				■ Blue	Continuous	—— Default	Color_5			
DefPoints				■ White	Continuous	—— Default	Color_7			
Title Block				■ White	Continuous	—— Default	Color_7			
Vports				■ White	Continuous	—— Default	Color_7			
Xref				■ White	Continuous	—— Default	Color_7			

Figure 19–32

The results are illustrated in Figure 19–33, with all layers associated with Body frozen in the upper right and lower right viewports. This is the reason "Body*" was entered as the name of the layer to freeze. Notice, in the previous prompt sequence, that the command automatically switched to Paper Space to select the viewports and then back to Model Space to complete the command.

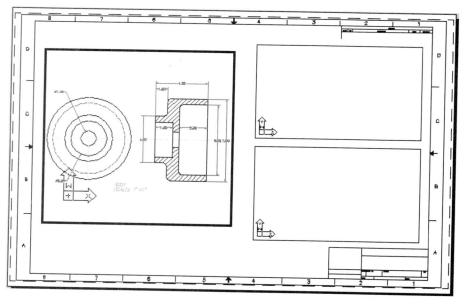

Figure 19–33

With the large viewport at the left still active, as in Figure 19–33, issue the ZOOM command along with the XP option to scale the image at a factor of 1XP, which is another way of saying full size. The Viewports toolbar, shown in Figure 19–34, provides an alternate method for zooming to Paper Space scale factors—simply choose the desired scale from the drop-down list box.

Command: **Z** *(For ZOOM)*
Specify corner of window, enter a scale factor (nX or nXP), or [All/Center/Dynamic/
   Extents/Previous/Scale/Window] <real time>: **1XP**

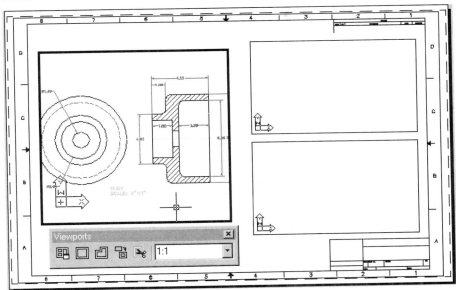

Figure 19–34

If the image appears too large for the viewport, adjust the size of the viewport using grips.

Next, make the upper right viewport current and use the External Reference dialog box to attach Bearing.Dwg to this viewport. Again, the image appears in all viewports, illustrated in Figure 19–35. Use the VPLAYER command to freeze all layers beginning with Bearing in the large viewport on the left and in the lower right viewport.

Command: **VPLAYER**
Enter an option [?/Freeze/Thaw/Reset/Newfrz/Vpvisdflt]: **F** *(For Freeze)*
Enter layer name(s) to freeze: **Bearing*** *(For all layers that begin with Bearing)*
Enter an option [All/Select/Current] <Current>: **S** *(For Select)*
Switching to Paper space.
Select objects: *(Pick the left viewport in Figure 19–35)*
Select objects: *(Pick the lower right viewport in Figure 19–35)*
Select objects: *(Press ENTER to continue)*
Switching to Model space.
Enter an option [?/Freeze/Thaw/Reset/Newfrz/Vpvisdflt]: *(Press ENTER to exit this
   command)*

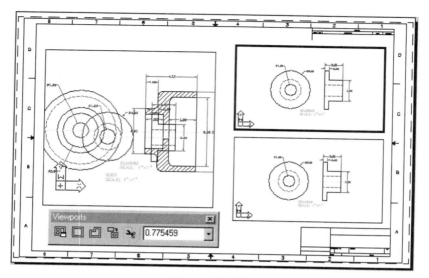

Figure 19–35

With the upper right viewport still active, as in Figure 19–35, use the Viewports toolbar or issue the ZOOM command along with the XP option to scale the image at a factor of 1XP or full size. The results are illustrated in Figure 19–36.

Command: **Z** *(For ZOOM)*
Specify corner of window, enter a scale factor (nX or nXP), or [All/Center/Dynamic/Extents/Previous/Scale/Window] <real time>: **1XP**

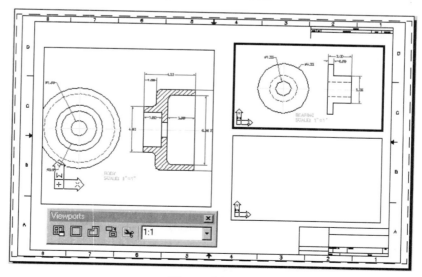

Figure 19–36

The lower right viewport is now made current and the External Reference dialog box is used to attach Bushing.Dwg to this viewport. The image again appears in all viewports, illustrated in Figure 19–37. The VPLAYER command is used to freeze all layers beginning with Bushing in the large viewport on the left and in the upper right viewport.

Command: **VPLAYER**
Enter an option [?/Freeze/Thaw/Reset/Newfrz/Vpvisdflt]: **F** *(For Freeze)*
Enter layer name(s) to freeze: **Bushing*** *(For all layers that begin with Bushing)*
Enter an option [All/Select/Current] <Current>: **S** *(For Select)*
Switching to Paper space.
Select objects: *(Pick the left viewport in Figure 19–37)*
Select objects: *(Pick the upper right viewport in Figure 19–37)*
Select objects: *(Press ENTER to continue)*
Switching to Model space.
Enter an option [?/Freeze/Thaw/Reset/Newfrz/Vpvisdflt]: *(Press ENTER to exit this command)*

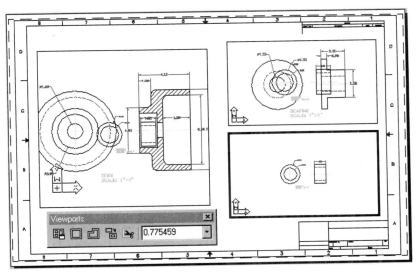

Figure 19–37

With the lower right viewport still active, as in Figure 19–37, use the Viewports toolbar or issue the ZOOM command along with the XP option to scale the image at a factor of 2XP or twice its normal size. The results are illustrated in Figure 19–38.

Command: **Z** *(For ZOOM)*
Specify corner of window, enter a scale factor (nX or nXP), or [All/Center/Dynamic/ Extents/Previous/Scale/Window] <real time>: **2XP**

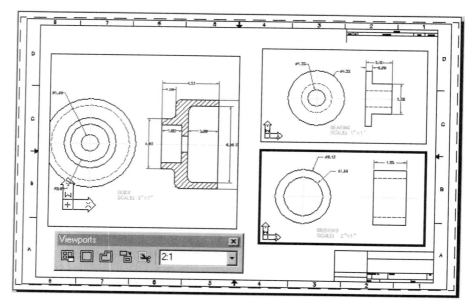

Figure 19–38

If necessary, switch back to Paper Space and use grips to adjust all viewports so that the entire image is displayed inside each viewport. The results are illustrated in Figure 19–39.

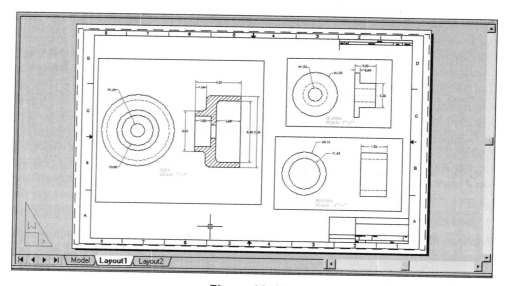

Figure 19–39

The detail sheet is almost ready to be plotted. Switch to floating Model Space and make the upper right viewport active. Activate the Layer Properties Manager dialog box to observe the layer states, shown in Figure 19–40. Notice, in the upper right viewport, how layers that begin with Bearing are thawed and visible while all layers that begin with Body and Bushing are frozen in the viewport. This is the result of using the VPLAYER command.

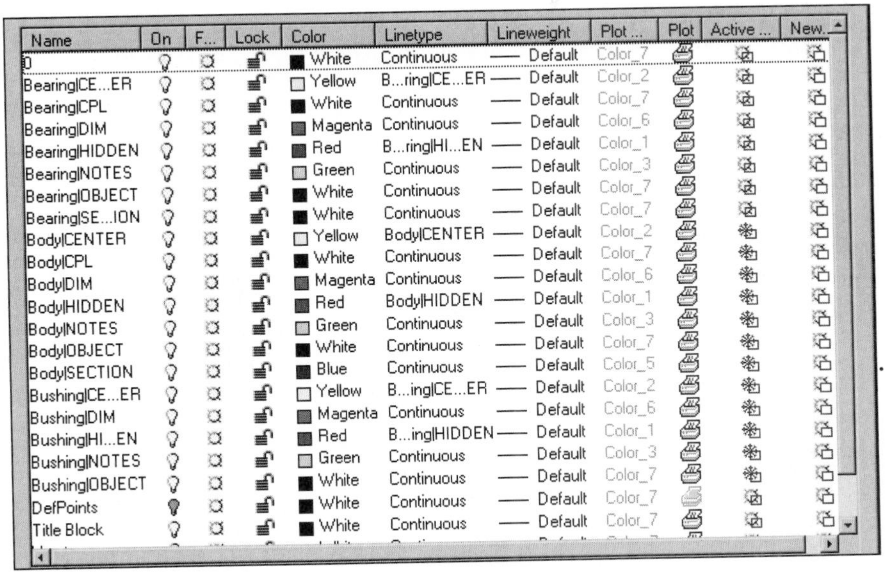

Figure 19–40

While inside the Layer Properties Manager dialog box, right-click one of the layer names in the dialog box to bring up the shortcut menu in Figure 19–41. Click on Select All to select all layers.

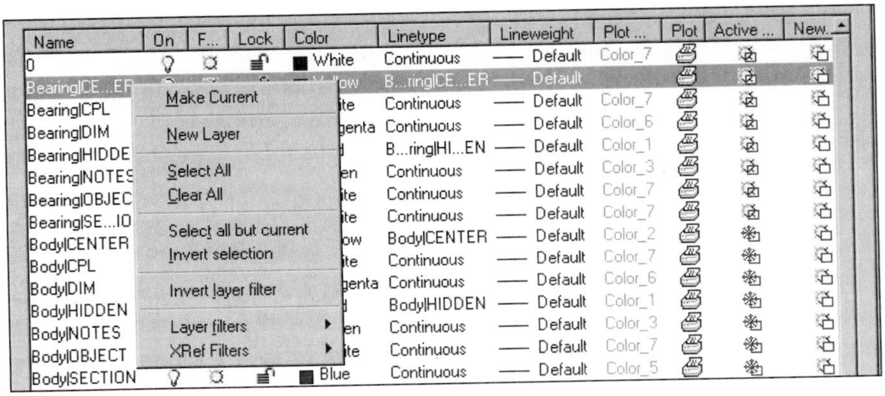

Figure 19–41

With all layers highlighted as in Figure 19–42, click on the icon to Freeze/Thaw in new viewports. Now if any new viewports are created for extra details, the images of the Body, Bearing, and Bushing will not appear in the new viewports. This is yet another productivity tool available for Paper Space layout through the Layer Properties Manager dialog box. This same effect can be accomplished through the VPLAYER command and the following sequence:

Command: **VPLAYER**
Enter an option [?/Freeze/Thaw/Reset/Newfrz/Vpvisdflt]: **V** *(For Vpvisdflt)*
Enter layer name(s) to change viewport visibility: **Body*,Bearing*,Bushing***
Enter a viewport visibility option [Frozen/Thawed] <Thawed>: **F** *(For Frozen)*
Enter an option [?/Freeze/Thaw/Reset/Newfrz/Vpvisdflt]: *(Press ENTER to exit this command)*

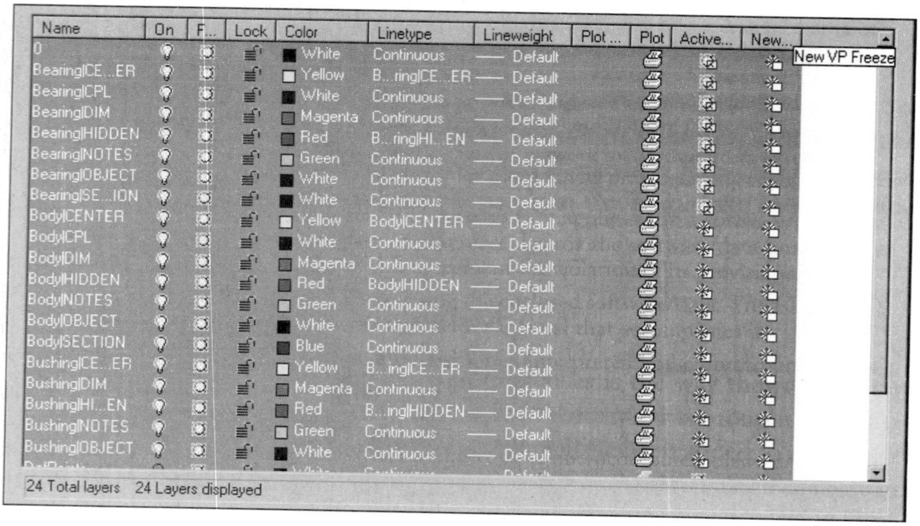

Figure 19–42

Note: This detail sheet is complete. If more details need to be added, the Xref layer must be thawed in new viewports through the Layer Properties Manager dialog box.

The Vpvisdflt option used above stands for Viewport Visibility Default and allows for layers to be frozen or thawed in any new viewport.

Finally, turn off the Vports layer to view the details.

LAYER 0 IN MODEL SPACE

When you construct objects in Model Space, Layer 0 is to remain neutral. In other words, it is recommended that you not draw any geometry or objects such as lines, circles, or arcs on Layer 0. Still, you might accidentally construct on Layer 0 and get away with it in Model Space, but not in Paper Space, especially when you create a detail sheet consisting of numerous drawing that have been attached through the XREF command.

Figure 19–43 consists of a typical layout in Paper Space; two views of different drawings have been laid out. Activating the Xref Manager dialog box shows that both drawings were attached to the viewports in Figure 19–44; namely Flange1 and Guide_5A. Also, both drawings were drawn on Layer 0. Before you use the ZOOM command with the XP option, certain layers need to be frozen in designated viewports, which will leave both drawings displayed in their own viewports.

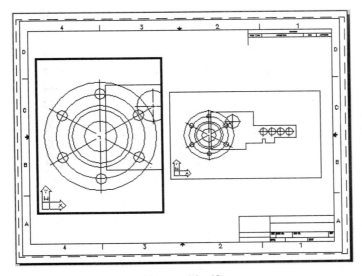

Figure 19–44

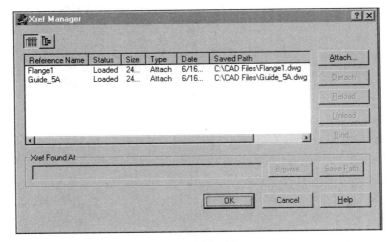

Figure 19–44

When the Layer Properties Manager dialog box opens, observe the layers that appear in the list in Figure 19–45. Instead of the familiar layers of Flange1|0 and Guide_A5|0, all that displays is Layer 0. There is no way to freeze the Layer 0 in the left viewport without affecting the layers in the right viewport. For this reason it is highly recommended that you keep all geometry off Layer 0. At this point, each drawing must be opened, layers assigned, and geometry changed to the new layers. Finally, each External Reference needs to be reattached.

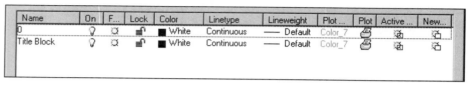

Figure 19–45

NOTES TO THE STUDENT AND INSTRUCTOR

Two tutorial exercises have been designed around the topic of Paper Space. As with all tutorials, you will tend to follow the steps very closely, taking care not to make a mistake. This is to be expected. However, most individuals rush through the tutorials to get the correct solution only to forget the steps used to complete the tutorial. This is also to be expected.

Of all tutorial exercises in this book, the Paper Space tutorials will probably pose the greatest challenge. Certain operations have to be performed in paper space while other operations require floating model space. For example, it was already illustrated that viewports constructed with the MVIEW command must be accomplished in Paper Space; Scaling an image to Paper Space units using the XP option of the ZOOM command must be accomplished in floating Model Space, and so on.

It is recommended to both student and instructor that all tutorial exercises and especially those tutorials that deal with Paper Space, be performed two or even three times. Completing the tutorial the first time will give you the confidence that it can be done; however, you may not understand all of the steps involved. Completing the tutorial a second time will allow you to focus on where certain operations are performed and why things behave the way they do. This still may not be enough. Completing the tutorial exercise a third time will allow you to anticipate each step and have a better idea of which operation to perform in each step.

This recommendation should be exercised with all tutorials; however, it should especially be practiced when you work on the following Paper Space tutorials. For applying the concepts of paper space to other drawing files, problems have been set up at the end of this chapter that take advantage of previously completed drawings from other chapters. Use Paper Space as the layout mechanism for each of the problems.

TUTORIAL EXERCISE: HOUSING2.DWG

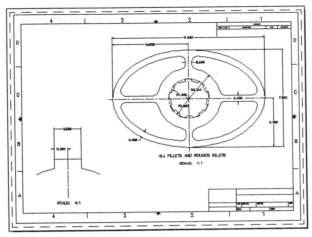

Figure 19–46

Purpose

This tutorial is designed to lay out two views of the same drawing at different scales on the same sheet using the layout mode (Paper Space environment) in Figure 19–46. This drawing is provided on the CD under the name of Housing2.Dwg.

System Settings

Keep the default drawing limits at (0.0000,0.0000) for the lower left corner and (12.0000,9.0000) for the upper right corner. Use the Drawing Units Dialog box (activated through the UNITS command) to change the number of decimal places past the zero from four units to three units.

Layers

In addition to the current layers of the drawing, create the following layers:

Name	Color	Linetype
Border	White	Continuous
Detail-cen	Yellow	Center
Detail-dim	Yellow	Continuous
Vports	Green	Continuous

Suggested Commands

Open the drawing Housing2.Dwg. Enter the layout mode (Paper Space environment), insert a "C" size title block, and create two viewports: one large and the other small. As the image of Housing2 appears in both, use the ZOOM command to scale the first image to Paper Space units at a factor of 1XP or full size. Then scale the detail using a scale factor of 4XP, which is four times its original size. Freeze the Dim layer in the detail viewport while leaving it intact in the large viewport. Create a new dimension style called DETAIL and change the dimension scale to reflect the current scale of the detail image inside the viewport. Place two dimensions inside this viewport. Then Freeze the Detail-dim layer in the large viewport. Plot the drawing out at a scale of 1=1.

Whenever possible, substitute the appropriate command alias in place of the full AutoCAD command in each tutorial step; for example, use "CP" for the COPY command, "L" for the LINE command, and so on. The complete listing of all command aliases is located in Table 1–2.

STEP I

Before beginning the layout of Housing2 and including one detail, create the following new layers, illustrated in Figure 19–47, in the Layer Properties Manager dialog box: Border, Detail-cen, Detail-dim, and Vports. Here is an explanation of the purpose of these layers:

Name	On	Freeze...	Lock	Color	Linetype	Lineweight	Plot Style	Plot
0	♀	☼	🔓	■ White	CONTINUOUS	—— Default	Style_1	🖨
CEN	♀	☼	🔓	☐ Yellow	CENTER	—— Default	Style_1	🖨
DefPoints	♀	☼	🔓	■ White	CONTINUOUS	—— Default	Style_1	🖨
DIM	♀	☼	🔓	■ M...nta	CONTINUOUS	—— Default	Style_1	🖨
OBJECT	♀	☼	🔓	■ White	CONTINUOUS	—— Default	Style_1	🖨
Border	♀	☼	🔓	■ White	CONTINUOUS	—— Default	Normal	🖨
Detail-cen	♀	☼	🔓	☐ Yellow	CENTER	—— Default	Normal	🖨
Detail-dim	♀	☼	🔓	☐ Yellow	CONTINUOUS	—— Default	Normal	🖨
Vports	♀	☼	🔓	☐ Green	CONTINUOUS	—— Default	Normal	🖨

Figure 19–47

BORDER

This layer will hold the title block information that is inserted in Paper Space.

DETAIL-CEN

This layer is designed specifically for the detail that will accompany the main image of Housing2. A special centerline will need to be drawn in the detail. However, a centerline will already be visible in the detail on the Cen layer. Paper Space allows for layers to be visible in certain viewports and hidden, frozen, or invisible in others. Once the Cen layer is frozen only in the detail viewport, a new centerline will be drawn in the detail viewport on the Detail-cen layer.

DETAIL-DIM

This layer is specifically designed to hold any dimension drawn in the detail viewport.

Again, as new dimensions are drawn in the detail viewport, they must be frozen, or turned off, in the main Housing2 viewport. This is easy to accomplish while you are in the Paper Space environment.

VPORTS

This layer is designed to hold all viewport information as each is drawn. Viewports need to be visible for sizing them with grips if the image inside the viewport is too large or small in relation to the image. Once the image has been properly scaled in the viewport and all information is visible in the viewport, the Layer Properties Manager dialog box is used to turn off, freeze, or set to no-plot the Vports layer. In this way, only the images and title block will plot out.

STEP 2

Enter layout mode by selecting a layout tab at the bottom of the drawing screen or by clicking on the MODEL/PAPER toggle button located in the bottom Status bar area in Figure 19–48.

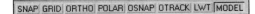

Figure 19–48

The Page Setup dialog box in Figure 19–49 is displayed. Change the layout name in the edit box to "Housing2 and Detail." With the Layout Setting tab selected, choose a "C" size sheet of paper from the drop down list box. Be sure the current plotter is the DesignJet 750C Plus. Verify that the plotting scale is 1:1 and click OK to accept the settings.

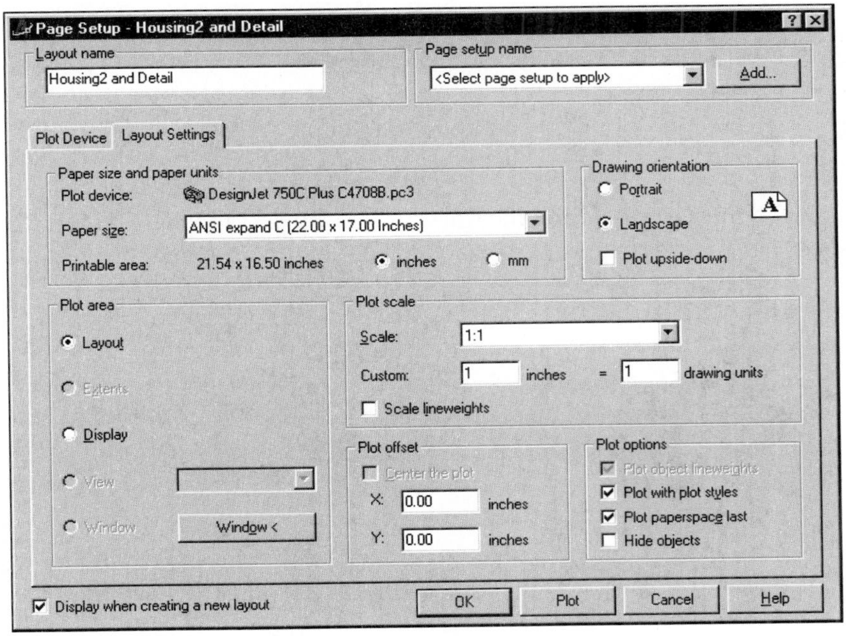

Figure 19–49

You will be returned to the screen in Paper Space and provided a view of your new drawing sheet layout (see Figure 19–50). The "C" size sheet is shown with a dashed line representing the printable area. You have also been provided with a floating viewport, which allows you to see the housing.

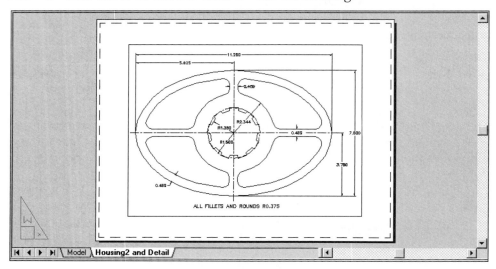

Figure 19–50

STEP 3

Insert the ANSI-C title block on the Border layer, using the INSERT command or the AutoCAD DesignCenter. Your display should be similar to Figure 19–51.

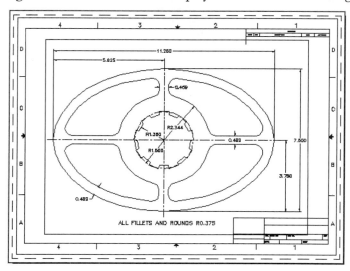

Figure 19–51

STEP 4

To make room for a second viewport that will contain a detail, resize the viewport using grips. Your drawing should appear similar to Figure 19–52. Next, create an additional viewport inside of Paper Space using the MVIEW command. This can be entered at the command prompt, or by choosing Viewports > from the View pull-down menu and then 1 Viewport, or by selecting from the Viewports toolbar, as in Figure 19–53. The size of each viewport is not critical at this stage. Use the illustration in Figure 19–54 as a guide for how large to construct the new viewport.

Command: **MV** *(For MVIEW)*
Specify corner of viewport or
[ON/OFF/Fit/Hideplot/Lock/Object/
    Polygonal/Restore/2/3/4] <Fit>: *(Pick
    at "A" in Figure 19–54)*
Specify opposite corner: *(Pick at "B" in
    Figure 19–54)*
Regenerating model.

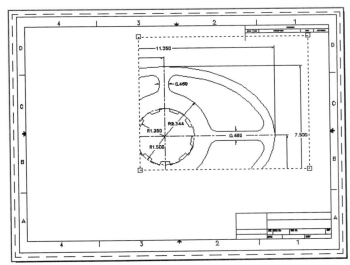

Figure 19–52

Do not be concerned if the viewports overlap each other. This is acceptable inside the Paper Space environment. After creating these viewports, notice that the original image of the drawing of Housing2 reappears inside both viewports. The large viewport will contain the entire image of Housing2; the smaller viewport will contain an enlargement of one circular gear tooth, which will be dimensioned inside the detail viewport.

Single Viewport

Figure 19–53

(A)
(B)
11.250
0.459
R2.344
R1.250
R1.500
0.480
7.500

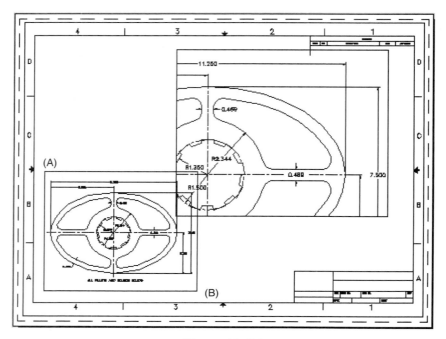

Figure 19–54

Before continuing to the next step, change both viewports to the Vports layer; their color should now be green.

STEP 5

Switch to floating Model Space by entering MS at the command prompt, by clicking on the PAPER/MODEL toggle button located in the status bar shown in Figure 19–55, or by double-clicking a viewport.

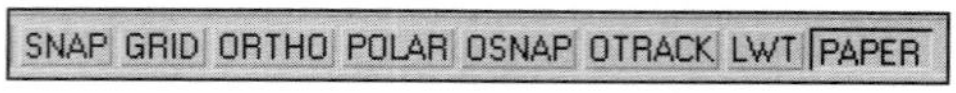

SNAP GRID ORTHO POLAR OSNAP OTRACK LWT PAPER

Figure 19–55

Command: **MS** *(For MSPACE)*
(Switches to Floating model space)

Notice that the User Coordinate System icon is present in both viewports. However, only one viewport is now active. Click inside the large viewport in Figure 19–56 to make this the active viewport. Notice how the viewport outline appears thicker to signify that it is the current viewport. The image inside of this viewport will be scaled full size in relation to Paper Space units. While in the large viewport, use the Viewports toolbar or issue the ZOOM command to scale the image inside the viewport to Paper Space units. Since the image is designed to be scaled at full size, enter a value of 1:1 in the toolbar or 1XP for the ZOOM command scale factor.

Command: **Z** *(For ZOOM)*
Specify corner of window, enter a scale factor (nX or nXP), or [All/Center/ Dynamic/Extents/Previous/Scale/ Window] <real time>: **1XP**

The image inside the large viewport is now scaled full size as in Figure 19–56. If necessary, switch to Paper Space and use grips to size the viewport to the image.

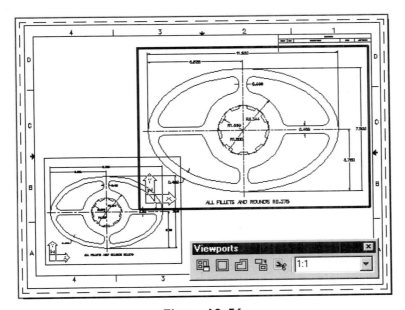

Figure 19–56

STEP 6

Be sure you are in floating Model Space; now move to the smaller viewport and make it active by clicking inside it, as in Figure 19–57. Again, this viewport outline will appear thicker, signifying that it is the current viewport. This image needs to be enlarged to four times its original size for viewing the gear tooth to be dimensioned. Before increasing the size, use the PAN command to center the tooth area that is to be enlarged in the viewport. With this small viewport active, use the Viewports toolbar or issue the ZOOM command to scale the image inside the small viewport to four times the Paper Space units. Since this image is designed to be viewed at four times its normal size, enter a value of 4:1 in the toolbar or 4XP for the ZOOM command scale factor.

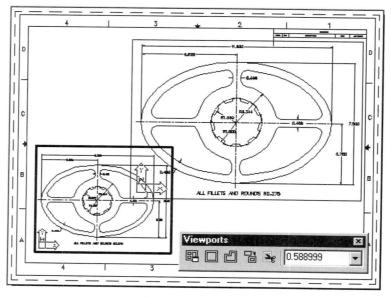

Figure 19–57

Command: **Z** *(For ZOOM)*
Specify corner of window, enter a scale factor (nX or nXP), or [All/Center/Dynamic/Extents/Previous/Scale/Window] <real time>: **4XP**

Notice that the image increases in size. Use the PAN command to locate the gear tooth located at the 90° position (see Figure 19–58). If necessary, switch to Paper Space and use grips to size the viewport to the image.

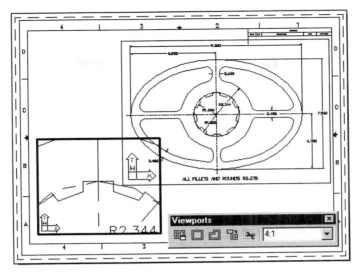

Figure 19–58

STEP 7

Before continuing with the next operation, be sure floating Model Space is active and the small viewport containing the detail is the current viewport. Activate the Layer Properties Manager dialog box and use Figure 19–59 as a guide to freeze the CEN layer in the active viewport. This will freeze the centerlines located only in the detail viewport; the centerlines located in the large viewport will remain visible (see Figure 19–60).

Name	On	Freeze...	Lock	Color	Linetype	Lineweight	Plot Style	Plot	Active ...	New ...
0	♀	✿	⬤	■ White	CONTINUOUS	—— Default	Style_1	🖨	🗗	🗗
Border	♀	✿	⬤	■ White	CONTINUOUS	—— Default	Normal	🖨	🗗	🗗
CEN	♀	✿	⬤	White	CENTER	—— Default	Style_1	🖨	🗗	🗗
DefPoints	♀	✿	⬤	■ White	CONTINUOUS	—— Default	Style_1	🖨	🗗	🗗
Detail-cen	♀	✿	⬤	☐ Yellow	CENTER	—— Default	Normal	🖨	🗗	🗗
Detail-dim	♀	✿	⬤	☐ Yellow	CONTINUOUS	—— Default	Normal	🖨	🗗	🗗
DIM	♀	✿	⬤	■ White	CONTINUOUS	—— Default	Style_1	🖨	🗗	🗗
OBJECT	♀	✿	⬤	■ White	CONTINUOUS	—— Default	Style_1	🖨	🗗	🗗
Title Block	♀	✿	⬤	■ White	CONTINUOUS	—— Default	Normal	🖨	🗗	🗗
Vports	♀	✿	⬤	☐ Green	CONTINUOUS	—— Default	Normal	🖨	🗗	🗗

Figure 19–59

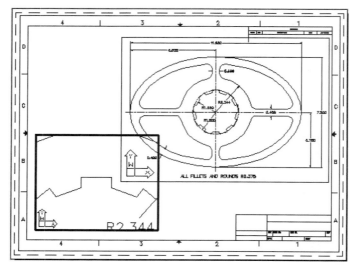

Figure 19–60

STEP 8

Make Detail-cen the new current layer
using the Layer Properties Manager dialog box shown in Figure 19–61.

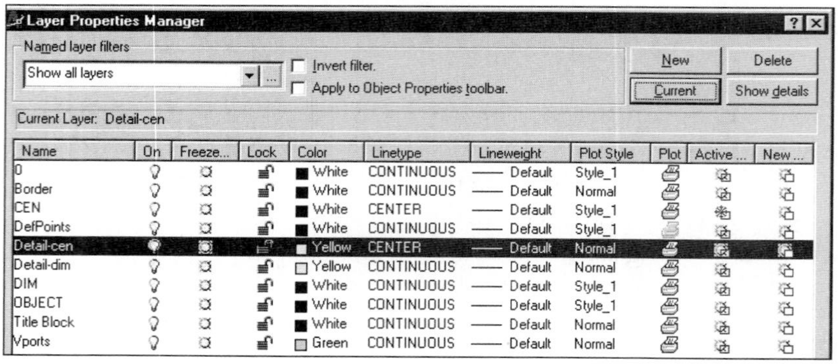

Figure 19–61

Next, construct a centerline in the detail
view from the midpoint of the gear tooth
at "A" a distance of 0.75 units in the 270
direction using a polar coordinate or the
Polar Tracking mode in Figure 19–62.

Command: **L** *(For LINE)*
Specify first point: **Mid**
of *(Pick the midpoint of the line at "A")*
Specify next point or [Undo]:
  @0.75<270
Specify next point or [Undo]: *(Press
  ENTER to exit this command)*

By default, the PSLTSCALE system variable is turned on and allows Paper Space to control the linetype scale depending on the scale of the image inside the viewport. If PSLTSCALE were turned off, the line would resemble the continuous linetype. Because the scale of the viewport increased by a factor of four, the linetype would be too large to display its dashes.

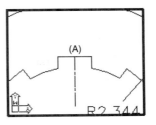

Figure 19–62

STEP 9

Before you begin to dimension inside the viewport containing the detail of the gear tooth, a few items need to be prepared.

First, make Detail-dim the new current layer as in Figure 19–63.

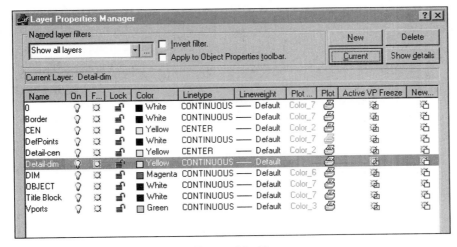

Figure 19–63

Next, create a new dimension style: display the Dimension Style Manager dialog box using the DDIM command (D is the command's alias). Click the New button to display the Create New Dimension Style dialog box shown in Figure 19–64.

In the New Style Name edit box, change the name to DETAIL and click on the Continue button. This will create a new dimension style called DETAIL from the information contained in the STANDARD dimension style.

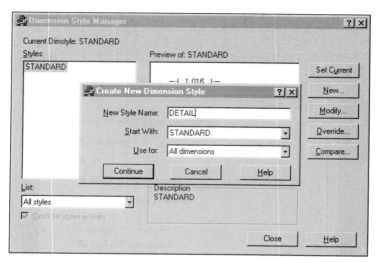

Figure 19–64

One other change needs to be made. Click on the Fit tab in the New Dimension Style: dialog box in Figure 19–65. In the Scale for Dimension Features area, select the radio button for Scale dimensions to layout (paperspace). This will au-tomatically make the dimension settings relative to the scale of the Paper Space viewport and ensure all dimension text, arrowheads, and other dimension settings will be identical in appearance to the dimensions located in the main viewport.

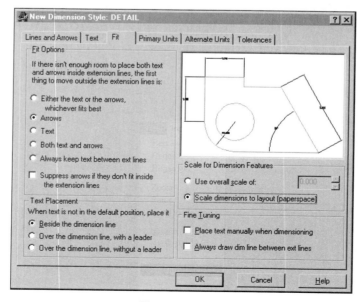

Figure 19–65

Save this change to the DETAIL dimension style by clicking OK as in Figure 19–65. In the Dimension Style Manager dialog box, click the Set Current button to make DETAIL the current dimension style (see Figure 19–66), and then click the Close button.

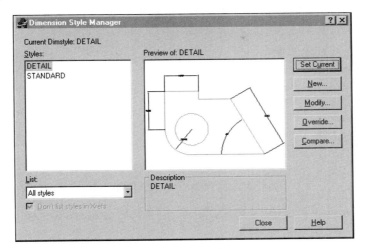

Figure 19–66

STEP 10

With Detail-dim as the current layer and with the enlargement of the gear tooth as the current viewport in floating Model Space, place a horizontal dimension from the corner of the gear tooth to the endpoint of the centerline in Figure 19–67.

Command: **DLI** *(For DIMLINEAR)*
Specify first extension line origin or
    <select object>: **End**
of *(Pick the endpoint of the line at "A")*
Specify second extension line origin: **End**
of *(Pick the endpoint of the line at "B")*
Specify dimension line location or
[Mtext/Text/Angle/Horizontal/Vertical/
    Rotated]: *(Pick a point at "C")*
Dimension text = 0.250

With the linear dimension placed, add a baseline dimension using the DIMBASELINE command in Figure 19–67.

Command: **DBA** *(For DIMBASELINE)*
Specify a second extension line origin or
    [Undo/Select] <Select>): **End**
of *(Pick the endpoint of the line at "D")*
Dimension text = 0.500
Specify a second extension line origin or
    [Undo/Select] <Select>): *(Press ENTER
    to exit this mode)*
Select base dimension: *(Press ENTER to exit
    this command)*

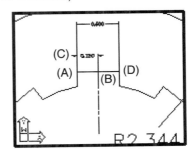

Figure 19–67

STEP 11

Notice how the original dimension of R2.344 still appears inside the small viewport. Just as the Cen layer was frozen in this viewport back in step 7, take this opportunity to freeze the Dim layer only in the small viewport through the Layer Properties Manager dialog box shown in Figure 19–68. Remember that this operation freezes the layer only in the small viewport while keeping the Dim layer visible in the large viewport.

Name	On	F...	Lock	Color	Linetype	Lineweight	Plot ...	Plot	Active VP Freeze	New...
0	♀	☼	🔓	■ White	CONTINUOUS	—— Default	Color_7	🖨	❄	❄
Border	♀	☼	🔓	■ White	CONTINUOUS	—— Default	Color_7	🖨	❄	❄
CEN	♀	☼	🔓	□ Yellow	CENTER	—— Default	Color_2	🖨	❄	❄
DefPoints	♀	☼	🔓	■ White	CONTINUOUS	—— Default	Color_7	🖨	❄	❄
Detail-cen	♀	☼	🔓	□ Yellow	CENTER	—— Default	Color_2	🖨	❄	❄
Detail-dim	♀	☼	🔓	□ Green	CONTINUOUS	—— Default	Color_3	🖨	❄	❄
DIM	♀	☼	🔓	□ Magenta	CONTINUOUS	—— Default		🖨	❄	❄
OBJECT	♀	☼	🔓	■ White	CONTINUOUS	—— Default	Color_7	🖨	❄	❄
Title Block	♀	☼	🔓	■ White	CONTINUOUS	—— Default	Color_7	🖨	❄	❄
Vports	♀	☼	🔓	□ Green	CONTINUOUS	—— Default	Color_3	🖨	❄	❄

Figure 19–68

STEP 12

Before you complete the layout of this drawing, a few layers need to be frozen in the main viewport holding Housing2. Because a centerline was added to the detail on the Detail-cen layer and two dimensions were added to the detail in the Detail-dim layer, the images of all objects drawn on these layers is present in the main viewport (see Figure 19–69). To fix this, first be sure the main viewport is active by double clicking inside it anywhere. Then activate the Layer Properties Manager dialog box illustrated in Figure 19–70 and freeze the Detail-cen and Detail-cen layers in the active viewport. Notice that both dimensions and centerline disappear from the main viewport.

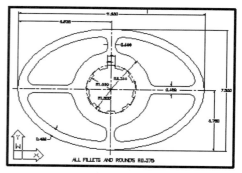

Figure 19–69

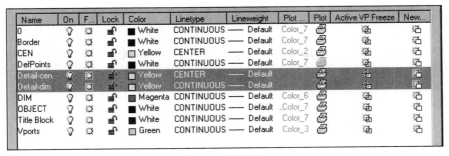

Figure 19–70

STEP 13

Complete the layout of the drawing by switching to Paper Space and adjusting all viewports using grips until the desired effect is achieved, as in Figure 19–71.

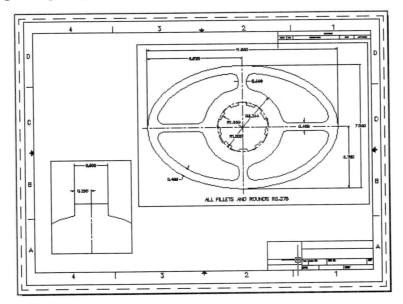

Figure 19–71

STEP 14

Turn off the Vports layer. Scale notes need to be added to both views. This should be accomplished in Paper Space with the current layer set to Dim. Use the MTEXT or DTEXT commands to place the note "SCALE: 4:1" under the detail view and the note "SCALE: 1:1" under the main view of Housing2 in Figure 19–72. Use a text height of 0.18 units. The completed drawing is illustrated in Figure 19–73 and can be plotted out at a scale of 1=1.

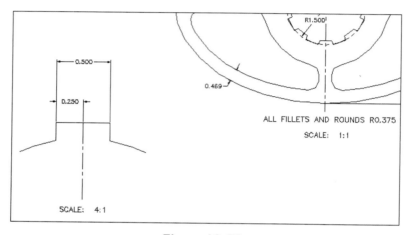

Figure 19–72

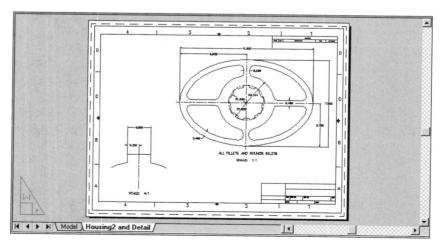

Figure 19–73

TUTORIAL EXERCISE: ARCH_DETAILS.DWG

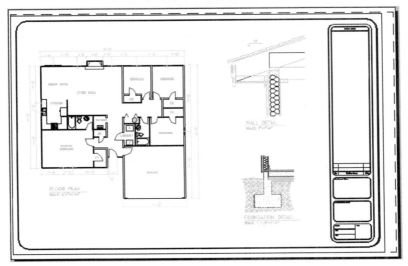

Figure 19–74

Purpose:

This tutorial exercise is designed to lay out three architectural details displayed in Figure 19–74 in the layout mode (Paper Space environment). The three details consist of a floor plan, cornice detail, and foundation detail. External references will also be used to assist in the control of layers.

System Settings:

Begin a new drawing called "Arch_Details." Use the Drawing Units Dialog box (activated through the UNITS command) to change the system of units to Architectural. Keep all default values for the units of the drawing.

Layers:

Create the following layers with the format:

Name	Color	Linetype
Xref	White	Continuous
Title_block	White	Continuous
Viewport	Green	Continuous

Suggested Commands:

A new drawing will be started using layout mode (Paper Space environment) and three viewports will be created through the MVIEW command. Three architectural details provided on the CD will then be attached to each viewport through the External Reference dialog box. The VPLAYER command will be used to freeze layers in certain viewports while keeping them visible in others. Each detail will be scaled to the Paper Space viewport through the XP option of the ZOOM command. This will enable the drawing to be plotted out at a scale factor of 1=1 because the scaling was already performed in each viewport.

Whenever possible, substitute the appropriate command alias in place of the full AutoCAD command in each tutorial step; for example, use "CP" for the COPY command, "L" for the LINE command, and so on. The complete listing of all command aliases is located in Table 1–2.

tutorial **EXERCISE**

Phase I—Individual Drawing Preparation

While three drawings have been provided to step you through this tutorial, it is important to be aware of certain settings inside each drawing based on the scale of the drawing. These usually affect the linetype scale, dimension scale, and individual text height for notes. All three drawings will be explained in the steps that follow.

STEP 1

Open the drawing Floor.Dwg shown in Figure 19–75. This drawing will be plotted out at a scale of 1/4"=1'-0".

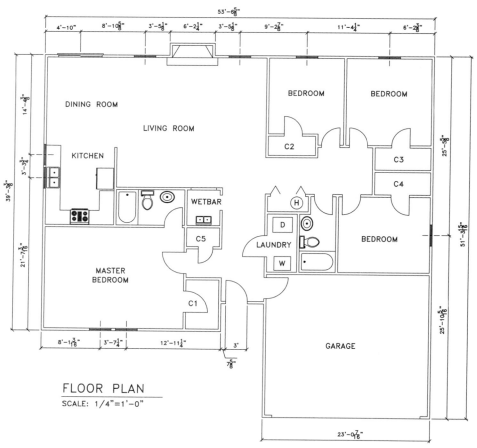

FLOOR PLAN
SCALE: 1/4"=1'-0"

Figure 19–75

Issue the DDIM command and click on the Modify button to display the settings for the Architectural dimension style. Select the Fit tab and notice the overall scale value is set to 48, which is the multiplier found when 1' or 12 is divided by 1/4 or 0.25 units. All dimension parameters such as text and arrow height will be multiplied by a value of 48. Also take note of the layers assigned to this drawing (see Figure 19–76). Close this drawing when you are finished viewing the layers.

Layer name	State	Color	Linetype
0	On	7 (white)	CONTINUOUS
CENTER	On	2 (yellow)	CENTER2
COUNTERTOP	On	7 (white)	CONTINUOUS
DEFPOINTS	On	7 (white)	CONTINUOUS
DIM	On	2 (yellow)	CONTINUOUS
DOORS	On	1 (red)	CONTINUOUS
FIREPLACE	On	1 (red)	CONTINUOUS
FRAMING	On	3 (green)	CONTINUOUS
NOTES	On	4 (cyan)	CONTINUOUS
SYMBOLS	On	7 (white)	CONTINUOUS
TEXT	On	4 (cyan)	CONTINUOUS
WALLS	On	7 (white)	CONTINUOUS
WINDOWS	On	6 (magenta)	CONTINUOUS

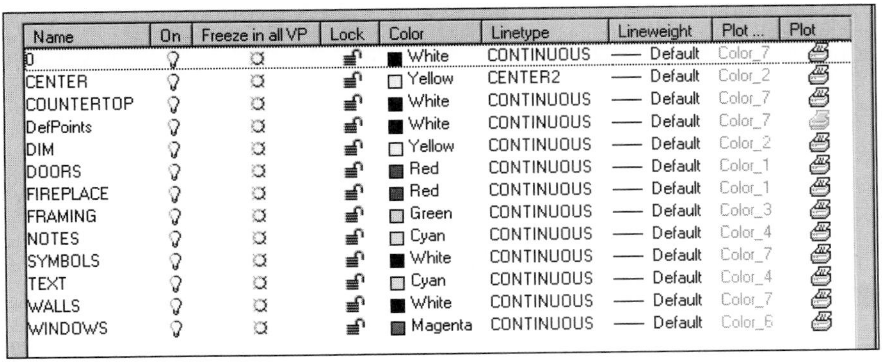

Figure 19–76

STEP 2

Open the drawing Cornice.Dwg (see Figure 19–77). This drawing will be plotted out at a scale of 3"=1'-0".

Issue the DDIM command and click on the Modify button to display the settings for the Standard dimension style. Select the Fit tab and notice the overall scale value is set to 4, which is the multiplier found when 1' or 12 is divided by 3. All dimension parameters such as text and arrow height will be multiplied by a value of 4. Also take note of the layers assigned to this drawing (see Figure 19–78). Close this drawing when you are finished viewing the layers.

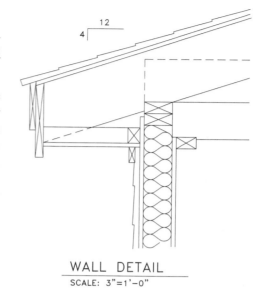

WALL DETAIL
SCALE: 3"=1'-0"

Figure 19–77

Layer name	State	Color	Linetype
0	On	7 (white)	CONTINUOUS
FRAMING	On	3 (green)	CONTINUOUS
HIDDEN	On	1 (red)	HIDDEN
INSULATION	On	7 (white)	CONTINUOUS
NOTES	On	4 (cyan)	CONTINUOUS
SECTION	On	2 (yellow)	CONTINUOUS

Name	On	Freeze in all VP	Lock	Color	Linetype	Lineweight	Plot ...	Plot
0	♀	✿	⬤	■ White	CONTINUOUS	— Default	Color_7	🖶
FRAMING	♀	✿	⬤	□ Green	CONTINUOUS	— Default	Color_3	🖶
HIDDEN	♀	✿	⬤	■ Red	HIDDEN	— Default	Color_1	🖶
INSULATION	♀	✿	⬤	■ White	CONTINUOUS	— Default	Color_7	🖶
NOTES	♀	✿	⬤	□ Cyan	CONTINUOUS	— Default	Color_4	🖶
SECTION	♀	✿	⬤	□ Yellow	CONTINUOUS	— Default	Color_2	🖶

Figure 19–78

STEP 3

Open the drawing Fndation.Dwg. (see Figure 19–79). This drawing will be plotted out at a scale of 1 1/2"=1'-0".

Issue the DDIM command and click on the Modify button to display the settings for the Standard dimension style. Select the Fit tab and notice the overall scale value is set to 8, which is the multiplier found when 1' or 12 is divided by 1 1/2 or 1.50 units. All dimension parameters such as text and arrow height will be multiplied by 8. Also take note of the layers assigned to this drawing (see Figure 19–80). Close this drawing when you are finished viewing the layers.

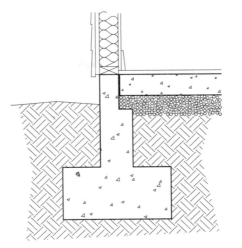

FOUNDATION DETAIL
SCALE: 1 1/2"=1'-0"

Figure 19–79

Layer name	State	Color	Linetype
0	On	7 (white)	CONTINUOUS
CONCRETE	On	2 (yellow)	CONTINUOUS
CONSTRUCTION	On	1 (red)	CONTINUOUS
EARTH	On	3 (green)	CONTINUOUS
FRAMING	On	3 (green)	CONTINUOUS
GRAVEL	On	4 (cyan)	CONTINUOUS
INSULATION	On	7 (white)	CONTINUOUS
NOTES	On	4 (cyan)	CONTINUOUS

Name	On	Freeze in all VP	Lock	Color	Linetype	Lineweight	Plot ...	Plot
0	♀	☼	≝	■ White	CONTINUOUS	—— Default	Color_7	🖨
CONCRETE	♀	☼	≝	☐ Yellow	CONTINUOUS	—— Default	Color_2	🖨
CONSTRUCTION	♀	☼	≝	■ Red	CONTINUOUS	—— Default	Color_1	🖨
EARTH	♀	☼	≝	☐ Green	CONTINUOUS	—— Default	Color_3	🖨
FRAMING	♀	☼	≝	☐ Green	CONTINUOUS	—— Default	Color_3	🖨
GRAVEL	♀	☼	≝	☐ Cyan	CONTINUOUS	—— Default	Color_4	🖨
INSULATION	♀	☼	≝	■ White	CONTINUOUS	—— Default	Color_7	🖨
NOTES	♀	☼	≝	☐ Cyan	CONTINUOUS	—— Default	Color_4	🖨

Figure 19–80

Phase II—Drawing Layout Setup

STEP 4

All three details that will make up the drawing Arch_Details.Dwg are designed to fit on a "D" size sheet of paper. Use the Layout Wizard to setup the drawing from scratch. To begin creating a layout, enter LAYOUTWIZARD at the command prompt, choose it from the Insert pull down menu under Layout > Layout Wizard, or select it from the Tools pull down menu under Wizards > Create Layout to begin creating a layout. Type "Detail Sheet" as in Figure 19–81 for the name of the layout, and select the Next button.

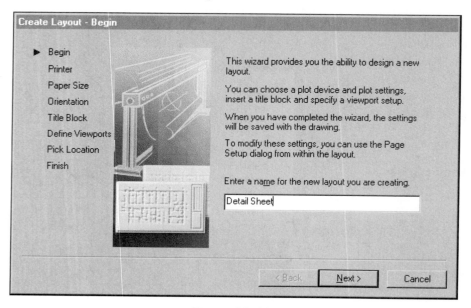

Figure 19–81

In the Create Layout - Printer dialog box, select a DesignJet 750C plotter which will support a "D" size paper. In Figure 19–82, the Create Layout - Paper Size dialog box, select the ARCH expand D sheet size and pick the Next button. Choose Landscape in the Create Layout - Orientation dialog box. In Figure 19–83, the Create Layout - Title Block dialog box, select Architectural Title Block.Dwg as the title block to insert. To insert the drawing as a block rather than attach it as an external reference, verify that the Block radio button is selected in the Type area.

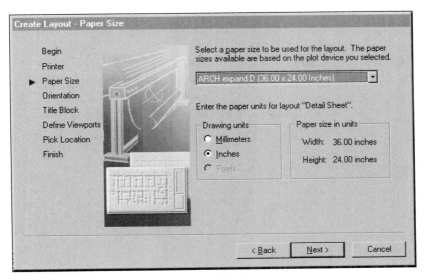

Figure 19–82

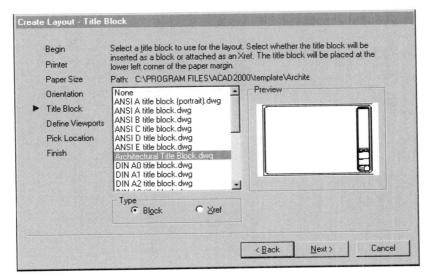

Figure 19–83

In Figure 19–84, the Create Layout - Define Layouts dialog box, you have several options for creating viewports. Select Single as the viewport setup, with a scale of 1/4" = 1'0". This viewport will be used for attaching the Floor.Dwg. The MVIEW command will be used later to create the other two viewports we will need.

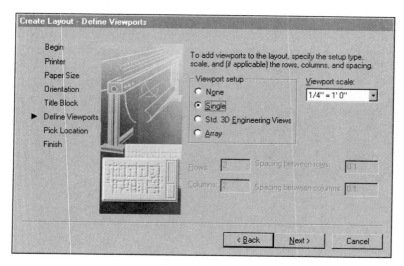

Figure 19–84

In the Create Layout - Pick Location dialog box click the Select Location button and select the viewport corners using the command prompt sequence shown below and Figure 19–85 as a guide. Click the Finish button in the last dialog box to complete the layout wizard.

Specify first corner: **1,3.75** (At "A")
Specify opposite corner: **18,20.50** (At "B")

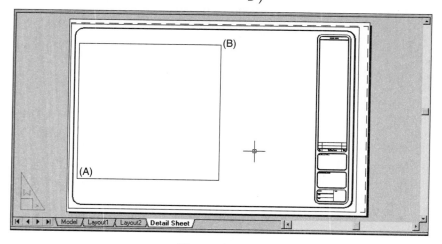

Figure 19–85

STEP 5

Notice three layers that were created earlier in the Layer Properties Manager dialog box in Figure 19–86: a layer called Title_block for all title block information, a layer called Viewport for the viewports used to aid in the layout of the drawing, and the Xref layer for external references. If any of these layers are missing, create them now in the Layer Properties Manager dialog box. Use the PROPERTIES command to move the title block and viewport created by the wizard to their respective layers.

Name	On	Freeze in all VP	Lock	Color	Linetype	Lineweight	Plot ...	Plot
0	♀	☼	🔓	■ White	Continuous	—— Default	Color_7	🖨
Title_block	♀	☼	🔓	■ White	Continuous	—— Default	Color_7	🖨
Viewport	♀	☼	🔓	■ White	Continuous	—— Default	Color_7	🖨
Xref	♀	☼	🔓	■ White	Continuous	—— Default	Color_7	🖨

Figure 19–86

STEP 6

In this step, first make Viewport the current layer. All viewports will be drawn on this layer. Then, use the MVIEW command to make two additional viewports on the Viewport layer. These are illustrated in Figure 19–87. One of these viewports will hold the file Cornice.Dwg, the other the Fndation.Dwg.

Command: **MV** *(For MVIEW)*
Specify corner of viewport or
[ON/OFF/Fit/Hideplot/Lock/Object/
    Polygonal/Restore/2/3/4] <Fit>:
    19.00,11.50 *(At "A")*
Specify opposite corner: **28.50,21.50** *(At "B")*

Regenerating model.
Command: **MV** *(For MVIEW)*
Specify corner of viewport or
[ON/OFF/Fit/Hideplot/Lock/Object/
    Polygonal/Restore/2/3/4] <Fit>:
    19.00,1.25 *(At "C")*
Specify opposite corner: **28.50,10.50** *(At "D")*
Regenerating model.

Although the two new viewports were created with coordinates, in practice, the viewports are drawn to any size and scaled once the images are placed inside each.

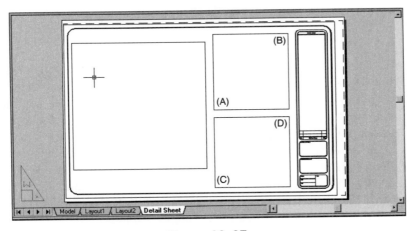

Figure 19–87

Phase III—Drawing Attachment
STEP 7

The Xref layer should have already been created; make this layer the new current layer. In the Status area, click on the PAPER button; this should switch to MODEL, illustrated in Figure 19–88, floating Model Space. Make the large viewport active by clicking anywhere inside it as in Figure 19–89.

SNAP	GRID	ORTHO	POLAR	OSNAP	OTRACK	LWT	MODEL

Figure 19–88

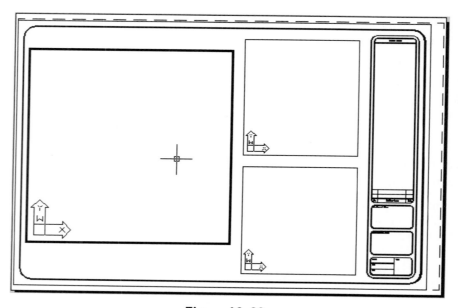

Figure 19–89

STEP 8

Use the XREF command and click on the Attach button to attach the file Floor.Dwg. Pick an insertion point on the screen for this external reference (see Figure 19–90); keep all remaining default values. Use the PAN command to center the floor plan in the viewport as shown in Figure 19–91. Examine all layers of the drawing through the Layer Properties Manager dialog box shown in Figure 19–92. Notice how numerous layers associated with the floor plan begin with the name "FLOOR." As a drawing is attached through the XREF command, all layers in the drawing are dependent on the original file name. This is AutoCAD's way of keeping track of all layers that belong to an external reference file. Notice how the file name and layer name are separated with the vertical bar (|) called the pipe symbol.

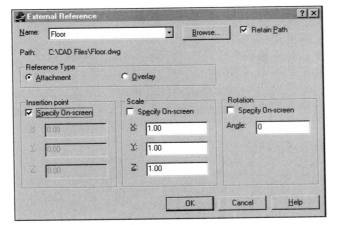

Figure 19–90

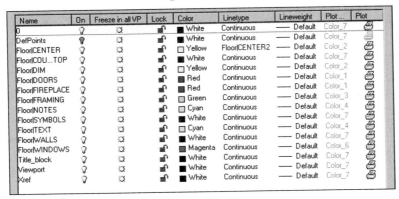

Figure 19–91

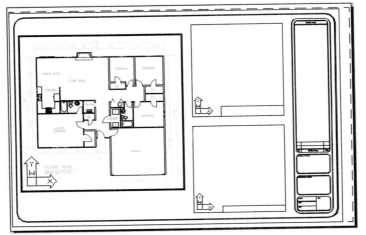

Figure 19–92

STEP 9

From the previous step, notice also how the floor plan is visible in the other two viewports. Activate each of these other viewports and perform a ZOOM-Extents in each. Your display should appear similar to Figure 19–93.

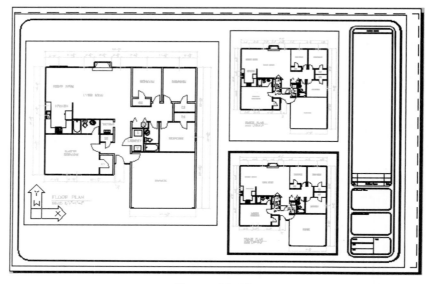

Figure 19–93

STEP 10

One of the advantages of using external references to arrange details in numerous viewports is that it is easy to freeze layers on certain viewports while keeping layers visible in others. Use the VPLAYER (Viewport Layer) command to accomplish this. Freeze all layers that begin with "Floor." As you proceed, this command automatically switches to Paper Space to select the viewports for freezing layers; the command switches back to Model Space when completed. The results of this operation are illustrated in Figure 19–94.

Command: **VPLAYER**
Enter an option [?/Freeze/Thaw/Reset/
  Newfrz/Vpvisdflt]: **F** (For Freeze)

Enter layer name(s) to freeze: **Floor***
  (For all layers that begin with Floor)
Enter an option [All/Select/Current]
  <Current>: **S** *(For Select)*
Switching to Paper space.
Select objects: *(Pick the upper right
  viewport in Figure 19–93)*
Select objects: *(Pick the lower right
  viewport in Figure 19–93)*
Select objects: *(Press ENTER to continue)*
Switching to Model space.
Enter an option [?/Freeze/Thaw/Reset/
  Newfrz/Vpvisdflt]: *(Press ENTER to exit
  this command and freeze all layers
  beginning with Floor)*

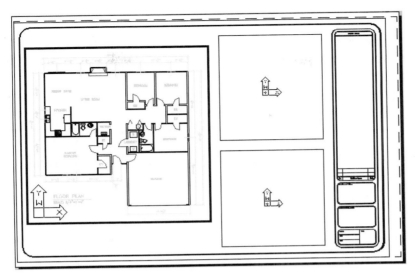

Figure 19–94

STEP 11

Be sure the large viewport, which contains the floor plan, is active. The floor plan is to be plotted out at a scale of 1/4"=1'-0". Normally, you would use the ZOOM command and enter a value of 1/48XP to properly scale the image of the floor plan to reflect Paper Space units. This is not required, due to the scaling of this viewport during utilization of the layout wizard earlier (see Figure 19–84). To verify that the image of the floor plan is properly scaled, open the Viewports toolbar. The toolbar displays the scale of the active viewport in Figure 19–95.

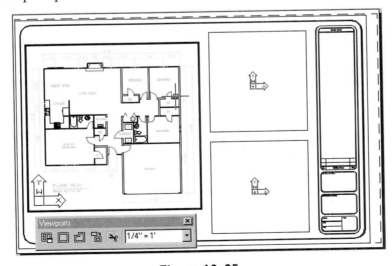

Figure 19–95

STEP 12

Check to see that you are in floating Model Space (the MODEL button should be displayed in the status area), and activate the upper right viewport. Then use the XREF command to attach the file Cornice.Dwg inside this viewport in Figure 19–96. It is important to attach this drawing file in the middle of the upper right viewport. The reason is that when you scale the image to size using the ZOOM command, the image or a portion of the image will be visible in the viewport, enabling you to pan the image into position. If this image is inserted at 0,0 and the image scaled, it may not be visible in the viewport. Panning the image may take a long time; images have even been known to be lost in floating Model Space because they were inserted at 0,0. As in all cases involving floating Model Space, notice the Cornice, although small, is also visible in the other two viewports.

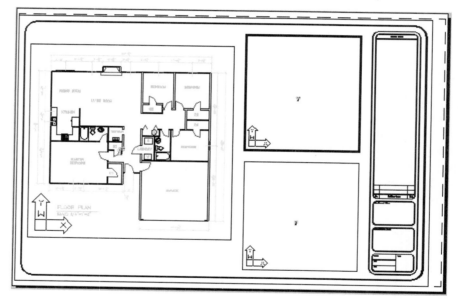

Figure 19–96

STEP 13

As with the floor plan, use the VPLAYER (Viewport Layer) command to freeze all layers that begin with "Cornice" in the left and lower right viewports. The results are illustrated in Figure 19–97.

Command: **VPLAYER**
Enter an option [?/Freeze/Thaw/Reset/
   Newfrz/Vpvisdflt]: **F** *(For Freeze)*
Enter layer name(s) to freeze: **Cornice***
   (For all layers that begin with Cornice)
Enter an option [All/Select/Current]
   <Current>: **S** *(For Select)*

Switching to Paper space.
Select objects: *(Pick the large left viewport in Figure 19–97)*
Select objects: *(Pick the lower right viewport in Figure 19–97)*
Select objects: *(Press ENTER to continue)*
Switching to Model space.
Enter an option [?/Freeze/Thaw/Reset/
   Newfrz/Vpvisdflt]: *(Press ENTER to exit this command and freeze all layers beginning with Cornice)*

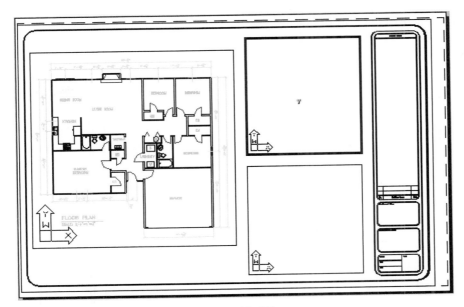

Figure 19–97

STEP 14

Be sure the upper right viewport is still active. Although the image of the Cornice is visible, it is not necessarily at the proper scale. The cornice is to be plotted out at a scale of 3"=1'-0". Use the Viewports toolbar and select the scale from the drop-down list box or use the ZOOM command and enter a value of 1/4XP to properly scale the image of the cornice detail to reflect Paper Space units (see Figure 19–98).

If the image does not display in its entirety in the viewport, you can use the PAN command to pan the image across the viewport until it is positioned properly. You can easily do this, especially since this is still the active viewport in floating Model Space. You could also click on the MODEL button in the Status area to switch back to PAPER, click on the edge of the viewport to expose grips, and stretch the viewport to the desired size.

Command: **Z** *(For ZOOM)*
Specify corner of window, enter a scale
    factor (nX or nXP), or [All/Center/
    Dynamic/Extents/Previous/Scale/
    Window] <real time>: **1/4XP**

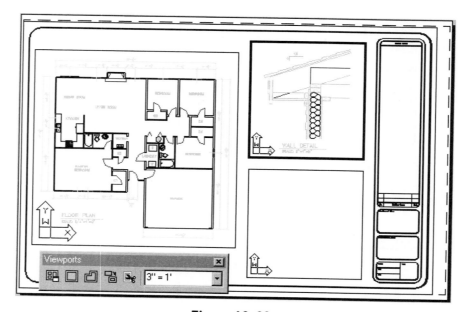

Figure 19–98

STEP 15

Again, check to see that you are in floating Model Space (the MODEL button should be displayed in the status area), and activate the lower right viewport. Then use the XREF command to attach the file Fndation.Dwg inside this viewport as in Figure 19–99. As with the Cornice, attach the Fndation drawing file in the middle of the lower right viewport. As in all cases involving floating Model Space, notice that the file Fndation, although small, is also visible in the other two viewports.

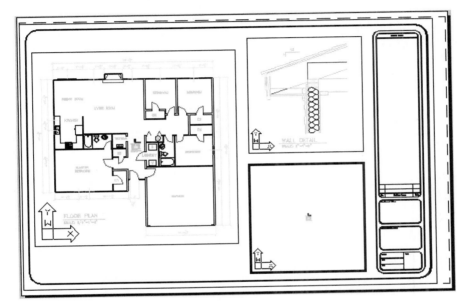

Figure 19–99

STEP 16

Use the VPLAYER (Viewport Layer) command to freeze all layers that begin with "Fndation" in the left and upper right viewports. The results are illustrated in Figure 19–100.

Command: **VPLAYER**
Enter an option [?/Freeze/Thaw/Reset/
  Newfrz/Vpvisdflt]: **F** *(For Freeze)*
Enter layer name(s) to freeze:
  Fndation* *(For all layers that begin with Fndation)*
Enter an option [All/Select/Current]
  <Current>: **S** *(For Select)*

Switching to Paper space.
Select objects: *(Pick the large left viewport in Figure 19–100)*
Select objects: *(Pick the upper right viewport in Figure 19–100)*
Select objects: *(Press ENTER to continue)*
Switching to Model space.
Enter an option [?/Freeze/Thaw/Reset/
  Newfrz/Vpvisdflt]: *(Press ENTER to exit this command and freeze all layers beginning with Fndation)*

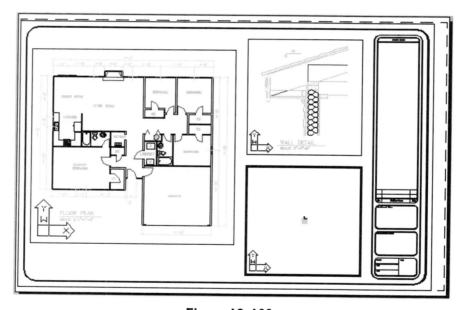

Figure 19–100

STEP 17

Check to see that the lower left viewport is active. Although the image of the foundation is visible, it is not necessarily at the proper scale. The foundation is to be plotted out at a scale of 1 1/2"=1'-0". Use the Viewports toolbar to select the scale (1:8) or use the ZOOM command and enter a value of 1/8XP to properly scale the image of the foundation detail to reflect Paper Space units (see Figure 19–101).

Command: **Z** *(For ZOOM)*
Specify corner of window, enter a scale factor (nX or nXP), or [All/Center/Dynamic/Extents/Previous/Scale/Window] <real time>: **1/8XP**

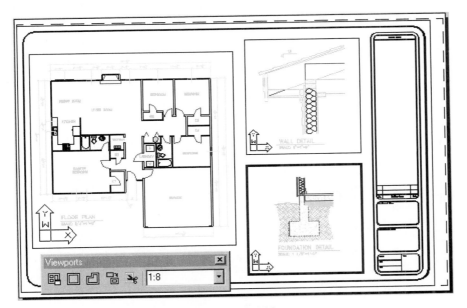

Figure 19–101

If the image does not display in its entirety in the viewport, you can use the PAN command to pan the image across the viewport until it is positioned properly. You can do this easily, especially since this is still the active viewport in floating Model Space. You could also click on the MODEL button in the Status area to switch back to PAPER, click on the edge of the viewport to expose grips, and stretch the viewport to the desired size.

STEP 18

All views are now properly positioned with images frozen in certain viewports and visible in others. If, however, another viewport were to be constructed through the MVIEW command, all three images would appear in the new viewport. To prevent existing images from displaying in any new viewports, use the VPLAYER command along with the Vpvisdflt option. This options stands for Viewport Visibility Default. If certain layers are identified through this option and the layers are frozen, they will not appear in any new viewport created.

Command: **VPLAYER**
Enter an option [?/Freeze/Thaw/Reset/
  Newfrz/Vpvisdflt]: **V** (For Vpvisdflt)

Enter layer name(s) to change viewport
  visibility:
  Floor*,Cornice*,Fndation*
Enter a viewport visibility option [Frozen/
  Thawed] <Thawed>: **F** (For Frozen)
Enter an option [?/Freeze/Thaw/Reset/
  Newfrz/Vpvisdflt]: (Press ENTER to exit
  this command)

Yet another way to freeze all layers in any new viewports is to activate the Layer Properties Manager dialog box shown in Figure 19–102. First, right-click a layer name and choose Select All from the shortcut menu. With all layers selected, click on one of the icons in the "New VP Freeze" column. This changes all sun symbols to the snowflake (see Figure 19–102).

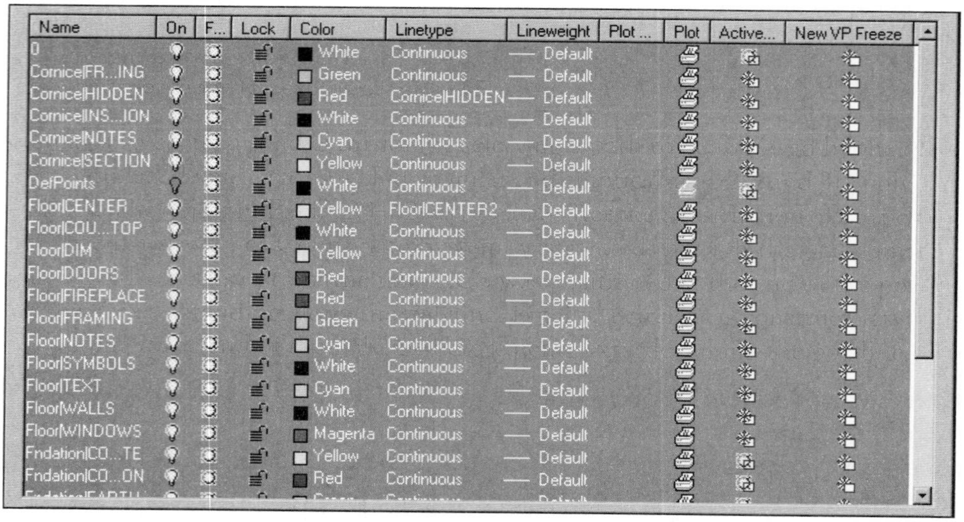

Figure 19–102

STEP 19

Turn off the Vports layer to display only the three details. Be sure the PAPER button is displayed in the Status bar shown in Figure 19–103. The drawing of the three details can now be plotted out at to a scale of 1=1 because the ZOOM XP was used to scale each detail to the specific viewport it occupies. The completed layout is illustrated in Figure 19–104.

| SNAP | GRID | ORTHO | POLAR | OSNAP | OTRACK | LWT | PAPER |

Figure 19–103

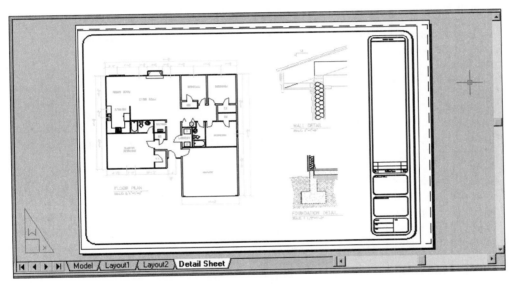

Figure 19–104

PROBLEMS FOR CHAPTER 19

Directions for Chapter 19 Problems:

Arrange Problems 13–1 and 13–2 from Chapter 13, "Analyzing 2D Drawings" on the same sheet in Paper Space. Create layers called Vports and Xref. Use ANSI_C.Dwg as the title block inside Paper Space. Scale each image inside its floating Model Space viewport at a scale of 1XP or full size.

Arrange Problems 13–3 and 13–4 from Chapter 13, "Analyzing 2D Drawings" on the same sheet in Paper Space. Create layers called Vports and Xref. Use ANSI_C.Dwg as the title block inside Paper Space. Scale Problem 13–3 inside its floating Model Space viewport at a scale of 1/4XP; scale metric Problem 13–4 inside its floating Model Space viewport at a scale of 1/25.4XP.

Arrange Problems 13–5 and 13–6 from Chapter 13, "Analyzing 2D Drawings" on the same sheet in Paper Space. Create layers called Vports and Xref. Use ANSI_C.Dwg as the title block inside Paper Space. Scale each image inside its floating Model Space viewport at a scale of 1XP or full size.

Arrange Problems 13–7 and 13–8 from Chapter 13, "Analyzing 2D Drawings" on the same sheet in Paper Space. Create layers called Vports and Xref. Use ANSI_C.Dwg as the title block inside Paper Space. Scale each image inside its floating Model Space viewport at a scale of 1XP or full size.

Arrange Problems 13–9 and 13–10 from Chapter 13, "Analyzing 2D Drawings" on the same sheet in Paper Space. Create layers called Vports and Xref. Use ANSI_C.Dwg as the title block inside Paper Space. Scale metric Problem 13–9 inside its floating Model Space viewport at a scale of 1/25.4XP; scale Problem 13–104 inside its floating Model Space viewport at a scale of 1/2XP.

Arrange Problems 13–11 and 13–12 from Chapter 13, "Analyzing 2D Drawings" on the same sheet in Paper Space. Create layers called Vports and Xref. Use ANSI_D.Dwg as the title block inside Paper Space. Scale each image inside its floating Model Space viewport at a scale of 1XP or full size.

Arrange Problems 13–13 and 13–14 from Chapter 13, "Analyzing 2D Drawings" on the same sheet in Paper Space. Create layers called Vports and Xref. Use ANSI_D.Dwg as the title block inside Paper Space. Scale each image inside its floating Model Space viewport at a scale of 1XP or full size.

Arrange Problems 13–15 and 13–16 from Chapter 13, "Analyzing 2D Drawings" on the same sheet in Paper Space. Create layers called Vports and Xref. Use ANSI_D.Dwg as the title block inside Paper Space. Scale metric Problem 13–15 inside its floating Model Space viewport at a scale of 1/25.4XP; scale Problem 13–16 inside its floating Model Space viewport at a scale of 1XP.

Arrange Problems 13–17 and 13–18 from Chapter 13, "Analyzing 2D Drawings" on the same sheet in Paper Space. Create layers called Vports and Xref. Use ANSI_D.Dwg as the title block inside Paper Space. Scale each image inside its floating Model Space viewport at a scale of 1XP or full size.

Arrange Problems 13–19 and 13–20 from Chapter 13, "Analyzing 2D Drawings" on the same sheet in Paper Space. Create layers called Vports and Xref. Use ANSI_D.Dwg as the title block inside Paper Space. Scale Problem 13–19 inside its floating Model Space viewport at a scale of 1XP; scale metric Problem 13–20 inside its floating Model Space viewport at a scale of 1/25.4XP.

Arrange Problems 13–21 and 13–22 from Chapter 13, "Analyzing 2D Drawings" on the same sheet in Paper Space. Create layers called Vports and Xref. Use ANSI_D.Dwg as the title block inside Paper Space. Scale each image inside its floating Model Space viewport at a scale of 1XP or full size.

Arrange Problems 13–23 and 13–24 from Chapter 13, "Analyzing 2D Drawings" on the same sheet in Paper Space. Create layers called Vports and Xref. Use ANSI_D.Dwg as the title block inside Paper Space. Scale each image inside its floating Model Space viewport at a scale of 1XP or full size.

Arrange Problems 13–25 and 13–26 from Chapter 13, "Analyzing 2D Drawings" on the same sheet in Paper Space. Create layers called Vports and Xref. Use ANSI_D.Dwg as the title block inside Paper Space. Scale each image inside its floating Model Space viewport at a scale of 1XP or full size.

Arrange Problems 13–27 and 13–28 from Chapter 13, "Analyzing 2D Drawings" on the same sheet in Paper Space. Create layers called Vports and Xref. Use ANSI_D.Dwg as the title block inside Paper Space. Scale each image inside its floating Model Space viewport at a scale of 1XP or full size.

Arrange Problems 13–29 and 13–30 from Chapter 13, "Analyzing 2D Drawings" on the same sheet in Paper Space. Create layers called Vports and Xref. Use ANSI_D.Dwg as the title block inside Paper Space. Scale Problem 13–29 inside its floating Model Space viewport at a scale of 1/2XP; scale metric Problem 13–30 inside its floating Model Space viewport at a scale of 1/4XP.

Arrange Problem 13–31 from Chapter 13, "Analyzing 2D Drawings" in Paper Space. Use ArchEng.Dwg as the title block inside Paper Space. Scale Problem 13–31 inside its floating Model Space viewport at a scale of 1/48XP (1/4" = 1'-0").

Arrange Problem 13–32 from Chapter 13, "Analyzing 2D Drawings" in Paper Space. Use ArchEng.Dwg as the title block inside Paper Space. Scale Problem 13–32 inside its floating Model Space viewport at a scale of 1/4XP (3" = 1'-0").

problem EXERCISE

Arrange Problem 13–33 from Chapter 13, "Analyzing 2D Drawings" in Paper Space. Use ArchEng.Dwg as the title block inside Paper Space. Scale Problem 13–33 inside its floating Model Space viewport at a scale of 1/200XP.

Arrange Problem 13–34 from Chapter 13, "Analyzing 2D Drawings" in Paper Space. Use ArchEng.Dwg as the title block inside Paper Space. Scale Problem 13–34 inside its floating Model Space viewport at a scale of 1/24XP (1/2" = 1'-0").

Arrange Problem 13–35 from Chapter 13, "Analyzing 2D Drawings" in Paper Space. Use ArchEng.Dwg as the title block inside Paper Space. Scale Problem 13–35 inside its floating Model Space viewport at a scale of 1/24XP (1/2" = 1'-0").

Solid Modeling Fundamentals

It is said that humans see, hear, and exist in a 3D world. Why not draw in three dimensions, and visualize the object before placing objects on the computer screen? Part of learning the art of visualization is the study and construction of models in 3D in order to obtain as accurate as possible an image of an object undergoing design. Solid models are more informationally correct than wireframe models because a solid object may be analyzed through the calculation of such items as mass properties, center of gravity, surface area, moments of inertia, and much more. The solid model starts the true design process by defining objects as a series of primitives. Boxes, cubes, cylinders, spheres, and wedges are all examples of primitives. These building blocks are then joined together or subtracted from each other through certain modifying commands. Fillets and chamfers may be created to give the solid model a more realistic appearance and functionality. Two-dimensional views may be extracted from the solid model along with a cross section of the model. First, the methods of representing drawings will be reviewed and discussed briefly.

ORTHOGRAPHIC PROJECTION

The heart of any engineering drawing is the ability to break up the design into the three main views, Front, Top, and Right side, representing what is called orthographic or multiview projection (see Figure 20–1). The engineer or designer is then required to interpret the views and their dimensions, to paint a mental picture of a graphic version of the object as if it were already made or constructed.

Although most engineers have the skill to convert the multiview drawing to a picture in their minds, a vast majority of us would be confused by the numerous hidden lines and centerlines of a drawing and their meaning to the overall design. We need some type of picture to help us interpret the multiview drawing and get a feel for what the part looks like, including the functionality of the part. This may be the major advantage of constructing an object in 3D.

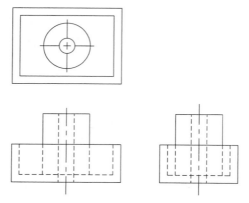

Figure 20–1

ISOMETRIC DRAWINGS

The isometric drawing is the easiest 3D representation to produce. It is based on tilting two axes at 30° angles along with a vertical line to represent the height of the object. As easy as the isometric is to produce, it is also one of the most inaccurate methods of producing a 3D image.

Figure 20–2 shows a view aligned to show the top of the object, Front view, and Right Side view. If we wished to see a different view of the object in isometric, a new drawing would have to be produced from scratch. This is the major disadvantage of using isometric drawings. Still, because of ease in drawing, isometrics remain popular in many school and technical drafting rooms.

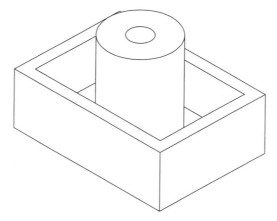

Figure 20–2

3D WIREFRAME MODELS

As shown in Figure 20–3, wireframes are fundamental types of models. However, as lines intersect from the front and back of the object, it is sometimes difficult to interpret the true design of the wireframe.

Through a user-definable coordinate system, previously defined as the UCS, a new coordinate system can be defined along any plane on which to place objects. Depth can be controlled by a number of aids, especially XYZ point filters. In this method, values are temporarily saved for later use inside a command. The values can be retrieved and a coordinate value entered to complete the command.

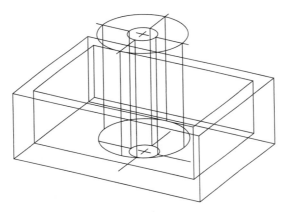

Figure 20–3

3D SURFACED MODELS

Surfacing picks up where the wireframe model leaves off. Because of the complexity of the wireframe and the number of intersecting lines, surfaces are applied to the wireframe. The surfaces are in the form of opaque objects, called 3D faces. As always, 3D faces are placed with Osnap options to assist in point selection. A hidden line removal is performed to view the model without the interference of other objects. The VPOINT command is used to view the model in different 3D positions to make sure all sides of the model have been surfaced. The format of the 3DFACE command is outlined with the following prompts and Figure 20–4.

Command: **3DFACE**
Specify first point or [Invisible]: *(Select the endpoint at "A")*
Specify second point or [Invisible]: *(Select the endpoint at "B")*
Specify third point or [Invisible] <exit>: *(Select the endpoint at "C")*
Specify fourth point or [Invisible] <create three-sided face>: *(Select the endpoint at "D")*
Specify third point or [Invisible] <exit>: *(Press* ENTER *to exit this command)*

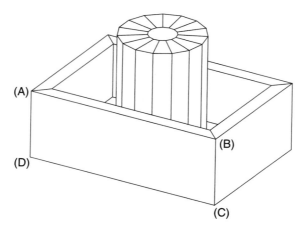

Figure 20–4

SOLID MODELS

The creation of a solid model remains the most exact way to represent a model in 3D. It is also the most versatile representation of an object. Solid models may be viewed as identical to wireframe and surfaced construction models. Wireframe models and surfaced models may be analyzed by distance readings and the identification of key points of the model. Key orthographic views such as Front, Top, and Right Side views may be extracted from wireframe models (see Figure 20–5). Surfaced models may be imported into shading packages for increased visualization (see Figure 20–6). Solid models do all of these operations and more. This is because the solid model, as it is called, is a solid representation of the actual object. From cylinders to slabs, wedges to boxes, all objects that go into the creation of a solid model have volume. This allows a model to be constructed of what are referred to as primitives. These primitives are then merged into one using addition and subtraction operations. What remains is the most versatile of 3D drawings. This method of creating models will be the main focus of this chapter.

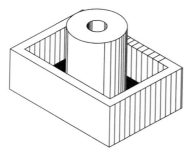

Figure 20–5

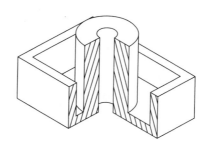

Figure 20–6

SOLID MODELING BASICS

All objects, no matter how simple or complex, are composed of simple geometric shapes or primitives. The shapes range from boxes to cylinders to cones and so on. Solid modeling allows for the creation of these primitives. Once created, the shapes are either merged or subtracted to form the final object. Follow the next series of steps to form the object in Figure 20–7.

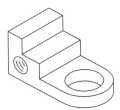

Figure 20–7

Begin the process of solid modeling by constructing a solid slab that will represent the base of the object. You can do this by using the BOX command. Supply the length, width, and height of the box, and the result is a solid slab, as shown in Figure 20–8A.

Next, construct a cylinder using the CYLINDER command. This cylinder will eventually form the curved end of the object. One of the advantages of constructing a solid model is the ability to merge primitives together to form composite solids. Using the UNION command, combine the cylinder and box to form the complete base of the object in Figure 20–8B.

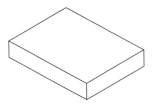

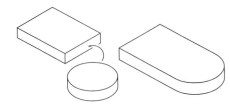

Figure 20–8A   **Figure 20–8B**

As the object progresses, create another solid box and move it into position on top of the base. There, use the UNION command to join this new block with the base. As you add new shapes during this process, they all become part of the same solid (see Figure 20–8C).

Yet another box is created and moved into position. However, instead of being combined with the block, this new box is removed from the solid. This process is called subtraction and, when complete, creates the step shown in Figure 20–8D. The AutoCAD command used during this process is SUBTRACT.

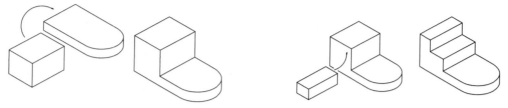

Figure 20–8C   **Figure 20–8D**

You can form holes in a similar fashion. First, use the CYLINDER command to create a cylinder the diameter and depth of the desired hole. Next, move the cylinder into the solid and then subtract it using the SUBTRACT command. Again, the object in Figure 20–8E represents a solid object.

Using existing AutoCAD tools such as User Coordinate Systems, create another cylinder with the CYLINDER command. It too is moved into position where it is subtracted through the SUBTRACT command. The complete solid model of the object is shown in Figure 20–8F (right).

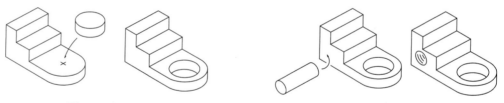

Figure 20–8E   **Figure 20–8F**

Advantages of constructing a solid model out of an object come in many forms. Profiles of different surfaces of the solid model may be taken. Section views of solids may be automatically formed and crosshatched, as in Figure 20–9.

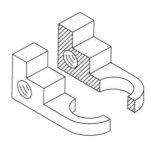

Figure 20–9

A very important analysis tool is the MASSPROP command, which is short for mass property. Information such as the mass and volume of the solid object may be calculated along with centroids and moments of inertia, components that are used in computer aided engineering (CAE) and in finite element analysis (FEA) of the model. See Figure 20–10.

Mass: 223.2 gm
Volume: 28.39 cu cm (Err: 3.397)

Bounding box: X:-1.501—6.001 cm
Y:-0.0009944—2.75 cm
Z:-3.251—3.251 cm

Centroid: X: 2.36 cm
Y: 0.7067 cm
Z: 0.07384 cm

Moments of inertia: X: 612.2 gm sq cm (Err: 76.16)
Y: 2667 gm sq cm (Err: 442.6)
Z: 2444 gm sq cm (Err: 383.2)

Figure 20–10

THE UCS COMMAND

Two-dimensional computer-aided design still remains the most popular form of representing drawings for most applications. However, in applications such as manufacturing, 3D models are becoming increasingly popular for creating rapid prototype models or for creating tool paths from the 3D model. To assist in this creation process, a series of User Coordinate Systems is used to create construction planes where features such as holes and slots are located. In Figure 20–11, a model of a box is displayed along with the User Coordinate System icon. The presence of the "W" signifies the World Coordinate System or home position of drawing. Figure 20–12 identifies the directions of the three User Coordinate System axes. Notice that as X and Y are identified in the User Coordinate System icon, the positive Z direction is straight up; a negative Z direction is down. The UCS command is used to create different user-defined coordinate systems. The command sequence follows along with all options; the UCS command and options can also be selected from the UCS toolbar illustrated in Figure 20–13.

Command: **UCS**
Current ucs name: *WORLD*
Enter an option [New/Move/orthoGraphic/Prev/Restore/Save/Del/Apply/?/World]
    <World>: **N** *(For New)*
Specify origin of new UCS or [ZAxis/3point/OBject/Face/View/X/Y/Z] <0,0,0>:

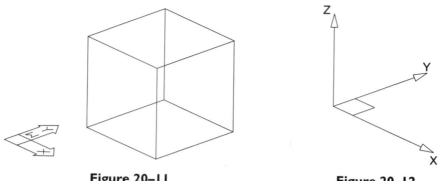

Figure 20–11   **Figure 20–12**

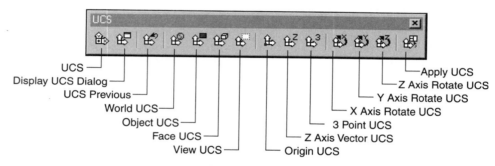

Figure 20–13

The following options are available when you use the UCS command:

> **New**—Identifies a new User Coordinate System using one of the following:
> **ZAxis**—From two points defining the Z axis
> **3point**—By 3 points
> **Object**—In relation to an object selected
> **Face**—From the selected face of a solid object
> **View**—By the current display
> **X/Y/Z**—By rotation along the X, Y, or Z axis
> **<0,0,0>**—By identifying origin (0,0,0) at any identified point
> **Move**—Identifies a new User Coordinate System by shifting origin in the XY plane or changing the Z coordinate
> **orthoGraphic**—Identifies new User Coordinate System from 6 preset systems (Top/Bottom/Front/Back/Left/Right)
> **Prev**—Sets the User Coordinate System icon to the previously defined User Coordinate System

Restore—Restores a previously saved User Coordinate System

Save—Saves the position of a User Coordinate System under a unique name given by the CAD operator

Del—Deletes a User Coordinate System from the database of the current drawing

Apply—Sets the current User Coordinate System setting to a specific viewport(s)

?—Lists all previously saved User Coordinate Systems

World—Switches to the World Coordinate System from any previously defined User Coordinate System

THE UCS-NEW—<0,0,0> DEFAULT OPTION

The New option of the ucs command defines a new User Coordinate System by one of several offered methods. The default method is moving the current UCS to a new 0,0,0 position while leaving the direction of the X, Y, and Z axes unchanged. The command sequence for using the <0,0,0> option is as follows:

Command: **UCS**
Current ucs name: *WORLD*
Enter an option [New/Move/orthoGraphic/Prev/Restore/Save/Del/Apply/?/World]
    <World>: **N** *(for New)*
Specify origin of new UCS or [ZAxis/3point/OBject/Face/View/X/Y/Z] <0,0,0>:
    (Identify a point for the new origin)

Figure 20–14A is a sample model with the current coordinate system being the World Coordinate System. Follow Figures 20–14B and 20–14C to define a new User Coordinate System using the <0,0,0> option.

Using the New option of the ucs command, identify a new origin point for 0,0,0 at "A" as shown in Figure 20–14B. This should move the User Coordinate System icon to the point that you specify. Once the icon has moved, the "W" is removed from the icon. If the icon remains in its previous location and does not move, use the UCSICON command with the Origin option to snap the icon to its new origin point. The "plus" that appears at the intersection of the X and Y axes on the icon indicates that the icon is displayed at 0,0,0.

Command: **UCS**
Current ucs name: *WORLD*
Enter an option [New/Move/orthoGraphic/Prev/Restore/Save/Del/Apply/?/World]
    <World>: **N** *(for New)*
Specify origin of new UCS or [ZAxis/3point/OBject/Face/View/X/Y/Z] <0,0,0>: **End**
of *(Select the endpoint of the line at "A")*

Objects may now be drawn parallel to this new coordinate system. Remember that 0,0,0 is identified by the "plus" on the User Coordinate System icon. See Figure 20–14C.

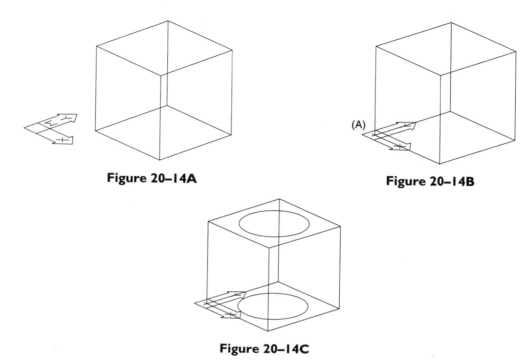

Figure 20–14A

Figure 20–14B

Figure 20–14C

THE UCS-NEW—3POINT OPTION

Use the 3point option of the UCS command to specify a new User Coordinate System by identifying an origin and new directions of its positive X and Y axes. Figure 20–15A shows a 3D cube in the World Coordinate System. To construct objects on the front panel, first define a new User Coordinate System parallel to the front. Use the following command sequence to accomplish this task.

Command: **UCS**
Current ucs name: *WORLD*
Enter an option [New/Move/orthoGraphic/Prev/Restore/Save/Del/Apply/?/World]
    <World>: **N** *(for New)*
Specify origin of new UCS or [ZAxis/3point/OBject/Face/View/X/Y/Z] <0,0,0>: **3** *(For 3point)*
Specify new origin point <0,0,0>: **End**
of *(Select the endpoint of the model at "A" as shown in Figure 20–15B)*
Specify point on positive portion of X-axis <>: **End**
of *(Select the endpoint of the model at "B")*

Specify point on positive-Y portion of the UCS XY plane <>: **End**
of *(Select the endpoint of the model at "C")*

With the User Coordinate System standing straight up, or aligned with the front of
the cube, any type of object may be constructed along this plane as in Figure 20–15C.

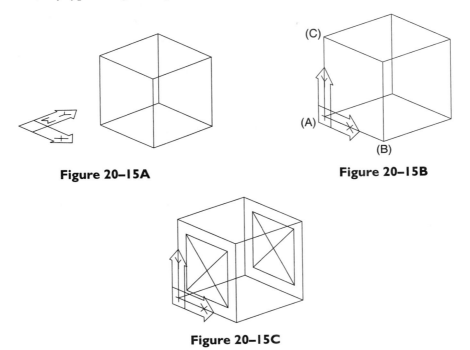

Figure 20–15A   **Figure 20–15B**

Figure 20–15C

The 3point method of defining a new User Coordinate System is quite useful in the
example shown in Figure 20–16, where a UCS needs to be aligned with the inclined
plane. Use the intersection at "A" as the center of the new UCS, the intersection at
"B" as the direction of the positive X axis, and the intersection at "C" as the direction
of the positive Y axis.

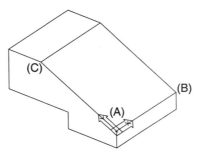

Figure 20–16

THE UCS–NEW—X/Y/Z ROTATION OPTIONS

Using the X/Y/Z rotation options will rotate the current user coordinate around the specific axis. Once a letter is selected as the pivot, a prompt appears asking for the rotation angle about the pivot axis. The right-hand rule is used to determine the positive direction of rotation around an axis. Think of the right hand gripping the pivot axis with the thumb pointing in the positive X, Y, or Z direction. The curling of the fingers on the right hand determines the direction of rotation. All positive rotations occur in the counter-clockwise directions. In Figure 20–17A, "A" shows a rotation about the X-axis, "B" shows a rotation about the Y-axis; and "C" shows a rotation about the Z-axis.

Given the same cube shown in Figure 20–17B in the World Coordinate System, the X option of the UCS command will be used to stand the icon straight up by entering a 90° rotation value, as in the following prompt sequence.

Command: **UCS**
Current ucs name: *NO NAME*
Enter an option [New/Move/orthoGraphic/Prev/Restore/Save/Del/Apply/?/World]
   <World>: **N** *(for New)*
Specify origin of new UCS or [ZAxis/3point/OBject/Face/View/X/Y/Z] <0,0,0>: **X**
Specify rotation angle about X axis <90>: *(Press* ENTER *to accept 90° of rotation)*

The X-axis is used as the pivot of rotation; entering a value of 90° rotates the icon the desired degrees in the counterclockwise direction as in Figure 20–17C.

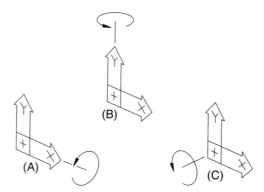

Figure 20–17A

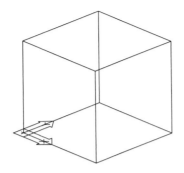

Figure 20–17B

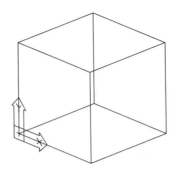

Figure 20–17C

THE UCS—NEW—OBJECT OPTION

Given the 3D cube in Figure 20–18A, another option for defining a new User Coordinate System is by selecting an object and having the User Coordinate System align to that object. Use the following command sequence and Figure 20–18B to accomplish this.

Command: **UCS**
Current ucs name: *WORLD*
Enter an option [New/Move/orthoGraphic/Prev/Restore/Save/Del/Apply/?/World]
    <World>: **N** *(for New)*
Specify origin of new UCS or [ZAxis/3point/OBject/Face/View/X/Y/Z] <0,0,0>: **OB**
    (for the Object option)
Select object to align UCS: *(Select the circle in Figure 20–18B)*

Selecting the circle in Figure 20–18B aligns the User Coordinate System to the object selected. Objects determine the alignment of the User Coordinate System in many ways. In the case of the circle, the center of the circle becomes the origin of the User Coordinate System. Where the circle was selected becomes the point through which the positive X axis aligns.

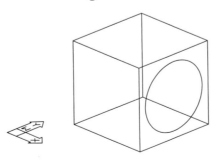

Figure 20–18A

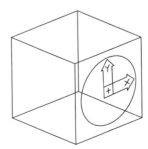

Figure 20–18B

THE UCS—NEWFACE—OPTION

The Face option of the UCS command allows you to establish a new coordinate system aligned to the selected face of a solid object. Given the current User Coordinate System in Figure 20–19A and a solid box, follow the command sequence below to align the User Coordinate System with the Face option.

Command: **UCS**
Current ucs name: *NO NAME*
Enter an option [New/Move/orthoGraphic/Prev/Restore/Save/Del/Apply/?/World] <World>: **N** *(for New)*
Specify origin of new UCS or [ZAxis/3point/OBject/Face/View/X/Y/Z] <0,0,0>: **F** *(for Face)*
Select face of solid object: *(Select the edge of the model at point "A" in Figure 20–19B)*
Enter an option [Next/Xflip/Yflip] <accept>: **N** *(the Next option relocates the UCS to the other face shared by the edge; see Figure 20–19C)*
Enter an option [Next/Xflip/Yflip] <accept>: **N** *(the Next option returns the UCS to the original face in Figure 20–19B)*
Enter an option [Next/Xflip/Yflip] <accept>: **X** *(the Xflip option rotates the UCS about the X axis 180° in Figure 20–19D)*
Enter an option [Next/Xflip/Yflip] <accept>: **X** *(the Xflip option rotates the UCS back to its location in Figure 20–19B)*
Enter an option [Next/Xflip/Yflip] <accept>: **Y** *(the Yflip option rotates the UCS about the Y axis 180° in Figure 20–19E)*
Enter an option [Next/Xflip/Yflip] <accept>: **Y** *(the Yflip option rotates the UCS back to its location in Figure 20–19B)*
Enter an option [Next/Xflip/Yflip] <accept>: *(Press ENTER to accept the UCS position)*

The results are displayed in Figure 20–19F with the User Coordinate System aligned to the right face. To select a face in this command you can click within the edges of a face or on one of its edges. Where you click directly affects the location of the origin. Use of the Next, Xflip, and Yflip options, which were demonstrated in the command sequence, were not required to obtain the results in Figure 20–19F.

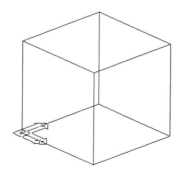

Figure 20–19A

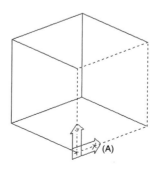

Figure 20–19B

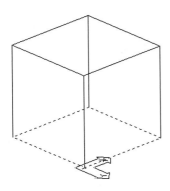

Figure 20–19C

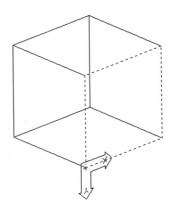

Figure 20–19D

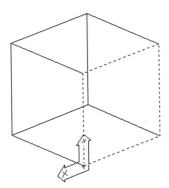

Figure 20–19E

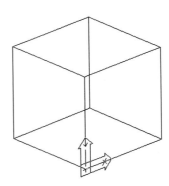

Figure 20–19F

THE UCS–NEW—VIEW OPTION

The View option of the UCS command allows you to establish a new coordinate system where the XY plane is perpendicular to the current screen viewing direction; in other words, it is parallel to the display screen. Given the current User Coordinate System in Figure 20–20A, follow the prompt below along with Figure 20–20B to align the User Coordinate System with the View option.

Command: **UCS**
Current ucs name: *NO NAME*
Enter an option [New/Move/orthoGraphic/Prev/Restore/Save/Del/Apply/?/World]
    <World>: **N** *(for New)*
Specify origin of new UCS or [ZAxis/3point/OBject/Face/View/X/Y/Z] <0,0,0>: **V** *(for View)*

The results are displayed in Figure 20–20B with the User Coordinate System aligned parallel to the display screen.

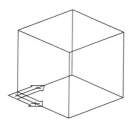

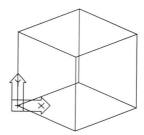

Figure 20–20A   **Figure 20–20B**

APPLICATIONS OF ROTATING THE UCS

Follow the next series of steps to perform numerous rotations of the User Coordinate System in the example of the cube shown in Figure 20–21A.

Begin rotating the User Coordinate System along the X-axis at a rotation angle of 90° using the following prompt sequence. This will align the User Coordinate System with the front of the cube shown in Figure 20–21B.

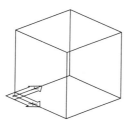

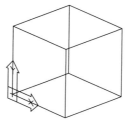

Figure 20–21A   **Figure 20–21B**

Command: **UCS**
Current ucs name: *NO NAME*
Enter an option [New/Move/orthoGraphic/Prev/Restore/Save/Del/Apply/?/World]
    <World>: **N** *(for New)*
Specify origin of new UCS or [ZAxis/3point/OBject/Face/View/X/Y/Z] <0,0,0>: **X**
Specify rotation angle about X axis <90>: *(Press ENTER to accept 90° of rotation)*

Next, align the User Coordinate System with one of the sides of the cube by performing the rotation using the Y axis as the pivot axis (see Figure 20–21C). The positive angle rotates the icon in the counterclockwise direction.

Command: **UCS**
Current ucs name: *NO NAME*
Enter an option [New/Move/orthoGraphic/Prev/Restore/Save/Del/Apply/?/World]
   <World>: **N** *(for New)*
Specify origin of new UCS or [ZAxis/3point/OBject/Face/View/X/Y/Z] <0,0,0>: **Y**
Specify rotation angle about Y axis <90>: *(Press ENTER to accept 90° of rotation)*

Next, rotate the User Coordinate System using the Z axis as the pivot axis. The degree of rotation entered at 45° tilts the User Coordinate System shown in Figure 20–21D.

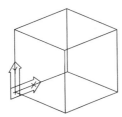

Figure 20–21C

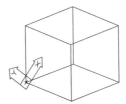

Figure 20–21D

Command: **UCS**
Current ucs name: *NO NAME*
Enter an option [New/Move/orthoGraphic/Prev/Restore/Save/Del/Apply/?/World]
   <World>: **N** *(for New)*
Specify origin of new UCS or [ZAxis/3point/OBject/Face/View/X/Y/Z] <0,0,0>: **Z**
Specify rotation angle about Z axis <90>: **45**

To tilt the UCS icon along the Z axis pointing down at a 45° angle, enter a rotation angle of -90°. The results are shown in Figure 20–21E.

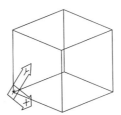

Figure 20–21E

Command: **UCS**
Current ucs name: *NO NAME*
Enter an option [New/Move/orthoGraphic/Prev/Restore/Save/Del/Apply/?/World]
   <World>: **N** *(for New)*
Specify origin of new UCS or [ZAxis/3point/OBject/Face/View/X/Y/Z] <0,0,0>: **Z**
Specify rotation angle about Z axis <90>: **-90**

USING THE UCS DIALOG BOX

It is considered good practice to assign a name to a User Coordinate System once it has been created. Once numerous User Coordinate Systems have been defined in a drawing, using their name instead of recreating each coordinate system easily restores them. You can accomplish this easily using the Restore option of the UCS command. Another method is to choose Named UCS... from the Tools pull-down menu illustrated in Figure 20–22A. This displays the UCS dialog box in Figure 20–22B with the Named UCSs tab selected. All User Coordinate Systems defined in the drawing are listed here. To make one of these coordinate systems current, highlight the desired UCS name and select the Set Current button shown in Figure 20–22B. This dialog box tab provides a quick method of restoring previously defined coordinate systems without entering them at the keyboard.

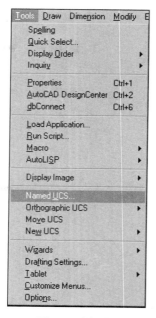

Figure 20–22A

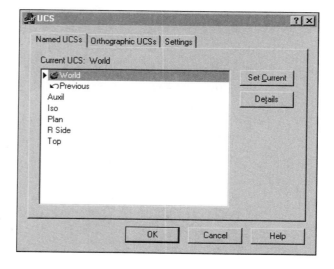

Figure 20–22B

Choosing Preset... from the Orthographic UCS cascading menu on the Tools pull-down menu (see Figure 20–23A) also displays the UCS dialog box, but with the Orthographic UCSs tab selected, as shown in Figure 20–23B. Use this tab to automatically align the User Coordinate System icon to sides of an object such as Front view, Top view, Back view, Right Side view, Left Side view, and Bottom view.

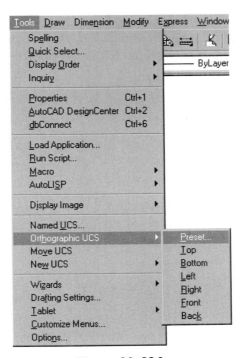

Figure 20–23A

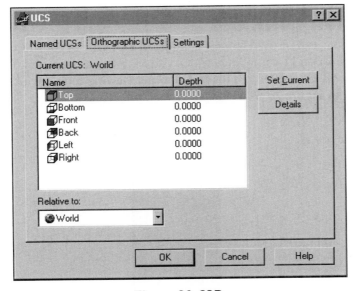

Figure 20–23B

The Settings tab of the UCS dialog box in Figure 20–24 controls the setting of UCS system variables. In this tab, you can control such things as whether the UCS icon is displayed at the origin point, in the lower left corner of the screen, or if it is displayed at all.

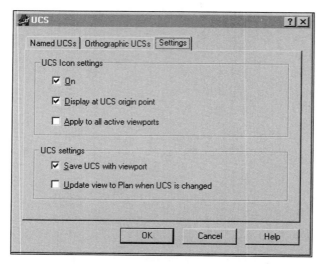

Figure 20–24

VIEWING 3D MODELS
THE 3DORBIT COMMAND

Various methods can be used to view a model in 3D. One of the more efficient ways is through the 3DORBIT command, which can be selected from the View pulldown menu shown in Figure 20–25. This and all other viewing tools are located in the 3D Orbit toolbar in Figure 20–26.

Figure 20–25

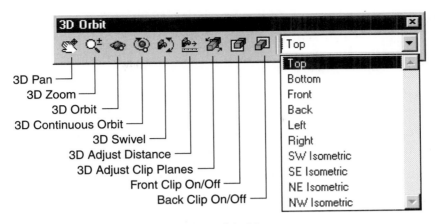

Figure 20–26

Choosing this command displays your model similar to Figure 20–27. Use the large circle to guide your model through a series of dynamic rotation maneuvers. Notice also that the User Coordinate System has changed to three arrows to better define the X, Y, and Z axes. For instance, moving your cursor outside the large circle at "A" allows you to dynamically rotate your model only in a circular direction. Moving your cursor to either of the circle quadrant identifiers at "B" and "C" allows you to rotate your model horizontally. Moving your cursor to either of the circle quadrant identifiers at "D" and "E" allows you to rotate your model vertically. Moving your cursor inside the large circle at "F" allows you to dynamically rotate your model to any viewing position, whether above or below.

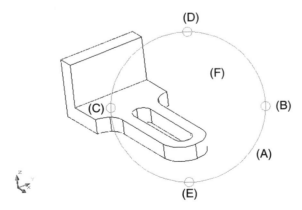

Figure 20–27

Other components of the 3D Orbit toolbar:

3D Continuous Orbit—Allows you to view an object in a continuous orbit motion.

3D Swivel—Allows you to view an object with a motion that is similar to looking through a camera viewfinder.

3D Adjust Distance—Allows you to view an object closer or farther away.

3D Adjust Clip Planes—Allows you to view an object after removing a portion of that object from the view with a front or back clipping plane.

Front Clip On/Off—Allows you to turn the front clipping plane on or off.

Back Clip On/Off—Allows you to turn the back clipping plane on or off.

THE VPOINT COMMAND

The object in Figure 20–28 represents the plan view of a 3D model already created. Unfortunately, it becomes very difficult to understand what the model looks like in its present condition. Use the VPOINT command to view a wireframe, surfaced or solid model in three dimensions. The following is a typical prompt sequence for the VPOINT command:

Command: **VPOINT**
Switching to the WCS
Current view direction: VIEWDIR=0.0000,0.0000,1.0000
Specify a view point or [Rotate] <display compass and tripod>:

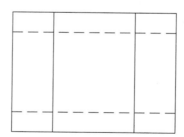

Figure 20–28

The VPOINT command stands for "View point." A point is identified in 3D space. This point becomes a location where the model is viewed. The point may be entered at the keyboard as an X,Y,Z coordinate or selected with the aid of a 2D compass (see Figure 20–32). The object shown in Figure 20–29 is a typical result of the use of the VPOINT command.

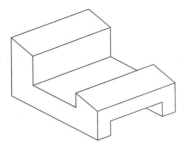

Figure 20–29

Numerous ways exist to select the VPOINT command. Choosing 3D Views from the View pull-down menu exposes the 3D Views cascading menu in Figure 20–30. Choose from numerous viewpoint options. Items listed as Top, Bottom, SW Isometric, and NE Isometric are standard, stored view points used to view a 3D model from these different locations. They provide a fast way to view a model in a primary or isometric view.

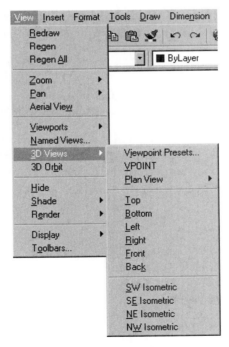

Figure 20–30

A View toolbar is also available as shown in Figure 20–31. It contains options similar to the 3D Views cascading menu on the View pull-down menu. The View toolbar has the extra advantage of having a picture in the form of an icon that guides you through which viewpoint to pick for the desired viewing effect.

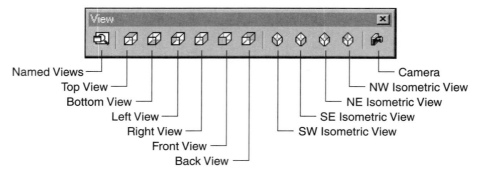

Figure 20–31

Besides selecting from the pull-down menu or toolbar, one method of defining a viewpoint is through the use of an X,Y,Z coordinate. In the following command sequence, a new viewing point is established 1 unit in the positive X direction, 1 unit in the negative Y direction, and 1 unit in the positive Z direction.

Command: **VPOINT**
Switching to the WCS
Current view direction: VIEWDIR=0.0000,0.0000,1.0000
Specify a view point or [Rotate] <display compass and tripod>: **1,-1,1**

Another method of defining a viewing point is to define two angular axes by rotation. The first angle defines the viewpoint in the X-Y axis. However, this is only 2D. The second angle defines the viewpoint from the X-Y axis. This tilts the viewing point up for a positive angle or down for a negative angle.

Command: **VPOINT**
Switching to the WCS
Current view direction: VIEWDIR=1.0000,-1.0000,1.0000
Specify a view point or [Rotate] <display compass and tripod>: **R** *(For Rotate)*
Enter angle in XY plane from X axis <315>: **45**
Enter angle from XY plane <35>: **30**

If the first command prompt is followed by pressing ENTER, a graphic image consisting of a compass and tripods appear in Figure 20–32. Although the compass appears two-dimensional, it enables you to pick a viewpoint depending on how you read the compass. The intersection of the horizontal and vertical lines forms the North Pole of the compass. The inner circle forms the equator and the outer circle forms the South Pole. The following examples illustrate numerous viewing points.

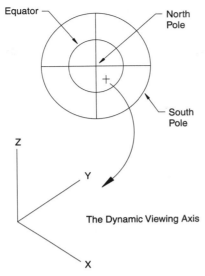

Figure 20–32

Command: **VPOINT**
Switching to the WCS
Current view direction: VIEWDIR=1.0000,-1.0000,1.0000
Specify a view point or [Rotate] <display compass and tripod>: *(Press* ENTER *to display the viewpoint compass in Figure 20–32)*

VIEWING ALONG THE EQUATOR

To obtain the results in Figure 20–33, use the VPOINT command and mark a point at "A" to view the Front view, mark a point at "B" to view the Top view, and mark a point at "C" to view the Right Side view. Coordinates could also have been entered to achieve the same results:

Command: **VPOINT**
Switching to the WCS
Current view direction: VIEWDIR=1.0000,-1.0000,1.0000
Specify a view point or [Rotate] <display compass and tripod>: **0,-1,0** *(At "A")*

Command: **VPOINT**
Switching to the WCS
Current view direction: VIEWDIR=1.0000,-1.0000,1.0000
Specify a view point or [Rotate] <display compass and tripod>: **0,0,1** *(At "B")*

Command: **VPOINT**
Switching to the WCS
Current view direction: VIEWDIR=1.0000,-1.0000,1.0000
Specify a view point or [Rotate] <display compass and tripod>: **1,0,0** *(At "C")*

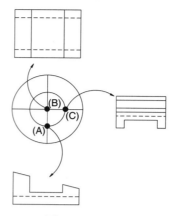

Figure 20–33

VIEWING NEAR THE NORTH POLE

Picking the four points shown in Figure 20–34 results in the different viewing points for the object. Since all points are inside the inner circle, the results are aerial views, or views from above. Remember that the inner circle symbolizes the Equator. Depending on which quadrant you select, you will look up at the object from the right corner, left corner, or either of the rear corners.

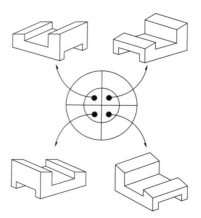

Figure 20–34

VIEWING NEAR THE SOUTH POLE

Picking the four points shown in Figure 20–35 results in underground views, or viewing the object from underneath. This is true because all points selected lie between the small and large circles. Again, remember that the small circle symbolizes the Equator, while the large circle symbolizes the South Pole.

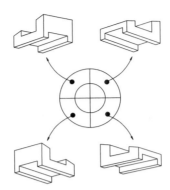

Figure 20–35

USING THE VIEWPOINT PRESETS DIALOG BOX

Choosing Viewpoint Presets… from the 3D Views cascading menu on the View pull-down menu, shown in Figure 20–36, displays the Viewpoint Presets dialog box shown in Figure 20–37. Use this dialog box to help define a new point of viewing a 3D model. The square image with a circle in the center allows you to define a viewing point in the XY plane. The semicircular-shaped image allows you to define a viewing point from the XY plane. The combination of both directions forms the 3D viewpoint.

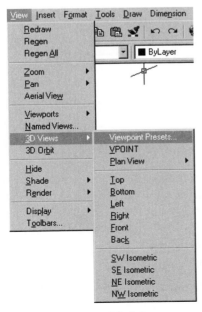

Figure 20–36

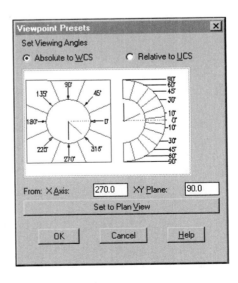

Figure 20–37

When selecting a viewing point in the X-Y axis, you have two options for selecting the desired point of view. When you pick outside the circle on 0, 90, 180, etc., as in "A" in Figure 20–38A, the viewpoint snaps from one angle to another in increments of 45°. The resulting angle is displayed at the bottom of the square. If you want a more detailed view point in the X-Y axis, pick a point inside and near the center of the circle at "B." The resulting angle will display all values in between the 45° increment from before. If you already know the angle in the X-Y axis, you may enter it in the X Axis edit box.

Figure 20–38A illustrates setting a viewing angle in the XY plane. Figure 20–38B shows the second half of the Viewpoint Presets dialog box, which defines an angle from the XY plane. Selecting a point in between both semicircles at "A" snaps the viewpoint in 10°, 30°, 45°, 60°, and 90° increments. If you want a more detailed selection, pick the viewing point inside the smaller semicircle and near its center at "B." This will allow you to select an angle different from the five default values listed above. An alternate method of selecting an angle from the XY plane would be to place the value in the XY Plane edit box.

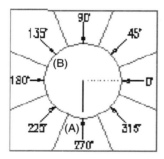

Figure 20–38A

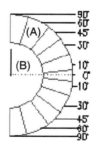

Figure 20–38B

HOW VIEW POINT CAN AFFECT THE UCS ICON

The User Coordinate System icon may take on different forms depending upon how it is viewed using the VPOINT command. Viewing the icon in plan view or in an aerial view displays the image in Figure 20–39A. The presence of the "+" sign indicates the icon is located in the current viewport at the current UCS origin. The UCS icon must be totally displayed on the screen; if the icon is unable to be totally shown, the "+" sign is removed and the icon locates itself in the lower left corner of the display screen.

When you view a model from below, or in the negative Z direction, the User Coordinate System icon takes the form of the image in Figure 20–39B. The absence of the two lines near the intersection of the X and Y axes acts as a visual cue alerting you that you are either viewing the model from below or from the icon's back side.

Figure 20–39A

Figure 20–39B

When you work with multiple viewports, defining the UCS in one viewport displays the normal icon in Figure 20–39A; yet in an adjacent viewport, the icon may disappear and be replaced with the image in Figure 20–39C. This is commonly referred to as the "broken pencil" icon; it appears when the current UCS is viewed orthogonal or perpendicular to the viewing angle in a particular viewport. This icon also warns you that drawing lines or other objects may not be obvious in the present viewport.

In all previous examples (Figures 20–39A through 20–39C), the UCS icon has been displayed while you work in the Model Space environment. The icon displayed in Figure 20–39D represents the Paper Space environment. This environment is strictly two-dimensional and is used as a layout tool for arranging views and displaying title block information.

Figure 20–39C

Figure 20–39D

SELECTING SOLID MODELING COMMANDS

Figures 20–40 and 20–41 show two of the more popular locations for accessing solid modeling commands. In Figure 20–40, solid modeling commands may be chosen from the Draw pull-down menu. Choosing Solids > displays four groupings of commands. The first grouping displays Box, Sphere, Cylinder, Cone, Wedge, and Torus, which are considered the building blocks of the solid model and are used to construct basic primitives. The second grouping displays the EXTRUDE and REVOLVE commands, which are additional ways to construct solid models. Polyline outlines or circles may be extruded to a thickness that you designate. You may also revolve other polyline outlines about an object representing a centerline of rotation. The third grouping of solid modeling commands enables you to slice the solid model in half, cut the solid model into what is called a section, and perform an interference check where two solids are constructed near each other but must not touch or intersect. The last grouping, Setup >, displays three commands designed to extract the orthographic views from the solid model. The three commands are SOLPROF, SOLVIEW, and SOLDRAW. Notice also in Figure 20–41 the complete grouping of solid modeling commands on a toolbar. This is another convenient way to select solid modeling commands.

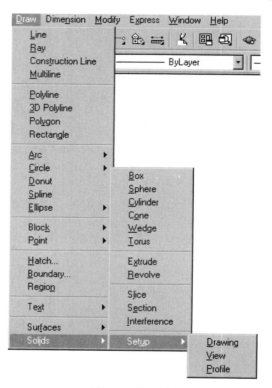

Figure 20–40

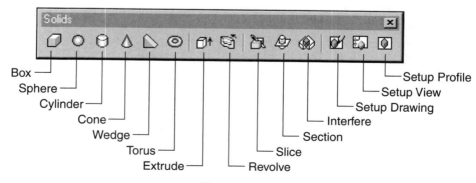

Figure 20–41

USING THE BOX COMMAND

Use the BOX command to create a 3D box. One corner of the box is located along with a diagonal corner. A height is assigned to complete the definition of the box. If you know all three dimensions, you may construct the solid box by entering values for its length, width, and height. You may construct a cube if all three dimensions are the same value. Figure 20–42 shows examples of solid boxes.

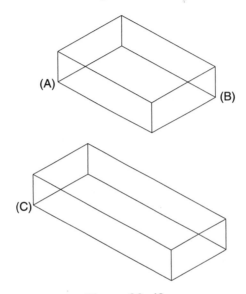

Figure 20–42

Command: **BOX**
Specify corner of box or [CEnter] <0,0,0>: *(Pick a point at "A")*
Specify corner or [Cube/Length]: *(Pick a point at "B")*
Specify height: **1**

Command: **BOX**
Specify corner of box or [CEnter] <0,0,0>: *(Pick a point at "C")*
Specify corner or [Cube/Length]: **L** *(To define the length of the box)*
Specify length: **5**
Specify width: **2**
Specify height: **1**

USING THE CONE COMMAND

Use the CONE command and Figure 20–43 to construct a cone of specified height with the base of the cone either circular or elliptical.

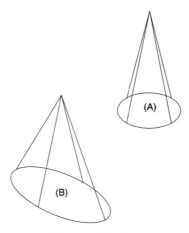

Figure 20–43

Command: **CONE**
Current wire frame density: ISOLINES=4
Specify center point for base of cone or [Elliptical] <0,0,0>: *(Pick a point designating the center of the cone at "A")*
Specify radius for base of cone or [Diameter]: **1**
Specify height of cone or [Apex]: **4**

Command: **CONE**
Current wire frame density: ISOLINES=4
Specify center point for base of cone or [Elliptical] <0,0,0>: **E** *(For an elliptical base)*
Specify axis endpoint of ellipse for base of cone or [Center]: **C** *(To define the center of the ellipse)*
Specify center point of ellipse for base of cone <0,0,0>: *(Pick a point designating the center of the elliptical base at "B" in Figure 20–43)*
Specify axis endpoint of ellipse for base of cone: **@2<0**
Specify length of other axis for base of cone: **@1<90**
Specify height of cone or [Apex]: **4**

USING THE WEDGE COMMAND

Yet another solid primitive is the wedge that consists of a box that has been diagonally cut. Use the WEDGE command to create a wedge with the base parallel to the current User Coordinate System and the slope of the wedge constructed along the X axis. Use the following prompts and Figure 20–44 as examples of how to use this command.

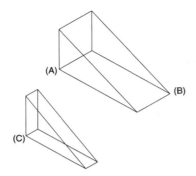

Figure 20–44

Command: **WEDGE**
Specify first corner of wedge or [CEnter] <0,0,0>: *(Pick a point at "A")*
Specify corner or [Cube/Length]: *(Pick a point at "B")*
Specify height: **3**

Command: **WEDGE**
Specify first corner of wedge or [CEnter] <0,0,0>: *(Pick a point at "C")*
Specify corner or [Cube/Length]: **L** *(To define the length of the wedge)*
Specify length: **5**
Specify width: **1**
Specify height: **3**

USING THE CYLINDER COMMAND

The CYLINDER command is similar to the CONE command except that the cylinder is drawn without a taper. The central axis of the cylinder is along the Z axis of the current User Coordinate System. Use this command and Figure 20–45 to construct a cylinder with either a circular or elliptical base.

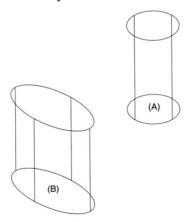

Figure 20–45

Command: **CYLINDER**
Current wire frame density: ISOLINES=4
Specify center point for base of cylinder or [Elliptical] <0,0,0>: *(Pick a point at "A" to specify the center of the cylinder)*
Specify radius for base of cylinder or [Diameter]: **D** *(For Diameter)*
Specify diameter for base of cylinder: **2**
Specify height of cylinder or [Center of other end]: **4**

Command: **CYLINDER**
Current wire frame density: ISOLINES=4
Specify center point for base of cylinder or [Elliptical] <0,0,0>: **E** *(For an elliptical base)*
Specify axis endpoint of ellipse for base of cylinder or [Center]: **C** *(To define the center of the ellipse)*
Specify center point of ellipse for base of cylinder <0,0,0>: *(Pick a point at "B" designating the center of the elliptical base)*
Specify axis endpoint of ellipse for base of cylinder: **@2<0**
Specify length of other axis for base of cylinder: **@1<90**
Specify height of cylinder or [Center of other end]: **4**

USING THE SPHERE COMMAND

Use the SPHERE command to construct a sphere by defining the center of the sphere along with a radius or diameter. As with the cylinder, the central axis of a sphere is along the Z axis of the current User Coordinate System. In Figure 20–46, the sphere is constructed in wireframe mode and does not look like much of a sphere. Perform a hidden line removal using the HIDE command to get a better view of the sphere.

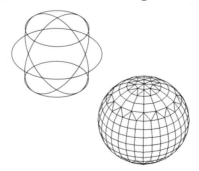

Figure 20–46

Command: **SPHERE**
Current wire frame density: ISOLINES=4
Specify center of sphere <0,0,0>: *(Identify the center of the sphere through coordinate entry or by picking)*
Specify radius of sphere or [Diameter]: **D** *(For Diameter)*
Specify diameter: **4**

Command: **HIDE**

THE TORUS COMMAND

A torus is formed when a circle is revolved about a line in the same plane as the circle. In other words, a torus is similar to a 3D donut. The torus may be constructed with either the radius or diameter method. When you use the radius method, two radius values must be used to define the torus; one for the radius of the tube and the other for the radius from the center of the torus to the center of the tube. Use two diameter values when specifying a torus by diameter. Once you construct the torus, it lies parallel to the current User Coordinate System. Use the following prompts to construct a torus similar to Figure 20–47.

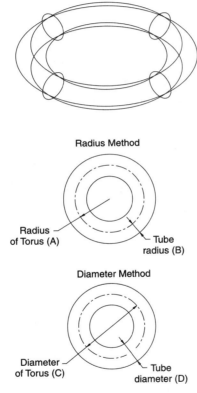

Figure 20–47

Command: **TORUS**
Current wire frame density: ISOLINES=4
Specify center of torus <0,0,0>: *(Identify the center of the torus through coordinate entry or by picking)*
Specify radius of torus or [Diameter]: **5** *(For the radius of the torus at "A")*
Specify radius or tube or [Diameter]: **1** *(For the radius of the tube at "B")*

Command: **TORUS**

Current wire frame density: ISOLINES=4

Specify center of torus <0,0,0>: *(Identify the center of the torus through coordinate entry or by picking)*

Specify radius of torus or [Diameter]: **D** *(For diameter of the torus at "C")*

Specify diameter: **4**

Specify radius or tube or [Diameter]: **D** *(For diameter of the tube at "D")*

Specify diameter: **2**

USING BOOLEAN OPERATIONS

To combine one or more primitives to form a common solid, a system is available to illustrate the relationship between the individual items that make up the solid model. This system is called a Boolean operation. Boolean operations must act on at least a pair of primitives, regions, or solids. These operations in the form of commands are located in the Modify pull-down menu under Solids Editing. They can also be obtained through the Solids Editing toolbar. Boolean operations allow you to add two or more objects together, subtract a single or group of objects from another, or find the overlapping volume—in other words, to form the solid common to both primitives. Displayed in Figure 20–48 are the UNION, SUBTRACT, and INTERSECT commands that you use to perform the Boolean operations previously explained.

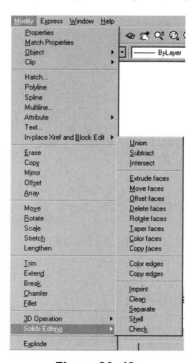

Figure 20–48

In Figure 20–49A, a cylinder has been constructed along with a square block. Depending on which Boolean operation you use, the results could be quite different. In Figure 20–49B, both the square slab and cylinder are considered one solid object. This is the purpose of the UNION command: to join or unite two solid primitives into one. Figure 20–49C illustrates the intersection of the two solid primitives or the area that both solids have in common. This solid is obtained through the INTERSECT command. Figure 20–49D shows the results of removing or subtracting the cylinder from the square slab—a hole is formed inside the square slab as a result of using the SUBTRACT command. All Boolean operation commands can work on numerous solid primitives, that is, if you subtract numerous cylinders from a slab, you can subtract all cylinders at the same time.

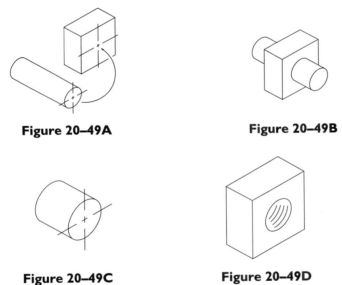

Figure 20–49A   **Figure 20–49B**

Figure 20–49C   **Figure 20–49D**

THE UNION COMMAND

This construction operation joins two or more selected solid objects together into a single solid object. See Figure 20–50.

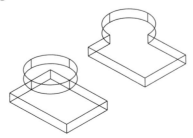

Figure 20–50

Command: **UNION**
Select objects: *(Pick the box and cylinder)*
Select objects: *(Press ENTER to perform the union operation)*

THE SUBTRACT COMMAND

Use this command to subtract one or more solid objects from a source object. See Figure 20–51.

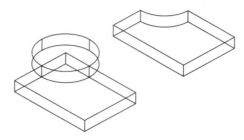

Figure 20–51

Command: **SU** *(For SUBTRACT)*
Select solids and regions to subtract from...
Select objects: *(Pick the box)*
Select objects: *(Press ENTER to continue with this command)*
Select solids and regions to subtract...
Select objects: *(Pick the cylinder)*
Select objects: *(Press ENTER to perform the subtraction operation)*

THE INTERSECT COMMAND

Use this command to find the solid common to a group of selected solid objects. See Figure 20–52.

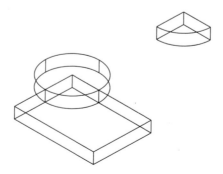

Figure 20–52

Command: **IN** *(For INTERSECT)*
Select objects: *(Pick the box and cylinder)*
Select objects: *(Press ENTER to perform the intersection operation)*

3D APPLICATIONS OF UNIONING SOLIDS

Figure 20–53 shows an object consisting of one horizontal solid box, two vertical solid boxes, and two extruded semi-circular shapes. All primitives have been positioned with the MOVE command. To join all solid primitives into one solid object, use the UNION command. The order of selection of these solids for this command is not important. Use the following prompts and Figure 20–53.

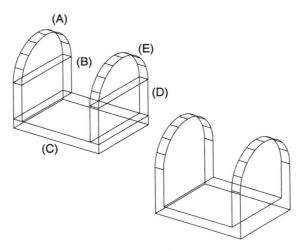

Figure 20–53

Command: **UNION**
Select objects: *(Select the solid extrusion at "A")*
Select objects: *(Select the vertical solid box at "B")*
Select objects: *(Select the horizontal solid box at "C")*
Select objects: *(Select the vertical solid box at "D")*
Select objects: *(Select the solid extrusion at "E")*
Select objects: *(Press ENTER to perform the union operation)*

3D APPLICATIONS OF MOVING SOLIDS

Using the same problem from the previous example, let us now add a hole in the center of the base. The cylinder was already created through the CYLINDER command. It now needs to be moved to the exact center of the base. You may use the MOVE command along with the OSNAP-Tracking mode to accomplish this. Tracking mode will automatically activate the Ortho mode when it is in use.

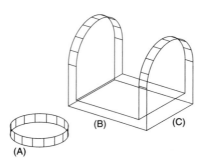

Figure 20-54

Command: **M** *(For MOVE)*
Select objects: *(Select the cylinder at "A")*
Select objects: *(Press ENTER to continue with this command)*
Specify base point or displacement: **Cen**
of *(Select the bottom of the cylinder at "A")*
Specify second point of displacement or <use first point as displacement>: **TK** *(To activate tracking mode)*
First tracking point: **Mid**
of *(Select the midpoint of the bottom of the base at "B")*
Next point (Press ENTER to end tracking): **Mid**
of *(Select the midpoint of the bottom of the base at "C")*
Next point (Press ENTER to end tracking): *(Press ENTER to end tracking and perform the move operation)*

3D APPLICATIONS OF SUBTRACTING SOLIDS

Now that the solid cylinder is in position, use the SUBTRACT command to remove the cylinder from the base of the main solid and create a hole in the base. See Figure 20–55.

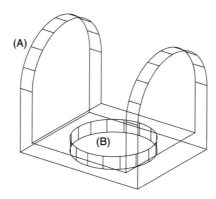

Figure 20-55

Command: **SU** *(For SUBTRACT)*
Select solids and regions to subtract from...
Select objects: *(Select the main solid as source at "A")*
Select objects: *(Press ENTER to continue with this command)*
Select solids and regions to subtract...
Select objects: *(Select the cylinder at "B")*
Select objects: *(Press ENTER to perform the subtraction operation)*

3D APPLICATIONS OF ALIGNING SOLIDS

Two more holes need to be added to the vertical sides of the solid object. A cylinder was already constructed; however, it is in the vertical position. This object needs to be rotated and moved into position. The ALIGN command would be a good command to use in this situation. A series of source and destination points guides the placement of one object on another. The object rotates and moves into position. Use the following command prompt sequence and Figure 20–56 for the ALIGN command.

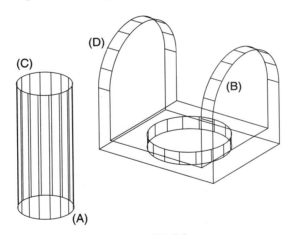

Figure 20–56

Command: **AL** *(For ALIGN)*
Select objects: *(Select the cylinder)*
Select objects: *(Press ENTER to continue with this command)*
Specify first source point: **Cen**
of *(Select the bottom of the cylinder at "A")*
Specify first destination point: **Cen**
of *(Select the outside the main solid at "B")*
Specify second source point: **Cen**
of *(Select the top of the cylinder at "C")*
Specify second destination point: **Cen**
of *(Select the outside the main solid at "D")*

Specify third source point or <continue>: *(Press* ENTER *to continue)*
Scale objects based on alignment points? [Yes/No] <N>: *(Press* ENTER *to accept the default value and perform the align operation)*

The cylinder is removed to create holes in each of the vertical sides of the solid object through the SUBTRACT command. The completed object is shown in Figure 20–57.

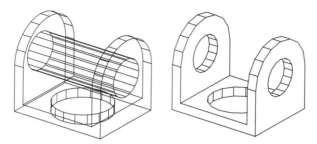

Figure 20–57

A BETTER UNDERSTANDING OF INTERSECTIONS

One of the most misunderstood yet powerful Boolean operations is the Intersection. In Figure 20–58, the object at "C" represents the finished model. Look at the sequence beginning at "A" to see how to prepare the solid primitives for an Intersection operation. Two separate 3DSOLID objects are created at "A." One object represents a block that has been filleted along with the placement of a hole drilled through. The other object represents the U-shaped extrusion. With both objects modeled, they are moved on top of each other at "B." The OSNAP-Midpoint mode was used to accomplish this. Finally the INTERSECT command is used to find the common volume shared by both objects; namely the illustration at "C."

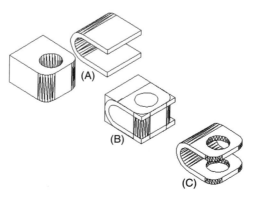

Figure 20–58

The object in Figure 20–59 is another example of how the INTERSECT command is applied to a solid model. For the results at "B" to be obtained of a cylinder that has numerous cuts, the cylinder is first extruded as a separate model. Then the cuts are made in another model at "A". Again, both models are moved together and then the INTERSECT command used to achieve the results at "B." Before undertaking any solid model, first analyze how the model is to be constructed. Using intersections can create dramatic results, which would normally require numerous union and subtraction operations.

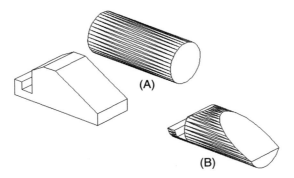

(A)

(B)

Figure 20–59

CREATING SOLID EXTRUSIONS

The EXTRUDE command creates a solid by extrusion. Only polylines and circles may be extruded. Once objects are polylines, use the following prompts to construct a solid extrusion of the object in Figure 20–60. For the height of the extrusion, you can enter a positive numeric value or you can determine the distance by picking two points on the display screen.

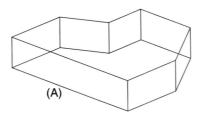

(A)

Figure 20–60

Command: **EXT** *(For EXTRUDE)*
Current wire frame density: ISOLINES=4
Select objects: *(Select the polyline object at "A" in Figure 20–60)*
Select objects: *(Press* ENTER *to continue with this command)*
Specify height of extrusion or [Path]: **1.00**
Specify angle of taper for extrusion <0>: *(Press* ENTER *to accept the default and perform the extrusion operation)*

You may create an optional taper along with the extrusion by entering an angle value for the prompt, "Specify angle of taper for extrusion".

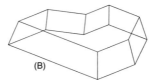

(B)

Figure 20–61

Command: **EXT** *(For EXTRUDE)*
Current wire frame density: ISOLINES=4
Select objects: *(Select the polyline object at "B" in Figure 20–61)*
Select objects: *(Press* ENTER *to continue with this command)*
Specify height of extrusion or [Path]: **1.00**
Specify angle of taper for extrusion <0>: **15**

You may also create a solid extrusion by selecting a path to be followed by the object being extruded. Typical paths include regular and elliptical arcs, 2D and 3D polylines, or splines. The object in Figure 20–62 was created after the construction of the polyline path. Then, a new User Coordinate System was created through the UCS command along with the Z axis option; the new UCS was positioned at the end of the polyline with the Z axis extending along the polyline. Finally, a circle was constructed with its center point at the end of the polyline before being extruded along the polyline path.

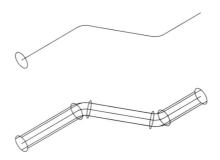

Figure 20–62

Command: **EXT** *(For EXTRUDE)*
Current wire frame density: ISOLINES=4
Select objects: *(Select the small circle as the object to extrude)*
Select objects: *(Press ENTER to continue)*
Specify height of extrusion or [Path]: **P** *(For Path)*
Select extrusion path: *(Select the polyline object representing the path)*

CREATING REVOLVED SOLIDS

The REVOLVE command creates a solid by revolving an object about an axis of revolution. Only polylines, polygons, circles, ellipses, and 3D polylines may be revolved. If a group of objects are not in the form of polylines, group them together using the PEDIT command. The polyline must form a closed shape. The resulting image in Figure 20–63 represents a solid object.

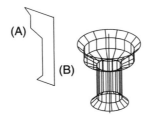

Figure 20–63

Command: **REV** *(For REVOLVE)*
Current wire frame density: ISOLINES=4
Select objects: *(Select profile "A" as the object to revolve)*
Select objects: *(Press ENTER to continue with this command)*
Specify start point for axis of revolution or
Define axis by [Object/X (axis)/Y (axis)]: **O** *(For Object)*
Select an object: *(Select line "B")*
Specify angle of revolution <360>: *(Press ENTER to accept the default and perform the revolving operation)*

A practical application of this type of solid would be to first construct an additional solid consisting of a cylinder using the CYLINDER command. Be sure this solid is larger in diameter than the revolved solid. Existing OSNAP options are fully supported in solid modeling. Use the Center option of OSNAP along with the MOVE command to position the revolved solid inside the cylinder.

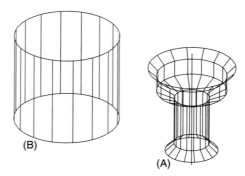

Figure 20–64

Command: **M** *(For MOVE)*
Select objects: *(Select the revolved solid in Figure 20–64)*
Select objects: *(Press ENTER to continue with this command)*
Specify base point or displacement: **Cen**
of *(Select the center of the revolved solid at "A")*
Specify second point of displacement or <use first point as displacement>: **Cen**
of *(Select the center of the cylinder at "B")*

Then, once the revolved solid is positioned inside the cylinder, use the SUBTRACT command to subtract the revolved solid from the cylinder (see Figure 20–65). Use the HIDE command to perform a hidden line removal at "B" to check that the solid is correct (this would be difficult to interpret in wireframe mode).

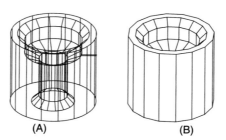

Figure 20–65

USING THE ISOLINES AND FACETRES COMMANDS

Tessellation refers to the lines that are displayed on any curved surface to help you in visualizing the surface shown in Figure 20–66. Tessellation lines are automatically formed when you construct solid primitives such as cylinders and cones. These lines are also calculated when you perform solid modeling operations such as SUBTRACT and UNION.

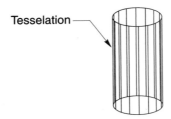

Figure 20–66

THE ISOLINES COMMAND

The number of tessellation lines per curved object is controlled by the command called ISOLINES. By default, this variable is set to a value of 4. Figure 20–67 shows the results of setting this variable to other values such as 8 and 12. The more lines used to describe a curved surface, the more accurate the surface will look; however, it will take longer to process hidden line removals using the HIDE command and Boolean operation commands such as UNION and SUBTRACT.

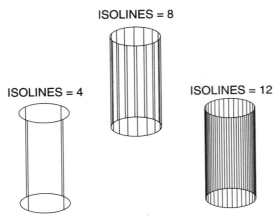

Figure 20–67

THE FACETRES COMMAND

As shown in Figure 20–67, tessellation lines on cylinders can affect wireframe models. When you perform hidden line removals on these objects, the results are displayed in Figure 20–68. The cylinder with FACETRES set to 0.50 processes much more quickly than the cylinder with FACETRES set to 1, because there are fewer surfaces to process in such operations as hidden line removals. However, the image with FACETRES set to 1 shows a more defined circle. The default value for FACETRES is 0.50, which seems adequate for most applications.

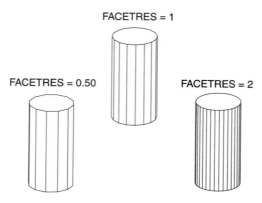

Figure 20–68

FILLETING SOLID MODELS

Filleting of complex objects is easily handled with the FILLET command. This is the same command as the one used to create a 2D fillet. Use the following prompt sequence and Figure 20–69.

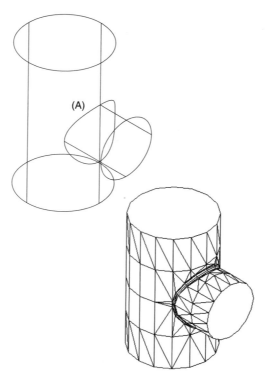

Figure 20–69

Command: **F** *(For FILLET)*
Current settings: Mode = TRIM, Radius = 0.5000
Select first object or [Polyline/Radius/Trim]: *(Select the edge at "A" which represents the intersection of both cylinders)*
Enter fillet radius <0.5000>: **0.25**
Select an edge or [Chain/Radius]: *(Press* ENTER *to perform the fillet operation)*
I edge(s) selected for fillet.

A group of objects with a series of edges may be filleted with the Chain option of the FILLET command. Use the following prompt sequence and Figure 20–70.

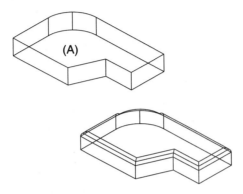

Figure 20–70

Command: **F** *(For FILLET)*
Current settings: Mode = TRIM, Radius = 0.5000
Select first object or [Polyline/Radius/Trim]: *(Select the edge at "A")*
Enter fillet radius <0.5000>: *(Press* ENTER *to accept the default value)*
Select an edge or [Chain/Radius]: **C** *(For chain mode)*
Select an edge chain or [Edge/Radius]: *(Select all edges that form the top of the plate)*
Select an edge chain or [Edge/Radius]: *(Press* ENTER *when finished selecting all edges to perform the fillet operation)*
7 edge(s) selected for fillet.

CHAMFERING SOLID MODELS

Just as the FILLET command uses the Chain mode to group a series of edges together, the CHAMFER command uses the Loop option to do the same type of operation. Use the following prompt sequence and Figure 20–71.

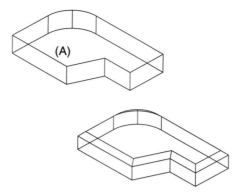

Figure 20–71

Command: **CHA** *(For CHAMFER)*
(TRIM mode) Current chamfer Dist1 = 0.5000, Dist2 = 0.5000
Select first line or [Polyline/Distance/Angle/Trim/Method]: *(Pick the edge at "A")*
Base surface selection... *(The top surface should be highlighted. If not, continue with the next series of prompts)*
Enter surface selection option [Next/OK (current)] <OK>: **N** *(To select the next surface; the top surface should highlight)*
Enter surface selection option [Next/OK (current)] <OK>: *(Press ENTER to accept the top surface)*
Specify base surface chamfer distance <0.5000>: *(Press ENTER to accept the default)*
Specify other surface chamfer distance <0.5000>: *(Press ENTER to accept the default)*
Select an edge or [Loop]: **L** *(To loop all edges together into one)*
Select an edge loop or [Edge]: *(Pick any edge)*
Select an edge loop or [Edge]: *(Press ENTER to perform the chamfering operation)*

MODELING TECHNIQUES USING VIEWPORTS

Using viewports can be very important and beneficial when laying out a solid model. Use the VPORTS command to lay out the display screen in a number of individually different display areas. Select this command by choosing Viewports from the View pull-down menu shown in Figure 20–72. A cascading menu appears with the different types of viewports available. Choosing New Viewports... displays the Viewports dialog box shown in Figure 20–73. A number of viewport configurations may be selected that will convert the display screen to the VPORT configuration desired. Selecting 3D in the Setup area automatically sets up 3D view points for the viewports.

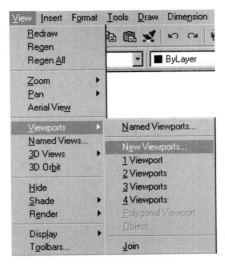

Figure 20–72

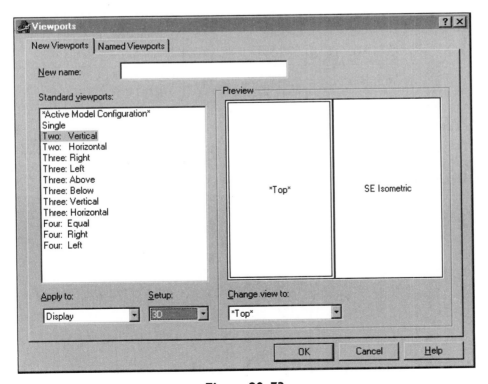

Figure 20–73

Figure 20–74 shows the results of the dialog box settings and the way viewports can be used in the construction of a 3D solid model. The image in the right viewport represents the solid model viewed in three dimensions. The image on the left represents the model in 2D mode. The images are viewed based on the World Coordinate System. This gives you two viewports to help visualize the construction of the solid model.

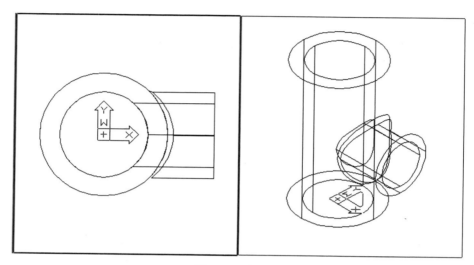

Figure 20–74

OBTAINING MASS PROPERTIES

Use the MASSPROP command to calculate the mass properties of a selected solid (see Figure 20–75A). All calculations in Figure 20–75B are based on the current position of the User Coordinate System.

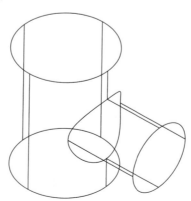

Figure 20–75A

Command: **MASSPROP**
Select objects: *(Select the model)*

```
————————————————SOLIDS————————————————
```

Mass:	26.8978
Volume:	26.8978
Bounding box:	X: 5.8821 — 11.6617
	Y: 3.1262 — 6.6855
	Z: 0.0000 — 5.0000
Centroid:	X: 8.2360
	Y: 4.9058
	Z: 2.5000
Moments of inertia:	
	X: 895.3560
	Y: 2108.3669
	Z: 2569.2755
Products of inertia:	
	XY: 1086.7870
	YZ: 329.8900
	ZX: 553.8236
Radii of gyration:	
	X: 5.7695
	Y: 8.8535
	Z: 9.7734

Principal moments and X-Y-Z directions about centroid:
 I: 79.8905 along [1.0000 0.0000 0.0000]
 J: 115.7440 along [0.0000 1.0000 0.0000]
 K: 97.4092 along [0.0000 0.0000 1.0000]

Figure 20–75B

THE INTERFERE COMMAND

Use this command to find any interference shared by a series of solids. As an interference is identified, a solid may be created out of the common volume similar to Figure 20–76.

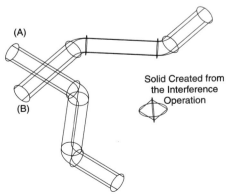

(A)

(B)

Solid Created from
the Interference
Operation

Figure 20–76

Command: **INTERFERE**
Select first set of solids: *(Select solid "A")*
Select objects: I found
Select objects: *(Press* ENTER *to continue)*
Select second set of solids: *(Select solid "B")*
Select objects: I found
Select objects: *(Press* ENTER *to continue)*
Comparing I solid against I solid.
Interfering solids (first set): I
                              (second set): I
Interfering pairs: I
Create interference solids? [Yes/No] <N>: **Y**

CUTTING SECTIONS FROM SOLID MODELS

THE SECTION COMMAND

Use this command to create a cross section of a selected solid object similar to Figure 20–77. Note that section lines are not used to define the surfaces cut by the operation. You have control over how the section will be created; in Figure 20–77, the location of the User Coordinate System defines the cutting plane.

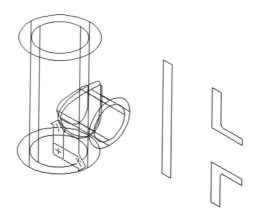

Figure 20–77

Command: **SEC** *(For SECTION)*
Select objects: *(Select the object in Figure 20–77)*
Select objects: *(Press* ENTER *to continue with this command)*
Specify first point on Section plane by [Object/Zaxis/View/XY/YZ/ZX/3points]
   <3points>: **XY**
Specify a point on the XY-plane <0,0,0>: *(Press* ENTER *to perform the sectioning operation)*

SLICING SOLID MODELS

THE SLICE COMMAND

This command is similar to the SECTION command except that the solid model is actually cut or sliced at a location that you define; in the example in Figure 20–78, this location is defined by the User Coordinate System. Before the slice is made, you also have the option of keeping the desired half while discarding the other.

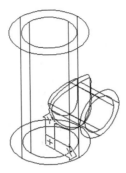

Figure 20–78

Command: **SL** *(For SLICE)*
Select objects: *(Select the solid object in Figure 20–78)*
Select objects: *(Press ENTER to continue with this command)*
Specify first point on slicing plane by [Object/Zaxis/View/XY/YZ/ZX/3points]
    <3points>: **XY**
Specify a point on XY-plane <0,0,0>: *(Press ENTER to accept this default value)*
Specify a point on desired side of the plane or [keep Both sides]: **B** *(To keep both sides)*

The MOVE command was used to separate both halves in Figure 20–79.

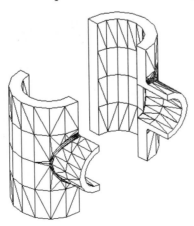

Figure 20–79

SHADING SOLID MODELS

Various shading modes are available to help you better visualize the solid model you are constructing. You should use shading modes as often as possible. Access the seven shading modes by choosing Shade from the View pull-down menu in Figure 20–80A. There is also a Shade toolbar dedicated to all shading modes, shown in Figure 20–80B. Each one of these modes is explained in the next series of paragraphs.

Figure 20–80A

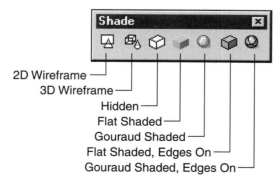

Figure 20–80B

2D WIREFRAME

This shading mode, illustrated in Figure 20–81, displays a wireframe image of a solid model. Note that the User Coordinate System icon is displayed flat or in 2D. This is how the command name is derived.

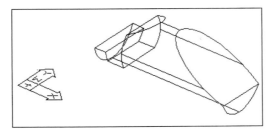

Figure 20–81

3D WIREFRAME

The 3D Wireframe mode is identical to the 2D mode regarding the wireframe mode. Notice how the User Coordinate System in Figure 20–82 has switched to a better defined coordinate system icon that illustrates the directions of the X, Y, and Z axes.

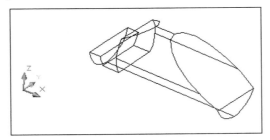

Figure 20–82

HIDDEN

The Hidden mode performs a hidden line removal, which is illustrated by the model in Figure 20–83. Only those surfaces in front of your viewing plane are visible.

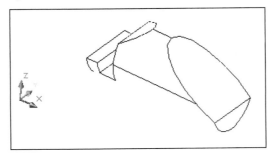

Figure 20–83

FLAT SHADE

You can achieve more dramatic viewing results by performing a Flat Shade of a solid model in Figure 20–84. Here the model is shaded based on the current color and screen background. The system variable FACETRES controls the smoothness of this shaded image.

Figure 20–84

GOURAUD SHADED

This shade mode provides the highest quality because it not only applies color to the model but also smoothes all edges as well (see Figure 20–85).

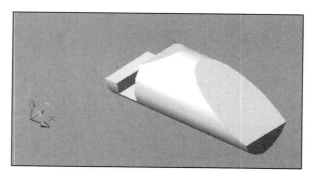

Figure 20–85

FLAT SHADED, EDGES ON

At times, you need to stay in flat shaded mode yet still select edges of the model for performing other operations. To better select edges of a shaded model, turn the edges on using this mode. The results are illustrated in Figure 20–86, with the wireframe lines showing through the shaded model.

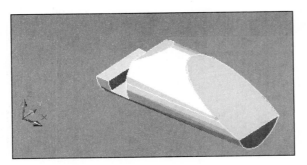

Figure 20–86

GOURAUD SHADED, EDGES ON

As with the flat shaded mode, you can have the wireframe edges show through a Gouraud Shaded model, as in Figure 20–87.

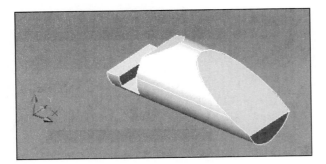

Figure 20–87

In addition, when you perform such commands as 3DORBIT, the current shade mode remains persistent. This means that if you are in Gouraud Shaded mode and you rotate your model using 3DORBIT, the model remains shaded through the rotation operation.

TUTORIAL EXERCISE: BPLATE.DWG

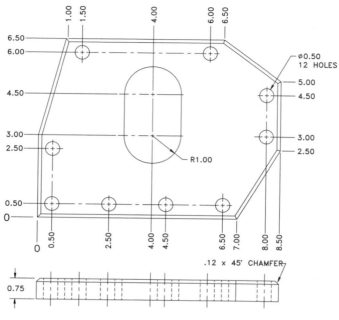

Figure 20–88

Purpose

The purpose of this tutorial is to produce a solid model of the Bplate shown in Figure 20–88.

System Settings

Keep the current limits settings of (0,0) for the lower left corner and (12,9) for the upper right corner. Change the number of decimal places past the zero from four to two using the Drawing Units dialog box. Keep all remaining system settings.

Layers

Special layers do not have to be created for this tutorial exercise.

Suggested Commands

Begin this tutorial by constructing the profile of the Bplate using polylines. Add all circles; use the EXTRUDE command to extrude all objects at a thickness of 0.75 units. Chamfer the top edge of the plate using the CHAMFER command. Use the SUBTRACT command to subtract all cylinders from the Bplate forming the holes in the plate. Perform hidden line removals using the HIDE command and create shaded models using the SHADE command.

Whenever possible, substitute the appropriate command alias in place of the full AutoCAD command in each tutorial step. For example, use "CP" for the COPY command, "L" for the LINE command, and so on. The complete listing of all command aliases is located in Table 1–2.

STEP 1

Begin the Bplate by establishing a new coordinate system using the following UCS command. Define the origin at 2.00,1.50. Use the UCSICON command, if necessary, to update the User Coordinate System icon to the new coordinate system location on the display screen. Use the PLINE command to draw the profile of the Bplate. See Figure 20–89.

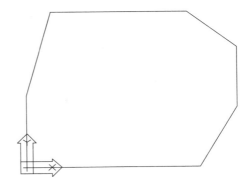

Figure 20–89

Command: **UCS**
Current ucs name: *WORLD*
Enter an option [New/Move/
    orthoGraphic/Prev/Restore/Save/Del/
    Apply/?/World] <World>: **M** *(For
    Move)*
Specify new origin point or [Zdepth]
<0,0,0>: **2.00,1.50**

Command: **UCSICON**
Enter an option [ON/OFF/All/Noorigin/
    ORigin] <ON>: **OR** *(For Origin)*

Command: **PL** *(For PLINE)*
Specify start point: **0,0**
Current line-width is 0.00
Specify next point or [Arc/Close/
    Halfwidth/Length/Undo/Width]:
    @7.00<0

Specify next point or [Arc/Close/
    Halfwidth/Length/Undo/Width]:
    @1.50,2.50
Specify next point or [Arc/Close/
    Halfwidth/Length/Undo/Width]:
    @2.50<90
Specify next point or [Arc/Close/
    Halfwidth/Length/Undo/Width]: **@-
    2.00,1.50**
Specify next point or [Arc/Close/
    Halfwidth/Length/Undo/Width]:
    @5.50<180
Specify next point or [Arc/Close/
    Halfwidth/Length/Undo/Width]: **@-
    1.00,-3.50**
Specify next point or [Arc/Close/
    Halfwidth/Length/Undo/Width]: **C** *(To
    close the polyline and return to the
    command prompt)*

STEP 2

Draw the nine circles of 0.50 diameter by placing one circle at "A" and copying the remaining circles to their desired locations (see Figure 20–90). Use the Multiple option of the COPY command to accomplish this.

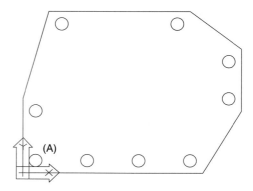

Figure 20–90

Command: **C** *(For CIRCLE)*
Specify center point for circle or [3P/2P/ Ttr (tan tan radius)]: **0.50,0.50**
Specify radius of circle or [Diameter]: **D** *(For Diameter)*
Specify diameter of circle: **0.50**

Command: **CP** *(For COPY)*
Select objects: **L** *(for the last circle)*
Select objects: *(Press ENTER to continue with this command)*
Specify base point or displacement, or [Multiple]: **M** *(For Multiple)*
Specify base point: **@** *(References the center of the 0.50 circle)*
Specify second point of displacement or <use first point as displacement>: **2.50,0.50**
Specify second point of displacement or <use first point as displacement>: **4.50,0.50**

Specify second point of displacement or <use first point as displacement>: **6.50,0.50**
Specify second point of displacement or <use first point as displacement>: **8.00,3.00**
Specify second point of displacement or <use first point as displacement>: **8.00,4.50**
Specify second point of displacement or <use first point as displacement>: **0.50,2.50**
Specify second point of displacement or <use first point as displacement>: **1.50,6.00**
Specify second point of displacement or <use first point as displacement>: **6.00,6.00**
Specify second point of displacement or <use first point as displacement>: *(Press ENTER to exit this command)*

STEP 3

Form the slot by placing two circles using the CIRCLE command followed by two lines drawn from the quadrants of the circles using the OSNAP-Quadrant mode. See Figure 20–91.

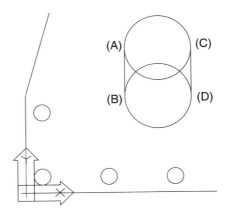

Figure 20–91

Command: **C** *(For CIRCLE)*
Specify center point for circle or [3P/2P/
   Ttr (tan tan radius)]: **4.00,3.00**
Specify radius of circle or [Diameter]
   <0.25>: **1.00**

Command: **C** *(For CIRCLE)*
Specify center point for circle or [3P/2P/
   Ttr (tan tan radius)]: **4.00,4.50**
Specify radius of circle or [Diameter]
   <1.00> *(Press* ENTER *to accept the
   default value)*

Command: **L** *(For LINE)*
Specify first point: **Qua**

of *(Select the quadrant of the circle at "A")*
Specify next point or [Undo]: **Qua**
of *(Select the quadrant of the circle at "B")*
Specify next point or [Undo]: *(Press*
   ENTER *to exit this command)*

Command: **L** *(For LINE)*
Specify first point: **Qua**
of *(Select the quadrant of the circle at "C")*
Specify next point or [Undo]: **Qua**
of *(Select the quadrant of the circle at "D")*
Specify next point or [Undo]: *(Press*
   ENTER *to exit this command)*

STEP 4

Use the TRIM command to trim away any unnecessary arcs to form the slot. See Figure 20–92.

Command: **TR** *(For TRIM)*
Current settings: Projection=UCS
   Edge=None
Select cutting edges ...
Select objects: *(Select both vertical lines at "A" and "B")*
Select objects: *(Press ENTER to continue with this command)*
Select object to trim or [Project/Edge/ Undo]: *(Select the circle at "C")*
Select object to trim or [Project/Edge/ Undo]: *(Select the circle at "D")*

Select object to trim or [Project/Edge/ Undo]: *(Press ENTER to exit this command).*

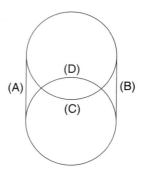

Figure 20–92

STEP 5

The next step illustrates the use of the EXTRUDE command to give the plate a thickness of 0.75 units. However, this command only operates on circles and polylines. Currently, all objects may be extruded except for the two arcs and lines representing the slot (see Figure 20–93). Use the PEDIT command to convert these objects to a single polyline. Choose this command under Polyline > on the Modify pull-down menu.

Command: **PE** *(For PEDIT)*
Select polyline: *(Select the bottom arc at "A")*
Object selected is not a polyline
Do you want to turn it into one? <Y>
   (Press ENTER to accept this default value)
Enter an option [Close/Join/Width/Edit vertex/Fit/Spline/Decurve/Ltype gen/ Undo]: **J** *(For Join)*
Select objects: *(Select the objects labeled "B," "C," and "D")*

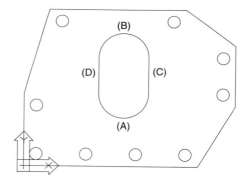

Figure 20–93

Select objects: *(Press ENTER to perform the joining operation)*
3 segments added to polyline
Enter an option [Open/Join/Width/Edit vertex/Fit/Spline/Decurve/Ltype gen/ Undo]: *(Press ENTER to exit this command)*

STEP 6

Use the EXTRUDE command to give Bplate thickness. Enter this command from the keyboard or choose Solids from the Draw pull-down menu and then Extrude. Use the All option to select all circles and polyline outlines. Enter a value of 0.75 as the height of the extrusion. Keep the default value for the extrusion taper angle. When finished with this operation, turn off the UCS icon with the UCSICON command. See Figure 20–94.

Command: **EXT** (For EXTRUDE)
Current wire frame density: ISOLINES=4
Select objects: **All**
Select objects: (Press ENTER to continue with this command)
Specify height of extrusion or [Path]:
    0.75

Specify angle of taper for extrusion <0>:
    (Press ENTER to execute the extrusion operation)

Command: **UCSICON**
Enter an option [ON/OFF/All/Noorigin/ORigin] <ON>: **Off**

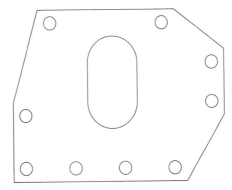

Figure 20–94

STEP 7

Use the VPOINT command to view the plate in three dimensions using the following prompt sequence and Figure 20–95. Or choose 3D Views from the View pull-down menu and then SE Isometric to achieve the same results. Next, use the CHAMFER command to place a chamfer of 0.12 units along the top edge of the plate. The top surface may not highlight when you are prompted for the base surface; use the Next option until the top surface is highlighted. After entering the chamfer distances, use the Loop option to create the chamfer along the entire edge without picking each individual edge along the top.

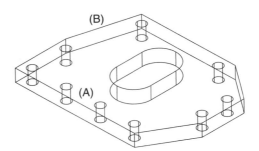

Figure 20–95

Command: **VPOINT**
Switching to the WCS
Current view direction:
    VIEWDIR=0.00,0.00,1.00

Specify a view point or [Rotate] <display compass and tripod>: **1,-1,1**

Command: **CHA** (For CHAMFER)

(TRIM mode) Current chamfer Dist1 = 0.50, Dist2 = 0.50

Select first line or [Polyline/Distance/ Angle/Trim/Method]: (Select the line at "A")

Base surface selection...

Enter surface selection option [Next/OK (current)] <OK>: **N** (For the next surface, which should be the top surface)

Enter surface selection option [Next/OK (current)] <OK>: (Press ENTER to accept the top surface as the base)

Specify base surface chamfer distance <0.5000>: **0.12**

Specify other surface chamfer distance <0.5000>: **0.12**

Select an edge or [Loop]: **L** (For loop)

Select an edge loop or [Edge]: (Select the edge along the top at "B")

Select an edge loop or [Edge]: (Press ENTER to perform the chamfer operation)

STEP 8

Because all of the objects have been extruded a distance of 0.75 units, the holes and slot are considered individual solid objects that do not yet belong to the base. Use the SUBTRACT command to subtract the holes and slot from the base plate (see Figure 20–96). This operation will resemble the drilling of holes and the milling for the slot. Choose this command from the Solids Editing cascading menu on the Modify pull-down menu.

Command: **SU** (For SUBTRACT)

Select solids and regions to subtract from...

Select objects: (Select the base of the Bplate along any edge)

Select objects: (Press ENTER to continue with this command)

Select solids and regions to subtract...

Select objects: (Select every hole and the slot)

Select objects: (Press ENTER to perform the subtraction operation)

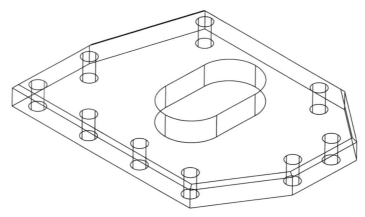

Figure 20–96

STEP 9

To see the object with all hidden edges removed, use the HIDE command. Choose this command from the View pull-down menu. See Figure 20–97.

Command: **HI** *(For HIDE)*
Regenerating model.

Regenerating the display will return the model to its wireframe mode.

Command: **RE** *(For REGEN)*
Regenerating model.

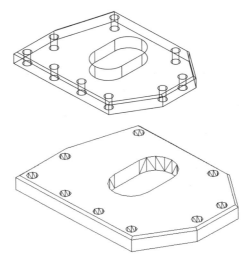

Figure 20–97

STEP 10

From Figure 20–97 in the previous step, the circles and slot look somewhat irregular after you perform the HIDE operation. To smooth out the circles and arcs representing the slot, use the FACETRES command. This stands for facet resolution and it controls the density at which circles and arcs are displayed. Set the facet resolution to a new value of 1, perform a HIDE operation, and obs results. See Figure 20–98.

Command: **FACETRES**
Enter new value for FACETRES
   <0.5000>: **1**

Command: **HI** *(For HIDE)*
Regenerating model.

Notice the circles are better defined due to the increased density. Regenerating the display will return the model to its wireframe mode.

Command: **RE** *(For REGEN)*
Regenerating model.

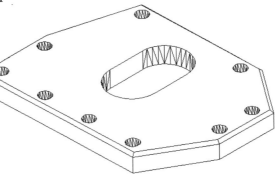

Figure 20–98

STEP 11

FACETRES controls the density of circles and arcs when you perform hidden line removals. The ISOLINES command controls the density of lines that represent circles and arcs in the wireframe mode. These lines are called tessellation lines. Change the value of ISOLINES from 4 to a new value of 15, regenerate the screen, and observe the results. See Figure 20–99.

Command: **ISOLINES**
Enter new value for ISOLINES <4>: **15**

Command: **RE** *(For REGEN)*
Regenerating model.

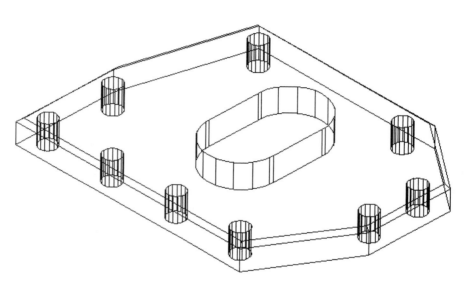

Figure 20–99

STEP 12

Use the SHADE command to produce a shaded image consisting of hidden line removal in addition to the shading being performed in the original color of the model. There are no features to control lights, materials, or shadows. This shading mechanism is merely a way to get a fast shaded image of a model before you prepare it to be rendered. See Figure 20–100.

Command: **SHA** *(For SHADE)*
Regenerating model.

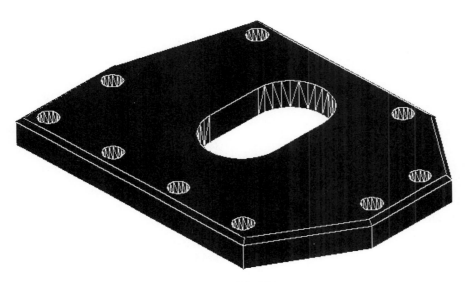

Figure 20–100

TUTORIAL EXERCISE: GUIDE.DWG

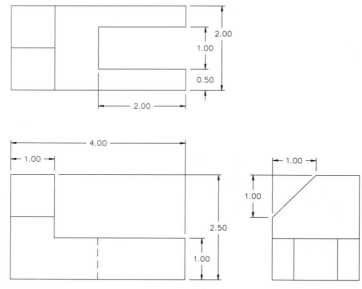

Figure 20–101

Purpose

This tutorial exercise is designed to produce a 3D solid model of the Guide from the information supplied in the orthographic drawing in Figure 20–101.

System Settings

Use the Drawing Units dialog box, keep the system of measurement set to decimal but change the number of decimal places past the zero from a value of 4 to 2. Leave the current limits of (0,0) by (12,9) as the default setting.

Layers

Special layers do not have to be created for this tutorial exercise.

Suggested Commands

Begin this drawing by constructing solid primitives of all components of the Guide using the BOX and WEDGE commands. Move the components into position and begin merging solids using the UNION command. To form the rectangular slot, move the solid box into position and use the SUBTRACT command to subtract the rectangle from the solid, thus forming the slot. Do the same procedure for the wedge. Perform a hidden line removal and view the solid.

Whenever possible, substitute the appropriate command alias in place of the full AutoCAD command in each tutorial step. For example, use "CP" for the COPY command, "L" for the LINE command, and so on. The complete listing of all command aliases is located in Table 1–2.

STEP 1

Begin this tutorial by constructing a solid box 4 units long by 2 units wide and 1 unit in height using the BOX command. See Figure 20–102. Begin this box at absolute coordinate 4.00,5.50. This slab will represent the base of the guide.

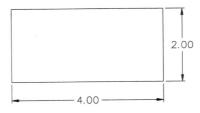

Command: **BOX**
Specify corner of box or [CEenter]
    <0,0,0>: **4.00,5.50**
Specify corner or [Cube/Length]: **L** *(For Length)*
Specify length: **4.00**
Specify width: **2.00**
Specify height: **1.00**

Figure 20–102

STEP 2

Construct a solid box 1 unit long by 2 units wide and 1.5 units in height using the BOX command (see Figure 20–103). Begin this box at absolute coordinate 2.00,1.50. This slab will represent the vertical column of the guide.

Command: **BOX**
Specify corner of box or [CEenter]
    <0,0,0>: **2.00,1.50**
Specify corner or [Cube/Length]: **L** *(For Length)*
Specify length: **1.00**
Specify width: **2.00**
Specify height: **1.50**

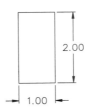

Figure 20–103

STEP 3

Construct a solid box 2 units long by 1 unit wide and 1 unit in height using the BOX command (see Figure 20–104). Begin this box at absolute coordinate 5.50,1.50. This slab will represent the rectangular slot made into the slab that will be subtracted at a later time.

Command: **BOX**
Specify corner of box or [CEenter]
    <0,0,0>: **5.50,1.50**
Specify corner or [Cube/Length]: **L** *(For Length)*
Specify length: **2.00**
Specify width: **1.00**
Specify height: **1.00**

Figure 20–104

STEP 4

Use the WEDGE command to draw a wedge 1 unit in length, 1 unit wide, and 1 unit in height (see Figure 20–105). Begin this primitive at absolute coordinate 9.50,1.50. This wedge will be subtracted from the vertical column to form the inclined surface.

Command: **WEDGE**
Specify first corner of wedge or
    [CEenter] <0,0,0>: **9.50,1.50**
Specify corner or [Cube/Length]: **L** *(For Length)*
Specify length: **1.00**
Specify width: **1.00**
Specify height: **1.00**

Figure 20–105

STEP 5

Use the VPOINT command to view the four solid primitives in three dimensions (see Figure 20–106). Use a new view point of 1,-1,1. You can use a preset view point by choosing 3D Views from the View pull-down menu and then SE Isometric. Then use the MOVE command to move the vertical column at "A" to the top of the base at "B."

Command: **VPOINT**
Switching to the WCS
Current view direction:
   VIEWDIR=0.00,0.00,1.00
Specify a view point or [Rotate] <display compass and tripod>: 1,-1,1

Command: **M** *(For MOVE)*
Select objects: *(Select the solid box at "A")*
Select objects: *(Press ENTER to continue with this command)*
Specify base point or displacement: **End** of *(Select the endpoint of the solid at "A")*
Specify second point of displacement or <use first point as displacement>: **End** of *(Select the endpoint of the base at "B")*

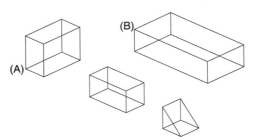

Figure 20–106

STEP 6

Use the UNION command to join the base and vertical column into one object. See Figure 20–107.

Command: **UNION**
Select objects: *(Select the base at "A" and column at "B")*
Select objects: *(Press ENTER to perform the union)*

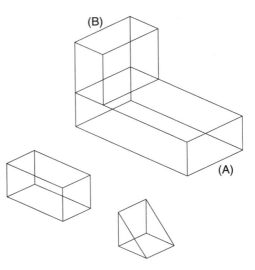

Figure 20–107

STEP 7

Use the MOVE command to position the rectangle from its midpoint at "A" to the midpoint of the base at "B" in Figure 20–108. In a moment, the small rectangle will be subtracted to form the rectangular slot in the base.

Command: **M** *(For MOVE)*
Select objects: *(Select box "A")*
Select objects: *(Press ENTER to continue with this command)*
Specify base point or displacement: **Mid** of *(Select the midpoint of the rectangle at "A")*
Specify second point of displacement or <use first point as displacement>: **Mid** of *(Select the midpoint of the base at "B")*

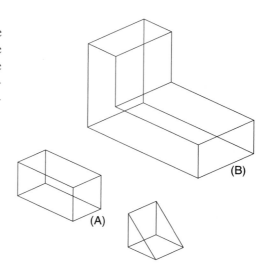

Figure 20–108

STEP 8

Use the SUBTRACT command to subtract the small box from the base of the solid object. See Figure 20–109.

Command: **SU** *(For SUBTRACT)*
Select solids and regions to subtract from...
Select objects: *(Select solid object "A")*
Select objects: *(Press ENTER to continue with this command)*
Select solids and regions to subtract...
Select objects: *(Select box "B" to subtract)*
Select objects: *(Press ENTER to perform the subtraction operation)*

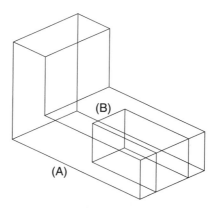

Figure 20–109

STEP 9

Use the ALIGN command to match points along a source object (wedge) with points along a destination object (guide). The selection of three sets of points will guide the placement and rotation of the wedge into the guide. See Figure 20–110.

Command: **AL** *(For ALIGN)*
Select objects: *(Select the wedge)*
Select objects: *(Press ENTER to continue with this command)*
Specify first source point: **End**
of *(Pick the endpoint of the wedge at "A")*
Specify first destination point: **End**
of *(Pick the endpoint of the guide at "B")*
Specify second source point: **End**
of *(Pick the endpoint of the wedge at "C")*
Specify second destination point: **End**
of *(Pick the endpoint of the guide at "D")*
Specify third source point or <continue>:
    End
of *(Pick the endpoint of the wedge at "E")*
Specify third destination point: **End**
of *(Pick the endpoint of the guide at "F")*

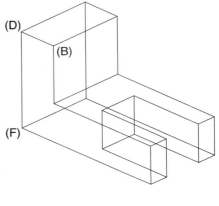
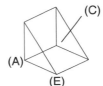

Figure 20–110

STEP 10

Use the SUBTRACT command to subtract the wedge from the guide, as shown in Figure 20–111.

Command: **SU** *(For SUBTRACT)*
Select solids and regions to subtract from...
Select objects: *(Select guide "A")*
Select objects: *(Press ENTER to continue with this command)*
Select solids and regions to subtract...
Select objects: *(Select the wedge at "B")*
Select objects: *(Press ENTER to perform the subtraction operation)*

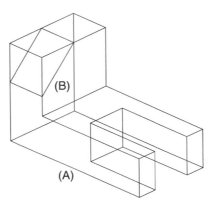

Figure 20–111

STEP 11

An alternate method of creating the inclined surface in the guide is to use the CHAMFER command on the vertical column of the guide. See Figure 20–112.

Command: **CHA** *(For CHAMFER)*
(TRIM mode) Current chamfer Dist1 = 0.50, Dist2 = 0.50
Select first line or [Polyline/Distance/ Angle/Trim/Method]: *(Select the line at "A")*
Base surface selection...
Enter surface selection option [Next/OK (current)] <OK>: *(Press ENTER to accept the base surface)*
Specify base surface chamfer distance <0.50>: 1
Specify other surface chamfer distance <0.50>: 1
Select an edge or [Loop]: *(Select the edge at "A")*
Select an edge or [Loop]: *(Press ENTER to perform the chamfer operation)*

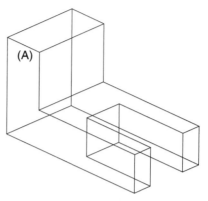

Figure 20–112

STEP 12

Use the HIDE command to perform a hidden line removal on all surfaces of the object. The results are shown in Figure 20–113.

Command: **HI** *(For HIDE)*
Regenerating model.

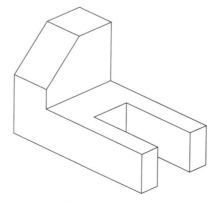

Figure 20–113

TUTORIAL EXERCISE: LEVER.DWG

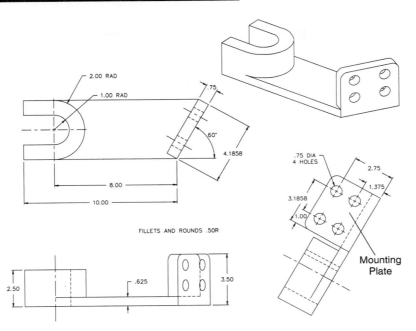

FILLETS AND ROUNDS .50R

Mounting Plate

Figure 20–114

Purpose

This tutorial is designed to use the UCS command to construct a 3D solid model of the Lever in Figure 20–114.

System Settings

Begin a new drawing called "Lever." Use the Drawing Units dialog box to change the number of decimal places past the zero from four to two. Keep all default values for the UNITS command. Using the LIMITS command, keep (0,0) for the lower left corner and change the upper right corner from (12,9) to (15.50,9.50).

Layers

Create the following layer with the format:

Name	Color	Linetype
Model	Cyan	Continuous

Suggested Commands

Begin the layout of this problem by constructing the plan view of the Lever. Use the EXTRUDE command to create the height of the individual components that make up the lever. Position the UCS to construct the mounting plate.

Whenever possible, substitute the appropriate command alias in place of the full AutoCAD command in each tutorial step. For example, use "CP" for the COPY command, "L" for the LINE command, and so on. The complete listing of all command aliases is located in Table 1–2.

STEP 1

Begin this drawing by making the Model layer current. Then construct two circles of radius values 1.00 and 2.00 using the CIRCLE command and 4.00,5.00 as the center of both circles (see Figure 20–115). Use the @ symbol to identify the last known point (4.00,5.00) as the center of the circle.

Command: **C** *(For CIRCLE)*
Specify center point for circle or [3P/2P/
   Ttr (tan tan radius)]: **4.00,5.00**
Specify radius of circle or [Diameter]:
   1.00

Command: **C** *(For CIRCLE)*
Specify center point for circle or [3P/2P/
   Ttr (tan tan radius)]: **@** *(To identify the
   last point)*
Specify radius of circle or [Diameter]
   <1.00>: **2.00**

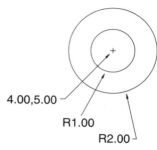

Figure 20–115

STEP 2

Draw a vertical line 2 units to the left of the center of the circles and 4 units long in the 270 direction, or use the Direct Distance mode to construct the line in the 270 direction. In the command sequence below, use OSNAP-Tracking to accomplish this. Turn on OSNAP, OTRACK, and ORTHO at the status bar. Set Quadrant as a running object snap.

Command: **L** *(For LINE)*
Specify first point: *(Move the cursor over
   point "A" to acquire the quadrant for
   tracking, then move the cursor over point
   "B" to acquire the quadrant for tracking.
   As you move the cursor to point "C," two
   dotted lines will appear; pick a point at
   this intersection for the starting point of*

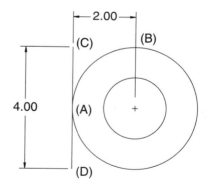

Figure 20–116

*your line. Move the cursor toward point
"D" to use Direct Distance mode for the
next prompt.)*
Specify next point or [Undo]: **4**
Specify next point or [Undo]: *(Press
   ENTER to exit this command)*

STEP 3

Draw four horizontal lines from quadrant points on the two circles a distance of 2 units. Polar coordinates or the Direct Distance mode may be used to accomplish this. These lines should intersect with the vertical line drawn in the previous step. See also Figure 20–117.

Command: **L** *(For LINE)*
Specify first point: **Qua**
of *(Select the quadrant of the large circle at "A")*
Specify next point or [Undo]: **@2<180**
Specify next point or [Undo]: *(Press ENTER to exit this command)*

Repeat the above procedure and draw three more lines from points "B," "C," and "D" in Figure 20–117. Using the same polar coordinate value for the length, namely, @2<180. An alternate method would be to use the Multiple option of the COPY command to duplicate the three remaining lines.

Figure 20–117

STEP 4

Use the TRIM command, select the two dashed lines in Figure 20–118 as cutting edges, and trim the left side of the large circle.

Command: **TR** *(For TRIM)*
Current settings: Projection=UCS Edge=None
Select cutting edges ...
Select objects: *(Select the two dashed lines in Figure 20–118)*
Select objects: *(Press ENTER to continue with this command)*
Select object to trim or [Project/Edge/Undo]: *(Pick the circle at "A")*

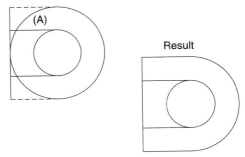

Result

Figure 20–118

Select object to trim or [Project/Edge/Undo]: *(Press ENTER to exit this command)*

STEP 5

Use the TRIM command again and select the two dashed lines in Figure 20–119 as cutting edges. Select the left side of the small circle and the middle of the vertical line as the objects to trim.

Command: **TR** *(For TRIM)*
Current settings: Projection=UCS
    Edge=None
Select cutting edges ...
Select objects: *(Select the two dashed lines in Figure 20–119)*
Select objects: *(Press ENTER to continue with this command)*
Select object to trim or [Project/Edge/Undo]: *(Select the vertical line at "A")*

Select object to trim or [Project/Edge/Undo]: *(Select the small circle at "B")*
Select object to trim or [Project/Edge/Undo]: *(Press ENTER to exit this command)*

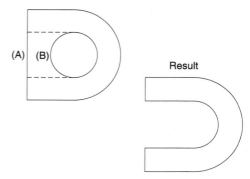

Figure 20–119

STEP 6

Complete a partial plan view by using the LINE command and Figure 20–120 to draw the four lines. Use OSNAP-Endpoint in combination with polar coordinates.

Command: **L** *(For LINE)*
Specify first point: **End**
of *(Select the endpoint of the line or arc labeled "Start")*
Specify next point or [Undo]: **@8.00<0**

Specify next point or [Undo]: **@4.1858<60**
Specify next point or [Undo]: **@0.75<150**
Specify next point or [Undo]: **End**
of *(Select the endpoint of the line or arc labeled "End")*
Specify next point or [Undo]: *(Press ENTER to exit this command)*

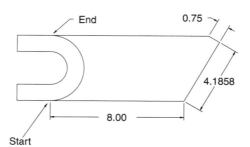

Figure 20–120

STEP 7

View the model in 3D using the VPOINT command. Use a viewing position of 1,-1,1 or choose 3D Views from the View pull-down menu and then SE Isometric. See Figure 20–121.

Command: **VPOINT**
Current view direction:
   VIEWDIR=0.00,0.00,1.00
Specify a view point or [Rotate] <display compass and tripod>: **1,-1,1**

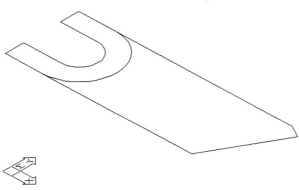

Figure 20–121

STEP 8

Move the four lines that represent the base away from the U-shaped figure as shown in Figure 20–122. Since both figures are at different heights, they will be treated as different objects when you extrude.

Command: **M** *(For MOVE)*
Select objects: *(Select all four lines labeled "A," "B," "C," and "D")*
Select objects: *(Press ENTER to continue with this command)*
Specify base point or displacement: **End** of *(Pick the endpoint near "A")*
Specify second point of displacement or <use first point as displacement>: **@2.00<0**

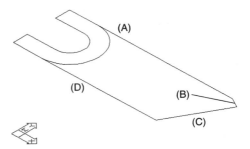

Figure 20–122

STEP 9

Copy arc "A" to the endpoint of the line at "C." This will allow the shape represented by the lines to be closed through the placement of the arc. See Figure 20–123.

Command: **CP** *(For COPY)*
Select objects: *(Select arc "A")*
Select objects: *(Press ENTER to continue with this command)*
Specify base point or displacement, or [Multiple]: **End**
of *(Pick the endpoint of the arc at "B")*
Specify second point of displacement or <use first point as displacement>: **End**
of *(Pick the endpoint of the arc at "C")*

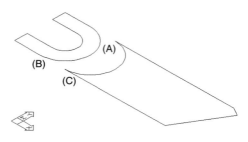

Figure 20–123

STEP 10

Convert both shapes to polylines using the PEDIT command. Refer also to Figure 20–124.

Command: **PE** *(For PEDIT)*
Select polyline: *(Select the arc at "A")*
Object selected is not a polyline
Do you want to turn it into one? <Y> *(Press ENTER to accept the default)*
Enter an option [Close/Join/Width/Edit vertex/Fit/Spline/Decurve/Ltype gen/Undo]: **J** *(For Join)*
Select objects: *(Select all arc and line segments that make up the outline of the U-shaped item)*
Select objects: *(Press ENTER to perform the join operation)*
7 segments added to polyline
Enter an option [Open/Join/Width/Edit vertex/Fit/Spline/Decurve/Ltype gen/Undo]: *(Press ENTER to exit this command)*

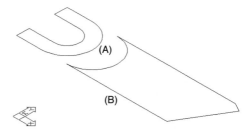

Figure 20–124

Command: **PE** *(For PEDIT)*
Select polyline: *(Select the line at "B")*
Object selected is not a polyline
Do you want to turn it into one? <Y>
　　(Press ENTER to accept the default)
Enter an option [Close/Join/Width/Edit
　　vertex/Fit/Spline/Decurve/Ltype gen/
　　Undo]: **J** *(For Join)*
Select objects: *(Select the other line and arc
　　segments that make up the outline of the
　　other shape)*

Select objects: *(Press ENTER to perform the
　　join operation)*
4 segments added to polyline
Enter an option [Open/Join/Width/Edit
　　vertex/Fit/Spline/Decurve/Ltype gen/
　　Undo]: *(Press ENTER to exit this
　　command)*

STEP 11

Extrude the U-shaped figure to a height
of 2.50 units. Extrude the other shape to
a height of 0.625 units. Perform both operations using the EXTRUDE command and
Figure 20–125.

Command: **EXT** *(For EXTRUDE)*
Current wire frame density: ISOLINES=4
Select objects: *(Select the U-shaped figure)*
Select objects: *(Press ENTER to continue
　　with this command)*
Specify height of extrusion or [Path]:
　　2.50
Specify angle of taper for extrusion <0>:
　　*(Press ENTER to perform the extrusion
　　operation)*

Command: **EXT** *(For EXTRUDE)*
Current wire frame density: ISOLINES=4
Select objects: *(Select the other shape at
　　"A")*
Select objects: *(Press ENTER to continue
　　with this command)*
Specify height of extrusion or [Path]:
　　0.625
Specify angle of taper for extrusion <0>:
　　*(Press ENTER to perform the extrusion
　　operation)*

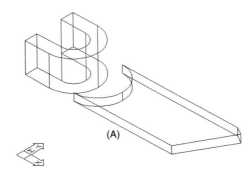

(A)

Figure 20–125

STEP 12

With both shapes extruded, connect both shapes together from the quadrant of one shape to the quadrant of the other shape using the MOVE command. See Figure 20–126.

Command: **M** *(For MOVE)*
Select objects: *(Select shape "A")*
Select objects: *(Press ENTER to continue with this command)*
Specify base point or displacement: **Qua**
of *(Pick the bottom quadrant of the arc at "B")*
Specify second point of displacement or <use first point as displacement>:
Qua
of *(Pick the bottom quadrant of the arc at "C")*

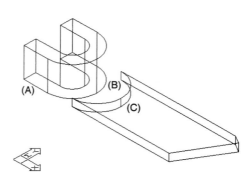

Figure 20–126

STEP 13

Join both shapes into one using the UNION command, as shown in Figure 20–127.

Command: **UNION**
Select objects: *(Select shape "A" and "B")*
Select objects: *(Press ENTER to perform the union operation)*

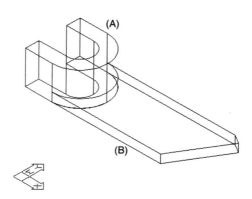

Figure 20–127

STEP 14

Define a new origin for the User Coordinate System by placing it at the endpoint at "A" in Figure 20–128. If the icon does not move to the new location, use the UCSICON command and the Origin option to update the position of the User Coordinate System icon.

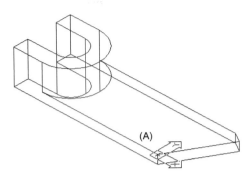

Command: **UCS**
Current ucs name: *WORLD*
Enter an option [New/Move/
    orthoGraphic/Prev/Restore/Save/Del/
    Apply/?/World] <World>: **M** *(For
    Move)*
Specify new origin point or [Zdepth]
    <0,0,0>: **End**
of *(Pick the line at "A" to find the endpoint)*

Figure 20–128

Command: **UCSICON**
Enter an option [ON/OFF/All/Noorigin/
    ORigin] <ON>: **OR** *(For Origin)*

STEP 15

Before beginning the construction of the mounting plate consisting of the four holes, rotate the UCS icon 60° about the Z axis and then 90° about the X axis. See Figure 20–129.

Command: **UCS**
Current ucs name: *NO NAME*
Enter an option [New/Move/
    orthoGraphic/Prev/Restore/Save/Del/
    Apply/?/World] <World>: **N** *(For New)*
Specify origin of new UCS or [ZAxis/
    3point/OBject/Face/View/X/Y/Z]
    <0,0,0>: **Z**

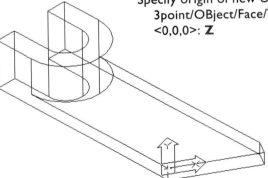

Figure 20–129

Specify rotation angle about Z axis <90>:
60

Command: **UCS**
Current ucs name: *NO NAME*
Enter an option [New/Move/
   orthoGraphic/Prev/Restore/Save/Del/
   Apply/?/World] <World>: **N** *(For New)*

Specify origin of new UCS or [ZAxis/
   3point/OBject/Face/View/X/Y/Z]
   <0,0,0>: **X**
Specify rotation angle about X axis <90>:
   (Press ENTER to accept default)

STEP 16

Construct a polyline representing the
front face of the mounting plate (see Figure 20–130). Polar coordinates or the Direct Distance mode may be used to accomplish this step. Round off the two top
corners of the mounting plate using the
FILLET command.

Command: **PL** *(For PLINE)*
Specify start point: **0,0,0**
Current line-width is 0.00
Specify next point or [Arc/Close/
   Halfwidth/Length/Undo/Width]:
   @3.50<90
Specify next point or [Arc/Close/
   Halfwidth/Length/Undo/Width]:
   @4.1858<0
Specify next point or [Arc/Close/
   Halfwidth/Length/Undo/Width]:
   @3.50<270
Specify next point or [Arc/Close/
   Halfwidth/Length/Undo/Width]: **C** *(To
   close the polyline and exit the command)*

Command: **F** *(For FILLET)*
Current settings: Mode = TRIM Radius
   = 0.50
Select first object or [Polyline/Radius/
   Trim]: *(Pick the polyline at "A")*
Select second object: *(Pick the pline at
   "B")*

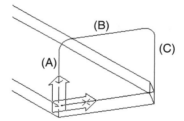

Figure 20–130

Repeat the FILLET command to round off
the other corner.

Command: **F** *(For FILLET)*
Current settings: Mode = TRIM Radius
   = 0.50
Select first object or [Polyline/Radius/
   Trim]: *(Pick the polyline at "B")*
Select second object: *(Pick the pline at
   "C")*

STEP 17

Extrude the face of the mounting plate back a distance of -0.75 units to give it thickness using the EXTRUDE command. Once the mounting plate has been extruded, join the plate to the Lever using the UNION command. See Figure 20–131.

Command: **EXT** *(For EXTRUDE)*
Current wire frame density: ISOLINES=4
Select objects: *(Select the face of the mounting plate)*

Select objects: *(Press ENTER to continue with this command)*
Specify height of extrusion or [Path]: **- 0.75**
Specify angle of taper for extrusion <0>: *(Press ENTER to perform the extrusion operation)*

Command: **UNION**
Select objects: *(Select the mounting plate and the Lever)*
Select objects: *(Press ENTER to perform the union operation)*

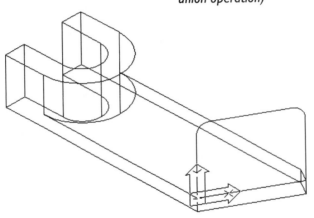

Figure 20–131

STEP 18

Lay out one of the holes into the mounting plate. Use the OSNAP-From option to establish a reference point that will allow you to construct the cylinder -1.00 units in the X axis and 1.375 units along the Y axis. See Figure 20–132.

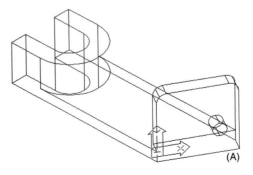

(A)

Figure 20–132

Command: **CYLINDER**
Current wire frame density: ISOLINES=4
Specify center point for base of cylinder
    or [Elliptical] <0,0,0>: **From**
Base point: **End**
of *(Pick the endpoint at "A")*

<Offset>: **@-1.00,1.375**
Specify radius for base of cylinder or
    [Diameter]: **D** *(For Diameter)*
Specify diameter for base of cylinder:
    0.75
Specify height of cylinder or [Center of
    other end]: **-0.75**

STEP 19

Use the ARRAY command to create a rectangular pattern consisting of two rows and two columns. The distance between the rows will be 1.375 units and the distance between columns will be -2.1858 units. Then subtract the four cylinders from the Lever to complete the object using the SUBTRACT command. See Figure 20–133.

Command: **AR** *(For ARRAY)*
Select objects: *(Select the cylinder just created)*
Select objects: *(Press ENTER to continue with this command)*
Enter the type of array [Rectangular/ Polar] <R>: **R** *(For rectangular)*
Enter the number of rows (—) <1>: **2**

Enter the number of columns (|||) <1>: **2**
Enter the distance between rows or
    specify unit cell (—): **1.375**
Specify the distance between columns
    (||||): **-2.1858**

Command: **SU** *(For SUBTRACT)*
Select solids and regions to subtract
    from...
Select objects: *(Select the object at "A")*
Select objects: *(Press ENTER to continue with command)*
Select solids and regions to subtract...
Select objects: *(Select the four individual cylinders just created through the ARRAY command)*
Select objects: *(Press ENTER to perform the subtraction operation)*

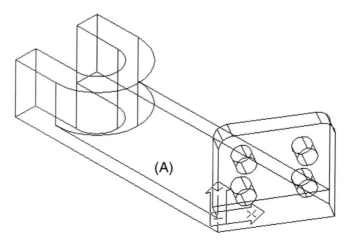

(A)

Figure 20–133

STEP 20

Set FACETRES to a value of 2 and perform a hidden line removal to get a better look at the model. See Figure 20–134.

Command: **FACETRES**
Enter new value for FACETRES
   <0.5000>: **2**

Command: **HI** *(For HIDE)*

Perform a REGEN to convert the model
   back to its wireframe state.

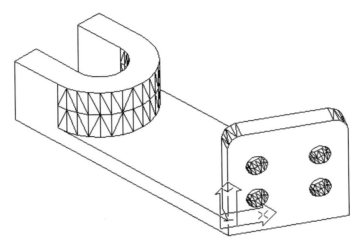

Figure 20–134

STEP 21

Return the User Coordinate System icon to its home position by using the World option of the UCS command. Then use the MASSPROP command to calculate the mass properties of the lever. See Figure 20–135. Variances may occur in a few of the data fields when using this command.

Command: **UCS**
Current ucs name: *NO NAME*

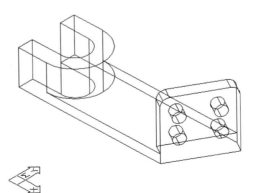

Figure 20–135

Enter an option [New/Move/
orthoGraphic/Prev/Restore/Save/Del/
Apply/?/World] <World>: (Press ENTER
to accept the default as WORLD)

Command: **MASSPROP**
Select objects: (Select anywhere along the
model)
Select objects: (Press ENTER to perform the
mass property calculation)

———————————————SOLIDS———————————————

Mass:	50.22
Volume:	50.22
Bounding box:	X: 4.00–16.09
	Y: 3.00–7.00
	Z: 0.00–4.13
Centroid:	X: 9.71
	Y: 5.03
	Z: 1.12
Moments of inertia:	X: 1461.23
	Y: 5552.24
	Z: 6786.64
Products of inertia:	XY: 2469.71
	YZ: 281.33
	ZX: 571.53
Radii of gyration:	X: 5.39
	Y: 10.51
	Z: 11.62

Principal moments and X-Y-Z directions about centroid:
I: 124.10 along [1.00 0.02 0.04]
J: 756.73 along [-0.02 1.00 -0.08]
K: 782.03 along [-0.04 0.08 1.00]

TUTORIAL EXERCISE: COLLAR.DWG

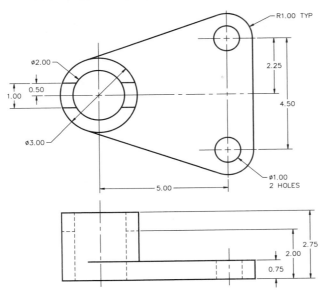

Figure 20–136

Purpose

This tutorial is designed to construct a solid model of the Collar using the dimensions in Figure 20–136.

System Settings

Use the current limits set to 0,0 for the lower left corner and (12,9) for the upper right corner. Change the number of decimal places from four to two using the Drawing Units dialog box. Snap and Grid values may remain as set by the default.

Layers

Create the following layer:

Name	Color	Linetype
Model	Cyan	Continuous

Suggested Commands

Begin this tutorial by laying out the Collar in plan view and drawing the basic shape out-lined in the Top view. Convert the objects to a polyline and extrude the objects to form a solid. Draw a cylinder and combine this object with the base. Add another cylinder and then subtract it to from the large hole through the model. Add two small cylinders and subtract them from the base to form the smaller holes. Construct a solid box, use the MOVE command to move the box into position, and subtract it to form the slot across the large cylinder.

Whenever possible, substitute the appropriate command alias in place of the full AutoCAD command in each tutorial step. For example, use "CP" for the COPY command, "L" for the LINE command, and so on. The complete listing of all command aliases is located in Table 1–2.

STEP I

Begin the Collar by setting the Model layer current. Then draw the three circles shown in Figure 20–137 using the CIRCLE command. Place the center of the circle at "A" at 0,0. Perform a ZOOM-All after all three circles have been constructed. The center marks identifying the centers of the circles are used for illustrative purposes and do not need to be placed in the drawing.

Command: **C** *(For CIRCLE)*
Specify center point for circle or [3P/2P/ Ttr (tan tan radius)]: **0,0**
Specify radius of circle or [Diameter]: **D** *(For Diameter)*
Specify diameter of circle: **3.00**

Command: **C** *(For CIRCLE)*
Specify center point for circle or [3P/2P/ Ttr (tan tan radius)]: **5.00,2.25**
Specify radius of circle or [Diameter] <1.50>: **1.00**

Command: **C** *(For CIRCLE)*
Specify center point for circle or [3P/2P/ Ttr (tan tan radius)]: **5.00,-2.25**
Specify radius of circle or [Diameter] <1.00>: *(Press ENTER to accept default)*

Command: **Z** *(For ZOOM)*
Specify corner of window, enter a scale factor (nX or nXP), or [All/Center/ Dynamic/Extents/Previous/Scale/ Window] <real time>: **All**

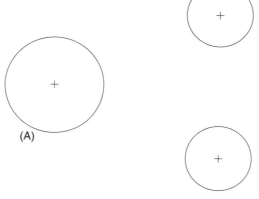

(A)

Figure 20–137

STEP 2

Draw lines tangent to the three arcs using the LINE command and the OSNAP-Tangent mode. Also see Figure 20–138.

Command: **L** *(For LINE)*
Specify first point: **Tan**
to *(Select the circle at "A")*

Specify next point or [Undo]: **Tan**
to *(Select the circle at "B")*
To point: *(Press the ENTER key to exit the command)*

Repeat the above procedure to draw lines from "C" to "D" and "E" to "F".

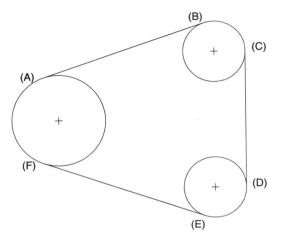

Figure 20–138

STEP 3

Use the TRIM command to trim the circles. When prompted to select the cutting edge object, press ENTER; this will make cutting edges out of all objects in the drawing. See Figure 20–139.

Command: **TR** *(For TRIM)*
Current settings: Projection=UCS
   Edge=None
Select cutting edges ...
Select objects: *(Press* ENTER *to create cutting edges out of all objects)*

Select object to trim or [Project/Edge/
   Undo]: *(Select the circle at "A")*
Select object to trim or [Project/Edge/
   Undo]: *(Select the circle at "B")*
Select object to trim or [Project/Edge/
   Undo]: *(Select the circle at "C")*
Select object to trim or [Project/Edge/
   Undo]: *(Press* ENTER *to exit this command)*

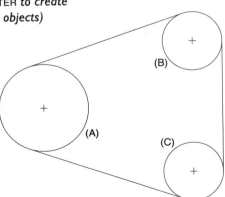

Figure 20–139

STEP 4

Prepare to construct the bracket by viewing the object in 3D using the VPOINT command and the coordinates 1,-1,1. You may also select a view point by choosing 3D Views from the View pulldown menu and then SE Isometric. See Figure 20–140.

Command: **VPOINT**
Current view direction:
   VIEWDIR=0.00,0.00,1.00
Specify a view point or [Rotate] <display
   compass and tripod>: **1,-1,1**

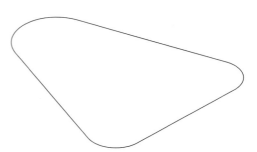

Figure 20–140

STEP 5

Convert all objects to a polyline using the Join option of the PEDIT command. See Figure 20–141.

Command: **PE** *(For PEDIT)*
Select polyline: *(Select the arc at "A")*
Object selected is not a polyline
Do you want to turn it into one? <Y>
   (Press ENTER *to accept the default value)*
Enter an option [Close/Join/Width/Edit
   vertex/Fit/Spline/Decurve/Ltype gen/
   Undo]: **J** *(For Join)*
Select objects: *(Select the three lines and
   remaining two arcs shown in Figure 20–
   141)*
Select objects: *(Press* ENTER *to perform the
   joining operation)*
5 segments added to polyline
Enter an option [Open/Join/Width/Edit
   vertex/Fit/Spline/Decurve/Ltype gen/
   Undo]: *(Press* ENTER *to exit this
   command)*

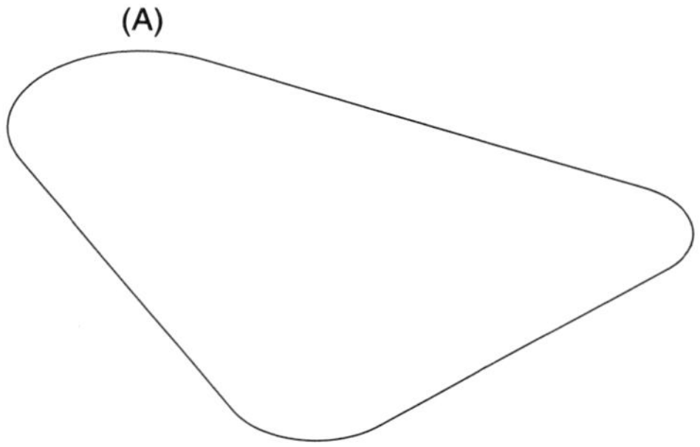

(A)

Figure 20–141

STEP 6

Use the EXTRUDE command to extrude the base to a thickness of 0.75 units. See Figure 20–142.

Command: **EXT** *(For EXTRUDE)*
Current wire frame density: ISOLINES=4
Select objects: *(Select the polyline)*
Select objects: *(Press ENTER to continue with this command)*
Specify height of extrusion or [Path]: **0.75**
Specify angle of taper for extrusion <0>: *(Press ENTER to perform the extrusion operation)*

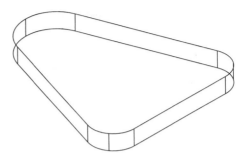

Figure 20–142

STEP 7

Create a cylinder using the CYLINDER command. Begin the center point of the cylinder at 0,0,0 with a diameter of 3.00 units and a height of 2.75 units. You may have to perform a ZOOM-All to display the entire model. See Figure 20–143.

Command: **CYLINDER**
Current wire frame density: ISOLINES=4
Specify center point for base of cylinder or [Elliptical] <0,0,0>: *(Press ENTER to accept the default of 0,0,0)*
Specify radius for base of cylinder or [Diameter]: **D** *(For Diameter)*
Specify diameter for base of cylinder: **3**
Specify height of cylinder or [Center of other end]: **2.75**

Command: **Z** *(For ZOOM)*
Specify corner of window, enter a scale factor (nX or nXP), or [All/Center/Dynamic/Extents/Previous/Scale/Window] <real time>: **All**

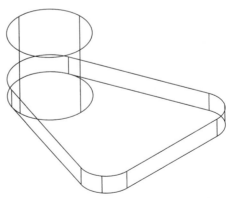

Figure 20–143

STEP 8

Merge the cylinder just created with the extruded base using the UNION command. See Figure 20–144.

Command: **UNION**
Select objects: *(Select the extruded base and cylinder)*
Select objects: *(Press ENTER to perform the union operation)*

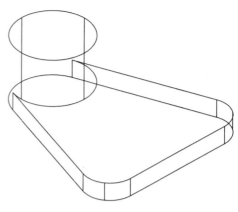

Figure 20–144

STEP 9

Use the CYLINDER command to create a 2.00-unit diameter cylinder representing a through hole, as shown in Figure 20–145. The height of the cylinder is 2.75 units with the center point at 0,0,0.

Command: **CYLINDER**
Current wire frame density: ISOLINES=4
Specify center point for base of cylinder or [Elliptical] <0,0,0>: *(Press ENTER to accept the default of 0,0,0)*
Specify radius for base of cylinder or [Diameter]: **D** *(For Diameter)*
Specify diameter for base of cylinder: **2.00**
Specify height of cylinder or [Center of other end]: **2.75**

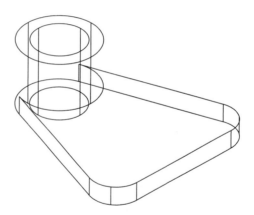

Figure 20–145

STEP 10

To cut the hole through the outer cylinder, use the SUBTRACT command. Select the base as the source object; select the inner cylinder as the object to subtract.

Use the HIDE command to view the results after performing a hidden line removal. Your display should appear similar to Figure 20–146.

Command: **SU** *(For SUBTRACT)*
Select solids and regions to subtract
  from…
Select objects: *(Select the base of the
  Collar)*
Select objects: *(Press* ENTER *to continue
  with this command)*
Select solids and regions to subtract…
Select objects: *(Select the 2.00 diameter
  cylinder just created)*
Select objects: *(Press* ENTER *to perform the
  subtraction operation)*

Command: **HI** *(For HIDE)*
Command: **RE** *(For REGEN)*
  (To return to wireframe mode)

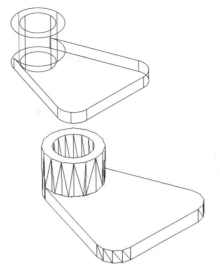

Figure 20–146

STEP 11

Begin placing the two small drill holes in
the base using the CYLINDER command.
Use the OSNAP-Center mode to place
each cylinder at the center of arcs "A" and
"B" in Figure 20–147.

Command: **CYLINDER**
Current wire frame density: ISOLINES=4
Specify center point for base of cylinder
  or [Elliptical] <0,0,0>: **Cen**
of *(Pick the bottom edge of the arc at "A")*
Specify radius for base of cylinder or
  [Diameter]: **D** *(For Diameter)*
Specify diameter for base of cylinder:
  1.00
Specify height of cylinder or [Center of
  other end]: **0.75**

Command: **CYLINDER**
Current wire frame density: ISOLINES=4
Specify center point for base of cylinder
  or [Elliptical] <0,0,0>: **Cen**
of *(Pick the bottom edge of the arc at "B")*
Specify radius for base of cylinder or
  [Diameter]: **D** *(For Diameter)*

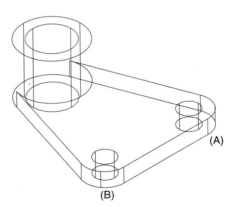

Figure 20–147

Specify diameter for base of cylinder:
  1.00
Specify height of cylinder or [Center of
  other end]: **0.75**

STEP 12

Subtract both 1.00-diameter cylinders from the base of the model using the SUB-TRACT command. Use the HIDE command to view the results after performing a hidden line removal. Your display should appear similar to Figure 20–148.

Command: **SU** (For SUBTRACT)
Select solids and regions to subtract from...
Select objects: (Pick the base of the Collar)
Select objects: (Press ENTER to continue with this command)
Select solids and regions to subtract...
Select objects: (Select both 1.00 diameter cylinders at the right)
Select objects: (Press ENTER to perform the subtraction operation)

Command: **HI** (For HIDE)
Command: **RE** (For REGEN)
    (To return to wireframe mode)

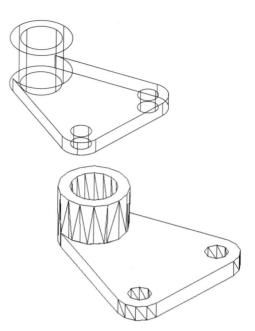

Figure 20–148

STEP 13

Begin constructing the rectangular slot that will pass through the two cylinders (see Figure 20–149). Use the BOX command to accomplish this. Locate the center of the box at 0,0,0 and make the box 4 units long, 1 unit wide, and 0.75 units high. Then move the box to the top of the cylinder. Use the geometry calculator to select the base point of displacement at the top of the box with the MEE function. After you locate two endpoints, the function will calculate the midpoint.

Command: **BOX**
Specify corner of box or [CEnter]
    <0,0,0>: **CE** (For the center of the box)

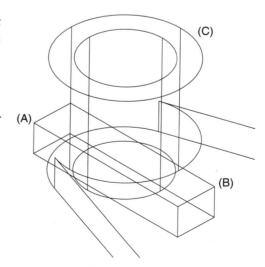

Figure 20–149

Specify center of box <0,0,0>: *(Press ENTER to accept the default center value of 0,0,0).*
Specify corner or [Cube/Length]: **L**
Specify length: **4**
Specify width: **1**
Specify height: **0.75**

Command: **M** *(For MOVE)*
Select objects: *(Select the box just constructed)*
Select objects: *(Press ENTER to continue with this command)*

Specify base point or displacement: **'CAL** *(To activate the geometry calculator)*
Initializing...
>> Expression: **MEE**
>> Select one endpoint for MEE: *(Pick a point on one corner of the box at "A")*
>> Select another endpoint for MEE: *(Pick a point on the opposite corner of the box at "B")*
Specify second point of displacement or <use first point as displacement>: **Cen**
of *(Pick the edge of the cylinder at "C")*

STEP 14

Use the SUBTRACT command to subtract the rectangular box from the solid model, as in Figure 20–150.

Command: **SU** *(For SUBTRACT)*
Select solids and regions to subtract from...
Select objects: *(Select the model at "A")*
Select objects: *(Press ENTER to continue with this command)*

Select solids and regions to subtract...
Select objects: *(Select the rectangular box at "B")*

Select objects: *(Press ENTER to perform the subtraction operation)*

The completed solid mode should appear similar to Figure 20–151.

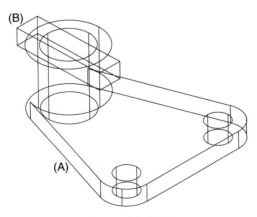

Figure 20–150

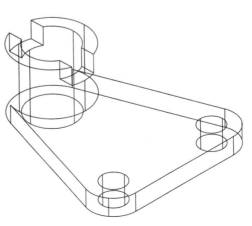

Figure 20–151

STEP 15

Change the facet resolution to a higher value. Then perform a hidden line removal to see the appearance of the model shown in Figure 20–152.

Command: **FACETRES**
Enter new value for FACETRES
  <0.5000>: **1**

Command: **HI** (For HIDE)

Command: **RE** (For REGEN)
  (To return to wireframe mode)

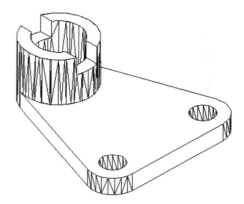

Figure 20–152

STEP 16

Important information may be extracted from the solid model to be used for design and analysis. The following calculations are obtained when you use the MASSPROP command. The following properties are calculated by this command: Mass, Volume, Bounding Box, Centroid, Moments of Inertia, Products of Inertia, Radii of Gyration, and Principal Mo-

ments about Centroid. Variances may occur in a few of the data fields when using this command.

Command: **MASSPROP**
Select objects: (Select the model of the Collar)
Select objects: (Press ENTER to perform the mass property calculation)

————————SOLIDS————————

Mass:	29.11
Volume:	29.11
Bounding box:	X: -1.50–6.00
	Y: -3.25–3.25
	Z: 0.00–2.75
Centroid:	X: 2.36
	Y: 0.00
	Z: 0.69
Moments of inertia:	X: 82.92
	Y: 319.72
	Z: 349.84
Products of inertia:	XY: 0.00
	YZ: 0.00
	ZX: 25.72
Radii of gyration:	X: 1.69
	Y: 3.31
	Z: 3.47

Principal moments and X-Y-Z directions about centroid:
  I: 65.09 along [0.98 0.00 -0.17]
  J: 144.18 along [0.00 1.00 0.00]
  K: 192.12 along [0.17 0.00 0.98]

PROBLEMS FOR CHAPTER 20

Directions for Problems 20–1a through 20–1i

1. Create a 3D solid model of each object on layer "Model."

2. When completed, calculate the volume of the solid model using the MASSPROP command.

PROBLEM 20–1A

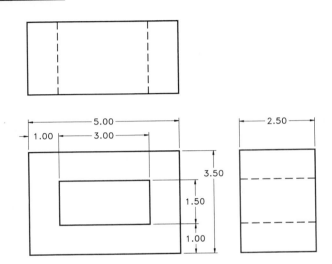

PROBLEM 20–1B

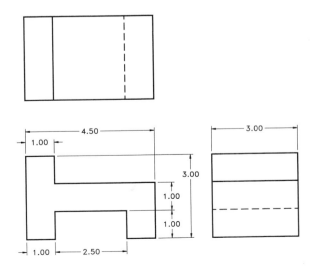

PROBLEM 20–1C

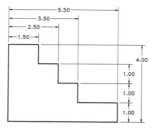

PROBLEM 20–1D

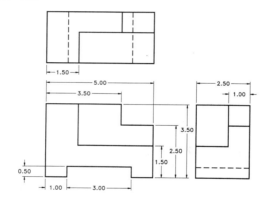

PROBLEM 20–1E

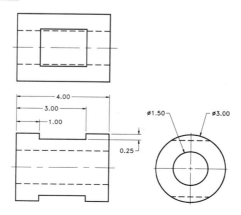

PROBLEM 20–1F

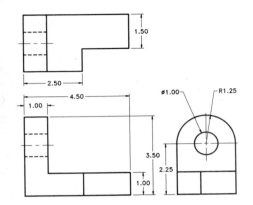

PROBLEM 20–1G

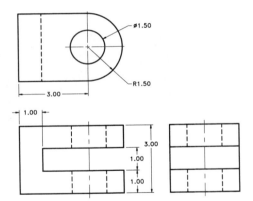

PROBLEM 20–1H

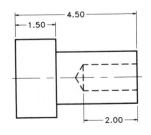

PROBLEM 20–11

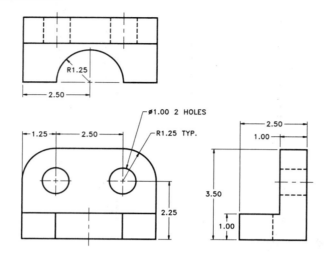

R1.25

2.50

Ø1.00 2 HOLES

1.25 2.50 R1.25 TYP.

2.50

1.00

3.50

1.00

2.25

Directions for Problems 20–2 through 20–30

1. Create a 3D solid model of each object on Layer "Model."

2. When completed, calculate the volume of the solid model using the MASSPROP command.

PROBLEM 20–2

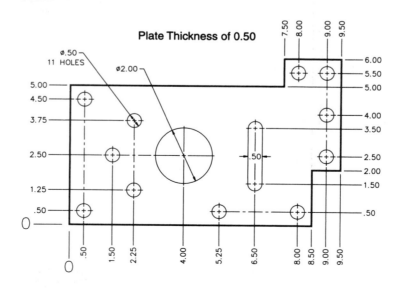

Plate Thickness of 0.50

Ø.50
11 HOLES

Ø2.00

.50

PROBLEM 20-3

Plate Thickness of 0.75

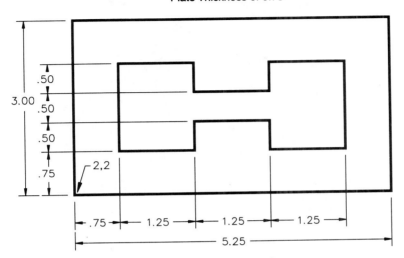

PROBLEM 20-4

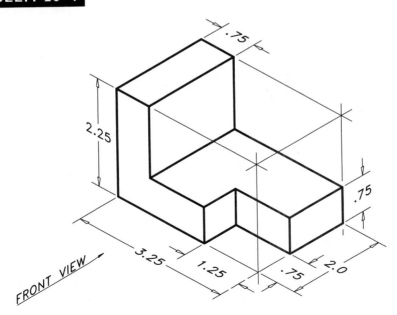

FRONT VIEW

PROBLEM 20–5

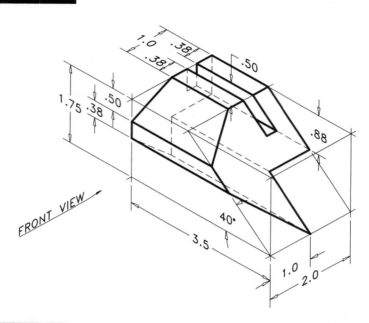

FRONT VIEW

1.0 .38
.38
.50
1.75 .38
.50
.88
40°
3.5
1.0
2.0

PROBLEM 20–6

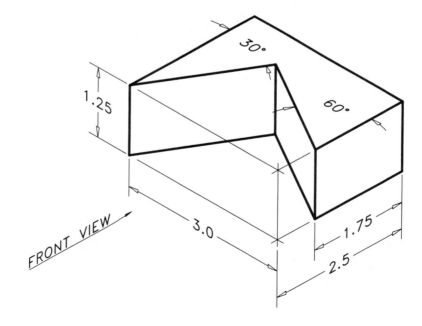

30°
60°
1.25
FRONT VIEW
3.0
1.75
2.5

PROBLEM 20–7

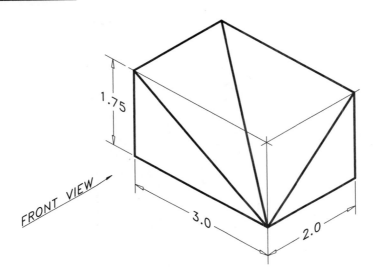

FRONT VIEW

1.75

3.0

2.0

PROBLEM 20–8

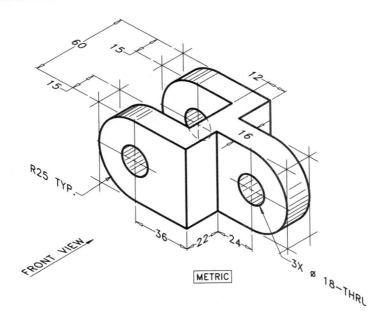

60

15

15

12

16

R25 TYP.

36

22

24

FRONT VIEW

METRIC

3X ø 18–THRU

PROBLEM 20–9

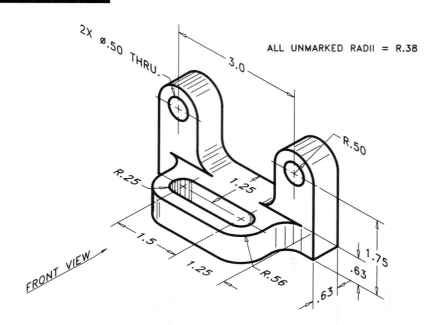

2X ⌀.50 THRU.

ALL UNMARKED RADII = R.38

3.0

R.50

R.25

1.25

R.56

1.5

1.25

1.75

.63

.63

FRONT VIEW

PROBLEM 20–10

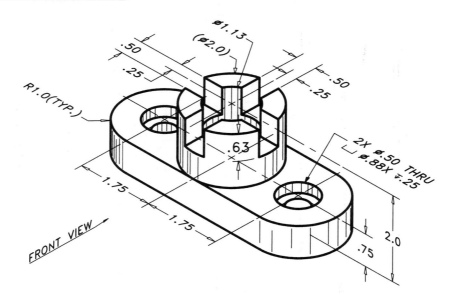

⌀1.13

(⌀2.0)

.50

.25

.50

.25

R1.0(TYP.)

.63

2X ⌀.50 THRU
⌴ ⌀.88X ▽.25

1.75

1.75

2.0

.75

FRONT VIEW

PROBLEM 20-11

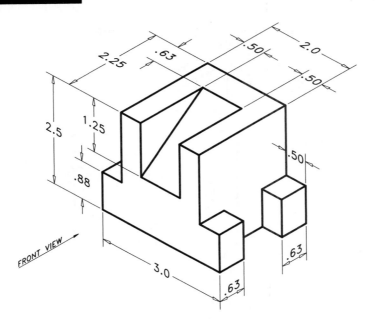

FRONT VIEW

PROBLEM 20-12

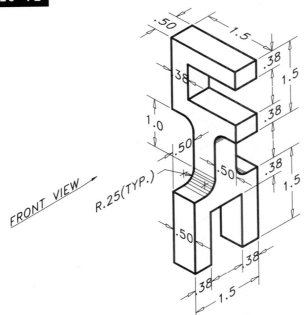

FRONT VIEW

PROBLEM 20–13

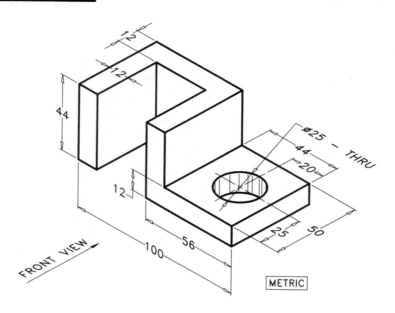

12

12

44

12

ø25 – THRU

44

20

25

50

56

100

FRONT VIEW

METRIC

PROBLEM 20–14

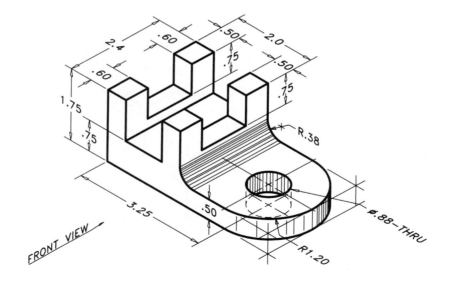

2.4

.60

.60

.50

2.0

.75

.50

.75

1.75

.75

R.38

3.25

.50

R1.20

ø.88–THRU

FRONT VIEW

PROBLEM 20–15

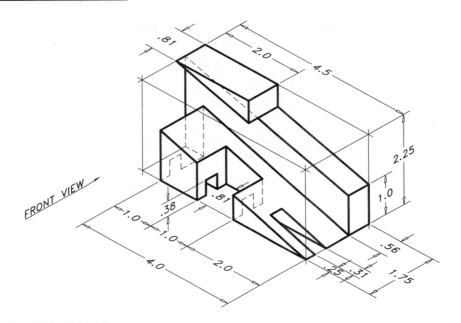

FRONT VIEW

PROBLEM 20–16

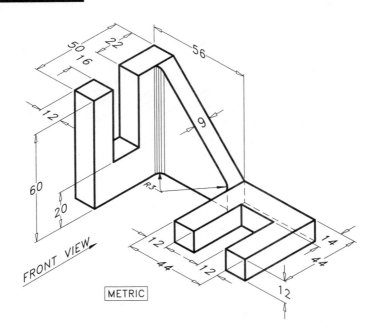

FRONT VIEW

METRIC

problem **EXERCISE**

PROBLEM 20–17

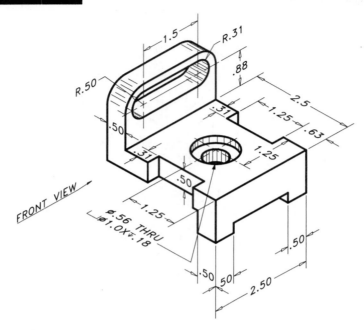

FRONT VIEW

PROBLEM 20–18

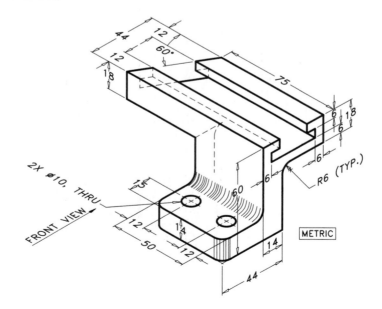

FRONT VIEW

METRIC

PROBLEM 20-19

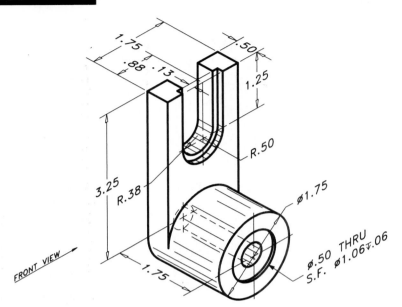

FRONT VIEW

1.75
.88 · .13
.50
1.25
3.25
R.38
R.50
Ø1.75
Ø.50 THRU
S.F. Ø1.06⊽.06
1.75

PROBLEM 20-20

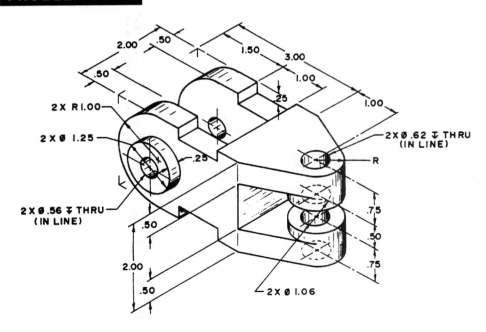

2.00
.50
1.50
3.00
.50
1.00
.25
2 X R1.00
2 X Ø 1.25
.25
2 X Ø.62 ⊥ THRU
(IN LINE)
R
2 X Ø.56 ⊥ THRU
(IN LINE)
.50
.75
.50
.75
2.00
.50
2 X Ø 1.06
1.00

PROBLEM 20–21

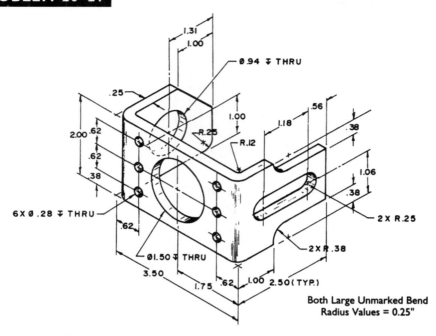

Ø.94 ⯑ THRU

.25

1.31
1.00

2.00 .62
.62
.38

R.25
R.12

1.00

1.18 .56
.38

1.06
.38

6X Ø.28 ⯑ THRU

.62

Ø1.50 ⯑ THRU

3.50
1.75 .62 1.00 2.50 (TYP.)

2 X R.25

2 X R.38

Both Large Unmarked Bend
Radius Values = 0.25"

PROBLEM 20–22

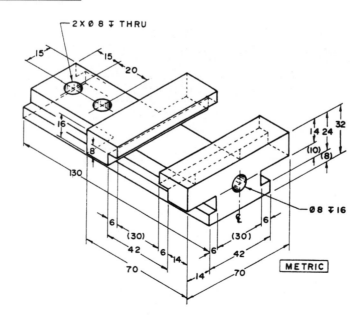

2 X Ø 8 ⯑ THRU

15 15 20

16
8

130

6
(30)
42
70

14 24 32
(10) (8)

14

Ø 8 ⯑ 16

6 6
(30)
42
6 14
14 70

METRIC

PROBLEM 20–23

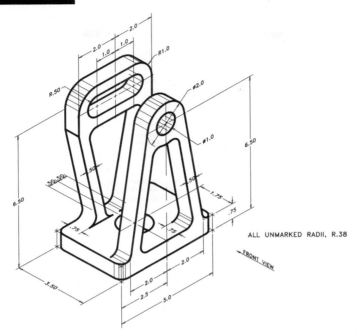

ALL UNMARKED RADII, R.38

FRONT VIEW

PROBLEM 20–24

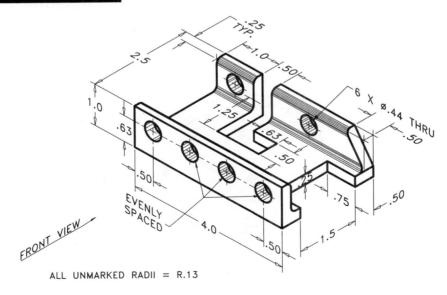

FRONT VIEW

ALL UNMARKED RADII = R.13

PROBLEM 20–25

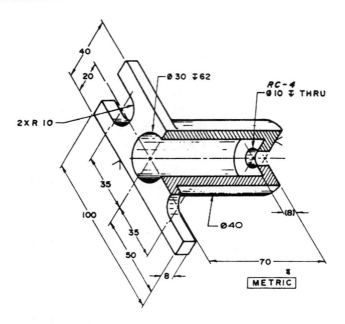

40
20
2 X R 10
Ø 30 ⌻62
RC-4
Ø 10 ⌻ THRU
35
100
35
50
8
Ø40
(8)
70
METRIC

PROBLEM 20–26

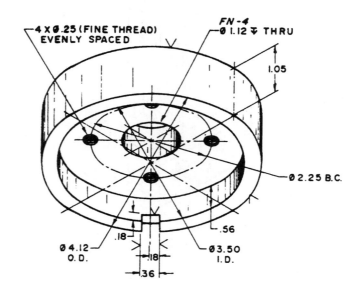

4 X Ø.25 (FINE THREAD)
EVENLY SPACED
FN-4
Ø 1.12 ⌻ THRU
1.05
Ø 2.25 B.C.
.56
.18
Ø 4.12
O. D.
.18
Ø 3.50
I. D.
.36

PROBLEM 20–27

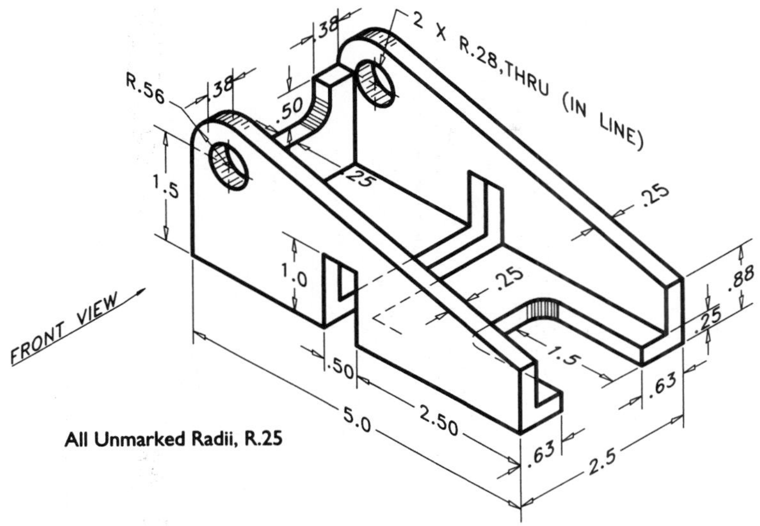
2 X R.28,THRU (IN LINE)
R.56
.38
.38
.50
1.5
.25
.25
.25
1.0
.88
.25
1.5
FRONT VIEW
.50
2.50
.63
5.0
.63
2.5

All Unmarked Radii, R.25

PROBLEM 20–28

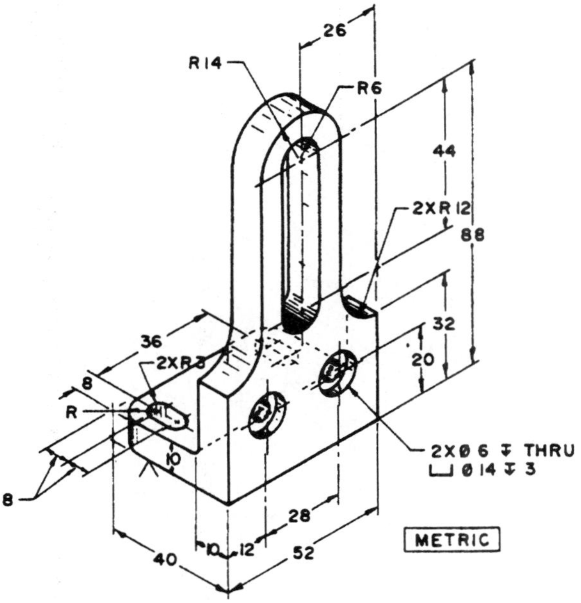
26
R14
R6
44
2XR12
88
32
20
36
2XR3
8
R
10
8
2XØ6 ↧ THRU
⌴ Ø14 ↧ 3
28
40
10
12
52
METRIC

Note: Use a Fillet Radius of 2 units for the inside unmarked radius.

PROBLEM 20-29

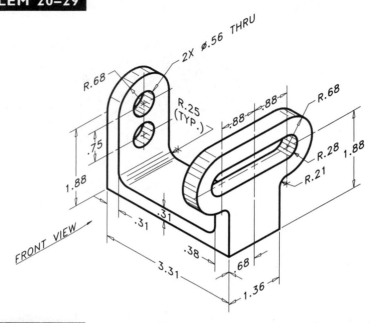

2X Ø.56 THRU

R.68

R.25 (TYP.)

R.68

.88 .88 .88

R.28

R.21

1.88

.75

1.88

.31

.31

.38

.68

3.31

1.36

FRONT VIEW

PROBLEM 20-30

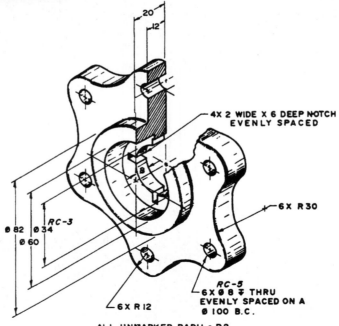

20

12

4X 2 WIDE X 6 DEEP NOTCH
EVENLY SPACED

6X R30

RC-3

Ø 82
Ø 34
Ø 60

RC-5
6X Ø 8 ⊥ THRU
EVENLY SPACED ON A
Ø 100 B.C.

6X R12

ALL UNMARKED RADII = R2

METRIC

CHAPTER 21

Editing Solid Models

THE ALIGN COMMAND

Once a solid model is created, various methods are available to edit the model. The three methods discussed in this chapter are the ALIGN, ROTATE3D, and SOLIDEDIT commands. Various small drawings, which can be opened, will be used to illustrate these commands.

The first method discussed is the use of the ALIGN command. Open the drawing 21_Align3D.Dwg. The objects in Figure 21–1 need to be positioned or aligned to form the object in "A." At this point, it is unclear at what angle the objects are currently rotated. When you use the ALIGN command, it is not necessary to know this information. Rather, you line up source points with destination points. When the three points are identified, the object moves and aligns to the three points. The first destination point acts as a base point to which the object being aligned will lock.

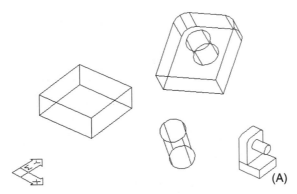
(A)

Figure 21–1

Follow the prompt sequence below and the illustration in Figure 21–2 for aligning the hole plate with the bottom base. The first destination point acts as a base point to which the cylinder locates.

Command: **HI** *(For HIDE)*
Command: **AL** *(For ALIGN)*
Select objects: *(Select the object with the hole)*
Select objects: *(Press* ENTER *to continue)*
Specify first source point: *(Select the endpoint at "A")*
Specify first destination point: *(Select the endpoint at "B")*
Specify second source point: *(Select the endpoint at "C")*
Specify second destination point: *(Select the endpoint at "D")*
Specify third source point or <continue>: *(Select the endpoint at "E")*
Specify third destination point: *(Select the endpoint at "F")*

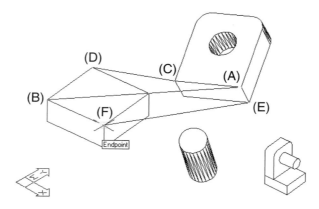

Figure 21–2

The results of the previous step are illustrated in Figure 21–3. Next, align the cylinder with the hole. Circular shapes need only two sets of source and destination points for the shapes to be properly aligned.

Command: **AL** *(For ALIGN)*
Select objects: *(Select the cylinder)*
Select objects: *(Press* ENTER *to continue)*
Specify first source point: *(Select the top center of the cylinder at "A")*
Specify first destination point: *(Select the outer center of the hole at "B")*
Specify second source point: *(Select the bottom center of the cylinder at "C")*
Specify second destination point: *(Select the inner center of the hole at "D")*
Specify third source point or <continue>: *(Press* ENTER *to continue)*
Scale objects based on alignment points? [Yes/No] <N>: *(Press* ENTER *to accept the default and align the cylinder with the hole)*

The completed 3D model is illustrated in Figure 21–4. The ALIGN command provides an easy means of putting solid objects together to form assembly models.

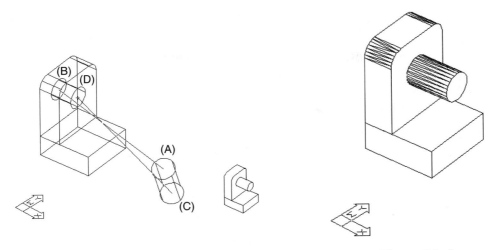

Figure 21–3

Figure 21–4

You are not limited to using the ALIGN command on 3D objects. There are 2D applications of the command, as shown in Figure 21–5. Open the drawing 21_Align2D.Dwg. Here the edge of the triangle needs to the aligned with the edge of the rectangle. Follow the prompt sequence for performing this task. Again, pay attention to the importance of the first destination point in aligning the object.

Command: **AL** *(For ALIGN)*
Select objects: *(Select the triangle)*
Select objects: *(Press* ENTER *to continue)*
Specify first source point: *(Select the endpoint at "A")*
Specify first destination point: *(Select the endpoint at "B")*
Specify second source point: *(Select the endpoint at "C")*
Specify second destination point: *(Select the endpoint at "D")*
Specify third source point or <continue>: *(Press* ENTER *to continue)*
Scale objects based on alignment points? [Yes/No] <N>: *(Press* ENTER *to accept the default and align the triangle with the rectangle)*

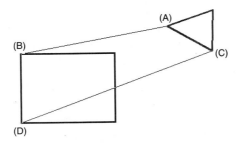

Figure 21–5

The results are illustrated in 21–6. Notice that the second destination point in this example does not stretch the object. Rather, it provides direction in aligning the triangle with the edge of the rectangle.

Open 21_Align2D.Dwg again. Answering Yes to the prompt "Scale objects based on alignment points?" displays the results in Figure 21–7. Here the triangle is scaled in size based on the two destination points of the rectangle.

Command: **AL** *(For ALIGN)*
Select objects: *(Select the triangle)*
Select objects: *(Press ENTER to continue)*
Specify first source point: *(Select the endpoint at "A" in Figure 21-5)*
Specify first destination point: *(Select the endpoint at "B" in Figure 21-5)*
Specify second source point: *(Select the endpoint at "C" in Figure 21-5)*
Specify second destination point: *(Select the endpoint at "D" in Figure 21-5)*
Specify third source point or <continue>: *(Press ENTER to continue)*
Scale objects based on alignment points? [Yes/No] <N>: **Y** *(To scale the edge of the
    triangle with the edge of the rectangle)*

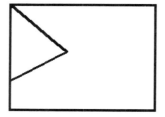

Figure 21–6

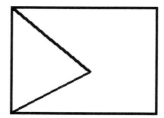

Figure 21–7

THE ROTATE3D COMMAND

Another way to position objects in 3D is by rotating them into position using the ROTATE3D command. A thorough understanding of the User Coordinate System is a must in operating this command. Open the drawing 21_Rot3D.Dwg. In Figure 21–8, a base containing a slot needs to be joined with the two rectangular boxes to form a back and side.

First select the box at "A" as the object to rotate in 3D. You will be prompted to define the axis of rotation. This axis will serve as a pivot point where the rotation occurs. In Figure 21–8, the axis of rotation is the Y axis from the current position of the User Coordinate System icon. Entering a positive angle of 90° will rotate the box in the counterclockwise direction. Negative angles rotate in the clockwise direction.

Command: **ROTATE3D**
Current positive angle: ANGDIR=counterclockwise ANGBASE=0
Select objects: *(Select box "A")*
Select objects: *(Press* ENTER *to continue)*
Specify first point on axis or define axis by
[Object/Last/View/Xaxis/Yaxis/Zaxis/2points]: **Y** *(For Yaxis)*
Specify a point on the Y axis <0,0,0>: *(Select the endpoint at "A")*
Specify rotation angle or [Reference]: **90**

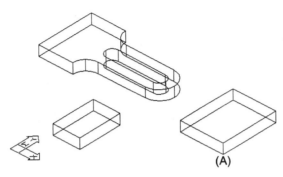

Figure 21–8

Next, the second box in Figure 21–9 is rotated 90° with the X axis as the axis of rotation.

Command: **ROTATE3D**
Current positive angle: ANGDIR=counterclockwise ANGBASE=0
Select objects: *(Select box "A")*
Select objects: *(Press* ENTER *to continue)*
Specify first point on axis or define axis by
[Object/Last/View/Xaxis/Yaxis/Zaxis/2points]: **X** *(For Xaxis)*
Specify a point on the X axis <0,0,0>: *(Select the endpoint at "A")*
Specify rotation angle or [Reference]: **90**

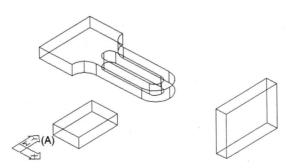

Figure 21–9

The results of these operations are illustrated in Figure 21–10. Once the boxes are rotated to the correct angles, they are moved into position through Object Snap modes. Box "A" is moved from the endpoint of the corner at "A" to the endpoint of the corner at "C." Box "B" is moved from the endpoint of the corner at "B" to the endpoint of the corner at "C." Once moved, they are then joined to the model through the UNION command. The results are illustrated in Figure 21–11, with a hidden line removal operation applied to the model.

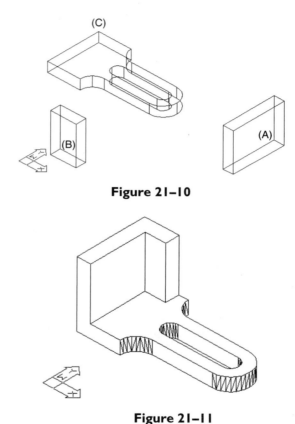

Figure 21–10

Figure 21–11

THE SOLIDEDIT COMMAND

Once features such as holes, slots, and extrusions are constructed in a solid model, the time may come to make changes to these features. This is the function of the SOLIDEDIT command. Select this command from the Solids Editing toolbar shown in Figure 21–12 or choose Solids Editing from the Modify pull-down menu shown in Figure 21–13. Both menus are arranged in three groupings; namely Face, Edge, and Body editing. These groupings will be discussed in the pages that follow. Also, using the

Solids Editing toolbar or pull-down menu to perform the following operations will eliminate a number of steps that will otherwise be necessary if the command is entered at the command prompt.

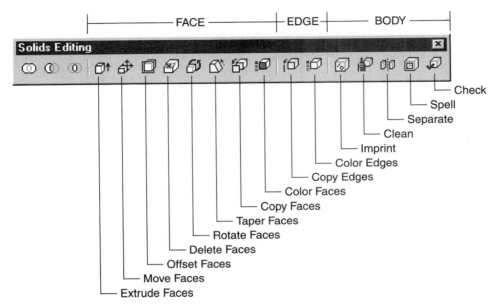

Figure 21–12

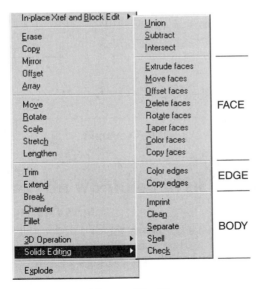

Figure 21–13

EXTRUDING (FACE EDITING)

Faces may be lengthened or shortened through the Extrude option of the SOLIDEDIT command. A positive distance extrudes the face in the direction of its normal. A negative distance extrudes the face in the opposite direction. Open the drawing 21_ExtrudeDwg. In Figure 21–14, the highlighted face at "A" needs to be decreased in height.

Command: **SOLIDEDIT**
Solids editing automatic checking: SOLIDCHECK=1
Enter a solids editing option [Face/Edge/Body/Undo/eXit] <eXit>: **F** *(For Face)*
Enter a face editing option
[Extrude/Move/Rotate/Offset/Taper/Delete/Copy/coLor/Undo/eXit] <eXit>: **E** *(For Extrude)*
Select faces or [Undo/Remove]: *(Select the face inside the area represented by "A")*
Select faces or [Undo/Remove/ALL]: *(Press* ENTER *to continue)*
Specify height of extrusion or [Path]: **-10.00**
Specify angle of taper for extrusion <0>: *(Press* ENTER*)*
Solid validation started.
Solid validation completed.
Enter a face editing option
[Extrude/Move/Rotate/Offset/Taper/Delete/Copy/coLor/Undo/eXit] <eXit>: *(Press* ENTER*)*
Solids editing automatic checking: SOLIDCHECK=1
Enter a solids editing option [Face/Edge/Body/Undo/eXit] <eXit>: *(Press* ENTER*)*

The result is illustrated at "B."

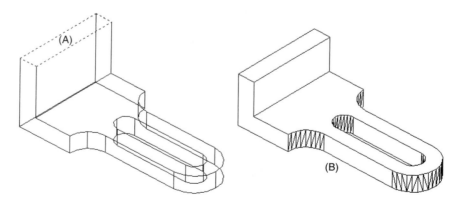

Figure 21–14

MOVING (FACE EDITING)

Open the drawing 21_Move.Dwg. The object in Figure 21–15 illustrates two intersecting cylinders. The two horizontal cylinders need to be moved 1 unit up from their current location. The cylinders are first selected at "A" and "B" through the SOLIDEDIT command along with the Move option.

Command: **SOLIDEDIT**
Solids editing automatic checking: SOLIDCHECK=1
Enter a solids editing option [Face/Edge/Body/Undo/eXit] <eXit>: **F** *(For Face)*
Enter a face editing option
[Extrude/Move/Rotate/Offset/Taper/Delete/Copy/coLor/Undo/eXit] <eXit>: **M** *(For Move)*
Select faces or [Undo/Remove]: *(Select both highlighted faces at "A" and "B")*
Select faces or [Undo/Remove/ALL]: *(Press* ENTER *to continue)*
Specify a base point or displacement: *(Pick the center of the bottom vertical cylinder)*
Specify a second point of displacement: **@0,0,1**
Solid validation started.
Solid validation completed.
Enter a face editing option
[Extrude/Move/Rotate/Offset/Taper/Delete/Copy/coLor/Undo/eXit] <eXit>: *(Press ENTER)*
Solids editing automatic checking: SOLIDCHECK=1
Enter a solids editing option [Face/Edge/Body/Undo/eXit] <eXit>: *(Press* ENTER*)*

The results are illustrated at "C."

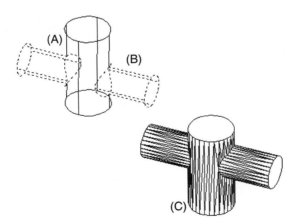

Figure 21–15

ROTATING (FACE EDITING)

Open the drawing 21_Rotate.Dwg. In Figure 21–16, the triangular extrusion needs to be rotated 45° in the clockwise direction. Use the Rotate Face option of the SOLIDEDIT command to accomplish this. You must select all top faces of the triangular wedge at "A," "B," and "C." You will also have to remove the face that makes up the top of the rectangular base at "D" before proceeding.

Command: **SOLIDEDIT**
Solids editing automatic checking: SOLIDCHECK=1
Enter a solids editing option [Face/Edge/Body/Undo/eXit] <eXit>: **F** *(For Face)*
Enter a face editing option
[Extrude/Move/Rotate/Offset/Taper/Delete/Copy/coLor/Undo/eXit] <eXit>: **R** *(For Rotate)*
Select faces or [Undo/Remove]: *(Select all top faces that make up the triangular extrusion at "A", "B", and "C")*
Select faces or [Undo/Remove/ALL]: **R** *(For Remove)*
Remove faces or [Undo/Add/ALL]: *(Select the face at "D" to remove)*
Remove faces or [Undo/Add/ALL]: *(Press ENTER to continue)*
Specify an axis point or [Axis by object/View/Xaxis/Yaxis/Zaxis] <2points>: **Z** *(For Zaxis)*
Specify the origin of the rotation <0,0,0>: *(Select the endpoint at "E")*
Specify a rotation angle or [Reference]: **-45** *(To rotate the triangular extrusion 45° in the clockwise direction)*
Solid validation started.
Solid validation completed.
Enter a face editing option
[Extrude/Move/Rotate/Offset/Taper/Delete/Copy/coLor/Undo/eXit] <eXit>: *(Press ENTER)*
Solids editing automatic checking: SOLIDCHECK=1
Enter a solids editing option [Face/Edge/Body/Undo/eXit] <eXit>: *(Press ENTER)*

The results are illustrated at "F."

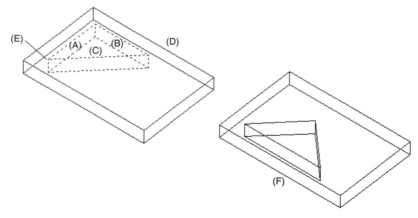

Figure 21–16

OFFSETTING (FACE EDITING)

Open the drawing 21_Offset.Dwg. In the image in Figure 21–17 the holes need to be increased or decreased in size. Use the Offset Face option of the SOLIDEDIT command to increase or decrease the size of selected faces. Using positive values actually increases the volume of the solid. Therefore, the feature being offset gets smaller, similar to the illustration at "B." Entering negative values reduces the volume of the solid; this means the feature being offset gets larger, as in the figures at "C." Study the prompt sequence and Figure 21–17 for the mechanics of this command option.

Command: **SOLIDEDIT**
Solids editing automatic checking: SOLIDCHECK=1
Enter a solids editing option [Face/Edge/Body/Undo/eXit] <eXit>: **F** *(For Face)*
Enter a face editing option
[Extrude/Move/Rotate/Offset/Taper/Delete/Copy/coLor/Undo/eXit] <eXit>: **O** *(For Offset)*
Select faces or [Undo/Remove]: *(Select the two holes at "D" and "E")*
Select faces or [Undo/Remove/ALL]: **R** *(For Remove)*
Remove faces or [Undo/Add/ALL]: *(Select the bottom or top edges of the object at "F" and "G")*
Remove faces or [Undo/Add/ALL]: *(Press ENTER to continue)*
Specify the offset distance: **0.50**
Solid validation started.
Solid validation completed.
Enter a face editing option
[Extrude/Move/Rotate/Offset/Taper/Delete/Copy/coLor/Undo/eXit] <eXit>: *(Press ENTER)*
Solids editing automatic checking: SOLIDCHECK=1
Enter a solids editing option [Face/Edge/Body/Undo/eXit] <eXit>: *(Press ENTER)*

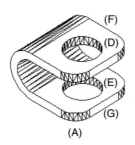
(A)

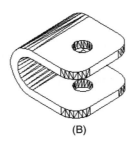
(B)

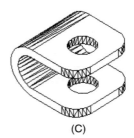
(C)

Figure 21–17

TAPERING (FACE EDITING)

Open the drawing 21_Taper.Dwg. The object at "A" in Figure 21–18 represents a solid box that needs to have tapers applied to all sides. Using the Taper Face option of the SOLIDEDIT command allows you to accomplish this task. Entering a positive angle moves the location of the second point into the part as shown at "F." Entering a negative angle moves the location of the second point away from the part as shown at "G."

Command: **SOLIDEDIT**
Solids editing automatic checking: SOLIDCHECK=1
Enter a solids editing option [Face/Edge/Body/Undo/eXit] <eXit>: **F** *(For Face)*
Enter a face editing option
[Extrude/Move/Rotate/Offset/Taper/Delete/Copy/coLor/Undo/eXit] <eXit>: **T** *(For Taper)*
Select faces or [Undo/Remove]: *(Select faces "A" through "D")*
Select faces or [Undo/Remove/ALL]: *(Press* ENTER *to continue)*
Specify the base point: *(Select the endpoint at "E")*
Specify another point along the axis of tapering: *(Select the endpoint at "B")*
Specify the taper angle: **10** *(For the angle of the taper)*
Solid validation started.
Solid validation completed.
Enter a face editing option
[Extrude/Move/Rotate/Offset/Taper/Delete/Copy/coLor/Undo/eXit] <eXit>: *(Press* ENTER*)*
Solids editing automatic checking: SOLIDCHECK=1
Enter a solids editing option [Face/Edge/Body/Undo/eXit] <eXit>: *(Press* ENTER*)*

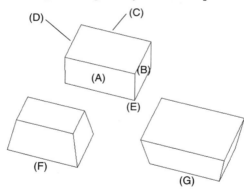

Figure 21–18

DELETING (FACE EDITING)

Faces can be erased through the Delete Face option of the SOLIDEDIT command. Open the drawing 21_Delete.Dwg. In Figure 21–19, select the hole at "A" as the face to erase. Since the top face of the plate also highlights, you will have to deselect it using the Remove option.

Command: **SOLIDEDIT**
Solids editing automatic checking: SOLIDCHECK=1
Enter a solids editing option [Face/Edge/Body/Undo/eXit] <eXit>: **F** *(For Face)*
Enter a face editing option
[Extrude/Move/Rotate/Offset/Taper/Delete/Copy/coLor/Undo/eXit] <eXit>: **D** *(For Delete)*

Select faces or [Undo/Remove]: *(Select the hole at "A")*
Select faces or [Undo/Remove/ALL]: **R** *(For Remove)*
Remove faces or [Undo/Add/ALL]: *(Select the edge of the top face at "B")*
Remove faces or [Undo/Add/ALL: *(Press* ENTER *to continue)*
Solid validation started.
Solid validation completed.
Enter a face editing option
[Extrude/Move/Rotate/Offset/Taper/Delete/Copy/coLor/Undo/eXit] <eXit>: *(Press*
    ENTER*)*
Solids editing automatic checking: SOLIDCHECK=1
Enter a solids editing option [Face/Edge/Body/Undo/eXit] <eXit>: *(Press* ENTER*)*

The results are illustrated in "C."

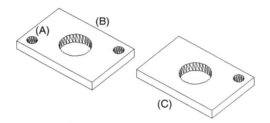

Figure 21–19

COPYING (FACE EDITING)

You can copy a face for use in the creation of another solid model using the Copy Face option of the SOLIDEDIT command. Open the drawing 21_Copy.Dwg. In Figure 21–20, the face at "A" is selected to copy. Notice that all objects making up the face, such as the rectangle and circles, are highlighted. You will have to remove the face at "B." Picking a base point and second point copies the face at "C." The rectangle and circles are considered a single object called a region. The EXTRUDE command is used on this region to produce the results illustrated at "D."

Command: **SOLIDEDIT**
Solids editing automatic checking: SOLIDCHECK=1
Enter a solids editing option [Face/Edge/Body/Undo/eXit] <eXit>: **F** *(For Face)*
Enter a face editing option
[Extrude/Move/Rotate/Offset/Taper/Delete/Copy/coLor/Undo/eXit] <eXit>: **C** *(For
    Copy)*
Select faces or [Undo/Remove]: *(Select the bottom face at "A")*
Select faces or [Undo/Remove/ALL]: **R** *(For Remove)*
Remove faces or [Undo/Add/ALL]: *(Select the edge of the face at "B" to be deselected)*
Remove faces or [Undo/Add/ALL]: *(Press* ENTER *to continue)*
Specify a base point or displacement: *(Pick a point to copy from)*
Specify a second point of displacement: *(Pick a point to copy to)*

Enter a face editing option
[Extrude/Move/Rotate/Offset/Taper/Delete/Copy/coLor/Undo/eXit] <eXit>: *(Press*
   ENTER*)*
Solids editing automatic checking: SOLIDCHECK=1
Enter a solids editing option [Face/Edge/Body/Undo/eXit] <eXit>: *(Press* ENTER*)*

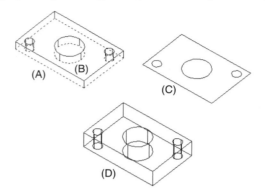

Figure 21–20

IMPRINTING (BODY EDITING)

An interesting means of adding construction geometry to a solid model is though the process of imprinting. Open the drawing 21_Imprint.Dwg. In the imprinting illustration in Figure 21–21 at "A," a rectangle along with slot is already modeled in 3D. A line was constructed from the midpoints of the rectangle at "A." Next, the SOLIDEDIT command was run on the model and the line.

Command: **SOLIDEDIT**
Solids editing automatic checking: SOLIDCHECK=1
Enter a solids editing option [Face/Edge/Body/Undo/eXit] <eXit>: **B** *(For Body)*
Enter a body editing option
[Imprint/seParate solids/Shell/cLean/Check/Undo/eXit] <eXit>: **I** *(For Imprint)*
Select a 3D solid: *(Select the solid model)*
Select an object to imprint: *(Select line "B")*
Delete the source object <N>: **Y** *(For Yes)*
Select an object to imprint: *(Press* ENTER *to perform the imprint operation)*
Enter a body editing option
[Imprint/seParate solids/Shell/cLean/Check/Undo/eXit] <eXit>: *(Press* ENTER*)*
Solids editing automatic checking: SOLIDCHECK=1
Enter a solids editing option [Face/Edge/Body/Undo/eXit] <eXit>: *(Press* ENTER*)*

The segments of the lines that come in contact with the part remain on the part's surface. However, these are no longer line segments; rather these lines now belong to the part. The lines actually separate the top surface into two faces.

Use the Extrude Face option of the SOLIDEDIT command on one of the newly created faces at "C" and increase the face in height by an extra 1.50 units. The results are illustrated in "D," with the face being given a taller height of. Splitting the top surface into two faces is possible through the Imprint option of the SOLIDEDIT command.

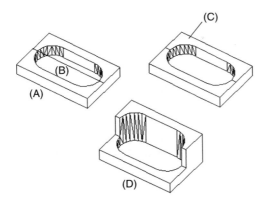

Figure 21–21

SEPARATING SOLIDS (BODY EDITING)

Editing solid models can at times lead to confusing results. Open the drawing 21_Separate.Dwg. In Figure 21–22, the model at "A" is about to be sliced in half with a thin box created at the center of the circle and spanning the depth of the rectangular shelf.

Use the SUBTRACT command and subtract the rectangular box from the solid object. When you subtract the box and list the solid at "B," both halves of the object highlight even though they appear separate. To convert the single solid model to two separate models, use the SOLIDEDIT command followed by the Body option and then use the Separate option.

Command: **SOLIDEDIT**
Solids editing automatic checking: SOLIDCHECK=1
Enter a solids editing option [Face/Edge/Body/Undo/eXit] <eXit>: **B** *(For Body)*
Enter a body editing option
[Imprint/seParate solids/Shell/cLean/Check/Undo/eXit] <eXit>: **P** *(For Separate)*
Select a 3D solid: *(Pick the solid model at "C")*
Enter a body editing option
[Imprint/seParate solids/Shell/cLean/Check/Undo/eXit] <eXit>: *(Press ENTER)*
Solids editing automatic checking: SOLIDCHECK=1
Enter a solids editing option [Face/Edge/Body/Undo/eXit] <eXit>: *(Press ENTER)*

This action separates the single model into two. When you select the model in "C," only one half highlights.

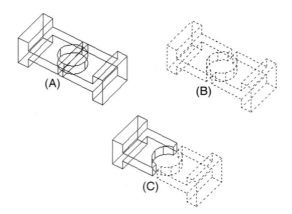

Figure 21–22

SHELLING (BODY EDITING)

Shelling is the process of constructing a thin wall inside or outside of a solid model. Positive thickness will produce the thin wall inside; negative values for thickness will produce the thin wall outside. This wall thickness remains constant throughout the entire model. Faces may be removed during the shelling operation to create an opening. Open the drawing 21_Shell.Dwg. For clarity, it is important to first rotate your model at "A" into a position where the face to be removed is visible. Follow this action by performing a hidden line removal with the HIDE command at "B" in Figure 21–23. Now the Shell option of SOLIDEDIT can be used with predictable results. An additional note: only one shell is permitted in a model.

Command: **SOLIDEDIT**
Solids editing automatic checking: SOLIDCHECK=1
Enter a solids editing option [Face/Edge/Body/Undo/eXit] <eXit>: **B** *(For Body)*
Enter a body editing option
[Imprint/seParate solids/Shell/cLean/Check/Undo/eXit] <eXit>: **S** *(For Shell)*
Select a 3D solid: *(Select the solid model at "B")*
Remove faces or [Undo/Add/ALL]: *(Pick a point at "C")*
Remove faces or [Undo/Add/ALL]: *(Press ENTER to continue)*
Enter the shell offset distance: **0.20**
Solid validation started.
Solid validation completed.
Enter a body editing option
[Imprint/seParate solids/Shell/cLean/Check/Undo/eXit] <eXit>: *(Press ENTER)*
Solids editing automatic checking: SOLIDCHECK=1
Enter a solids editing option [Face/Edge/Body/Undo/eXit] <eXit>: *(Press ENTER)*

The results are illustrated at "D."

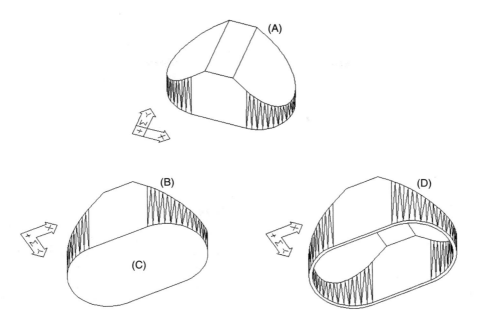

Figure 21–23

CLEANING (BODY EDITING)

When imprinted lines that form faces are not used, they can be deleted from a model by the Clean option when the SOLIDEDIT command is used on a Body. Open the drawing 21_Clean.Dwg. The object at "A" in Figure 21–24 illustrates lines originally constructed on the top of the solid model. These lines were then imprinted at "B." During the imprint process, the source lines are deleted. Since these lines now belong to the model, the Clean option is used remove the lines previously imprinted. The results are illustrated at "C."

Command: **SOLIDEDIT**
Solids editing automatic checking: SOLIDCHECK=1
Enter a solids editing option [Face/Edge/Body/Undo/eXit] <eXit>: **B** *(For Body)*
Enter a body editing option
[Imprint/seParate solids/Shell/cLean/Check/Undo/eXit] <eXit>: **L** *(For Clean)*
Select a 3D solid: *(Select the solid model at "B")*
Enter a body editing option
[Imprint/seParate solids/Shell/cLean/Check/Undo/eXit] <eXit>: *(Press* ENTER*)*
Solids editing automatic checking: SOLIDCHECK=1
Enter a solids editing option [Face/Edge/Body/Undo/eXit] <eXit>: *(Press* ENTER*)*

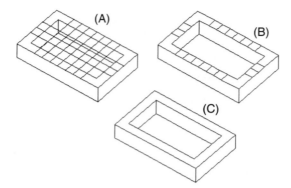

Figure 21–24

Additional options of the SOLIDEDIT command include the ability to apply a color to a selected face or edge of a solid model. You can even copy the edge of a model to be used for construction purposes on other models.

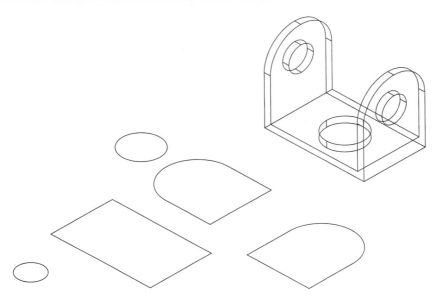

TUTORIAL EXERCISE: 21_ROTATE3D.DWG

Figure 21–25

Purpose

This tutorial exercise is designed to produce 3D objects and then use the ROTATE3D command to rotate the objects into position before creating unions or performing subtractions. See Figure 21–25.

System Settings

The drawing units, limits, grid, and snap values are already set for this drawing. Check to see that the following Object Snap modes are already set: Endpoint, Extension, Intersection, Center. Also check to see that the POLAR, OSNAP, and OTRACK modes are turned on. These are located in the bottom status bar.

Layers

Make sure the current layer is Model.

Suggested Commands

Open the drawing file 21_ROTATE3D.Dwg. All 2D objects have been converted to polyline objects. The EXTRUDE command is used to create a thickness of 0.50 units for most objects. Then the ROTATE3D command is used to rotate the sides to the proper angle before they are moved into place and joined together. The cylinders are also rotated and moved into place before being subtracted from the base and sides to create holes.

Whenever possible, substitute the appropriate command alias in place of the full AutoCAD command in each tutorial step. For example, use "CP" for the COPY command, "L" for the LINE command, and so on. The complete listing of all command aliases is located in Chapter 1, Table 1–2.

STEP 1

Extrude the base, two side panels, and large circle to a height of 0.50 units, as shown in Figure 21–26. Then extrude the small circle to a height of 8 units. This represents the total length of the base and will be used to cut a hole through the two side panels. Use the EXTRUDE command for both operations.

Command: **EXT** *(For EXTRUDE)*
Current wire frame density: ISOLINES=4
Select objects: *(Select the base at "A", the two side panels at "B" and "C", and the large circle at "D")*
Select objects: *(Press ENTER to continue)*
Specify height of extrusion or [Path]: **0.50**
Specify angle of taper for extrusion <0>: *(Press ENTER to perform the extrude operation)*

Command: **EXT** *(For EXTRUDE)*
Current wire frame density: ISOLINES=4
Select objects: *(Select the small circle at "E")*
Select objects: *(Press ENTER to continue)*
Specify height of extrusion or [Path]: **8.00**
Specify angle of taper for extrusion <0>: *(Press ENTER to perform the extrude operation)*

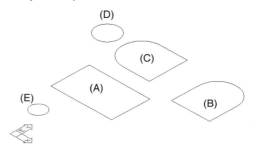

Figure 21–26

STEP 2

Use the MOVE command to position the large cylinder in the center of the base. You may also use the geometry calculator and the MEE function to center the cylinder in the base. See Figure 21–27.

Command: **M** *(For MOVE)*
Select objects: *(Select the large cylinder at "A")*
Select objects: *(Press ENTER to continue)*
Specify base point or displacement: *(Pick the top face of the large cylinder at "A" to identify its center)*
Specify second point of displacement or <use first point as displacement>: **'CAL**
Initializing...>> Expression: **MEE** *(For Midpoint and two Endpoints)*

>> Select one endpoint for MEE: *(Pick a point along the top corner of the base at "B")*
>> Select another endpoint for MEE: *(Pick a point along the opposite corner of the base at "C")*
(10.6466 0.320423 0.5)

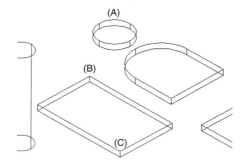

Figure 21–27

STEP 3

Begin rotating the remaining shapes into place using the ROTATE3D command (see Figure 21–28). Being careful to observe the current position of the User Coordinate System, rotate one of the side panels at an angle of 90° using the Y axis as the axis of rotation.

Command: **ROTATE3D**
Current positive angle:
 ANGDIR=counterclockwise
 ANGBASE=0
Select objects: *(Select the side panel labeled "A")*
Select objects: *(Press ENTER to continue)*
Specify first point on axis or define axis by [Object/Last/View/Xaxis/Yaxis/Zaxis/2points]: **Y** *(For Y axis)*
Specify a point on the Y axis <0,0,0>: *(Pick the endpoint of the side panel at "B")*
Specify rotation angle or [Reference]: **90**

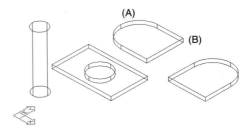

Figure 21–28

STEP 4

Position the side panel just rotated on the top of the base using the MOVE command. See Figure 21–29.

Command: **M** *(For MOVE)*
Select objects: *(Select the side panel at "A")*
Select objects: *(Press ENTER to continue)*
Specify base point or displacement: *(Pick the endpoint of the side panel at "A")*
Specify second point of displacement or <use first point as displacement>: *(Pick the endpoint of the base at "B")*

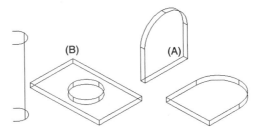

Figure 21–29

STEP 5

Now rotate the second side panel into position. This will take two steps to accomplish. First rotate the panel 90° along the X axis. See Figure 21–30.

Command: **ROTATE3D**
Current positive angle:
    ANGDIR=counterclockwise
    ANGBASE=0
Select objects: *(Select the second side panel)*
Select objects: *(Press* ENTER *to continue)*
Specify first point on axis or define axis by [Object/Last/View/Xaxis/Yaxis/Zaxis/2points]: **X** *(For X axis)*
Specify a point on the X axis <0,0,0>: *(Pick the endpoint of the side panel at "A")*
Specify rotation angle or [Reference]: **90**

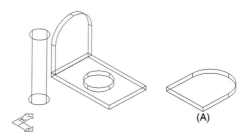

Figure 21–30

STEP 6

Now rotate the same panel 90° along the Z Axis. See Figure 21–31.

Command: **ROTATE3D**
Current positive angle:
    ANGDIR=counterclockwise
    ANGBASE=0
Select objects: *(Select the same side panel)*
Select objects: *(Press* ENTER *to continue)*
Specify first point on axis or define axis by [Object/Last/View/Xaxis/Yaxis/Zaxis/2points]: **Z** *(For Z axis)*
Specify a point on the Z axis <0,0,0>: *(Pick the endpoint of the side panel at "A")*
Specify rotation angle or [Reference]: **90**

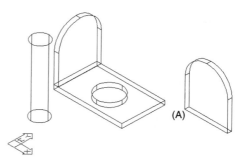

Figure 21–31

STEP 7

Finally, position the panel along the top of the base using the MOVE command. See Figure 21–32.

Command: **M** *(For MOVE)*
Select objects: *(Select the side panel at "A")*
Select objects: *(Press ENTER to continue)*
Specify base point or displacement: *(Pick the endpoint of the side panel at "A")*
Specify second point of displacement or <use first point as displacement>: *(Pick the top endpoint of the base at "B")*

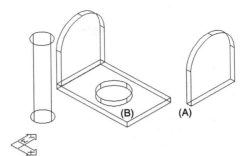

Figure 21–32

STEP 8

Rotate the long cylinder 90° using the Y axis as the axis of rotation, as shown in Figure 21–33.

Command: **ROTATE3D**
Current positive angle:
   ANGDIR=counterclockwise
   ANGBASE=0
Select objects: *(Select the long cylinder)*
Select objects: *(Press ENTER to continue)*
Specify first point on axis or define axis by [Object/Last/View/Xaxis/Yaxis/Zaxis/2points]: **Y** *(For Y axis)*
Specify a point on the Y axis <0,0,0>: *(Pick the edge of the cylinder at "A" to identify its center)*
Specify rotation angle or [Reference]: **90**

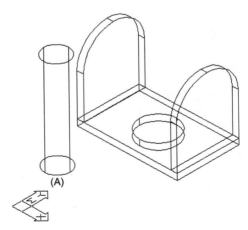

Figure 21–33

STEP 9

Move the long cylinder into position; it should span the complete length of the object and should be lined up with the center of the side panels. See Figure 21–34.

Command: **M** *(For MOVE)*
Select objects: *(Select the long cylinder)*
Select objects: *(Press ENTER to continue)*
Specify base point or displacement: *(Pick the edge of the cylinder at "A" to identify its center)*
Specify second point of displacement or <use first point as displacement>: *(Pick the outer edge of the side panel at "B" to identify its center)*

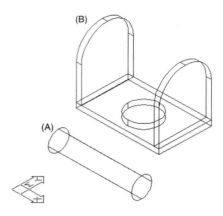

Figure 21–34

STEP 10

Finally, use the SUBTRACT command to subtract the two cylinders from the object; the side panels will automatically be unioned to the base when you complete this command. See Figure 21–35.

Command: **SU** *(For SUBTRACT)*
Select solids and regions to subtract from ..
Select objects: *(Select the two side panels and the base)*
Select objects: *(Press ENTER to continue)*
Select solids and regions to subtract ..
Select objects: *(Select the two cylinders to subtract)*
Select objects: *(Press ENTER to perform the union/subtraction operations)*

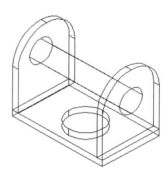

Figure 21–35

The completed object is displayed in Figure 21–36.

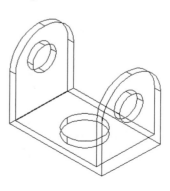

Figure 21–36

TUTORIAL EXERCISE: STEP_HOUSING.DWG

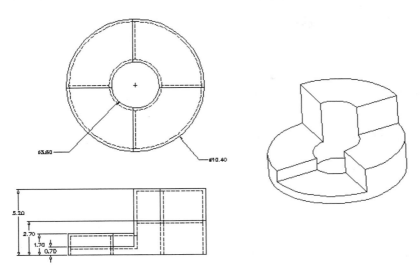

Figure 21–37

Purpose

This tutorial exercise is designed to create the solid model in Figure 21–37 using the SOLIDEDIT command.

System Settings

Use the Drawing Units dialog box to change the number of decimal places from four to two. Keep the default settings for limits.

Layers

Create a layer called Model for this tutorial exercise.

Suggested Commands

Two circles will be drawn and extruded to form two cylinders. The smaller cylinder will then be subtracted from the larger cylinder to create a hole. Construction lines will be drawn from each quadrant of the large circle.

These lines will then be imprinted on the model. This will enable the four faces to be extruded separately through the SOLIDEDIT command. After this is performed, the model will be rotated to display the base surface and a thin wall will be created through the Shell option of the SOLIDEDIT command. Finally, the model will be prepared for a rendering operation with the placement of a light and assignment of a material.

Whenever possible, substitute the appropriate command alias in place of the full AutoCAD command in each tutorial step. For example, use "CP" for the COPY command, "L" for the LINE command, and so on. The complete listing of all command aliases is located in Chapter 1, Table 1–2.

STEP 1

Draw one circle at a 5.00 diameter and another at a 2.00 diameter as in Figure 21–38.

Command: **C** *(For CIRCLE)*
Specify center point for circle or [3P/2P/ Ttr (tan tan radius)]: **7.00,4.00**
Specify radius of circle or [Diameter]: **D** *(For Diameter)*
Specify diameter of circle: **5.00**
Command: **C** *(For CIRCLE)*
Specify center point for circle or [3P/2P/ Ttr (tan tan radius)]: **@** *(To locate the last point)*
Specify radius of circle or [Diameter] <2.50>: **D** *(For Diameter)*
Specify diameter of circle <5.00>: **2.00**

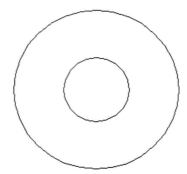

Figure 21–38

STEP 2

View the model from the South West position using the -VIEW command. You could also activate this position by choosing 3D Views from the View pull-down menu and then SW Isometric. The results are illustrated in Figure 21–39.

Command: **-VIEW**
Enter an option [?/Orthographic/Delete/ Restore/Save/Ucs/Window]: **Swiso** *(For South West Isometric)*
Regenerating model.

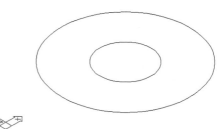

Figure 21–39

STEP 3

Extrude both circles to a height of 0.50 units using the EXTRUDE command (see Figure 21–40).

Command: **EXT** *(For EXTRUDE)*
Current wire frame density: ISOLINES=4
Select objects: *(Select both circles)*
Select objects: *(Press ENTER to continue)*
Specify height of extrusion or [Path]:
   0.50
Specify angle of taper for extrusion <0>:
   (Press ENTER to perform the extrusion operation)

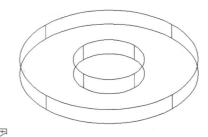

Figure 21–40

STEP 4

Subtract the small cylinder from the large cylinder to form a single solid model. Perform a hidden line removal using the HIDE command. Your model should appear similar to Figure 21–41.

Command: **SU** *(For SUBTRACT)*
Select solids and regions to subtract from ..
Select objects: *(Select the large cylinder)*
Select objects: *(Press ENTER to continue)*
Select solids and regions to subtract ..
Select objects: *(Select the small cylinder)*
Select objects: *(Press ENTER to perform the subtraction operation)*
Command: **HI** *(For HIDE)*
Regenerating model.

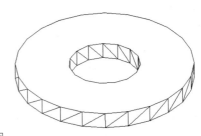

Figure 21–41

STEP 5

Draw lines at the top of the large cylinder in Figure 21–42. Use the OSNAP-Quadrant mode to accomplish this task.

Command: **L** *(For LINE)*
Specify first point: **Qua**
of *(Pick the quadrant of the model at "A")*
Specify next point or [Undo]: **Qua**
of *(Pick the quadrant of the model at "B")*
Specify next point or [Undo]: *(Press ENTER to exit this command)*
Command: **L** *(For LINE)*
Specify first point: **Qua**
of *(Pick the quadrant of the model at "C")*
Specify next point or [Undo]: **Qua**
of *(Pick the quadrant of the model at "D")*
Specify next point or [Undo]: *(Press ENTER to exit this command)*

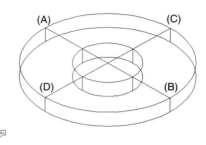

Figure 21–42

STEP 6

Use the Imprint option of the SOLIDEDIT command to create faces at the top of the model in Figure 21–43.

Command: **SOLIDEDIT**
Solids editing automatic checking: SOLIDCHECK=1
Enter a solids editing option [Face/Edge/Body/Undo/eXit] <eXit>: **B** *(For Body)*
Enter a body editing option
[Imprint/seParate solids/Shell/cLean/Check/Undo/eXit] <eXit>: **I** *(For Imprint)*
Select a 3D solid: *(Select the model)*
Select an object to imprint: *(Select line "A")*
Delete the source object <N>: *(Press ENTER)*
Select an object to imprint: *(Select line "B")*
Delete the source object <N>: *(Press ENTER)*

Select an object to imprint: *(Press ENTER to end imprint selection)*
Enter a body editing option
[Imprint/seParate solids/Shell/cLean/Check/Undo/eXit] <eXit>: *(Press ENTER)*
Solids editing automatic checking: SOLIDCHECK=1
Enter a solids editing option [Face/Edge/Body/Undo/eXit] <eXit>: *(Press ENTER)*

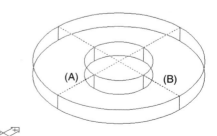

Figure 21–43

STEP 7

Use the face formed by the imprinted lines in Figure 21–44A and extrude the face at a height of 2.50 units.

Command: **SOLIDEDIT**
Solids editing automatic checking:
   SOLIDCHECK=1
Enter a solids editing option [Face/Edge/
   Body/Undo/eXit] <eXit>: **F** *(For Face)*
Enter a face editing option
[Extrude/Move/Rotate/Offset/Taper/
   Delete/Copy/coLor/Undo/eXit]
   <eXit>: **E** *(For Extrude)*
Select faces or [Undo/Remove]: *(Pick the
   face at A" in Figure 21–44A)*
Select faces or [Undo/Remove/ALL]:
   (Press ENTER to continue)
Specify height of extrusion or [Path]:
   2.50
Specify angle of taper for extrusion <0>:
   (Press ENTER to continue)
Solid validation started.
Solid validation completed.
Enter a face editing option
[Extrude/Move/Rotate/Offset/Taper/
   Delete/Copy/coLor/Undo/eXit]
   <eXit>: *(Press ENTER)*
Solids editing automatic checking:
   SOLIDCHECK=1

Enter a solids editing option [Face/Edge/
   Body/Undo/eXit] <eXit>: *(Press ENTER)*

Your model should appear similar to Figure 21–44B.

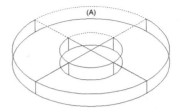

Figure 21–44A

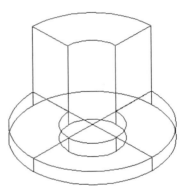

Figure 21–44B

STEP 8

All imprinted lines have disappeared and we need to extrude more faces. Use the SOLIDEDIT command and imprint the highlighted lines in Figure 21–45. Use the same prompt sequence as back in Step 6 to perform this operation.

Command: **SOLIDEDIT**
Solids editing automatic checking:
   SOLIDCHECK=1
Enter a solids editing option [Face/Edge/
   Body/Undo/eXit] <eXit>: **B** *(For Body)*
Enter a body editing option
[Imprint/seParate solids/Shell/cLean/
   Check/Undo/eXit] <eXit>: **I** *(For
   Imprint)*
Select a 3D solid: *(Select the model)*
Select an object to imprint: *(Select line
   "A")*
Delete the source object <N>: *(Press
   ENTER)*
Select an object to imprint: *(Select line
   "B")*
Delete the source object <N>: *(Press
   ENTER)*
Select an object to imprint: *(Press ENTER
   to end imprint selection)*
Enter a body editing option
[Imprint/seParate solids/Shell/cLean/
   Check/Undo/eXit] <eXit>: *(Press
   ENTER)*
Solids editing automatic checking:
   SOLIDCHECK=1
Enter a solids editing option [Face/Edge/
   Body/Undo/eXit] <eXit>: *(Press ENTER)*

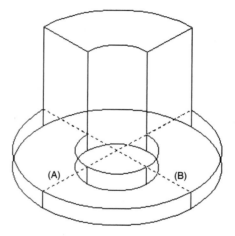

Figure 21–45

STEP 9

Extrude the highlighted face in Figure 21–46A using a negative distance of -0.25 units. This will drop the face down that distance in Figure 21–46B.

Command: **SOLIDEDIT**
Solids editing automatic checking:
    SOLIDCHECK=1
Enter a solids editing option [Face/Edge/
    Body/Undo/eXit] <eXit>: **F** *(For Face)*
Enter a face editing option
[Extrude/Move/Rotate/Offset/Taper/
    Delete/Copy/coLor/Undo/eXit]
    <eXit>: **E** *(For Extrude)*
Select faces or [Undo/Remove]: *(Pick the
    face at A" in Figure 21-46A)*
Select faces or [Undo/Remove/ALL]:
    (Press ENTER *to continue)*
Specify height of extrusion or [Path]: **-
0.25**
Specify angle of taper for extrusion <0>:
    (Press ENTER *to continue)*
Solid validation started.
Solid validation completed.
Enter a face editing option
[Extrude/Move/Rotate/Offset/Taper/
    Delete/Copy/coLor/Undo/eXit]
    <eXit>: *(Press* ENTER*)*
Solids editing automatic checking:
    SOLIDCHECK=1
Enter a solids editing option [Face/Edge/
    Body/Undo/eXit] <eXit>: *(Press* ENTER*)*

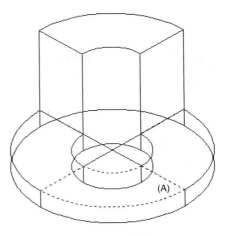

Figure 21–46A

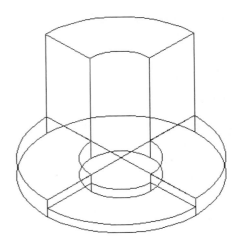

Figure 21–46B

STEP 10

Extrude the face in Figure 21–47A to a height of 0.50 using the SOLIDEDIT command. Your display should appear similar to Figure 21–47B.

Command: **SOLIDEDIT**
Solids editing automatic checking:
SOLIDCHECK=1
Enter a solids editing option [Face/Edge/
Body/Undo/eXit] <eXit>: **F** *(For Face)*
Enter a face editing option
[Extrude/Move/Rotate/Offset/Taper/
Delete/Copy/coLor/Undo/eXit]
<eXit>: **E** *(For Extrude)*
Select faces or [Undo/Remove]: *(Select
the face at "A" in Figure 21-47A)*
Select faces or [Undo/Remove/ALL]:
(Press ENTER to continue)
Specify height of extrusion or [Path]:
0.50
Specify angle of taper for extrusion <0>:
(Press ENTER)
Solid validation started.
Solid validation completed.
Enter a face editing option
[Extrude/Move/Rotate/Offset/Taper/
Delete/Copy/coLor/Undo/eXit]
<eXit>: *(Press ENTER)*
Solids editing automatic checking:
SOLIDCHECK=1
Enter a solids editing option [Face/Edge/
Body/Undo/eXit] <eXit>: *(Press ENTER)*

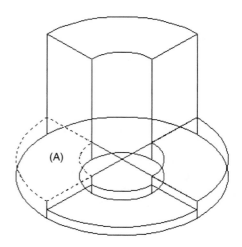

Figure 21–47A

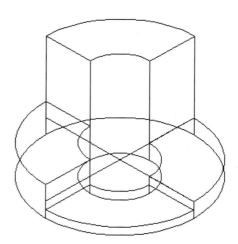

Figure 21–47B

STEP 11

Extrude the face in Figure 21–48A to a height of 1.00 using the SOLIDEDIT command. Your display should appear similar to Figure 21–48B.

Command: **SOLIDEDIT**
Solids editing automatic checking:
 SOLIDCHECK=1
Enter a solids editing option [Face/Edge/
 Body/Undo/eXit] <eXit>: **F** *(For Face)*
Enter a face editing option
[Extrude/Move/Rotate/Offset/Taper/
 Delete/Copy/coLor/Undo/eXit]
 <eXit>: **E** *(For Extrude)*
Select faces or [Undo/Remove]: *(Select
 the face at "A" in Figure 21-48A)*
Select faces or [Undo/Remove/ALL]:
 (Press ENTER *to continue)*
Specify height of extrusion or [Path]:
 1.00
Specify angle of taper for extrusion <0>:
 (Press ENTER*)*
Solid validation started.
Solid validation completed.
Enter a face editing option
[Extrude/Move/Rotate/Offset/Taper/
 Delete/Copy/coLor/Undo/eXit]
 <eXit>: *(Press* ENTER*)*
Solids editing automatic checking:
 SOLIDCHECK=1
Enter a solids editing option [Face/Edge/
 Body/Undo/eXit] <eXit>: *(Press* ENTER*)*

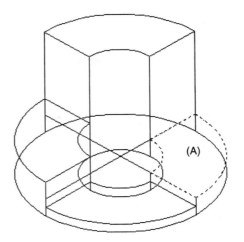

Figure 21–48A

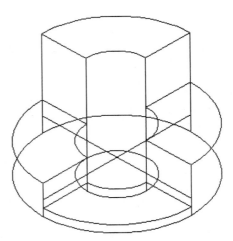

Figure 21–48B

STEP 12

Erase the two lines originally used to construct the imprints. Change the Facet resolution to 5 using the FACETRES command. Perform a hidden line removal using the HIDE command. Your display should appear similar to Figure 21–49.

Command: **FACETRES**
Enter new value for FACETRES
  <0.5000>: **5.00**
Command: **HI** (For HIDE)
Regenerating model.

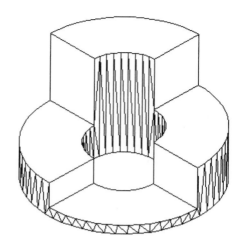

Figure 21–49

STEP 13

Before continuing with changes, save current position of your model as a named view. Entering the VIEW command displays the View dialog box in Figure 21–50. Clicking on the New button displays the New View dialog box. Enter "SW Iso" as the name of the view.

Click the OK button in the New View dialog box to create the view. Click the OK button in the main View dialog box to return to your drawing. Now this view can be called up no matter how you change the display of the model.

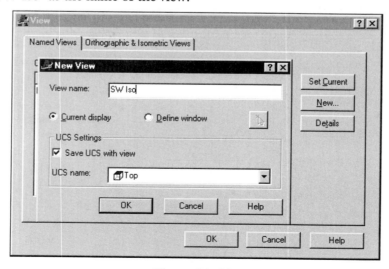

Figure 21–50

STEP 14

We will be creating a hollow shell of your model. To do this, you need to view the model from below. Entering VP at the command prompt displays the Viewpoint Presets dialog box in Figure 21–51A. Click in the right image until the XY

Plane box reads -60°. You could also enter -60 in the XY Plane: window. This will allow you to create a viewing position underneath your model. Clicking the OK button displays your model as in Figure 21–51B.

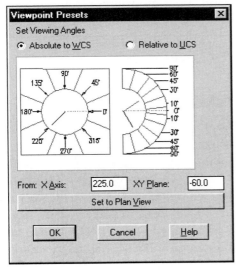

Figure 21–51A

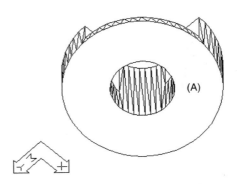

(A)

Figure 21–51B

STEP 15

The SOLIDEDIT command uses the Shell option to create a thin wall on the inside of your model. However, we want to have the bottom of the object open. You will remove the bottom face before assigning the shell distance. It is always best to remove this face when your model is displayed with its hidden lines removed (performing a hidden line removal.) Pick the face back at "A" in Figure 21-51B.

Command: **SOLIDEDIT**
Solids editing automatic checking:
 SOLIDCHECK=1
Enter a solids editing option [Face/Edge/
 Body/Undo/eXit] <eXit>: **B** (For Body)
Enter a body editing option
[Imprint/seParate solids/Shell/cLean/
 Check/Undo/eXit] <eXit>: **S** (For Shell)
Select a 3D solid: (Select the edge of the
 model)
Remove faces or [Undo/Add/ALL]: (Select
 the face at "A" in Figure 21–51B)
Remove faces or [Undo/Add/ALL]: (Press
 ENTER to continue)
Enter the shell offset distance: **0.20**
Solid validation started.
Solid validation completed.
Enter a body editing option

[Imprint/seParate solids/Shell/cLean/
 Check/Undo/eXit] <eXit>: (Press
 ENTER)
Solids editing automatic checking:
 SOLIDCHECK=1
Enter a solids editing option [Face/Edge/
 Body/Undo/eXit] <eXit>: (Press ENTER)

Perform another hidden line removal to view the shell results using the HIDE command. Your display should appear similar to Figure 21–52.

Command: **HI** (For HIDE)
Regenerating model.

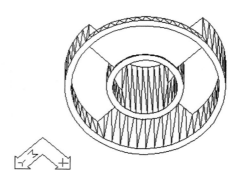

Figure 21–52

STEP 16

Use the VIEW command to display the View dialog box. Set the "SW Iso" view current in Figure 21–53A. Your display should appear similar to Figure 21–53B.

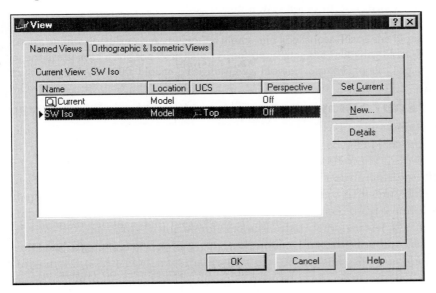

Figure 21–53A

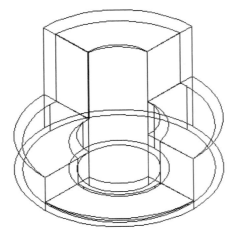

Figure 21–53B

STEP 17

For special effects, the next series of steps will demonstrate how easy it is to render this object. A light will be placed, material attached, and the object rendered in a photo realistic fashion. First enter the LIGHT command. This displays the Lights dialog box shown in Figure 21–54A. Click the New button to display the New Point Light dialog box. Enter the new light name as LT1 in Figure 21–54B. Then click on the Modify < button; this takes you back to your drawing. You need to position the light 8 units above the base of your model using a technique called XYZ point filters. Follow the command

prompt sequence and dialog boxes to accomplish this task.

Command: **LIGHT**
(The Lights dialog box appears in Figure 21-54A. Click on the New... button. This displays the New Point Light dialog box in Figure 21-54B. To locate the light 8 units above the bottom hole, click on the Modify< button and follow the prompts below)
Enter light location <current>: **.XY**
of **Cen**
of *(Pick the center of the bottom hole at "A" in Figure 21–54C)*
(need Z): **8.00**

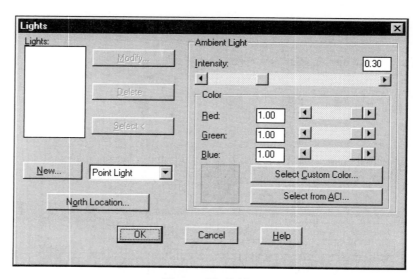

Figure 21–54A

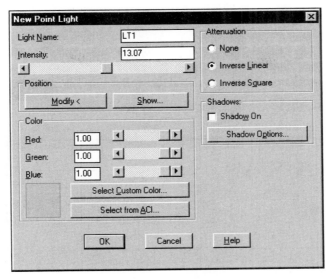

Figure 21–54B

When finished entering the height value of 8.00, click the OK button to dismiss the New Point Light dialog box. Click the OK button to dismiss the main Lights dialog box and return to your drawing.

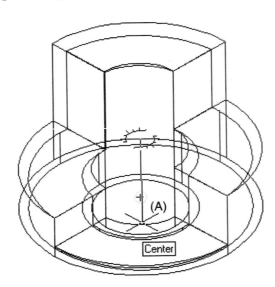

Figure 21–54C

STEP 18

Activate the Materials dialog box by entering the RMAT command. Notice there are no materials currently defined. Click on the Materials Library button in Figure 21–55A.

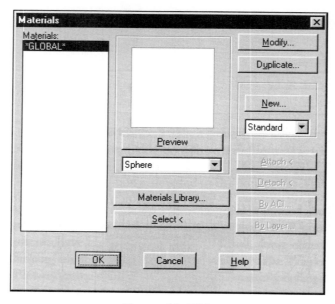

Figure 21–55A

This activates the Materials Library dialog box. Find the material labeled "BLUE GLASS" and select it. Click on the <- Import button in Figure 21–55B to import it into the current drawing. When finished, click the OK button to return to the Materials dialog box.

In the Materials dialog box, click on the Attach< button in Figure 21–55C. This will return you to the model. Pick the model anywhere to assign the material. When the Materials dialog box reappears, click the OK button to dismiss this dialog box and return to you drawing.

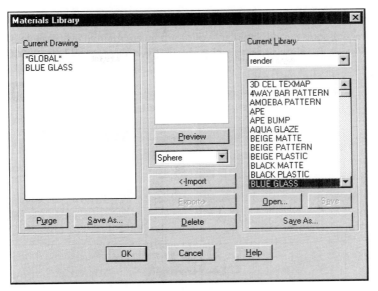

Figure 21–55B

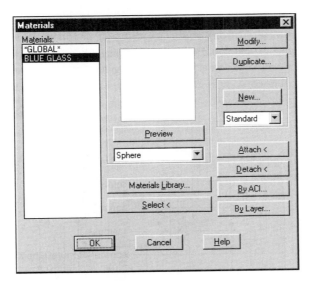

Figure 21–55C

STEP 19

Activate the Render dialog box by entering the RENDER command. As shown in Figure 21–56A, change the Rendering Type to Photo Real; keep all other settings. Click the Render button to render the image in the BLUE GLASS material type. The final image of the object should appear similar to Figure 21–56B.

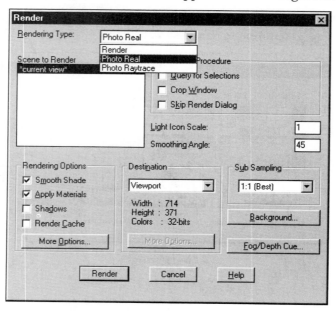

Figure 21–56A

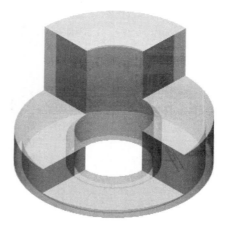

Figure 21–56B

Creating Orthographic Views from a Solid Model

CHAPTER 22

THE SOLVIEW AND SOLDRAW COMMANDS

One of the advantages of creating a 3D image in the form of a solid model is the ability to use the data of the solid model numerous times for other purposes. The purpose of this chapter is to generate 2D orthographic views from the solid model. Two commands will be used to perform this operation: SOLVIEW and SOLDRAW. The SOLVIEW command is used to create a layout of the orthographic views. The SOLDRAW command draws the views on specific layers (this includes the drawing of hidden lines to show invisible features). Included in the layout of 2D orthographic views will be the creation of isometric, section, and auxiliary views of a drawing.

Once you create the solid model, you can lay out and draw its orthogonal views using the commands located in the Draw pull-down menu under Solids and then Setup. There are three commands that deal with the ability to draw, view, or profile orthogonal views. Only two areas will be discussed in this section, namely the ability to lay out 2D views and then the ability to construct a drawing of these views.

Choosing Solids from the Draw pull-down menu and then Setup, as shown in Figure 22–1, exposes the Drawing (SOLDRAW), View (SOLVIEW), and Profile (SOLPROF) commands. Only SOLDRAW and SOLVIEW will be discussed in this chapter. The Solids toolbar in Figure 22–2 also exposes these two commands.

Once the SOLVIEW command is entered, the display screen automatically switches to the first layout or Paper Space environment. Using SOLVIEW will lay out a view based on responses to a series of prompts, depending on the type of view you want to create. Usually, the initial view that serves as the starting point for other orthogonal views is based on the current user coordinate system. This needs to be determined before you begin this command. Once an initial view is created, it is very easy to create Ortho, Section, Isometric, and even Auxiliary views.

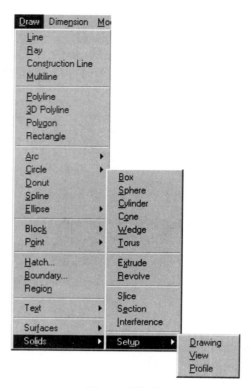

Figure 22–1

Figure 22–2

As SOLVIEW is used as a layout tool, the images of these views are taken from the original solid model. In other words, after you lay out a view, it does not contain any 2D features, such as hidden lines. As shown in Figure 22–1, clicking on Drawing activates the SOLDRAW command, which is used to draw the view once it has been laid out through the SOLVIEW command.

Before using the SOLVIEW command, study the illustration of the solid model in Figure 22–3. In particular, pay close attention to the position of the User Coordinate System icon. The current position of the User Coordinate System will begin the creation of the first or base view of the 2D drawing. In addition, before you start the SOLVIEW command, remember to load the Hidden linetype. This will automatically assign this linetype to any new layer that requires hidden lines for the drawing mode. If the linetype is not loaded at this point, it must be manually assigned to each layer that contains hidden lines through the Layer Properties Manager dialog box.

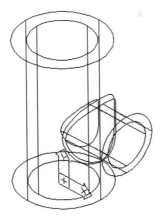

Figure 22–3

Activating SOLVIEW automatically switches the display to the layout or Paper Space environment. Since this is the first view to be laid out, the UCS option will be used to base the view on the current User Coordinate System. A scale value may be entered for the view. For the View Center, click anywhere on the screen and notice the view being constructed. You can pick numerous times on the screen until the view is in a desired location. When completed, press ENTER to place the view. The next series of prompts enables you to construct a viewport around the view. It is very important to make this viewport large enough for dimensions to fit inside. Once the view is given a name, it is laid out similar to Figure 22–4.

Command: **SOLVIEW**
Enter an option [Ucs/Ortho/Auxiliary/Section]: **U** *(For Ucs)*
Enter an option [Named/World/?/Current] <Current>: *(Press ENTER)*
Enter view scale <1.0000>: *(Press ENTER)*
Specify view center: *(Pick a point on the screen to display the view at its center; keep picking until the view is in the desired location)*
Specify view center <specify viewport>: *(Press ENTER to place the view)*
Specify first corner of viewport: *(Pick a point at "A")*
Specify opposite corner of viewport: *(Pick a point at "B")*
Enter view name: **FRONT**
UCSVIEW = 1 UCS will be saved with view
Enter an option [Ucs/Ortho/Auxiliary/Section]: *(Press ENTER to exit this command)*

Once the view has been laid out through the SOLVIEW command, use SOLDRAW to actually draw the view in two dimensions. If the Hidden linetype was loaded prior to using the SOLVIEW command, hidden lines will automatically be assigned to layers that contain hidden line information. The result of using the SOLDRAW command is shown in Figure 22–5.

Command: **SOLDRAW**
(If in Model Space, you a switched to Paper Space)
Select viewports to draw:
Select objects: *(Pick anywhere on the viewport in Figure 22–5)*
Select objects: *(Press ENTER to perform the Soldraw operation)*
One solid selected.

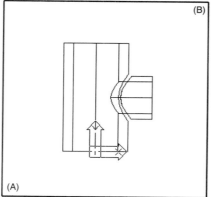

Figure 22–4

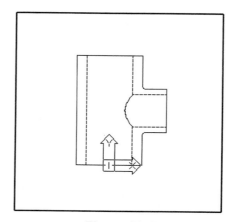

Figure 22–5

The use of layers in 2D-view layout is so important that when you run the SOLVIEW command, the layers shown in Figure 22–6 are created automatically. With the exception of Model and 0, the layers that begin with "FRONT" and the VPORTS layer were all created by the SOLVIEW command. The FRONT-DIM layer is designed to hold dimension information for the Front view. FRONT-HID holds all hidden lines information for the Front view; FRONT-VIS holds all visible line information for the Front view. All Paper Space viewports are placed on the VPORTS layer. The Model layer has automatically been frozen in the current viewport to show the visible and hidden lines.

Name	On	Freeze..	Lock	Color	Linetype	Lineweight	Plot Style	Plot	Active ...	New ...
0	♀	☼	⬛	■ White	CONTINUOUS	—— Default	Color_7	🖨	🔘	🔘
FRONT-DIM	♀	☼	⬛	■ White	CONTINUOUS	—— Default	Color_7	🖨	🔘	🔘
FRONT-HID	♀	☼	⬛	■ White	HIDDEN	—— Default	Color_7	🖨	🔘	🔘
FRONT-VIS	♀	☼	⬛	■ White	CONTINUOUS	—— Default	Color_7	🖨	🔘	🔘
MODEL	♀	☼	⬛	■ White	CONTINUOUS	—— Default	Color_7	🖨	🔘	🔘
VPORTS	♀	☼	⬛	■ White	CONTINUOUS	—— Default	Color_7	🖨	🔘	🔘

Figure 22–6

In order for the view shown in Figure 22–7 to be dimensioned, three operations must be performed. First, double-click inside the Front view to be sure it is the current floating Model Space viewport. Next, make FRONT-DIM the current layer. Finally, set the User Coordinate System to the current view using the View option. The UCS icon should be similar to the illustration in Figure 22–7. Now add all dimensions to the view using conventional dimensioning commands with the aid of Object snap modes. When you work on adding dimensions to another view, the same three operations must be made in the new view: make the viewport active by double-clicking inside it, make the layer holding the dimension information the current layer, and update the UCS to the current view with the View option.

When you draw the views using the SOLDRAW command and then add the dimensions, switching back to the solid model by clicking on the Model tab displays the image shown in Figure 22–8. In addition to the solid model of the object, the constructed view and dimensions are also displayed. All drawn views from Paper Space will display in the model. Use the Layer Properties Manager dialog box along with the Freeze option to freeze all drawing-related layers to isolate the solid model.

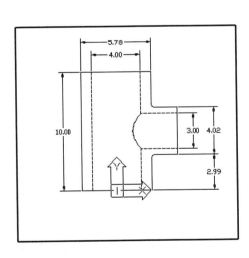

Figure 22–7

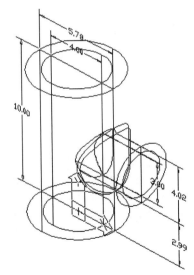

Figure 22–8

CREATING ORTHOGRAPHIC VIEWS

Once the first view is created, orthographic views can easily be created with the Ortho option of the SOLVIEW command. Open the drawing 22_Ortho.Dwg. Notice that you are in a layout and a Front view is already created. Follow the command prompt sequence below to create two orthographic views. When finished, your drawing should appear similar to Figure 22–9.

Command: **SOLVIEW**
Enter an option [Ucs/Ortho/Auxiliary/Section]: **O** *(For Ortho)*
Specify side of viewport to project: *(Select the midpoint at "A")*
Specify view center: *(Pick a point above the front view to locate the top view)*
Specify view center <specify viewport>: *(Press* ENTER *to place the view)*
Specify first corner of viewport: *(Pick a point at "B")*
Specify opposite corner of viewport: *(Pick a point at "C")*
Enter view name: **TOP**
UCSVIEW = I UCS will be saved with view
Enter an option [Ucs/Ortho/Auxiliary/Section]: **O** *(For Ortho)*
Specify side of viewport to project: *(Select the midpoint at "D")*
Specify view center: *(Pick a point to the right of the front view to locate the right side view)*
Specify view center <specify viewport>: *(Press* ENTER *to place the view)*
Specify first corner of viewport: *(Pick a point at "E")*
Specify opposite corner of viewport: *(Pick a point at "F")*
Enter view name: **R_SIDE**
UCSVIEW = I UCS will be saved with view
Enter an option [Ucs/Ortho/Auxiliary/Section]: *(Press* ENTER *to exit this command)*

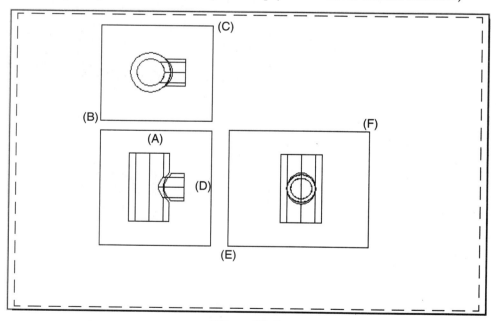

Figure 22–9

Running the SOLDRAW command on the three views displays the image as in Figure 22–10. Notice the appearance of the hidden lines in all views. The VPORTS layer is turned off to display only the three views.

Command: **SOLDRAW**
Select viewports to draw:
Select objects: *(Select the three viewports that contain the front, top and right side view information)*
Select objects: *(Press* ENTER *to perform the Soldraw operation)*

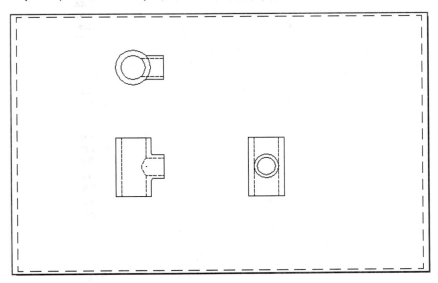

Figure 22–10

CREATING AN AUXILIARY VIEW

The true size and shape of the surface containing the large counterbored hole cannot be drawn with orthographic views alone. An auxiliary view must be used to document this information. Open the drawing 22_Auxiliary.Dwg. From the 3D model in Figure 22–11, create a Front view based on the current User Coordinate System and using the SOLVIEW command, as shown in Figure 22–12.

Command: **SOLVIEW**
Enter an option [Ucs/Ortho/Auxiliary/Section]: **U** *(For Ucs)*
Enter an option [Named/World/?/Current] <Current>: *(Press* ENTER*)*
Enter view scale <1.0000>: *(Press* ENTER*)*
Specify view center: *(Pick a point to locate the view)*
Specify view center <specify viewport>: *(Press* ENTER *to place the view)*
Specify first corner of viewport: *(Pick a point at "A")*
Specify opposite corner of viewport: *(Pick a point at "B")*
Enter view name: **FRONT**
UCSVIEW = 1 UCS will be saved with view
Enter an option [Ucs/Ortho/Auxiliary/Section]: *(Press* ENTER *to exit this command)*

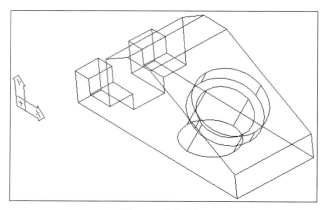

Figure 22–11

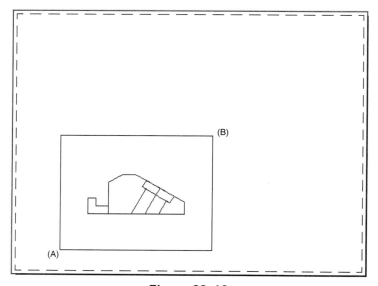

Figure 22–12

Begin the process of constructing an auxiliary view in Figure 22–13. Click in the endpoints at "A" and "B" to establish the edge of the surface to view. Pick a point at "C" as the side from which to view the auxiliary view. Notice how the Paper Space icon tilts perpendicular to the edge of the auxiliary view. Pick a location for the auxiliary view and establish a viewport. The result is illustrated in Figure 22–14.

Command: **SOLVIEW**
Enter an option [Ucs/Ortho/Auxiliary/Section]: **A** *(For Auxiliary)*
Specify first point of inclined plane: **End**
of *(Pick the endpoint at "A" in Figure 22–13)*
Specify second point of inclined plane: **End**

of *(Pick the endpoint at "B" in Figure 22–13)*
Specify side to view from: *(Pick a point inside of the viewport at "C" in Figure 22–13)*
Specify view center: *(Pick a point to locate the view)*
Specify view center <specify viewport>: *(Press* ENTER *to place the view)*
Specify first corner of viewport: *(Pick a point at "D" in Figure 22–14)*
Specify opposite corner of viewport: *(Pick a point at "E" in Figure 22–14)*
Enter view name: **AUXILIARY**
UCSVIEW = 1 UCS will be saved with view
Enter an option [Ucs/Ortho/Auxiliary/Section]: *(Press* ENTER *to exit this command)*

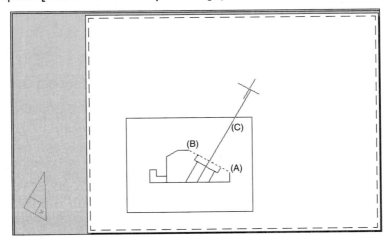

Figure 22–13

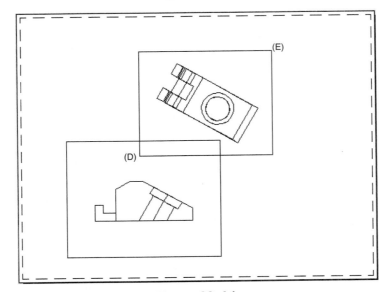

Figure 22–14

Run the SOLDRAW command and turn off the VPORTS layer. The finished result is illustrated in Figure 22–15. Hidden lines display only because this linetype was previously loaded.

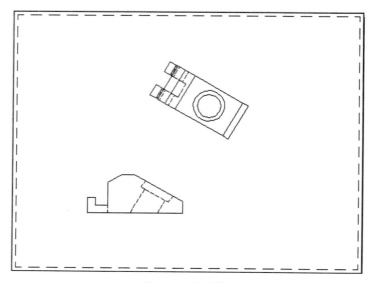

Figure 22–15

CREATING A SECTION VIEW

The SOLVIEW command is versatile enough to create a full section view of an object. Before you use the Section option, it is highly recommended that you turn off POLAR, OSNAP, and OTRACK. This will ensure predictable results when you cut the section. Open the drawing 22_Section.Dwg. From the model in Figure 22–16, create a Top view based on the current User Coordinate System as in Figure 22–17.

Command: **SOLVIEW**
Regenerating layout.
Enter an option [Ucs/Ortho/Auxiliary/Section]: **U** *(For Ucs)*
Enter an option [Named/World/?/Current] <Current>: *(Press ENTER)*
Enter view scale <1.0000>: *(Press ENTER)*
Specify view center: *(Pick a point to locate the view)*
Specify view center <specify viewport>: *(Press ENTER to place the view)*
Specify first corner of viewport: *(Pick a point at "A")*
Specify opposite corner of viewport: *(Pick a point at "B")*
Enter view name: **TOP**
UCSVIEW = 1 UCS will be saved with view
Enter an option [Ucs/Ortho/Auxiliary/Section]: *(Press ENTER to exit this command)*

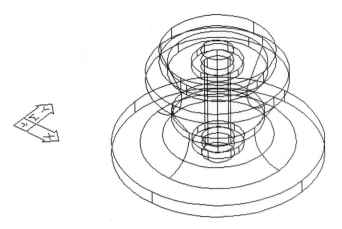

Figure 22–16

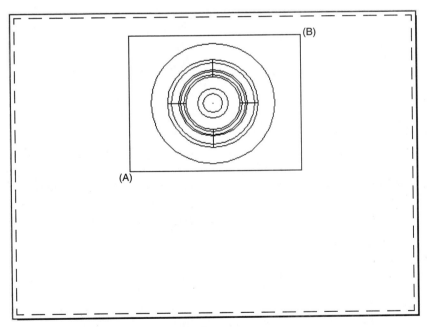

Figure 22–17

Begin the process of creating the section. You must first establish the cutting plane line in the Top view as in Figure 22–18. After the cutting plane line is drawn, you select the side from which to view the section. You then locate the section view, similar to the placement of an orthographic view.

Command: **SOLVIEW**
Enter an option [Ucs/Ortho/Auxiliary/Section]: **S** *(For Section)*
Specify first point of cutting plane: **Qua**
of *(Pick a point at "A" in Figure 22–18)*
Specify second point of cutting plane: *(With Ortho on, pick a point at "B" in Figure 22–18)*
Specify side to view from: *(Pick a point inside of the viewport at "C" in Figure 22–18)*
Enter view scale <1.0000>: *(Press* ENTER*)*
Specify view center: *(Pick a point below the top view to locate the view)*
Specify view center <specify viewport>: *(Press* ENTER *to place the view)*
Specify first corner of viewport: *(Pick a point at "D" in Figure 22–19)*
Specify opposite corner of viewport: *(Pick a point at "E" in Figure 22–19)*
Enter view name: **FRONT_SECTION**
UCSVIEW = 1 UCS will be saved with view
Enter an option [Ucs/Ortho/Auxiliary/Section]: *(Press* ENTER *to continue)*

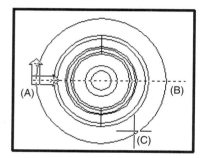

Figure 22–18

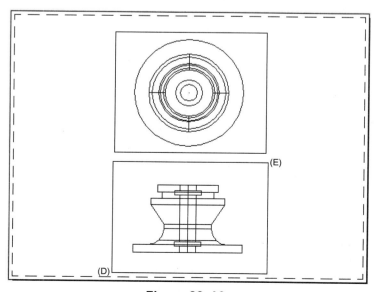

Figure 22–19

Running the SOLDRAW command on the viewports results in the image in Figure 22–20. You can also activate the viewport displaying the section view and use the HATCHEDIT command to edit the hatch pattern. In Figure 22–21, the hatch pattern scale was increased to a value of 2.00 and the viewports turned off.

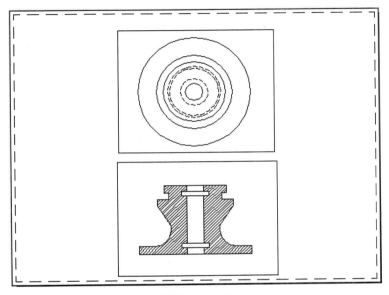

Figure 22–20

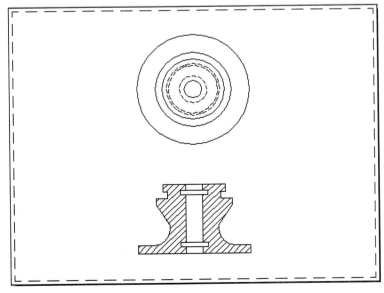

Figure 22–21

CREATING AN ISOMETRIC VIEW

Once orthographic, section, and auxiliary views are projected, you also have an opportunity to project an isometric view based on the current position of the 3D model. This type of projection relies entirely on the position of the User Coordinate System. Open the drawing 22_Iso.Dwg. This 3D model should appear similar to Figure 22–22. To prepare this image to be projected as an isometric view, first define a new User Coordinate System based on the current view. See the prompt sequence below to accomplish this task. Your image should appear similar to Figure 22–23.

Command: **UCS**
Current ucs name: *WORLD*
Enter an option [New/Move/orthoGraphic/Prev/Restore/Save/Del/Apply/?/World]
<World>: **N** *(For New)*
Specify origin of new UCS or [ZAxis/3point/OBject/Face/View/X/Y/Z] <0,0,0>: **V** *(For View)*

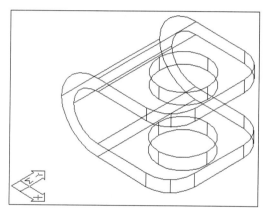

Figure 22–22

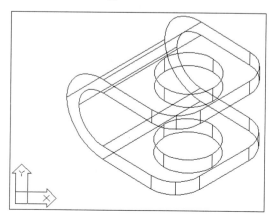

Figure 22–23

Next, run the SOLVIEW command based on the current UCS. Locate the view in the sample layout and construct a viewport around the isometric in Figure 22–24. Since dimensions are placed in the orthographic view drawings and not on an isometric, you can tighten up on the size of the viewport.

Command: **SOLVIEW**
Regenerating layout.
Enter an option [Ucs/Ortho/Auxiliary/Section]: **U** *(For Ucs)*
Enter an option [Named/World/?/Current] <Current>: *(Press* ENTER*)*
Enter view scale <1.0000>: *(Press* ENTER*)*
Specify view center: *(Pick a point to locate the view)*
Specify view center <specify viewport>: *(Press* ENTER *to place the view)*
Specify first corner of viewport: *(Pick a point at "A")*
Specify opposite corner of viewport: *(Pick a point at "B")*
Enter view name: **ISO**
UCSVIEW = 1 UCS will be saved with view
Enter an option [Ucs/Ortho/Auxiliary/Section]: *(Press* ENTER *to exit this command)*

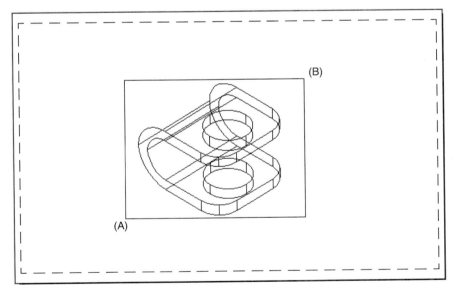

Figure 22–24

Running the SOLDRAW command on the isometric results in visible lines as well as hidden lines being displayed in Figure 22–25. Since a layer exists called ISO-HID that contains the hidden lines for the isometric drawing, use the Layer Properties Manager dialog box to turn off this layer. This results in the image shown in Figure 22–26.

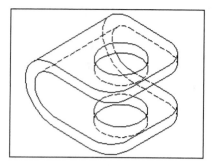

Figure 22–25

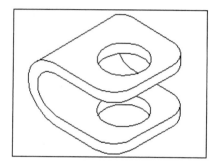

Figure 22–26

TUTORIAL EXERCISE: COLUMN.DWG

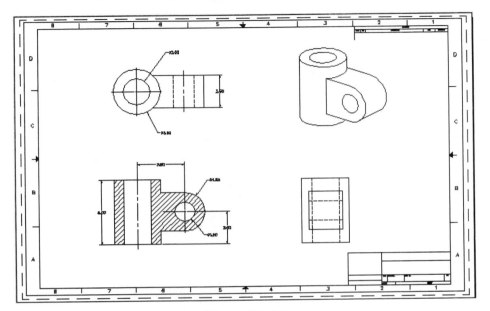

Figure 22–27

Purpose

This tutorial exercise is designed to generate orthographic views of the solid model "22_Column" using the SOLVIEW and SOLDRAW commands. See Figure 22–27.

System Settings

Open the drawing file 22_Column.Dwg. Keep all remaining default settings, including units and limits.

Layers

The following layers are already created:

Name	Color	Linetype
Model	Cyan	Continuous

Suggested Commands

Use the SOLVIEW command to lay out the Front, Top, Right Side, and Isometric views of the column. Next, draw the views in two dimensions using the SOLDRAW command. Dimensioning techniques will be explained. This exercise switches you from Model Space to Paper Space to accomplish the necessary tasks in building the solid model and extracting the views.

Whenever possible, substitute the appropriate command alias in place of the full AutoCAD command in each tutorial step. For example, use "CP" for the COPY command, "L" for the LINE command, and so on. The complete listing of all command aliases is located in Chapter 1, Table 1–2.

Phase I—Drawing Preparation

STEP 1

The Hidden linetype should be pre-loaded before you extract any views. Check to see that the Hidden linetype is loaded. Also, turn off POLAR, OSNAP, and OTRACK in the Status bar before using the SOLVIEW command. See Figure 22–28.

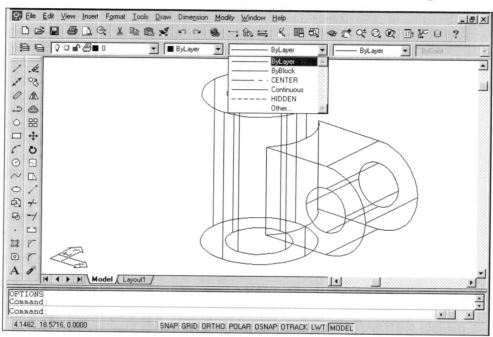

Figure 22–28

STEP 2

Activate the Options dialog box and click on the Display tab. In the Layout elements area, remove the check from the box for Create viewport in new layouts, as shown in Figure 22–29. This will turn off the automatic creation of a viewport when you first use the SOLVIEW command.

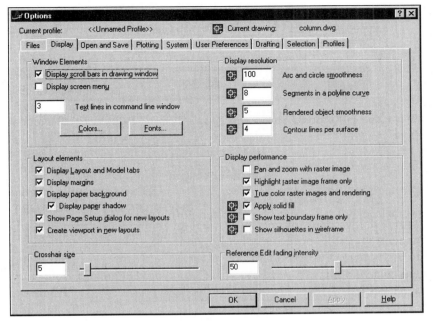

Figure 22–29

Phase II—Extracting the Orthographic Views

With the completed model of the Column, begin the extraction process by first noting the position of the User Coordinate System icon in Figure 22–30. This position will be used to extract the first view, the Top view. Once the Top view is created, the Front and Right Side views follow.

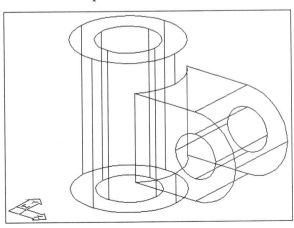

Figure 22–30

STEP 3

Use the SOLVIEW command to begin the extraction process. When activating this view, you are automatically switched to the Paper Space environment, shown in Figure 22–31. You will build an initial view based on the current position of the User Coordinate System. Once you locate the view on the drawing, you can change the scale of the view, give the view a name, and construct a viewport around the view. When constructing the viewport, make it large enough to accommodate the dimensions that need to be added later. See Figure 22–31 and the following prompts.

Command: **SOLVIEW**
Enter an option [Ucs/Ortho/Auxiliary/ Section]: **U** *(For Ucs)*

Enter an option [Named/World/?/ Current] <Current>: *(Press ENTER)*
Enter view scale <1.0000>: *(Press ENTER)*
Specify view center: *(Locate the view approximately in the upper left corner of the drawing sheet)*
Specify view center <specify viewport>: *(Press ENTER to place the view)*
Specify first corner of viewport: *(Pick a point approximately at "A")*
Specify opposite corner of viewport: *(Pick a point approximately at "B")*
Enter view name: **TOP**
UCSVIEW = 1   UCS will be saved with view
Enter an option [Ucs/Ortho/Auxiliary/ Section]: *(Press ENTER to exit this command)*

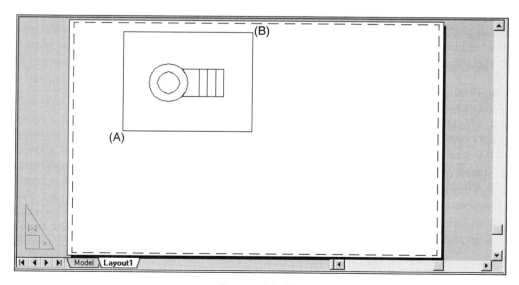

Figure 22–31

STEP 4

Before continuing with the SOLVIEW command to extract the other views, view the layers at this point through the Layer Properties Manager dialog box in Figure 22–32. Four layers were automatically created as a result of laying out the first view through SOLVIEW: TOP-DIM, TOP-HID, TOP-VIS, and VPORTS. TOP-DIM will hold all dimension information located in the Top view. TOP-HID will hold all hidden line information for the Top view, and TOP-VIS will hold all visible line information in the Top view. The VPORTS layer will hold all viewports created. These viewports will be turned off in the last step so as to isolate the views. These four layers are automatically visible in the current viewport and automatically frozen in other viewports; the presence of the snowflake symbol states this. Now return to the SOLVIEW command and lay out the remaining orthographic views.

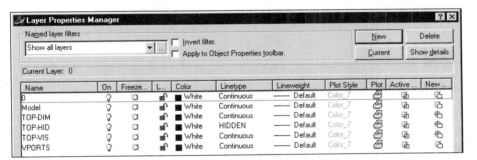

Figure 22–32

STEP 5

Use SOLVIEW and the Section option to create the Front view in full section based on the Top view already placed. Establish the cutting plane of the section from the Quadrant of the cylinder at "A" to the midpoint of the line at "B" in Figure 22–33A. Pick a point approximately at "C" to identify the side from which to view the section. As the new view is being located, Ortho mode is automatically turned on to keep both views lined up with each other in the same orientation. Create a viewport by following the prompts to clip a first corner and other corner. Again, you will add dimensions to this view later on, so make the viewport a good size. Name the view "FRONT" and continue by laying out the R_SIDE view, using the same prompt sequence. Create a small viewport because no dimensions will be placed on this view. When finished, your drawing should look similar to Figure 22–33B.

Command: **SOLVIEW**
Enter an option [Ucs/Ortho/Auxiliary/
    Section]: **S** *(For Section)*
Specify first point of cutting plane: **Qua**

of (Pick the quadrant at "A")

Specify second point of cutting plane: **Mid**

of (Pick the midpoint at "B")

Specify side to view from: (Pick a point at "C")

Enter view scale <1.0000>: (Press ENTER)

Specify view center: (Pick a location below the TOP view)

Specify view center <specify viewport>: (Press ENTER to place the view)

Specify first corner of viewport: (Pick a point at "D")

Specify opposite corner of viewport: (Pick a point at "E")

Enter view name: **FRONT**

UCSVIEW = 1 UCS will be saved with view

Enter an option [Ucs/Ortho/Auxiliary/ Section]: **O** (For Ortho)

Specify side of viewport to project: (Pick the viewport at "F")

Specify view center: (Pick a point to the right of the FRONT view)

Specify view center <specify viewport>: (Press ENTER to place the view)

Specify first corner of viewport: (Pick a point at "G")

Specify opposite corner of viewport: (Pick a point at "H")

Enter view name: **R_SIDE**

UCSVIEW = 1 UCS will be saved with view

Enter an option [Ucs/Ortho/Auxiliary/ Section]: (Press ENTER to exit this command)

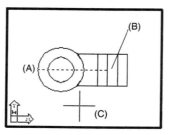

Figure 22–33A

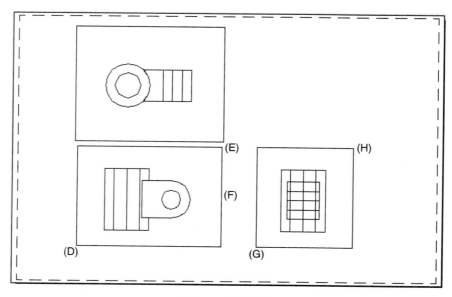

Figure 22–33B

STEP 6

Before creating the isometric view, click on the MODEL button while still in Layout1 to switch to PAPER (Paper Space). To create a projection of an isometric view, return to the solid model by clicking on the Model tab. Then use the ucs command to create a new User Coordinate System based on the current view. See Figure 22–34. Also save this User Coordinate System as ISO using the Save option of the ucs command. This makes ISO the current User Coordinate System.

(While in Layout1, click on the MODEL button in the Status bar to switch to PAPER)
(Click on the Model tab)

Command: **UCS**
Current ucs name: *WORLD*
Enter an option [New/Move/
   orthoGraphic/Prev/Restore/Save/Del/
   Apply/?/World]
<World>: **N** *(For New)*
Specify origin of new UCS or [ZAxis/
   3point/OBject/Face/View/X/Y/Z]
   <0,0,0>: **V** *(For View)*
Command: **UCS**
Current ucs name: *NO NAME*
Enter an option [New/Move/
   orthoGraphic/Prev/Restore/Save/Del/
   Apply/?/World]
<World>: **S** *(For Save)*
Enter name to save current UCS or [?]:
   ISO

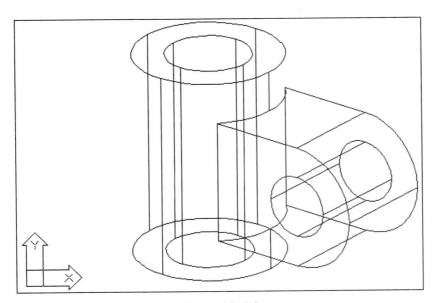

Figure 22–34

STEP 7

Return to Paper Space by clicking on the Layout1 tab. Issue the SOLVIEW command and use the UCS option to lay out the isometric view of the column based on the current User Coordinate System. See Figure 22–35.

(Click on the Layout1 tab)
Command: SOLVIEW
Enter an option [Ucs/Ortho/Auxiliary/
   Section]: **U** *(For Ucs)*
Enter an option [Named/World/?/
   Current] <Current>: *(Press ENTER)*
Enter view scale <1.0000>: *(Press ENTER)*
Specify view center: *(Pick a point to locate
   the view)*
Specify view center <specify viewport>:
   (Press ENTER to place the view)
Specify first corner of viewport: *(Pick a
   point at "A")*
Specify opposite corner of viewport:
   (Pick a point at "B")

Enter view name: **ISO**
UCSVIEW = 1 UCS will be saved with
   view
Enter an option [Ucs/Ortho/Auxiliary/
   Section]: *(Press ENTER to exit this
   command)*

All views have been successfully laid out. However, none of the orthographic views shows hidden features defined by hidden lines. Also, the Front view in full section does not show section lines to illustrate which surfaces were cut by the imaginary cutting plane line. The SOLVIEW command is used only as a layout tool. To actually draw the view complete with visible, object, and section lines, use the SOLDRAW command. This will be covered in Phase III of this project.

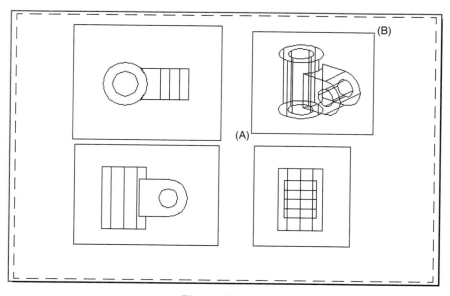

Figure 22–35

PHASE III—Using SOLDRAW to Construct the 2D Views

STEP 8

The SOLDRAW command is designed to follow the SOLVIEW command. SOLVIEW is used as a layout tool; it allows you to position the Top, Front, R_side, and Isometric views in the Paper Space environment. SOLDRAW is used to draw object, hidden, and section lines. Dimensions will be placed in the drawing during another phase. See Figure 22–36.

> Command: **SOLDRAW**
> Select viewports to draw..
> Select objects: **All** (*To select all viewports*)
> Select objects: (*Press ENTER to draw the views*)

Figure 22–36

STEP 9

Notice that the isometric view in Figure 22–36 has been drawn complete with hidden lines; these lines need to be made invisible in the isometric view. To accomplish this, first double-click inside the viewport holding the isometric view.

Then activate the Layer Properties Manager dialog box, shown in Figure 22–37A, and freeze the layer containing the hidden lines, ISO-HID. This will freeze the hidden lines in only the Isometric view, shown in Figure 22–37B.

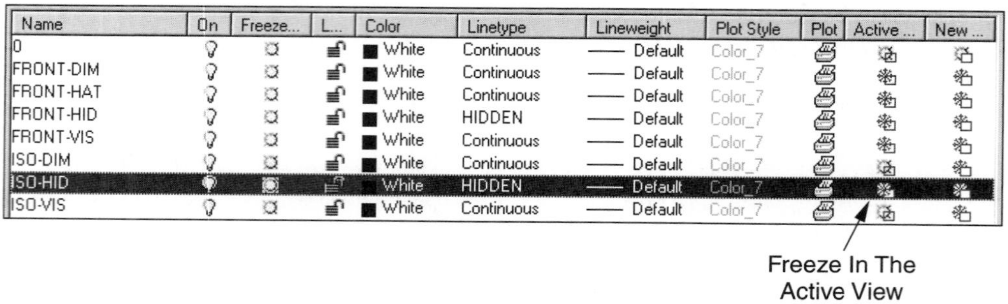

Name	On	Freeze...	L...	Color	Linetype	Lineweight	Plot Style	Plot	Active ...	New ...
0	♀	☼	🔓	■ White	Continuous	—— Default	Color_7	🖨	ⓥ	ⓥ
FRONT-DIM	♀	☼	🔓	■ White	Continuous	—— Default	Color_7	🖨	❄	❄
FRONT-HAT	♀	☼	🔓	■ White	Continuous	—— Default	Color_7	🖨	❄	❄
FRONT-HID	♀	☼	🔓	■ White	HIDDEN	—— Default	Color_7	🖨	❄	❄
FRONT-VIS	♀	☼	🔓	■ White	Continuous	—— Default	Color_7	🖨	❄	❄
ISO-DIM	♀	☼	🔓	■ White	Continuous	—— Default	Color_7	🖨	ⓥ	❄
ISO-HID	♀	☼	🔒	■ White	HIDDEN	—— Default	Color_7	🖨	❄	❄
ISO-VIS	♀	☼	🔓	■ White	Continuous	—— Default	Color_7	🖨	ⓥ	❄

Freeze In The
Active View

Figure 22–37A

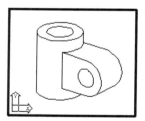

Figure 22–37B

STEP 10

Activate the Front viewport holding the section view information by clicking inside that viewport. The crosshatching looks too dense and needs to be scaled up in size through the Hatchedit dialog box in Figure 22–38A.

In the Scale edit box of the Hatchedit dialog box, change the hatch pattern scale to a new value of 2 units. Click the OK button to change the scale of the hatch pattern. The Front view should appear similar to Figure 22–38B.

Command: **HE** *(For HATCHEDIT)*
Select hatch object: *(Select the hatch object located in the viewport in Figure 22–38B)*

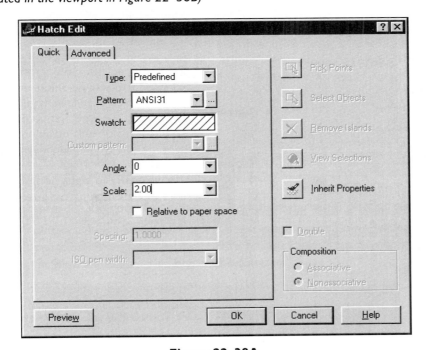

Figure 22–38A

Figure 22–38B

PHASE IV—Adding Dimensions to the Views

STEP 11

Begin adding dimensions to the Front view; do this while still in floating Model Space. First, click in the viewport containing the Front view information. Then, issue the UCS command along with the View option. This will line up the User Coordinate System icon parallel to the display screen, enabling you to place the dimensions as in a normal 2D drawing. Once the icon has been changed, make the FRONT-DIM layer current and turn off the FRONT-HAT layer, as in Figure 22–39. As you place the dimensions in the Front view, they should not appear in other viewports. The SOLVIEW command automatically creates dimension layers and freezes the layers in the viewports in which they do not apply.

(Click in the viewport at "A" in Figure 22–39, which contains the Front view)
Command: **UCS**
Current ucs name: *NO NAME*
Enter an option [New/Move/
 orthoGraphic/Prev/Restore/Save/Del/
 Apply/?/World]
<World>: **N** *(For New)*
Specify origin of new UCS or [ZAxis/
 3point/OBject/Face/View/X/Y/Z]
 <0,0,0>: **V** *(For View)*

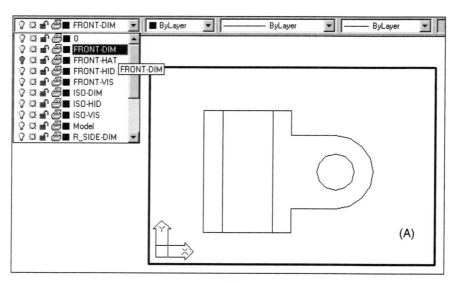

Figure 22–39

STEP 12

Draw a series of centerlines on the FRONT-DIM layer. To accomplish this, use the DIMCENTER command and pick the edge of the arc at "A." Then turn ORTHO mode on and construct a vertical centerline from the midpoint at "B" to the approximate point at "C," as in Figure 22–40.

Command: **DCE** *(For DIMCENTER)*
Select arc or circle: *(Select the edge of the arc at "A")*

Command: **L** *(For LINE)*
Specify first point: **Mid**
of *(Select the line at "B")*
Specify next point or [Undo]: *(Pick a point at "C")*
Specify next point or [Undo]: *(Press ENTER to exit this command)*

Use grips to extend the line in Figure 22–40. Then change the highlighted lines in the figure to the Center linetype (this linetype was also previously loaded).

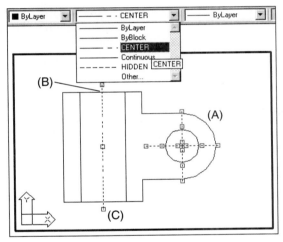

Figure 22–40

STEP 13

Begin adding dimensions to the Front view. Before performing this operation, turn on OSNAP. Set the dimension center mark control (DIMCEN) to 0. This will prevent the placement of duplicate center marks when you place diameter and radius dimensions. Check that the Endpoint and Intersection modes are set, because numerous intersections will be re-

quired in dimensioning. With the Front viewport active and the User Coordinate System set to the current view, begin placing the linear, radius, and diameter dimensions, using Figure 22–41A as a guide.

Command: **DIMCEN**
Enter new value for DIMCEN <-0.0900>:
  0
Command: **DLI** *(For DIMLINEAR)*

Specify first extension line origin or
   <select object>: (Pick the endpoint at
   "A")
Specify second extension line origin: (Pick
   the endpoint at "B")
Specify dimension line location or
[Mtext/Text/Angle/Horizontal/Vertical/
   Rotated]: (Pick at "C")
Dimension text = 5.00
Command: **DLI** (For DIMLINEAR)
Specify first extension line origin or
   <select object>: (Pick the endpoint at
   "D")
Specify second extension line origin: (Pick
   the endpoint at "E")
Specify dimension line location or
[Mtext/Text/Angle/Horizontal/Vertical/
   Rotated]: (Pick at "F")
Dimension text = 2.50
Command: **DLI** (For DIMLINEAR)

Specify first extension line origin or
   <select object>: (Pick the endpoint at
   "G")
Specify second extension line origin: (Pick
   the endpoint at "H")
Specify dimension line location or
[Mtext/Text/Angle/Horizontal/Vertical/
   Rotated]: (Pick at "I")
Dimension text = 3.50
Command: **DRA** (For DIMRADIUS)
Select arc or circle: (Select the edge of the
   arc at "J")
Dimension text = 1.50
Specify dimension line location or [Mtext/
   Text/Angle]: (Pick at "K")
Command: **DDI** (For DIMDIAMETER)
Select arc or circle: (Select the edge of the
   circle at "L")
Dimension text = 1.50
Specify dimension line location or [Mtext/
   Text/Angle]: (Pick at "M")

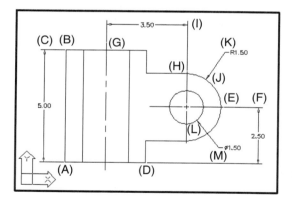

Figure 22–41A

Before continuing to the next step, activate the Layer Control box shown in Figure 22–41B and turn the FRONT-HAT layer back on.

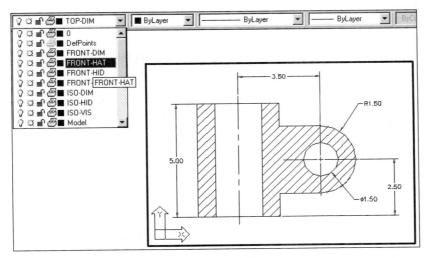

Figure 22–41B

STEP 14

Switch back to Paper Space by double-clicking in the area at "A" in Figure 22–42, which can be anywhere outside the viewport. Then pick the viewport; it will highlight and grips will appear. Right-click to display the shortcut menu. Choose Display Locked and then Yes to lock the viewport. This will prevent any accidental zooming of the image inside the Paper Space viewport. When finished, press ESC twice to remove the object highlight and grips from the screen.

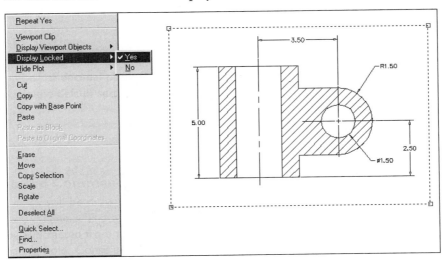

Figure 22–42

STEP 15

Activate the Top view by double-clicking inside the top viewport. Issue the UCS command along with the View option. This will position the User Coordinate System icon parallel to the display screen. This will also enable the dimensions to be placed as in a normal 2D drawing. Once the icon has been aligned to the view, make the TOP-DIM layer current as in Figure 22–43. As the dimensions are being placed, they will not appear in other viewports. The SOLVIEW command automatically creates dimension layers and freezes the layers in the viewports in which they do not apply. Begin adding dimensions to the Top view; do this while still in Model Space.

(Click in the viewport at "A" in Figure 22–43, which contains the Top view)
Command: **UCS**
Current ucs name: *NO NAME*
Enter an option [New/Move/
    orthoGraphic/Prev/Restore/Save/Del/
    Apply/?/World]
<World>: **N** *(For New)*
Specify origin of new UCS or [ZAxis/
    3point/OBject/Face/View/X/Y/Z]
    <0,0,0>: **V** *(For View)*

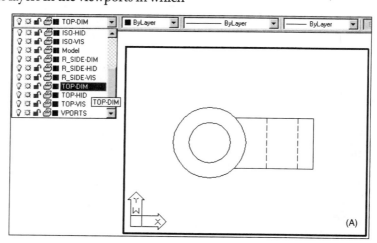

Figure 22–43

STEP 16

Draw a series of centerlines on the TOP-DIM layer. First reset the center mark setting DIMCEN to a value of -0.09 units. Use the DIMCENTER command and pick the edge of the circle at "A." Then turn ORTHO mode on and construct a vertical centerline from the midpoint at "B" to the approximate point at "C," as in Figure 22–44, with assistance from the Geometry Calculator and the MEE (Midpoint-Endpoint-Endpoint) option.

Command: **DIMCEN**
Enter new value for DIMCEN <0.0000>:
    -0.09

Command: **DCE** *(For DIMCENTER)*
Select arc or circle: *(Select the edge of the arc at "A")*
Command: **L** *(For LINE)*
Specify first point: **'CAL**
Initializing...>> Expression: **MEE**
>> Select one endpoint for MEE: *(Select the endpoint of the hidden line at "B")*
>> Select another endpoint for MEE: *(Select the endpoint of the hidden line at "C")*
(11.3822 5.55593 0.0)
Specify next point or [Undo]: *(Pick a point approximately at "D")*
Specify next point or [Undo]: *(Press ENTER to exit this command)*

Use grips to stretch the line to "E" in Figure 22–44. Then change the highlighted lines in the figure to the Center linetype.

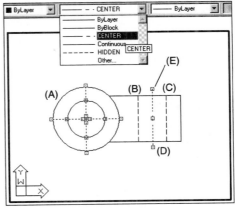

Figure 22–44

STEP 17

Turn off the setting that controls the placement of a center mark when placing diameter and radius dimensions. Begin adding linear, radius, and diameter dimensions to the Top view. The OSNAP Endpoint and Intersection modes should still be active and running. See Figure 22–45.

Command: **DIMCEN**
Enter new value for DIMCEN <-0.0900>: **0**
Command: **DLI** *(For DIMLINEAR)*
Specify first extension line origin or <select object>: *(Select the endpoint at "A")*
Specify second extension line origin: *(Select the endpoint at "B")*
Specify dimension line location or [Mtext/Text/Angle/Horizontal/Vertical/Rotated]: *(Pick at "C")*
Dimension text = 2.50
Command: **DDI** *(For DIMDIAMETER)*
Select arc or circle: *(Select the edge of the circle at "D")*

Dimension text = 2.00
Specify dimension line location or [Mtext/Text/Angle]: *(Pick at "E")*
Command: **DDI** *(For DIMDIAMETER)*
Select arc or circle: *(Select the edge of the circle at "F")*
Dimension text = 3.50
Specify dimension line location or [Mtext/Text/Angle]: *(Pick at "G")*

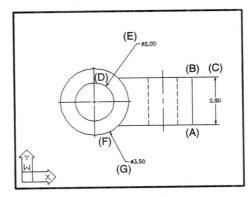

Figure 22–45

STEP 18

Switch back to Paper Space by double-clicking in the area at "A" in Figure 22–46. Then pick the viewport; it will highlight and grips will appear. Right-click to display the shortcut menu. Choose Display Locked and then Yes to lock the viewport.

This will prevent any accidental zooming of the image inside the Paper Space viewport. When finished, press ESC twice to remove the object highlight and grips from the screen.

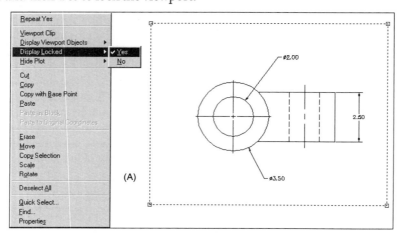

Figure 22–46

STEP 19

Activate the Layer Control box and turn off the VPORTS layer in Figure 22–47.

This will display only the views without the rectangular viewports present.

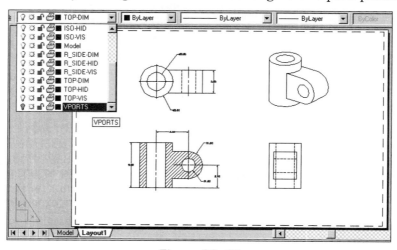

Figure 22–47

STEP 20

An ANSI-D title block can be inserted in the completed drawing in Figure 22–48.

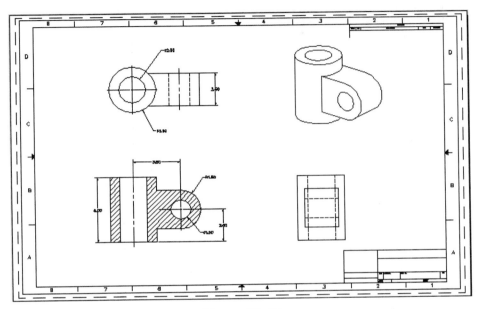

Figure 22–48

PROBLEMS FOR CHAPTER 22

1. Using the Tutorial Exercise of Column.Dwg as a guide, create an engineering drawing of Problems 20–1 through 20–30 consisting of Front, Top, Right Side, and Isometric views.

2. Properly dimension the engineering drawing.

INDEX